Electric Energy Systems
Analysis and Operation

Second Edition

Electric Power Engineering Series

Series Editor Leonard L. Grigsby

Electromechanical Systems, Electric Machines, and Applied Mechatronics

Sergey E. Lyshevski

Power Quality

C. Sankaran

Power System Operations and Electricity Markets

Fred I. Denny and David E. Dismukes

Electric Machines

Charles A. Gross

Electric Energy Systems

Analysis and Operation

Antonio Gómez-Expósito, Antonio J. Conejo, and Claudio A. Cañizares

The Induction Machines Design Handbook, Second Edition

Ion Boldea and Syed A. Nasar

Linear Synchronous Motors

Transportation and Automation Systems, Second Edition

Jacek F. Gieras, Zbigniew J. Piech, and Bronislaw Tomczuk

Electric Power Generation, Transmission, and Distribution, Third Edition

Leonard L. Grigsby

Computational Methods for Electric Power Systems, Third Edition

Mariesa L. Crow

Electric Energy Systems

Analysis and Operation, Second Edition

Antonio Gómez-Expósito, Antonio J. Conejo, and Claudio A. Cañizares

For more information about this series, please visit: https://www.crcpress.com/Electric-Power-Engineering-Series/book-series/CRCELEPOWENG

Electric Energy Systems
Analysis and Operation

Second Edition

Edited by

Antonio Gómez-Expósito
Antonio J. Conejo
Claudio A. Cañizares

CRC Press
Taylor & Francis Group
Boca Raton London New York

CRC Press is an imprint of the
Taylor & Francis Group, an **informa** business

CRC Press
Taylor & Francis Group
6000 Broken Sound Parkway NW, Suite 300
Boca Raton, FL 33487-2742

First issued in paperback 2020

ISBN-13: 978-1-138-72479-2 (hbk)
ISBN-13: 978-0-367-73427-5 (pbk)

Library of Congress Cataloging-in-Publication Data

Names: Gómez-Expósito, Antonio, editor. | Conejo, Antonio J., editor. |
Cañizares, Claudio A., editor.
Title: Electric energy systems : analysis and operation / edited by Antonio
Gómez-Expósito, Antonio J. Conejo, Claudio A. Cañizares.
Description: Second edition. | Boca Raton : Taylor & Francis, CRC Press,
2018. | Includes bibliographical references and index.
Identifiers: LCCN 2018006938| ISBN 9781138724792 (hardback : alk. paper) |
ISBN 9781315192246 (ebook)
Subjects: LCSH: Electric power systems.
Classification: LCC TK1001 .E347 2018 | DDC 621.319--dc23
LC record available at https://lccn.loc.gov/2018006938

Visit the Taylor & Francis Web site at
http://www.taylorandfrancis.com

and the CRC Press Web site at
http://www.crcpress.com

To our families

Contents

List of Figures

List of Tables

Preface

Since the first edition of this book saw the light, several electric energy systems trends have gained momentum or consolidated. The most relevant ones include:

1. Exponential growth of renewable energy (RE) generation, mainly solar and wind, which currently accounts worldwide for nearly 25% of the electricity production. Unlike in the past, this added generation's power output is essentially uncontrollable by the transmission system operator, which notices its effect as a reduction of the "net demand." In addition, as the share of RE increases, the global rotating inertia and synchronizing power are being proportionally reduced.

2. Deployment of an increasingly higher number of High-Voltage DC (HVDC) lines, overhead for long distance transmission of large amounts of energy, but especially submarine for the connection of off-shore RE to the bulk system and for a variety of grid interconnections, or underground when overhead lines are not adequate. Overall, electrical cables are expected to increase their share in power transmission and distribution networks.

3. Digitalization of primary and secondary substations, along with cheaper and ubiquitous information and communication systems, which can potentially provide between two to three orders of magnitude more information than in the past to the Energy Management System (EMS). One of the most promising technologies that can influence the way power systems are operated is phasor measurement units (PMU), also called synchrophasors.

4. Quick transformation of previously passive and inelastic customers into so-called "prosumers," capable of producing and/or storing some of their energy needs, being also responsive to electricity price signals thanks to the widespread adoption of smart meters.

In fact, many experts believe electric energy systems are going through a true revolution or changing of the guard, which is replacing the centralized, fossil fuel-based paradigm of the twentieth century by a partly distributed, smarter and decarbonized system.

Although the book structure has been kept intact, 9 out of the 12 chapters have been updated to a greater or lesser extent in the second edition, keeping in mind the wide spectrum of potential readers (from academia to industry) while, as stated in the first edition, trying to avoid an encyclopedic or very specialized approach. The reader is referred to the Preface of the first edition for a chapter-by-chapter detailed overview of the book contents. A brief description of the chapters in the second edition is provided next.

Chapter 1 provides an up-to-date overview of the structure and functioning of power systems, including their technical, economic, and regulatory layers. Issues such as environmental effects and climate change, the integration of renewables, including distributed energy resources, and associated regulatory challenges, are considered.

In Chapter 2, devoted to the steady-state linear models of power system components, the section on insulated cables has been considerably enlarged, providing expressions for series and shunt impedance calculations, as well as a description of earthing techniques, that are important in this context.

Chapter 3 covers the classical and well-established load flow problem. Even though modest algorithmic and modeling improvements have been made in recent years, particularly when solving radial distribution feeders, their industrial or practical relevance is still uncertain. For this reason, this chapter remains as in the previous edition.

Chapter 4 now includes a whole section devoted to the inclusion of PMUs in state estimation. Furthermore, the last section on parameter and topology errors has been enlarged by incorporating the recently introduced implicit models.

Chapter 5 addresses the economic operation of power systems including economic dispatch, unit commitment and market operation. Regarding market operation, the chapter provides an introduction to market clearing, producer self-scheduling and offering, and consumer energy procurement.

Chapter 6 provides an analysis of system security with a focus on the optimal power flow and the security constrained optimal power flow problems. A detailed description of system security concepts with emphasis on the transmission network is also provided.

The second half of the book begins with Chapter 7, dealing with unbalanced and nonlinear models of power components. In the first edition, this chapter already considered relevant power electronics building blocks, such as the Voltage Source Converter (VSC), which is becoming the workhorse of the smart grid paradigm. As a consequence, Chapter 7 remains unaltered in the second edition.

The book has been revised to include a description of the prevalent RE generation technologies and associate controls in Chapter 9, and their impact on power system stability in Chapter 10. It is important to mention that, even though RE plants have evolved quickly, the technologies being deployed are now mature and thus no significant changes are expected on this front in the midterm. Furthermore, Chapters 9 and 10 focus on large plants connected at the transmission system level, as opposed to distributed generation being deployed at increasing rates at the distribution system level, especially solar power generation with battery storage systems, which are beyond the scope of the present book.

Chapters 11 revisits the load flow problem under nonsinusoidal and unbalanced conditions. As in the linear balanced case (Chapter 3), no changes have been made to this chapter, which already presented sufficient algorithmic details in the first edition, at least for the nonspecialized reader.

Chapter 12 gives a comprehensive overview of electromagnetic transients in power systems, including modeling, switching transients, overvoltage transients, and analytical and computational analysis techniques. As in Chapter 2, the section on insulated cables has been significantly expanded.

Antonio Gómez-Expósito
Seville, Spain

Antonio J. Conejo
Columbus, OH, USA

Claudio A. Cañizares
Waterloo, ON, Canada

Preface to the First Edition

The purpose of this book is to merge, update, and extend the material in both classical power system analysis books and power economics books, within the framework of currently restructured electric energy systems. This effort is clearly needed to address the operations and planning problems in nowadays unbundled generation, transmission, and distribution systems.

In addressing the aforementioned issues, we realized that the challenges were significant. First of all, we had to provide added value to the reference books on the topic. Fortunately, achieving this goal was facilitated by the economic and technical revolution that the electric sector has been undergoing during the last two decades. This revolution has dramatically changed or made obsolete many important concepts as they are considered in existing books. Another difference between this and earlier books lies in the addition of some advanced chapters, which are usually covered in specialized monographs, as well as in the deeper treatment of certain classical topics.

The second and perhaps more difficult challenge was the need to avoid an *encyclopedic* approach, both in scope and content. A book coauthored by 24 researchers, each one writing on his own area of expertise, can easily degenerate into a voluminous collection of disconnected papers, which might only be useful for a minority of specialists. Being aware of this risk, the authors have made a significant effort to begin with the basic principles, paying attention to the topics any power engineer should know, including many solved examples, and directing the reader to other chapters if necessary. Some redundant material has been intentionally left out in order not to distract the reader's attention by cross referencing. An added advantage of the approach used here is that many chapters become self-sufficient for those readers with a certain background in power systems who simply desire to keep themselves updated.

Our objective was to keep the spectrum of readers to whom this book is directed fairly broad. On the one hand, instructors and undergraduate students of engineering schools can use it as a textbook, with the material being possibly fully covered in two terms. Depending on the particular context, the instructor may have to pick up only a subset of chapters, discarding those that are covered in other subjects of the curriculum. This may be the case of protections, overvoltages, synchronous machines, and so on. On the other hand, considering the advanced level of certain chapters, and the inclusion, for the first time in a textbook of this breadth and depth, of the regulatory issues, which are changing the electric sector, this volume can also serve as a handbook for graduate students and practicing professionals who lack the time to search for original sources (papers and technical reports). These readers will surely welcome the large number of references at the end of each chapter, which will allow them to acquire a deeper knowledge on the topics of their own interest.

It is assumed here that the reader has a minimum background in algebra (matrices, complex numbers, etc.), calculus (linear differential equations, Laplace and Fourier transform, etc.), physics (electromagnetic fields, rotating mass dynamics, etc.), circuits (nodal equations, three-phase circuits, etc.) and, if possible, electric machines and microeconomics. This is usually the case of those undergraduate students who enroll for the first time in a course on power system analysis.

Owing to space limitations, the book is mainly focused on the operation of generation and transmission systems, although part of the material (e.g., certain component models, three-phase and harmonic load flows, reliability indices, and protections) is of application also in the analysis of distribution networks. For the same reason, the long-term planning problem has not been explicitly dealt with; nevertheless, several chapters and parts of others (e.g., load flow, generation scheduling, security, reliability, and stability) present essential tools for network expansion studies, design and comparison of alternatives, and so on, which are directly linked with the short-term planning problem.

The book has been organized in 12 chapters and three appendices, which could have been arranged in several ways. A possibility could have been to begin with the most classical chapters (load flow, frequency regulation, economic dispatch, short circuits, and transient stability), and then continue with what could be labeled advanced topics (market issues, state estimation, electromagnetic transients, harmonics, etc.). However, it is difficult in many cases to place the division between basic and advanced material properly speaking, particularly in those chapters specifically dealing with the operation of generation and transmission systems. The scheme adopted in this book is based instead on the different working regimes of the power system, which are crucial to determine the techniques and tools needed to study the time scales involved.

The five chapters following the introductory one cover what is essentially the balanced sinusoidal steady-state regime, which, rigorously speaking, should be called quasi-steady-state because of the slow but continuous variation of the loads. In this context, mainly related to the real-time operation of power systems, phasors, and complex power constitute the basic tools on which the different analytical and computational methods are built. On the other hand, the last six chapters are devoted to the transient and nonsinusoidal states of a power system, including both balanced and unbalanced conditions. The system under transients originated by faults is first dealt with, followed by the slower electromechanical oscillations, to end with the faster electromagnetic transients.

Chapter 1 makes an original presentation of what power systems have been in the past and what they have become nowadays, from the technical, economical, and regulatory points of view. It constitutes by itself very valuable material to be disseminated among those young students who erroneously believe that professional challenges can only be found in computer engineering and communications.

Besides conventional components, Chapter 2 briefly deals with cables and asynchronous machine modeling, of renewed interest in view of the growth of distributed generation. An introduction to load forecasting techniques is also presented.

In Chapter 3, which is devoted to the classic power flow or load flow problem, the section on large-scale systems, complemented by Appendix A, stands out. The reader may find interesting the discussion about the simplifications behind the fast decoupled load flow. Power flow and voltage regulating devices are presented, starting from a common framework, in an original manner.

Chapter 4 provides many more details than it is usual in textbooks about advanced topics related to state estimation, like nonquadratic estimators and topology error identification.

Chapter 5 starts with a rigorous and general treatment of the economic dispatch problem, paying special attention to transmission loss coefficients, and finishes with the formulation of the optimization problems currently faced by producers, consumers, and other agents of electricity markets.

The presence of Chapter 6, entirely devoted to the operation of the transmission subsystem, is new in textbooks of this nature, but we believe it is fundamental to provide a comprehensive view of all the problems and tasks involved at this level. The new paradigm under which transmission networks are operated, based on open and nondiscriminatory access, and the resulting challenges are presented and discussed in this chapter.

The second part of the book starts with Chapter 7, which is devoted to a general treatment of three-phase linear and nonlinear models of power system components, including power electronics components, such as filters, voltage-source converters, etc. Several of the models described here are used in the following chapters.

Chapter 8 comprises two closely related subjects, namely fault analysis, including a brief reference to grounding systems, and protections. More attention than usual is paid to the matrix-based systematic analysis of short circuits in large-scale systems.

Automatic regulation and control of voltages and frequency is dealt with in Chapter 9 from a broader perspective, beginning with local or primary control strategies and ending with the region-wide secondary and tertiary control schemes. These controls and associated concepts, such as hierarchical and wide-area control, are presented and discussed within the context of practical grid requirements in Europe and North America, and in view of their role in competitive electricity markets, particularly in relation to ancillary services.

Power system stability analysis is discussed in Chapter 10 in a novel and up-to-date manner. This chapter assumes that the reader is familiar with basic system element models and controls, particularly those associated with the synchronous machine, as discussed in Chapters 7 and 9 and Appendix C. Each stability subtopic, that is, angle, voltage, and frequency stability, is first defined and the main concepts and analysis techniques are then explained using basic and simple system models. This is followed by a discussion of practical applications, analysis tools and measures for stability improvement, closing with a brief description of a real instability event, which is a unique feature of this book with respect to other textbooks in the topic.

Chapters 11 addresses again the power flow problem but under nonsinusoidal and unbalanced conditions. This chapter presents advanced topics whose relevance is steadily increasing, given the growing portion and size of electronic converters connected to power systems.

Ignored or superficially covered in most textbooks, in Chapter 12 analytical and computational techniques for the study of electromagnetic transients are explained in detail, as well as some related applications, like propagation and limitation of overvoltages.

The book closes with three appendices covering the solution of large-scale sparse systems of linear equations (Appendix A), the fundamentals of optimization (Appendix B), and the modeling of induction and synchronous machines (Appendix C).

The analytical and computational techniques covered in this book have been selected taking into account that the speed of response in the analysis of very large nonlinear systems is often crucial for the results to be of practical use. In this sense, electric energy systems constitute a unique case, because of their size, complexity, and strict control requirements.

We are thankful to all the colleagues, students, and institutions who have helped us in the complex endeavor of editing and partly writing this book. We hope that this book will be helpful for the new generations of power engineering students and professionals, which is the sole motivation for this project.

Antonio Gómez Expósito
Sevilla, Spain

Antonio J. Conejo
Ciudad Real, Spain

Claudio A. Cañizares
Waterloo, Canada

Editors

Antonio Gómez-Expósito received a six-year industrial engineer degree, major in electrical engineering, in 1982, and a PhD in 1985, both with honors from the University of Seville, Spain. He is currently the Endesa Red chair professor at the University of Seville, where he chaired the Department of Electrical Engineering for nearly 12 years and the Post-Graduate Program on Electrical Energy Systems. In the late 1980s he created, and still leads, the Power Engineering group, currently comprising about 30 researchers, including 20 doctors. He was also a visiting professor in California and Canada.

Professor Gómez-Expósito has coauthored several textbooks and monographs about circuit theory and power system analysis, of which the one published by Marcel Dekker, *Power System State Estimation: Theory and Implementation*, stands out. He is also the coauthor of over 300 technical publications, nearly one half of them in indexed journals. He has been the principal investigator or participated in over 80 research and development projects, both publicly and privately funded, most of them developed in close cooperation with the major national and European companies of the electrical sector. In 2012, he led the creation of the spin-off company Ingelectus.

Professor Gómez-Expósito has served on the editorial board of the *IEEE Transactions on Power Systems* and other journals, and has also belonged to the scientific or technical committees of the major European conferences related to power systems in the last 20 years. Among other recognitions, he received in 2011 the "Juan López Peñalver" Research and Technology Transfer Award, granted by the Government of Andalusia, and the Outstanding Engineer Award, granted by the Spanish Chapter of the IEEE/PES in 2010. He is a fellow of the IEEE and a member of the Royal Sevillian Academy of Sciences.

Antonio J. Conejo, professor at the Ohio State University, OH, USA, received a BS from University P. Comillas, Madrid, Spain, an MS from Massachusetts Institute of Technology, USA, and a PhD from the Royal Institute of Technology, Sweden. He has published over 190 papers in SCI journals and is the author or coauthor of books published by Springer, John Wiley, McGraw-Hill, and CRC. He has been the principal investigator of many research projects financed by public agencies and the power industry and has supervised 20 PhD theses. He is an IEEE Fellow and a former editor-in-chief of the *IEEE Transactions on Power Systems*, the flagship journal of the power engineering profession.

Claudio A. Cañizares is a full professor and the Hydro One endowed chair at the Electrical and Computer Engineering (E&CE) Department of the University of Waterloo, where he has held various academic and administrative positions since 1993. He received the electrical engineer degree from the Escuela Politécnica Nacional (EPN) in Quito-Ecuador in 1984, where he held different teaching and administrative positions between 1983 and 1993, and his MSc (1988) and PhD (1991) in electrical engineering are from the University of Wisconsin-Madison. His research activities focus on the study of stability, control, optimization, modeling, simulation, and computational issues in large as well as small grids and energy systems in the context of competitive energy markets and smart grids. In these areas, he has led or been an integral part of many grants and contracts from government agencies and private companies, and has collaborated with various industry and university researchers in Canada and abroad, supervising/co-supervising a large number of research fellows and graduate students. He has authored/co-authored many highly cited journal and conference papers, as well as several technical reports, book chapters, disclosures and patents, and has been invited to make multiple keynote speeches, seminars, and presentations at numerous institutions and conferences worldwide. He is a fellow of the IEEE, as well as a fellow of the Royal Society of Canada, where he is currently the director of the Applied Science and Engineering Division of the Academy of Science, and a fellow of the Canadian Academy of Engineering. He is also the recipient of the 2017 IEEE Power & Energy Society (PES) Outstanding Power Engineering Educator Award, the 2016 IEEE Canada Electric Power Medal, and of various IEEE PES Technical Council and Committee awards and recognitions, holding leadership positions in several IEEE-PES Technical Committees, Working Groups, and Task Forces.

Contributors

Michel Rivier Abbad
Instituto de Investigación Tecnológica
Escuela Técnica Superior de Ingeniería ICAI
Universidad Pontificia Comillas
Madrid, Spain

Ali Abur
Northeastern University
Boston, Massachusetts

Enrique Acha
Tampere University of Technology
Tampere, Finland

Fernando L. Alvarado
University of Wisconsin
Madison, Wisconsin

Göran Andersson
ETH
Zurich, Switzerland

Carlos Álvarez Bel
Polytechnic University of Valencia
Valencia, Spain

Claudio A. Cañizares
University of Waterloo
Waterloo, Ontario, Canada

José Cidrás
Universidad de Vigo
Vigo, Spain

Antonio J. Conejo
Department of Electrical and Computer
   Engineering
The Ohio State University
Columbus, Ohio

Francisco D. Galiana
McGill University
Montréal, Québec, Canada

Julio García-Mayordomo
Universidad Politécnica de Madrid
Madrid, Spain

Antonio Gómez-Expósito
Universidad de Sevilla
Seville, Spain

José R. Martí
University of British Columbia
Vancouver, British Columbia, Canada

José L. Martínez-Ramos
Department of Electrical Engineering
Universidad de Sevilla
Seville, Spain

Juan A. Martínez-Velasco
Universitat Politécnica de Catalunya
Barcelona, Spain

José F. Miñambres
University of the Basque Country UPV/EHU
Bilbao, Spain

Antonio F. Otero
Universidad de Vigo
Vigo, Spain

Ignacio J. Pérez-Arriaga
Sloan School of Management
Massachussets Institute of Technology
Boston, Massachusetts

and

Instituto de Investigación Tecnológica
Escuela Técnica Superior de Ingeniería ICAI
Universidad Pontificia Comillas
Madrid, Spain

Luis Rouco
Universidad Pontificia Comillas de Madrid
Madrid, Spain

Hugh Rudnick
Department of Electrical Engineering
Pontificia Universidad Catolica de Chile
Santiago, Chile

Julio Usaola
Universidad Carlos III de Madrid
Madrid, Spain

Wilsun Xu
University of Alberta
Edmonton, Alberta, Canada

1 Electric Energy Systems
An Overview

*Ignacio J. Pérez-Arriaga, Hugh Rudnick, and
Michel Rivier Abbad*

CONTENTS

1.1 A FIRST VISION

1.1.1 THE ENERGY CHALLENGES IN MODERN TIMES

Energy is a fundamental ingredient of modern society and its supply impacts directly on the social and economic development of nations. Economic growth and energy consumption go hand in hand. The development and quality of our life and our work are totally dependent on a continuous, abundant, economic and environmental-friendly energy supply. Coal, oil and natural gas have been the traditional basic energy sources, and this reliance on energy for economic growth has historically implied dependence on third parties for energy supply, with geopolitical connotations arising, as these energy resources have not been generally in places where high consumption has developed. Energy has transformed itself in a new form of international political power, utilized by owners of energy resources (mainly oil and natural gas). At some time, concerns aroused on the decline of volumes of oil and natural gas (coal remains an abundant resource) and on the consequent energy price increases in the medium to long term. However, the advances in exploration and drilling technologies, and the development of shale gas, have minimized this fear. It is actually the climate change which has driven the major changes on this regard. Greenhouse emissions are heavily penalizing the traditional resources, being renewable sources (mainly wind but fundamentally solar) seen as the future main energy resources, as the backbone of a decarbonized abundant energy supply. Strong geopolitical impacts may be expected from such a transformation.

Within that framework, electricity has become a favorite form of energy usage at the consumer end, with coal, oil, gas, uranium, hydro, wind, solar irradiation, and other basic resources used to generate electricity. With its versatility and controllability, instant availability and consumer-end cleanliness, electricity has become an indispensable, multipurpose form of energy. Its domestic use now extends far beyond the initial purpose, to which it owes its colloquial name ("light" or "lights"), and has become virtually irreplaceable in kitchens—for refrigerators, ovens, and cookers or ranges, and any number of other appliances—and in the rest of the house as well, for air conditioner, radio, television, computers, and the like. But electricity usage is even broader in the commercial and industrial domains: in addition to providing power for lighting and air conditioning, it drives motors with a host of applications: lifts, cranes, mills, pumps, compressors, lathes, or other machine tools, and so on and so forth: it is nearly impossible to imagine an industrial activity that does not use some sort of electricity. Being also electricity one of the energy forms where renewable sources can more economically an efficiently substitute traditional greenhouse emitting energy sources, electrification of other sectors such as the terrestrial transport or the heating might be expected. Thus, modern societies have become totally dependent on an abundant electricity supply and it is alike they will become even more dependent on the future.

1.1.2 Characteristics of Electricity

At first glance, electricity must appear to be a commodity much like any other on consumers' list of routine expenses. In fact, this may be the point of view that prompted the revolution that has rocked electric energy systems worldwide, as they have been engulfed in the wave of liberalization and deregulation that has changed so many other sectors of the economy [12,13,20]. And yet electricity is defined by a series of properties that distinguish it from other products, an argument often wielded in an attempt to prevent or at least limit the implementation of such changes in the electricity industry. The chief characteristic of electricity as a product that differentiates it from all others is that it is not susceptible, in practice, to being stored or inventoried. Electricity can, of course, be stored in batteries, but price, performance, and inconvenience makes this impractical up to now for handling the amounts of energy usually needed in the developed world. Therefore, electricity must be generated and transmitted as it is consumed, which means that electric systems are dynamic and highly complex, as well as immense. At any given time, these vast dynamic systems must strike a balance between generation and demand, and the disturbance caused by the failure of a single component may be transmitted across the entire system almost instantaneously. This sobering fact plays a decisive role in the structure, operation, and planning of electric energy systems, as discussed below.

Another peculiarity of electricity is its transmission: this is not a product that can be shipped in "packages" from its origin to destination by the most suitable medium at any given time. Electric power is transmitted over grids in which the pathway cannot be chosen at will, but is determined by Kirchhoff's laws, whereby current distribution depends on impedance in the lines and other elements through which electricity flows [4]. Except in very simple cases, all that can be said is that electric power flows into the system at one point and out of it at another, because ascribing the flow to any given path is extraordinarily complex and somewhat arbitrary. Moreover, according to these laws of physics, the alternative routes that form the grid are highly interdependent, so that any variation in a transmission facility may cause the instantaneous reconfiguration of power flows and that, in turn, may have a substantial effect on other facilities. All this renders the dynamic balance referred to in the preceding paragraph even more complex.

1.1.3 Electrical Energy Systems: The Biggest Industrial System Created by Humankind

Indeed, for all its apparent grandiloquence, the introductory sentence to this unit may be no exaggeration. The combination of the extreme convenience of utility and countless applications of electricity, on the one hand, and its particularities, on the other hand, has engendered these immense and sophisticated industrial systems. Their size has to do with their scope, as they are designed to carry electricity practically to any place inhabited by human beings from electric power stations located wherever a supply of primary energy—in the form of potential energy in moving water or any of several fuels—is most readily available. Carrying electric power from place of origin to place of consumption calls for transmission grids and distribution grids or networks that interconnect the entire system and enable it to work as an integrated whole. Their sophistication is a result of the complexity of the problem, determined by the characteristics discussed above: the apparently fragile dynamic equilibrium between generation and demand that must be permanently maintained is depicted in the highly regular patterns followed by the characteristic magnitudes involved—the value and frequency of voltage and currents as well as the waveform of these signals. Such regularity is achieved with complicated control systems that, based on the innumerable measurements that continuously monitor system performance, adapt its response to constantly changing conditions. A major share of these control tasks is performed by powerful computers in energy management centers running a host of management applications: some estimate demand at different grid buses several minutes, hours, days, or months in advance; other models determine the generation needed to meet this demand; yet other programs compute the flow in system lines and transformers and the voltage at grid buses under a number of assumptions on operating conditions or component failure, and determine the most suitable action to take in each case. Others study the dynamic behavior of the electric power system under various types of disturbance [9]. Some models not only attempt to determine the most suitable control measures to take when a problem arises, but also to anticipate their possible occurrence, modifying system operating conditions to reduce or eliminate its vulnerability to the most likely contingencies.

This, however, is not all: the economic aspect of the problem must also be borne in mind. The actors that make the system work may be private companies that logically attempt to maximize their earnings or public institutions that aim to minimize the cost of the service provided. In either case, the economic implications of the decisions made cannot be ignored, except, of course, where system safety is at stake. The system operates under normal conditions practically always, so there is sufficient time to make decisions that are not only safe, but also economically sound. Hence, when demand rises foreseeably during the day, power should be drawn from the facilities with unused capacity that can generate power most efficiently. The objective is to meet daily load curve needs with power generated at the lowest and least variable cost. This new dimension in the operation of electric energy systems is present in all timescales [8]: from the hourly dispatch of generating plant to the choice of which units should start-up and stop and when, including decisions on the use of hydroelectric reserve capacity, maintenance programing and investment in new facilities. It should moreover be stressed that all these decisions are made in a context of uncertainty: about the future demand to be met, plant availability, the prices of the various parameters involved in the production process, in particular, fuel, and even the regulatory legislation in effect when long-term decisions are to be implemented.

1.1.4 History

1.1.4.1 Technological Aspects

The first electric light systems, installed around 1870, consisted of individual dynamos that fed the electrical system—arc lamps—in place in a single residence. Thomas Edison discovered the incandescent light bulb around 1880 and authored the idea of increasing the scale of the process by using a single generator to feed many more bulbs. In 1882, Edison's first generator, driven by a steam turbine located on Pearl Street in lower Manhattan, successfully fed a direct current (DC) at a voltage of 100 V to around four hundred 80 W bulbs in office and residential buildings on the Wall Street. Shortly, thereafter London's 60 kW Holborn Viaduct station was commissioned, which also generated 100 V DC. This local generation and distribution scheme was quickly adopted, exclusively for lighting, in many urban and rural communities worldwide.

The invention of the transformer thanks to Gaulard, Gibbs, Zipernowsky, Bláthy and Déri in 1881–1884 revealed, in a process not exempt from controversy, the advantages of alternating current (AC), which made it possible to conveniently raise the voltage to reduce line losses and voltage drops over long transmission distances. Alternating, single-phase electric current was first transmitted in

1884 at a voltage of 18 kV. On August 24, 1891, three-phase current was first transmitted from the hydroelectric power station at Lauffen to the International Exposition at Frankfurt, 175 km away. Swiss engineer Charles Brown, who with his colleague and fellow countryman Walter Boveri founded the Brown–Boveri Company that very year, designed the three-phase AC generator and the oil-immersed transformer used in the station. In 1990, the Institute of Electrical and Electronic Engineers (IEEE) agreed to take August 24, 1891 as the date marking the beginning of the industrial use and transmission of AC.

The transmission capacity of AC lines increases in proportion to the square of the voltage, whereas the cost per unit of power transmitted declines in the same proportion. There was then an obvious motivation to surmount the technological barriers limiting the use of higher voltages. Voltages of up to 150 kV were in place by 1910 and the first 245 kV line was commissioned in 1922. The maximum voltage for AC has continued to climb ever since, as Figure 1.1 shows. And yet DC has also always been used, since it has advantages over AC in certain applications, such as electrical traction and especially electricity transmission in very long distance overhead lines, and in underground and submarine cables for which reactive power becomes a major issue if built in AC. The upward trend in maximum DC voltage throughout the twentieth century is also depicted in Figure 1.1.

The alternating voltage frequency to be used in these systems was another of the basic design parameters that had to be determined. Higher frequencies can accommodate more compact generating and consumption units, an advantage offset, however, by the steeper voltage drops in transmission and distribution lines that their use involves. Some countries such as the United States, Canada, Central American countries, and northernmost South American countries adopted a frequency of 60 Hz, while countries in the rest of South America, Europe, Asia, and Africa adopted a frequency of 50 Hz. The International Electrotechnical Commission was created in 1906 to standardize electrical facilities everywhere as far as possible [22]. It was, however, unable to standardize frequency, which continues to divide countries around the world into two different groups.

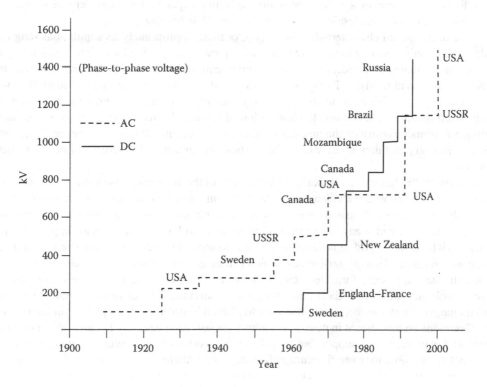

FIGURE 1.1   Maximum AC- and DC-rated voltages. (From Torá, J. L., *Transporte de la energía eléctrica*, Universidad Pontificia Comillas (ICAI-ICADE), Madrid, 1997. [1])

The advantages of interconnecting small electric energy systems soon became obvious. The reliability of each system was enhanced by the support received from the others in the event of emergencies. Reserve capacity could also be reduced, since each system would be able to draw from the total grid reserve capacity. With such interconnections, it was possible to deploy the generator units to be able to meet demand most economically at any given time; the advantage this affords is particularly relevant when peak demand time frames vary from one system to another and when the generation technology mix, hydroelectric and steam, for instance, likewise differs. In 1926, the English Parliament created the Central Electricity Board and commissioned it to build a high-voltage grid that would interconnect the 500 largest generation stations then in operation.

1.1.4.2 Organizational Aspects

What sort of organizational structure is in place in the sector responsible for planning, operating, and maintaining electric energy systems? Who makes the decisions in each case and under what criteria? The answers to these questions have evolved over time, largely to adapt to the conditioning factors imposed by technological development, but also depending on prevailing economic theory. As mentioned above, the first industrial applications of electricity were strictly local, with a generator feeding a series of light bulbs in the surrounding area. Whole hosts of individual systems sprang up under private or public, usually municipal, initiative, primarily to provide urban lighting and, somewhat later, to drive electric motors for many different purposes. The vertically integrated electric utility, which generates, transmits, distributes, and supplies electricity, evolved naturally and was the predominant model in most countries until very recently. The enormous growth of electricity consumption, the huge economies of scale in electricity generation, and the increase in the transmission capacity of high-voltage lines drove the development of transmission grids, often under state protection, to interconnect individual systems, giving rise to literally nationwide systems. Technical specialization and the huge volume of resources required to build large power stations led to the coexistence of local distribution companies, with scant or nil production capacity, and large vertically integrated utilities, which also sold wholesale electric power to small distributors.

Because of its special characteristics, electricity, or more appropriately its supply, has long been regarded to be a public service in most countries, an approach that justified state intervention to guarantee reasonable quality and price. In some cases, intervention consisted of nationalizing the electricity industry, such as in nearly all European countries until the 1990s. In others, electric utilities were subject to the legal provisions typically applied to monopolies, namely the requirement to meet certain minimum quality standards and the imposition of regulated prices to cover the costs incurred, including a reasonable return on the investment made. Until recently, this was the generally accepted model for industry regulation, in which the vertical integration of electric utilities was never questioned.

In the early 1990s, however, a radically different view of the electricity business began to take hold the world over. This approach challenged the vertically integrated structure of electric power suppliers. Densely interconnected transmission grids in most countries and even between countries now enable a generator located at any bus on the grid to compete with other operators to supply electricity virtually anywhere on the grid. It is therefore possible to separate strictly monopolistic grid activities from the generation and supply businesses, which can be conducted on a competitive market.

Under this new approach of the electricity business, electric power system operation and planning acquire a whole new dimension. Each generator decides individually when and how much to produce, how to manage the water in their dams, and how to plan and implement their plant maintenance programs. Decisions on investment in new power plants are not made centrally by anybody or company responsible for guaranteeing supply, but by private investors seeking a return on their investment, who are not responsible for overall security of supply. Nevertheless, after years of liberalized experience in several parts of the world, security of supply concerns have led the administration in some countries to apply a certain degree of intervention in generation expansion planning, through the launch of centralized auctions for energy supply with the goal of assuring enough power plant investments. The distribution business has not been significantly impacted by this new regulatory framework, except that it must be unbundled from supply, which is now competitive. Transmission, on the contrary, has been the object of a major overhaul, for its crucial importance in determining the competitive conditions under which wholesale market actors must operate.

Although technological and economic factors are ever present, the role played by regulation grows in importance with the geographic and political scope of electrical systems, particularly under competitive market conditions. In regional or supranational electric energy systems, for instance, rules have had to be established in the virtual absence of regulatory provisions for the operation of international markets. The European Union's Internal Electricity Market is paradigmatic in this regard: it covers 27 countries, 25 in the European Union including Norway and Switzerland. Other regional markets, in different stages of implementation, include the Australian national market, which encompasses several states in that country; Mercosur, servicing Argentina, Brazil, Paraguay, and Uruguay; the Central American Electricity Market; and the Regional Transmission Organizations in the United States that link several different but centrally managed electric utilities.

The motivation for establishing these regional markets is essentially economic: lower costs to maintain system safety and the advantage of mutually beneficial transactions among the different systems. Interconnecting whole electric energy systems poses interesting technological problems such as cooperation to maintain a common frequency across the entire system, abiding by trade arrangements stipulated between the various countries, support in emergency situations, global analysis and control of certain grid stability phenomena [6], or management of grid restrictions derived from international trade, that had been essentially solved or kept under control in the context of vertically integrated electric utilities through well-established rules for support in emergencies in a climate of cooperation, scant competition, and limited trade.

These technical problems have become more acute and their complexity has grown with the need to accommodate economic and regulatory considerations in the recent context of open competition. The proliferation of international transactions conducted in a completely decentralized manner by individual players—buyers and sellers entitled to access the regional grid as a whole—has complicated matters even further. In addition to these technological problems, other issues must also be addressed, such as harmonizing different national regulations, organizing and designing operating rules for regional markets, determining the transmission tolls to be applied in international transactions, pursuing economic efficiency in the allocation of limited grid capacity, and solving technical restrictions or proposing suitable regulatory mechanisms to ensure efficient transmission grid expansion. The integration of large amounts of renewable generation, most of them almost zero production cost based, is challenging the initial designs of electricity markets. The volatility of the energy market price will increase substantially; the flexibility of production will play a major role, energy price could be expected to be close to zero many hours. Market designs will need to be adapted to these new characteristics, ensuring market outcomes will still be attractive to new investments, especially those required to satisfy new electricity systems' needs.

1.1.5 ENVIRONMENTAL IMPACT AND CLIMATE CHANGE

In addition to ongoing technological development and the winds of change blowing in the global economy, a factor of increasing weight in the electricity industry, as in all other human activities, is the growing awareness of the importance of the natural environment. There is a widespread belief that one of the major challenges facing humanity today is the design of a model for sustainable development, defined to be development that meets the needs of the present without compromising the ability of future generations to meet their own needs. Besides such weighty issues as the enormous social and economic inequalities between peoples or the existence of a growth model that can hardly be extended to the entire world population, other questions, such as the intense use of the known energy resources and their adverse impact on the environment, problems that relate directly to electric energy systems, also come under the umbrella of sustainable development. For these reasons, environmental impact is a factor of increasing relevance and importance that conditions the present operation and development of these systems and will indisputably have an even more intense effect on the industry in the future.

A growing concern in modern society is the issue of climate change and the impact of greenhouse gas emissions. There is wide scientific agreement that the rising levels of greenhouse gases, that trap heat in the atmosphere, are conditioning a significant climate change by warming the atmosphere and ocean. The temperature of the earth has gone up about 0.85°C over the period 1880 to 2012, according to the Intergovernmental Panel on Climate Change Fifth Assessment Report. This has conditioned problems that manifest as extreme weather and climate events such as heat waves, droughts; intense tropical cyclone activity, heavy precipitation events and floods, rising sea levels; melting glaciers, and so on. Scientists project that this climate change will pose severe risks to human health, food production, agriculture, freshwater supplies, coastline cities, and natural resources in general. Generation is arguably the line of business in electric energy systems that produces the greatest environmental impact, in particular, with regard to fossil fuel plant emissions and the production of moderately and highly radioactive waste. As far as combustion is concerned, coal- and oil-fired steam plants vie with the transport industry for first place in the emission of both carbon dioxide (CO_2)—associated with climate change—and sulfur dioxide (SO_2)—the former related to the formation of tropospheric ozone and both responsible for acid rain. Carbon dioxide is an inevitable by-product of the combustion of organic material, NO_x comes from the nitrogen in the air and SO_2 from the sulfur in coal and oil. Other environmental effects of conventional steam power stations include the emission of particles and heavy metals, the generation of solid waste such as fly ash and slag, the heating of river, reservoir, or sea water to cover. To combat climate change, countries worldwide are favoring the development of renewable forms of generation, wind and solar being the highest growing forms. Carbon capture technologies are being researched as well as carbon pricing instruments such as carbon taxes and emission trading systems. The creation of a new market mechanism to trade carbon credits—which will replace the existing markets under the Kyoto Protocol have been discussed under the climate negotiations and the rules and procedures are expected to be agreed only by 2020.

In any event, it must be borne in mind that even generation facilities that use renewable energy and are considered to be the most environment-friendly technologies have an adverse impact. The most numerous, namely hydroelectric power plants, which have existed ever since electric power was first industrialized, change the surroundings radically. Some of its ill-effects are alteration of hydrology, disturbance of habitats, or even transformation of the microclimate, not to mention the risk of accidents that can spell vast ecological and human disaster. Other more recent technologies also have adverse consequences: wind, the disturbance of natural habitats and noise; solar, land occupancy, and the pollution inherent in the manufacture of the components required for the cells, and more specifically, the heavy metals present in their waste products; the use of biomass has the same drawbacks as conventional steam plants, although the effect is less intense, no SO_2 is emitted and, if properly managed, it is neutral with respect to CO_2 emissions. In fact, all electricity generation activities have one feature in common, namely the occupation of land and visual impact, but the area involved and the (not necessarily proportional) extent of social rejection vary considerably with technology and specific local conditions. Public opposition to power projects is currently a major barrier for their further deployment. In effect, power infrastructure development affects the population directly involved, with costs and benefits that are not necessarily distributed in a fair way.

In a similar vein, the huge overhead lines that carry electric power across plains, mountain ranges, valleys, and coasts and circle large cities, have at least a visual impact on the environment, which is being taken more and more seriously. Less visible but indubitably present are the electromagnetic fields that go hand in hand with the physics of electricity, although their potential effects on people, fauna, and flora, still under examination, seems to be rather harmless. Such considerations have important consequences, since environmental permits and rights of way constitute strong constraints on the expansion of the transmission grid. As a result, the grid is operating closer and closer to its maximum capacity, occasioning new technical problems, relating to its dynamic behavior, for instance, which logically have economic consequences. In some cases, alternative solutions are available, albeit at a higher cost, such as running underground lines in densely populated areas.

But the question is not solely one of establishing the magnitude of the environmental impact of the electricity industry or of the awareness that minimizing this impact generally entails increased system costs. The question, rather, is whether this impact should be considered when deciding how to best allocate society's scant resources. In a free market, the tool for resource allocation is product price; in this case, of the various power options. Nonetheless, the general opinion, among both the public at large and governmental authorities at the various levels, is that energy prices do not cover all the types of impact discussed above. This is what is known as a market failure or externality, defined to be the consequences of some productive or consumption processes for other economic agents that are not fully accounted for by production or consumption costs. The existence of such externalities, also called external costs, therefore leads to an undue allocation of resources in the economy, preventing the market from properly and efficiently allocating resources on the grounds of their price. Indeed, since account is not taken of these external costs, the price of energy is lower, and therefore consumption and environmental impact are higher, than they would be if total power costs were efficiently allocated. The existence of externalities, if not taken into consideration, also leads to the choice of more highly polluting power technologies than if allocation were optimum. To correct this market failure and reach optimum allocation, such costs must be internalized, building them into the price in a way that the economic agents can include them in their decision-making and ensure an optimum outcome for society as a whole. See the above discussion on mechanisms to deal with emissions issues.

1.2   THE TECHNOLOGICAL ENVIRONMENT

1.2.1   Electric Power System Structure [4,5,9,10]

Electric energy systems have developed along more or less the same lines in all countries, converging toward a very similar structure and configuration [5]. This is hardly surprising when account is taken of the very specific characteristics of the product they sell. As mentioned earlier, electricity generation, transmission, distribution, and supply are inevitably conditioned by the fact that generation and demand must be in instantaneous and permanent balance. The relevance of technical factors in maintaining such large-scale systems in dynamic equilibrium cannot be overlooked. A disturbance anywhere on the system may endanger the overall dynamic balance, with adverse consequences for the supply of electricity across vast areas, even whole regions of a country or an entire country [10]. It is perhaps for this reason that the existence of sophisticated real-time control, supervision, and monitoring systems, together with the protection facilities is what, from the technical standpoint, chiefly differentiates the configuration and structure of electric energy systems from other industrial activities. The functions typical of any industry, such as production, shipping, planning, and organization, are also highly specialized in the electricity industry.

The organization of the electricity industry, like any other, is divided into production centers generating plant; transmission (equivalent to transport or shipping in other industries); the high-voltage grid; distribution; the low-voltage grid or network; and consumption (also termed supply in some contexts), in addition to the associated protection and control systems. More formally, system configuration and structure are as depicted in Figure 1.2. Production centers generate electricity at voltages of several kilovolts, typically from 6 to 20 kV, and immediately transform this power to voltages of hundreds of kilovolts: 132, 220, 400, 500, and 700 kV are relatively common values to optimize long-distance transmission over electric lines to the areas where consumption is most intense. Raising the voltage makes it possible to transmit large amounts of electric power, the entire output from a nuclear powered generator for instance, over long distances using reasonably inexpensive cable technology with minimum line losses. The transmission grid interconnects all the major production and consumption centers, generally forming a very dense web to guarantee high reliability, with alternative pathways for the supply of electric power in the event of failure in a few of the lines. These

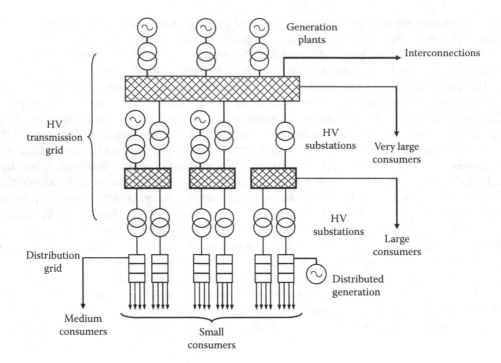

FIGURE 1.2 Electric power system configuration and structure.

electric power transmission highways are interconnected at communication nodes known as electric substations; more local (covering regions of a country) grids are spun out from the stations at a somewhat lower voltage, 132, 66, or 45 kV in Spain, for instance, and in turn feed local distribution networks, which bring electric power to consumers at less hazardous voltages, adapted to consumer needs: 20,000, 15,000, 6600, 380, or 220 V. Successive substations step the working voltage down in several phases and centralize the measuring and protection devices for the entire transmission grid. The configuration of these distribution grids is usually radial, with tentacles stretching out to even the most remote consumption points. As the lines are split up at each step, the grids carry less and less power and consequently can operate at lower voltages. Consumers connect to the voltage level best suited to their power needs, in accordance with the basic principle that the lower the voltage, the smaller the power capacity. This means that highly energy-intensive businesses, iron and steel plants and mills, aluminum plants, railways, and the like, connect directly to the high-voltage grid; other major consumers, for example, large factories receive power at a somewhat lower voltage and small consumers like households, retailers, small factories are connected to the low-voltage network. On the basis of a more or less reciprocal principle, generating stations with a very small output feed their electric power directly into the distribution network instead of connecting to the high-voltage grid. Such generators, which usually run small hydroelectric, photovoltaic, wind, combined heat and power, or other types of modular power stations engaging in distributed generation, are sometimes grouped under a single category for regulatory purposes.

This traditional worldwide-extended structure of power systems is being challenged by the steeply increase of distributed generation. The support of the use of renewable energy and CHP for environmental reasons, the encouragement of self-consumption practices (not always properly founded), the savings in network investments and in network related energy losses, the increasing concern on resilience issues, together with technological advances leading to lower wind and solar generation, microturbine or fuel cell costs, is prompting spectacular growth in decentralized generation. This together with the large progresses on battery storage systems, the slow but sure penetration of electric vehicles, the cheap possibilities of performing demand response (DR) actions by controlling the consumption of electrical devices, is stirring up the role of consumers, which may become active players delivering generation and services to the system. Consumers may become also producers, being "prosumer" a widely use name to describe this transformation. The growth of distributed generation may be more intense in developing countries, where traditional electric infrastructure is still highly insufficient.

The points below focus on these chief components of electric energy systems: consumption, production, transmission, distribution, and protection and control.

1.2.2 Consumption and Demand Management

1.2.2.1 Demand Growth

Electricity demand has undergone high, sustained growth since the beginning. The creation of standards for the electricity "product," voltage, frequency, current, paved the way for the enormous boom in electricity consumption. This, in turn, laid the ground for the standardization of electrically powered fixtures and facilities—from light bulbs and motors to PCs—dramatically lowering manufacturing costs and enhancing product versatility, making it possible to use a given electrically powered item virtually anywhere. Electric power consumption is one of the clearest indicators of a country's industrial development and closely parallels GDP growth. As noted earlier, there are scarcely any production processes or sectors involved in creating wealth that do not require electricity. But electric power consumption has also been used as a measure of social development. Electricity consumption per capita and especially the degree of electrification in a country, that is, the percentage of the population living in electrified homes, provide a clear indication of the standard of living [17]. This is not surprising, since such basics as lighting, supply of potable water, refrigerators, and other household appliances depend on access to electricity. The curves in Figures 1.3 through 1.5 relate the growth in electricity consumption to other basic indicators, such as GDP, population, or energy consumption. The growth rate is obviously higher in countries with low baseline levels of electric power consumption and high economic growth.

Electrification rates and electricity consumption per capita vary widely from one area of the world to another, as Table 1.1 eloquently illustrates [2]. One-third of the Earth's 6 million inhabitants have no electricity.

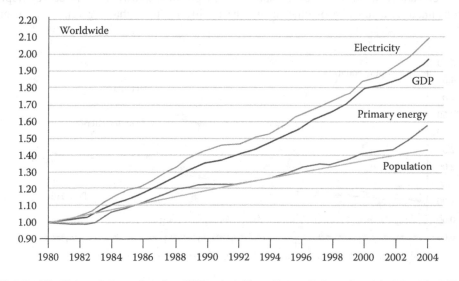

FIGURE 1.3 World growth rate referred to 1980 value. (From Energy Information Administration, U.S. Government; U.S. Department of Agriculture. [18])

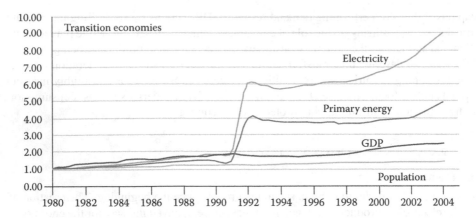

FIGURE 1.4 Transition economies' growth rate referred to 1980 value. (From Energy Information Administration, U.S. Government; U.S. Department of Agriculture. [18])

But the growth in electricity consumption is not limited to developing countries: it has definitely steadied, but is certainly not flat in developed countries. While the industrial world's consumer mentality may partly be driving such growth, it is nonetheless true that new uses are continuously found for electric power. The generalized use of air conditioning in these countries is an obvious example and one that has brought a radical change in seasonal consumption curves, as explained below.

1.2.2.2 Demand Profiles

Consumption is characterized by a variety of items from the technical standpoint. The two most important items are power and energy. Power, measured in watts (W), is the energy (Wh) required per unit of time. Power, therefore, is the instantaneous energy consumed. Since electric power is not stored, electric facilities must be designed to withstand the maximum instantaneous energy consumed, in other words, to withstand the maximum power load in the system throughout the consumption cycle. Therefore, not only the total electric capacity needed, but the demand profile over time is especially relevant to characterize consumption. Such profiles, known as load curves, represent power consumed as a function of time. It may be readily deduced that a given value of energy

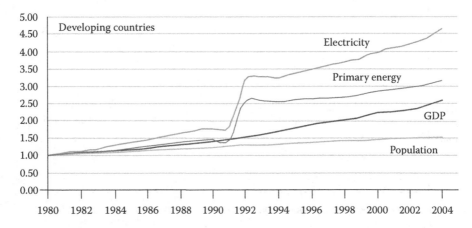

FIGURE 1.5 Developing countries' growth rate referred to 1980 value. (From Energy Information Administration, U.S. Government; U.S. Department of Agriculture. [18])

TABLE 1.1

Electricity Consumption Per Capita (kWh), 1980–2004

Region	1980	1983	1986	1989	1992	1995	1998	2001	2004
North America	9792	9863	10,677	12,035	12,115	12,716	13,258	13,392	13,715
Latin America	796	860	1005	1048	1090	1209	1331	1356	1495
Europe	4111	4240	4711	5005	4983	5192	5520	5854	6138
Former Soviet Union	4584	4881	5045	5351	4848	3989	3777	4003	4266
Asia & Oceania	476	514	587	697	793	926	1009	1112	1329
Middle East	599	807	929	1064	1137	1364	1563	1744	2027
Africa	375	392	443	457	450	467	477	496	542
World	1663	1698	1827	1980	1986	2062	2150	2242	2421

Source: Energy Information Administration, U.S. Government [16].

consumed may have a number of related load profiles. Some may be flat, indicating very constant electricity consumption over time, while others may have one or several very steep valleys or peaks, denoting very variable demand. An aluminum plant working around the clock 365 days a year and a factory operating at full capacity only during the daytime on weekdays would exemplify these two types of profiles. Load profiles commonly generate repetitive patterns over time. Thus, for instance, weekday demand is normally very uniform, as is the weekly load during a given season. Therefore, depending on the timescale considered, the load profile to be used may be daily, weekly, monthly, seasonal, yearly, or even multiyearly. Load profiles also have economic relevance, as will be seen in the discussion below: for any given demand level, it is less expensive to cover a flat than a spiked load profile. For this reason, load curves constitute one of the most relevant parameters considered in the methods used to set tariffs.

Summing all the individual consumption curves for an electric power system yields the total daily, weekly, monthly, seasonal, yearly, and multiyearly load curves, each with a characteristic and highly significant power profile. Figures 1.6 through 1.8 show load curves for a South American electric energy system, specifically central Chile (www.cdec-sic.cl), with indication of energy supplied by run of river hydro (pass), reservoir hydro (dam), and thermal generation. There are very clear peaks and valleys in each, denoting cyclical, maximum, and minimum demands. Demand forecasting is an essential problem to solve in foreseeing the conditions under which the system will be operating in the short, medium, and long term. Normal procedure is to base the prediction on

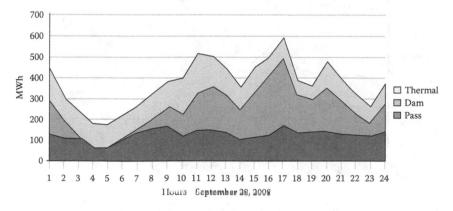

FIGURE 1.6 Hourly power load for a South American system.

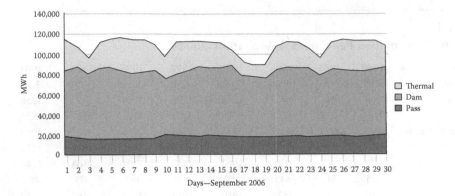

FIGURE 1.7   Monthly power load for a South American system.

historical data adjusted to take account of factors affecting the expected load. The most important of these factors include temperature, since many electrical devices are used for space heating or cooling; number of working days, to account for the difference in consumption on business days and holidays; and economic growth, in view of the above-mentioned close relationship between economic activity and electricity consumption. Therefore, consumption at any given time can be reasonably well predicted from time series data corrected for foreseeable variations in growth, working days, and temperature, and taking account as well of special events that may have a substantial effect on demand.

Aggregate electricity consumption can also be represented as a monotonic load curve, which is particularly useful in certain applications and studies. Such curves represent the length of time that demand exceeds a given load. The thick line in Figure 1.9 is the approximate monotonic load curve for a Canadian utility, which generates with hydro energy (www.gcpud.org): the abscissa values represent time in hours and the ordinate values demand in megawatts. Therefore, each point on the curve indicates the total hours during the year that demand exceeded a given value. In the example, the load was in excess of 15,000 MW for a total of 1200 h in 2005. The load monotone can be plotted directly from the chronological load curve by ranking demand in descending order. The integral of the load monotone represents the energy consumed in the time frame considered. It will be noted, however,

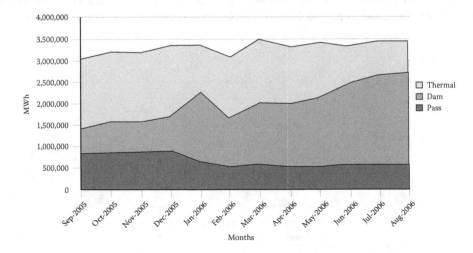

FIGURE 1.8   Yearly power load for a South American system.

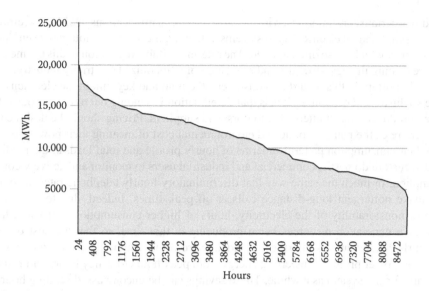

FIGURE 1.9 Monotonic load curve for a Canadian utility. (From Grant County PUD, http://www.gcpud.org/energy.htm.)

that whereas a given load curve can have only one load monotone, the opposite is not true. Although the chronological information contained in load curves is lost in monotonic curves, the latter are widely used for their simplicity. Probabilistic monotone curves are commonly used in prospective studies, which are based on demand forecasts subject to some degree of uncertainty; in this case the x-axis values represent the likelihood that demand will exceed a given value. As in the case of chronological demand profiles, monotonic load curves can be plotted for daily, weekly, monthly, seasonal, yearly, or multiyearly consumption.

In addition to the power and energy properties discussed at length in the foregoing paragraphs, consumption is characterized by other technical factors. Account must be taken, for instance, of the fact that while real power and energy are consumed in the system, reactive power is also either generated or consumed, usually the latter, since inductive motors, which consume reactive power, generally predominate. This gives rise to a power factor less than unity, which penalizes consumption as far as the tariff charged is concerned, because it entails the circulation of unproductive current and ohmic dissipation, and line capacity saturation with it. Moreover, consumption may depend on supply conditions (voltage, frequency) be static or dynamic, or vary with connection time due to heating or other effects. All of this must be taken into account in load modeling.

1.2.2.3 Demand-Side Management

Sustained electricity consumption growth exists despite substantial improvement in the efficiency of most equipment and processes using electric power, which reduces the input power in kWh needed to attain a given result. More and more voices are being raised in defense of the need to rationalize the consumption of electricity and all other forms of energy. Aware of the environmental impact of such consumption and the vast amount of natural resources that are literally going up in smoke, such voices rightly call for intergenerational solidarity to be able to bequeath to coming generations an ecologically acceptable planet whose energy resources have not been depleted. Hence the importance of demand-side management (DSM), a term coined in the United States to mean all the techniques and actions geared to rationalizing the consumption of electric power. The aim, on one hand, is more efficient use of existing consumption to reduce the enormous investment involved in the construction of new stations and the substantial cost of producing electricity and, on the other hand, just energy savings by cutting down certain consumptions, with the same beneficial implications. DSM is therefore becoming an active component of electric energy systems, reflecting the attempt to internalize environmental costs, for instance, which are so often ignored. The role and suitable regulation of this business is one of the challenges facing the new structure and regulation of a liberalized electricity industry.

It may be important in this regard for consumers, the final and key link in the electricity chain, to receive the sophisticated economic signals that deregulation is sending out to the various other players involved producers, transmitters, distributors, and suppliers. Pricing should be designed to make consumers aware of the real economic and environmental cost of meeting their power needs, taking account of their consumption patterns in terms of hourly profile and total load. In the medium term, this should accustom domestic, commercial and industrial users to monitor and actively control electric consumption, in much the same way that discriminatory hourly telephone rates encourage customers to make nonurgent long-distance calls at off-peak times. Indeed, due to the, in practical terms, almost nonstorability of the electricity, hours of higher consumption will entail larger and more expensive generation resources being producing at that time, so that the cost of delivering energy is hourly dependent, linked to the demand time profile. Consuming the same amount of energy with a different power profile, transferring energy from peak hours to valley hours will result in economic savings for the system as a whole. These savings can be encompassed sending hourly differentiating prices to consumers, incentivizing them to transfer some of their consumption from peak hours to valley hours. This is called peak shaving. A roll-out program to substitute millions of old consumptions meters, unable to discriminate the time profile of consumers demand, by smarter meters, able to monitor the hourly behavior of the demand is in place in many places and is critical to foster a proper DR to prices. Similarly, customers will voluntarily not only transfer energy from peak hours to valley hours but also reduce electricity consumption by foregoing the most superfluous applications at times when higher prices signal that expensive resources are being deployed or that the margin between the demand and supply of electric power is narrow. The capacity of demand to respond to pricing is generally characterized by a parameter termed price elasticity of demand. This is defined to be the percentage variation in consumption of electricity or any other product in response to a unit variation in the price. Electricity demand is characterized, generally speaking, by scant short-term elasticity; in other words, the reaction to changes in price are small, although this assertion is more accurate for some types of consumer than others. Such limited elasticity is arguably due to the mentality prevailing until very recently in the electricity industry: continuity of supply was regarded to be a nearly sacred duty, to be fulfilled at any price. Consumers who were identified, indeed, as subscribers rather than customers were merely passive recipients of the service provided. Moreover, as discussed previously, the deployment of distributed generation and storable devices, the advances in communications technology, the progress on load controllability, is changing consumers role radically (towards a prosumer role), being now also a potential provider of services (reserves, flexibility, etc.) to the system. In order to manage such a dispersed an enormous volume of small prosumers, aggregators (a role normally assumed by retailers) is emerging as a figure devoted to group and manage the provision of all these small individual services up to a volume large enough to be valuable to the system. DSM will become as relevant as other areas, such as generation.

1.2.2.4 Service Quality

Electric power consumption may be very sensitive to the technical properties of the supply of electricity. Many devices malfunction or simply do not operate at all, unless the voltage wave is perfectly sinusoidal and its frequency and magnitude are constant and stable over time. The precision, quality, features, and performance of electrical devices depend on the quality of the current that powers them. Problems may also arise in almost any type of electrical device when the supply voltage is too low or too high (overvoltage). Computer, motor, and household appliance performance may suffer or these devices may even fail altogether when the supply voltage swings up or down. Most electrically powered equipment, especially a particularly expensive equipment or any equipment regarded to be vital for the proper and safe operation of all kinds of processes, is fitted with protection systems—fuses, circuit breakers and switches, protection relays—to prevent damage caused by voltage fluctuations outside an acceptable range. Thus, for instance, the motors that drive the cooling pumps in nuclear power plants are fitted with under and overvoltage protection that may even trip systems that cause plant shutdown, given the vital role of these motors in safe plant operation. Finally, outages, whether short or long, are clearly detrimental to service quality. For example, unstored information representing hours of work on a PC may be lost because of an untimely power outage. But power failures can cause even greater harm in industries such as foundries or in chemical or mechanical processes whose interruption may entail huge losses.

In developed countries, where the universal supply of electricity is guaranteed, attention increasingly focuses on quality, as in any other commercial product. Consumption and consumers have become more demanding in this regard and electricity industry regulation authorities assiduously include quality standards in laws and regulations. Designing the proper signals to suitably combine efficiency with high-quality service is one of the major challenges facing the new regulatory system.

The factors that basically characterize quality of electricity service are set out briefly below:

- *Supply outages*: Supply interruptions may have serious consequences for consumers. The duration of such interruptions may be very short—in which case they are called micro-outages, often caused by the reconnection of switches after a short circuit—or long. Normally, the harm caused increases nonlinearly with the duration of the outage.
- *Voltage drops*: Momentary dip in supply voltage caused by system short circuits or failures, lasting only until the fault is cleared, or due to the start-up of nearby motors with high input demand when first switched on, occasioning voltage drops in the supply network. Some devices are particularly sensitive to these drops, particularly motors whose electromagnetic torque varies with the square of the supply voltage.
- *Voltage wave harmonics*: Deviations from the fundamental frequency of the voltage sine wave due to the saturation of ferromagnetic materials, in system transformers or generators, for instance, or to the loads themselves; these deviations may also have adverse effects on consumer appliances.
- *Flicker*: Low-frequency fluctuations in voltage amplitude normally due to certain types of loads. Arc furnaces and electronic devices with thyristors usually cause flicker, which is detrimental to the proper operation of devices connected to the network. The solution to this problem is complex, since it depends not on the supplier but on system loads.
- *Overvoltage*: Voltage increases caused by short circuits, faults, lightning, or any other event, potentially causing severe damage to consumer appliances.

Finally, it should be added that electric power consumption may vary broadly with temperature or contingencies. What must be borne in mind in this regard is, as mentioned earlier, that this demand must be met instantaneously and therefore the electric power supply system, power stations, transmission, distribution, must be designed to be able to detect and respond immediately to such variations. The system must be fitted with sophisticated measurement, control and supervisory equipment, and must have reserve generating capacity ready to go into production at all times. But most users flipping switches in their homes or workplaces to turn on the lights or start-up an appliance or tool are blithely unaware of the host of systems, services, and processes needed to provide that service.

1.2.3 GENERATION: CONVENTIONAL AND RENEWABLES

1.2.3.1 Different Generation Technologies

The electricity required to meet these consumption needs is generated in production centers commonly called power plants or stations, where a source of primary energy is converted into electric power with clearly defined characteristics. Specifically, these facilities generate a three-phase, sinusoidal voltage system, with a strictly standardized and controlled wave frequency and amplitude.

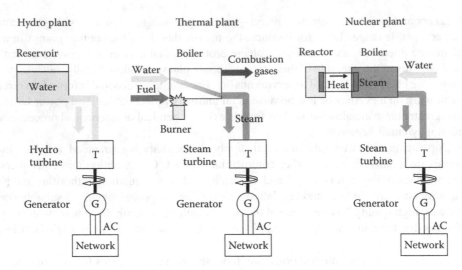

FIGURE 1.10 Hydroelectric, thermal, and nuclear plants.

There are many generation technologies, usually associated with the fuel used. Conventional power stations are divided into hydroelectric, thermal, and nuclear plants, as shown in Figure 1.10.

The primary source of energy used in hydroelectric stations is water, which is expressed, energetically speaking, in terms of flow rate and height or "head." Hydroelectric energy is converted by a hydraulic turbine into mechanical energy, characterized by the torque and speed of the shaft coupled to the electric generator. In other words, hydraulic energy is converted into electrical energy in the generator, producing voltage and current in the machine terminals. Because of the source of primary energy used, hydroelectric stations produce less atmospheric pollution than other conventional generation technologies. Another advantage to this type of stations, in addition to the cost of the fuel and lack of pollution, is their connection and disconnection flexibility, making them highly suitable regulating stations to adjust production to demand needs. Nonetheless, they are costly to build, and ensuring a steady supply of water normally involves flooding vast areas. And finally, their operation is contingent upon a highly random factor—rainfall in the area where they are sited.

The various types of hydroelectric stations can be grouped under three main categories, which are distinguished in system operation:

1. Conventional hydroelectric stations such as depicted in Figure 1.10 are the most common type; their characteristics are as described in the preceding paragraph.
2. Run-of-the-river plants have no storage capacity and consequently cannot be deployed to use water resources for opportunistic generation; for this reason, they are not used as regulating stations.
3. Pumping power stations have a raised reservoir to which they can pump water when electric power is cheaper, and then dump it on to a turbine when it is more cost-effective to do so. They can be regarded to be an efficient means of storing energy, but not electricity per se.

Steam or thermal power stations, such as depicted in Figure 1.10, in which the primary energy is provided by a fossil fuel (coal, fuel-oil, or gas), are, respectively, termed coal-fired, oil-fired, or gas-fired stations. The operating principle behind these stations is basically as follows:

1. The fuel is burned in a boiler to produce high-pressure steam.
2. High-pressure steam is converted in the steam turbine into mechanical energy.
3. Mechanical energy, as in hydroelectric plants, is converted into electric power by the generator.

The thermal efficiency of steam power stations, which convert thermal to mechanical to electrical energy, depends primarily on the calorific value of the fuel used. In any event, the highest efficiency reached is never over 45%. Owing to the heat inertia of the boiler, approximately 7 h, these stations cannot be readily connected and disconnected, that is, they are less flexible in this respect than hydro-electric plants. In light of this, start or stop studies are conducted on steam power plants to establish operating orders and they are sometimes placed in standby operation, without generating any power whatsoever. Although fuel may be subject to variations in price, in most countries a constant supply is regarded to be routinely available. Therefore, such stations can be used for regulation, subject only to their connection inertia.

There are two types of steam plant technologies that use gas as a fuel, as shown in Figure 1.11. On the one hand, there are gas turbine plants where, like in jet engines, gas combustion in high-pressure air feeds a turbine that produces mechanical energy, in turn absorbed by an AC generator. And on the other hand, there are combined cycle or combined cycle gas turbine (CCGT) plants, which, as today's technology of choice, merit further comment. The operation of these stations, as may be inferred from their name, involves two types of cycles. In the primary cycle a compressor attached to the shaft of a gas turbine absorbs air at atmospheric pressure, compresses it, and guides it to a combustion chamber where the gas that triggers combustion is likewise injected. The resulting gas expands in the turbine blades to produce mechanical energy. The gas expelled from the turbine, which is still at a high temperature, is used to heat a water vapor circuit where the latent heat in the gas is converted into mechanical energy in a steam turbine. Finally, electricity is generated by one or two AC generators connected to a single common shaft or two separate shafts, one for each cycle. Thanks to the latest advances in ceramics, the materials used to protect the blades from high temperatures, performance in these cycles is substantially higher than in open gas or conventional steam turbine cycles, with thermal efficiency values of up to 60% in some facilities. This, together with a considerable reduction in polluting emissions, a high degree of modularity and reasonable investment costs, make CCGT one of the most competitive generation technologies available.

Nuclear power plants (see Figure 1.10), also known as atomic power plants, consist essentially of a nuclear reactor that produces vast amounts of heat with the atomic fission of the uranium. This heat is

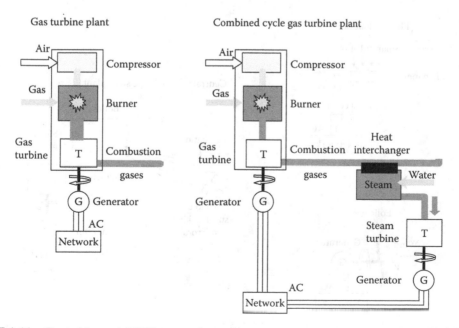

FIGURE 1.11   Gas turbine and CCGT power plants.

transferred to a fluid, carbon dioxide, liquid sodium, or water, and carried to a heat exchanger where it is transferred to a water circuit. Like in steam stations, the rest of the process involves transforming the steam produced into mechanical energy in a steam turbine and then into electric power with an AC generator. There are two drawbacks to the use of nuclear power plants which are difficult to solve, and which have made them socially unacceptable in some countries: the magnitude of the catastrophe in the event of an accident, no matter how low the risk, and the problem of eliminating radioactive waste. In light of these difficulties, some countries have imposed a moratorium on the construction of nuclear power plants. From the standpoint of system operation, nuclear power stations are always base plants, rarely used for regulation because of the inherent hazards in changing the cooling conditions in the nuclear reactor.

In electric power grids, most production presently takes place in the so-called conventional stations, described in the foregoing discussion. There are, however, other types of power stations that are gradually acquiring significance in some areas and countries. These are often called alternative plants, characterized by their limited environmental impact and the use of renewable sources of energy: wind, solar, biomass, and CHP (combined heat and power or "cogeneration") plants, depicted in Figures 1.12 and 1.13.

Of all these technologies, the one that have undergone most spectacular growth in recent years are wind and solar energy: in fact, CCGT, solar and wind technologies account for very nearly all the new medium- or short-term generation plant. Wind farms may be fitted with synchronous AC generators, such as the ones used in other types of power stations, or asynchronous facilities, which accommodate small variations in speed when the torque fluctuates, to reduce equipment wear due to variations in wind speed. In asynchronous stations, capacitors are needed to generate the reactive power consumed by the induction machinery. These stations may be connected to the grid directly or indirectly, through a rectifier, an inverter, and a filter. While the generation of DC makes it possible to work at variable speeds, this comes at a cost, in addition to line loss and reliability issues, although the reactive power generated can be controlled by power electronics.

The source of solar energy is abundant and may represent in the future, for sure, one of the main sources of energy for greenhouse emissions to be controlled. After a large period of very small growth due to a scantly developed and very expensive technology, promising improvements have been

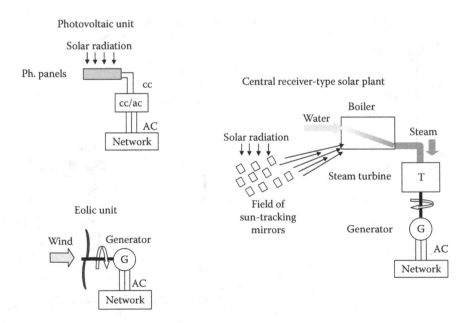

FIGURE 1.12   Solar and wind (eolic) power stations.

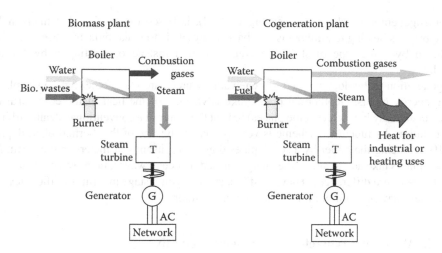

FIGURE 1.13 Biomass and cogeneration power plants.

achieved leading to more affordable solar technology. Still "major scale-up in the decades ahead will depend on the solar industry's ability to overcome several major hurdles with respect to cost, the availability of technology and materials to support very large-scale expansion, and successful integration at large scale into existing electric systems," as stated in [15].

Up to now the installation of photovoltaic cells, which convert solar energy directly to DC current for storage in batteries, has been largely deployed in some countries such as Germany, mainly as self-generation facilities, typically deployed in building and house roofs, usually fostered by direct incentives program or by a friendly regulation. Other technologies, solar steam power stations, use solar radiation to heat a fluid and generate electricity thermodynamically. Few of them have been commissioned for commercial operation worldwide. Different functional schemes can be found:

- Parabolic trough stations use parabolic collectors to focus radiation on pipes and heat the oil they carry. This oil then releases heat in a steam turbine cycle in stations using solar energy alone, or in a variety of cycles in hybrid plants.
- Central receiver or solar power tower stations, such as shown in Figure 1.12, have a field of heliostats, sun-tracking mirrors, that focus radiation on any of a number of types of receivers, normally located in a tower where heat is accumulated for subsequent use in any kind of power cycle.
- Solar dish generation is similar to the central receiver design but on a smaller scale, in which each module has its own "dishes" or parabolic disks and its own receiver. They generally use Stirling or gas turbine cycles; the principal advantage is their modularity.

Biomass generation (see Figure 1.13), which means obtaining energy from biological resources—energy crops (also called biomass feedstocks), livestock waste, or forestry residue, and so on—uses a resource available in nearly any habitat and perhaps for that reason is gaining popularity in developing countries such as India. The two basic approaches taken in this technology are:

1. Direct combustion in specific furnaces to produce steam subsequently used in a turbine cycle, like in conventional steam power stations.
2. Gasification of the organic matter to obtain a combustible gas, usually with a high methane content, generally used to feed an internal combustion engine or gas turbine coupled to an electric generator. Matter can be gasified with physical–chemical or anaerobic biological processes.

CHP or cogeneration technology (see Figure 1.13), is based on the fact that many industrial plants have process heating requirements: the basic principle is to make industrial use of the surplus heat produced by some type of steam generation system instead of wasting it by cooling the return fluid.

Other generation technologies based on renewable sources are the geothermal power plants and the marine energy plants. Geothermal plants take advantage of the heat of the core earth (which may be especially easy in reach in some places) to heat the water of a conventional steam turbine thermal power station. Its functional scheme is very similar to the one of the thermal plant depicted in Figure 1.10, replacing the boiler by a set of pipes digging thousands of meters in the earth. Marine based power plants take advantage of the energy carried by ocean waves, tides, sea currents, salinity, and ocean temperature differences. They are still at an embryonic stage in terms of efficiency and relevance but they may play a more important role in the future.

1.2.3.2 The Why's and Wherefore's of a Generation Mix

The existence of such a wide variety of technologies in most countries can be justified in a number of ways. First, there is a purely economic justification deriving from the load curve. The range of fixed investment costs to build a station and the operating costs to generate electricity vary widely from one technology to another. Nuclear power plants, for instance, call for very high investment, but have comparatively low operating costs due to the price of the fuel, in this case uranium, and the efficiency of the energy conversion process, making nuclear power an attractive technology from the standpoint discussed here for the arm of the demand curve that covers 8760 h in the year. The other extreme is gas turbine technology, which has the highest operating but the lowest investment costs, making it a very attractive type of generation to cover demand peaks, that is, a relatively small number of hours per year. Conventional steam stations fall in between these two extremes. Obviously, the assumptions on which economic analyses are based and which justify the coexistence of different technologies always involve some degree of uncertainty in connection with the shape of the future demand curve, fuel costs, specific operation of each generating station, capital costs, regulatory decisions, market prices (as appropriate), and so on.

Not only economic arguments, but also political and environmental strategy weigh heavily in the reasons for deploying a technology mix in electricity generation. Ensuring a supply of fuel as independent as possible of political and economic crises, be they international, such as the oil price crisis, or domestic, such as a miners' strike, entails the implementation of a diversification strategy. Moreover, the internalization of environmental costs and medium- and long-term environmental sustainability go hand in hand with regulatory measures to encourage the use of production technologies with lesser environmental impact.

Today, most electricity generation takes place in large production centers scattered across a country, often at long distances from the major consumption centers. It seems natural to build stations close to the source of fuel—mines and ports for coal, refineries for fuel-oil, regasifiers and pipelines for gas-fired stations, rivers with a heavy flow or head for hydroelectric stations—as well as the coast or rivers, since water is a vital coolant in large steam plants. Attempts are usually made to site large power stations at a substantial distance from densely populated areas due to issues such as pollution and the adverse social reaction to nuclear power plants. The transmission grid is responsible for carrying the electricity generated to the consumption centers. The enormous size of modern electric power plants is a result of the lower unit costs obtained by increasing their size to the present dimensions. This effect, known as economies of scale, has weighed, for instance, in the decision to build nuclear plants with an installed capacity of up to 1000 MW, or 500 MW or even larger coal- or oil-fired steam stations, since they are more competitive than smaller plants using the same technologies. The appearance of CCGT technology has reversed this trend since, as such plants are much more modular, they can be much smaller and still be competitive. The next few decades may see a dramatic rise in distributed generation, with power stations located much closer to consumers, supported by regulatory measures encouraging diversification, energy savings, such as in CHP, and reduction of environmental impact. The generation of electric power in large plants is characterized, economically, by very heavy investments, amortized in the very long term (25 or 30 years), after several years of construction (5, 10, or even more in the case of nuclear plants or large-scale hydroelectric stations). The high financial risk that this entails can be assumed by state-owned entities or private initiative if there is a sufficient governmental guarantee to ensure the recovery of investment and operating costs through regulated tariffs. The appearance of CCGT technology has changed the economic context substantially by significantly reducing risk: these stations are more flexible, modular and competitive, smaller, and therefore quicker to build. These issues have greatly facilitated private investment, in the wake of the recent regulatory changes to introduce free competition in the electricity industry.

The deployment of renewable source-based generation may condition also the generation mix discussed in this section. The production of wind and solar generation units will depend on the actual wind blowing and solar irradiation happening in that instant and in that geographical location. They are often described as intermittent generation sources, since their production cannot be managed. Fuels of conventional generation technologies can be purchased and stored, enabling its use for production whenever required, whereas the production of such renewable kind of generation is determined by external, noncontrollable, factors. The rest of the generation mix should adapt its production to accommodate wind and solar generation production. This enhances the need of flexible generation technologies able to respond in short term to changes, sometimes very steeply, in wind and solar generation based production. Therefore, the penetration of very large amounts of such generation technologies will condition the generation technology mix required to encompass efficiency and technical restrictions. Storage facilities may play a critical role in the future generation mix since they will break the nonstorability paradigm of the electricity and will be able to provide the increasing flexibility needs. But still investment cost of such kind of facilities must decline substantially for a large deployment of them being an efficient solution.

1.2.4 Transmission

The transmission grid connects large and geographically scattered production centers to demand hubs, generally located near cities and industrial areas, maintaining the electric power system fully interconnected and in synchronic operation. The long-distance transmission of huge amounts of power necessitates operating at high voltages to reduce circulating current intensity and, therefore, line loss. The transmission grid is the backbone of the electric power system, interconnecting all its neuralgic centers. Its key role in the dynamic equilibrium between production and consumption determines its typically web-like structure, in which every station on the grid is backed up by all the others to avert the consequences of possible failures. Ideally, the system should operate as though all generation and all demand were connected to a single bus. It is fitted with sophisticated measurement, protection and control equipment, so overall system operation is not compromised by faults, that is, short circuits, lightning, dispatch errors, or equipment failure. The transmission grid has acquired particular relevance in the new regulatory context that encourages competition, since it is the wholesale market facilitator, the meeting point for market players, as discussed below. The growth of transmission grid capacity, together with the development of connectivity between transmission grids, both within and across national boundaries, has paved the way for regional or international scale electricity markets. Although AC technology is the one in place in all existing power system since almost the origin of power systems due to the role of AC transformers, advances in power electronics are paving the way towards the use of high-voltage DC (HVDC) grids. HVDC technology fits better the transferring of large amounts of energy through large distances, appropriate to integrate for instance very large renewable projects in extended regional electricity systems [21]. HVDC technology is being used today for underground and submarine cables to avoid reactive power related problems linked to the use of AC in those cases.

1.2.4.1 Power Lines

Transmission grid lines consist of aluminum cables with a steel core that rest on towers. Line design is based on both mechanical and electrical considerations. The towers must be sturdy enough to bear the weight of the cables and withstand the voltage in the cables while maintaining the minimum safety distance between cables, between the cables and the towers, and between the cables and the ground. A very visible assembly of insulators attaches the cables to the towers. Since each insulator can accommodate voltage of 12–18 kV, 400 kV lines need on the order of 20–25 such links in the insulation chain. Sometimes two lines run along a parallel route, sharing the same towers: this is known as a double circuit, an example of which is illustrated in Figure 1.14.

Electrically, the section of the cables determines the maximum current intensity they can transmit and therefore it determines the transmission capacity of the line. The greater the intensity, the greater the line losses due to the Joule effect, higher conductor temperature, and greater cable expansion and lengthening with the concomitant shorter distance to the ground and greater risk of discharge. To reduce so-called corona discharge (rupture of the insulation capacity of the air around the cables due to high electrical fields, occasioning line losses and electromagnetic disturbance that may cause interference in communications systems) each phase of the line is generally divided into two, three, or more cables, giving rise to duplex or triplex cables. One of the most important line parameters, inductance, depends largely on the relative geometric position of the three phases on the tower. Moreover, the lines provoke a capacitive effect with the earth that fixes the value of their capacitance to ground. Consequently, the inductive effect predominates in lines that carry power close to their capacity limit, which consume reactive energy, whereas the capacitive effect prevails and the lines generate reactive energy when they carry small amounts of power, typically at night. Some transmission lines run underground, mostly in urban grids where the operating voltage is lower and only very rarely in the case of very high-voltage circuits. High-voltage underground systems involve the deployment of rather expensive technology, since the very short distance between the line and the ground necessitates the installation of very heavy duty insulators. These lines have a much more pronounced capacitive effect than overhead lines.

In a meshed system such as the transmission grid, energy flows are distributed across the lines depending on their impedance, in accordance with Kirchhoff's laws. The long distances and large scale of the power transmitted may reduce the grid's ability to maintain system operation, favoring the appearance of instability detrimental to the dynamic equilibrium between generation and demand. This may reduce line transmission capacity to less than its natural thermal limit.

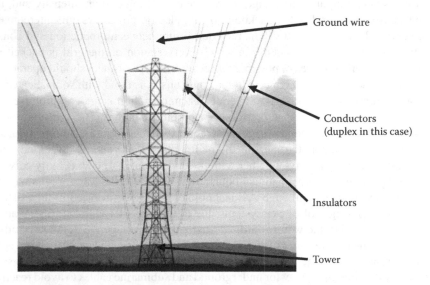

FIGURE 1.14 Double-circuit line (400 kV).

As noted above, for reasons of environmental impact, it is increasingly difficult to expand and reinforce the transmission system, translating into a growing need to make optimum use of existing facilities. This represents an important challenge, since it entails narrowing safety margins and perfecting protection, measurement, and control logic. With the development of power electronics, new devices have become available; they attempt to increase actual line capacity and steer current flow towards the lines with the smallest loads. Such devices are known as flexible alternating current transmission systems (FACTS).

1.2.4.2 Substations

Substations constitute the second fundamental component of the transmission grid. They have three chief functions: they are the interconnection buses for lines, the transformation nodes that feed the distribution networks that reach consumers, and the centers where system measurement, protection, interruption, and dispatch equipment are sited. Typically, several high-voltage lines feed into the substation, which steps the voltage down and sends the resulting current over the outgoing (lower voltage) transmission or distribution lines. Materially, the substation is structured around thick bars to which the various lines connect. Circuit opening and closing facilities ensure the connection and disconnection operations needed for dispatch, configuration changes, or the isolation of failed lines or other elements. There is a wide variety of substation configurations. Busbar numbers and arrangement (single, split, double, or triple bar substations, with or without transfer bars, or ring-shaped) and the number of circuit breaker and dispatch devices per outgoing or incoming line determine the configuration type. Increasing the number of such devices increases substation costs but enhances safety, preventing such anomalies as momentary downstream outages due to simple dispatching operations.

The most representative technological facility in substations is the transformer, which raises or lowers voltage. Transformation is performed electromagnetically with two sets of (high and low voltage) coils wound around a ferromagnetic core. The entire assembly is immersed in a vat of oil to ensure optimum conductor insulation. These are very large scale, expensive, heavy, high-performance facilities, with a very low failure rate. Many transformers are involved in system voltage control: in these, the windings are fitted with taps that allow for slight modifications in the turns ratio and therefore voltage step-up or step-down. In some transformers, regulation can be performed when charged, while in others it may not. Figure 1.15 shows several substation transformers. Other substation components include line breaker and switching devices. As noted earlier, substations are the interconnection buses on the grid, where the connections between the various elements are made or severed. This function, which is natural and foreseeable in normal operation, is crucial in the event of failure. Indeed, system must absolutely be protected from the short circuits occurring in lines or substation bars, since they trigger the circulation of very strong currents that could damage cables and equipment. A fault must then be cleared, that is, the overcurrent canceled, as soon as possible and isolated to repair the damaged component; otherwise, the system as a whole may be endangered. The most sophisticated line breakers are automatic circuit breakers, which are able to open a circuit when overcurrents occur. The protection devices detect overcurrents and, applying appropriate logic, decide which lines must be opened to clear the fault. Constructively speaking, there are many types of such breakers, ranging from compressed air (or pneumatic) or magnetic blowout breakers for small power and voltage to circuit breakers immersed in oil or sulfuric hexafluoride devices for systems with very high voltage and capacity. One special feature of these mechanisms is their ability to open twice in immediate succession. Since many faults have a very short duration because the cause of the outage disappears spontaneously, if due to a false contact or one that is burnt out by the current flow for instance, the system usually attempts to reconnect the circuit breaker automatically, in case the cause of the fault has in fact been eliminated. If not, the breaker will reopen. It should be noted here that due to breaker construction design, it is not usually possible to ascertain from plain sight whether a breaker is open (= off) or closed.

FIGURE 1.15   Substation power transformers.

Once the fault is cleared and identified, the damaged area must be electrically isolated to reconnect the rest of the elements initially shutdown by the circuit breaker. This is done with local disconnectors, used to open or close a line when the current is negligible. Their function, therefore, is not to cut off the current, but simply to visibly isolate a section of line or a device, machine, substation bar, or any other element so it can be handled for repair or maintenance in the total assurance that it is not charged. The operator closes the circuit manually after confirming that the circuit breaker has worked properly and removed voltage from the entire area in question. There are several different types of disconnectors: rotary, sliding, column rotary, and pantographic.

Finally, the line breakers used in grid dispatching have a break capacity on the order of the nominal intensity of the current in the circuit or line they are designed to open or close. Consequently, they do not open in the event of short circuiting. Air-break switches, automatic air switches, automatic gas circuit breakers, magnetic blowout switches, and oil or hexafluoride switches are some of the protection devices used for this purpose.

Today some substations are entirely immersed in hexafluoride. Although more expensive, this arrangement makes it possible to considerably shorten the distance between bars, conductors, and cables, and is particularly attractive for urban environments where square footage is costly. Such substations are, moreover, extremely safe.

1.2.5 DISTRIBUTION

Lower voltage networks branch off the high-voltage grid from the substations in multiple directions to carry electric power to even the most secluded areas. The structure of this network, generically called the distribution grid or network, is very different from transmission grid structure. The upper or regional level, which actually forms a part of the transmission grid, has an open web or loop configuration and operates at somewhat lower but still very high voltages, typically 132, 66, and 45 kV. The substations fed by this part of the grid step the voltage down to 20, 15, or 6.6 kV, splitting power off into the distribution network per se, which is the part of the system that supplies power to the final consumer. The structure of this network may vary, but its operation is always radial. The substations normally house circuit breakers that protect the feeders, that is, lines running to other transformer stations where the voltage is stepped down again to supply low-voltage power, which may be 380, 220, 127, or 110 V, depending on the country, to residential customers, wholesalers, and retailers and the like. Consumers connect into the system at the voltage level best suited to the scale of consumption. In rural areas, the distribution networks are generally radial and consist of less expensive overhead lines, because load density is not very high, the reliability required is lower due to the smaller number of users, and space is not an issue. One problem encountered is that reliability declines as the distance from the substation grows. For this reason, the network is sometimes designed to provide downstream emergency supply in the event of failures. The voltage drop problems that also arise in these networks are solved by placing taps on the transformers and the capacitor banks that supply reactive power.

Distribution networks in urban areas, which are characterized by high load densities concentrated in small areas, generally run underground. Owing to the larger number of users, reliability requirements are stricter. While it is more costly to lay and repair underground lines, the distances involved are much shorter than in rural networks. Urban system structure is usually meshed for greater reliability, but by and large these networks operate radially, with circuit breakers normally open, for reasons of cost and ease of operation. Distribution networks, which comprise thousands of kilometers of wiring, are subject to more frequent failure than the transmission grid and their structure is less redundant; this means that most of the supply outages that affect the final consumer originate in the distribution network. In terms of investment, they account for a large share of total system costs and normally call for an investment several times higher than the transmission grid. The spectacular growth in distributed generation and the increasing active role of the demand (towards a prosumer figure), as discussed previously in this chapter, is challenging the traditional structure of power systems worldwide. It entails profound changes in the functions, planning and operation of transmission grids and distribution networks and the economic management of electric energy systems. From a technical point of view, distribution networks are changing their passive role, in which energy flows always from upstream higher voltage nodes where energy is injected towards the end-consumers nodes where energy is retrieved, into a much more dynamic and active role in which energy is injected at any point in the network and energy may flow in any direction, upstream or downstream. Control, visibility and monitoring of devices connected to the grid, of distributed generation production, of the state of the network (flows, voltages, topology), become relevant at the distribution grid level, as it is already at transmission grid level. The concept of "smart grid" has been adopted to describe the new functionalities distribution networks should comply with to accommodate to all these unstoppable path towards a much more decentralized and flexible structure of power systems. Supported by the impressive progresses on Information and Communication Technologies (ITC), the deployment of smart grids is starting.

1.2.6 CONTROL AND PROTECTION

This review of the chief technological aspects of electric energy systems concludes with a brief description of control and protection systems and equipment. The role and importance of these elements in sustaining system operations have been stressed repeatedly in the foregoing discussion. In view of the wide variety of such systems, the following discussion will be limited to a mere enumeration of the devices involved. They are organized by levels or layers.

On the first level, the elements that comprise the system backbone, generating stations, high-voltage grids, large substations, are centrally monitored and controlled from a control center that supervises the system status in real time (generating plant, line flows, voltage levels, voltage wave frequency, and the like) by remotely transmitted and duly processed measurements. This supervisory and control system goes by the name of SCADA, acronym for supervisory control and data acquisition. These control centers—there may be one for the entire country, or several, scaled by order of importance and coordinated—strive to ensure system safety and may transmit instructions to generating stations to produce real or reactive power, order grid dispatching operations, change transformer taps or connect capacitor banks. Such instructions are based on system data, interpreted by operators on the grounds of their experience or with the support of sophisticated models that analyze operating conditions and determine line flows or bus voltages under different hypothetical contingencies.

The control systems installed in production plants constitute the second level of operation. The two most important such systems are speed and voltage regulators. Speed regulators maintain the instantaneous balance between generation and consumption in the system as a whole. Generating plant must respond immediately to any increase or decrease in demand. Similarly, the chance tripping of a unit in operation at any given time, where nuclear power is involved, this may mean up to 1000 MW, occasions an instantaneous imbalance between power generated and consumed that must be compensated for by immediately replacing the failed unit. When the power generated differs from system load, the surplus power or power shortage is stored or withdrawn, respectively, from the kinetic energy stored in rotating machines. Speeding up or slowing down these facilities provokes a change in revolutions per minute in the AC generators or the frequency of the wavelength generated. Such parameter changes automatically activate the respective steam, water, or gas-driven valve to modify plant generation accordingly. This is called primary regulation of load–frequency control.

A power shortage caused by a power station failure, for instance, prompts a joint response across the entire interconnected system (all the countries synchronously connected to the country where the shortage occurred are involved), which prevents system frequency from falling further, but is unable to reestablish it exactly to the nominal value. Nor do the power exchanges with neighboring systems sustain their predetermined values due to the flows required to maintain the frequency. A second control loop known as AGC or automatic generation control reestablishes frequency to the nominal value and the exchange operations to their initial values. This constitutes what is known as secondary regulation, which is also usually automatic, and does not involve all generators, in particular, none located on neighboring systems. The extra generation required is redistributed among the stations chosen for this purpose. This also regenerates the primary reserve capacity, to ensure continued operation and prevent system standstill, resulting from units reaching their limit capacity. Finally, tertiary regulation may also be implemented. At this level, in which supervision is not automatic, the control center may change long-term dispatching instructions to enhance economic efficiency and restore the so-called secondary reserve capacity, in much the same way that secondary control restores the primary reserve capacity. It will be noted that secondary and tertiary regulation form a part of the higher control level referred to above, but they have been described here for greater clarity.

Power stations are fitted with a second control loop related to system voltage. System voltage must be kept within certain allowable margins to ensure system safety and guarantee that the power delivered is of a reasonable quality. The voltage level of an electric power system is closely related to the balance of reactive power. High reactive consumption, either by charged lines or inductive motors, tends to depress system voltage, whereas a supply of reactive power, from uncharged lines or capacitor banks for instance, tends to raise system voltage. For these reasons, power stations, which are able to produce or consume reactive power at will with their (synchronous) AC generators, are ideal candidates to monitor and correct dangerous voltage fluctuations. The voltage regulator measures voltages at generator terminals or selected points of the system, compares the measurement with a reference value and adjusts the AC generator excitation current accordingly, which controls the reactive energy supplied or absorbed by the unit.

Power stations, naturally, are fitted with protection systems that prevent potential damage. The AC generator, pumps, turbines, and any other vital component are equipped with the respective measuring systems, tripping relays, and alarms. The approach is as discussed above for substations: the protection relays must detect and locate faults, the automatic circuit breakers must clear them and the disconnectors must isolate the failure to be able to reestablish service in the rest of the system while the fault is repaired. Protection relays must be sensitive enough to detect the fault, selective to minimize the impact of clearance, able to respond quickly for protection to be effective, and reliable, that is, neither tripping operations unnecessarily nor failing to act in critical situations. They must also be robust, since they operate under widely varying adverse circumstances, and must be able to operate independently and automatically, even in the absence of electricity.

1.3 THE ECONOMIC ENVIRONMENT

1.3.1 THE ELECTRIC SECTOR AND ECONOMIC ACTIVITY

The economic end of electric power system management is extremely complex due, among others, to the breadth of the task, which covers financial, pricing, social, business, and environmental factors, not to mention investment planning and system operation, these are last closely related to the technological aspects of such systems. And all of these issues must be handled within the regulatory and legal context prevailing in each country. This, naturally, significantly conditions not only the approach and the margins within which each of these activities may be conducted, but also determines exactly who the decision-makers are. With the profound regulatory change underway in the industry in many areas of the world, any attempt at describing the economic environment should address both the traditional settings, still in place in many countries, as well as more liberalized situations. Instead of comparing the two approaches subject by subject, the following discussion first provides an overview of the well-established expansion planning and system operation functions in the traditional scenario, and then highlights the philosophy and changes introduced by the liberalized regulations.

The planning and actual operation of an electric power system are the result of a complex chain of decisions. The first link comprises long-term provisions—capacity expansion, fuel contracts; the second medium-term planning—hydroelectric management, facility maintenance programs; the third short-term specifications—generating unit connection, operating capacity in reserve; and the fourth actual system operation—generating unit dispatching, frequency regulation, response to possible emergency situations. Decision-making is supported by computing models fed by highly sophisticated data acquisition and communications systems. Today's resources make it possible, for instance, to precisely calculate the marginal cost of meeting demand (that is, the cost of one additional kWh) at any given point on the system at a given time, taking account of the entire chain of decisions referred above [11].

Decisions affecting electric power system expansion and operation should be guided by criteria of economic efficiency to minimize the cost of delivering an acceptable quality of supply to consumers or customers. Nonetheless, constant account must be taken of technical considerations to ensure the material feasibility of supplying electric power, arguably an issue much more vital in this than any other industry, given its specific characteristics. As discussed below, the importance of such considerations grows as the time lapsing between decision-making and implementation narrows down to real time, when the distinction between economic and technical factors blurs and no clear line can be drawn between them. Given the size, dimension, and complexity of the problem, the entire decision chain must be rationalized and organized. This is achieved by ranking the expansion and operating functions chronologically. In longer term decisions, for instance, where future uncertainty and economic criteria carry considerable weight, a rough approximation of system technological behavior suffices. Such decisions successively guide shorter term decision-making in which technical specifics are much more relevant, culminating in real-time operation, where system dynamics must be analyzed in full detail, millisecond by millisecond.

1.3.2 EXPANSION AND OPERATION IN THE TRADITIONAL CONTEXT

In this context, a government-controlled centralized coordinator is responsible for overall electric power system operation decisions, control, and monitoring. This body is likewise entrusted with the formulation of plans for system expansion, as regards the installation of both new generating capacity and transmission grid lines or facilities. It is also often responsible for implementing such plans if the grid is publicly owned.

The underlying criterion for the entire decision-making processes is maximization of social utility in the production and consumption of electric power. This involves two fundamental concerns. The first is to attempt to minimize the entire chain of costs incurred to provide service, including both investment and operating costs. Nonetheless, the attainment of an inexpensive service is not the only factor used to measure social utility. The quality of supply must also be satisfactory. Utility is low for both industrial and residential consumers where service is cheap but plagued by constant outages. A number of uncertainties (rainfall, real demand growth, generating, transmission, and distribution equipment failure) make it impossible to guarantee outage-free service in future scenarios. In other words, there will always be some likelihood of inability to service all the demand at all times, which is a measure of system reliability. It is equally clear, however, that such likelihood of failure can be minimized by investing in more facilities and operating more conservatively. Increased reliability entails higher costs. For this reason, the first criterion referred to above, cost minimization, must be qualified to accommodate the second criterion, which reflects system reliability. This can be built into the decision-making process in a variety of ways. One is to set a minimum reliability threshold based on past experience and social perception of the concept, measured in terms of the likelihood of an interruption of supply of electric power or some other similar indicator. A more soundly based method consists of attempting to quantify the financial harm caused by an interruption of service based on its utility to consumers. Building this factor into the cost minimization process as one more expense to be taken into account is actually a way of maximizing the social utility of the service. The inherent difficulty in this second approach lies in quantifying "utility," which may vary from one consumer or individual to another, and which there is no clear way to measure, although attempts have been made in this respect by conducting systematic surveys of consumers, specifically designed for this purpose.

Reliability is a factor that involves the entire decision-making process, from long to short term. Service interruption may be due to investment-related issues: if system-installed capacity is insufficient to cover demand, perhaps because demand growth has been unexpectedly steep, hydrological conditions particularly adverse, transmission capacity lacking or the implementation of new investment delayed; to operating problems: poor reservoir management, lack of immediate response to failures in units or lines due to a shortage of reserve capacity; or real-time system stability problems. In light of this, in nearly all the decision-making scenarios reviewed below, cost and reliability factors must be brought into balance in the decision reached. The term "adequacy" is generally used to describe long-term reliability and "safety" to refer to short-term operations.

There is no standard way to organize electric power system planning and operation. The solution to such a complex problem nearly always involves breaking it down into simpler problems and all the approaches taken to date involve, in one way or another, time-scaled hierarchical decomposition. Decision-making is ranked by time frame and the respective functions are scaled accordingly. The highest level comprises the longest term decisions, which tend to address strategic problems. The solutions adopted are then passed on to the lower levels, delimiting their scope of action. At successively lower levels of the scale, as functions approach real time, they must seek the optimum solution within the restrictions imposed by the problem posed as well as by the guidelines received from the higher levels. This structure, abstractly defined in this section, is described more specifically in the items below.

1.3.2.1 Long Term

First level decision-making takes a long-term approach, projecting anywhere from 2 or 3 to 10, 15, or more years into the future, to define generating plant and transmission and distribution grid investments. The process involves determining the type, dimensions and timing of new generation, and transmission facilities to be installed based on several parameters, namely demand growth forecasts, technological alternatives and costs, estimated fuel availability and price trends, reliability criteria adopted, environmental impact restrictions, diversification policies, and objectives relating to depen-dence on the foreign sector. Such distant horizons are necessary because the (very large) investments involved are justified on the grounds of the earnings over the service life of such facilities, which may be from 25 to 30 years in the case of steam power stations and much longer for hydroelectric plants.

With such distant horizons, uncertainty is obviously a key determining factor. A whole suite of scenarios must be addressed, the respective probabilistic assessments conducted and the most suitable criteria adopted, such as minimization of expected average costs, minimization of regret, or minimi-zation of risk, taken as the variance of cost distribution.

For the same reason, it makes no sense in this type of studies to evaluate the technical behavior of system operation in detail, since it is neither feasible nor sensible to seek accuracy in the evaluation of operating costs when the process involves much greater levels of financial uncertainty.

One of the chief requisites to the process is a good database, which must contain information such as updated data on technologies as well as historical series on demand, hydrology (rainfall), equip-ment failure rates, and so on. The long-term demand forecast built from these data assumes the form of a probability curve that determines system expansion needs. As indicated above, the hourly demand profile is as important as total demand in this regard, since the choice of technologies depends largely on this information. The next step is to determine the generating plant expansion required to meet demand under terms which, while respecting the different strategic criteria mentioned above, seek the option that minimizes the anticipated costs over the entire period considered. Such costs include the fixed costs of the investment chosen and the operating costs throughout the entire period, which will, obviously, depend on the type of investments made. The support of simulation and opti-mization models is often enlisted for such estimates. Because of the scale of the problem posed, decomposition analysis techniques generally used are those which run iteratively and alternatively between two modules, one specializing in expansion cost calculations and the other in operating costs, exchanging information between the two as necessary until the results converge.

Decisions to enlarge the transmission grid have traditionally depended on new generation plant investment needs and demand center growth. This is because considerably less investment and time were needed to build grid facilities than to build generating stations, although in countries where geog-raphy and distance call for very large-scale and costly transmission systems, this may not be wholly true. Once the new plants are sited and consumption and production growth rates are estimated, grid expansion is determined, comparing the necessary investment costs with the benefits afforded by the system: lower operating costs, smaller system losses, and greater reliability in demand coverage. The decision process is also impacted by safety criteria that ensure that supply will not be subject to interruptions due to grid reinforcements or other technical aspects, such as voltage or stability issues.

Expansion decisions are, naturally, dynamic over time, inasmuch as they must be periodically adjusted when real data on demand growth, technological innovations, or fuel purchase terms modify the assumptions underlying the initial expansion plans.

1.3.2.2 Medium Term

Once future investment is defined, medium- and long-term facility operating plans must be laid. For a one- to three-year horizon, depending on the system in question, such planning involves determining the best unit and grid maintenance cycle program, the most beneficial fuel purchase policy and the most efficient use of power plants subject to primary energy limitations, hydroelectric stations in par-ticular, or to yearly production restrictions for environmental reasons. Electricity generating plants are sophisticated systems with thousands of components that must be revised periodically to prevent major and sometimes hazardous failure and assure plant efficiency, from the technical standpoint. Operation of conventional steam plants is usually interrupted around 20 days a year for these pur-poses. Nuclear plants need to recharge their fuel (uranium bars) once every 18 months, so mainte-nance tasks are programed to concur with such plant shutdowns. Electric power lines and the components of the transmission and distribution grids located in substations also need upkeep, such as the replacement of faulty insulators or their cleaning to prevent loss of insulation power. Although technology to perform these tasks on live lines is more and more commonly available, most of these operations are conducted on deenergized facilities for obvious safety reasons, which involves disconnecting certain lines or parts of substations. This, in turn, requires careful maintenance program planning to interfere as little as possible with system operation.

Fuel management also calls for careful planning, sometimes far in advance. Once input needs are defined, fuel purchases must be planned, often in international markets, to buy at the most advantageous price, make shipping and storage arrangements, and take all other necessary logistical measures to ensure that stations do not run out of fuel. Negotiation of fuel contracts will impact the operation decisions of the system, not only because fuel prices constitutes a direct input to operation cost of plants, but also due to specific clauses agreed on the delivery of the fuel. For instance it is common to purchase large amounts of gas under a so-called "take or pay" kind of contract. With this kind of contract, the company either takes from the supplier the total agreed volume of gas purchased at a given price or pays the supplier a penalty, often close to that gas price. Once signed, it is therefore almost like a sunk cost for the plant owner, and this will condition the operation of the plant. This is particularly applicable to liquefied natural gas (LNG) contracts served through maritime transport, with limited gas storage capacity, that sometimes require to use all stored gas in a zero cost condition. Finally, the use of water resources in hydroelectric plants must also be planned, as if it, too, were a fuel. In fact, water can be seen as a cost-free fuel in limited supply. Therefore, its use must be scheduled in the manner most beneficial to the system. Run- of-the-river stations require no planning, but decisions are imperative where there is an option to either generate energy or store water for production at a later time. Depending on the size of the reservoir, such decisions may cover time frames ranging anywhere from a single day to several weeks, months, or even years for the largest reservoirs, the so-called carry-over storage reservoirs. Stations whose management and regulation extend over several months should be planned on a yearly or multiyearly basis. Since the logical aim of such planning is to attempt to replace the most expensive thermal production, this type of planning is usually termed hydrothermal coordination. A similar approach is taken for any other technology subject to restrictions on use that limit accumulated production in a given time frame, typically seasonal or annual. One example would be the existence of mandatory national fuel consumption quotas or yearly pollution limits.

1.3.2.3 Short Term

Short-term decision-making is referred to a weekly scale, that is, from a few days up to one month. It involves determining the production plan for hydroelectric and steam power stations on an hourly basis for each day of the week or month. This plan must abide, moreover, by the instructions received from the immediately higher decision level described in the preceding section in connection with maintenance action, weekly or monthly hydroelectric management, emissions plan, management of fuel quotas, and so on.

At this level, system details are extremely relevant, and account must be taken of aspects such as steam plant generating unit start-up and shutdown processes and costs; the hydrological restrictions in place in river basins; stations in tandem arrangement; demand chronology profiles, which call for accurate production monitoring; generating capacity to be held in reserve to respond immediately to fortuitous equipment failure, and so on.

The possibility of varying steam power station output is limited by the technical characteristics of their generating units. It takes an inactive station a certain minimum amount of time to recover operational status, which is primarily determined by the time needed to heat the boiler to a suitable temperature. This minimum lead time depends therefore on the cooling state of the boiler: in other words, the amount of time it has been off. The most conventional steam power plants may need from 8 to 10 h if the boiler is completely cold. Gas and CCGT plants are more flexible, with lead times of from 1 to 2 h or even only a few minutes for simple gas turbines. As a result, the cost of starting up a steam power station is significant and can be quantified as the price of the fuel that must be unproductively burnt to heat the boiler to the appropriate temperature. For this reason, even if demand declines substantially, it may not be cost-effective to disconnect certain steam stations at night, but rather to maintain a minimum production level. This level, known as the plant's minimum load, is relatively high, generally speaking, on the order of 30%–40% of the station's maximum output, due to boiler combustion stability requirements. Depending on the findings of cost-effectiveness studies, then, a decision must be made on whether it is more economically sound to start and stop the station everyday (daily start-up cycle) or shutdown on the weekends only (weekly cycle) or simply not shutdown ever, as in the case of nuclear stations. Sometimes it may be more efficient to keep the boiler hot without producing anything. Boiler thermal constants and their limits likewise impose constraints on how quickly steam plant output rates can be modified, which are known as upward or downward ramp constraints. All of this calls for careful planning of unit start-ups and shutdowns, a problem known as unit commitment.

This decision is also strongly influenced by weekly or monthly hydroelectric management, as well as system reserve capacity requirements. Hydroelectric stations are much more flexible in this respect, with practically zero lead time, no significant start-up costs and virtually no real limits to modulating generating capacity. The optimal scheduling of hydroelectric production to cover seasonal variations in demand takes account of a number of considerations: higher level decisions on the amount of water resources to be used in the week or month, the most cost-effective hourly distribution (hydrothermal coordination during the month or week) and the technical constraints on steam plant units. Hydroelectric generation and the respective use of water in reservoirs must likewise abide by possible restrictions imposed by water management for other purposes (irrigation, fauna, minimum reservoir and river flow levels, etc.) as well as other conditioning factors characteristic of water works and their configuration: canals, reservoir limits, reservoirs in tandem, or pipelines.

Even in this case, reliability criteria play a role in decision-making. Provision must be made for the immediate replacement of any plant in the system that may reasonably be expected to fail or for the ability to respond to transmission grid incidents. This translates into the start-up and connection of new units which, although unnecessary under normal conditions, would not otherwise be able to generate power for several hours if needed to cover emergencies. Finally, wind and solar irradiation should be forecasted as accurately as possible in the short term to properly predict wind farm and solar based productions. Indeed, they will condition not only the hourly dispatch of other generation units but also the production flexibility of other units required to cope with the intermittent nature of their production. Some units may be scheduled to start-up in order to assure being able to manage unforeseen or even foreseen decreases/increases of these kind of renewable production. Substantial progress has been achieved in the forecast of wind in a time scope of few hours, although forecast errors still become quite relevant as the time scope enlarge to a day or few days.

1.3.2.4 Real Time

Real-time operating functions are based essentially on safety criteria rather than on financial considerations. The economic component of the process is defined by higher level decisions, although sight should never be lost, as noted earlier, of the economic aspects of reliability. Supervision, control, and monitoring ensure the technical viability of the immense and dynamic electric power system, as described in more detail in the "control and protection" section included above in this chapter.

1.3.3 Expansion and Operation in the New Regulatory Context

The new electricity industry regulatory environment is bringing profound change to electric power system operating and planning habits. Industry liberalization has gone hand in hand with dramatic decentralization of planning and operating functions. System expansion, now focused on investment, and operation are the result of individual company decisions based on the maximization of business earnings, either under organized tendering or through private electric power supply contracts. Economic and financial risk and anticipated returns on investment, instead of the traditional cost minimization criteria, drive decision-making. The challenge for administrative and regulatory authorities is to design liberalized market rules that ensure that the strictly entrepreneurial behavior of each market player leads to overall minimization of system costs, reflected in the tariff charged to final consumers.

Real-time system operation, however, continues to be a centralized task. A central operator, usually called the system operator, oversees system safety under a supervision and control scheme that is essentially the same as described above for the traditional setting. Ensuring the real-time technical viability of the system calls for advanced coordination of all the available resources, which in turn requires absolute independence from the various actors' individual interests. Electric power system operation is viewed, therefore, from a wholly different vantage point. There are new functions, responsibilities, and ways to broach the decision-making process, and changes in the roles played by each of the agents involved. Electric utilities have had to reorganize to assume their new market functions. They are faced with the challenge of adapting to a new environment in which they must change many operating habits and where new duties and tasks are arising. Some of the most novel of these tasks have to do with tendering, landing contracts, and formulating yearly budgets—revenues less operating expenses—all within the framework of risk management policy.

1.3.3.1 Long Term

The new environment has completely revolutionized the approach taken to electricity generation planning. The liberalization and decentralization of investment decisions ascribe to each actor the responsibility for assessing the advisability of investing in new generating facilities, based on individual benefit–cost analysis. Hence, new investment is studied from the perspective of estimated revenues over a period of several years, in other words, of the evaluation of future market performance. Such analyses address factors associated with estimated fuel prices, demand levels, new third-party investment, market prices, and the like. And financial risk assessment plays a crucial role in such reviews, because it is the key to suitable investment financing. Rainfall and demand stochasticity, as well as the possible interplay between price scenarios and third-party expansion decisions, are aspects that have to be considered in this environment. Risk management constitutes one of the chief activities and drivers of the planning and operation of electric energy systems. The formulation of contracts to supply power or obtain financing, along with access to futures markets and options, are elements of immense importance in this environment.

While transmission grid planning continues to be centralized, the perspective has changed and today's decisions are subject to much greater uncertainty. The basic planning criterion, namely optimization of the social utility of electricity production and consumption, remains unvaried. But it no longer necessarily acquires the form of minimization of production costs, but rather maximization of individual market agents' profits: for consumers, electricity consumption utility minus electricity acquisition costs, and for producers, electricity sale revenues minus generating costs. Moreover, the uncertainty surrounding the planner's decisions has risen dramatically. Traditionally, the grid was planned based on prior decisions on expansion of generating capacity, information which in an environment of free competition is not compiled centrally or *a priori*, but rather is the result of business decisions made individually by the different players at any given time. Consequently, neither the amount nor the location of new generating plant is known for certain when the grid is being planned. Furthermore, renewable generation expansion and operation adds a new important uncertainty factor to be taken into account, especially given shorter construction times of solar and wind plants. Besides, the time that may lapse from when the decision to build a line is made until it becomes operational is growing longer, due in particular to difficulties arising in connection with the environment and territorial organization. As a result, construction lead times are often longer for the transmission grid than for the power stations themselves.

1.3.3.2   Medium Term

Of the various agents participating in the electricity market, several must try to optimize their medium- and short-term decisions: consumers, suppliers, and producers. In the system operation setting, however, producers play the lead role. The three medium-term aims pursued by generators are as follows:

1. Formulate medium-term economic forecasts: revenue projections and yearly budgets.
2. Provide support for the long-term functions mentioned above: contract management, determination of long-term business strategies, and investment assessment.
3. Provide support for short-term functions, in particular the formulation of bids on daily markets for electric power and ancillary services: hydroelectric output guidelines, valuation of water reserves, and guidelines for steam production subject to provisions on national coal quotas, environmental restrictions, or similar.

The new approach for designing models to support these decisions, which seeks optimization of each agent's own market earnings, calls in one way or another for building new theoretical microeconomic- and game theory-based concepts into the model. Markets are regarded to be dynamic elements that stabilize around certain points of equilibrium characterized by the various agents' production structures. Alternatively, they may be viewed as the result of a certain game in which the established rules ultimately impose a strategy to be followed by each agent in response to the reactions of all the others.

1.3.3.3   Short Term

Although there is a variety of ways to organize electricity markets, they all have a series of short-term (typically daily) markets, as well as intradaily markets and ancillary service markets, where the short-term output of generating stations is determined. Out of the three, the daily bid market is usually the most important in terms of trading volume. Consequently, short-term operation tends to revolve around the preparation of daily bids. This process is governed by higher level strategic decisions. As such, it is informed by the output guidelines deriving from analyses covering longer terms and its role consists of setting market prices on a day-to-day basis. The estimation of short-term prices, on this timescale the uncertainty associated with rainfall and unit availability is very small, is an important task, since it serves as a guide for decisions on the internalization of steam generating units costs and whether to draw on hydroelectric production. As a result of this and enlightened assumptions on the behavior of the competition, companies establish the bids, specifying quantity and price, with which they compete on the wholesale market. Being their production cost almost null, wind and solar based generation units should rely on their production forecasts to set their bids in those markets.

1.3.3.4   Real Time

As in the traditional setting, real-time operation is strongly influenced by safety considerations and has a similar structure, although considerable effort is generally made to differentiate and clearly value the various types of so-called ancillary services provided by each agent in this respect. And wherever possible, market mechanisms have been implemented to competitively decide who provides what service and at what price. Such mechanisms are often deployed in connection with operating, secondary or tertiary, capacity in reserve, as well at times with voltage control and even with system cold-starts. Primary frequency or output control continues to be a mandatory basic service at the system operator's disposal as an element vital to system safety.

In this competitive framework, companies must offer their services taking account of the costs incurred in their power stations to provide them, along with other considerations such as market opportunities.

1.4   THE REGULATORY AND SOCIAL ENVIRONMENT

1.4.1   TRADITIONAL REGULATION AND REGULATION OF COMPETITIVE MARKETS

In a nutshell, regulation can be defined as a "system that allows a government to formalize and institutionalize its commitments to protect consumers and investors" [3]. Depending on the development of the electricity industry in each country, and even in different regions of a country, the prevailing ideology, specific natural resources, and technological change, among other factors, the electric power industry has adopted different organizational and ownership (private or public, at the municipal, provincial, or national level) formulae.

Despite this diversity, ever since the electricity industry reached maturity and until very recently, regulation around the world was uniformly of the sort applied to a public service provided under monopoly arrangements: guaranteed franchise for the typically vertically integrated electricity utility and price regulation based on the costs incurred to provide the service. This is what will be referred to in the present discussion as the "traditional" regulatory approach. Under this scheme, the relations between different electric utilities were generally characterized by voluntary cooperation in a number of areas, such as joint management of frequency regulation or operating reserve capacity, exchange for reasons of economy or emergency, normally the latter, and third-party use of grids for current transmission or distribution under terms negotiated by the parties concerned.

This regulatory uniformity was altered in 1982 when Chile introduced an innovative approach, which separated the basic activities involved in the provision of electric power. Under the new arrangements, most of the industry was privatized and an organized, competitive (within rather strict limits) power pool or wholesale market was created, with centralized dispatching based on declared variable costs. All generators were paid a "system marginal price," and long-term contracts were instituted to offset price volatility. The new scheme also envisaged planning guidelines for generation, in which the state assumed a merely subsidiary role, and free access to the grid subject to payment of transmission tolls. Similar reforms were not introduced anywhere else place until 1990 when the electricity industry was much more radically transformed in England and Wales, and shortly thereafter in Argentina (1991) and Norway (1991). Since then, many other countries, including Colombia, Sweden, Finland, New Zealand, the Australian states of Victoria and New South Wales (later to give rise to the Australian national market), Central America, Peru, Ecuador, Bolivia, El Salvador, many states in the United States and Canada, country members of the European Economic Community (EEC), among others, have established or are in the process of establishing regulatory frameworks for a free electric power market. Certain elements of free competition have been introduced in traditional regulatory contexts by countries in Eastern Europe, as well as in Mexico, Malaysia, Philippines, Indonesia, Thailand, Japan, India, and Jamaica.

1.4.2   NEW REGULATORY ENVIRONMENT

1.4.2.1   Motivation

The regulatory change in the electricity industry, which forms a part of the present wave of economic liberalization affecting businesses such as air carriage, telecommunications, banking services, gas supply, and so on, has been possible thanks to a variety of factors. On the one hand, the development of the capacity to interconnect electric energy systems has led to an effective increase in the size of relevant potential markets, eliminating or reducing the possible economies of scale to be had in a single production unit. On the other hand, competitive generating technologies that can be built in shorter times are, at least initially, opening the recently created markets up to a flood of new entrants. In some countries, the determining factor has been dissatisfaction with the traditional approach due to its most common shortcomings: excess of governmental intervention, confusion over the state's dual role as owner and regulator, financial and technical management inefficiencies due to the lack of competition, or lack of investment capacity. Finally, technological advances in areas such as metering, communications, and information processing have paved the way for the advent of competition in the supply of electric power to the final consumer.

1.4.2.2 Fundamentals

The new electricity industry regulation is based on a fundamental premise: that it is possible to have a wholesale market for electric power open to all generators, those already existing and those that voluntarily enter the market, and all consumer entities [11]. The core of this wholesale market is typically a spot market for electricity [11], with respect or as an alternative to which medium- and long-term contracts of different types, and even organized markets for electricity derivatives, are established. The agents trading on such markets are generators, authorized consumers, and different categories of supplier companies, acting on behalf of noneligible consumer groups or eligible consumers, or simply as strict intermediaries. The new regulatory context must also address many other issues, such as: creation of a retail market enabling all consumers to exercise their right to choose a supplier; the mechanisms and institutions needed to coordinate organized markets and, especially, system technical operation; transmission grid and distribution network access, expansion and remuneration, as well as quality of supply and the establishment of transmission tolls to use them; or the design of the transition from a traditional to a competitive market, protecting consumers' and utilities' legitimate interests [7].

1.4.2.3 Requirements

In this environment, sight must not be lost of the technological and economic features of the electricity industry that condition the design of the regulatory provisions governing it, namely:

- The infrastructure needed for electric power generation, transmission, and distribution is costly, highly specific, and long-lasting.
- Electric power is essential to consumers, making public opinion highly sensitive to possible service outages or poor quality.
- Electric power is not economically storable in significant quantities and therefore production must be instantaneously adapted to demand.
- Real operation of an electric power system is the result of a complex chain of scaled decisions, as described in detail in the preceding section.
- The supply of electric power combines activities that clearly conform to natural monopoly requirements (transmission and distribution services or system operation) with others that can be conducted under competitive conditions (generation and supply).
- The organization and ownership structures of electric companies vary widely across different electric energy systems.

Account must be taken of the fact that the supply of electric power under competitive arrangements is subject to the existence of certain activities associated essentially with the transmission grid and distribution network, whose control entails absolute power over the electricity market. Consequently, these grid and network-associated activities must be absolutely and wholly independent of the competitive activities, namely production and supply. For this reason, and given that when most liberalization processes are introduced, the industry is dominated by vertically integrated utilities, that is, companies conducting all stages of the business from production to billing the final consumer, industry organization, and ownership structure nearly always has to be modified before competition mechanisms can be implemented. Naturally, after such reform, attention also needs to be paid to issues such as the horizontal concentration of production and supplier companies and the vertical integration between them, to ensure free and fair competition.

1.4.3 Nature of Electric Activities

A careful examination of the whole process of electric power supply to end users allows the identification of various activities of a very distinct technical and economic nature, and therefore,
capable of receiving a different regulatory treatment [15]. The classic division into generation,
transmission, and distribution is excessively simple and also has gross errors, such as integrating
the activity of "distribution" into one single category. As a minimum, distribution includes two
activities radically different in nature: on the one hand, there is the "distribution" service that permits energy to arrive physically from the transmission network to the end users and has the form
of a natural monopoly; and on the other hand, there is the "marketing" service of such energy
that is acquired in bulk and then retailed, and may thus be carried out in a competitive scenario.

Upon a first general classification, activities may be sorted into basic categories: production
(generation and ancillary services), network (transmission and distribution), transaction (wholesale
market, retail market, balancing market), and coordination (system and market operation), and
some other complementary classifications, such as measurement or billing. This breakdown
may be considered excessive; however, it is needed to start at the design level with such an analysis to set up an adequate regulatory framework. By way of example, planning the expansion of
the transmission network is an activity that must be regulated in some way or other, given its
characteristic of natural monopoly and its significant influence on the electricity market conditions. Conversely, once the characteristics and start-up date of a new transmission facility are
decided, its construction may be assigned by some contest process where price and quality
compete.

1.4.3.1 Unbundling of Activities

The number of activities indicated before does not necessarily imply a corresponding multiplicity of
entities for their execution [14]. As shall be seen below, there are synergies and transaction costs that
make it convenient in certain cases for a single entity to undertake several activities. It will also be
stated, however, that conflicts of interest may arise for an entity that is in charge of more than one
activity, when the execution of one such activity may benefit it against other agents in another activity
that is open to competition. Different separation levels may be applied, but it is necessary to adjust
them to each particular case. Basically, these activities may be separated into four types: accounting,
management, legal status (that is, different companies that may belong to same owners through a
trust), and ownership.

The basic rule on separation of activities under the new regulation is that a single entity cannot
execute regulated activities (for instance, distribution) and competitive activities (such as generation)
at the same time. The potential support to be provided by the regulated activity to the competitive one
is an evident advantage for the latter and such advantage is legally out of order. Similarly, the risk of
the competitive activity cannot be transferred to the regulated one, since it will definitely fall on consumers who do not have the option to choose.

An adequate transparency in the regulated activities also requires at least accounting separation
among the corresponding business units. Companies engaged in regulated activities are not permitted
to conduct diversified activities (that is, activities not related to electricity) or must at least be subject
to the authorization of the regulatory agency. Such authorization shall be initially based on the non-
existence of negative impacts on the regulated business which could ultimately be borne by consumers who do not have the option to choose.

The new regulatory framework design shall consider the various benefits and inconveniences
when assigning activities to entities and setting the separation levels, and shall also take into account
the specific characteristics of the actual system, particularly the initial business structure. Several
valid alternatives are generally possible, as shown by the diverse experiences undergone in countries
that have adopted the new electricity regulatory framework.

1.4.3.2   Generation Activities

Generation activities include ordinary and special power generation. Special power generation is typically understood as cogeneration and production technologies that use renewable resources, and regularly differ from the ordinary one in that it receives a more favorable treatment by way of compensation, priority in operation, among others, as is the usual case in a number of countries. The so-called ancillary or complementary services should also be included when they are provided by generators, contributing to an adequate level of quality and security of energy supply.

The ordinary generation is a nonregulated activity conducted in competitive conditions, with no entry restrictions and free access to the networks. Selling production may take place through different transaction processes, basically in the spot market or through contracts, as described below when referring to transaction activities. The special generation, from the regulatory point of view, bears no other difference from the ordinary generation. The fundamental reason to support the special generation is its lower environmental impact as compared to the ordinary generation. The failure to explicitly consider environmental costs in the prices of the electricity market today is offset by the use of different regulatory patterns so as to level off the playing field for all production technologies, enabling them to compete at arm's length conditions by tacitly or explicitly recognizing the total costs effectively incurred.

The following rules are included in the regulatory schemes currently in use or proposed to promote the special power generation: (a) obligation for companies that trade or distribute energy to offer them at administratively fixed prices; (b) a premium, whether prefixed or assigned by means of competitive mechanisms, per kWh produced with generators that are eligible for that purpose; (c) exemption from certain taxes, particularly those that are imposed on the production of energy; (d) assistance to investment or to R&D programs related to these technologies; (e) obligatory quotas for the purchase of energy from the special generation sector by qualified traders and consumers, thereby encouraging the creation of a parallel market for the purchase and sale of this energy; and (f) voluntary purchases of renewable energy by end users who pay an extra amount for it to finance the special generation activity. When choosing the adequate approach, efficiency, that is, the degree of target attainment versus the additional cost incurred, should be basically valued while avoiding interference with the market operation as much as possible. Decreasing costs of solar and wind technologies are permitting access without support mechanisms.

In addition to the production of energy, the groups engaged in generation contribute to the provision of other services that are essential in an efficient and safe supply of electricity: they provide reserves for future operations enabling them to act within different timescales and face the unavoidable mismatches between demand and the generation; they help regulate voltage in the electricity network in the different operating conditions or else they permit a prompt recovery of the service in the event of a general failure. The current tendency in the regulations of this group of additional generating activities, globally called ancillary services, is summed up in two basic views: (a) the use, whenever possible, of a market approach for the allocation and compensation of said services, or, otherwise, the direct application of regulation and (b) the assignation of the charges arising from the incurred costs to the agents that caused the demand.

1.4.3.3   Network Activities

As stated above, the electric power supply necessarily requires the use of networks which, given their technical and economic characteristics, have to be managed and regulated as a natural monopoly. Such a requirement is a basic condition for the new sector regulation.

Network activities include investment planning, construction, maintenance planning, actual maintenance, and operation. The investment planning process determines the commissioning date, location, capacity, and other characteristics of the new assets of a network. The maintenance planning process determines downtime periods for each line to make the repairs and actions required to keep them functional and reliable. Construction and maintenance are activities that may be executed by specialized companies, not necessarily electricity companies. Network operation is the management of energy flows within the network through actions executed directly in the physical transmission facilities and in coordination with actions executed in the production and use facilities. Networks may also take part in the provision of certain ancillary services, such as voltage regulation, that are usually directly regulated.

Both the planning of additions and the planning of maintenance for the transmission network have an impact on the manner in which activities are coordinated that, in turn, affect the electricity market. Consequently, the independence of the relevant responsible entity, typically the operator of the system, regarding market agents should be guaranteed. However, both planning activities also have obvious effects on the planning of the network construction and maintenance to be made by the transmission companies. Even though it cannot be denied that the synergies among the different network activities suggest that they should be managed by one single company, there are important regulatory reasons to separate the system operation from any transmission company.

On the other hand, the distribution networks do not have the problem of interference in market coordination. Thus, all network activities in a given area may be easily executed by the same company, namely, the local distributor. Within the regulatory purview, there are two other fundamental differences between the distribution activity and the transmission activity: (a) the great majority of end users are connected to the distribution network, which is the reason for the particular importance of quality in the services and (b) the high number of distribution facilities makes the individual regulatory treatment impossible, particularly regarding payment, and results in the use of global simplifying procedures. The new regulation on electricity networks may be reduced to three main aspects: access, investments, and prices, which are discussed below.

1.4.3.4 Transmission

Access: The systems that have adhered to the new regulation have implicit access to the transmission network for all agents authorized to take part in the wholesale market. The network capacity obviously imposes a physical limitation to access and there are several restriction management procedures to solve the potential conflict situations that may arise. These procedures extend from the application of nodal prices or zone prices in an organized wholesale market (thereby implicitly solving the network restriction) to auctions held to assign the limited capacity among the different agents, or to dispatch modifications by the system operator according to preestablished rules, and even to the previous assignment of long-term rights of use of the network, whether through an auction or based on participation in the construction of lines.

Investments: As stated in Section 1.3.3, the purpose of the new regulations at the design level is to obtain a network that maximizes the benefits added to producers and consumers, who should finally undertake the network costs. Certain explicit reliability criteria are generally used in the network planning instead of being fully included in the economic functions that will be optimized.

The approach that is mainly used is centralized planning, conferred to a specialized entity, subject to preestablished selection criteria regarding the best alternatives and always under a final administrative consent for each facility. Traditionally such an entity is a vertically integrated company, which is the system operator under the new regulations. Payment for the network is fixed by the regulator or for certain facilities it is determined directly from the construction and maintenance bidding processes. Such procedure may be open to the participation and proposal of interested parties, with the regulator's intervention preventing the possibility of an over investment. This approach may pose a difficulty in a liberalized environment, since, as aforesaid, it is very difficult to foresee the future development of power generation as it may also be affected by the network expansion.

A second approach makes the single transmission company fully responsible for the operation and planning of the network. In this case, the company would also be the operator of the system, who shall (a) inform the users of the foreseeable congestion or "remnant capacity" of the network at its different access nodes within a reasonable time limit, (b) ensure that the network complies with certain formally preestablished design and service standards, and (c) assume the expansion of the network facilities, if required, to respond to access requests, provided the standards continue to be met. Transmission payment would be preset by the regulator and should cover the costs of an efficient company that provides the service within the established conditions. This method does not guarantee, nor does it foster, an optimal expansion of the network.

A third approach is to leave the initiative of network reinforcement to the actual users thereof, who can weigh the contribution to investment costs required from them against the benefits arising from every possible reinforcement due to better access, less congestion, or loss reduction. The regulator assesses the public convenience of the proposed reinforcements and, if the evaluation is positive, it calls a bidding process for the execution of its construction and maintenance. The awarded transmission company is compensated according to its bidding terms, leaving the operation of the facility to the system's operator. This procedure is as market-oriented as permitted by the network regulation; however, its administration is complex and relies mainly on the availability of adequate network prices to promote an appropriate location of its agents.

Prices: Since transmission activities in the network are regulated, the prices applied in this field must allow sufficient funds to cover the entire costs (equity or feasibility criteria). It is also fundamental that the agents receive adequate economic signals (efficiency criteria) regarding their location within the network, both in the short run, for a correct market operation considering losses and possible congestions, and in the long run, to promote a correct location of future producing or consuming agents. And prices should not be discriminatory. Four cost concepts are often mentioned under the title of prices in the transmission network that need to be distinguished and treated correctly: network infrastructure costs, Ohm losses, congestions, and ancillary services. The only relevant network costs are investment and facility maintenance costs and, in practice, none of them is related to the electrical use of transmission assets. Losses take place in the network but are actually related to production costs; the same occurs with excess reprograming costs that may be derived from the existing congestion or other restrictions related to the network. As stated above, ancillary services are mainly a generation activity and, as such, they must be regulated.

Losses and congestion in the network give rise to economic signals that may be seen each time as modifications of the market price. Thus, the single market price becomes a nodal price, that is, a different price in each network node that adequately conveys the economic impact of the different locations of generators and consumers. If the use of a single market price is preferred, this should not lead one to dismiss the economic signals of losses and restrictions. Losses that are imputable to each agent, either at a marginal value or at a mean value, may have a correcting effect on the prices of supplies or, preferably, on the actual amounts produced or generated, so that agents may absorb the losses in their supplies to the wholesale market. The economic treatment of congestion has already been covered in the section on access to the network.

The application of nodal prices instead of a single market price gives rise to a surplus that may be used to cover a part of the network costs, usually not over 20%. In any case, whether a single market price or nodal prices are applied, the issue of assigning all or a great portion of the transmission network costs to the users has to be solved. Very different methods have been used or proposed to carry out such distribution. The most popular one, especially in countries that have well-developed networks and no such large distances to cover between generation and demand, is simply the "postage stamp" method consisting of a uniform charge per kWh that is injected into or withdrawn from the network, or per installed kWh, independent from the location in the network. When reinforcement of the short-term location signals in the network has been deemed convenient in relation to losses and restrictions, procedures have been implemented that lead to quantifying the electric use of the network made by each agent of the network, or the economic benefit each user gets from such network, or else, each one's responsibility in the development of the existing network. In the context of several interconnected systems, with electricity regulations that are generally different but allow for transactions between their respective agents, a basic regulatory error has to be avoided when determining the network toll to be applied to two agents involved in a transaction, who are located in different systems. In the United States, such error is called "pancaking" and means that charges are applied to the transaction adding the tolls charged by the electricity grids that they should have to cross to implement such transaction. It should be noted that in this situation, the toll amount critically depends upon the territorial structure, whether electricity companies or countries, which has little to do with the true costs imposed by such a transaction on the electric network of the group of systems. Failing the coordination level required to establish a single regional toll to cover the global network cost, a reasonable and easy-to-apply approach could be that each agent only pays the network charge corresponding to the network infrastructure in his country, as a single right to connect to the regional network, with a separate charge for losses and restrictions considered jointly. Coordinated procedures could also be established to assign the costs of losses and the management of restrictions that affect international transactions.

1.4.3.5 Distribution

Access: The distribution activity is regulated, and distributing companies have an obligation to supply in the area where they have been granted an explicit or implicit territorial license. Accordingly, any consumer located in this area is entitled to be connected to the network and receive the supply under the quality conditions legally established for such purpose.

The great majority of the systems that have adopted the new regulation offer free access, that is, freedom to choose a trading company, to the end users, who shall generally be connected to a certain grid. It should be noted that the change of trading company in no way modifies the rights and duties of a consumer regarding the distribution network to which the consumer is physically connected.

The foregoing does not mean that any consumer may demand special connection requirements from the distributor if the consumer is not obliged at least to cover the excess costs incurred. Regulation of the intakes is usually detailed and conditioned upon the rules established by local administrations and seeks to find the fair middle point between two sides. On the one hand, it may require the distributor to provide a universal network service in its area, with no other cost than the one acknowledged by the rules of this activity; on the other hand, it may impose some kind of economic restraint on consumers in an attempt to prevent demands from becoming excessive and unreasonable. The problem of access to distribution worsens with the existence of users of these networks, other than the end users. They are the generators, typically those of small to medium capacity, who are connected to the distribution network and to other distributors who are generally small and whose supply comes from another upstream distributor.

Investments: As in the case of transmission, the object here is also to obtain an "optimal" network to provide the consumer with a more satisfactory balance between cost of electricity and quality of service. However, distribution requires a specific approach, since the large number of utilities makes it more difficult to apply an individual treatment and requires global solutions [13].

On the basis of distribution regulation is the compensation procedure, as it needs to allow for a return on capital invested that is consistent with the risk involved in this activity, while accurately promoting service quality and loss reduction. Simultaneously, one should avoid the tendency to use the effectively incurred costs as a basis, since they cannot be verified or justified in detail. The general tendency is to use procedures such as price cap or revenue cap, which determine the trajectory of distribution rates or total revenues for the distributor in a number of years, usually four or five, until the regulator conducts a new revision. Rates or revenues are established from an analysis of the most adequate remuneration for the available network and the anticipated operating and expansion costs of the network over the considered period of time.

Several procedures are used to determine these costs. Thus, for example, a starting point may be a regulation based on a yardstick competition between similar distributors. On the basis of a cost database that includes the most significant characteristics of companies, advanced statistical techniques permit the establishment of different types of comparison among them and the adequate level of compensation for any additional distributor that may be considered. Another approach would be based on "model companies" or "benchmark networks." A much greater analysis of the distribution activity is required here, although it is also possible to get closer to the conditions that warrant the compensation level for each company, particularly when compared against others. Reference models design perfectly adjusted networks and business organizations, the costs of which, with timely adjustments to adapt to the actual conditions, serve as a basis to set the remuneration for each distributor. During the period until the next review, other factors may also be added to adapt the compensation to the growth of the market. Having such a reference model in the network, this approach can conveniently represent the levels of losses and quality of service in an explicit way. Consequently, remuneration is consistent with the losses and quality preestablished by the regulator for each area, and economic incentives for the improvement of performance; and sanctions for noncompliance may more easily be created based on the historical performance of distributors in both aspects. The enforcement of obligatory design and operation standards that have been formally put into force for the network allows the assurance that investments, even though not optimal, comply at least with some minimum quality criteria.

Prices: Since distribution is a regulated activity, its prices should permit to cover the total costs involved in this activity, which are basically related to investment, operation, and maintenance. The distribution network is not relevant for the actual activities of market coordination and system operation; therefore, when setting distribution prices, it is important to ensure that the user of such network, mainly the end user, receives an adequate economic indication of its contribution to the network costs and losses. Today this can be only achieved approximately, as for most consumers, the installed measurement and billing system facilities only consider the use of energy within extended periods of time. The most frequent approach used to establish distribution tolls is simply the allocation of the regulated costs of this activity among the users of the network, and the only discrimination refers to the level of connection voltage and contracted power. Users connected in each voltage level will only take part in the costs incurred at their level and above. Since distribution networks are largely designed to cover peak demands, it is important to estimate the contributing factor of each consumer to the peak demand. For those consumers that fail to have meters with an adequate time discriminating capacity, standardized load profiles, which reflect the usual characteristics of the different kinds of consumers, should be applied.

Ohm losses in the distribution network affect the charges to be paid by consumers in at least two basic ways. On the one hand, in calculating network tolls, the demand of each consumer at each voltage level is already affected by its corresponding loss factor. On the other hand, excluding payment of the network expenses, the energy consumption charge should be applied to this value as increased by the losses inflicted on the system rather than to the actual consumption at the consumer's facilities.

1.4.3.6 Transaction Activities

Risk management is a key aspect when considering transaction activities in any of its modalities. For generators, risk management consists in weighing the opportunity to wait for selling energy in the uncertain spot market versus acquiring sales commitments through different kinds of medium- and long-term agreements in preestablished quantities, prices, and terms. For consumers who have access to the wholesale market, risk management is symmetrical to that mentioned above. In the case of trading companies that deal with consumers who have the option to choose, the managed risk shows two sides: on the one hand, the acquisition of energy in the wholesale market either at the spot price or through agreements, and, on the other hand, the sale of energy at rates freely negotiated with consumers. For trading companies dealing with consumers who do not have the option to choose, the managed risk level depends critically on its regulation: in an extreme scenario, the regulator permits a total pass-through of the wholesale price for energy at the regulated rates, fully annulling the risk of the transaction; in another extreme scenario, the regulated rate is established *a priori*, based on some estimation of the medium market price, probably subject to some subsequent adjustments, with the trader fully absorbing the risk in the purchase price. Reasonable regulation methods are between the extremes, limiting the risk for the trading company, but not completely, so that there is an incentive for taking an active part in the wholesale market by making the best possible use of the available transaction means.

1.4.3.6.1 Transactions in the Wholesale Market Context

In the wholesale market [12], generators, authorized consumers (generally starting from the larger ones), and trading entities of any kind (which trade with consumers at a regulated rate, which trade with clients who are able to choose their supplier, dealers, and brokers) may freely carry out transactions among each other, either at the spot market or by means of a contract. The relevant rules often establish restrictions on transactions between agents, sometimes just transitory, usually intended to prevent positions, leading to the abuse of dominating position when there is a strong horizontal concentration or vertical integration. The current trend worldwide is toward a total liberalization of the transaction means, for as much as it is possible.

Even though an organized market is not essential at the design level, all competitive electricity markets have established some kind of organized market with standardized transactions and generally with an anonymous (that is, nonbilateral) matching of production and demand offers. Such an organized market normally includes a spot market, with a daily horizon and hourly or semihourly matching intervals that serve as a reference for other transactions. In the more developed electricity markets, whenever the volatility and competitive conditions permit and there is sufficient contracting volume, organized electricity markets of derivatives or futures have arisen, providing the agents with a more flexible means of contract to manage the risks involved. In the organized spot market, generators usually offer their energy from 1 day to 24 h later. These production offers are matched to the demand offers by making use of very different proceedings that are based on economic criteria. Demand is simply estimated by any independent entity in systems where the demanding agents cannot supply. Each generator gets paid after 1 h for the energy it produces at the system marginal price, which is basically the price of the marginal offer at that hour. When the installed capacity of a given technology is well adapted to the system demand and to the rest of all the generating system, such payment for the energy at the marginal price of the system allows the recovery of the fixed and variable production costs [15]. New challenges are arising with higher penetration of zero or near zero marginal technologies such as wind and solar, creating conditions that may drive to a redesign of liberalized competitive markets.

The uncertainty about the spot price results in the use by both the generators and the demanding agents of several safeguard mechanisms against risk, among which bilateral agreements on spot price differences can be highlighted, which are solely economic in nature and are ignored when establishing a priority order when matching supply and demand offers. Other contract variables that are only authorized in some systems are the physical bilateral agreement that enables a sales agent to supply to a specific buyer, without making use of the mechanism of spot-market offers. This seems to be the current tendency: the coexistence of a voluntary spot market and physical bilateral agreements. However, the question whether the hypothetical freedom of action provided by these agreements compensates for the apparent regulatory and organized complexity of managing another type of transaction is still being discussed. International exchanges are a particular transaction case in the wholesale market where the novelty is the treatment to be given to foreign agents who are generally subject to different regulatory frameworks that are open to competition at another level.

The practical differences in the regulation of international exchanges arise from the reciprocity demands existing in the regulatory treatment between one country and another, since distribution among systems of the economic benefits provided by interconnections may critically depend on such reciprocity. Consistency within the multinational regulatory context where exchanges occur is essential. One extreme would be the total absence of regulatory integration, leading to transactions being subject to discretionary and negotiated access conditions to the network with no restriction on opportunistic behavior based on the position within the network or in the monopoly where certain transactions take place. The other extreme is one of a strong regulatory integration, which, as a minimum, guarantees access to all networks under regulated, transparent, and nondiscriminatory conditions. In this case, the rules adopted shall reach a minimum level of consistency in their design and application of access charges, that may avoid, for example, the repeated payment of network tolls in all of the systems supposedly affected by a transaction (the above-mentioned "pancaking") and be closest to the concept of a regional access toll, that would exist if the aggregate of all systems should constitute a single system [15].

1.4.3.6.2 Transactions in the Context of the Retail Market

At the retail level, consumers who are unable to choose their supplier have to purchase their energy at the regulated rate from the assigned trading company, which is typically closely related to the distributor to which they are physically connected. In most regulations, the company may be the same with separate accounting. Depending on the specific regulation, and normally for a transitory period, consumers who have the right to choose may be permitted to continue to purchase the energy from their original trading company at the corresponding regulated rate. Trading to consumers who do not have the option to choose, or to those who being able to choose are permitted to continue to pay at a regulated rate, is a regulated activity, which is paid for according to acknowledged costs and subject to standards of quality in the services provided to the client. Consumers with the right to choose may turn to any trading company to contract their electricity supply at a price that is freely negotiated between them. Such a price shall include the regulated rates for the transmission services and the distribution network to which they are connected, as well as other regulated charges that are applicable in each actual regulation and are adequately settled by the trading company. The price for the trading service and for the energy is freely set by the trading company. Consequently, the sale to consumers who have the right to choose is a nonregulated activity basically intended to manage risk. Indeed it is a business with very high monetary cash flow and a reduced margin of benefits based on the presumably high competition in the activity, which largely depends on adequate wholesale purchase management and its adjustment to negotiated supply contracts with qualified consumers.

1.4.3.7 Ancillary Activities

It is generally the market operator who settles the economic transactions performed in the different markets they manage. The operator may also be in charge of settling other transactions and concepts related to the wholesale market, such as ancillary services, losses, technical restrictions, balances of final deviations, and even, in some systems, bilateral agreements among agents. The settlement of regulated activities, such as transmission, distribution, or other issues, such as assistance to the special generation, is in general entrusted to the management, or the specialized and independent entity to which it delegates such activity.

Traditionally, measurement of consumption and the corresponding billing were an integral part of the trading activity while they formed an inseparable set with distribution. Under the new regulation it is possible that these activities take place independently by expert companies competing for the provision of these services. This is also the case with the installation of intakes for end users, an activity that is usually considered a part of distribution, but may be, and in fact often is, performed by competing independent installers. It is possible that in the future other activities may be identified, which can logically be executed separately, or new activities may appear, for instance within the purview of the revived philosophy of "multiutility," where trading companies offer other services apart from electricity.

1.4.3.8 Coordination Activities

1.4.3.8.1 Market Operation

Theoretically, agents could deal among themselves solely through any kind of bilateral relationship and in any period of time, as they may freely negotiate. However, as stated above, virtually all of the electricity systems that have adopted the new regulation have created at least one organized spot market, with very different formats but managed by an independent agency which in this document has been called the market operator (and corresponds to the usual English term "power exchange"). Often the same entity manages other organized markets that complement the spot market, either within a time limit still more reduced ("regulatory" or "intradaily" markets destined to make adjustments in generation or demand through competitive resources) or otherwise extending the time from weeks to years, that is, organized markets for forward electricity agreements. Since offer matching in all these markets takes place anonymously, the market operator must act as a clearing house. Even though every system has a specific entity that formally acts as a market operator, there is no problem, in principle, that other organizations may compete in this activity, both by offering electricity derivatives and by partially matching production and demand offers in the short term. The market operator must be seen as a "facilitator" of transactions rather than the agent of a specific electricity-related activity.

The most characteristic duty of the market operator is the management of a spot market. This market plays such a relevant role that its design is a constant central subject matter of the new regulation. A characteristic of spot-market models in the first systems that were open to competition was generally the passive role of demand, which basically took the form of an agent who accepted prices, who was obliged to cross this market to purchase or sell energy and used algorithmic optimization procedures, similar to those of the daily generation dispatch or weekly programing within the traditional regulatory framework, to match production and demand offers. Conversely, the tendency in the most recent spot market models is to (a) let production and demand offers act at arm's length conditions, both in quantity and in price, allowing agents to use physical bilateral agreements as an alternative to using of the spot market; (b) make matches in more real time, either by reducing the matching time limit in the main market, or through a succession or "zoom" of markets, or by giving priority to the final regulation market where production or consumption deviations by agents are valued in respect of scheduled amounts; and (c) simplify the form of offer submission as well as the corresponding matching process to make it more transparent, transferring the complex processes involved to the preparation of offers by the agents.

1.4.3.8.2   System Operation

The system operation is an activity intended to guarantee the functioning of the electricity system under safe conditions and in a way that is consistent with the production and consumption decisions made by market agents [6]. Strictly speaking, there is a classic coordination activity in every electricity system, the starting point of which in the new regulation is the result of matching offers and physical bilateral agreements, instead of the result of traditional proceedings aimed to minimize production costs. Such activity determines the actual production method by generators and the network operating instructions for transmission companies, so as to implement the market results provided by the market operator and the agents, if any, by way of physical bilateral contracts and considering the available technical restrictions, especially all those relating to safety in the system.

From its privileged position, the system operator, as a connoisseur of the functioning and technical limitations of the electricity system, shall be responsible for applying technical access criteria to networks and keeping the agents of the system informed of the foreseeable conditions of use of the same in the short, medium, and long term. Given the nature of the electric energy system, its operation is an activity that has to be performed in a centralized fashion, subject to regulation, regarding service costs, operation viewpoints, and control of actions. An essential aspect is that of the independence of the entity in charge of the system operation, the system operator, to ensure a nondiscriminatory treatment to agents, for example, in the application of technical restrictions to the transactions conducted in the market. The aspects relating to the independence of the system operator are reinforced when, as is reasonable due to the synergies existing among the different activities, it is entrusted with other tasks such as management of ancillary services or expansion planning or maintenance of the transmission network. And, as stated when discussing the network activities, a conflict of interest arises when the system operator also acts as a transmitter.

1.4.4   The Integration of Renewables

As already mentioned, the dramatic cost reduction of renewable technologies (mainly wind and solar) coupled to a compromise of the developed world to reduce its greenhouse emissions, has implied a growing penetration of variable renewable technologies into power systems, that pose both operational and regulatory challenges. The growing need is to incorporate flexibility into the systems to cope with the intermittent condition of wind plus the limited timeframe of solar generation, without compromising system reliability and maintaining affordable costs to consumers. Operational changes arise as intermittent wind couples to variable solar PV resources that are only present during the day, conditioning faster changes in supply and requiring other generation technologies to provide the complementary resources, responding within short time intervals. System operators have to assess the required reserves to respond to those changes. While nuclear and coal thermal do not have the flexibility to respond rapidly to those changes, gas and diesel turbines as well as hydro reservoirs may perform that role. Otherwise, storage technologies, mainly pumped storage, can provide rapid response to changes in supply. The different impacts of the integration of renewables into other generation technologies may be acknowledged, and new regulations and planning processes may be implemented to attenuate or compensate impacts. The creation of ancillary markets and the identification and remuneration of sources of flexibility help the system operator to face that integration and maintain adequate and secure supply.

1.4.5 PRACTICAL ASPECTS OF REGULATION [14]

This section outlines certain basic regulatory concepts, together with the fundamentals and requirements of the new electricity industry regulation. As noted earlier, this subject has been addressed in greater depth in other modules. However, no overview of this environment would be complete without at least a brief mention of certain general aspects that must be dealt with to practically implement the new legislation.

1.4.5.1 Transition to Competition [7]

A change from the traditional to the new regulatory model must define not only the final aim pursued, but also how to reach it. Obviously, the key conditioning factor is the baseline state of the electricity industry to be transformed—its organizational and ownership structure. When ownership is private, two types of possible difficulties may appear in connection with implementation of the new regulations. On the one hand, the probable maladjustment of existing generating plants to both demand and modern production technologies. This gives rise to what are known as "stranded costs" when the regulatory framework changes, since the market valuation of the maladjusted generating assets is normally lower than that acknowledge in the former traditional framework. On the other hand, the degree of business concentration and vertical integration may not be acceptable. What is relatively simple to solve when utilities are publicly owned, by passing the bill on to the taxpayer, is more difficult when private interests are involved. One possible strategy in the changeover to new arrangements consists of reaching an agreement with company shareholders whereby the government makes concessions respecting the first difficulty in exchange for greater company flexibility in solving the second.

1.4.5.2 Stranded Benefits

"Stranded benefits" is the term used by some authors to mean the social benefits forfeited when certain public goods are no longer produced in the move from a traditional to a competitive electricity industry. The public goods involved can be divided into three categories:

1. Protection for consumers unable to meet the real cost of electricity.
2. Environmental protection, since the cost of the environmental impact of electric power supply is not internalized in the market price.

3. Others, such as the diversification of production technologies, or R&D activities with no clear short-term impact on competitive business.

1.4.5.3 Environmental Costs

The substantial environmental impact of electric energy systems must be specifically addressed by the legislation, since the market will not be able to suitably deal with this aspect until environmental costs are included in the price of electricity. It is for this reason that the different regulatory frameworks must adopt measures to correct market failings in a number of areas:

- Encouragement of less polluting production technologies through specific regulatory mechanisms.
- Implementation of programs for energy savings and electric power DSM.
- Imposition of pollution limits and pollution-related excise taxes in electricity generation in place of real internalization of environmental costs.

1.4.5.4 Structural Aspects

Regulatory theory and experience to date show that, even where regulatory development is appropriate for each activity, if business structure is not suitable or consumer choice is not guaranteed, the new legislation will fail. A transitional period for the gradual adaptation of structures is, obviously, acceptable, but the basic requirements to ensure the existence of competition, which is essential to the new regulatory environment, are at least as follows:

- A limit to horizontal concentration, since competition needs well-matched rivalry; this entails establishing a minimum number and a maximum size of utility companies, relative to the volume of the geographically relevant market.
- A limit to vertical integration to prevent situations of privilege existing in one area of the business from being used to the detriment of competition in another: for example, the vertical integration of two activities, one conducted under regulated monopoly arrangements and the other on the free market.
- Consumer freedom of choice of supplier and producer and supplier access to the wholesale market.

1.4.5.5 Security of Supply in Generation

In the systems governed by the new regulatory provisions, production companies are generally under no obligation whatsoever to supply electric power, while there is no centralized planning of generating resources, which is left, rather, to private initiative. For distributors, the obligation is limited to being connected to the grid and providing network service and customer support. Supplier companies typically commit to purchasing electricity at the price applicable to their wholesale market operations. The question posed, therefore, is whether the market, of its own accord, will provide satisfactory security of supply at the power generation level or if some additional regulatory mechanism needs to be introduced. No international consensus has been reached in this regard, with countries opting for one alternative or the other. This is arguably the issue of greatest importance still waiting for a solution under the new regulatory scheme.

1.4.5.6 Independent Regulatory Body

This new regulatory model calls for an independent and specialized body whose chief function is to ensure fair competition and settle differences arising around market operation. This regulatory body should preferably be independent of government to prevent political aims from interfering with industry regulation, and to attain higher staff specialization, provide for greater transparency in the action taken, and ensure greater regulatory stability.

1.4.6 Environmental and Social Restrictions

Further than the control of the environmental costs created by electric power systems, particular emphasis must be placed on how the development of these systems affects society and especially the people involved. Fifty years ago, power systems were received as drivers of socioeconomic growth, but attitudes have changed and people are more concerned about impacts on the environment and sustainable development. Thus, public opposition to power projects is currently a major barrier for their further deployment. In effect, power infrastructure development affects the population directly involved, with costs and benefits that are not necessarily distributed in a fair way. Population living close to a new thermal power generation plant may not be satisfied with the physical impact of the plant nor share the benefits that it may provide to people living elsewhere. This is also true for transmission lines that cover thousands of kilometers and impact diverse and separate communities, which often oppose their development, as they do not see the benefits they carry. It is no uncommon to have large generation plants with long interconnections that benefit the metropolis or large industry, but impose local costs that are not necessarily compensated.

An impact that may create abrupt social reactions is the displacement of communities due to the impacts of new power plant construction. This is especially so in the construction of hydroelectric dams, that historically has originated substantial social and political conflicts. While remote urban population benefits from these dams the people that are displaced, as well as those living in the areas where the displaced population are relocated, may suffer social and economic deterioration. The construction process of the plants, with thousands of workers transiently moving into the area is a common condition for the development of social, housing, health and safety problems.

The need to create conditions that share benefits with communities and stakeholders affected in a transparent manner is central to power system expansion, succeeding in reducing social conflicts. Social issues should be considered from the initial steps of system planning, long before they are constructed. The international experience reports the utilization of different instruments: participation in decisions of infrastructure location, shared property of installations, economic compensations, economic stimulus to local community projects, interregional cooperation, modernization of roads during construction, reducing visibility impacts, after the fact monitoring of projects, among others. Social groups that learn early about the power system projects and engage and participate in the planning processes are less likely to conflict with the investors and the government.

1.4.7 Universal Access to Energy

There is a rather large difference between the challenges regarding electrification in countries with a mature electric infrastructure and in developing economies. In most countries with a mature electric infrastructure access to electricity is taken for granted, though utilities are grappling with the challenge to connect a variety of distributed generators to existing distribution grids and nevertheless maintaining security of supply. The efforts of power utilities and other stakeholders in countries in the developing world are directed to provide access to reliable electricity services in yet un-electrified or poorly electrified areas, and to cope with the affordability problems of poor communities. The potential benefits of electrification, apart from the benefits related to the improvement of the living conditions, include socioeconomic benefits, sociopolitical benefits, and environmental benefits.

Access to electricity is fundamental and it is the connection needed to bond the third world into the global economy, particularly for continents like Africa and Asia. In effect, as the World Bank indicates, about 1.2 billion people do not have access to electricity worldwide. The bank indicates that, although 1.8 billion people obtained connections to electricity between 1990 and 2010, the rate was only slightly ahead of the population growth of 1.6 billion over the same period. Electricity expansion growth will have to double to meet a 100% access target by 2030, and getting there would require an additional $45 billion invested in access every year, five times the current annual level.

The challenge is particularly important in rural electrification, which will not take place on its own, unless there is a clear direction or support by the government, either in creating the institutional framework for it to take place or in supplying the necessary basic funding or subsidies for it to develop. It is important that this takes place based on clear social economic objectives. Rural electrification is not just one of supplying rural households with basic electricity, but looking at the economic impact that electricity may have on poor communities.

1.4.8 THE TRENDS IN REGULATION: INTERNATIONAL EXPERIENCES [17]

The innovations that were started in the 1980s in the electricity industry, where the basic activities of electricity supply were separated and competitive wholesale markets were created, have extended worldwide. Experience in the introduction of wholesale competition has been mixed and good and bad reform processes have taken place. The positive deregulation of Australia and the Nord Pool countries (Denmark, Finland, Norway, and Sweden) have stimulated new reforms and improvements of existing ones. The California reform 2000–2001 fiasco was a hard blow for changes being formulated elsewhere. Thus, supporters and opponents of the changes have arisen, arguing to either extend competition or to retract reform. Nevertheless, the ongoing tendency continues to be toward deregulation and introduction of competition, with the new processes learning from the successes and failures of previous regulatory schemes.

Countries in advanced stages of reform have continued implementing and refining competition, for example introducing full retail competition by opening their domestic residential markets. In some of the countries, such as Sweden, a household consumer may even buy electricity from suppliers in any Nordic country. While load management was seen as a tool to defer investment in preform monopolistic energy supply, demand price response is now searched for as a means to face energy shortages in competitive markets.

As new efficient thermal generation technologies develop through the combined cycle natural gas plants, electricity markets have become intermingled with gas markets. The need to make coherent arrangements in both markets is seen as a necessity. In this new context, electric transmission expansion needs to be assessed in direct relation to the development of gas duct networks.

Energy security has become an issue worldwide, given the dependence of many large world economies on fossil fuel supply from third countries, and the political use that is being made of energy resources. This is coupled to an increase in prices for oil and gas, given escalating global demand for fuels and stretched supply chains. The energy security issue has become so relevant that the European Union defined a policy for energy security. The policy identifies the need to diversify Europe's energy sources, entering into strategic partnerships of consumers with major potential suppliers, and to strengthen coordination and cooperation amongst countries in the internal energy market.

Coupled with the above developments, electricity trade is crossing international boundaries and are merging to form multinational markets. The merging of trading structures implies the need to harmonize regulations as well as operational practices. Of particular interest is the need to harmonize tariffs for the use of transmission networks for international trade, with open access to wires being extended across boundaries. Multinational energy integration also requires the creation of supranational institutions to coordinate system operation as well as for conflict resolution. The global aim of enforcing wholesale competition must be always kept in mind.

An issue that has been a continuous source of discussion since the first reforms took place is: how to achieve an adequate expansion of supply? The supply adequacy objective aims at ensuring an optimal level of overall generation capacity to respond to growing demand and an optimal mix of generation technologies to ensure continuity of supply. The discussion has centered on the needed economic signals for new installed capacity, fundamental in markets with significant growth and highly subject to supply shocks (for example, droughts). Essentially three regulatory paths have developed worldwide to provide these economic signals:

1. *Capacity payments*: This method remunerates investment in generation by its contribution to peak capacity, related to its technology and availability, independent of its energy contribution. Besides paying for consumed energy, consumers pay an uplift proportional to their maximum demand.
2. *Capacity markets*: Capacity markets have developed based on a contract reliability base and financial options. Distributors are requested to contract a product "capacity," assuring a certain minimal level of long-term reliability.
3. *Energy only markets*: Generators and consumers interact through energy spot prices, without restrictions, resultant from a bid-based dispatch. Remuneration of installed capacity comes from the difference between the resultant market energy spot price and the production variable costs.

Finally, while deregulation aimed at promoting competition in generation and supply markets, regulations were introduced to deal with the monopolistic segments of the electricity chain. Different forms of incentive-based regulation of transmission and distribution tariffs have been introduced, with governments and regulators taking the lead, and successfully stimulating cost reductions that have benefited the final consumer. However, the risk that frequently arises is the potential for governmental and regulatory intervention in these markets, with the excuse to protect customers from monopoly or achieve new social benefits. The boundary between minimum government intervention to correct market failures and direct full intervention to orientate markets is easily crossed. Energy security, subsidy of renewables, nuclear power, and transmission expansion are just a few examples where governments have increasingly intervened the market to orientate certain social changes, particularly when markets have faced difficulties in energy supply.

1.5 MODELING REQUIREMENTS OF MODERN ELECTRIC ENERGY SYSTEMS

As electric energy systems grow in size and complexity, the need to adequately model their operation and expansion becomes a growing challenge for engineers. The alternative operational conditions and contingencies that affect systems, make system operation a multidimensional mathematical problem, that not only needs to consider technical issues but also economic ones. On the other hand, the design of future system expansions often requires to consider multiple stochastic conditions that imply handling thousands, if not million, variables and options. The building of mathematical models for generation plants, system controls, networks, loads, and other equipment is a challenge. Data collection and handling to feed those models are also demanding tasks. There are so many elements that interact in a diversity of time frameworks in major scale networks that the challenge keeps growing.

Thus, since the birth of computers and digital processing in the 1940s, they have been used extensively in electric energy system studies. Digital models and calculation tools have been developed in a wide range of applications, with engineers in the field trying the latest state-of-the-art advances to assess and simulate system behavior better. Computer algorithms have been developed for, at least, the following areas of analysis, several dealt with in later chapters:

1. Load flow analysis (power flow and voltage analysis and control)
2. Short circuit analysis
3. System stability (angular and voltage stability)
4. Electromagnetic transients
5. Economic operation (economic dispatch, unit commitment, hydrothermal coordination, optimal load flow)
6. Protection coordination
7. Expansion planning
8. Reliability analysis [19]

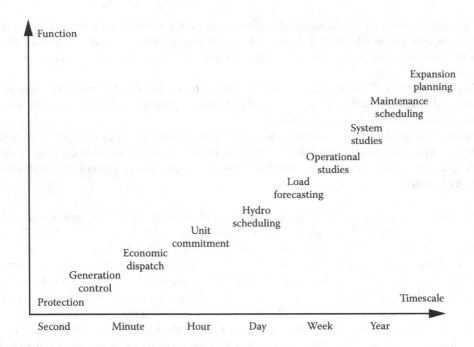

FIGURE 1.16   Timescales and computer studies.

9. Risk assessment
10. Load forecasting
11. State estimation

Even with increasingly powerful computer technologies (processing times and digital storage), the difficulty to study energy systems makes it impossible to handle simultaneously steady-state and dynamic conditions, so that a study time segmentation is often made, depending on time constants. Computer analysis is made separately of problems dealing with steady state, mid-term, and long-term time constants, as illustrated in Figure 1.16.

The study of electric energy system engineering problems involves the development of special computer programs that require mathematical and operational research algorithms (linear and non-linear, real and integer, continuous and discrete, deterministic, and stochastic) to study steady-state conditions, and eigenvalue and step-by-step integration algorithms to study dynamic conditions.

Another dimension of computer use in electric energy systems is in the online control of equipment and networks. Data collection, online monitoring, and control make direct use of computers to operate complex networks. Man–machine interface then becomes a main need, to provide analysts with the minimum and sufficient results to take actions in real-time applications.

1.6   FUTURE CHALLENGES AND PROSPECTS

This section is not intended to be a lengthy or detailed discussion of the prospects for change in the electricity industry, presently one of the most dynamic sectors of the economy. Rather, it contains a necessarily incomplete but hopefully representative annotated list of technological, economic, or regulatory developments that are expected to acquire relevance in the electricity industry in the years to come. Some of these may change the industry radically in the future. In most of the areas, the enormous potential for change lies in the interactions between technology, economics, and regulation.

- *Distributed generation*: As discussed in the chapter, distributed generation, demand management, and self-consumption may radically change the way power systems are understood at the time of writing, challenging seriously the business models of traditional utilities in this sector. Major changes could be expected in this sense.
- *Off-grid rural electrification*: The most suitable solutions for electric service to the over 2 billion people who presently lack access to electricity may often be small isolated grids or individual systems, using suitable distributed generation technologies.
- *Environmental and strategic considerations*: Environmental restrictions and the progressive internalization of costs deriving from environmental impact, together with long-term approaches to security of supply, will have a gradual and significant effect on future investment in new production resources. Measures originated in climate change concerns will growingly shape future power system expansion. Specific markets for "green electricity" and pollutant emissions will tend to sprout up everywhere.
- *Use of the electricity grid for telecommunications*: Recent technological developments make it possible to use local distribution networks for the high-speed transmission of information. Electric utilities may therefore provide competitive Internet services as well, very likely, as many others yet to be defined, such as remote metering of electricity consumption or DSM.
- *Multiutilities*: As some electric utilities did in the past, today's companies are beginning to offer additional services (gas, water, or telecommunications distribution) in a single package to take advantage of the synergies existing in these various lines of business.
- *Storage devices*: Development of affordable batteries or hydrogen based storage technologies will represent a paramount change in power systems. These technolgies will become specially relevant in the context of the expected very large penetration of renewables to comply with the decarbonization goals commited by the international comunity.
- *Superconductors*: Although the use of these components in electric power is still limited to the manufacture of large electromagnets and experimental facilities, superconductivity may change the future design of large transmission equipment, particularly in and around large cities.
- *FACTS*: The difficulties encountered to enlarge transmission grid capacity and the technical and economic problems caused by loop flows, especially in multinational markets, will further the development of electronic devices to control grid flows to optimize the use of the individual transmission capacity of each grid. Their universal use will change the traditional approach to transmission grid supervision and control.
- *HVDC grids*: DC technology may enlarge significantly its role thanks to its clear advantages, provided costs continue to decline and some technical problems such as the DC breaker capabilities are overcome.
- *Technical and economic management of regional markets*: The creation of multinational electricity markets, consisting of the coordinated operation of organized competitive markets or energy exchanges in parallel with countless bilateral transactions, while leaving technical management of security of operation in the hands of a series of independent system operators, poses complex organizational problems, which must be suitably solved if both system safety and economic efficiency are to be suitably guaranteed. A paradigmatic example of such a problem is the coordinated management of grid restrictions. New organizational schemes and communications and information systems, along with coordination models and algorithms, will have to be developed in response to this challenge. Important progresses have been already achieved in this sense in regional markets such as the European one or the Central American one.
- *Electricity trading in the digital economy*: The liberalization of an industry as important as electric power, together with the development of e-commerce, is already leading to the spectacular development of electricity trading on the Internet, with products ranging from

long-term contracts to online purchases and including risk insurance, to cover climate risk, for instance. Blockchain technology can solve data management problems.

* *Clean development mechanisms*: Clean development mechanisms have created an international emission market under the Kyoto Protocol. The market allows industrialized countries committing to reduce greenhouse gas emissions to invest in emission reducing projects in developing countries. Emission reductions are paid to new generators that can demonstrate that through their investment they are replacing other generating facilities that would emit CO_2.

REFERENCES

1. From Torá, J. L., *Transporte de la energía eléctrica*, Universidad Pontificia Comillas (ICAI-ICADE), Madrid, 1997.
2. The World Bank, *Energy Services for the World's Poor: Energy and Development Report 2000*, ESMAP Report, Washington DC, 2000.
3. Tenenbaum, B., *The Real World of Power Sector Regulation*, The World Bank, Viewpoint, June 1995.
4. Grainger, J. J. and Stevenson, J. W. D., *Power System Analysis*, McGraw-Hill, New York, 1996.
5. Elgerd, O., *Electric Energy Systems Theory: An Introduction*, McGraw-Hill, New York, 1982.
6. Kundur, P., *Power System Stability and Control*, McGraw-Hill, New York, 1994.
7. Hunt, S., *Making Competition Work in Electricity*, Wiley Finance, New York, 2002.
8. Wood, A. J. and Wollenberg B. F., *Power Generation, Operation and Control* (2nd ed.), John Wiley & Sons, New York, 1996.
9. Saadat, H., *Power System Analysis*, McGraw-Hill, New York, 1999.
10. Bergen, A. R., and Vittal, V., *Power Systems Analysis*, Prentice Hall, 1999.
11. Schweppe, F. C., Caramanis, M. C., Tabors, R. D., and Bohn, R. E., *Spot Pricing of Electricity*, Kluwer Academic Publishers, Norwell, MA, 1988.
12. Stoft, S., *Power System Economics: Designing Markets for Electricity*, Wiley-IEEE Press, New York, 2002.
13. Rothwell, G., and Gómez, T., *Electricity Economics: Regulation and Deregulation*, Wiley-IEEE Press, Piscataway, NJ, 2003.
14. Newbery, D. M., *Privatization, Restructuring, and Regulation of Network Utilities*, The MIT Press, Cambridge, MA, 2000.
15. Pérez-Arriaga, I. J., *Regulation of the Power Sector*, Springer, London, 2013.
16. http://www.eia.doe.gov
17. https://www.iea.org/
18. http://www.ers.usda.gov
19. http://www.nerc.com
20. http://stoft.com/
21. http://www.cigre.org/
22. http://www.iec.ch/

2 Steady-State Single-Phase Models of Power System Components

Antonio F. Otero and José Cidrás

CONTENTS

2.1 MODELING ELECTRICAL NETWORKS

Any type of analysis and simulation of an electric energy system has to be performed by means of a mathematical model which, based on a set of equations with diverse complexity degrees, allows us to know the behavior of these systems under different conditions. As it was mentioned in the previous chapter, the size and sophistication of electric energy systems are very high. Therefore, the claim of an accurate and valid model for all operating conditions can be computationally very expensive or even unaffordable.

In this chapter, the single-phase equivalent models for the basic elements of the electric energy systems under sinusoidal steady-state conditions and with symmetrical loads are described. In particular, elements considered are: transmission lines, transformers, synchronous machines, asynchronous machines, and electric loads.

2.2 PER UNIT VALUES

The analysis of electric energy systems can be numerically simplified using the *per unit values* normalization technique. The *per unit value* (V_{pu}) is obtained dividing the physical quantity (V) by a defined *reference value* (V_B):

$$V_{pu} = \frac{V}{V_B} \tag{2.1}$$

In electrical systems, the following four basic quantities are involved: voltage ($\mathcal{V}$), power ($\mathcal{S}$), current ($\mathcal{I}$), and impedance ($\mathcal{Z}$). The simplifying capability of the *per unit values* is determined by a suitable choice of the reference values in a way that the basic modular electrical relationships of a three-phase system are met:

$$Z = \frac{V}{\sqrt{3} \cdot I}; \quad S = \sqrt{3} \cdot V \cdot I \tag{2.2}$$

where S is the three-phase apparent power, V the phase-to-phase voltage, I the phase current, and Z the phase impedance (assuming a star-connected equivalent system).

Therefore, choosing two of the four reference values (typically apparent power and voltage), the remaining two values can be calculated. For instance, in a three-phase system, choosing a phase-to-phase reference voltage V_B and a three-phase reference power S_B, the reference current and impedance are

$$Z_B = \frac{V_B^2}{S_B}; \quad I_B = \frac{S_B}{\sqrt{3}V_B} \tag{2.3}$$

Every system quantity can be expressed in per unit values dividing it by the corresponding reference value:

$$\mathcal{V}_{pu} = \frac{\mathcal{V}}{V_B}; \quad \mathcal{I}_{pu} = \frac{\mathcal{I}}{I_B}; \quad \mathcal{S}_{pu} = \frac{\mathcal{S}}{S_B}; \quad \mathcal{Z}_{pu} = \frac{\mathcal{Z}}{Z_B} \tag{2.4}$$

The analysis of the electric system expressed in per unit values is more clarifying and simpler. This is especially true for the voltages. If reference voltages are chosen matching the rated voltages, all the per unit voltage values will be around unity. For any value other than 1, we will have a direct and immediate perception of the voltage variation level compared to the rated value.

Moreover, using per unit values the $\sqrt{3}$ multiplying factor of three-phase Equations 2.2 disappears

$$Z_{pu} = \frac{V_{pu}}{I_{pu}}; \quad S_{pu} = V_{pu} \cdot I_{pu} \tag{2.5}$$

Another interesting advantage of the *per unit values* is achieved in the case of electrical power transformers in systems with different voltage levels. If reference voltages at both sides are chosen matching the rated voltage ratio, the transformer ratio becomes "1" in per unit values. In addition, the transformer short-circuit impedance (usually expressed in percent value), when divided by 100, will get the same value as the per unit short-circuit impedance of the transformer referred to its rated voltage and power.

It is not unusual to find in the same system electrical elements expressed in per unit or percent values referred to different reference quantities. For instance, transformers and electrical machines internal impedances are usually shown referring to their different rated values. Then, a change of reference has to be performed in the way that a pu quantity $\mathcal{V}_{pu,1}$ referred to a reference value V_{B1} can be referred to another different value V_{B2} as follows:

$$\mathcal{V}_{pu,2} = \mathcal{V}_{pu,1} \cdot \frac{V_{B1}}{V_{B2}} \tag{2.6}$$

EXAMPLE 2.1

Let us apply the *per unit values* technique to the system of Figure 2.1. It contains three different voltage level sections.

First, a reasonable reference power of $S_B = 100$ MVA for all the system is considered. Then, the reference voltage for one of the sections has to be chosen. For example, $V_{B_{S1}} = 220$ kV at the left side of the T1 transformer. From here, the two other reference voltages are determined applying the rated transformer ratios and from Equation 2.3, the reference impedances and currents are obtained (see Table 2.1).

The pu values of the elements of the equivalent circuit (see Figure 2.2) have to be calculated. Because of the rated values of the T1 transformer are equal to the chosen reference values, its reactance in pu is $X_{12_{pu}} = 0.11$ pu straightaway. However, the T2 transformer has a different rated power, so the following change of reference has to be performed:

$$X_{34_{pu}} = 0.08 \cdot \frac{132^2}{20} : \frac{132^2}{100} = 0.4 \text{ pu} \tag{2.7}$$

The impedance of line 2–3 given in Ω has to be divided by the corresponding reference impedance:

$$R_{23_{pu}} = \frac{2}{174.24} = 0.0115 \text{ pu}; \quad X_{23_{pu}} = \frac{6}{174.24} = 0.0344 \text{ pu} \tag{2.8}$$

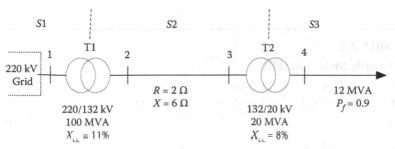

FIGURE 2.1   One-line diagram of the example system.

TABLE 2.1

Example Reference Values

	S1	**S2**	**S3**	
		Section		
S_B		100		MVA
V_B	220	132	20	kV
Z_B	484	174.24	4	Ω
I_B	262.4	437.4	2886.7	A

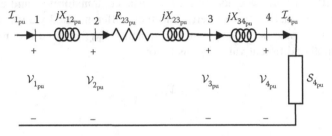

FIGURE 2.2 Equivalent circuit in *per unit values.*

Finally, the load at point 4

$$\mathcal{S}_{4_{pu}} = \frac{12}{100} \underline{/a\cos(0.9)} = 0.12\underline{/25.8}\,\text{pu} \tag{2.9}$$

Assuming the given load voltage is $V_4 = 20\underline{/0}$ kV, then $V_{4_{pu}} = 1\underline{/0}$ pu, all the voltages and currents at any system point can be easily obtained from the pu circuit of Figure 2.2 (see Table 2.2).

2.3 TRANSMISSION LINES

Electric power lines are an essential part of electric systems whose role is the transmission of the electric energy between two points in a safe, reliable, and efficient way. Its performance is characterized by four basic electrical parameters that are uniformly distributed along the whole length of the line: resistance, inductance, capacitance, and conductance.

In most situations, three-phase electric lines are unbalanced. For long lines, the unbalance degree can be reduced by means of a phase transposition procedure. For shorter lines, the unbalance is small

TABLE 2.2

Example System Results

	1	**2**	**3**	**4**	
V_{pu}	$1.031\underline{/3.23}$	$1.025\underline{/2.59}$	$1.022\underline{/2.42}$	$1\underline{/0}$	pu
I_{pu}		$0.12\underline{/-25.8}$			pu
V	$226.9\underline{/3.23}$	$135.3\underline{/2.59}$	$134.9\underline{/2.42}$	$20\underline{/0}$	kV
I	$31.49\underline{/-25.8}$	$52.49\underline{/-25.8}$		$346.4\underline{/-25.8}$	A

enough to be negligible. Anyway, in most practical cases, the analysis of an electric line can be performed by means of a simple single-phase equivalent circuit that will be described later in this chapter.

Considering a sinusoidal steady state at an angular frequency ω, the line parameters are usually grouped into a serial impedance $\mathcal{Z}'$ and a shunt admittance $\mathcal{Y}'$ given by:

$$\mathcal{Z}' = R' + j\omega L' \quad (\Omega/\text{km}) \tag{2.10}$$

$$\mathcal{Y}' = G' + j\omega C' \quad (\Omega^{-1}/\text{km}) \tag{2.11}$$

where $R'(\Omega/\text{km})$, $L'(H/\text{km})$, $C'(F/\text{km})$, and $G'(\Omega^{-1}/\text{km})$ are the electrical parameters per unit of length and per phase.

2.3.1 Electrical Parameters of Lines

2.3.1.1 Resistance

The electrical resistance of a solid homogeneous conductor for direct current can be obtained as:

$$R_{dc} = \rho \frac{l}{S} \tag{2.12}$$

being l its length, S its cross-sectional area, and ρ the resistivity of the material (Table 2.3).

In practice, the above expression is not useful. Typical conductors used in transmission lines are not strictly solids but made from a number of stranded wires and, in many cases, combining more than one material. Furthermore, conventional electric systems are circulated by alternating currents (50 or 60 Hz) and an increment of resistance due to *skin effect* has to be considered. For the extremely low frequencies used in electric energy systems this phenomenon is quite small and very difficult to be analytically evaluated. Fortunately, conductor a.c. resistances can be obtained from manufacturer data sheets or standards.

Another important factor that affects the resistance of a conductor is the temperature. For conductor materials, the resistivity increases with temperature according to an approximately linear relationship. Therefore, resistances R_1 and R_2 at two different temperatures θ_1 and θ_2 are related by the following expression:

$$R_2 = R_1(1 + \alpha(\theta_2 - \theta_1)) \tag{2.13}$$

where α is a coefficient depending on the conductor material (Table 2.3).

A suitable design involves the evaluation of the conductors resistance at the maximum allowable working temperature. For insulated cables, this temperature is limited by the insulator temperature rating (i.e., 90°C for EPR and XLPE). For overhead lines with bare conductors, the maximum design temperature is mainly determined by the maximum line sag and the admissible clearances to ground. Depending on the countries and their regulations this maximum temperature will be in the range of 50–90°C.

TABLE 2.3

Typical Values of Resistivity and Temperature Coefficient

	ρ $(\Omega \cdot \text{mm}^2/\text{m})$	α $\left(\frac{1}{C}\right)$
Copper	0.017	3.9×10^{-3}
Aluminum	0.028	4.0×10^{-3}

In bundled conductors lines where the phases are built with more than one conductor (usually, 2, 3, or 4), the per-phase resistance of the line is given by

$$R' = \frac{R_c}{n_x} \tag{2.14}$$

where R_c is the resistance of an individual conductor and n_x the number of conductors per phase.

2.3.1.2 Inductance

An electric current flowing through a conductor generates a magnetic field of concentric circular lines around the conductor. If the current, and therefore the field, varies with time, a voltage (emf) will be induced in any electrical circuit linked by the magnetic field flux ϕ according Faraday's Law:

$$v(t) = \frac{d\phi}{dt} \tag{2.15}$$

The inductance L is the constant coefficient that relates the magnetic flux with the current and it only depends on the geometry of the circuits:

$$L = \frac{\phi}{\mathcal{I}} \tag{2.16}$$

In three-phase electric lines, this phenomenon can be represented by means of the self-inductances for each phase and mutual inductances between any two phases. Considering a generic phase x consisting of n_x equal conductors of radius r, the self-inductance coefficient is calculated by the following expression:

$$L_{xx} = \frac{\mu_0}{2\pi}\left(\frac{k_r}{4n_x} + \ln\frac{d_\infty}{rmg_x}\right) \tag{2.17}$$

where rmg_x is the geometric mean radius of the phase x and represents the geometric mean of the distances between all the conductors belonging to this phase x:

$$rmg_x = \sqrt[n_x n_x]{\prod_{i=1}^{n_x}\prod_{j=1}^{n_x} d_{x_i x_j}} \tag{2.18}$$

For cylindrical conductors, in the above expression any ii factor representing the distance from a conductor to itself is equal to its radius, $d_{x_i x_i} = r_i$.

The k_r coefficient depends on the shape and the structure of the conductors, being $k_r = 1$ for cylindrical solid conductors. For any other type of conductor, this coefficient should be either calculated or taken from manufacturer data. Table 2.4 shows the k_r values for concentrically stranded conductors with two or more layers of homogeneous cylindrical wires. In practice, $k_r = 1$ is commonly used.

TABLE 2.4

Values of k_r for Stranded Conductors

	Number of Layers				
	2	3	4	5	6
No. conductors	7	19	37	61	91
k_r	1.281	1.108	1.056	1.035	1.025

Considering another phase y consisting of n_y conductors, the mutual inductance between both phases (x, y) is

$$L_{xy} = \frac{\mu_0}{2\pi} \ln \frac{d_\infty}{dmg_{xy}} \tag{2.19}$$

where dmg_{xy} is the geometric mean distance between the two phases x and y and is calculated from

$$dmg_{xy} = \sqrt[n_x n_y]{\prod_{i=1}^{n_x} \prod_{j=1}^{n_y} d_{x_i y_j}} \tag{2.20}$$

The d_∞ distance mathematically represents the integration upper limit for the calculation of the magnetic fluxes linking every conductor. If we assume that the sum of the phase currents is zero, this distance has to be a value large enough (theoretically infinite) to make the sum of the magnetic fields generated by the three phases neglectable. However, if the sum of the phase currents is not null, there will be a return conductor (real or fictitious) and d_∞ will express the distance to it (i.e., ground return). Its value depends on the ground characteristics and can be calculated from more or less complex equations according to the operating conditions of the line [1,2].

The constant $\mu_0 = 4\pi 10^{-7} H/m$ is the permeability of vacuum (magnetic constant) and inductances are obtained in the same units.

In a three-phase line, the inductive phenomenon can be written in a matrix form:

$$\begin{pmatrix} \phi_1 \\ \phi_2 \\ \phi_3 \end{pmatrix} = \begin{pmatrix} L_{11} & L_{12} & L_{13} \\ L_{21} & L_{22} & L_{23} \\ L_{31} & L_{32} & L_{33} \end{pmatrix} \cdot \begin{pmatrix} \mathcal{I}_1 \\ \mathcal{I}_2 \\ \mathcal{I}_3 \end{pmatrix} \tag{2.21}$$

If the three phases of the line are identical ($rmg_1 = rmg_2 = rmg_3 = rmg$) and equally separated ($dmg_{12} = dmg_{13} = dmg_{23} = dmg$), the inductance matrix is perfectly balanced being $L_{11} = L_{22} = L_{33} = L_s$ and $L_{12} = L_{13} = L_{23} = L_m$. Moreover, if the condition $\mathcal{I}_1 + \mathcal{I}_2 + \mathcal{I}_3 = 0$ is fulfilled, a decoupled inductance per phase can be obtained as:

$$L' = L_s - L_m = \frac{\mu_0}{2\pi} \left(\frac{k_r}{4n_x} + \ln \frac{dmg}{rmg} \right) \tag{2.22}$$

In practice, real lines are never perfectly balanced. However, if a transposition of the phases is considered as a normal situation, the inductance per phase can be obtained with a similar expression with

$$rmg = \sqrt[3]{rmg_1 \cdot rmg_2 \cdot rmg_3} \tag{2.23}$$

and

$$dmg = \sqrt[3]{dmg_{12} \cdot dmg_{13} \cdot dmg_{23}} \tag{2.24}$$

2.3.1.3 Capacitance

The capacitance of power lines is associated with the existence of the electric field generated by the conductors.

In three-phase power lines, the relationship between potentials and electric charges in the different phases is expressed in terms of the denominated self and mutual *coefficients of potential*. Assuming the earth to behave as a flat conductor at zero potential, the traditional method of images can be used to analyze the electric field generated by the conductors and to get the *coefficients of potential*.

For a generic phase x consisting of n_x conductors, the self-coefficient of potential is given by

$$P_{xx} = \frac{1}{2\pi\varepsilon_0} \ln \frac{hmg_x}{rmg_x} \tag{2.25}$$

where hmg_x is the geometric mean of the distances from the x-phase conductors to their mirror images with respect to the ground plane. The mutual coefficient of potential between the phase x and another phase y with n_y conductors is

$$P_{xy} = \frac{1}{2\pi\varepsilon_0} \ln \frac{hmg_{xy}}{dmg_{xy}} \tag{2.26}$$

where hmg_{xy} is the geometric mean of the distances from the x-phase conductors to the mirror images with respect to ground of the y-phase conductors.

The constant $\varepsilon_0 \approx 8.854 \times 10^{-12}$ F/m is the vacuum permittivity and capacitances will have the same units.

In a three-phase line, the relationship between coefficients of potential and electric charges can be written in a matrix form as

$$\begin{pmatrix} V_1 \\ V_2 \\ V_3 \end{pmatrix} = \begin{pmatrix} P_{11} & P_{12} & P_{13} \\ P_{21} & P_{22} & P_{23} \\ P_{31} & P_{32} & P_{33} \end{pmatrix} \cdot \begin{pmatrix} Q_1 \\ Q_2 \\ Q_3 \end{pmatrix} \tag{2.27}$$

Under balanced conditions, the above relationship can be simplified into a decoupled per-phase coefficient given by

$$P' = \frac{1}{2\pi\varepsilon_0} \ln \frac{dmg \cdot hmg_s}{rmg \cdot hmg_m} \tag{2.28}$$

where

$$hmg_m = \sqrt[3]{hmg_{12} \cdot hmg_{13} \cdot hmg_{23}} \tag{2.29}$$

and

$$hmg_s = \sqrt[3]{hmg_1 \cdot hmg_2 \cdot hmg_3} \tag{2.30}$$

The per-phase capacitance is just the inverse value:

$$C' = \frac{2\pi\varepsilon_0}{\ln(dmg \cdot hmg_s)/(rmg \cdot hmg_m)} \tag{2.31}$$

Given that in typical overhead lines $hmg_s \approx hmg_m$, the following simplified expression is commonly used

$$C' = \frac{2\pi\varepsilon_0}{\ln dmg/rmg} \tag{2.32}$$

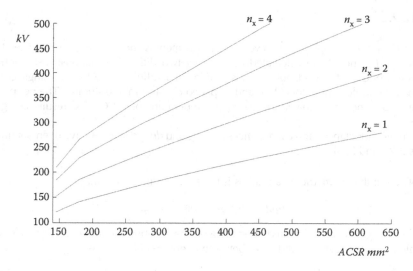

FIGURE 2.3 Typical values for corona critical voltage (phase to phase).

2.3.1.4 Conductance

The conductance G' represents the active power losses P_g produced by the current flowing through the insulating elements and the air due to their nonideal insulating condition:

$$G' = \frac{P_g}{V^2} \tag{2.33}$$

The *conductance* is analytically difficult to evaluate because it strongly depends on the highly random environmental conditions and the surface state of the conductors.

In overhead transmission lines the major part of these losses comes from the so-called *corona effect* which also generates other undesirable effects as audible noise and radio interferences.

This phenomenon occurs when the phase-to-ground voltage is high enough to provoke the electric field on the surface of conductors to exceed the value of the dielectric strength of air. Therefore, there is a *critical voltage* value above which the *corona effect* begins and it should not be overcome.

The way to reduce the *corona effect* (and therefore to raise the *critical voltage*) is by decreasing the value of the electric field on the surface of the conductors. This can be done using higher diameter conductors or more than one conductor per phase (bundle conductors). Figure 2.3 shows typical values of the critical voltage in standard conditions for different size conductors and for 1 to 4 bundle conductors per phase.

Typical values for overhead line parameters are given in Table 2.5.

TABLE 2.5

Typical Values for Overhead Line Parameters (Per Circuit)

	n_x	R' (Ω/km)	L' (mH/km)	$X'_{50\,Hz}$ (Ω/km)	C' (nF/km)
66 kV	1	0.1–0.2	1.1–1.2	0.34–0.38	10
132 kV	1	0.06–0.1	1.1–1.2	0.34–0.38	9
	2	0.03–0.05	0.8	0.25	14
220–400 kV	1	0.05–0.08	1.2–1.4	0.38–0.44	9
	2	0.025–0.04	0.9	0.28	12
	4	0.012–0.02	0.7	0.22	16

EXAMPLE 2.2

A double circuit transmission line with vertical arrangement is shown in Figure 2.4. The electrical parameters R', L', and C' are calculated considering two different situations: first, each circuit independently (S-C) and second, both circuits coupled in parallel (D-C). Also, one single conductor per-phase and bundle conductors (2, 3, and 4 spaced $d = 40$ cm) are studied. The per-phase a.c. resistance is obtained at a maximum working temperature of 70°C. The results are given in Table 2.6.

To calculate inductances and capacitances the rmg and dmg distances have been obtained with Equations 2.23 and 2.24.

- S-C case: the geometric mean radius is the same for the three phases:

$$rmg_1 = rmg_2 = rmg_3 = rmg_b \qquad (2.34)$$

where rmg_b can be calculated depending on the number of conductors per phase as indicated in Figure 2.4. The distances between every two phases are

$$dmg_{12} = d_{12}; \quad dmg_{13} = d_{13}; \quad dmg_{23} = d_{23} \qquad (2.35)$$

which can be obtained from the figure.

- D-C case: for each phase i

$$rmg_i = \sqrt{rmg_b \cdot d_{ii'}} \qquad (2.36)$$

and between every two phases i and j

$$dmg_{ij} = \sqrt[4]{d_{ij} \cdot d_{ij'} \cdot d_{i'j} \cdot d_{i'j'}} \qquad (2.37)$$

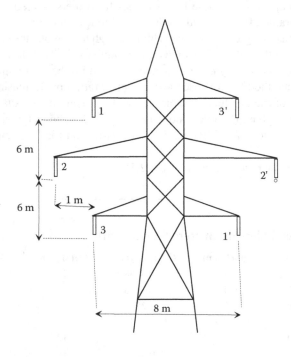

$$V_r = 220 \text{ kV}$$
$$402\text{-Al1}/52\text{-ST1A}$$
$$r = 13.85 \text{ mm}$$
$$R_{ac,70°C} = 0.091 \ \Omega/\text{km}$$

	rmg_b
⊙	r
⊙ ⊙ (d)	$\sqrt{r \cdot d}$
⊙ ⊙ ⊙	$\sqrt[3]{r \cdot d^2}$
⊙ ⊙ ⊙ ⊙	$\sqrt[4]{r \cdot d^3 \sqrt{2}}$

FIGURE 2.4 Example line.

TABLE 2.6

Parameters of the Example Line

	n_x	R' (Ω/km)	L' (mH/km)	C' (nF/km)
S-C	1	0.091	1.31	8.81
	2	0.054	0.95	12.01
	3	0.030	0.83	13.70
	4	0.023	0.75	15.02
D-C	1	0.045	0.65	18.4
	2	0.023	0.46	25.5
	3	0.015	0.39	29.2
	4	0.011	0.36	32.4

2.3.2 CABLES

It is becoming more and more difficult to build new overhead power lines due to the increasing social concerns about their environmental and visual impact. Therefore, especially in densely populated areas and in zones of special ecological protection, the solution came from the use of underground power lines based on insulated cables. Furthermore, the need to interconnect the electrical systems of different areas has brought a submarine electrical links proliferation. The high capacitance values of the H.V. insulated cables operating on a.c., present technical restrictions that limit the feasible length of such facilities to a few tens of kilometers. For longer distances, when an overhead line is not possible, the only solution is to resort to an underground or submarine d.c. operated line (HVDC link).

Regarding the electrical modeling of insulated cables, some differences compared to overhead lines have to be taken into account.

High-voltage cables are constituted by a series of successive layers of different materials and with different purposes. The three layers that mainly determine the electrical response of the cable are

- *Conductor*: It is the core layer usually made from copper or aluminum through which the electric current flows.
- *Insulation*: It prevents direct contact between the conductor and any other object, while allowing a proper heat dissipation. A good insulating material should provide a high electrical resistance withstanding the dielectric stresses. The insulation of modern highvoltage cables is based on synthetic polymers such as cross-linked polyethylene (XLPE) or ethyl-ene-propylene (EPR).
- *Screen*: It is a conductive layer whose function is to obtain a symmetrical radial distribution of the electric field into the insulation layer. In high-voltage cables the screen must be grounded at one or several points, becoming then an equipotential surface.

Also, with the purpose of mechanical, physical or chemical protection some other layers such as sheaths or armors can be present.

2.3.2.1 Impedance

If the cable screens are not taken into account, the previous equations for per-phase resistance and inductance developed for overhead lines are still valid, being the per-phase impedance $\mathcal{Z} = R + j\omega L$.

However, the metallic screens of insulated cables are under the influence of electromagnetic induction generated by the conductor currents. Depending on how the earthing of the screens is made, induced voltages and currents may appear on them. The induced currents flowing through the screens generate additional Joule losses and new magnetic fields affecting the effective inductance of the line.

Assuming a perfectly balanced line where the phase currents $\Sigma \mathcal{I}_i = 0$ and the screen currents $\Sigma \mathcal{I}_{s_i} = 0$, the electrical behavior of screens can be expressed in terms of a mutual inductance between conductors and screens (M), a self-inductance of every screen (L_s) and an electrical resistance of every screen (R_s). The mutual inductance M and the screen self-inductance L_s are equal in practice and they can be calculated as:

$$M \approx L_s = \frac{\mu_0}{2\pi} \ln \frac{dmg}{rmg_s} \tag{2.38}$$

where $rmg_s = (r_{s_e} + r_{s_i})/2$ with r_{s_e} and r_{s_i} being the external and internal screen radius respectively. The screen resistance R_s depends on its material and its cross-sectional area.

In these conditions, the following relationships between currents and voltage drops for every phase can be written:

$$\Delta \mathcal{V} = \mathcal{Z} \cdot \mathcal{I} + jX_m \cdot \mathcal{I}_s \tag{2.39}$$

$$\Delta \mathcal{V}_s = \mathcal{Z}_s \cdot \mathcal{I}_s + jX_m \cdot \mathcal{I} \tag{2.40}$$

where $\Delta \mathcal{V}$ and $\Delta \mathcal{V}_s$ are the voltage drops across the phase conductor and across the screen respectively and $\mathcal{Z}_s = R_s + jX_s$ the screen impedance with $X_s = X_m = \omega L_s$.

When earthing the screens at only one side (*Single-Point*), the induced currents cannot flow through them ($\mathcal{I}_s = 0$). In this case, a screen-to-ground voltage $\Delta \mathcal{V}_s = X_m \cdot \mathcal{I}$ at the open side will be induced. This could create a hazardous situation for people or equipment that must be evaluated and, if necessary, controlled.

However, if the screens are earthed at both sides (*Double-Point*), we can assume that $\Delta \mathcal{V}_s = 0$ and then, the screens are traversed by currents given by:

$$\mathcal{I}_s = -\frac{X_m}{\mathcal{Z}_s} \cdot \mathcal{I} \tag{2.41}$$

which provoke the following Joule losses:

$$P_s = I_s^2 \cdot R_s \tag{2.42}$$

The abovementioned losses can be expressed as a function of the conductor current $\mathcal{I}$ in terms of an increase of the effective per-phase resistance:

$$\Delta R = \frac{R_s \cdot X_m^2}{R_s^2 + X_s^2} \tag{2.43}$$

In the same way, the effect of the screen currents over the inductance is the equivalent to a decrease of the per-phase inductance given by

$$\Delta L = -\frac{L_s \cdot X_m^2}{R_s^2 + X_s^2} \tag{2.44}$$

The elimination of screen losses while maintaining a low induced voltage can be achieved by using a *Cross-Bonding* earthing scheme. The cable screens are sectionalized and transposed between the three phases to obtain a sum of induced effects close to zero.

TABLE 2.7

Typical Values of Cable Parameters

	R' (Ω/km)	L' (mH/km)	$X'_{50\,Hz}$ (Ω/km)	C' (nF/km)
66 kV	0.02 – 0.2	0.3 – 0.6	0.1 – 0.2	200 – 400
132 kV	0.01 – 0.1	0.3 – 0.6	0.1 – 0.2	150 – 300
220–400 kV	0.01 – 0.05	0.4 – 0.6	0.12 – 0.2	130 – 220

2.3.2.2 Capacitance

The problem of the radial electric field in an insulated cable has a well-known solution. Due to the presence of the screen, there is no electric field interaction between the different phases. So, every cable constitutes a conventional cylindrical capacitor with the following capacitance:

$$C = \frac{2\pi\varepsilon}{\ln(r_{s_i})/(r_c)} \qquad (2.45)$$

where $\varepsilon = \varepsilon_r\varepsilon_0$ is the dielectric permittivity of the insulating material (typically $\varepsilon_r \approx 2.3 - 3$), r_{s_i} is the internal radius of the screen and r_c is the radius of the conductor core (external and internal radius of the insulation layer, respectively).

If the insulating material was ideal, the current flowing through it would be perfectly capacitive (a $\pi/2$ *rad* phase shift with the voltage) and no losses will be expected in this capacitor. In practice, this phase shift is slightly smaller because a certain amount of dielectric losses due to leakage and polarization currents are dissipated as heat. Consequently, the phase shift between current and voltage is $\pi/2 - \delta$ where δ is the so-called *dielectric loss angle* which is an attribute of the insulating material ($\approx 0.001 - 0.005$).

The dielectric losses in a cable can be expressed as a function of the dielectric losses angle as:

$$P_e = V^2\omega C \tan\delta \qquad (2.46)$$

where V is the conductor-to-ground voltage and C is the capacitance of the cable (Table 2.7).

EXAMPLE 2.3

A three-phase underground line with a trefoil layout is shown in Figure 2.5. Every phase is made from an insulated cable with characteristics given below.

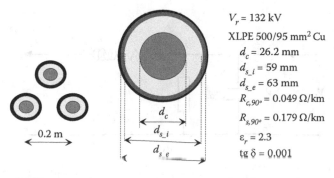

V_r = 132 kV

XLPE 500/95 mm^2 Cu

d_c = 26.2 mm

d_{s_i} = 59 mm

d_{s_e} = 63 mm

$R_{c,90°}$ = 0.049 Ω/km

$R_{s,90°}$ = 0.179 Ω/km

ε_r = 2.3

tg δ = 0.001

0.2 m

FIGURE 2.5   Insulated cables example.

TABLE 2.8

Parameters of the Example Line

Screen Earthing	R' (Ω/km)	L' (mH/km)	C' (nF/km)
Single-point	0.049	0.59	68.5
Double-point	0.103	0.48	68.5

First, *rmg*, *rmg$_s$*, and *dmg* can be directly extracted from the figure data:

$$rmg = \frac{d_c}{2} = 0.0131\text{ m}; \quad rmg_s = \frac{r_{s_e} + r_{s_i}}{2} = 0.0305\text{ m}; \quad dmg = 0.2\text{ m}$$

Both *single-point* and *double-point* screen earthing schemes are considered. To determine the effect of screen currents the following variables must be calculated

$$L_s = M = 0.376\text{ mH/km}; \quad \Delta R = 0.0543\ \Omega/\text{km}; \quad \Delta L = -0.114\text{ mH/km}$$

and the results for the electrical parameters are shown in Table 2.8.

Let us assume the line is carrying a load current $I = 500$ A. For the *single-point* earthing case, no current can flow through the screens but at the open side a voltage to ground $\Delta V_s = X_m \cdot I = 59.1$ V/km is induced. However, for the *double-point* earthing case, $\Delta V_s = 0$ but an induced current as high as $I_s = 275.4$ A according to Equation 2.41 could flow along the screens.

2.3.3 SINUSOIDAL STEADY-STATE MODEL OF THE LINE

In a sinusoidal steady-state and under perfectly balanced conditions, the electrical performance of the lines can be analyzed by means of an equivalent single-phase circuit.

Given that the electrical parameters of the line are uniformly distributed along its overall length, a more appropriate representation of the line must be an infinite sequence of differential elements of length dx, as shown in Figure 2.6, where $R'(\Omega\text{ km})$, $L'(\text{H/km})$, $C'(\text{F/km})$, and $G'(\Omega^{-1}/\text{km})$ are the electrical parameters per unit of length and per phase. This system obeys the following differential equations (wave equation):

$$\frac{d^2 \mathcal{V}_x}{dx^2} = \gamma^2 \cdot \mathcal{V}_x; \quad \frac{d^2 \mathcal{I}_x}{dx^2} = \gamma^2 \cdot \mathcal{I}_x \tag{2.47}$$

where γ is the propagation constant given by

$$\gamma = \sqrt{\mathcal{Z}' \cdot \mathcal{Y}'} \tag{2.48}$$

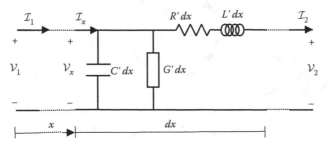

FIGURE 2.6 Differential element of the line.

$\mathcal{Z}' = R' + j\omega L'$ and $\mathcal{Y}' = G' + j\omega C'$ are the serial impedance and the shunt admittance, per phase and per unit length.

Considering as boundary conditions the voltage and the current at the end side of line ($\mathcal{V}_2$ and $\mathcal{I}_2$), the solution for the above equations for the voltage $\mathcal{V}_x$ and the current $\mathcal{I}_x$ at any point located at a distance x from the start side of the line is

$$\mathcal{V}_x = \mathcal{V}_2 \cosh(\gamma(l - x)) + \mathcal{Z}_c \mathcal{I}_2 \sinh(\gamma(l - x)) \tag{2.49}$$

$$\mathcal{I}_x = \mathcal{I}_2 \cosh(\gamma(l - x)) + \frac{\mathcal{V}_2}{\mathcal{Z}_c} \sinh(\gamma(l - x)) \tag{2.50}$$

where l is the total length of the line and $\mathcal{Z}_c = \sqrt{\mathcal{Z}'/\mathcal{Y}'}$ is the so-called *characteristic impedance* of the line (in Ω).

Then, the voltage and the current at the start side of the line can be written as

$$\mathcal{V}_1 = \mathcal{A} \cdot \mathcal{V}_2 + \mathcal{B} \cdot \mathcal{I}_2; \quad \mathcal{I}_1 = \mathcal{C} \cdot \mathcal{V}_2 + \mathcal{D} \cdot \mathcal{I}_2 \tag{2.51}$$

with

$$\mathcal{A} = \mathcal{D} = \cosh(\gamma l); \quad \mathcal{B} = \mathcal{Z}_c \sinh(\gamma l); \quad \mathcal{C} = \frac{\sinh(\gamma l)}{\mathcal{Z}_c} \tag{2.52}$$

There is a Π-equivalent circuit (see Figure 2.7) that satisfies the above distributed parameter equations of the line, with:

$$\mathcal{Z}_\pi = \mathcal{Z}_c \sinh(\gamma l); \quad \mathcal{Y}_\pi = \frac{1}{\mathcal{Z}_c} \tanh\left(\frac{\gamma l}{2}\right) \tag{2.53}$$

For typical overhead lines operating at 50 Hz, γ is in the order of 10^{-3} km^{-1}. Consequently, for line lengths up to 150–200 km, it is clearly acceptable to approximate $\sinh(\gamma l) \approx \gamma l$ and $\tanh(\gamma l/2) \approx \gamma l/2$, yielding

$$\mathcal{Z}_\pi \approx \mathcal{Z}_c \gamma l = \mathcal{Z}' l = \mathcal{Z} \tag{2.54}$$

$$\mathcal{Y}_\pi \approx \frac{1}{\mathcal{Z}_c} \frac{\gamma l}{2} = \frac{\mathcal{Y}' l}{2} = \frac{\mathcal{Y}}{2} \tag{2.55}$$

These are the lumped parameters of the π equivalent model of the line.

For very short overhead lines ($l < 50$ km), the transversal current flowing by the shunt admittance is small enough to be neglected, and the circuit is simplified to a series impedance $\mathcal{Z}$. Nevertheless, this approximation cannot be acceptable for high-voltage insulated cables due to their much higher capacitance.

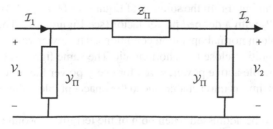

FIGURE 2.7 Equivalent Π circuit of the line.

2.3.4 Electrical Performance of Lines

From the lumped parameters Π-model of the line and assuming $G = 0$, the following equations for the voltages and currents at both sides of the line can be written as

$$S = 3 \cdot \mathcal{V}_2 \cdot \mathcal{I}_2^*; \quad \mathcal{I}_2 = \left(\frac{S}{3 \cdot \mathcal{V}_2}\right)^* \tag{2.56}$$

$$\mathcal{V}_1 = \mathcal{Z}\left(\mathcal{I}_2 + \mathcal{V}_2 \cdot \frac{\mathcal{Y}}{2}\right) + \mathcal{V}_2 \tag{2.57}$$

A practical use of the above equations is the calculation of the voltage regulation and the losses in an electric line. The voltage regulation is usually defined by a percent of the rated voltage V_r as:

$$\Delta V = \frac{V_1 - V_2}{V_r} \cdot 100 \tag{2.58}$$

where $\mathcal{V}_1 = V_1 \angle \delta$ and $\mathcal{V}_2 = V_2 \angle 0$.

The Joule losses can be obtained from

$$P_j = 3 \cdot I_{12}^2 \cdot R \tag{2.59}$$

with $\mathcal{I}_{12} = \mathcal{I}_2 + \mathcal{V}_2 \cdot \mathcal{Y}/2$.

If in the previous equations, the current is expressed in terms of the voltages, the following relations for the active and reactive powers are obtained

$$P = \frac{V_1 V_2}{Z} \cos(\beta - \delta) - \frac{V_2^2}{Z} \cos(\beta) \tag{2.60}$$

$$Q = \frac{V_1 V_2}{Z} \sin(\beta - \delta) - \frac{V_2^2}{Z} \sin(\beta) + \frac{Y V_2^2}{2} \tag{2.61}$$

where $\mathcal{Z} = Z \angle \beta$ and $Y = \omega C$.

The application of the above equations to different operating conditions provides some interesting conclusions.

One of the simplest study cases is when the line feeds a load $S_2 = P_2 + jQ_2$ at the end side at a known voltage $\mathcal{V}_2 = V_2 \underline{/0}$ (taken as angle reference). In such conditions, $\mathcal{V}_1$ can be directly calculated from Equation 2.57. Figure 2.8 shows how V_1 voltage (and also ΔV) increases with the transmitted power P_2, as expected. The lower the inductive power factor, the higher the voltage drop is. Also, for a no-load situation, a slight negative voltage drop due to the capacitive behavior of the open-end line (*Ferranti effect*) is observed. If the line voltage drop is limited to a maximum value (i.e., 10% in figure), a maximum transferable power is determined for every power factor.

A different case is when the line is supplying P_2 to a load at the end terminal while keeping a constant voltage $\mathcal{V}_1$ at the beginning. From the solution of Equations 60 and 61, two different voltages V_2 could deliver the power P_2 with a defined power factor (see Figure 2.9). The higher V_2 value is the only reasonably admissible in normal operation conditions. The lower value corresponds to an unacceptable load delivery condition close to a short circuit. The same figure shows how there is a theoretical maximum transmittable active power, given for every power factor, on the rightmost point of the graph. In practice, this limit is unreachable due to the unacceptable voltage drop with a high probability of voltage collapse.

Finally, a third case can be addressed when both of the terminal absolute voltages of the line are kept constant. Equation 2.60 shows how the active power P_2 depends on the phase shift δ between

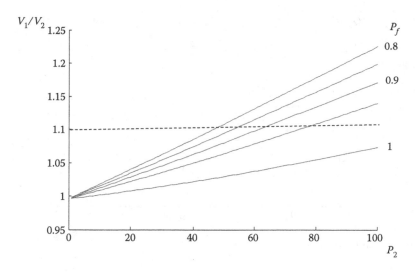

FIGURE 2.8 V_1 versus P_2 for different power factors.

voltages $\mathcal{V}_1$ and $\mathcal{V}_2$ (Figure 2.10). The maximum power has a limit given by

$$P = \frac{V_1 V_2}{Z} - \frac{V_2^2}{Z}\cos(\beta) \qquad (2.62)$$

when $\beta = \delta$.

For typical lines, the impedance argument β is relatively close to $\pi/2$, but such high values for the voltage angle difference δ are unacceptable due to stability problems. Figure 2.10 shows how reactive power flowing in the opposite direction increases very fast with the power load. This should imply the need for a compensation device connected at the receiving end capable of injecting a so high reactive power. A more practical limit of power could be considered if a reasonable maximum angle $\delta = \pi/6$ is taken.

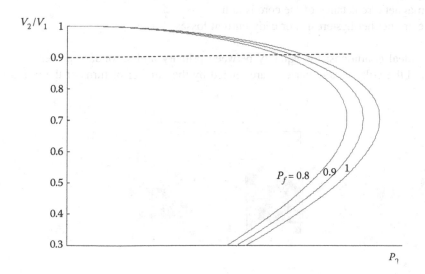

FIGURE 2.9 V_2 versus P_2 for different power factors.

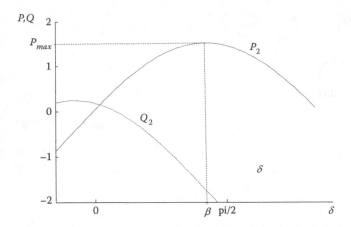

FIGURE 2.10 P_2 and Q_2 versus δ.

2.4 POWER TRANSFORMERS

Another fundamental element in electric networks is the power transformer, which is used to raise, lower, or regulate the voltage levels in a network. In this section, the electric model of different transformers used in electric networks will be presented.

2.4.1 SINGLE-PHASE TRANSFORMER

A single-phase power transformer essentially consists of two coils wound around a ferromagnetic material core, as shown schematically in Figure 2.11.

The ideal single-phase transformer that responds to the electric circuit of Figure 2.12 is defined by the following assumptions:

1. The resistance of the windings is zero.
2. The magnetic flux is totally confined within the magnetic core, that is, there is no leakage flux.
3. The magnetic reluctance of the core is null.
4. There are neither hysteresis nor eddy-current losses.

In these ideal conditions, the equality between primary and secondary powers is observed as $S_p = S_s$, and the voltages and currents are related by the number of turns of the primary, N_p, and

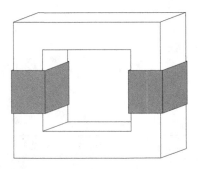

FIGURE 2.11 Single-phase transformer.

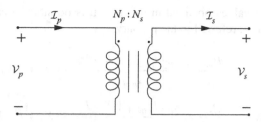

FIGURE 2.12 Electric circuit of the ideal single-phase transformer.

secondary, N_s, windings:

$$\frac{V_p}{V_s} = \frac{N_p}{N_s} = t; \qquad \frac{I_p}{I_s} = \left(\frac{N_s}{N_p}\right)^* = \left(\frac{N_s}{N_p}\right) = \frac{1}{t}$$

However, when a voltage is applied to one of the windings of a real transformer and the other winding is connected to a load, we can see that the active and reactive powers in both windings are not the same, and the ratio between primary and secondary voltages does not exactly match the turns ratio N_p/N_s. This raises the necessity of representing the transformer using a model closer to reality like the one shown in Figure 2.13.

In this circuit, to model the ohmic losses and the leakage flux corresponding to both windings, the series resistances R_p and R_s and the reactances X_p and X_s, respectively, are included. The branch made up of the resistance R_{Fe} and the reactance X_m represents the losses originating in its own ferromagnetic core due to eddy-currents and hysteresis phenomena as well as the fact that the reluctance of the magnetic circuit is not completely zero.

To determine the model of the real transformer, it must undergo two types of tests:

1. Open-circuit test, which is based on feeding one of the windings, for example, the primary, with its rated voltage $V_{p,\text{rated}}$, while leaving the other, the secondary, open (no-load connected). The result is the existence of a small current I_0, called magnetizing current, with a clear inductive character, a secondary voltage $V_{s,0}$ close to the rated voltage, a very small active power P_0, and a small reactive power Q_0. From the measured values, the following parameters can be deduced:

$$\frac{V_{p,\text{rated}}}{V_{s,0}} \approx \frac{N_p}{N_s}; \qquad R_{Fe} \approx \frac{V_{p,\text{rated}}^2}{P_0}; \qquad X_m \approx \frac{V_{p,\text{rated}}^2}{Q_0}$$

2. Short-circuit test, which is done by shorting the secondary and feeding the primary with a voltage $V_{p,\text{sc}}$, established so that the rated current $I_{p,\text{rated}}$ circulates through the winding. The results of this test: a secondary current $I_{s,\text{sc}}$, nearly equal to the rated one $((I_{p,\text{rated}})/(I_{s,\text{sc}}) \approx (1)/(t))$, and a small active power P_{sc}, lower than the reactive

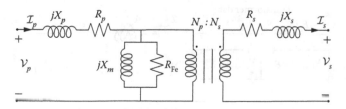

FIGURE 2.13 Electric circuit of the real single-phase transformer.

power Q_{sc}. From the values obtained in this test, it is possible to calculate the so-called short-circuit impedance, referred to the primary:

$$R_{sc} \approx \frac{P_{sc}}{I_{p,\text{rated}}^2} = \frac{R_p \cdot I_{p,\text{rated}}^2 + R_s \cdot I_{s,sc}^2}{I_{p,\text{rated}}^2} \approx R_p + t^2 \cdot R_s$$

$$X_{sc} \approx \frac{Q_{sc}}{I_{p,\text{rated}}^2} = \frac{X_p \cdot I_{p,\text{rated}}^2 + X_s \cdot I_{s,sc}^2}{I_{p,\text{rated}}^2} \approx X_p + t^2 \cdot X_s$$

It is normal practice to express the result of the short-circuit test in percent value as the ratio between the test voltage $\mathcal{V}_{p,sc}$ and the winding rated voltage $V_{p,\text{rated}}$, multiplied by 100, that is,

$$\mathcal{Z}_{sc,\%} = \frac{\mathcal{V}_{p,sc}}{V_{p,\text{rated}}} \cdot 100 \frac{\mathcal{I}_{p,sc}\mathcal{Z}_p + (N_p/N_s)\mathcal{I}_{s,\text{rated}}\mathcal{Z}_s}{((N_p)/(N_s))V_{s,\text{rated}}} \cdot 100$$

$$= \frac{\left[((N_s)/(N_p))^2 \mathcal{Z}_p + \mathcal{Z}_s \right] S_{s,\text{rated}}}{V_{s,\text{rated}}^2} \cdot 100 \qquad (2.63)$$

From the above equation, it can be deduced that $\mathcal{Z}_{sc,\%}$, when divided by 100, is the same as the per unit short-circuit impedance $\mathcal{Z}_{sc,pu}$, referred to the transformer voltage and power-rated values.

In operational conditions close to the rated ones, it is possible to neglect the magnetizing current as it is much lower than the load current, and, consequently, it is possible to eliminate the parallel branch of the circuit of Figure 2.13. If the result is converted to per unit values, the transformer model is reduced to the series circuit in Figure 2.14, where $R_{sc,pu}$ and $X_{sc,pu}$ are the resistance and the reactance (in per unit values), respectively, calculated in the short-circuit test.

From now on, unless otherwise stated, the variables and the parameters will be expressed in per unit values.

2.4.2 THREE-PHASE TRANSFORMER

Three-phase power transformers are an extension of the single-phase transformer, where increased number of windings can be connected in different ways (delta, wye). On the basis of the type of the ferromagnetic core, the three-phase transformer can be classified as follows:

- Bank of single-phase transformers, also called three-phase bank, made up of three single-phase transformers (Figure 2.15a).
- Three-leg transformer in which only one core with a leg for each phase is used (Figure 2.15b).
- Five-leg transformer whose core has five legs; three legs for each of the phases and two additional legs at its ends (Figure 2.15c).

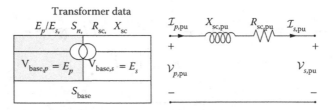

FIGURE 2.14 Per unit electric circuit of the real single-phase transformer.

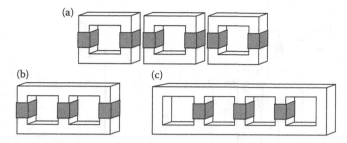

FIGURE 2.15 Three-phase transformer types depending on the iron core: (a) three-phase bank, (b) three legs, and (c) five legs.

In this section, only three-phase transformers under steady-state balanced conditions are considered. Consequently, their three-phase variables will respond to balanced sources and loads and therefore can be modeled and analyzed, whatever their typology, by means of three single-phase transformers connected adequately.

On the basis of the number of windings per core leg, we identify the following transformers:

- Transformers with two separate windings (primary–secondary).
- Transformers with three separate windings (primary–secondary–tertiary*).
- Transformers with a continuous winding. This category comprises the so-called reactances as well as the autotransformers, although owing to the existence of intermediate connections it is possible to define primary and secondary windings.

Three-phase transformers can also be classified by their winding connections, delta or wye, resulting in different transformer configurations: delta–wye, wye–wye, or delta–delta.

The way in which the primary and secondary windings are connected (delta or wye) gives rise to an angular displacement between the corresponding primary and the secondary voltages, also resulting in different voltage ratios in the transformation. That is, in three-phase transformers, unlike in single-phase transformers, there are winding ratios (primary–secondary–tertiary) both in magnitude and angle.

The angle ratios are normally expressed using a coefficient that indicates the phase shift between the primary and the secondary in units of 30°. This coefficient is sometimes called *vector group*. For example, if both the primary and the secondary windings are wye-connected, as indicated in Figure 2.16a, it can be seen how the angle displacement between the primary and the secondary voltages is 0°, and this is why this transformer is denominated Yy0. In this transformer, the voltage and the ratios can be mathematically expressed as (Figure 2.16b)

$$\frac{\mathcal{V}_A}{\mathcal{V}_a} = \frac{N_p}{N_s} \tag{2.64}$$

Another possible situation is when the primary is wye-connected and the secondary is delta-connected, as indicated in Figure 2.17a. From its phasor diagram (Figure 2.17b), keeping in mind that $\mathcal{V}_A/\mathcal{V}_{ab} = N_p/N_s$, the voltage ratio $\mathcal{V}_A/\mathcal{V}_a$ is immediately obtained

$$\frac{\mathcal{V}_A}{\mathcal{V}_a} = \frac{\mathcal{V}_A}{(\mathcal{V}_{ab})/(\sqrt{3}) \cdot e^{-j(\pi/6)}} = \frac{\mathcal{V}_A}{\mathcal{V}_{ab}} \cdot \sqrt{3} \cdot e^{j \cdot (\pi/6)} = \frac{N_p}{N_s} \cdot \sqrt{3} \cdot e^{j \cdot (\pi/6)} \tag{2.65}$$

---

* As a normal rule, the primary winding should be the one with the highest voltage, the secondary with second highest, and the tertiary, if it exists, with the lowest voltage.

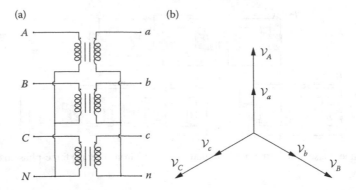

FIGURE 2.16 Connection for the Yy0 three-phase transformer and voltage phasor diagram: (a) connection scheme and (b) phasor diagram.

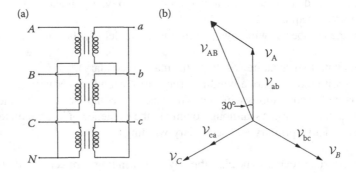

FIGURE 2.17 Connection for the Yd1 three-phase transformer and voltage phasor diagram: (a) connection scheme and (b) phasor diagram.

Namely, for this three-phase transformer the transformation ratio between the primary and the secondary is $N_p\sqrt{3}/N_s$, with an angle displacement of $30°$, which explains the denomination Yd1.

Combining the different connections, it is possible to obtain transformers with diverse vector groups (Yy6, Yd7, Yd5, Yd11, Dd6, Dd2, Dd8, Dd10, Dy7, Dy5, etc.).

On the basis of our discussion, it can be concluded that a three-phase transformer with a transformation ratio a (rated voltage ratio), and vector group h, responds to the model shown in Figure 2.18, where $\alpha = h \cdot \frac{\pi}{6}$.

A proper selection of the base values leads to the per unit circuit of the three-phase transformer shown in Figure 2.19.

On the other hand, the operating limits of the power transformer present the same characteristics as in the case of a line, making it simpler to analyze due to the lack of parallel capacitances.

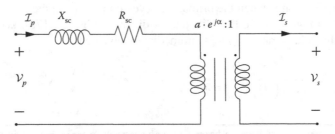

FIGURE 2.18 Single-phase equivalent circuit of a three-phase transformer.

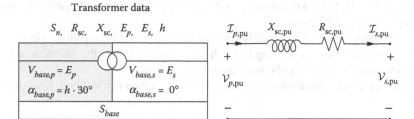

FIGURE 2.19 Single phase per unit equivalent circuit of a three-phase transformer.

EXAMPLE 2.4

Let us consider a Yd11, 240/11 kV transformer with a rated power of 50 MVA, a short-circuit reactance X_{sc} of 13.73%, and a resistance of 4%.

For a different base power of 100 MVA, the per unit model will be calculated. The rated transformer voltages are used as base voltages:

$$V_{base,p} = 240\,kV; \quad V_{base,s} = 11\,kV$$

$$\alpha_{base,p} = 30°; \quad \alpha_{base,s} = 0°$$

As the short-circuit impedance, given by X_{sc} and R_{sc} are referred to the rated power, a change of base has to be made

$$R_{sc,pu} + j \cdot X_{sc,pu} = (R_{sc} + j \cdot X_{sc}) \cdot \frac{S_{base}}{S_{rated}} = (0.04 + j \cdot 0.1373) \cdot \frac{100}{50} = 0.08 + j \cdot 0.2746 \quad (2.66)$$

The resulting single-phase equivalent circuit is similar to that of Figure 2.19.

2.4.3 THREE WINDING TRANSFORMERS

The three-winding transformers, as the name indicates, present three windings in each one of the phases (three coils per leg). Its equivalent single-phase model has three magnetic couplings, as shown in Figure 2.20a, where the three short-circuit impedances of the windings have been included and iron losses and magnetizing reactance have been neglected.

The different vector groups and connections in these types of transformers are treated in the same way as in the two-winding transformer. So, for example, in a Dy5d1 transformer, there is a phase

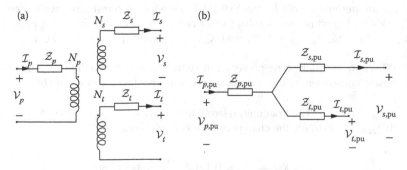

FIGURE 2.20 Single-phase equivalent circuit of a three-winding transformer.

displacement of 150° (Dy5) between primary and secondary, and another phase displacement of 30° (Dd1) between primary and tertiary.

If the three-winding transformer circuit is converted to per unit values having a single power base, and three power voltage bases according to the transformation ratios, the magnetic couplings disappear and a wye-shaped circuit is obtained, as in Figure 2.20b.

The impedances of the three-winding transformer are generally defined from the following tests:

- Short-circuit test, carried out with the secondary short circuited and the tertiary open, in such a way that connecting the primary to a $V_{p,sc}$ voltage, the rated current $I_{s,rated}$ circulates through the secondary. In these conditions, an impedance $Z_{ps,\%}$ is calculated as a ratio between the voltage $V_{p,sc}$ and the rated voltage $V_{p,rated}$, multiplied by 100 (in percent). From the winding transformation ratios and with a similar deduction to that of Equation 2.63, it is demonstrated that the per unit impedance so obtained ($Z_{ps,pu} = Z_{ps,\%}/100$) is equal to the impedance Z_{ps} in per unit when the base voltages are the rated voltages ($V_{p,rated}$ and $V_{s,rated}$) and the base power corresponds to that of the secondary winding $S_{s,rated}$ (normally the short-circuited winding).
- Short-circuit test, carried out with the tertiary short circuited and the secondary open, in such a way that the rated current circulates through the tertiary. From this test, the per unit $Z_{pt,pu}$ obtained is referred to rated base voltages ($V_{p,rated}$ and $V_{t,rated}$) and a base power equal to the rated tertiary one $S_{t,rated}$.
- Short-circuit test, carried out with the tertiary short circuited and the primary open, and fed by the secondary, from where the impedance $Z_{st,pu}$ is obtained. This impedance will be expressed, in this case, in the voltage rated bases ($V_{s,rated}$ and $V_{t,rated}$) and in the tertiary base power $S_{t,rated}$.

We may highlight that in the three-winding transformers, contrary to the two-winding case, it is normal for the windings to have different rated powers. This fact makes it necessary to change the bases in such a way that the three abovementioned impedances are expressed in the same base power. In these conditions, the three following equalities hold:

$$Z_{ps,pu} = Z_{p,pu} + Z_{s,pu}; \quad Z_{pt,pu} = Z_{p,pu} + Z_{t,pu}; \quad Z_{st,pu} = Z_{s,pu} + Z_{t,pu} \qquad (2.67)$$

constituting a three-equation system in three unknowns from which the three impedances $Z_{p,pu}$, $Z_{s,pu}$, and $Z_{t,pu}$ can be calculated.

EXAMPLE 2.5

Let us deduce the per unit model for the Yy0d11 three-winding transformer with rated voltages 139/65/13.8 kV and rated powers 40/40/13.33 MVA from which the following parameters are known: $X_{ps,\%} = 12.62\%$, $X_{pt,\%} = 6.41\%$, and $X_{st,\%} = 1.7381\%$ (as usually, the resistances are neglected).

$S_{base} = 100$ is considered the base power for the whole network, and the rated voltages will be the voltage bases; consequently, the per unit short-circuit reactance values will be

- $X_{ps} = X_{ps,\%}/100 = 0.1262$ per unit, referred to the power $S_{s,rated} = 40$ MVA. To express it with $S_{base} = 100$ MVA, the change of base is carried out:

$$X'_{ps} = X_{ps} \cdot \frac{S_{base}}{S_{s,rated}} = 0.1262 \cdot \frac{100}{40} = 0.315 \, pu$$

- $X_{pt} = X_{pt,\%}/100 = 0.0641$ per unit, referred to the power $S_{t,rated} = 13.33$ MVA. To express it with $S_{base} = 100$ MVA, the change of base is carried out

$$X'_{pt} = X_{pt} \cdot \frac{S_{base}}{S_{t,rated}} = 0.0641 \cdot \frac{100}{13.33} = 0.48 \, \text{pu}$$

- $X_{st} = X_{st,\%}/100 = 0.01738$ per unit, referred to the power $S_{t,rated} = 13.33$ MVA. To express it with $S_{base} = 100$ MVA, the change of base is carried out:

$$X'_{st} = X_{st} \cdot \frac{S_{base}}{S_{t,rated}} = 0.017381 \cdot \frac{100}{13.33} = 0.12 \, \text{pu} \tag{2.68}$$

From the new short-circuit impedance values and from solving the system Equation 2.67, the per unit reactances of the single-phase equivalent model for the three-winding transformer are obtained (wye-shaped circuit in Figure 2.20).

$$X_p = 0.3375 \, \text{pu}; \quad X_s = -0.0225 \, \text{pu}; \quad X_t = 0.1425 \, \text{pu}$$

2.4.4 REGULATING TRANSFORMERS

A type of power transformer that is of special interest in electric network operation is the regulating transformer. From the point of view of the variable to be controlled, these transformers can be of two types:

1. Voltage-magnitude regulating transformer
2. Phase-angle regulating transformer

This book does not discuss the manufacturing technology of these transformers in depth, but describes their electrical behavior and models.

2.4.4.1 Voltage-Magnitude Regulating Transformers

The simplest scheme consists of a transformer with one of its windings having different changing taps corresponding to different number of turns. In this way, the transformer modifies the V_s voltage magnitude and the V_p supposed constant by changing the number of turns of N_p (high-voltage side) and therefore changing the turns ratio N_p/N_s. In this manner, a tap-changing regulating transformer is obtained, which responds to the circuit shown in Figure 2.21.

From the equivalent circuit of Figure 2.21, when converted to per unit values (with their voltage bases maintaining the rated transformation ratio N_p/N_s), the model in Figure 2.22 is obtained. Here, $Z_{sc,pu}$ is the per unit short-circuit impedance of the transformer, which is equal to that obtained without tap modification, and a represents the regulation between primary and secondary voltages, so that $a = 1 + \Delta N_p/N_p$.

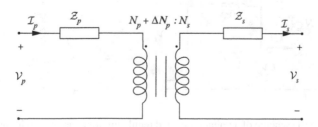

FIGURE 2.21 Tap-changing regulating transformer circuit.

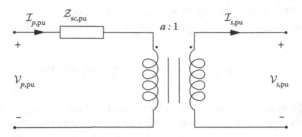

FIGURE 2.22 Per unit tap-changing regulating transformer circuit.

For the generic case of a transformer, with magnitude regulation it is possible to obtain an equivalent circuit without coupling, resulting in a simpler scheme. Considering the nodal analysis of the circuit in Figure 2.22, the following matrix equation is obtained:

$$\begin{pmatrix} \mathcal{I}_{p,pu} \\ \mathcal{I}_{s,pu} \end{pmatrix} = \begin{pmatrix} \mathcal{Y} & -a\mathcal{Y} \\ -a\mathcal{Y} & a^2\mathcal{Y} \end{pmatrix} \cdot \begin{pmatrix} \mathcal{V}_{p,pu} \\ \mathcal{V}_{s,pu} \end{pmatrix} \tag{2.69}$$

where $\mathcal{Y} = 1/\mathcal{Z}_{sc,pu}$.

From the above equations, the equivalent circuit in Figure 2.23 is easily obtained.

Another device that enables the regulation of the voltage magnitude is based on a voltage-injection scheme, like the one shown in Figure 2.24a, where the $\mathcal{V}_s$ voltage is obtained as the sum of $\mathcal{V}_p$ and a voltage proportional to itself, $k \cdot \mathcal{V}_p$:

$$\mathcal{V}_s = \mathcal{V}_p + k \cdot \mathcal{V}_p = (k+1) \cdot \mathcal{V}_p$$

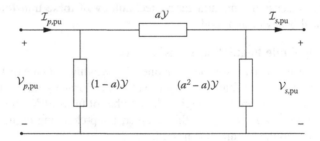

FIGURE 2.23 π model of the circuit in Figure 2.22.

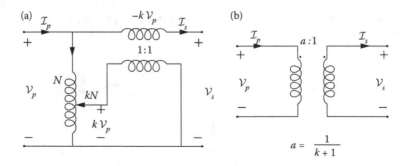

FIGURE 2.24 Connection scheme of a voltage-injected regulating transformer and its single-phase equivalent circuit: (a) connection scheme and (b) equivalent circuit.

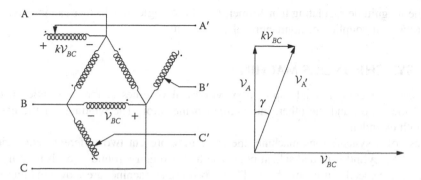

FIGURE 2.25 Phase-angle regulating transformer.

From the above equation, the equivalent circuit in Figure 2.24b is deduced. These regulating transformers are connected at the end of a line, with the aim of regulating the voltage magnitude in that node. A treatment similar to the one used for the tap-changing transformer can be applied to this model to obtain an equivalent circuit.

2.4.4.2 Phase-Angle Regulating Transformer

Similar to the magnitude regulating transformer just described, it is possible to have a transformer for the regulation of the phase-angle displacement between the two ends. The scheme of this type of regulating transformer is shown in Figure 2.25, where it is possible to appreciate how a 90° out-of-phase voltage is injected into each phase to be regulated. The result can be observed in the phasor diagram, where the V_A voltage is shifted by the injected voltage $k \cdot V_{BC}$, so that $V_{A'} = V_A + k \cdot V_{BC}$. The same occurs in the other two phases.

Owing to the fact that the values of k must be small

$$V_{A'} \approx V_A \cdot e^{-j \cdot \gamma}$$

where

$$\gamma \approx \mathrm{tg}\gamma = \frac{k \cdot V_{BC}}{V_A} = k \cdot \sqrt{3}$$

Consequently, the circuit in Figure 2.25 permits a phase shift of value γ between the V_A and the $V_{A'}$ voltages, their magnitudes remaining practically the same. The single-phase equivalent circuit of the phase-angle regulating transformer is shown in Figure 2.26, where the coupling transformation ratio is a complex number $\mathcal{T} = 1 \angle \gamma$, with

$$\frac{V_p}{V_s} = \mathcal{T} = 1 \angle \gamma; \qquad \frac{I_p}{I_s} = \frac{1}{\mathcal{T}^*} = 1 \angle \gamma$$

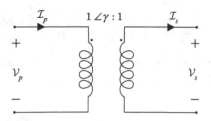

FIGURE 2.26 Single-phase circuit of the phase-angle regulating transformer.

Unlike the magnitude regulating transformer, the phase-angle transformer cannot have an equivalent circuit without coupling or nonreciprocal components.

2.5   THE SYNCHRONOUS MACHINE

The synchronous machine is made up of two cylindrical volumes of ferromagnetic material: one fixed, called the stator, and the other, mobile, called the rotor. The windings of the machine are located in both cylinders.

In the case of a synchronous machine, the rotor may present two different geometric shapes, namely, round (or cylindrical) and salient pole. The first geometry, round rotor, is used in machines that rotate at higher speeds (turbo-machine). The two different machines are shown in Figure 2.27. All the theoretical explanations given in this section will initially refer to the two-pole round-rotor machine (the reader is referred to Appendix C, where a more detailed and generic analysis of rotating machines is performed).

2.5.1   ROUND-ROTOR SYNCHRONOUS MACHINE

In the synchronous machine, the rotor winding is fed by a d.c. current (there are also machines with a rotor with permanent magnets) in such a way that a symmetrical magnetic flux, which encircles the stator, is created. Owing to the rotor rotation, voltages are induced in the stator windings, which can be used to feed electric loads. To obtain a balanced three-phase voltage generation in the stator windings, these are geometrically distributed every $120°$.

To facilitate the representations and explanations, it is customary to show the magnetic variables of the machine (fluxes, flux density phasor, and magnetomotive force) with the flux density phasor $\vec{B}$ as reference, which indicates the magnetic flux direction perpendicular to the coil surface. Figure 2.28 shows the references established in a coil, assuming all the variables considered d.c. have a positive value.

Figure 2.29 shows the phasor $\vec{B}_r$ created by the rotor coil when fed by the so-called excitation current.

For constant rotation speed, Ω, and according to the electromagnetic relations, the stator winding-induced voltages generated by the rotor flux can be expressed as

$$e_{aa'} = \sqrt{2}E \cdot \sin(\Omega t)$$

$$e_{bb'} = \sqrt{2}E \cdot \sin(\Omega t - 120°) \tag{2.70}$$

$$e_{cc'} = \sqrt{2}E \cdot \sin(\Omega t - 240°)$$

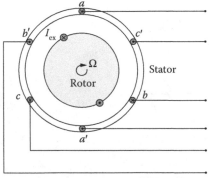
Round rotor

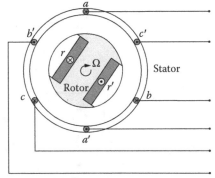
Salient-pole rotor

FIGURE 2.27   Types of synchronous machines (two poles).

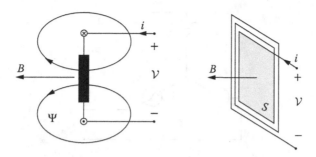

FIGURE 2.28 Variables and references in an electromagnetic circuit.

where E is the rms value of the voltage in each winding, which depends on the rotation speed Ω and the excitation current I_{ex}, which circulates through the rotor winding.

To obtain voltages at the industrial frequency of 50 Hz, with a two-pole machine, the rotor rotation speed has to be 50 rd/s (50 Hz = 50 rps = 3000 rpm), that is, $\Omega = 2\pi 50$ (rd/s). If the machine has p pairs of poles, the 50 Hz voltages are obtained with a rotation speed of $\Omega = 2\pi 50/p$ (rd/s) = $3000/p$ (rpm). Consequently, the relation between electrical frequency $\omega(f)$ and mechanical rotation speed Ω, is $\omega = 2\pi f = \Omega \cdot p$.

On the other hand, if all the a', b', and c' terminals of the stator windings are joined, a wye configuration is obtained. In the same way, if the a' terminals are joined with the b, b' with the c, and c' with the a, a delta configuration is obtained, with the phase-to-phase voltages available at the machine terminals.

From now on, unless otherwise stated, a wye configuration will always be assumed for the generator. In the case of a delta configuration, a delta—wye transformation (available in any electric engineering book) can be used.

Thus, a no-load synchronous machine can be modeled using a balanced three-phase system of instantaneous voltages e_a, e_b, and e_c, and associated phasors $\mathcal{E}_a$, $\mathcal{E}_b$, and $\mathcal{E}_c$, with the following equations:

$$e_a = \sqrt{2}E \cdot \sin(\omega t)$$

$$e_b = \sqrt{2}E \cdot \sin(\omega t - 120°) \tag{2.71}$$

$$e_c = \sqrt{2}E \cdot \sin(\omega t - 240°)$$

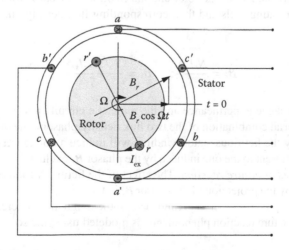

FIGURE 2.29 Magnetic flux density phasor created by the rotor in the synchronous machine.

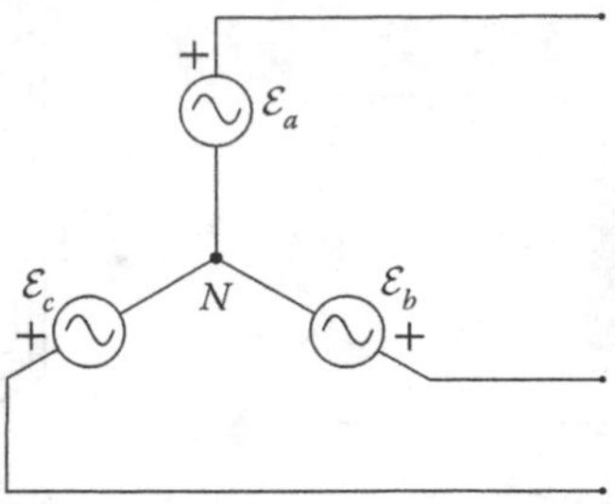

FIGURE 2.30   Electric circuit of a no-load synchronous machine.

If saturations effects are not considered, the voltage magnitude E can be expressed as $E = k \cdot \omega \cdot I_{ex}$, where k is a constant and I_{ex} is the rotor-winding excitation current. This last expression for the rms voltage E is very important to realize the voltage regulation capability of the synchronous machine, because it shows how, to regulate the voltage at the machine terminals, supposing a constant frequency, it is possible to modify the excitation current.

Figure 2.30 shows the steady-state electric circuit of a synchronous machine under no-load conditions, solely made up of a three-phase voltage system (ideal three-phase source).

Till now, the no-load synchronous machine has been described. However, if an electric load is connected to the machine terminals and the voltages are considered as having a frequency f, the currents i_a, i_b, and i_c will flow through the stator windings, also with the same f frequency.

These currents will present a phase displacement φ in relation to the corresponding voltages induced by the rotor e_a, e_b, and e_c:

$$i_a = \sqrt{2}I \cdot \sin(\omega t - \varphi)$$

$$i_b = \sqrt{2}I \cdot \sin(\omega t - 120° - \varphi) \qquad (2.72)$$

$$i_c = \sqrt{2}I \cdot \sin(\omega t - 240° - \varphi)$$

The stator currents, due to the windings geometric arrangement, create a flux density phasor $\vec{B}_s$ of constant value, which rotates at the same speed as the rotor, or, in other words, at the same frequency of the currents that created it. In this way, from the point of view of the magnetic couplings, the windings of the machine can be represented as two rotating coils: the coil of the rotor $(r - r')$, which generates the phasor $\vec{B}_r$, and a fictitious coil in the stator $(s - s')$, which generates the phasor $\vec{B}_s$. Both coils are fed by d.c. currents: I_{ex} for the rotor and the fictitious current I_s for the stator. Figure 2.31 shows the two defined rotating coils and their corresponding flux density phasors, which are calculated from

$$\vec{B}_r = \frac{N_r I_r \Lambda}{S}; \quad \vec{B}_s = \frac{3\sqrt{2}N_s I_s \Lambda}{S}$$

where Λ and S are the magnetic permeance and the magnetic circuit cross section.

Therefore, the phasorial combination of the two flux density phasors, the one created by the rotor and the other created by the fictitious stator winding, will induce a voltage in each one of the stator windings that will be different to the one induced by the phasor $\vec{B}_r$ exclusively (of magnitude E). This phenomenon is known as *armature reaction*. The effect of the fictitious winding phasor $s - s'$ on the voltage E is calculated by the projection of the phasor $\vec{B}_s$ in the direction of the phasor $\vec{B}_r$. In this way, the flux of the flux density $B_s \sin(\varphi)$ is superimposed on the magnetic flux created by the phasor $\vec{B}_r$. Mathematically, the armature reaction phenomenon is modeled using the so-called armature reaction reactance X_{ri}.

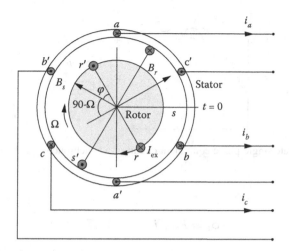

FIGURE 2.31 Flux density phasors in a two-pole synchronous machine.

On the other hand, there is a portion of the flux created by the phasor $\vec{B}_s$ not embracing the rotor winding, which is modeled using a reactance X_p. The combination of this last reactance with an armature reaction reactance X_{ri} creates the so-called synchronous reactance X_s of the machine.

Apart from the effects of the magnetic phenomena, the machine also has resistive phenomena due to the losses in the electric conductors of the coils, which are modeled using series resistances. Consequently, the steady-state model of a synchronous machine is represented by the circuit shown in Figure 2.32, where, as usual, the resistances have been neglected due to their reduced value, compared with the reactances.

On the other hand, the balanced three-phase circuit in Figure 2.32 can be studied by resorting to the equivalent single-phase circuit in Figure 2.33, whose parameters are

$\mathcal{E}$: Internal voltage, function of the excitation current and the rotor rotating speed, whose rms value is obtained from the no-load test of the machine.

X_s: Synchronous reactance, obtained from the short-circuit test and which models all the fluxes created by the stator.

$\mathcal{I}$: Current delivered by the machine.

$\mathcal{V}$: Voltage at the machine terminals or output voltage.

From Figure 2.33, the expressions for active and reactive powers delivered by the machine to any load connected to its terminals are easily established, considered as a generator, and modeled

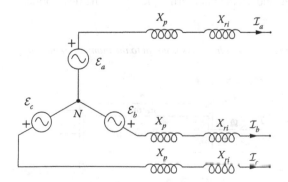

FIGURE 2.32 Three-phase circuit of an on-load synchronous machine.

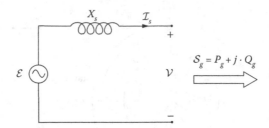

FIGURE 2.33 Steady-state single-phase equivalent circuit of a synchronous machine.

in per unit values:

$$S_g = P_g + j \cdot Q_g = \mathcal{V} \cdot \mathcal{I}^* \tag{2.73}$$

where S_g is the apparent power at the machine terminals, P_g is the active power, and Q_g is the reactive power.

Considering the phasor voltages $\mathcal{V} = V\angle 0°$ and $\mathcal{E} = E\angle \delta$ and substituting the value of the current $\mathcal{I} = (\mathcal{E} - \mathcal{V})/jX_s$ in Equation 2.73, two expressions are obtained, similar to the ones defined for the series model of a line:

$$P_g = \frac{EV}{X_s} \cdot \sin \delta \tag{2.74}$$

$$Q_g = \frac{EV}{X_s} \cdot \cos \delta - \frac{V^2}{X_s} \tag{2.75}$$

Also, the rotor of the machine moves, thanks to the energy (power) provided by the prime mover (turbine). If the power delivered by this prime mover is called P_m, the power flow can be illustrated as shown in Figure 2.34.

Applying the theorem of the power conservation to Figure 2.34, it is deduced that $P_m = P_g + P_{losses}$, where P_{losses} represents the losses that are solely mechanical if the electric resistances of the windings are not considered. The former expressions are also valid when the machine is working as a motor. In this case, the machine under steady-state conditions absorbs electric power and delivers it to a mechanical load.

From the abovementioned power flow, it can be concluded that the aspects related to reactive power are only linked to the electrical parts of the machine (electric circuit), while for the active power there is a relation between the prime mover and the electric circuit. This relation between the active powers of the generator and the prime mover mechanical power is established, for any working condition, from the elementary principle in mechanics applied to masses with rotating movement:

The sum of torques applied to a rotating mass is equal to the moment of inertia of the mass times its angular acceleration.

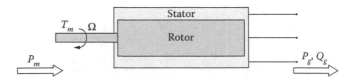

FIGURE 2.34 Representation of a synchronous machine.

Applying this rule to the powers of the machine and considering a generator-working condition, the following dynamic equation is obtained:

$$P_m - P_g - P_{losses} = M \cdot \frac{d\Omega}{dt} + D \cdot (\Omega - \Omega_s) \tag{2.76}$$

where M is a machine constant related with the total (generator plus turbine) inertia, the number of poles, and the rotor synchronous speed, Ω is the rotor angular speed, Ω_s is the rotor angular synchronous speed, and D is the so-called damping constant of the machine (defined in Chapter 10).

2.5.2 Working Conditions of the Synchronous Machine

The synchronous machine is a reversible device and therefore can work as a generator and as a motor. The difference between both working conditions is established from the sign of the mechanical power. In both working conditions, the reactive power can be consumed or delivered by the machine, as shown in Equation 2.75, in such a way that

$$E \cdot \cos \delta > V \quad \rightarrow \quad Q_g > 0$$
$$E \cdot \cos \delta < V \quad \rightarrow \quad Q_g < 0$$

Consequently, the synchronous machine can work in the four quadrants of the $P-Q$ axis as presented in Figure 2.35.

Synchronous generators represent nearly all the electric energy production in the network. For this reason, these generators are the basic devices that maintain the network frequency and, to a great extent, the node voltages. Subsequently, synchronous generators have frequency controls ($P-f$ control) as well as controls for the magnitude of the generated voltage ($Q-V$ control). A more detailed explanation of the generator controllers can be found in Chapter 9.

According to the $Q-V$ and $P-f$ regulators available, the steady-state operation conditions of a synchronous generator seen from its output terminals are the same as those of an element with specified active power and rms voltage (PV node type, see Chapter 3). In the case of the machine working as a motor, the model is defined in terms of specified active and reactive power (PQ node type).

On the other hand, the synchronous generator, as well as any other electric device, presents technological and economic restrictions in its operation, among which the following can be cited:

- *Mechanical power limits*: $P_{m,max}$, which is established by the maximum power that can extracted from the turbine, and $P_{m,min}$, the minimum power that the turbine can deliver because of economic or technological reasons.
- *Stator thermal limit*: I_{max}, is the maximum current that can circulate through the stator windings. Considering a constant voltage at the machine terminals, the thermal limit can be established as a maximum apparent power S_{max}.

Generator $\delta > 0$ $E \cdot \cos \delta < V$	Generator $\delta > 0$ $E \cdot \cos \delta > V$
Motor $\delta < 0$ $E \cdot \cos \delta < V$	Motor $\delta < 0$ $E \cdot \cos \delta > V$

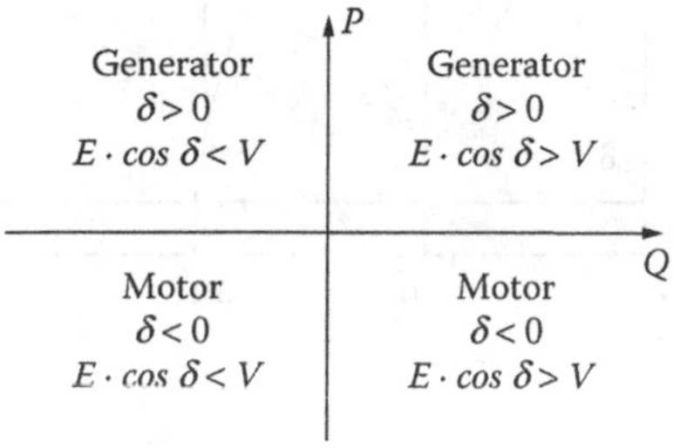

FIGURE 2.35 Operating conditions of the synchronous machine.

- *Maximum internal voltage limit*: E_{max}, is defined by the maximum excitation current that can circulate through the rotor windings or by the maximum insulation voltage.
- *Stability limit*: δ_{max}, is established from machine stability considerations that are thoroughly explained in Chapter 10. However, in the absence of other criteria, the upper value for this limit can be established as 90°, corresponding to the maximum active power the generator can deliver to the network.

EXAMPLE 2.6

Next, with a numerical example, the loci of the synchronous generator limits will be determined graphically and analytically, and, subsequently, its operation zone. The graphical representation is made in a P–Q coordinates system. Generator data are

- Rated voltage at the terminals $V = 11$ kV.
- Synchronous reactance $X_s = 1.2$ pu.
- Turbine rated power $P_m = 40$ MW.
- Rated apparent power of the generator $S_{rated} = 50$ MVA.

From the generator data, its operation limits, expressed in per unit values and referred to $V_{base} = 11$ kV and $S_{base} = 50$ MVA, are

- *Maximum active power*: $P_{m,max}$ = rated mechanical power of the turbine, in this case $40/50 = 0.8$ per unit. In Figure 2.36, the loci of this power is the straight line $P_m = 0.8$ per unit.

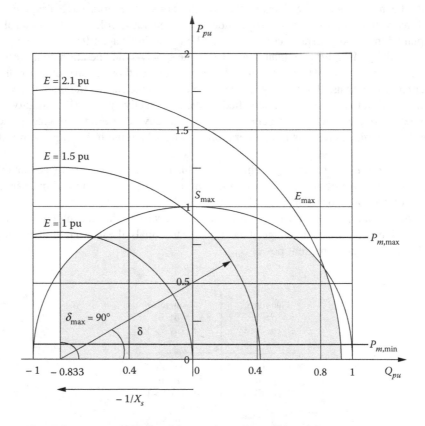

FIGURE 2.36 Operation zones and limits of the synchronous machine.

- *Minimum active power*: $P_{m,min}$ = minimum mechanical power necessary for the correct operation of the system (estimated 10% of the rated power). In Figure 2.36, the loci of this power is the straight line $P_m = 0.08$ per unit.
- *Maximum stator current*: I_{max}, whose value for $V = 1.0$ per unit coincides with the maximum apparent power of the generator, S_{max}, estimated to be equal to the rated power, $S_{max} = S_{rated} = 1$ per unit. Therefore, the loci is defined by the points satisfying $P^2 + Q^2 = S^2_{max}$, which correspond to a circumference with radius S_{max} and is centered in the point $(P = 0, Q = 0)$.
- *Maximum rotor current*: This value can be expressed in terms of the maximum internal voltage, E_{max}. In this way, always working in per unit values, the loci for E_{max}, combining Equations 74 and 75, is calculated with the equation

$$P^2 + \left(Q + \frac{V^2}{X_s}\right)^2 = \frac{E^2_{max}V^2}{X^2_s}$$

In Figure 2.36, this is the circumference with radius E_{max}/X_s and centered in the point $(P = 0, Q = -1/X_s)$.
- *Steady-state stability limit*: This characteristic limit of the synchronous machine is defined as a phase displacement value of $\delta = 90°$, between the internal voltage, $\mathcal{E}$, and the terminal voltage, $\mathcal{V} = V\angle 0°$. In this situation, the generator's reactive power is $Q = -1/X_s$. Consequently, the loci of this limit is the straight line $Q_{min} = -1/X_s$.

With the abovementioned limits, the operation zone of the electrical generator is defined by the shaded area in Figure 2.36.

2.5.3 Salient-Pole Rotor Synchronous Generator

In addition to the round-rotor synchronous machine (already described), another type of machine called the salient-pole rotor will be analyzed and modeled in this section. With this aim, the salient-pole machine is shown in Figure 2.37, where, now, instead of representing the rotor and stator flux density phasors, magnetomotive phasors are used, which, unlike the flux density phasors, do not depend on the magnetic circuit geometry.

On observing the salient-pole machine, it can easily be seen that it does not present a complete symmetry, as occurs in the round-rotor machine. Consequently, the magnetic permeance corresponding to the F_s phasor is a function (assumed sinusoidal) of the rotor position, which allows a machine model with sinusoidal parameters to be established.

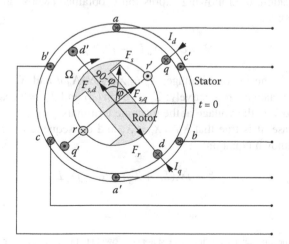

FIGURE 2.37 Salient-pole rotor synchronous generator.

However, it is possible to define in the machine two axes of symmetry in the longitudinal (direct-axis) and transversal (quadrature-axis) directions to the rotor pole axis ($d - d'$ and $q - q'$ axes in Figure 2.37), which allow the modeling of the machine with constant parameters.* This analysis is based on the breakdown of the F_s magnetomotive force into the two abovementioned axes ($F_{s,d}$ and $F_{s,q}$).

In this way, from the electrical and magnetic point of view, the salient-pole machine can be represented by three coils:

1. The rotor coil ($r - r'$) circulated by the d.c. current I_r, which generates F_r and the phasor B_r.
2. The fictitious stator coil ($d - d'$) circulated by a d.c. current I_d, which originates $F_{s,d}$ and $B_{s,d}$.
3. The fictitious stator coil ($q - q'$) circulated by a d.c. current I_q, which originates $F_{s,q}$ and $B_{s,q}$.

The mentioned magnetic variables can be expressed as

$$F_r = N_r I_r; \quad F_{s,d} = N_s I_d; \quad F_{s,q} = N_s I_q \tag{2.77}$$

$$B_r = \frac{N_r I_r \Lambda_d}{S_d}; \quad B_{s,d} = \frac{N_s I_d \Lambda_d}{S_d}; \quad B_{s,q} = \frac{N_s I_q \Lambda_q}{S_q} \tag{2.78}$$

where Λ_d and S_d (Λ_q and S_q) are the permeance and the cross section of the magnetic circuit through which the flux associated with the magnetomotive force $F_{s,d}$ ($F_{s,q}$) flows.

From the abovementioned equations, it is possible to obtain the magnetic flux linkages of the windings ($d - d'$) and ($q - q'$) as

$$\psi_d = (N_s I_d - N_r I_r)\Lambda_d$$
$$\psi_q = N_s I_q \Lambda_q$$

According to Lenz's Law, the induced voltages in the fixed stator coils can be calculated. For phase a this yields.[†]:

$$u_{a,q} = (-N_s^2 I_d + N_r N_s I_r)\Lambda_d \omega \sin(\omega t)$$

$$u_{a,d} = N_s^2 I_q \Lambda_q \omega \sin(\omega t - 90°)$$

If the leakage flux is added to the former equations, modeled as a leakage reactance X_p, and expressed in phasor notation, the following expression is obtained (whose graphical representation is shown in Figure 2.38):

$$\mathcal{V} = \mathcal{E} - jX_d \mathcal{I}_d - jX_q \mathcal{I}_q \tag{2.79}$$

where $\mathcal{E} = N_s N_r \Lambda_d I_r \omega$ is the internal voltage; $X_d = X_p + N_s^2 \Lambda_d \omega$ and $X_q = X_p + N_s^2 \Lambda_q \omega$ are the direct and quadrature reactances, respectively; $\mathcal{I}_d = jI \sin(\varphi)$ and $\mathcal{I}_q = I \cos(\varphi)$ are the direct and quadrature currents; and $\mathcal{V}$ is the voltage at the machine terminals.

In the round-rotor case, it is true that $X_d = X_q = X_s$ and consequently the above Equation 2.79 reduces to the already-known equation:

$$\mathcal{V} = \mathcal{E} - jX_s(\mathcal{I}_d + \mathcal{I}_q) = \mathcal{E} - jX_s \mathcal{I} \tag{2.80}$$

---

* The Park transformation is a mathematical transformation which breaks down the electrical and magnetic variables into direct- and quadrature-axis quantities.

[†] The voltages are related to the flux density existing in the stator coils owing to the fluxes of the mobile fictitious coils, and not to the voltages in the fictitious coils which will be d.c. magnitudes.

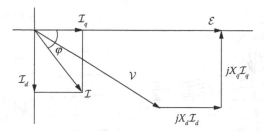

FIGURE 2.38 Phasor diagram of the salient-pole synchronous machine.

2.6 THE INDUCTION MACHINE

The induction machine is an electrical machine in which the stator windings are fed through a three-phase voltage source, while the rotor windings are short circuited and are circulated by currents induced by the stator (the reader is referred to Appendix C, where a generic rotating machine is considered).

In balanced steady-state conditions, the induction machine has an analog behavior to that of a transformer [3] and hence a transformer model can be used to represent this machine (Section 2.4). Thus, a simplified version of the equivalent single-phase circuit for wye-connected stator windings, assuming that the rotor parameters and variables are referred to the stator, can be seen in Figure 2.39.

The per unit values of the variables and parameters arising in that circuit are

P and Q: Active and reactive powers at the machine terminals.
V_s and $\mathcal{I}_s$: Voltage and current at the generator terminals (stator).
$\mathcal{I}_r$: Rotor current referred to the stator.
$Z_{sc} = R_{sc} + jX_{sc}$: Short-circuit impedance comprising that of the stator plus that of the rotor referred to the stator.
X_m: Per-phase reactance modeling the magnetization of the ferromagnetic core.
s: Slip of the machine obtained as $s = (\omega_s - \omega)/\omega_s$, where ω_s is the synchronous speed in rad/s.
P_m: Mechanical power exchanged by the machine with the outer world, which can be expressed from the electric parameters as

$$P_m = I_r^2 \cdot R_r \cdot \frac{1 - s}{s}$$

where R_r is the rotor resistance referred to the stator.

Typically, the steady-state operation of the induction machine is represented by the mechanical power-slip and reactive power-slip characteristics, according to the source voltage V_s. Figure 2.40 shows typical curves for different voltages.

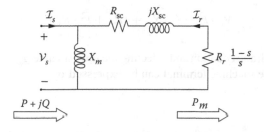

FIGURE 2.39 Approximate single-phase equivalent circuit of an induction machine.

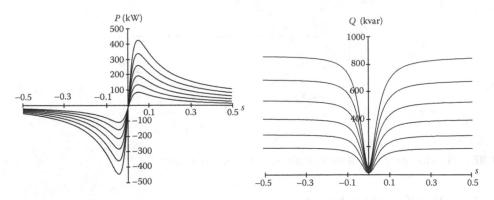

FIGURE 2.40   *P–s* and *Q–s* characteristics of an induction machine.

In the *P–s* graph in Figure 2.40, it can be observed that for slips greater than 0 (sub-synchronous rotor speeds), the mechanical power is positive (the active power at the terminals is consumed) and the machine is said to be working as a motor. Whereas, for negative slips (super-synchronous rotor speeds), the power is negative (the active power at the terminals is generated) and therefore it works as a generator. On the other hand, in the *Q–s* graph it can be observed that the reactive power is always positive (i.e., consumed). So, it can be concluded that induction machines need to be connected to a voltage source (electric network) that delivers the reactive power. Nevertheless, there are auto-excitation systems for these types of machines based on electronic devices that attempt to eliminate the need for the electric network, but they are out of the scope of this chapter.

Other interesting, extreme situations that can be observed in the graphs are the values for slips $s = 1$ (locked-rotor) and $s = 0$ (no-load). For the first case, locked-rotor, the mechanical power is zero and the reactive power presents its maximum value. Whereas for the machine under no-load conditions, the mechanical power is also zero and the reactive power is reduced to the one needed for the magnetizing reactance.

From the point of view of the steady-state equivalent circuit, as can be seen in Figure 2.39, the induction machine can be modeled with an equivalent admittance $\mathcal{Y}_{eq}$ defined by

$$\mathcal{Y}_{eq} = j(B_c - B_m) + \frac{1}{R_r(1 - s)/(s) + R_{sc} + jX_{sc}} \tag{2.81}$$

where B_m is the inverse of the magnetizing reactance (X_m), and B_c is the susceptance of the capacitor bank, which is located parallel to the machine terminals for the power factor compensation.

Another possible steady-state model of a synchronous machine is as a power source. This model is obtained from the relations between the powers in Figure 2.39 [4]

$$V_s^2 \cdot I_r^2 = I_r^2 \cdot X_{sc} + \left(P_m + I_r^2 \cdot R_{sc}\right)^2$$

From the above bi-quadratic equation and selecting the lowest value, $I_r^2 = f(V_s, P_m)$ is obtained. In this way, the powers at the machine terminal can be expressed by

$$P = P_m + f(V_s, P_m) \cdot R_{sc}$$
$$Q = V_s^2 \cdot (B_c - B_m) + f(V_s, P_m) \cdot X_{sc}$$

If some practical approximations as $B_c = B_m$, $R_{sc} = 0$, and $V_s = 1$ are admitted, the powers can be formulated as

$$P \approx P_m; \quad Q \approx f(V_s, P_m) \cdot X_{sc} \approx \frac{P_m^2 X_{sc}}{V_s - X_{sc}} \approx P^2 X_{sc}$$

The operation limits of an induction machine, as in other types of machines and elements of electric networks, depend on technological and economic conditions. In this case, three operation limits are defined:

- *Thermal or stator current limit*: This limit is defined by the maximum current that can circulate through the stator windings of the machine, and is calculated as

$$I_s = \frac{\sqrt{P^2 + Q^2}}{V_s} < I_{s,\max}$$

- *Dielectric insulation or maximum feeding voltage limit*: $V_s < V_{s,\max}$.
- *Stability or magnetizing limit*: This limit can be graphically established, for a determined voltage, as the maximum value of the mechanical power that crosses at the corresponding P_m–s curve in Figure 2.40.

2.7 LOADS

In the previous sections, the electric loads have been considered individually, by means of concrete models. The static loads are modeled by impedances or power sources, whereas the dynamic loads refer to synchronous and induction motors. However, it is obvious that there is a higher diversity of electric energy-consuming devices that do not exactly respond to the above models. On the other hand, the majority of the studies on electric networks consider loads that combine and bring together different consuming devices (e.g., a substation feeding a particular area of a city where industries, shops, and residencial loads coexist). Consequently, using only a qualitative criterion, the electric loads can be classified as *particular* and *global*.

A particular load consists of only one consuming device and therefore responds to a specific model (motor, lamp, rectifier, furnace, etc.). However, a global load consists of several consuming devices with different characteristics or different operating conditions (electric switchyard, substation, group of motors, etc.). These kinds of loads are normally defined using the so-called *aggregated models*. It is obvious that the most complex aggregation is the one in which the devices are of different types and present diverse operating conditions.

Electric loads can be classified in other possible ways, some of them are as follows:

- According to their mathematical function: *linear* and *nonlinear*
- According to the type of variables considered: *electrical*, *electro mechanical*, *thermo electrical*, *environmental*, *temporal*, and so on
- According to their modeling: *deterministic* and *random*

The modeling of an aggregated electric load is analyzed in this section, and the reader can learn more about the particular load models in other chapters of this book or in the references. The most frequent modeling of the loads as impedances, currents, and power sources are described initially. Then, functional models, which attempt to relate the load with voltage, feeding frequency, and so on, are studied. This type of model is interesting for the dynamic analysis of the networks and for the load management issue [5,6]. Finally, the load demand-predictive models are analyzed, which are focused on the load characterization to estimate its future values.

2.7.1 STEADY-STATE MODELS

The modeling of a load as a "negative" power source has been traditionally used for representing the large consumptions of electric substations. Thus, it is clearly a model of consumer aggregation. The active and reactive power values are obtained through measurements taken in the field and are normally represented by the so-called *daily curve*, where kW and kvar consumptions are shown at hourly intervals. One example of this type of representation is shown in Figure 2.41. In addition, to be useful for aggregated loads, there may be specific devices that admit the modeling as a power source, as happens with induction motors or controlled synchronous motors.

The model that defines a load as an *impedance* is useful for aggregated consumptions in MV and LV distribution networks, rarely used in the analysis of transmission networks, except in some particular cases where the nodal equations are linearized. As in the previous model, that of the impedance is used for some specific loads such as incandescent lamps, electric heaters, and so on.

The current source models are less frequent in the aggregated loads, and their main use is in the harmonic analysis of electric networks, with the aim of carrying out an iterative calculation process. For this reason, this model is generally not very useful in practice. Nevertheless, there are some devices that respond to this model, as in the case of controlled electronic rectifiers.

In most cases, the loads cannot be simulated using only one type of model, so it is more convenient to use a model with different types. The models for this type of load are generally called *generic steady-state models* and are defined to emphasize the aggregated character of a load. One of the many ways to express a generic consumption is using the expression [5]

$$P = K_p \cdot V^\alpha; \quad Q = K_q \cdot V^\beta$$

where P and Q are the active and reactive load consumptions, K_p and K_q are the active and reactive power coefficients, respectively, defined from the powers of an operating point, V is the rms voltage at the consumption node, and α and β are the active and reactive power exponents in that order.

It can be observed that $\alpha = 0$ and $\beta = 0$ leads to a power source; for $\alpha = 2$ and $\beta = 2$, the model is an admittance and for $\alpha = 1$ and $\beta = 1$, a current source model is obtained. Experience shows that this type of model behaves adequately for per unit values of $0.90 < V < 1.1$ and 48.5 Hz $< f < 51.5$ Hz. The typical exponent values, obtained from practice, move within the intervals $1.5 \leq \alpha \leq 2.0$ and $3.0 \leq \beta \leq 4.0$.

Figure 2.42 shows the characteristics of a generic consumption for different voltages and exponents.

2.7.2 FUNCTIONAL MODELS

It is evident that the described steady-state models establish a partial functional model. So, an impedance or a power source is sensitive to the variations of the feeding voltage, although in an opposite manner. For an impedance, a voltage increase means a current increase, whereas for a power source it

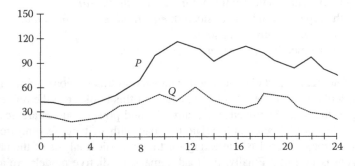

FIGURE 2.41   Daily load curve of an entire substation.

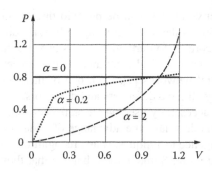
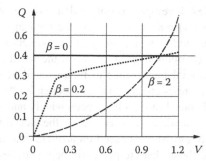

FIGURE 2.42 Representation of generic consumptions.

means a decrease. Thus, a functional model, unless the load type is known, is not limited *a priori* to a linear expression [5–7].

In general, a steady-state functional model responds to a nonlinear function, as $P = p(V, \omega)$ and $Q = q(V, \omega)$ [8], which can be linearized applying the Taylor method:

$$P = P_0 + \sum K_i \cdot \Delta V^i + \sum K_i' \cdot \Delta \omega^i \tag{2.82}$$

$$Q = Q_0 + \sum L_i \cdot \Delta V^i + \sum L_i' \cdot \Delta \omega^i \tag{2.83}$$

2.7.3 PREDICTIVE MODELS

The tendency to deregulate and privatize the electric sector in some countries has increased the interest in demand-predictive models. In this way, the predictive models that were normally used for network planning are now being applied to obtain very short-term (minutes) and short-term (hours) consumption models. The consequence is a growth in the use of predictive techniques due to the emergence of a highly competitive electric market.

In this section, some of the commonly used methods in the prediction of electric demand will be briefly studied. To do this, the demand singularities from a qualitative point of view are described. These singularities make the load prediction a problem with a difficult solution.

The randomness of electric consumption is well known due to its uncontrollability and its dependence on variables outside the electric network. Furthermore, the electric usage is not centered in only one activity but is present in nearly all social activities (traction, lighting, shows, services, thermal loads, etc.). Consequently, the randomness along with the extension of electric consumption make the demand characterization a complex problem. However, it is possible to appreciate certain specific demand behavior that can help to understand and characterize it:

- *Demand depending on the time of the day*: As can be observed from any daily curve, there is a variation in consumption during the day. In this way, it is known that the hours of least consumption are at night due to lower social activity, and the hours of highest consumption are during daytime.
- *Demand depending on the type of day*: Owing to the strong relation between consumption and economic activities, there are different levels of consumption subject to the type of day (working days or holidays).
- *Demand depending on the type of society or area determined*: For the same type of day, the daily consumption curves are different, related to the activity in the area. In this way, the consumption of an industrial area is clearly very different from that of a residential, agricultural, or public services area.
- *Demand depending on weather conditions*: For the same area and type of day, the consumptions are strongly influenced by the weather (temperature, wind, cloud, etc.).

The studies developed to obtain demand-predictive models can be put into three main groups: models based on *time series*, where the consumption is defined according to historic data; *causal or functional models* where the load is modeled according to particular factors or variables outside the network (exogenous), for example, the weather or the social area; and the most recent models based on artificial intelligence techniques and neural networks [9]. Each one of these groups has pros and cons, respectively, and, consequently, they all have defenders and detractors.

Time and causal models have been widely and successfully used where statistical techniques are employed to calculate them. However, the causal models have the advantage of understanding the demand behavior. Models based on artificial intelligence techniques, shape recognition, and neural networks are those that are presently being studied more. Some authors believe that these models need to be checked more exhaustively and that their procedures need to be systematized to become a reliable prediction technique.

With the aim to present some of the predictive models, a mathematical expression for a consumption y in a determined area z for each time instant t, can be

$$y(t) = f(z, C(t), t) \tag{2.84}$$

where the area z is defined by a set of activities (industries, domestic customers, services, etc.) and $C(t)$ is the set of variables that define the weather conditions (temperature, humidity, wind speed, cloud, etc.), which have different values for every time instant t.

Applying Equation 2.84 to the historic data necessary to obtain the predictive models, the following expression is obtained:

$$y_k = f(z, C_k, k) \tag{2.85}$$

where k represents time intervals in such a way that $t = k \cdot \tau$, where the τ interval is selected according to the desired time horizon (minutes, hours, days, or weeks).

It is obvious that the previous function cannot be interpreted as a deterministic relation between the present variables, due to the existence of other unquantifiable variables, but can be representative of random behavior. In short, the abovementioned function attempts to establish a causal relation between the consumption value and the weather conditions, the area, or the loads that define it, from a set of data and historic measurements.

2.7.3.1 Causal Predictive Models

Causal models try to establish a relation between electric demand and weather conditions, where a previous demand classification with respect to the time component is admitted to eliminate its effects on the model. The most basic demand classification is when the working days and the rest are considered separately and when it is possible to admit some type of periodicity in daily demand curve owing to the area activities [10,11]. In this way, causal models generally admit a demand $y(t)$ break down, to specific areas and days as in Reference 11

$$y(t) = y_d(t) + y_c(t) \tag{2.86}$$

where $y_d(t)$ is the deterministic component of the load that is obtained, for example, from the mean value of the hourly curve of the loads of a specific day, and $y_c(t)$ is the part of the consumption subject to the weather conditions.

Despite the existence of a large number of techniques to carry out the causal predictive model, the commonly used are based on linear regression [12,13]. The formula for the load forecast problem is of the type [10]

$$y_c(t) = a_0 + a_1 T + a_2 W + a_3 L + a_4 P \tag{2.87}$$

where a_0, a_1, a_2, a_3, and a_4 are coefficients to calculate and T, W, L, and R are the weather condition variables (temperature, wind speed, luminosity, and rainfall).

2.7.3.2 Time Series Predictive Models

These predictive models, based on the time series analysis, use the auto-regressive moving average (ARMA) [13,14]. The version that also uses exogenous variables (e.g., atmospheric temperature) is called ARMAX (auto-regressive moving average with exogenous variable) [15].

The time series methods are based on the fact that some periodicity in data exists (seasonal, daily, weekly, or hourly). The causes of this periodicity are of no interest.

Time series predictive models are defined using a $y(k)$ vector of observed data (measurements), which is a discrete nonsteady-state time series, where k is the sampling done using the least periodicity permitted (hours for a daily periodicity).

The application of the ARMA process to the $y(k)$ series requires the following steps: (a) transformation of the nonsteady series in a steady-state one $x(k)$, with constant mean and variance and (b) determination of ARMA parameters p and q and seasonal P and Q [13].

In this way, it can be concluded, according to the AR process, that the $x(k)$ values can be obtained from the previous values $x(k-1)$, $x(k-2)$, ..., $x(k-p)$ and from a random Gaussian variable $a(k)$ with a mean equal to zero and variance equal to σ^2. At the same time, due to the moving average process, the $x(k)$ values are also a linear combination of the previous values $a(k)$, $a(k-1)$, $a(k-2)$, ..., $a(k-q)$. The validity of the predictive model is determined by observing the difference between the two temporal series $y(k)$ and $x(k)$.

REFERENCES

1. J. R. Carson, Wave propagation in overhead wires with ground return, *Bell System Technical Journal*, 5, 1926, 539–554.
2. E. D. Sunde, *Earth Conduction Effects in Transmission Systems*, Dover Publications, New York, 1968.
3. P. C. Krause, *Analysis of Electric Machinery*, McGraw-Hill, New York, 1986.
4. A. Feijóo, and J. Cidrás, Modelling of wind farm in load flow analysis, *IEEE Transactions on Power Systems*, 15(1), 2000, 110–115.
5. IEEE Task Force Report, Load representation for dynamic performance analysis, *IEEE Transactions on Power Systems*, 8(2), 1993, 472–482.
6. C. Álvarez, R. P. Malhamé, and A. Gabaldón, A class of models for load management application and evaluation revisited, *IEEE Transactions on Power Systems*, 7(4), 1992, 1435–1443.
7. A. R. Bergen, D. J. Hill, and C. L. Marcot, Lyapunov function for multimachine power systems with generator flux decay and voltage dependent loads, *Electrical Power and Energy Systems*, 8(1), 1986, 2–10.
8. M. Khalifa, *High-Voltage Engineering*, Marcel Dekker Inc., New York, 1990.
9. H. Steinherz, C. E. Pedreira, and R. Castro, Neural networks for short-term load forecasting: A review and evaluation, *IEEE Transactions on Power Systems*, 16(1), 2001, 44–55.
10. M. J. H. Sterling, *Power System Control*, Peter Peregrinus Ltd., 1978.
11. J. Y. Fan, and J. D. McDonald, A real-time implementation of short-term load forecasting for distribution power systems, *IEEE Transactions on Power Systems*, 9(2), 1994, 988–994.
12. V. M. Vlahovic, and I. M. Vujosevic, Long-term forecasting: A critical review of direct-trend extrapolation methods, *Electrical Power and Energy Systems*, 9(1), 1987, 2–8.
13. D. Peña, *Estadística Modelos y métodos. Volúmen 2. Modelos Lineales y series Temporales*, Alianza Universidad Textos, Madrid, 1989 (2nd edition).
14. G. E. P. Box, and G. M. Jenkins, *Time Series Analysis: Forecasting and Control*, Holden-Day, 1970.
15. H. T. Yang, C. M. Huang, and C. L. Huang, Identification of ARMAX model for short load forecasting: An evolutionary programming aproach, *IEEE Transactions on Power Systems*, 11(1), 1996, 403–408.

3 Load Flow

Antonio Gómez-Expósito and Fernando L. Alvarado

CONTENTS

3.1 INTRODUCTION

The load flow or power flow problem consists of finding the steady-state operating point of an electric power system. More specifically, given the load demanded at consumption buses and the power supplied by generators, the aim is to obtain all bus voltages and complex power flowing through all network components.

Without doubt, the power flow solver is the most widely used application both in operating and in planning environments, either as a stand-alone tool or as a subroutine within more complex processes (stability analysis, optimization problems, training simulators, etc.).

During the daily grid operation, the load flow constitutes the basic tool for security analysis, by identifying unacceptable voltage deviations or potential component overloading, as a consequence of both natural load evolution and sudden structural changes. It also allows the planning engineer to simulate different future scenarios that may arise for a forecasted demand.

The power flow solution proceeds in two stages. The first and most critical one is aimed at finding the complex voltage at all buses, for which conventional linear circuit analysis techniques are not useful. This is a consequence of complex powers, rather than impedances and sources, being specified as binding constraints, leading to a set of nonlinear equations. The second step simply consists of computing the remaining magnitudes of interest, such as active and reactive power flows, ohmic losses, and so on, which is a trivial problem provided all bus voltages are available.

In this chapter, the most popular techniques for load flow computation are explained in detail and illustrated with examples. Furthermore, certain solution refinements, which are necessary to account for additional constraints imposed by a diversity of regulating devices, are addressed.

One of the keys to the terrific success attained by load flow solvers during the 1970s was the introduction of efficient numerical techniques to solve large and sparse equation systems. Such techniques, specifically developed by power engineers to allow execution of load flow algorithms on primitive computers, are discussed at the end of this chapter and reviewed in more detail in Appendix A.

3.2 NETWORK MODELING

As explained in the previous chapter, all power system components (lines, cables, transformers) interconnecting different buses can be represented through a two-port π model. Given the complex voltages of the terminal buses, this simple model allows sending and receiving power flows to be obtained and hence power losses.

However, when the analysis refers to the whole network, rather than to individual components, a more compact representation arises by resorting to the node or bus admittance matrix, which is obtained as follows.

Consider a generic bus i, as shown in Figure 3.1, connected by means of series admittances to a reduced subset of buses. In addition, a small shunt admittance directly connected to the neutral or ground node may exist, modeling the total charging current. The net current injected to the bus by generators or loads, $\mathcal{I}_i$, obeys Kirchhoff's current law, that is,

$$\mathcal{I}_i = \sum_{j \in i} y_{ij}(\mathcal{V}_i - \mathcal{V}_j) + y_{si}\mathcal{V}_i \tag{3.1}$$

where $\mathcal{V}_j$ denotes the complex voltage at bus j, and $j \in i$ refers to the set of buses $1, 2, \dots, m$ directly connected to bus i. Rearranging terms yields

$$\mathcal{I}_i = \left[\sum_{j \in i} y_{ij} + y_{si} \right] \mathcal{V}_i - \sum_{j \in i} y_{ij}\mathcal{V}_j \tag{3.2}$$

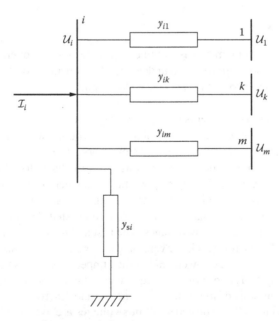

FIGURE 3.1 Elements connected to a generic bus i.

Repeating the above development for the whole set of n buses leads to the so-called nodal equations, which can be written in matrix form as follows:

$$\mathcal{I} = \mathcal{Y}\mathcal{V} \tag{3.3}$$

where $\mathcal{Y}$ is the $n \times n$ bus admittance matrix, and the elements of the column vectors $\mathcal{V}$ and $\mathcal{I}$ represent complex node voltages and net injected currents, respectively. The elements of matrix $\mathcal{Y}$ can be obtained by comparing the ith row of Equation 3.3

$$\mathcal{I}_i = \sum_{i=1}^{n} \mathcal{Y}_{ij}\mathcal{V}_j \quad i = 1, 2, \ldots, n \tag{3.4}$$

with Equation 3.2 leading to

$$\mathcal{Y}_{ii} = \left[\sum_{j \in i} y_{ij} + y_{si} \right]; \quad \mathcal{Y}_{ij} = -y_{ij} \tag{3.5}$$

In summary, the diagonal elements of $\mathcal{Y}$ are obtained by adding all admittances connected to the respective bus, whereas the off-diagonal elements are simply the negative admittance interconnecting the involved buses. In case there are several admittances in parallel, the equivalent admittance (i.e., the sum of such admittances) should be previously computed. The great majority of off-diagonal elements will be null, because a bus is directly connected on an average to only a few buses.

A minor disadvantage of the bus admittance concept lies in the fact that the shunt admittances of individual components, added to the respective diagonal, cannot be recovered, which slightly complicates the power flow expressions when the matrix elements, rather than the two-port parameters, are employed. This is also the case when power flows through individual lines in parallel are to be found.

3.3 PROBLEM FORMULATION

In addition to the n linear Equation 3.4, representing in compact form the network topology and components, the following power constraint should be enforced at each bus:

$$\mathcal{S}_i = \mathcal{S}_{Gi} - \mathcal{S}_{Li} = \mathcal{V}_i \mathcal{I}_i^* \tag{3.6}$$

where $\mathcal{S}_i$ is the net complex power injected to bus i, obtained in the general case as the difference between the complex power injected by generating elements, $\mathcal{S}_{Gi}$, and the complex power absorbed by loads, $\mathcal{S}_{Li}$. Eventually, $\mathcal{S}_{Li}$ may reflect the effect of other passive components not included in matrix $\mathcal{Y}$. The above expression, applied to all buses, can be written in matrix form as follows:

$$\mathcal{S} = \text{diag}(\mathcal{V})\mathcal{I}^* \tag{3.7}$$

where $\mathcal{S}$ is the column vector comprising bus complex powers and $\text{diag}(\mathcal{V})$ denotes a diagonal matrix whose elements are those of vector $\mathcal{V}$.

Given $\mathcal{Y}$, Equations 3.3 and 3.7 constitute a system of $2n$ complex equations in terms of the $3n$ complex unknowns $\mathcal{S}$, $\mathcal{V}$, and $\mathcal{I}$. In theory, knowing n of such unknowns, the resulting nonlinear system could be solved to obtain the remaining $2n$ variables. In practice, however, the complex nodal currents are seldom known or specified *a priori* in a power system. Therefore, they are usually removed from the unknown set by substitution of Equation 3.3 into Equation 3.7. This leads to the following nonlinear set of n complex equations:

$$S = \text{diag}(\mathcal{V})[\mathcal{Y}\mathcal{V}]^* \tag{3.8}$$

Expressing the complex power in terms of active and reactive power, $S = P + jQ$, and using rectangular coordinates for the elements of the admittance matrix, $\mathcal{Y} = G + jB$, the above expression yields

$$P + jQ = \text{diag}(\mathcal{V})[G - jB]\mathcal{V}^* \tag{3.9}$$

$$P_i + jQ_i = \mathcal{V}_i \sum_{j=1}^{n} (G_{ij} - jB_{ij})\mathcal{V}_j^* \quad i = 1, 2, \ldots, n \tag{3.10}$$

The most relevant methods described below to solve the load flow problem cannot directly work with the above equations, because the conjugate operator "$*$" prevents the application of derivatives in complex form. For this reason, it is customary to split them into $2n$ real equations. Usually, complex voltages are expressed in polar form, $\mathcal{V} = V \angle \theta$, leading to

$$P_i = V_i \sum_{j=1}^{n} V_j(G_{ij} \cos \theta_{ij} + B_{ij} \sin \theta_{ij}) \tag{3.11}$$

$$Q_i = V_i \sum_{j=1}^{n} V_j(G_{ij} \sin \theta_{ij} - B_{ij} \cos \theta_{ij}) \quad i = 1, 2, \ldots, n \tag{3.12}$$

Alternatively, expressing the voltages in rectangular coordinates $\mathcal{V} = V_r + jV_x$, yields

$$P_i = V_{ri} \sum_{j=1}^{n} (G_{ij}V_{rj} - B_{ij}V_{xj}) + V_{xi} \sum_{j=1}^{n} (G_{ij}V_{xj} + B_{ij}V_{rj}) \tag{3.13}$$

$$Q_i = V_{xi} \sum_{j=1}^{n} (G_{ij}V_{rj} - B_{ij}V_{xj}) - V_{ri} \sum_{j=1}^{n} (G_{ij}V_{xj} + B_{ij}V_{rj}) \quad i = 1, 2, \ldots, n \tag{3.14}$$

Unless otherwise indicated, polar coordinates will be assumed.

Each node provides two equations and four unknowns, which means that two variables per node should be specified to solve the resulting load flow equations. Depending on which variables are specified, two main types of buses can be distinguished:

- *Load or PQ buses*: Both active and reactive power absorbed by the sum of loads connected at the bus are specified. Assuming the power generated locally is null ($P_{Gi} = Q_{Gi} = 0$), this leads to the following bus constraints:

$$P_i = P_i^{\text{sp}} = -P_{Li}^{\text{sp}}; \quad Q_i = Q_i^{\text{sp}} = -Q_{Li}^{\text{sp}} \tag{3.15}$$

leaving the two voltage components, V_i and θ_i, as the remaining unknowns. A majority of buses, particularly at the lower voltage levels, belong to this type.
- *Generation or PV buses*: These are buses where the voltage regulator of a local generator keeps the voltage magnitude to a specified value (V_i^{sp}). Furthermore, the active power injected by the generator is specified according to certain economic criteria (see Chapter 6).

Taking into account the possible load demand, the resulting constraints are therefore

$$P_i = P_i^{\text{sp}} = P_{\text{G}i}^{\text{sp}} - P_{\text{L}i}^{\text{sp}}; \quad V_i = V_i^{\text{sp}} \tag{3.16}$$

leaving Q_i and θ_i as unknowns. A particular case of PV bus arises when a reactive power compensator (rotating or static), equipped with voltage regulator, is connected to a bus. In this case, $P_{\text{G}i}^{\text{sp}} = 0$. The presence of small generating units, not capable in general of performing voltage regulation, leads to PQ buses with appropriate (usually positive) specified values.

Nonetheless, if only those two types of buses were considered, all injected active powers should be specified *a priori*, which requires that ohmic losses be known in advance. However, power losses depend on the resulting power flows and cannot be accurately determined until the load flow itself is solved. Therefore, the active power of at least a generator should be left as an unknown. Fortunately, this extra unknown is compensated by the fact that, when performing steady-state AC analysis, the phase angle of an arbitrary phasor can be set arbitrarily to zero. This constitutes the phase angle reference for the remaining sinusoidal waveforms. For convenience, the voltage phasor of the generating bus whose active power remains unspecified is taken as reference for phase angles. This particular PV bus, known as the *slack* or *swing bus*, is usually chosen among those generating buses with largest capacity, frequently being in charge of frequency regulation duties. In summary, for the slack bus, the complex voltage is fully specified whereas both power components, active and reactive, belong to the set of unknowns.

It may be correctly argued that the slack bus concept is a mathematical artifact, without any direct link to the real world, as no bus in the system is explicitly in charge of providing all ohmic losses. Indeed, for very large systems, power losses may exceed by far the capacity of certain generators. However, if an estimation of power losses is available, which is usually the case, then all generators can share a fraction of those losses. This way, in addition to its own power, the slack bus will be responsible only for the power system imbalance, that is, the difference between the total load plus actual losses and the total specified generation, leading to a power flow profile closely resembling actual operation. Unlike ohmic losses, the resulting system imbalance may be positive or negative. In Reference 1, instead of determining *a priori* which bus plays the role of slack bus, it is selected on the fly during the load flow computation process in such a way that the system power imbalance is minimized.

Let n_L be the number of PQ buses. Then, the number of PV buses, excluding the slack bus, will be $n_\text{G} = n - n_\text{L} - 1$. Without the loss of generality, it will be assumed that the first n_L buses correspond with PQ buses, followed by ordinary PV buses and then the slack bus. Following this classification of buses, the load flow equations in polar form are

$$P_i^{\text{sp}} = V_i \sum_{j=1}^{n} V_j(G_{ij} \cos \theta_{ij} + B_{ij} \sin \theta_{ij}) \quad i = 1, 2, \ldots, n_\text{L} + n_\text{G} \tag{3.17}$$

$$Q_i^{\text{sp}} = V_i \sum_{j=1}^{n} V_j(G_{ij} \sin \theta_{ij} - B_{ij} \cos \theta_{ij}) \quad i = 1, 2, \ldots, n_\text{L} \tag{3.18}$$

Solving the load flow consists of finding the set of phase angles θ_i, $i = 1, 2, \ldots, n_\text{L} + n_\text{G}$, and the set of voltage magnitudes V_i, $i = 1, 2, \ldots, n_\text{L}$, satisfying the $2n_\text{L} + n_\text{G}$ Equations 3.17 and 3.18.

Specifying the complex voltage of the slack bus and freeing its complex power imply simply that the respective pair of equations will be ignored during the load flow process. Such equations will be useful afterward to obtain the slack complex power.

In the same way, the n_G Equation 3.12 excluded from Equation 3.18 will provide subsequently the reactive power required by each generator to keep its voltage to the target value. As the reactive power capability of generators is bounded, it is necessary to verify that none of the limits are violated, which complicates and slows down the solution process (this issue will be addressed in more detail in Section 3.7.1).

As the resulting equation system is nonlinear, its solution necessarily involves an iterative process, for which adequate initial values should be given to the state variables. Although finding suitable initial values may not be trivial in the general case, the so-called flat start is generally the best choice for the load flow problem. It consists of setting $\theta_i^0 = 0$ for every bus and $V_i^0 = 1$ pu for PQ buses, reflecting the fact that voltage magnitudes lie normally within a relatively narrow band around 1 pu while phase angle differences between adjacent buses are also quite small.

Sometimes, the solution of a previous run can be a good starting point if changes to the load or generation pattern have been small. This may be particularly useful for security assessment purposes involving the analysis of a long sequence of cases, with each one differing from the base case in a single topological change (see Chapter 6). Experience shows, however, that using arbitrary initial values, apparently closer to the true solution than the flat start profile, leads almost always to a larger number of iterations.

Once Equations 3.17 and 3.18 are solved, any other desired magnitude can be easily computed (commercial packages provide a long list of output files or graphical representations containing all the information specified in advance by the user).

When there is a single element (line or transformer) connected between buses i and j, the power flows leaving bus i can be obtained from

$$P_{ij} = V_i V_j (G_{ij} \cos \theta_{ij} + B_{ij} \sin \theta_{ij}) - G_{ij} V_i^2 \tag{3.19}$$

$$Q_{ij} = V_i V_j (G_{ij} \sin \theta_{ij} - B_{ij} \cos \theta_{ij}) + V_i^2 (B_{ij} - b_{s,ij}) \tag{3.20}$$

where $b_{s,ij}$ denotes the shunt susceptance associated to the respective π model (power flows leaving bus j can be easily obtained by properly exchanging subindices). In case there are several components connected in parallel between buses i and j, it is necessary to resort to the individual parameters associated to each π model (as explained in Chapter 2), since the elements of the bus admittance matrix represent aggregated values.

Similarly, total network losses (both active and reactive) can be computed either by adding the power injections at all buses, provided the complex power of the slack bus is available, or by adding the losses corresponding to each individual component. The second alternative is the only choice when losses corresponding to a specified area or subsystem are needed.

EXAMPLE 3.1

Consider the three-bus network shown in Figure 3.2 in which bus 1 is the slack bus ($\theta_1 = 0$), bus 2 is a PQ bus, and bus 3 is a PV bus. Data corresponding to this system, on a 100-MVA base, are listed in the table below (shunt admittances are considered negligible):

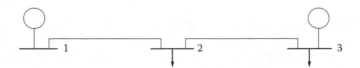

FIGURE 3.2 Three-bus illustrative system.

Bus	V	P_G	P_L	Q_G	Q_L
1	1.1	—	0	—	0
2	—	0	1	0	0.4
3	1.05	0.6	0.2	—	0.05
Series impedance 1–2:			$z_{12} = 0.03 + 0.3j$		
Series impedance 2–3:			$z_{23} = 0.06 + 0.2j$		

First, to simplify the computations below, it is convenient to build the bus admittance matrix:

$$y = \begin{bmatrix} 0.3300 - 3.3003j & -0.3300 + 3.3003j & 0 \\ -0.3300 + 3.3003j & 1.7062 - 7.8875j & -1.3761 + 4.5872j \\ 0 & -1.3761 + 4.5872j & 1.3761 - 4.5872j \end{bmatrix}$$

The unknowns in this case are θ_2, V_2 for bus 2 and θ_3 for bus 3, whereas the specified values are

$$P_2^{sp} = P_{G2} - P_{L2} = -1$$
$$Q_2^{sp} = Q_{G2} - Q_{L2} = -0.4$$
$$P_3^{sp} = P_{G3} - P_{L3} = 0.4$$

The nonlinear functions relating data with unknowns in polar form are as follows:

$$P_2^{sp} - V_2(1.7V_2 - 0.363\cos\theta_2 + 3.63\sin\theta_2 - 1.445\cos\theta_{23} + 4.816\sin\theta_{23}) = 0$$
$$Q_2^{sp} - V_2(7.887V_2 - 3.63\cos\theta_2 - 0.363\sin\theta_2 - 4.816\cos\theta_{23} - 1.445\sin\theta_{23}) = 0 \qquad (3.21)$$
$$P_3^{sp} - 1.05(1.445 - 1.376V_2\cos\theta_{32} + 4.587V_2\sin\theta_{32}) = 0$$

Solving the load flow consists of finding the solution to the above equations. Afterward, active and reactive power injected at bus 1, reactive power injected at bus 3, as well as the power flows through each line and associated losses, can be computed.

3.4 SIMPLE ITERATIVE METHODS

Historically, owing to the reduced computing power and amount of memory available in primitive computers, simple methods iterating on a bus each time were adopted. The common feature of such methods is that just a single row has to be sequentially manipulated, rather than entire admittance [2,3] or impedance matrices [4–6]. They are direct application to the nonlinear case of well-known linear equation solvers. Even though their practical interest is questionable nowadays, some of the most popular and simple algorithms, still offered in commercial packages, will be presented below.

3.4.1 GAUSS–SEIDEL METHOD

This scheme sequentially sweeps each node, updating its complex voltage in terms of the voltages of neighbor buses.

In general, finding the vector x satisfying the nonlinear system

$$f(x) = 0 \qquad (3.22)$$

can be reformulated like a fixed-point problem,

$$x = F(x)$$

whose solution, starting from the initial value x^0, is iteratively obtained through the sequence:

$$x_i^{k+1} = F_i(x_1^{k+1}, \ldots, x_{i-1}^{k+1}, x_i^k, \ldots, x_n^k) \quad i = 1, 2, \ldots, n \tag{3.23}$$

Fresh values of variables already updated ($i = 1, 2, \ldots, i-1$) are used when computing x_i.

Focusing now on the load flow problem, among the several ways Equation 3.10 can be rewritten, the following one has proved to be the most effective:

$$\mathcal{V}_i^{k+1} = \frac{1}{\mathcal{Y}_{ii}} \left[\frac{P_i^{sp} - jQ_i^{sp}}{(\mathcal{V}_i^k)^*} - \sum_{j=1}^{i-1} \mathcal{Y}_{ij}\mathcal{V}_j^{k+1} - \sum_{j=i+1}^{n} \mathcal{Y}_{ij}\mathcal{V}_j^k \right] \quad i = 1, 2, \ldots, n-1 \tag{3.24}$$

The iterative process is stopped when the condition

$$\max_i |\mathcal{V}_i^{k+1} - \mathcal{V}_i^k| \le \varepsilon \tag{3.25}$$

is satisfied, where ε is a sufficiently small threshold (e.g., 0.0001).

Although the computational effort per iteration is moderate, the convergence of this method is linear, which means that the tolerance decreases more or less linearly with the number of iterations (and tends to increase as the dimension n of the system increases). This poses an important limitation for large systems, as the total computational cost, and hence solution time, increases considerably as larger systems are solved. The number of iterations can be significantly reduced, sometimes to over one half, using an acceleration factor α,

$$[\mathcal{V}_i^{k+1}]^{acc} = \mathcal{V}_i^k + \alpha(\mathcal{V}_i^{k+1} - \mathcal{V}_i^k)$$

whose value should not exceed two to avoid divergence. Empirically determined optimal values of α lie between 1.4 and 1.6.

However, Equation 3.24 cannot be directly applied to PV buses for two reasons: (1) Q_i^{sp} is unknown for those buses and (2) the resulting voltage magnitude after each iteration will differ from the specified value. The method usually employed to circumvent the first problem consists of replacing Q_i^{sp} by the value computed with the best available voltages. The second limitation is averted by scaling the estimated voltage so that the phase angle is updated but the specified voltage magnitude is retained:

$$[\mathcal{V}_i^{k+1}]^{corr} = \frac{V_i^{sp} \, \mathcal{V}_i^{k+1}}{V_i^{k+1}}$$

This corrective mechanism should not be prematurely implemented not to deteriorate the global convergence of the process, inherently very poor.

Perhaps, the only practical application of the Gauss–Seidel method nowadays consists of its use as a starter of the Newton–Raphson method, only in those rare cases in which the latter does not converge from the flat start profile.

EXAMPLE 3.2

The system of 3.1 will be solved by the Gauss–Seidel method.

Taking initially $Q_3 = 0$ and flat start, the voltages obtained after applying expression 3.24 twice to buses 2 and 3 are

$$V_2 = 0.9803\angle{-3.369}; \quad V_3 = 1.05\angle{-2.086}$$

The following table presents the final results obtained as well as the number of iterations for two different stopping criteria.

Max $|\Delta V_i|$	\multicolumn{2}{c}{10^{-4}}		\multicolumn{2}{c}{10^{-5}}	
Iterations	\multicolumn{2}{c}{25}		\multicolumn{2}{c}{33}	
Bus	V	θ		V	θ
2	0.9931	-9.256		0.9931	-9.262
3	1.05	-5.479		1.05	-5.490

The poor convergence of the method is clearly observed, accelerating factors being nearly of no help in this case. The convergence rate would be even worse if bus 3 was of the PQ type (33 and 42 iterations, respectively).

3.4.2 IMPEDANCE MATRIX METHODS

The inverse of the bus admittance matrix

$$\mathcal{Z} = \mathcal{Y}^{-1},$$

known as the bus impedance matrix, finds application in the context of fault analysis (see Chapter 8). When the network is weakly connected to ground (very small shunt admittances), matrix $\mathcal{Y}$ is nearly singular and $\mathcal{Z}$ is numerically ill-defined. This problem is prevented by eliminating the slack bus and working with the resulting reduced matrices, as explained in what follows. Let V_r and $\mathcal{I}_r$ be the vectors obtained by removing the slack bus variables. Then, Equation 3.3 can be rewritten as

$$\mathcal{I}_r = \mathcal{Y}_r V_r + \mathcal{Y}_s V_s$$

where $\mathcal{Y}_r$ is the admittance matrix obtained by removing the row and column of the slack bus, $\mathcal{Y}_s$ is the eliminated column, and V_s is the slack bus voltage. Rearranging terms leads to

$$V_r = \mathcal{Z}_r[\mathcal{I}_r - \mathcal{Y}_s V_s] \tag{3.26}$$

where $\mathcal{Z}_r = \mathcal{Y}_r^{-1}$ is the reduced bus impedance matrix. Starting from a set of initial voltages V_r^0, bus currents are obtained from

$$\mathcal{I}_i = (P_i^{sp} - jQ_i^{sp})/V_i^* \quad i = 1, 2, \ldots, n - 1$$

Then they are substituted into Equation 3.26, and the process is repeated until convergence is obtained.

This basic procedure allows several refinements to be implemented, all of them leading to much better convergence than that of the Gauss–Seidel method. The price paid for the improved convergence comes from matrix $\mathcal{Z}_r$ being full, which means that the cost per iteration grows with n^2. If the sparsity techniques discussed in Appendix A are adopted, this burden is significantly alleviated by resorting to the triangular factorization of $\mathcal{Y}_r$, instead of using its explicit inverse. In spite of that, this category of methods is not deemed competitive nowadays, except perhaps when radial distribution networks are solved (see Section 3.8.2).

3.5 NEWTON–RAPHSON METHOD

This method successively improves unknown values through first-order approximations of the involved nonlinear functions. Retaining the first two terms in the Taylor series expansion of

Equation 3.22 around x^k yields

$$f(x) \cong f(x^k) + F(x^k)(x^{k+1} - x^k) = 0 \tag{3.27}$$

where $F = \partial f / \partial x$ is the Jacobian matrix of $f(x)$. Then, starting from the initial value x^0, corrections Δx^k are obtained by solving the linear equation system:

$$-F(x^k)\Delta x^k = f(x^k) \tag{3.28}$$

and updated values x^{k+1} from

$$x^{k+1} = x^k + \Delta x^k \tag{3.29}$$

The iterative process is stopped when

$$\max_i |f_i(x^k)| \le \varepsilon$$

for a sufficiently small ε. For values of x^0 close to the solution, the Newton–Raphson method converges quadratically (however, when it diverges, it does so also quadratically). Irrespective of the network size, starting from the flat voltage profile, it takes from three to five iterations to attain convergence [7], but this figure can be significantly increased when the solution adjustments considered in Section 3.7 are considered.

Unlike the simpler methods described above, which can be implemented in complex form, the need to carry out derivatives in the presence of the conjugate operator requires that the system of equations be decomposed into its real components. Hence, depending on how complex voltages are expressed, the polar or the rectangular load flow version is obtained, the former being by far the most popular.

3.5.1 POLAR FORMULATION

In this case, vector x comprises the following $2n_L + n_G$ elements:

$$x = [\theta | V]^T = [\theta_1, \theta_2, \dots, \theta_{n-1} | V_1, V_2, \dots, V_{n_L}]^T$$

and the respective nonlinear functions can be expressed, for every bus, as the difference (residual) between the specified power and the power computed with the most recent x value, that is,

$$f(x) = [\Delta P | \Delta Q]^T = [\Delta P_1, \Delta P_2, \dots, \Delta P_{n-1} | \Delta Q_1, \Delta Q_2, \dots, \Delta Q_{n_L}]^T$$

where

$$\Delta P_i = P_i^{sp} - V_i \sum_{j=1}^{n} V_j(G_{ij} \cos \theta_{ij} + B_{ij} \sin \theta_{ij}) \quad i = 1, 2, \dots, n-1 \tag{3.30}$$

$$\Delta Q_i = Q_i^{sp} - V_i \sum_{j=1}^{n} V_j(G_{ij} \sin \theta_{ij} - B_{ij} \cos \theta_{ij}) \quad i = 1, 2, \dots, n_L \tag{3.31}$$

On the basis of the above notation and dividing the Jacobian in blocks corresponding to those of the residual and unknown vectors, Equation 3.28, when applied to the load flow problem, becomes [8,9]:

$$\begin{bmatrix} H & N \\ M & L \end{bmatrix}^k \begin{bmatrix} \Delta\theta \\ \Delta V/V \end{bmatrix}^k = \begin{bmatrix} \Delta P \\ \Delta Q \end{bmatrix}^k \tag{3.32}$$

and Equation 3.29

$$\begin{bmatrix} \theta \\ V \end{bmatrix}^{k+1} = \begin{bmatrix} \theta \\ V \end{bmatrix}^{k} + \begin{bmatrix} \Delta\theta \\ \Delta V \end{bmatrix}^{k} \tag{3.33}$$

The use of $\Delta V/V$ instead of ΔV does not affect numerically the algorithm but makes the Jacobian matrix more symmetrical (note that, structurally the Jacobian is fully symmetric, but not numerically). Keeping in mind that

$$\frac{-\partial(f_i^{\mathrm{sp}} - f_i)}{\partial x_j} = \frac{\partial f_i}{\partial x_j}$$

where f is indistinctly P or Q and x refers to V or θ, the Jacobian elements are obtained according to their definitions as follows:

$$\begin{aligned} H_{ij} &= \frac{\partial P_i}{\partial \theta_j}; & N_{ij} &= \frac{V_j}{} \frac{\partial P_i}{\partial V_j} \\ M_{ij} &= \frac{\partial Q_i}{\partial \theta_j}; & L_{ij} &= \frac{V_j}{} \frac{\partial Q_i}{\partial V_j} \end{aligned} \tag{3.34}$$

The resulting expressions are collected in Table 3.1. There are many common terms among the Jacobian expressions and those of the residual or mismatch vectors ΔP and ΔQ, which should be taken into account to save computational effort. This way, the Jacobian is nearly a by-product of the computation of the power mismatch vectors.

Solving the load flow problem by means of the Newton–Raphson method consists of the following steps:

1. Initialize the state vector with the flat voltage profile or with the solution of a previous case.
2. Compute $[\Delta P|\Delta Q]$ and the Jacobian elements. If all components of the mismatch vector are in absolute value lower than ε, then stop. Otherwise, continue.
3. Obtain $[\Delta\theta|\Delta V/V]$ by solving the system of Equation 3.32, according to the sparsity techniques explained at the end of this chapter and in Appendix A.
4. Update $[\theta|V]$ by means of Equation 3.33 and go back to step 2.

For each PV bus, an equation is removed from the above system, which is one of the advantages of the polar formulation.

TABLE 3.1

Expressions Corresponding to the Jacobian Elements in Polar Form

For $i \neq j$

$$H_{ij} = L_{ij} = V_i V_j (G_{ij} \sin \theta_{ij} - B_{ij} \cos \theta_{ij})$$
$$N_{ij} = -M_{ij} = V_i V_j (G_{ij} \cos \theta_{ij} + B_{ij} \sin \theta_{ij})$$

For $i = j$

$$H_{ii} = -Q_i - B_{ii} V_i^2 \qquad L_{ii} = Q_i - B_{ii} V_i^2$$
$$N_{ii} = P_i + G_{ii} V_i^2 \qquad M_{ii} = P_i - G_{ii} V_i^2$$

In many cases, convergence is improved if ΔQ is replaced by $\Delta Q/V$, which is simply achieved dividing the respective row by V_i. This way, the only nonlinear term with respect to V_i in $\Delta Q_i/V_i$ is Q_i^{sp}/V_i, which is relatively small compared with the others.

Even though for clarity of presentation the Jacobian has been divided into block submatrices, in practice both rows and columns of every PQ bus are gathered together.

EXAMPLE 3.3

The Newton–Raphson method in polar form will be illustrated on the three-bus system of Figure 3.2. Starting from the preliminary analysis and expressions developed in Example 3.1, the equation system to solve at each iteration is

$$
\begin{bmatrix} H_{22} & H_{23} & N_{22} \\ H_{32} & H_{33} & N_{32} \\ \hline M_{22} & M_{23} & L_{22} \end{bmatrix}^k \begin{bmatrix} \Delta\theta_2 \\ \Delta\theta_3 \\ \hline \Delta V_2/V_2 \end{bmatrix}^k = \begin{bmatrix} \Delta P_2 \\ \Delta P_3 \\ \hline \Delta Q_2 \end{bmatrix}^k \tag{3.35}
$$

where the power residuals are given by Equation 3.21 and phase angles are in radians. Directly deriving those functions, or from Table 3.1, the Jacobian terms are obtained. For the diagonal blocks, H and L, the following expressions result:

$$
H_{22} = V_2(3.63\cos\theta_2 + 0.363\sin\theta_2 + 4.816\cos\theta_{23} + 1.445\sin\theta_{23})
$$
$$
H_{23} = V_2(-4.816\cos\theta_{23} - 1.445\sin\theta_{23})
$$
$$
H_{32} = 1.05(-4.587\cos\theta_{32} - 1.376\sin\theta_{32})
$$
$$
H_{33} = 1.05(4.587\cos\theta_{32} + 1.376\sin\theta_{32})
$$
$$
L_{22} = Q_2 + 7.887V_2^2
$$

whereas for the off-diagonal blocks, N and M:

$$
N_{22} = P_2 + 1.7062V_2^2
$$
$$
N_{32} = 1.05V_2(-1.376\cos\theta_{32} + 4.587\sin\theta_{32})
$$
$$
M_{22} = P_2 - 1.7062V_2^2
$$
$$
M_{23} = V_2(1.445\cos\theta_{23} - 4.816\sin\theta_{23})
$$

For the flat voltage profile, Equation 3.35 yields

$$
\begin{bmatrix} 8.446 & -4.816 & 1.6045 \\ -4.816 & 4.816 & -1.445 \\ -1.808 & 1.445 & 7.328 \end{bmatrix} \begin{bmatrix} \Delta\theta_2 \\ \Delta\theta_3 \\ \Delta V_2/V_2 \end{bmatrix}^{(1)} = \begin{bmatrix} -0.892 \\ 0.328 \\ 0.159 \end{bmatrix} \tag{3.36}
$$

the solution of which, expressing the angles in degrees for convenience, is

$$
[\Delta\theta_2 \quad \Delta\theta_3 \quad \Delta V_2/V_2]^{(1)} = [-9.0059 \quad -5.0982 \quad 0.0005]
$$

Next, the state vector is updated with the above increments, and the process is repeated. The table below presents the maximum residual component before solving the equation system at each iteration.

Iter.	Max $|\Delta P|, |\Delta Q|$	Component
1	0.892	ΔP_2
2	5.57×10^{-2}	ΔQ_2
3	4.77×10^{-4}	ΔQ_2
4	4.48×10^{-8}	ΔQ_2

From the above table it is clear that two or three iterations should be carried out depending on the convergence threshold being 10^{-3} or 10^{-4}. After three iterations, the following values result:

$$[\theta_2 \quad \theta_3 \quad V_2] = [-9.263 \quad -5.492 \quad 0.9930]$$

which are very close to those provided by the Gauss–Seidel method after 33 iterations.

With these values, it can be checked that the generator at bus 3 must provide 24.02 Mvar and that the slack bus power injection is $62.38\,\text{MW} + j37.72\,\text{Mvar}$. This means that ohmic losses amount to 2.38 MW (1.98% of the load demand).

The Jacobian matrix at the fourth iteration would be

$$\begin{bmatrix} 8.178 & -4.678 & 0.683 \\ -4.867 & 4.867 & -1.117 \\ -2.682 & 1.746 & 7.378 \end{bmatrix}$$

which, as can be seen, differs only slightly from that of Equation 3.36, corresponding to the first iteration, particularly regarding the diagonal blocks.

3.5.2 RECTANGULAR FORMULATION

The state vector x is now composed of the following $2n - 2$ elements:

$$x = [V_{r1}, V_{r2}, \ldots, V_{r(n-1)} | V_{x1}, V_{x2}, \ldots, V_{x(n-1)}]^{\text{T}}$$

The residual or power mismatch vectors are given by the expressions

$$\Delta P_i = P_i^{\text{sp}} - \left[V_{ri} \sum_{j=1}^{n} (G_{ij} V_{rj} - B_{ij} V_{xj}) + V_{xi} \sum_{j=1}^{n} (G_{ij} V_{xj} + B_{ij} V_{rj}) \right] \quad (3.37)$$
$$i = 1, 2, \ldots, n - 1$$

$$\Delta Q_i = Q_i^{\text{sp}} - \left[V_{xi} \sum_{j=1}^{n} (G_{ij} V_{rj} - B_{ij} V_{xj}) - V_{xi} \sum_{j=1}^{n} (G_{ij} V_{xj} + B_{ij} V_{rj}) \right] \quad (3.38)$$
$$i = 1, 2, \ldots, n_{\text{L}}$$

As clearly seen, the number of equations does not yet match the number of unknowns. The reason is that voltage magnitude constraints at PV buses

$$\Delta V_i^2 = (V_i^{\text{sp}})^2 - V_{ri}^2 - V_{xi}^2 = 0 \quad i = 1, 2, \ldots, n_{\text{G}} \quad (3.39)$$

should be additionally enforced.

TABLE 3.2

Expressions Corresponding to the Jacobian Elements in Rectangular Form

For $i \neq j$

$$S_{ij} = -W_{ij} = G_{ij}V_{ri} + B_{ij}V_{xi}$$
$$T_{ij} = U_{ij} = G_{ij}V_{xi} - B_{ij}V_{ri}$$
$$C_{ij} = D_{ij} = 0$$

For $i = j$

$$S_{ii} = I_{ri} + G_{ii}V_{ri} + B_{ii}V_{xi} \qquad U_{ii} = -I_{xi} - B_{ii}V_{ri} + G_{ii}V_{xi}$$
$$W_{ii} = I_{ri} - G_{ii}V_{ri} - B_{ii}V_{xi} \qquad T_{ii} = I_{xi} - B_{ii}V_{ri} + G_{ii}V_{xi}$$
$$C_{ii} = 2V_{ri} \qquad D_{ii} = 2V_{xi}$$

Therefore, in matrix form, the load flow equations in rectangular coordinates are

$$\begin{bmatrix} S & T \\ U & W \\ C & D \end{bmatrix}^k \begin{bmatrix} \Delta V_r \\ \Delta V_x \end{bmatrix}^k = \begin{bmatrix} \Delta P \\ \Delta Q \\ \Delta V^2 \end{bmatrix}^k \tag{3.40}$$

$$\begin{bmatrix} V_r \\ V_x \end{bmatrix}^{k+1} = \begin{bmatrix} V_r \\ V_x \end{bmatrix}^k + \begin{bmatrix} \Delta V_r \\ \Delta V_x \end{bmatrix}^k \tag{3.41}$$

where the Jacobian elements are those provided in Table 3.2. In that table, I_{ri} and I_{xi} refer, respectively, to the real and imaginary components of the net current injected at bus i, computed from the expression:

$$I_{ri} + jI_{xi} = \sum_{j=1}^{n} (G_{ij} + jB_{ij})(V_{rj} + jV_{xj}) \tag{3.42}$$

The iterative process consists of the same steps described for the polar formulation, except logically for the involved expressions.

A version of the rectangular formulation based on current rather than power mismatch vectors has been proposed in Reference 10. This version has been shown to be competitive with the polar formulation, specially for networks with a reduced number of PV buses. Indeed, resorting to current residuals offers certain advantages and leads to the most effective way of handling three-phase and harmonic load flows, as explained in Chapter 11.

3.6 FAST DECOUPLED LOAD FLOW

In spite of the sophisticated computational techniques discussed in Appendix A, execution times associated with the exact Newton–Raphson implementation described above can be unacceptable for certain online applications, specially those dealing with multiple cases and very large networks. Sometimes, speed of response is as important as accuracy, which justifies the efforts devoted in the 1970s to develop fast versions of the Newton–Raphson methodology.

The first and most evident simplification, considering the numerical values given in Example 3.3, consists of ignoring the Jacobian dependence on the current state. Note however that, as the Jacobian is essentially a by-product of the power mismatch computation, the major computational saving coming from the use of a constant Jacobian has to do with its triangular factorization. Furthermore, as the convergence rate slightly decreases, the extra iterations required frequently offset the saving achieved at each iteration.

The second and most important simplification arises by considering the weak coupling between active powers and voltage magnitudes, on the one hand, and reactive power and phase angles on the other [11,12], which translates into numerical values of matrices N and M in Equation 3.32 being significantly smaller than those of the diagonal blocks H and L. As can be checked from expressions given in Table 3.1, this is mainly due to two reasons: (a) phase angle differences between adjacent buses are rather small, implying that $\cos \theta_{ij} \approx 1$ and $\sin \theta_{ij} \approx 0$ and (b) for high-voltage transmission networks the ratio $r/x = g/b \ll 1$ (for 220 and 400 kV levels, this ratio lies between 1/5 and 1/10). Therefore, it is expected that the performance of decoupled models, that is, models ignoring the coupling between the active and the reactive subproblems, can be less satisfactory when solving strongly loaded systems and lower voltage systems (for 50-kV lines, the ratio $r/x \approx 1$, while it clearly exceeds unity for 20-kV lines). Note that the decoupling referred to in this paragraph is not so evident when rectangular coordinates are used [13], which partly explains why this formulation has not become so popular.

Among the several decoupled Newton–Raphson formulations proposed in the literature, by far the most successful is the so-called *fast decoupled load flow* (FDLF), published in 1974 [14]. In addition to zeroing matrices N and M, the following simplifying assumptions are made by this scheme:

1. The scaled mismatch vectors $\Delta P/V$, $\Delta Q/V$ are used, instead of ΔP, ΔQ.
2. On the basis of the facts that Q_i is usually lower than 1 pu and B_{ii} typically ranges between 20 and 50 pu, it is assumed that:

$$\cos \theta_{ij} \approx 1$$
$$G_{ij} \sin \theta_{ij} \ll B_{ij}$$
$$Q_i \ll B_{ii} V_i^2$$

3. In the active subproblem, voltage magnitudes are set to 1 pu, shunt reactors and capacitors are omitted in matrix H, including those of the π models, and voltage regulators of transformers are ignored (i.e., nominal turn ratios are assumed).
4. Regarding the reactive subproblem, phase shifting transformers are ignored in matrix L.

Keeping in mind those considerations, Equation 3.32 reduces to the following decoupled systems:

$$B' \, \Delta\theta = \frac{\Delta P}{V} \tag{3.43}$$

$$B'' \Delta V = \frac{\Delta Q}{V} \tag{3.44}$$

where matrices B' and B'', being constant, need to be built and factorized only once. Furthermore, experiments showed that ignoring line resistances in matrix B' benefits convergence to a certain extent. This way, the elements of matrices B' and B'' become

$$B'_{ij} = \frac{-1}{x_{ij}}; \quad B'_{ii} = \sum_{j \in i} \frac{1}{x_{ij}}$$

$$B''_{ij} = -B_{ij}; \quad B''_{ii} = -B_{ii}$$

where x_{ij} is the series reactance of the element linking buses i and j, B_{ij} is the imaginary component of the respective element of the bus admittance matrix, and $j \in i$ denotes the set of buses j adjacent to bus *i*. Note that B'' is a symmetric matrix, and also B' if the network does not contain phase shifters, which should be taken into account to save computational cost.

The iterative process consists of sequentially solving Equations 3.43 and 3.44, using each time the most recent values of θ and V, until both ΔP and ΔQ satisfy the convergence criterion. The convergence rate of the FDLF is roughly the same as that of the coupled version during the first iterations, but it slows down as the solution is approached. In any case, the extra iterations required are well offset by the fact that the cost per iteration can be between four and five times smaller than that of the standard Newton–Raphson method. This makes the FDLF the perfect tool in those applications involving a large number of load flow solutions (see Chapters 5 and 6).

As discussed above, the hypotheses on which the FDLF is founded can be questioned when the network is very loaded or the ratios r/x are high. In such cases, the FDLF may diverge or behave in an oscillatory manner, the fully coupled Newton–Raphson approach being preferable. In Reference 15, experimental results are presented suggesting that the FDLF shows a better convergence rate in difficult cases if line resistances are ignored in B'' rather than in B'. Since then, the standard FDLF is known as the "XB" version whereas that proposed in Reference 15 is called the "BX" version.

The good behavior shown by the FDLF is somewhat surprising, considering the large number of simplifications on which it is based. This has led researchers to look for a plausible theory capable of explaining the evidence provided by experimental results. In what follows, a summary of the most relevant arguments discussed in Reference 16 is presented.

Resorting to block arithmetic, the system Equation 3.32 can be decomposed into the following two subsystems

$$[H - NL^{-1}M]\Delta\theta = \Delta P - NL^{-1}\,\Delta Q \tag{3.45}$$

$$[L - MH^{-1}N]\Delta V = \Delta Q - MH^{-1}\,\Delta P \tag{3.46}$$

where the coupling terms have been shifted to the right side. Let us focus on the active subproblem, as similar conclusions are reached for the reactive one.

Unlike the coupled model, in which both ΔP^k and ΔQ^k are functions of vectors θ^k and V^k, obtained during the previous iteration, each half iteration of the FDLF makes use of values partly updated in the former half iteration. For instance, at step $k+1$, ΔP^{k+1} is a function of vectors θ^k and of the just updated V^{k+1}. Performing the Taylor expansion of ΔP^{k+1} around the point θ^k, V^k yields

$$\Delta P(\theta^k, V^{k+1}) \cong \Delta P(\theta^k, V^k) - N\,\Delta V^k$$

but keeping in mind that ΔV^k has just been obtained from the reactive subproblem

$$L\,\Delta V^k \cong \Delta Q(\theta^k, V^k)$$

the series approximation becomes

$$\Delta P(\theta^k, V^{k+1}) \cong \Delta P(\theta^k, V^k) - NL^{-1}\,\Delta Q(\theta^k, V^k)$$

Comparing the above expression with Equation 3.45, it can be concluded that the FDLF does not really ignore the coupling between both subproblems, but takes it into account implicitly by using variables that have been updated during the previous half iteration.

Regarding the coefficient matrix in Equation 3.45, it can be shown for flat voltage profile and radial networks [16] that

$$H - NL^{-1}M = B'$$

if resistances are ignored when building B'. The same happens for meshed networks with uniform r/x ratios. In the general case, the above expression is not exact (in fact, the left side matrix is full while B' is sparse), but the differences are quite small numerically. This explains why ignoring resistances leads generally to better results.

EXAMPLE 3.4

Let us apply the "XB" version of the FDLF to the network of Figure 3.2.

According to the simplifications discussed above, matrices B' and B'' are as follows (the reader should compare the coefficients below with those of the diagonal blocks of the Newton–Raphson Jacobian):

$$B' = \begin{bmatrix} 8.333 & -5 \\ -5 & 5 \end{bmatrix}; \quad B'' = [7.887]$$

The active power mismatch vector slightly differs from that of the Newton–Raphson method because of the division by V (in this particular case, only the component corresponding to bus 3 changes since $V_3 = 1.05$). The equation system arising during the first half iteration is given by

$$\begin{bmatrix} 8.333 & -5 \\ -5 & 5 \end{bmatrix} \begin{bmatrix} \Delta\theta_2 \\ \Delta\theta_3 \end{bmatrix}^{(1)} = \begin{bmatrix} -0.898 \\ 0.312 \end{bmatrix}$$

providing (in degrees)

$$[\Delta\theta_2 \quad \Delta\theta_3]^{(1)} = [-10.0736 \quad -6.4984]$$

After updating θ_2 and θ_3, the reactive power residual corresponding to the second half iteration is computed, yielding

$$\frac{\Delta Q_2}{V_2} = -5.96 \times 10^{-2}$$

This value is of the same order of magnitude as that of the largest residual arising during the second iteration of the Newton–Raphson method, corresponding also with ΔQ_2 (see the former example). The fact that updated phase angles have been used to obtain the reactive power mismatch vector roughly compensates for the minor approximations assumed by the FDLF.

As there is a single unknown, the voltage magnitude correction is obtained in this case just through a division ($\Delta V_2^{(1)} = -7.55 \times 10^{-3}$), leading to $V_2^{(2)} = 0.9924$. This value is significantly better than that obtained by the first iteration of the coupled version, which explains why the results provided by a full iteration of the FDLF are considered satisfactory for certain applications in which accuracy is not the major concern.

After iterating four times with P and three times with Q, all residual components are smaller than 0.0001, leading to the same results provided by the Newton–Raphson method.

3.7 INCLUSION OF REGULATING DEVICES AND ASSOCIATED LIMITS

The methodology described so far considers only the boundary constraints imposed by generation and load buses. In practice, however, the solution provided by a load flow solver should be adjusted to take into account additional restrictions arising from a diversity of regulating devices.

Such devices are intended to keep a given electrical magnitude y (power flow, voltage magnitude, etc.) as close as possible to a target value y^{sp}, for which they can act on a control variable x within specified limits ($x^{\min}$, $x^{\max}$). In turn, the specified value y^{sp} may refer to either a desired set point or a limit that has been exceeded.

The way those solution adjustments are implemented depends on the load flow methodology adopted. Existing procedures can be grouped into the following two categories [17]:

1. Both the state and mismatch vectors are dynamically adapted at each iteration to account for the additional equality constraints. Then, the structure of the linear system arising from the Newton–Raphson method becomes

$$
\begin{bmatrix} H & N & \partial P/\partial x \\ M & L & \partial Q/\partial x \\ \hline \partial y/\partial \theta & \partial y/\partial V & \partial y/\partial x \end{bmatrix}^k \begin{bmatrix} \Delta\theta \\ \Delta V \\ \hline \Delta x \end{bmatrix}^k = \begin{bmatrix} \Delta P \\ \Delta Q \\ \hline \Delta y \end{bmatrix}^k \qquad (3.47)
$$

where

$$
\Delta y^k = y^{\mathrm{sp}} - y^k
$$

represents the new components of the mismatch vector, and x comprises the set of associated control variables. If a variable in x hits a limit, then the regulation capability is lost, and the respective component of Δy is removed from the system of equations, along with x.

In those cases in which the magnitude y is already in the state vector, it is more convenient to reduce, rather than to enlarge, the model. This is done by setting $y = y^{\mathrm{sp}}$ and removing it from the unknown vector. The value of the control variable x is then computed at the end of the current iteration to check for limits (this is actually the way PV buses are handled in the polar formulation).

2. A local feedback process is implemented adjusting the control variable in proportion to the residual of the regulated magnitude. After each iteration, Δx is given by

$$
\Delta x^k = \alpha \Delta y^k \qquad (3.48)
$$

where the feedback gain α should be properly chosen not to slow down the convergence process (if very small) or to cause divergence (if very large). When both x and y explicitly appear in the linearized model, a convenient value for α is provided by computing the sensitivity between both variables. Alternatively, α can be empirically estimated by observing the corrections Δy achieved by Δx in former iterations.

This technique may be the right choice when there is an incentive to use constant Jacobian matrices throughout the iterative process, like in the FDLF. Note that the feedback process provides an approximation to the solution of Equation 3.47, obtained by setting to zero the new off-diagonal blocks and approximating the lower diagonal block by a diagonal matrix.

When the number of solution adjustments is large, the two techniques described above significantly increase the number of iterations, particularly if several interacting limits are reached. To prevent oscillatory behavior, it is advisable not to launch the adjustment process until the components of y are sufficiently accurate.

In the following sections, implementation details corresponding to the most common solution adjustments are provided.

3.7.1 SHUNT COMPENSATORS

In addition to conventional synchronous generators, other devices can be connected to a bus to regulate its voltage magnitude (or the one of an adjacent bus) by proper injection/absorption of reactive power. The list of such devices includes shunt capacitor banks and reactors, synchronous compensators, static var compensators (SVC), and STATic synchronous COMpensators (STATCOM), or a combination of them [18]. Irrespective of the technology adopted, the steady-state behavior of any shunt compensator at the fundamental frequency, when equipped with the appropriate voltage regulator, is essentially the same, except for the fact that conventional capacitors/reactors are limited to discrete steps.

In a majority of cases, the regulated voltage corresponds to the bus where the compensator is connected, leading to a conventional PV bus with $P^{sp} = 0$. No matter whether P^{sp} is null or not, the reactive power required to keep the voltage magnitude under control should be computed at each iteration and the specified limits checked out. When a limit, Q^{lim}, is violated the voltage magnitude can no longer be equal to V^{sp}, and the bus becomes, in fact, a PQ bus with $Q^{sp} = Q^{lim}$. Subsequently, if the voltage of such virtual PQ bus satisfies $V^k > V^{sp}$ with $Q^{sp} = Q^{max}$, or $V^k < V^{sp}$ with $Q^{sp} = Q^{min}$, then it becomes again a PV bus (this may happen because of the nonlinear interactions among nearby PV buses).

Sometimes, the voltage magnitude of an adjacent PQ bus is regulated, rather than that of the bus where the compensator is connected. This leads to an atypical situation in which three magnitudes (P, Q, and V) are specified at the regulated bus while a single magnitude (P) is kept constant at the compensator bus, but the total number of equations and unknowns remains balanced.

Actually, the dual situation is also very common, namely the possibility for a PQ bus to be transformed into a PV bus by the activation of a shunt compensator when its voltage magnitude is not within acceptable limits.

Implementing the bus-type switching process described above within the Newton–Raphson process is straightforward, as the Jacobian is updated at each iteration. In the polar version, switching bus i from PV to PQ type (or vice versa) consists simply of adding (removing) ΔQ_i and ΔV_i to the residual and unknown vectors, respectively. In the rectangular version, the size of the problem remains the same, as changing the bus-type amounts to replacing Equation 3.39 by Equation 3.38, or vice versa.

A little bit more tricky is the case in which the regulated bus j differs from the bus i where the reactive power is injected (remote control). In this case, when Q_i is within limits the residual vector should include ΔP_i, ΔP_j, and ΔQ_j, the unknowns being θ_i, θ_j, and V_i. Otherwise, both i and j are ordinary PQ buses. The complexity stems from the structural asymmetry of the Jacobian matrix arising by the elimination of the row corresponding to ΔQ_i and the column corresponding to ΔV_j, which somewhat complicates the solution of the linear equation system.

On the other hand, as the FDLF algorithm is based on the use of constant Jacobian matrices, it is recommended to use the sensitivity-based technique explained above, that leaves intact the matrix B'' each time a bus type is switched [14]. At each iteration, V^{sp} is corrected by means of Equation 3.48 so as to correct the excess (or deficit) of reactive power required to keep the updated voltage magnitude. For instance, if $Q_i^k > Q_i^{max}$, the following sequence of operations is performed:

$$\Delta Q_i = Q_i^{max} - Q_i^k$$
$$\Delta V_i = \alpha_i \, \Delta Q_i$$
$$[V_i^{k+1}]^{sp} = [V_i^k]^{sp} + \Delta V_i$$

A reasonable value for α_i can be obtained *a priori* by computing the sensitivity between V_i and Q_i in the reactive power subproblem. For this purpose, the following system is solved:

$$B_a'' S_i = e_i \tag{3.49}$$

where e_i is the ith column of the identity matrix, B_a'' is the matrix B'' augmented with the row and column corresponding with bus i, and S_i is the respective column of the sensitivity matrix (inverse of the Jacobian). Then, $\alpha_i = S_{ii}$ if V_i is regulated or $\alpha_i = S_{ji}$ when a remote voltage magnitude V_j is regulated.

When several interacting reactive power limits are violated, the feedback process leads usually to a large number of iterations, mainly because the sensitivity values so computed become less accurate.

In such cases, the following alternative technique, requiring only the diagonal values of B'' to be modified but not its structure, can be adopted: matrix B'' is formed as if all buses were of the PQ type (same structure as B'). Then, for each PV bus, ΔQ_i is set to zero, and an arbitrarily large number ρ is added to the respective diagonal of B'' (ρ should not be so large that ill conditioning of B'' is exacerbated). This way, $\Delta V_i \approx 0$ is obtained when the reactive subproblem is solved. If a reactive power limit is reached, the original diagonal value of B'' is simply restored. Note that, interpreting the solution of the incremental model Equation 3.44 as that of an associated DC resistive circuit, this technique is equivalent to short circuiting the respective node to ground by means of a large conductance, forcing the node voltage (voltage correction in the incremental model) to be null. The main computational overhead compared to the standard FDLF comes from the need to refactorize B'' each time at least a bus type is switched (see Appendix A).

EXAMPLE 3.5

According to Example 3.3, the reactive power that the generator at bus 3 has to inject to keep its voltage magnitude at 1.05 pu is 24.02 Mvar. To illustrate the feedback process discussed above, let us assume in this example that the maximum reactive power that can be supplied by generator 3 is 20 Mvar, which means that bus 3 should become a PQ bus during the iterative process.

The sensitivity of V_3 with respect to Q_3 is obtained in advance by solving Equation 3.49, which reduces in this case to

$$\begin{bmatrix} 7.887 & -4.587 \\ -4.587 & 4.587 \end{bmatrix} \begin{bmatrix} S_{23} \\ S_{33} \end{bmatrix} = \begin{bmatrix} 0 \\ 1 \end{bmatrix} \Rightarrow S_{33} = 0.521$$

Let us assume for simplicity that the corrective process is applied for the first time to the results provided by Example 3.4, that is, after the third reactive power semi-iteration (in practice, this process should begin earlier). Then, the new value V_3^{sp} for the next iteration is estimated from

$$V_3^{sp} = 1.05 + S_{33}(0.2 - 0.2402) = 1.029$$

Proceeding in this way, after two active power and three reactive power additional semi-iterations (for a convergence threshold 0.0001), the following solution is reached:

$$[\theta_2 \quad \theta_3 \quad V_2 \quad V_3] = [-9.32 \quad -5.29 \quad 0.979 \quad 1.0297]$$

requiring that generator 3 injects only 20 Mvar not 24 Mvar. As a consequence, the voltage at bus 2 might be too low, which may call for commissioning a local shunt compensator or the voltage set point at bus 1 to be raised.

3.7.2 REGULATING TRANSFORMERS

Either by means of the special arrangements described in Chapter 2 or by resorting to under-load tap changers, two types of regulating transformers can be considered, namely voltage magnitude regulators and phase shifters, which can be employed to control the voltage magnitude at one of their terminal buses or the power flow through them.

For the sake of generality, the ideal transformer with complex turn ratio shown in Figure 3.3, connected between buses k and j in series with an impedance, will be considered to develop the necessary incremental model. Then, ordinary regulating transformers are simply special cases of this more general situation.

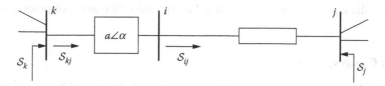

FIGURE 3.3   Generalized regulating transformer between buses k and j.

From the basic relationships characterizing the ideal transformer, the following expressions arise for the notation of Figure 3.3:

$$\mathcal{S}_{kj} = \mathcal{S}_{ij}; \quad V_i = aV_k; \quad \theta_i = \alpha + \theta_k \tag{3.50}$$

where the complex power flow $\mathcal{S}_{ij}$ is related with the complex voltages of buses i and j by means of Equations 3.19 and 3.20.

The 2 degrees of freedom provided by the complex turn ratio allow two magnitudes to be regulated, for instance, the active and reactive power flow through the transformer. In this case, the incremental model Equation 3.47 is of application with

$$\Delta x = [\Delta a \quad \Delta \alpha]^\mathrm{T}; \quad \Delta y = [\Delta P_{ij} \quad \Delta Q_{ij}]^\mathrm{T}$$

To compute the new Jacobian components, the following relationships, directly obtained from Equation 3.50, should be taken into account:

$$\frac{\partial f(\cdot)}{\partial V_k} = a \frac{\partial f(\cdot)}{\partial V_i}; \quad \frac{\partial f(\cdot)}{\partial a} = V_k \frac{\partial f(\cdot)}{\partial V_i}; \quad \frac{\partial f(\cdot)}{\partial \theta_k} = \frac{\partial f(\cdot)}{\partial \alpha} = \frac{\partial f(\cdot)}{\partial \theta_i} \tag{3.51}$$

where $f(\cdot)$ represents the power flows P_{ij} and Q_{ij}, related with V_i and θ_i through Equations 3.19 and 3.20, as well as the opposite power flows and the power injections at buses k and j.

By properly eliminating missing control variables and unregulated magnitudes, the generalized regulating transformer discussed so far leads immediately to the three particular cases usually found in practice, namely

- Tap changer regulating the reactive power flow ($Q_{ij} = Q_{ij}^{\mathrm{sp}}$). In this case, $\alpha = 0$ and a is the only control variable. The incremental model is obtained by removing ΔP_{ij} and $\Delta \alpha$ in Equation 3.47.
- Tap changer regulating the voltage magnitude ($V_k = V_k^{\mathrm{sp}}$). Again, a is the only control variable. The incremental model is obtained this time by removing ΔP_{ij}, ΔQ_{ij}, ΔV_k, and $\Delta \alpha$ in Equation 3.47. In other words, the new state variable a simply replaces V_k in the incremental model. If a reaches a limit, then ΔV_k is restored to the state vector [19].
- Phase shifter regulating the active power flow ($P_{ij} = P_{ij}^{\mathrm{sp}}$). In this case, $a = 1$ and α is the only control variable. The incremental model is obtained by removing ΔQ_{ij} and Δa in Equation 3.47.

In the FDLF, the need to keep the coefficient matrices constant makes the sensitivity-based feedback process an attractive choice. If $\Delta V_i^k = V_i^{\mathrm{sp}} - V_i^k$ represents the voltage magnitude error at iteration k, then the turn ratio a should be modified by the amount,

$$\Delta a^k = \alpha \, \Delta V_i^k$$

where $\alpha = \pm 1$ is an acceptable value, especially for transformers in radial configurations (the sign is positive when the tap changer is on the regulated bus side and negative otherwise). Irrespective of the method adopted, the tap value should be rounded at the end of the iterative process to the nearest discrete step.

3.7.3 SERIES COMPENSATORS

Another class of devices exists intended to modify the series impedance of transmission lines, either directly or by injecting a controllable voltage source with the help of an auxiliary series winding. Such devices, collectively known as series compensators, include [18] thyristor-controlled series capacitors (TCSC), thyristor-switched series capacitors (TSSC) as well as the series-connected module of the so-called unified power flow controllers (UPFC).

The most general situation arises when both the series resistance and the series reactance are compensated (note that modifying the series resistance implies the presence of a device capable of injecting/absorbing active power). For convenience, the series impedance of the compensated line, connected buses between i and j, will be expressed as follows:

$$z_{ij} = K_r r_{ij} + j K_x x_{ij} \tag{3.52}$$

where r_{ij} and x_{ij} denote the original line parameters and K_r, K_x, representing the respective compensation ratios, constitute the control variables. For instance, $K_r = 1$ and $K_x = 0.8$ means that the line resistance remains uncompensated while 20% of the line series reactance is compensated. If needed, the required compensating components, in ohms, can be expressed in terms of the compensation ratios:

$$r_c = (1 - K_r) r_{ij}; \quad x_c = (1 - K_x) x_{ij}$$

Note that both K_r and K_x are subject to operating limits, which depend on the particular technology adopted. For instance, for a conventional series capacitor, the following limits typically apply to K_x:

$$0.7 \le K_x \le 1$$

whereas much wider limits may apply to fast responding FACTS devices.

On the basis of Equation 3.52, the admittance matrix components appearing in Equations 3.19 and 3.20 can be expressed as nonlinear functions of K_r and K_x:

$$G_{ij} = \frac{-K_r r_{ij}}{K_r^2 r_{ij}^2 + K_x^2 x_{ij}^2}; \quad B_{ij} = \frac{K_x x_{ij}}{K_r^2 r_{ij}^2 + K_x^2 x_{ij}^2} \tag{3.53}$$

giving rise to the following derivatives

$$\begin{aligned}
\frac{\partial G_{ij}}{\partial K_r} &= r_{ij}(G_{ij}^2 - B_{ij}^2); & \frac{\partial G_{ij}}{\partial K_x} &= -2x_{ij}G_{ij}B_{ij} \\
\frac{\partial B_{ij}}{\partial K_x} &= x_{ij}(G_{ij}^2 - B_{ij}^2); & \frac{\partial B_{ij}}{\partial K_r} &= 2r_{ij}G_{ij}B_{ij}
\end{aligned} \tag{3.54}$$

In this case, the incremental model Equation 3.47 applies with

$$\Delta x = [\Delta K_r \quad \Delta K_x]^{\mathrm{T}}; \quad \Delta y = [\Delta P_{ij} \quad \Delta Q_{ij}]^{\mathrm{T}}$$

To obtain the new Jacobian elements, the following relationship should be taken into account:

$$\frac{\partial f(\cdot)}{\partial K} = \frac{\partial f(\cdot)}{\partial G_{ij}}\frac{\partial G_{ij}}{\partial K} + \frac{\partial f(\cdot)}{\partial B_{ij}}\frac{\partial B_{ij}}{\partial K} \tag{3.55}$$

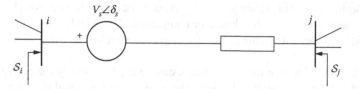

FIGURE 3.4   Voltage source model of a series compensator between buses i and j.

where K refers indistinctly to K_r or K_x, and $f(\cdot)$ represents the power flows P_{ij} and Q_{ij}, related with V_i and θ_i through Equations 3.19 and 3.20 , as well as the opposite power flows and the power injections at buses i and j. On the basis of the above relationship and on Equation 3.54, the required Jacobian components can be computed.

By properly adapting the generalized series compensator model developed so far, the following particular cases, of interest in practice, can be easily obtained:

- Series capacitors (several technological solutions are possible), intended to regulate the active power through the line ($P_{ij} = P_{ij}^{\text{sp}}$). In this case $K_r = 1$, K_x being the only control variable. The associated incremental model is obtained by removing ΔQ_{ij} and ΔK_r in Equation 3.47.
- Series voltage source (arising from the UPFC model), capable of independently regulating both the active and the reactive power flows. In the presence of this versatile compensator, the resulting line model, shown in Figure 3.4, can be related to that of the generic series compensator by means of

$$V_s \angle \delta_s = \frac{(K_r - 1)r_{ij} + j(K_x - 1)x_{ij}}{K_r r_{ij} + jK_x x_{ij}} (V_i - V_j) \qquad (3.56)$$

At each iteration, the above expression allows an equivalent voltage source to be computed in terms of the line compensation factors and complex voltages at the terminal buses. The incremental model corresponding to the generic series compensator remains valid so long as V_s lies within operating limits. In case V_s reaches its upper limit, V_s^{max}, then a degree of freedom is lost, and the two constants K_r and K_x are forced to satisfy Equation 3.56, with $V_s = V_s^{\text{max}}$. Consequently, either ΔP_{ij} or ΔQ_{ij} should be removed from the unknown vector to compensate for the extra constraint the two control variables must obey. In these circumstances, the control capability is due only to δ_s, the value of which determines to a large extent which power flow (active or reactive) is most affected by the presence of the voltage source. As explained in Reference 18, $\delta_s \approx \pm \theta_i$ makes the voltage source to behave much like a tap changer, whereas $\delta_s \approx \theta_i \pm 90°$ leads to a phase shifter (the active power involved in the source originates in the shunt converter of the UPFC, not shown in the figure). Note also that a series capacitor can be obtained if the voltage source phasor lags the line current by 90°.

3.7.4   Area Interchanges

When the load flow solution spans several interconnected areas, it is sometimes of interest to enforce that the total active power flowing out of a given area be equal to a predetermined amount, which is achieved by adjusting the set point of one or several generators within the area.

In the Newton–Raphson method, this constraint can be considered by adding the nonlinear equation

$$\Delta P_I^k = P_I^{\text{sp}} - \sum P_{ij}^k = 0$$

where P_I^{sp} is the desired power interchange, and P_{ij}^k refers to the power flows through the set of tie lines at iteration k. The additional constraint should be compensated obviously with a new unknown in the state vector, which can be, for instance, the active power supplied by the generator in charge of the regulation.

In the FDLF, the interchange error at iteration k can be used to modify the active power of regulating generators for the next iteration. This way, the total power generated within the area should be corrected by the amount

$$\Delta P_G = \alpha \, \Delta P_I^k$$

where $\alpha = 1$ is a reasonable value considering that ohmic losses in transmission networks are rather small. The total correction ΔP_G can then be assigned to a single generator or, more frequently, be pro-rated among a set of generators specified in advance.

3.8   SPECIAL-PURPOSE LOAD FLOWS

3.8.1   DC Load Flow

Even though both P and Q are nonlinear functions of V and θ, a reasonable linear approximation between P and θ can be found, leading to the so-called *DC load flow*. This model is based on the assumption that $V_i = 1$ at all buses, losing in this way the capability to track reactive power flows or any other voltage related data. With this hypothesis, the flow of active power, given by Equation 3.19, simplifies to

$$P_{ij} = G_{ij}(\cos \theta_{ij} - 1) + B_{ij} \sin \theta_{ij}$$

Considering that phase angle differences corresponding to adjacent buses are small ($\cos \theta_{ij} \approx 1$ and $\sin \theta_{ij} \approx \theta_i - \theta_j$) leads to

$$P_{ij} = B_{ij}(\theta_i - \theta_j)$$

The element B_{ij} equals minus the series susceptance between buses i and j, that is,

$$B_{ij} = \frac{x_{ij}}{r_{ij}^2 + x_{ij}^2} = \frac{1/x_{ij}}{1 + (r_{ij}/x_{ij})^2}$$

where r_{ij} and x_{ij} are the series resistance and reactance, respectively. For values of $r/x < 3$, typical in transmission networks, the error arising when B_{ij} is replaced by $1/x_{ij}$ is lower than 1%, yielding

$$P_{ij} = \frac{1}{x_{ij}}(\theta_i - \theta_j); \quad \theta_i - \theta_j = x_{ij}P_{ij} \tag{3.57}$$

Let A denote the branch-to-node incidence matrix, θ the vector of bus phase angles, both of them reduced by removing the slack row, X a diagonal matrix of branch reactances and P_f the vector of branch active power flows. Then the above expression can be written in matrix form as follows:

$$A^T\theta = XP_f \tag{3.58}$$

$$P_f = [X^{-1}A^T]\theta \tag{3.59}$$

Having ignored the branch resistances, the sum of all active powers is zero, which means that the slack power (or any other) is a linear combination of the remaining ones. If P denotes the vector of the net injected powers, except that of the slack bus, the Kirchhoff's "current" law applied to power flows leads to

$$P = AP_f \qquad (3.60)$$

Finally, eliminating branch power flows by means of Equation 3.59, the desired linear relationship between power injections and phase angles is obtained:

$$P = [AX^{-1}A^T]\theta = B\theta \qquad (3.61)$$

where matrix B has the same structure (sparse and symmetric) as that of the bus admittance matrix, its values being computed solely in terms of branch reactances. Note that matrix B is identical to matrix B' of the "XB" version of the FDLF. However, unlike the active subproblem of the FDLF, Equation 3.61 is not an incremental model. It can be easily checked that this linear model, derived from a full AC model, corresponds to that of a resistive-like DC circuit composed of reactances, in which active powers play the role of DC current injections and phase angles that of DC voltages, which explains its name.

Alternatively, phase angles can be removed, leading to a linear relationship between power flows and power injections:

$$P_f = [X^{-1}A^T B^{-1}]P = S_f P \qquad (3.62)$$

Under the DC load flow approximation, each element of matrix S_f provides the sensitivity between the respective power flow and power injection, assuming the slack bus fully compensates for changes in any bus injection. As such, matrix S_f is dense, but many of its elements relating electrically distant branch-bus pairs are negligible. The above expression is useful to quickly analyze power flow changes arising from branch or generator outages. Also, the so-called power transfer distribution factors (PTDF), which are essential to determine power transfer capabilities in competitive electricity markets, can be readily obtained from matrix S_f (see Chapter 6).

Although the DC model is lossless, actual power losses can be estimated in terms of active power flows by adding all terms of the form $R_{ij}P_{ij}^2$.

EXAMPLE 3.6

The DC load flow for the three-bus system of Figure 3.2 is obtained by solving the linear system:

$$\begin{bmatrix} 8.333 & -5 \\ -5 & 5 \end{bmatrix} \begin{bmatrix} \theta_2 \\ \theta_3 \end{bmatrix} = \begin{bmatrix} -1 \\ 0.4 \end{bmatrix}$$

While the coefficient matrix agrees with matrix B' in Example 3.4, the independent vector represents net injected powers rather than active power mismatches. The solution, in degrees, to the above system is

$$[\theta_2 \quad \theta_3] = [-10.314 \quad -5.731]$$

which is quite close to the exact AC solution. With those phase angles, branch power flows are obtained immediately (remember that phase angles are meant in radians, even though they are shown in angles for convenience):

$$P_{12} - \frac{1}{x_{12}}(\theta_1 \quad \theta_2) = 0.6\,\text{pu}; \qquad P_{32} - \frac{1}{x_{32}}(\theta_3 - \theta_2) = 0.4\,\text{pu}$$

Of course, in this trivial radial case, power flows can be directly deduced from power injections.

As power losses are neglected, the slack bus must provide just 60 MW. However, ignoring reactive power flows and assuming voltage magnitudes equal to 1 pu, ohmic losses can be estimated from

$$R_{12}P_{12}^2 + R_{32}P_{32}^2 = 0.0204 \text{ pu}$$

which compares quite well with the exact value (0.0238) obtained in Example 3.3.

3.8.2 RADIAL DISTRIBUTION NETWORKS

A majority of distribution networks are designed and built so that at least an alternative path is provided in case a feeder section fails for any reason, which leads to meshed configurations. However, owing to the geographical scope of such networks, reaching virtually any corner of a country, and the reduced number of customers each feeder serves, they are operated in a radial manner, in an attempt to reduce both the short-circuit power and the complexity of the switching and protection systems (see Chapter 8). This means that, at the distribution level, reliability (number and duration of interruptions) is clearly sacrificed for the sake of cost reduction.

On this kind of networks, characterized by high r/x ratios and comprising very short feeder sections, the performance of the FDLF is frequently poor, and the standard Newton–Raphson method gets in trouble when applied to certain ill-conditioned cases reported in the literature.

On the other hand, the particular configuration of such networks (radial topology, a single feeding point) is very suitable for the development of ad hoc procedures, which are simpler than but competitive with Jacobian-based algorithms. The method described below is an adaptation of that presented in Section 3.4.2 to radial networks [20] (only the single-phase version will be discussed, as the general three-phase case is dealt with in Chapter 11).

Consider a radial network with n buses, numbered from the head in such a way that each bus precedes its successors downstream. Starting with the flat start profile $\mathcal{V}^0$, the solution process consists of three steps, which are repeated until the complex voltages at two consecutive iterations are close enough

1. For the set of estimated voltages, obtain the net current drawn from each bus

$$\mathcal{I}_i^k = (\mathcal{S}_i^{\text{sp}}/\mathcal{V}_i^k)^* + y_{si}\mathcal{V}_i^k \quad i = n, n-1, \ldots, 2 \qquad (3.63)$$

where $\mathcal{S}_i^{\text{sp}}$ is the complex power absorbed by the local load at bus i and y_{si} is the shunt admittance, if any, connected to that bus.

2. Sweeping all tree branches in a backward manner (i.e., upstream), compute branch currents $\mathcal{I}_{ij}$ by means of Kirchhoff's current law

$$\mathcal{I}_{ij}^k = \mathcal{I}_j^k + \sum_{m \in j, m \neq i} \mathcal{I}_{jm} \quad j = n, n-1, \ldots, 2 \qquad (3.64)$$

where i and j are, respectively, the sending and receiving buses. In practice, steps 1 and 2 can be merged as a single one.

3. Sweeping the tree in the opposite direction (i.e., downstream), bus voltages are updated from the head by considering the respective branch voltage drops:

$$\mathcal{V}_j^{k+1} = \mathcal{V}_i^{k+1} - z_{ij}\mathcal{I}_{ij}^k \quad j = 2, 3, \ldots, n \qquad (3.65)$$

where z_{ij} is the series impedance of branch $i-j$, and the complex voltage of the feeding bus, $\mathcal{V}_1$, is specified (slack bus).

The main difference with respect to the method described in Section 3.4.2 is that, owing to the radial topology of the system, there is no need to explicitly build and handle any impedance or admittance matrix, a linked list being sufficient to perform the forward and backward sweeps.

A slightly more robust version arises by directly working with branch power flows, rather than branch currents [21]. For this purpose, Equation 3.63 is replaced by

$$\mathcal{S}_i^k = \mathcal{S}_i^{\text{sp}} + y_{si}^*(V_i^k)^2 \quad i = n, n-1, \ldots, 2 \tag{3.66}$$

Also, during the backward sweep, power losses of outgoing branches should be accounted for when computing branch power flows:

$$\mathcal{S}_{ij}^k = \mathcal{S}_j^k + \sum_{m \in j, m \neq i} \left[z_{jm} \frac{\mathcal{S}_{jm}^2}{V_m^2} + \mathcal{S}_{jm} \right] \quad j = n, n-1, \ldots, 2 \tag{3.67}$$

Then, the forward sweep updates voltages from

$$V_j^{k+1} = V_i^{k+1} - z_{ij} \left[\frac{\mathcal{S}_{ij}^k}{V_j^k} \right]^* \quad j = 2, 3, \ldots, n \tag{3.68}$$

3.9   LARGE-SCALE SYSTEMS

The Newton–Raphson method is used in the majority of modern power flow programs. The method requires, when applied to large systems, the use of very large matrices. As we have seen, the structure of the polar version of the Jacobian matrix has the following structure:

$$J = \begin{bmatrix} H & N \\ M & L \end{bmatrix} \tag{3.69}$$

The dimension of each Jacobian component is, in principle, $n \times n$ (where n is the number of nodes in the grid). As explained previously, certain rows and columns from the complete Jacobian are not used during actual computations. For instance, within the Jacobian component H (corresponding to derivatives of active power with respect to phase angles), the rows and columns corresponding to the reference node are eliminated, thus resulting in a smaller matrix. The same is true for the other three components of the Jacobian. Rows and columns corresponding to reactive power mismatches/residues in PV nodes, and columns corresponding to the respective voltage magnitudes are also omitted.

The most interesting property of these matrices is that, in addition to their large dimension for a system with many nodes, they are also quite sparse, that is, only a very small percentage of the total matrix entries have nonzero value. An inspection of the equations that define the Jacobian reveals that the structure (topology) of each of the four Jacobian components has the same structure as the admittance matrix $\mathcal{Y}$. This matrix has nonzero entries only when there is a direct connection between two nodes. This is best illustrated with a specific example. Consider the 20-node grid illustrated in Figure 3.5. Its graph and corresponding admittance matrix are illustrated in Figure 3.6.

The complete Jacobian (with rows for both P and Q and columns for both angle variables and voltage magnitudes) is illustrated in Figure 3.7. In this figure, the elements of H are represented by the symbol ∘, the ones for N with the symbol ◁, the ones for M with the symbol ▷, and the ones for L with the symbol ◇.

It is possible to improve the appearance and the computational efficiency of using this Jacobian if the rows and columns are reordered to group components that correspond to the same connections. The resulting new ordering of rows and columns is illustrated in Figure 3.8. In terms of individual

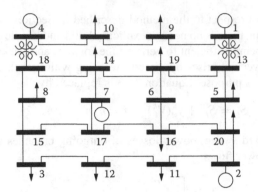

FIGURE 3.5 Single-line diagram for a 20-bus illustrative system. Thick lines represent busbars (electrical nodes), whereas thin lines stand for transmission lines or transformers. (From F. L. Alvarado et al., *Advances in Electric Power and Energy Conversion System Dynamics and Control* (C. T. Leondes, ed.), Control and Dynamic Systems, Vol. 41, Academic Press, Part 1. London, 1991, pp. 207–272.)

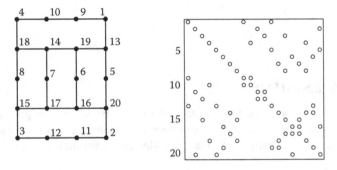

FIGURE 3.6 Graph and topology of the bus admittance matrix corresponding to the network of Figure 3.5. (From F. L. Alvarado et al., *Advances in Electric Power and Energy Conversion System Dynamics and Control* (C. T. Leondes, ed.), Control and Dynamic Systems, Vol. 41, Academic Press, Part 1. London, 1991, pp. 207–272.)

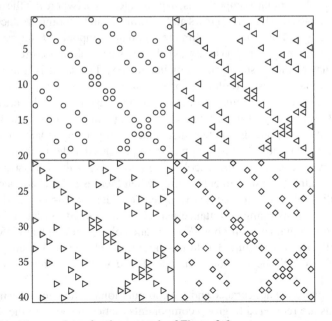

FIGURE 3.7 Full Jacobian topology for the network of Figure 3.6.

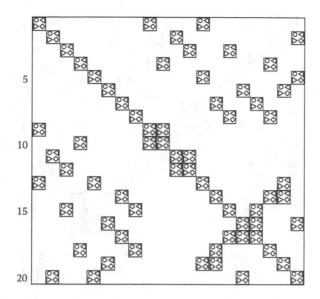

FIGURE 3.8 Restructured Jacobian topology.

entries, the dimension of this matrix is 40×40, but in terms of blocks, the dimension of this matrix is 20×20.

It is necessary to eliminate certain rows and columns from this structure. For the problem as stated, the following changes are required:

- The row and column corresponding to the reference node (node 4 in our example) is completely removed.
- For the other three generation nodes (nodes 1, 2, and 7), their corresponding terms are eliminated from the N, M, and L Jacobian submatrices.

The result is a matrix of dimension 35×35 (or 19×19 if you count blocks rather than individual matrix entries). This matrix is illustrated in Figure 3.9. Rows and columns have been removed based on the node type.

The most important step for an efficient solution involving large sparse matrices is the ordering of their rows and columns to reduce the number of nonzero entries in the matrix after it is factored. Because of its small dimension, it is difficult to illustrate the importance of this concept in the 20-bus example. In what follows, these notions will be illustrated using a network with 118 nodes.

There are many possible reordering methods, all of which reduce the number of entries in the factored Jacobian. The three main methods proposed by Tinnney and Hart [8] are

1. *A priori* ordering the nodes according to the valence or degree of the node (valence or degree refers to the number of connections between the node and the neighboring nodes or, equivalently, to the number of nonzero entries in the corresponding Jacobian row or column).
2. Dynamic ordering of the nodes according to their valence. In this case, the valence is defined as the number of connections between the nodes and its neighbors during each step of the elimination process. This method is also known as the minimum degree algorithm.
3. Dynamic ordering of the nodes according to valence, with the valence defined as the number of additional entries (fill-ins) that would occur if the node was eliminated at that step of the elimination process.

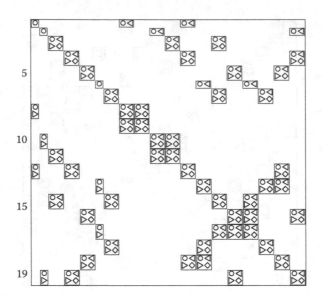

FIGURE 3.9 Topology of the reduced Jacobian.

Figure 3.10 illustrates the initial matrix structure corresponding to the 118-node example. Figure 3.11 illustrates the L and U factors if no ordering is done. The structure of the factors when the matrix has been ordered according to Tinney's method 2 is illustrated in Figure 3.12. The advantage of ordering is immediately evident, in spite of the relatively small dimension of this matrix. The impact and the differences among all three methods become even more evident in the case of larger matrices. The advantage of ordering is not restricted to a reduction in the

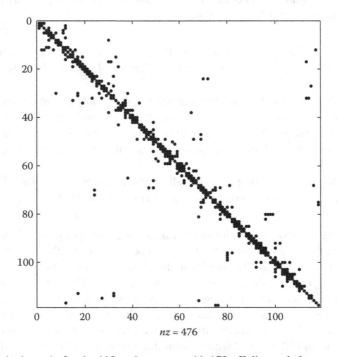

$nz = 476$

FIGURE 3.10 Original matrix for the 118-node system, with 179 off-diagonal elements in each half.

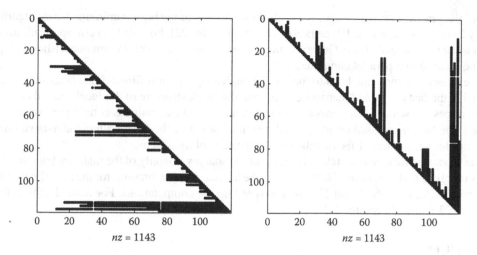

FIGURE 3.11 *L/U* factors of the 118-node matrix (without ordering). Each triangular factor contains 1025 off-diagonal elements, of which 846 are fill-ins.

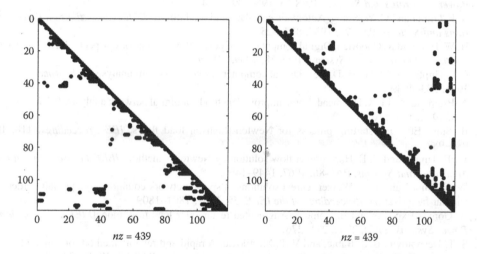

FIGURE 3.12 *L/U* factors of the 118-node matrix, when ordered according to Tinney's scheme 2. Each triangular factor contains 321 off-diagonal elements, of which 142 are *fill-ins*.

number of entries, but it is also reflected in a significant reduction in the computational effort required to obtain the factors.

Many variants of these basic three methods have been developed and implemented over the past 35 years. The purpose of these variants includes, among others, the following objectives:

- A reduction in the dependencies among operations and an increase in the available parallelism among the required computations
- An improvement in the speed of the ordering algorithms themselves by means of some approximations to the algorithms as stated
- Extensions to the case of unsymmetric matrices
- Proper consideration of the numerical aspects during the factorization of the matrix (a reduction in the accumulation of numerical error).

From the practical point of view, one of the best variants of the basic minimum degree algorithm (Tinney 2) is the so-called MMD method of George and Liu [22]. For most general applications, this variant is the recommended one. One exception to this recommendation is when one wishes to implement the method in a parallel computer.

The extension of these methods to the fast decoupled algorithm is direct if one takes into consideration that the fast decoupled matrices do not have the block structure of the Jacobian. All entries in these matrices are scalars. Furthermore, the factorization of these matrices is only performed once, and the same factors are used during several iterations of the method. This alters the relative importance of ordering, numerical factorization, and repeated solution steps.

More advanced methods can take advantage of the sparsity not only of the matrices but also of the vectors involved in the solution. They can also take advantage of techniques for the partial rather than complete rebuilding of the L and U factors to speed up the computations. For more details, refer to Reference 23 and to Appendix A.

REFERENCES

1. A. Gómez-Expósito, J. L. Martinez, and J. Riquelme, Slack bus selection to minimize the system power imbalance in load-flow studies, *IEEE Transactions on Power Systems*, 19(2), 2004, 987–995.
2. J. B. Ward and H. W. Hale, Digital computer solution of power flow problems, *IEEE Transactions on Power Apparatus and Systems*, PAS-75, 1956, 398–494.
3. A. F. Glimn and G. W. Stagg, Automatic calculation of load flows, *IEEE Transactions on Power Apparatus and Systems*, PAS-76, 1957, 817–825.
4. H. W. Hale and R. Goodrich, Digital computation of power flow—some new aspects, *IEEE Transactions on Power Apparatus and Systems*, PAS-78, 1959, 919–924.
5. P. P. Gupta and M. W. H. Davies, Digital computers in power system analysis, *IEE Proceedings*, 108, 1961, 383–404.
6. A. Brameller and J. K. Denmead, Some improved methods of digital network analysis, *IEE Proceedings*, 109, 1962, 109–116.
7. B. Stott, Effective starting process for Newton-Raphson load flows, *IEE Proceedings*, 118, 1971, 983–987.
8. W. F. Tinney and C. E. Hart, Power flow solution by Newton's method, *IEEE Transactions on Power Apparatus and Systems*, PAS-86, 1967, 1449–1460.
9. W. F. Tinney and J. W. Walker, Direct solutions of sparse network equations by optimally ordered triangular factorization, *Proceedings of the IEEE*, 55, 1967, 1801–1809.
10. A. Gómez-Expósito and E. Romero, Augmented rectangular load flow model, *IEEE Transactions on Power Systems*, 17(2), 2002, 271–276.
11. S. T. Despotovic, B. S. Babic, and V. P. Mastilovic, A rapid and reliable method for solving load flow problems, *IEEE Transactions on Power Apparatus and Systems*, PAS-90, 1971, 123–130.
12. B. Stott, Decoupled Newton load flow, *IEEE Transactions on Power Apparatus and Systems*, PAS-91, 1972, 1955–1959.
13. B. S. Babic, Decoupled load flow with variables in rectangular form, *IEE Proceedings*, Part C, 1983, 98–110.
14. B. Stott and O. Alsac, Fast decoupled load flow, *IEEE Transactions on Power Apparatus and Systems*, PAS-93, 1974, 859–869.
15. R. A. M. Van Amerongen, A general-purpose version of the fast decoupled load flow, *IEEE Transactions on Power Systems*, 4, 1989, 760–770.
16. A. Monticelli, A. García, and O. R. Saavedra, Fast decoupled load flow: Hypothesis, derivations, and testing, *IEEE Transactions on Power Systems*, 5(2), 1990, 556–564.
17. B. Stott, Review of load flow calculation methods, *Proceedings of the IEEE*, 62, 1974, 916–929.
18. N. Hingorani and L. Gyugyi, *Understanding FACTS: Concepts and Technology of Flexible AC Transmission Systems*, IEEE Press, New York, 2000.
19. N. M. Peterson and W. S. Meyer, Automatic adjustment of transformer and phase-shifter taps in the Newton power flow, *IEEE Transactions on Power Apparatus and Systems*, PAS-90, 1971, 103–108.
20. D. Shirmohammadi, H. W. Hong, A. Semlyen, and G. X. Luo, A compensation-based power flow method for weakly meshed distribution and transmission networks, *IEEE Transactions on Power Systems*, 3(2), 1988, 753–762.

21. G. X. Luo and A. Semlyen, Efficient load flow for large weakly meshed networks, *IEEE Transactions on Power Systems*, 5(4), 1990, 1309–1316.

22. A. George and J. W. H. Liu, The evolution of the minimum degree ordering algorithm, *SIAM Review*, 31, 1971, 1–19.

23. F. L. Alvarado, W. F. Tinney, and M. K. Enns, Sparsity in large-scale network computation, *Advances in Electric Power and Energy Conversion System Dynamics and Control* (C. T. Leondes, ed.), Control and Dynamic Systems, Vol. 41, Academic Press, Part 1. London, 1991, pp. 207–272.

4 State Estimation

Antonio Gómez-Expósito and Ali Abur

CONTENTS

4.1 INTRODUCTION

The renowned blackout that took place in the Northeast area of the United States in 1965 constituted a milestone in the way the increasingly complex electric energy systems were operated. The previous decades had seen how the primitive isolated systems began a systematic interconnection process, in an attempt to gain security (mutual help under contingencies) and to save money (exchange of cheap energy). Ironically, however, it was soon realized that operating such giant networks was anything but trivial, and that widespread blackouts could take place if the state of the network was not properly monitored and tracked in real time.

With the advent of the primitive digital computers, the first supervisory control and data acquisition (SCADA) systems were implemented in a diversity of complex industrial installations (railway systems, nuclear stations, etc.). As far as electric network control centers are concerned, such systems included several specialized functions, in addition to the typical monitoring tasks, like automatic generation control (AGC) and economic dispatch, for which the system frequency and generator power outputs had to be continuously supervised.

The 1965 blackout, and other major incidents that followed, pointed out the need to pay much more attention toward security issues, which called for more sophisticated SCADA environments. Thus, many more measurements, including line power flows, were captured at shorter intervals, and new computer functions, based on the load flow tool presented in Chapter 3, were developed. Eventually, those tools would allow the operators to properly assess the network losses, network security (unacceptable voltages and line flows), risk of instability, and so on.

However, initial attempts to obtain the network state through an online load flow were plagued by problems due to missing or inconsistent measurements. The concept of state estimation (SE) was first suggested by Fred Schweppe, who was an electrical engineering professor at MIT, to address these issues. SE had already been applied to smaller problems in control systems [1,2]. Computational developments that took place after this pioneering work, particularly those by Tinney and his colleagues [3], very soon made the SE a standard function in every control center.

State estimator provides a complete and reliable database, which is vital for the correct performance of all online application functions that are used in the operation and control of a power system, starting with security-related issues [4]. Solution provided by the SE is also essential for many of the off-line applications that are related to system planning (load forecasting, reliability, network expansion, etc.) as well as electricity markets. All these software tools and advances in computer architectures have led to the development of several generations of the so-called energy management systems (EMS), successors of the primitive SCADAs.

An SE comprises basically the following functions [5–7]:

1. *Measurement prefiltering*: A set of rudimentary consistency checks is implemented to detect and discard measurements that are clearly wrong (negative voltage magnitudes, out of range power flows, etc.).
2. *Topology processor*: On the basis of the status of switching devices and the physical layout of substations, electrical network model (electrical nodes, connectivity, nonenergized islands, etc.) is built.
3. *Observability analysis*: Determines if the system state can be obtained using the available measurements for the entire network. If only a subset of nodes is observable, then it identifies the observable islands.
4. *State estimation*: Computes the statistically optimal network state, or the state that best fits the remotely captured measurements for a given set of network parameters and network connectivity. As a by-product, it provides also the estimated measurements and associated covariance matrices.
5. *Bad data processor*: On the basis of certain statistical properties of the estimate, this function detects the presence of potential non-Gaussian errors in the set of measurements. If the

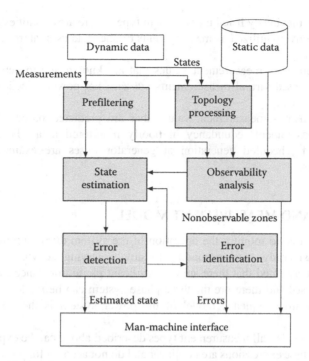

FIGURE 4.1   Building blocks of a state estimator.

redundancy is adequate, those undesirable measurements can be identified and removed. Modern estimators are also capable of detecting topology errors and estimating improved values of suspicious network parameters, such as transformer taps.

Figure 4.1 shows the functional relationships and data exchanges among these blocks.

In summary, the SE serves as a large-scale filter between the remote measurements and the high-level applications that constitute the EMS.

4.2   PROBLEM FORMULATION

The set of phasors representing the complex bus voltages of a power system is defined as the static state of that system. It is inherently assumed that the network topology along with the network parameters are known and are excluded from this definition of the state. This assumption will be relaxed later in Section 4.11 to enable detection of parameter errors.

The objective of static SE is to estimate the complex bus voltage phasors (the states) at every bus in a given power system. This is accomplished by processing the measurements of various line flows, bus injections, voltage and line current magnitudes as well as the information about the status of the circuit breakers (CBs), switches, transformer taps and the parameters of the transmission lines, transformers, and shunt capacitors and reactors. Measurement types, locations, and accuracies vary depending on the individual power system. Most commonly used measurement types are briefly described as follows:

1. Flows—real and reactive power flows measured at the terminal buses of transmission lines or transformers.
2. Injections—net real and reactive power injections at system buses.
3. Voltage magnitudes—measured at system buses.
4. Current magnitudes—ampere flows measured at the terminal buses of transmission lines or transformers.

In addition to these previously listed measurement types, there are quantities that are not actually measured but are effectively utilized as measurements by the state estimators. These are

1. Virtual measurements—measurement values that are known due to network constraints. Most commonly used virtual measurements are zero net power injections at buses with no load or generation.
2. Pseudomeasurements—measurement values that are predicted based on historical data to improve measurement redundancy in poorly monitored areas. Forecasted loads at load buses, and scheduled generation at generator buses are examples of this type of measurements.

4.3 NETWORK AND MEASUREMENT MODEL

Static SE function is used to monitor the operation of the system during normal operation, where the system is in quasi-steady state responding to slowly varying network load and generation. Hence, it is inherently assumed that there are no significant phase imbalances, and all transmission lines are fully transposed and therefore the three-phase system can be modeled by its single-phase equivalent circuit. Positive sequence model of the power system is then used in the problem formulation.

Given the network model, all measurement types described above can be expressed as a function of the system states. These expressions are nonlinear and do not account for possible errors due to the uncertainties in the network parameters, metering errors or biases, and noise that may be introduced through the telecommunication systems.

Consider a vector z containing the set of measurements that can be expressed in terms of the system states as follows:

$$z = \begin{bmatrix} z_1 \\ z_2 \\ \vdots \\ z_m \end{bmatrix} = \begin{bmatrix} h_1(x_1, x_2, \ldots, x_n) \\ h_2(x_1, x_2, \ldots, x_n) \\ \vdots \\ h_m(x_1, x_2, \ldots, x_n) \end{bmatrix} + \begin{bmatrix} e_1 \\ e_2 \\ \vdots \\ e_m \end{bmatrix} = h(x) + e \qquad (4.1)$$

where
$h^T = [h_1(x), h_2(x), \ldots, h_m(x)]$
$h_i(x)$ is the nonlinear function relating measurement i to the state vector x
$x^T = [x_1, x_2, \ldots, x_n]$ is the system state vector
$e^T = [e_1, e_2, \ldots, e_m]$ is the vector of measurement errors

To work with real as opposed to complex variables, the state vector is typically constructed in expanded form, where either the cartesian or the polar coordinates are used to represent each complex state variable (bus voltage) by two of its real components. Similarly, the measurement vector will contain real entries z_is, corresponding to real or reactive power flows and injections, or magnitude of voltage or current phasors. Indeed, the functions $h_i(x)$, which relate the states to the error free measurements, are also functions of the network model and parameters. However, based on the assumption that the network data are fully and accurately available, they are presented as functions of system states only.

It is customary, yet not always justifiable to make the following assumptions regarding the statistical properties of the measurement errors:

* Errors are distributed according to a normal distribution.
* Expected values of all errors are zero, that is, $E(e_i) = 0$, $i = 1, \ldots, m$.

- Errors are independent, that is, $E(e_i e_j) = 0$. Hence,

$$\text{Cov}(e) = E[e \cdot e^T] = R = \text{diag}\{\sigma_1^2, \sigma_2^2, \ldots, \sigma_m^2\}$$

The standard deviation σ_i of each measurement, i, is calculated to reflect the expected accuracy of the corresponding meter used. A formula used by the American Electric Power (AEP) Company's state estimator [8] is as follows:

$$\sigma_i = 0.0067 S_i + 0.0016 F S_i$$

where

$$S_i = \begin{cases} \sqrt{P_{km}^2 + Q_{km}^2} & \text{for flow } k - m \\ \sqrt{P_k^2 + Q_k^2} & \text{for injection at } k \\ |V_k| & \text{for voltage magnitude at } k \end{cases}$$

FS_i = Full scale deflection of the meter

Formulation of the SE problem is based on the concept of maximum likelihood estimation, which will be reviewed briefly below. Maximum likelihood estimator (MLE) of a random variable maximizes a likelihood function, which is defined based on assumptions of the problem formulation. Hence, the MLEs of different formulations will yield different results. The derivation below will consider the commonly used SE formulation that is based on the assumptions for the measurement errors stated previously.

The first assumption is that the errors are distributed according to a normal (or Gaussian) distribution. A random variable z is said to have a normal distribution if its probability density function (pdf), $f(z)$, is given as follows:

$$f(z) = \frac{1}{\sqrt{2\pi}\sigma} e^{-\frac{1}{2}(z-\mu)^2/\sigma^2}$$

where z is the random variable, μ the expected value (or mean) of $z = E(z)$, and σ the standard deviation of z.

A change of variable as shown below will yield a more common function referred to as the standard normal (or Gaussian) pdf, $\Phi(u)$. Since, all normal pdf's can be translated into $\Phi(u)$ via a simple change of variable, a single standardized pdf is all that needs to be calculated.

Let $u = (z - \mu)/\sigma$, then $E(u) = 0$, $\text{Var}(u) = 1.0$.

$$\Phi(u) = \frac{1}{\sqrt{2\pi}\sigma} e^{-u^2/2}$$

The second assumption can easily be incorporated into the pdf expression by setting the expected value of errors equal to zero.

The third assumption implies that the joint pdf of a set of m measurements can be obtained by simply taking the product of individual pdfs corresponding to each measurement. The resulting product function $f_m(z)$ given by

$$f_m(z) = f(z_1) f(z_2) \cdots f(z_m)$$

is called the *Likelihood Function* for the set of m measurements.

To simplify the arithmetic, the logarithm of the likelihood function, referred to as the log-likelihood function and denoted by $\mathcal{L}$, is used instead. The function $\mathcal{L}$ is expressed as

$$\mathcal{L} = \log f_m(z) = \sum_{i=1}^{m} \log f(z_i) = -\frac{1}{2} \sum_{i=1}^{m} \left(\frac{z_i - \mu_i}{\sigma_i} \right)^2 - \frac{m}{2} \log 2\pi - \sum_{i=1}^{m} \log \sigma_i$$

Note that maximizing $\mathcal{L}$ and $f_m(z)$ will yield the same optimal solution due to the monotonically increasing nature of the logarithmic function. Hence, the MLE of the state x can be found by maximizing the likelihood (or log-likelihood) function for a given set of observations $z_1, z_2, \ldots, z_m$. This is an optimization problem, which can be formulated as

$$\text{Maximize} \quad \log f_m(z)$$
$$\text{or}$$
$$\text{Minimize} \quad \sum_{i=1}^{m} \left(\frac{z_i - \mu_i}{\sigma_i} \right)^2$$

Let us define the residual of the measurement i as

$$r_i = z_i - E(z_i)$$

where $E(z_i) = h_i(x)$ and h_i is a nonlinear function relating the system state vector x to the ith measurement.

The reciprocal of the measurement variances can be thought of as weights assigned to individual measurements. High values are assigned for accurate measurements with small variance and low weights for measurements with large uncertainties. Let us denote these weights by $W_{ii} = \sigma_i^{-2}$, for measurement i.

Using the given notation and defined variables, formulation of the MLE can be rewritten as the following optimization problem:

$$\text{Minimize} \quad \sum_{i=1}^{m} W_{ii}\, r_i^2$$

$$\text{Subject to} \quad z_i = h_i(x) + r_i, \quad i = 1, \ldots, m$$

(4.2)

The solution of the above optimization problem is called the weighted least squares (WLS) estimator for x.

Note that the definition of the objective function in the given optimization problem directly follows from the choice of the likelihood function. Hence, variations in the choice of assumptions regarding the measurement errors' statistical properties will lead to different MLE formulations and solutions. Such other formulations using alternative assumptions are discussed in Section 4.9.

EXAMPLE 4.1

Consider the 3-bus system shown in Figure 4.2. Electrical data corresponding to this network are as follows:

| Line | | Resistance | Reactance | 1/2 Susceptance |
From Node	To Node	R (pu)	X (pu)	b_s (pu)
1	2	0.01	0.03	0.0
1	3	0.02	0.05	0.0
2	3	0.03	0.08	0.0

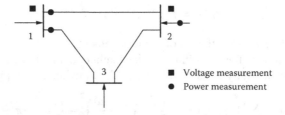

FIGURE 4.2   One-line diagram and measurement set for a 3-bus system.

The system is monitored through eight measurements, which means that $m = 8$ in Equation 4.1. Measurement values and associated weights are

Index, i	Type	Value (pu)	W_{ii} (pu)
1	p_{12}	0.888	15,625
2	p_{13}	1.173	15,625
3	p_2	−0.501	10,000
4	q_{12}	0.568	15,625
5	q_{13}	0.663	15,625
6	q_2	−0.286	10,000
7	V_1	1.006	62,500
8	V_2	0.968	62,500

In this case, the state vector x contains five elements ($n = 5$),

$$x^{\mathrm{T}} = [\theta_2, \theta_3, V_1, V_2, V_3]$$

$\theta_1 = 0$ being arbitrarily taken as phase-angle reference.

4.4   SOLUTION THROUGH THE NORMAL EQUATIONS

As we have just seen, obtaining the MLE estimate when measurement errors are independent and follow a normal distribution consists of solving the WLS problem defined by Equation 4.2, whose objective function can be rewritten as follows:

$$J(x) = [z - h(x)]^{\mathrm{T}} W [z - h(x)]$$

$$= \sum_{i=1}^{m} \frac{[z_i - h_i(x)]^2}{\sigma_i^2} \tag{4.3}$$

where $W = R^{-1}$.

At the solution, the following n first-order optimality conditions must be satisfied,

$$\frac{\partial J(x)}{\partial x} = 0 \Rightarrow H^{\mathrm{T}}(x) W [z - h(x)] = 0 \tag{4.4}$$

where

$$H(x) = \frac{\partial h(x)}{\partial x}$$

is the $m \times n$ Jacobian matrix of vector $h(x)$. The goal is to obtain the vector $\hat{x}$ satisfying the nonlinear Equation 4.4. Like in the load flow problem, the most effective way to do so lies in the application of the Newton–Raphson (NR) iterative process, that converges quadratically to the solution [2,3]. Neglecting those terms that contain the second-order derivatives of $h(x)$, the linear system comprising n equations that must be solved at each iteration is

$$G(x^k) \Delta x^k = H^{\mathrm{T}}(x^k) W [z - h(x^k)] \tag{4.5}$$

where x^k denotes the value of x at the kth iteration and

$$G(x) = H^{\mathrm{T}}(x) W H(x) \tag{4.6}$$

is known as the *gain matrix*. If H is of full rank, then the symmetric matrix G is positive definite and the system Equation 4.5 has a unique solution (see Section 4.6). After solving the equation system, the state vector is updated before repeating the process:

$$x^{k+1} = x^k + \Delta x^k$$

For simplicity of notation, ignoring the dependence of H on x, the iterative procedure based on the normal equations can be summarized as follows:

1. Initialize the state vector $x = x^0$ with the flat voltage profile ($V_i = 1$ pu, $\theta_i = 0$) and the iteration counter ($k = 0$). Perform the optimal ordering of the rows/columns of G, for which only its nonzero pattern is needed.
2. Compute the measurement residuals $\Delta z^k = z - h(x^k)$.
3. Obtain H and $G = H^T W H$.
4. Solve the linear system:

$$G \Delta x^k = H^T W \Delta z^k \tag{4.7}$$

taking advantage of the sparse structure of matrices H and G. This involves the Cholesky decomposition of the gain matrix ($G = U^T U$) and the subsequent forward/backward elimination process (see Appendix A).
5. Update the state vector ($x^{k+1} = x^k + \Delta x^k$) and the iteration counter ($k = k + 1$).
6. If any of the elements of Δx exceeds the specified convergence threshold then return to step 2. Otherwise, stop.

The Jacobian terms corresponding to power measurements (flows and injections) are collected in Table 4.1. Elements of the row corresponding to a voltage measurement (magnitude or phase angle) are all null except for the column associated with the measured variable, which is set as unity.

EXAMPLE 4.2

Let us consider the solution of the SE through the normal equations for the system of Example 4.1 (see Figure 4.2).

Starting from the flat voltage profile,

$$x^0 = \begin{matrix} \theta_2 \\ \theta_3 \\ V_1 \\ V_2 \\ V_3 \end{matrix} \begin{bmatrix} 0 \\ 0 \\ 1.0 \\ 1.0 \\ 1.0 \end{bmatrix}$$

and according to the expressions provided in Table 4.1, the resulting Jacobian is

$$H(x^0) = \begin{matrix} \partial p_{12} \\ \partial p_{13} \\ \partial p_2 \\ \partial q_{12} \\ \partial q_{13} \\ \partial q_2 \\ \partial V_1 \\ \partial V_2 \end{matrix} \begin{array}{ccc|ccc} \partial\theta_2 & \partial\theta_3 & & \partial V_1 & \partial V_2 & \partial V_3 \\ -30.0 & & & 10.0 & -10.0 & \\ & -17.2 & & 6.9 & & -6.9 \\ 40.9 & -10.9 & & -10.0 & 14.1 & -4.1 \\ \hline 10.0 & & & 30.0 & -30.0 & \\ & 6.9 & & 17.2 & & -17.2 \\ -14.1 & 4.1 & & -30.0 & 40.9 & -10.9 \\ & & & 1.0 & & \\ & & & & 1.0 & \end{array}$$

TABLE 4.1

Elements of H Corresponding to Power Injections and Flows

	Injection		Flow
$\dfrac{\partial P_i}{\partial V_i}$	$\sum_{j=1}^{N} V_j(G_{ij}\cos\theta_{ij} + B_{ij}\sin\theta_{ij}) + V_i G_{ii}$	$\dfrac{\partial P_{ij}}{\partial V_i}$	$V_j(G_{ij}\cos\theta_{ij} + B_{ij}\sin\theta_{ij}) - 2G_{ij}V_i$
$\dfrac{\partial P_i}{\partial V_j}$	$V_i(G_{ij}\cos\theta_{ij} + B_{ij}\sin\theta_{ij})$	$\dfrac{\partial P_{ij}}{\partial V_j}$	$V_i(G_{ij}\cos\theta_{ij} + B_{ij}\sin\theta_{ij})$
$\dfrac{\partial Q_i}{\partial V_i}$	$\sum_{j=1}^{N} V_j(G_{ij}\sin\theta_{ij} - B_{ij}\cos\theta_{ij}) - V_i B_{ii}$	$\dfrac{\partial Q_{ij}}{\partial V_i}$	$V_j(G_{ij}\sin\theta_{ij} - B_{ij}\cos\theta_{ij}) + 2V_i(B_{ij} - b_{ij}^p)$
$\dfrac{\partial Q_i}{\partial V_j}$	$V_i(G_{ij}\sin\theta_{ij} - B_{ij}\cos\theta_{ij})$	$\dfrac{\partial Q_{ij}}{\partial V_j}$	$V_i(G_{ij}\sin\theta_{ij} - B_{ij}\cos\theta_{ij})$
$\dfrac{\partial P_i}{\partial\theta_i}$	$\sum_{j=1}^{N} V_i V_j(-G_{ij}\sin\theta_{ij} + B_{ij}\cos\theta_{ij}) - V_i^2 B_{ii}$	$\dfrac{\partial P_{ij}}{\partial\theta_i}$	$V_i V_j(-G_{ij}\sin\theta_{ij} + B_{ij}\cos\theta_{ij})$
$\dfrac{\partial P_i}{\partial\theta_j}$	$V_i V_j(G_{ij}\sin\theta_{ij} - B_{ij}\cos\theta_{ij})$	$\dfrac{\partial P_{ij}}{\partial\theta_j}$	$V_i V_j(G_{ij}\sin\theta_{ij} - B_{ij}\cos\theta_{ij})$
$\dfrac{\partial Q_i}{\partial\theta_i}$	$\sum_{j=1}^{N} V_i V_j(G_{ij}\cos\theta_{ij} + B_{ij}\sin\theta_{ij}) - V_i^2 G_{ii}$	$\dfrac{\partial Q_{ij}}{\partial\theta_i}$	$V_i V_j(G_{ij}\cos\theta_{ij} + B_{ij}\sin\theta_{ij})$
$\dfrac{\partial Q_i}{\partial\theta_j}$	$-V_i V_j(G_{ij}\cos\theta_{ij} + B_{ij}\sin\theta_{ij})$	$\dfrac{\partial Q_{ij}}{\partial\theta_j}$	$-V_i V_j(G_{ij}\cos\theta_{ij} + B_{ij}\sin\theta_{ij})$

and the gain matrix

$$G(x^0) = 10^7 \begin{bmatrix} 3.4392 & -0.5068 & 0.0137 & & -0.0137 \\ -0.5068 & 0.6758 & -0.0137 & 0.0137 & 0.0000 \\ 0.0137 & -0.0137 & 3.1075 & -2.9324 & -0.1689 \\ & 0.0137 & -2.9324 & 3.4455 & -0.5068 \\ -0.0137 & 0.0000 & -0.1689 & -0.5068 & 0.6758 \end{bmatrix}$$

Then the iterative procedure described above is applied, the convergence threshold being 10^{-4}. The next table provides, for every iteration, the state vector increment Δx^k and the objective function (Equation 4.3):

Iteration, k	1	2	3
$\Delta\theta_2^k$	-2.1×10^{-2}	-5.52×10^{-4}	0
$\Delta\theta_2^k$	-4.52×10^{-2}	-2.69×10^{-3}	3.0×10^{-6}
ΔV_1^k	1.67×10^{-4}	-1.09×10^{-4}	-2.0×10^{-6}
ΔV_2^k	-2.53×10^{-2}	-1.06×10^{-4}	-2.0×10^{-6}
ΔV_3^k	-5.68×10^{-2}	1.15×10^{-3}	2.0×10^{-6}
Objective function			
$J(x^k)$	49×10^3	59.1	9.1

After this process, the estimate is

Node	$\hat{V}$ (pu)	$\hat{\theta}$ (°)
1	1.0001	0.0
2	0.9746	-1.25
3	0.9443	-2.75

Finally, the estimated measurements and associated residuals are

Index, i	Type	Measured (pu)	Estimated (pu)	Residual (pu)
1	p_{12}	0.888	0.893	−0.005
2	p_{13}	1.173	1.171	0.002
3	p_2	−0.501	−0.496	−0.005
4	q_{12}	0.568	0.558	0.01
5	q_{13}	0.663	0.668	−0.005
6	q_2	−0.286	−0.298	0.0123
7	V_1	1.006	1.000	−0.006
8	V_2	0.968	0.974	−0.006

It is worth noting that the coefficients of the gain matrix G after the third iteration

$$
G(x^3) = 10^7 \begin{bmatrix}
3.2086 & -0.4472 & -0.0698 & -0.0314 & 0.0038 \\
-0.4472 & 0.5955 & -0.0451 & 0.0045 & 0.0000 \\
-0.0698 & -0.0451 & 3.2011 & -2.8862 & -0.2160 \\
-0.0314 & 0.0045 & -2.8862 & 3.3105 & -0.4760 \\
0.0038 & 0.0000 & -0.2160 & -0.4760 & 0.6684
\end{bmatrix}
$$

are very close to those computed at flat start.

The more accurate the measurements the larger are the coefficients of W and, consequently, those of G (in this example they are in the order of 10^7). It can be easily proved that the estimated state does not change if the coefficients W_{ii} are normalized with a scalar, for instance, W_{ii}^{max}. However, it should be noted that this scaling will modify the objective function in the same proportion.

4.5 FAST DECOUPLED ESTIMATOR

As the SE is repetitively run in online environments and on networks that can be very large, any simplification that saves solution time is welcome. Like in the load flow tool, the most successful variants of the basic process described above have to do with certain approximations to the gain matrix.

In this regard, the most intuitive approach consists of not updating G at each iteration (NR with constant Jacobian), which saves one of the most expensive stages at each iteration, namely, its LU decomposition.

The most popular simplification, however, arises from the well-known decoupling between the active and the reactive subproblems, which leads to the so-called decoupled estimators [9,10]. Denoting the set of measurements related with active and reactive problems by z_a and z_r, respectively, and the phase angles and voltage magnitudes by x_a and x_r, the matrices involved in the solution process can be partitioned accordingly as follows:

$$
H = \begin{pmatrix} H_{aa} & H_{ar} \\ H_{ra} & H_{rr} \end{pmatrix} \quad W = \begin{pmatrix} W_a & 0 \\ 0 & W_r \end{pmatrix} \quad G = \begin{pmatrix} G_{aa} & G_{ar} \\ G_{ra} & G_{rr} \end{pmatrix}
$$

The numerical values corresponding to the off-diagonal blocks of matrix H, and consequently those of G, are significantly smaller than those of the diagonal blocks, which change very little during the iterative process. Resorting to the same simplifying assumptions as in the load flow case, and normalizing power measurements with the respective voltage magnitudes, constant (computed at flat voltage profile) and decoupled Jacobian and gain matrices are obtained

$$
G = \begin{pmatrix} G_{aa} & 0 \\ 0 & G_{rr} \end{pmatrix} \quad \begin{aligned} G_{aa} &= H_{aa}^T W_a H_{aa} \\ G_{rr} &= H_{rr}^T W_r H_{rr} \end{aligned}
$$

as well as an approximate decoupled right-hand side vector

$$\begin{pmatrix} T_a \\ T_r \end{pmatrix} = \begin{pmatrix} H_{aa}^T W_a \Delta z_a' \\ H_{rr}^T W_a \Delta z_r' \end{pmatrix} \qquad \begin{aligned} \Delta z_a' &= \Delta z_a/V \\ \Delta z_r' &= \Delta z_r/V \end{aligned}$$

Note that, while modifying G affects only the convergence speed, approximating the independent vector influences the estimate to a certain extent.

The WLS fast decoupled SE, comprising two sets of constant-coefficient, decoupled systems of equations is obtained, which involves the following iterative system:

1. Build and factorize G_{aa} and G_{rr}
2. Compute T_a
3. Solve $G_{aa}\Delta\theta = T_a$
4. Update phase angles, $\theta^{k+1} = \theta^k + \Delta\theta$
5. Compute T_r
6. Solve $G_{rr}\Delta V = T_r$
7. Update voltage magnitudes, $V^{k+1} = V^k + \Delta V$
8. Return to step 2 until convergence is reached

Computational saving arises from the following factors:

- Approximately, one half of the vector and matrix elements are not computed.
- Solving the two decoupled systems requires about one half of the operations involved in the solution of the full original system.
- Matrices G_{aa} and G_{rr} are built and factorized only once, which means that the forward/backward elimination process is only needed at each iteration.

EXAMPLE 4.3

The constant gain matrices for the system of Example 4.1 are

$$G_{aa} = 10^7 \begin{bmatrix} 3.837 & -0.5729 \\ -0.5729 & 0.7812 \end{bmatrix} \qquad G_{rr} = 10^7 \begin{bmatrix} 2.777 & -2.635 & -0.1357 \\ -2.635 & 3.090 & -0.4489 \\ -0.1357 & -0.4489 & 0.5846 \end{bmatrix}$$

It can be observed that the numerical values are very similar to those of the diagonal blocks of matrix G in Example 4.2.

The iterative process converges in 3.5 iterations (three active and four reactive problem solutions), to the state,

Node	$\hat{V}$ (pu)	$\hat{\theta}$ (°)
1	1.000	0.0
2	0.97438	−1.24
3	0.94401	−2.71

which differs slightly from that obtained in Example 4.2 owing to the approximations applied to the independent vector.

4.6 OBSERVABILITY ANALYSIS

Considering the measurement equations given by Equation 4.1, the aim of SE is to obtain the best possible solution for the vector x of n states, given the vector z of m measurements. Assuming that there are no measurement errors, that is, $e = 0$, at least an equal number of independent measurements as the number of states ($m \geq n$) will be needed to obtain a solution for x. Therefore, prior to estimating the system state, the measurement set should be analyzed to make sure that it contains at least n linearly independent measurements. This analysis is referred to as the observability test. The main purpose of this test is to check whether the state of the entire system can be estimated using the given set of measurements. If the test fails, then the analysis continues to identify all the observable islands. Observable islands are subsystems, which are observable on their own with respect to a chosen internal reference bus, making it possible to estimate all relevant quantities internal to these subsystems, such as line flows, voltages, and injections. Observable subsystems are interconnected by branches that are not observable, that is, line flows through these branches cannot be estimated using the given set of measurements.

Observability analysis can be formally described by considering the following linearized measurement model:

$$\Delta z = H \, \Delta x + e$$

The WLS estimate $\Delta \hat{x}$ will be given by

$$\Delta \hat{x} = (H^T R^{-1} H)^{-1} H^T R^{-1} \, \Delta z$$

A unique solution for Δx can be calculated if $(H^T R^{-1} H)$ is nonsingular or equivalently if H has full column rank, that is, rank $[H] = n$, where n is the total number of states.

As discussed in Section 4.5, there is weak coupling between the real power flow or injection measurements and the bus voltage magnitudes ($P - V$), and also between the reactive power flow or injection measurements and the bus voltage phase angles ($Q - \theta$). Thus, the above given linearized model can be decoupled as follows:

$$\Delta z_a = H_{aa} \Delta \theta + e_a$$
$$\Delta z_r = H_{rr} \Delta V + e_r$$

where the meaning of each term is explained above.

Using these decoupled models, the $P - \theta$ and $Q - V$ observabilities can be tested separately. In practice, most of the power flow and injection measurements come in pairs of real and reactive parts. So, the observability analysis is commonly carried out for $P - \theta$ model only. At the end of the analysis, each island has to contain at least one voltage magnitude measurement in order for it to be declared as an observable island. This condition can, however, be checked easily upon identifying the observable islands based on the decoupled model of $P - \theta$.

The last step in the observability analysis is the placement of pseudomeasurements in case the system is determined as unobservable for the given set of measurements. Pseudomeasurements can be assigned based on load forecasts, generation schedules, or some other source of information regarding the missing measurements. The measurement type as well as its location play a critical role in this choice, since the goal is to render the system fully observable by assigning a minimum of such pseudomeasurements.

Observability analysis can be carried out by either numerical or topological methods. Since both methods have been widely implemented in existing SE programs, they will be briefly reviewed.

4.6.1 NUMERICAL OBSERVABILITY ANALYSIS

Network observability is essentially determined by the type and location of the measurements. The operating state of the system or the actual parameters of the network components will not affect network observability, except for some degenerate cases leading to numerical cancelations in the SE equations.

Therefore, the measurement equations can be artificially simplified by assuming that all system branches have an impedance of $1.0j$ pu, and all bus voltage magnitudes are equal to 1.0 pu. Then the DC approximation for the real power flows along the system branches can be written as:

$$P_b = C\theta$$

where P_b is the vector of DC approximation for branch real power flows, C the reduced (excluding slack bus) branch-bus incidence matrix, and θ the vector of bus voltage phase angles.

Numerical observability analysis method is based on the observation that if all measurements are zero, none of the branch flows can be nonzero for a fully observable system [11,12]. If any of the branch flows are nonzero, then the corresponding branches will have to be unobservable. This requirement can be easily tested by using the DC approximation of the real part of the decoupled linear measurement model

$$H_{aa}\theta = z_a$$

leading to the following WLS estimate

$$\hat{\theta} = (H_{aa}^T H_{aa})^{-1} H_{aa}^T z_a$$

The observability requirement implies that choosing a null vector for z_a should yield an estimate $\hat{\theta}$ such that all estimated branch flows will be zero, that is, $P_b = C\hat{\theta} = 0$.

If $H_{aa}\hat{\theta} = 0$, but $P_b = C\hat{\theta} \neq 0$, then $\hat{\theta}$ will be called *an unobservable state* and the system will be declared unobservable. Those system branches i, with $P_b(i) \neq 0$, will be called *unobservable branches* of the system. Unobservable branches split the overall system into *observable islands*.

If a branch has no incident measurements, then the estimated state will be independent of the status (on/off) and parameters of this branch. Therefore, such branches can be disregarded when analyzing network observability. They are called the *irrelevant branches* of the system.

Let us again consider the decoupled linearized model

$$(H_{aa}^T H_{aa})\hat{\theta} = G_{aa}\hat{\theta} = 0$$

If G_{aa} is nonsingular, the system will be fully observable. If G_{aa} is singular, then row/column permutations can be used to reorder and partition the matrix as follows:

$$\begin{bmatrix} G_{11} & G_{12} \\ G_{21} & G_{22} \end{bmatrix} \begin{bmatrix} \hat{\theta}_a \\ \hat{\theta}_b \end{bmatrix} = \begin{bmatrix} 0 \\ 0 \end{bmatrix}$$

where G_{11} is a nonsingular submatrix within G_{aa}. By assigning arbitrary but distinct values to $\hat{\theta}_b$ entries as $\bar{\theta}_b$, one of many possible solutions for $\hat{\theta}_a$ can be obtained as

$$\hat{\theta}_a = -G_{11}^{-1} G_{12} \bar{\theta}_b$$

The branch flows corresponding to this solution $(\hat{\theta}_a, \bar{\theta}_b) = \hat{\theta}^*$ can then be found as

$$C\hat{\theta}^* = P_b^*$$

Those branches i with $P_b^*(i) \neq 0$ will be identified as the *unobservable branches*. The numerical observability analysis algorithm involves the following steps:

1. Removing all irrelevant branches.
2. Forming the decoupled linearized gain matrix for the $P - \theta$ estimation problem:

$$G_{aa} = H_{aa}^T R_a^{-1} H_{aa}$$

3. Declaring the system as fully observable if G_{aa} is found to be nonsingular. Else, identifying the unobservable branches using the above outlined procedure.
4. Removing the unobservable branches and all injections that are incident at these unobservable branches.
5. Returning to step 1.

EXAMPLE 4.4

Let us consider the network and measurement system of Figure 4.3.

There is one irrelevant branch in this example and that is branch 1–3. Note the lack of any injection or line flow measurements incident to this branch. The branch is removed from the system diagram prior to observability analysis.

$$G_{aa} = \begin{bmatrix} 2 & 0 & 0 & -3 & 0 & 1 \\ 0 & 2 & 0 & -3 & 0 & 1 \\ 0 & 0 & 1 & 0 & -1 & 0 \\ -3 & -3 & 0 & 9 & 0 & -3 \\ 0 & 0 & -1 & 0 & 2 & -1 \\ 1 & 1 & 0 & -3 & -1 & 2 \end{bmatrix} \quad H_{aa} = \begin{bmatrix} -1 & 1 & 0 & 0 & 0 & 0 \\ 0 & 0 & -1 & 0 & 1 & 0 \\ 0 & 0 & 0 & 0 & -1 & 1 \\ -1 & -1 & 0 & 3 & 0 & -1 \end{bmatrix}$$

Reordering the rows/columns of G_{aa} so that

$$\theta_a^{\mathsf{T}} = [\theta_1 \theta_2 \theta_3 \theta_5]$$
$$\theta_b^{\mathsf{T}} = [\theta_6 \theta_4]$$

the reordered matrix G_{aa}^{ord} will then be

$$G_{aa}^{ord} = \begin{bmatrix} 2 & 0 & 0 & 0 & 1 & -3 \\ 0 & 2 & 0 & 0 & 1 & -3 \\ 0 & 0 & 1 & -1 & 0 & 0 \\ 0 & 0 & -1 & 2 & -1 & 0 \\ 1 & 1 & 0 & -1 & 2 & -3 \\ -3 & -3 & 0 & 0 & -3 & 9 \end{bmatrix}$$

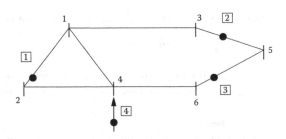

x Measurement number

FIGURE 4.3 Illustrative unobservable system.

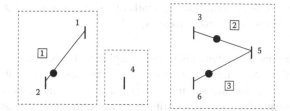

FIGURE 4.4 Resulting observable islands.

Let $\theta_6 = 1$, $\theta_4 = 0$, then solve for the rest as

$$\theta_1 = -0.5, \quad \theta_2 = -0.5, \quad \theta_3 = 1.0, \quad \theta_5 = 1$$

Calculating the resulting branch flows

Branch	Calculated Flow	Status
1–2	0	Observable
1–4	−0.5	Unobservable
2–4	−0.5	Unobservable
4–6	−0.5	Unobservable
3–5	0	Observable
5–6	0	Observable
1–3	−0.5	Unobservable

Removing all unobservable branches and the injection at bus 4 yields the network shown in Figure 4.4.

Therefore, three observable islands will be identified, that is, branch 1–2, branches 3–5, 5–6, and the isolated bus 4.

4.6.2 Topological Observability Analysis

Observability analysis aims to find at least n (number of system states) independent measurement equations to declare the system as observable. The linearized decoupled measurement equations can be expressed in terms of branch variables, that is, phase-angle differences between branch terminals, instead of the more common nodal variables, that is, bus phase angles. In such a formulation, it can be shown that there is a one-to-one correspondence between a set of n linearly independent columns in the measurement matrix equation and a network graph tree, formed based on assignment of measurements to branches. Measurements are assigned to branches according to the following rules [13–15]:

- If a branch flow is measured, the branch is assigned to its flow measurement.
- If an injection is measured at one of the terminal nodes of a branch, the branch will be assigned to that injection.
- Once a branch is assigned to a measurement, it cannot be assigned to any other measurement.

Topological observability method makes use of the measurement assignment rules to progressively build a network tree whose branches have distinct measurement assignments. The implementation of the method can take different forms due to the required heuristics for assigning injection measurements. Therefore, instead of presenting details of a single version of the implementation, the basic steps of the algorithm are outlined as follows:

1. First assign all the flow measurements to their respective branches.
2. Then assign injection measurements to respective branches in an arbitrary sequence. The aim of assigning an injection measurement to a branch is to enlarge the existing forest (several trees of subsystems) by merging existing trees through this newly assigned branch. If a new injection cannot accomplish this, it will be declared as a redundant measurement and will thus be disregarded.

After processing all the flows and injections, if a spanning tree (a tree containing all the system buses) cannot be found, then all the observable islands need to be identified. This is done iteratively as follows:

1. Discard those injections that have at least one incident branch, which does not form a loop with the branches of the already defined forest.
2. Update the forest accordingly and repeat step 1 until no more injections need to be removed.

The next example is used to illustrate the main steps involved in the implementation of the topological method.

EXAMPLE 4.5

Once again consider the network and measurement system of Figure 4.3 and proceed as follows:

- Assign measurements to branches: $1 \Rightarrow 1\text{--}2$, $2 \Rightarrow 3\text{--}5$, $3 \Rightarrow 5\text{--}6$, $4 \Rightarrow 4\text{--}6$.
- Branches 1–4 and 2–4 are those that do not form loops with the branches of the existing forest, which contains branches 3–5, 5–6, and 1–2. Therefore, the injection at bus 4 is discarded.
- The resulting observable islands are the same as those identified before (Figure 4.4).

4.7 ALTERNATIVE SOLUTION TECHNIQUES

Solving the WLS SE through the normal equations provides generally a satisfactory solution. However, in certain circumstances the use of normal equations may give rise to some numerical problems mainly related with the ill-conditioning of the gain matrix [16]. The ill-conditioning of a matrix is a measure of how close it is to a singular matrix, and it is quantified by means of the so-called condition number, κ. This number is unity for the identity matrix and tends to infinity for quasi-singular matrices. In practice, the condition number represents the extent to which the errors associated with the system coefficients, and those arising by round-off during the solution process are amplified. In certain cases, the iterative process may even not converge if the condition number is very high and the convergence threshold is very exigent.

It can be shown that $\kappa(H^\mathrm{T}H) = \kappa(H)^2$, which explains the added risk of explicitly using matrix G. Furthermore, some specific causes of ill-conditioning in the SE problem are as follows:

- Simultaneously using very large weights for virtual measurements and very low weights for pseudomeasurements.
- Very short and long lines simultaneously incident at the same bus.
- Large percentage of injection measurements.

EXAMPLE 4.6

The network in Example 4.1 will be modified by setting bus 2 as a null-injection node and increasing the load at bus 3 with the load removed from bus 2. The new measurements are

Index, i	Type	Value (pu)	W_{ii} (pu)
1	p_{12}	0.6390	15,625
2	p_{13}	1.4150	15,625
3	p_2	0	10^7
4	q_{12}	0.4247	15,625
5	q_{13}	0.8780	15,625
6	q_2	0	10^7
7	V_1	1.0065	62,500
8	V_2	0.9795	62,500

When the weight of the null injection at bus 2 is 1000 times larger than that of regular injection measurements, the condition number of the gain matrix is 8.056×10^5, against 1.525×10^3 for the matrix of Example 4.2, where only conventional measurements are used. After three iterations, the process converges to the following state:

Node	$\hat{V}$ (pu)	$\hat{\theta}$ (°)
1	0.9999	0.0
2	0.9860	−1.0498
3	0.9507	−3.9952

The above state leads to estimated powers injected at bus 2 being in the order of 10^{-4} pu. Using larger weights to model virtual measurements would lead to power injections that are closer to zero, at the cost of further deteriorating the gain matrix numerical conditioning.

Normalizing the coefficients W_{ii}, like in Example 4.2, has no effect on the condition number of G.

In this section, several alternative techniques are presented aiming at preventing the explicit use of G and somehow circumventing the other ill-conditioning sources described previously.

4.7.1 Orthogonal Factorization

To present this approach, it is preferable to scale the matrices and vectors in the normal Equation 4.5 so that the weighting matrix does not explicitly appear, which leads to

$$\underbrace{\tilde{H}^T \tilde{H}}_{G} \Delta x = \tilde{H}^T \Delta \tilde{z} \qquad (4.8)$$

where

$$\tilde{H} = W^{1/2} H \qquad (4.9)$$

$$\Delta \tilde{z} = W^{1/2} \Delta z \qquad (4.10)$$

The method is based on the orthogonal factorization of matrix $\tilde{H}$,

$$\tilde{H} = QR \qquad (4.11)$$

where Q is an $m \times m$ orthogonal matrix ($Q^T = Q^{-1}$) and R is an $m \times n$ trapezoidal matrix, that is,

$$R = \begin{pmatrix} U \\ 0 \end{pmatrix}$$

matrix U being, in the absence of round-off errors, the triangular matrix arising in the LU factorization of G.

Equation 4.11 is equivalent to

$$Q^T \tilde{H} = R$$

which is more amenable for the implementation of orthogonal factorization algorithms, such as Givens rotations. Those algorithms successively transform $\tilde{H}$ into R in a row-wise or column-wise manner.

Then, the normal equations can be transformed as follows:

$$\tilde{H}^T Q Q^T \tilde{H} \Delta x = \tilde{H}^T \Delta \tilde{z}$$

$$R^T R \Delta x = R^T Q^T \Delta \tilde{z}$$

$$U^T U \Delta x = U^T Q_n^T \Delta \tilde{z}$$

where Q_n represents the first n columns of Q. Furthermore, as U is a nonsingular matrix, the last equations leads to

$$U \Delta x = Q_n^T \Delta \tilde{z} \tag{4.12}$$

Computing Δx at each iteration by means of orthogonal transformations consists of the following steps [10,17,18]:

1. Perform the orthogonal factorization of $\tilde{H} = QR$.
2. Compute the independent vector $\Delta z_q = Q_n^T \Delta \tilde{z}$.
3. Obtain Δx by back substitution on $U \Delta x = \Delta z_q$.

As can be seen, there is no need to form the gain matrix G, whose condition number is much higher than that of H. On the other hand, owing to the way pivots are computed, the orthogonal factorization is numerically much more robust than the LU counterpart, which is particularly important when solving ill-conditioned systems. This explains why the simultaneous presence of very large and small measurement weights is better tolerated when this approach is adopted.

The price paid lies in the need to obtain the orthogonal matrix Q, which involves a notable computational cost. In practice, however, Q is not explicitly computed, but is expressed as the product of a number of elementary factors, each one provided by a Givens rotation.

The most popular implementation in commercial estimators is a hybrid method that avoids the use of Q. This technique consists of the following steps [19]:

1. Obtain U through orthogonal transformations of $\tilde{H}$. In this process, there is no need to store the components of Q.
2. Compute the independent vector $\Delta z_h = \tilde{H}^T \Delta \tilde{z}$.
3. Obtain Δx by solving the system $U^T U \Delta x = \Delta z_h$.

The technique is hybrid in the sense that normal equations are solved at step 3, but U is obtained by transforming H rather than G.

4.7.2 EQUALITY-CONSTRAINED SOLUTION

A more robust approach to modeling virtual measurements is by treating them as equality constraints of the WLS optimization problem, rather than as very accurate measurements. This leads to the following problem [20]:

$$\text{Minimize} \quad J(x) = \frac{1}{2}[z - h(x)]^T W[z - h(x)]$$

$$\text{Subject to} \quad c(x) = 0 \tag{4.13}$$

where $c(x) = 0$ represents the nonlinear constraints. This way, very large weights are eliminated from matrix W.

To solve this problem, the Lagrangian function is built,

$$\mathcal{L} = J(x) - \lambda^T c(x) \tag{4.14}$$

and the corresponding first-order optimality conditions are obtained,

$$\begin{aligned}
\partial \mathcal{L}(x)\partial x = 0 &\quad\Rightarrow\quad H^T W[z - h(x)] + C^T \lambda = 0 \\
\partial \mathcal{L}(x)\partial \lambda = 0 &\quad\Rightarrow\quad c(x) = 0
\end{aligned} \tag{4.15}$$

where C is the Jacobian of $c(x)$. The resulting nonlinear system is solved iteratively by solving the following linear system:

$$\begin{bmatrix} H^T W H & C^T \\ C & 0 \end{bmatrix} \begin{bmatrix} \Delta x \\ -\lambda \end{bmatrix} = \begin{bmatrix} H^T W \Delta z^k \\ -c(x^k) \end{bmatrix} \tag{4.16}$$

In this manner, even though G is computed, one of the main sources of ill-conditioning for this matrix vanishes. The price paid lies in the fact that the coefficient matrix in Equation 4.16 is no longer positive definite, which calls for more complex LU factorization schemes, capable of performing row pivoting to simultaneously preserve both numerical stability and sparsity of a matrix that becomes unsymmetrical.

Before ending this section, it is worth pointing out the relationship between the equality-constrained formulation and the classical one considering virtual measurements as very accurate measurements. This is obtained easily by writing $J(x)$ in such a way that both types of measurements, real and virtual, appear separately,

$$J(x) = \frac{1}{2}[z - h(x)]^T W[z - h(x)] + \rho c(x)^T c(x) \tag{4.17}$$

where ρ is a weighting coefficient much larger than the elements of W. Comparing the above expression with the Lagrangian function (Equation 4.14), the following relationship is obtained in the optimal solution,

$$\lambda = -\rho \, c(\hat{x})$$

That is, as the weighting factor ρ tends to infinity, the residuals of the virtual measurements tend to zero, but their product is a finite value given by the Lagrange multipliers. Indeed, the use of the Lagrange multipliers significantly enhances the numerical conditioning of the problem.

4.7.3 HACHTEL'S MATRIX-BASED SOLUTION

The basic idea behind this technique, also called the "tableau" method, is to retain the residuals as explicit variables of the problem [21]. This leads to the following optimization problem:

$$\text{Minimize} \quad J(x) = \frac{1}{2} r^{\mathrm{T}} W r$$
$$\text{Subject to} \quad c(x) = 0$$
$$r - z + h(x) = 0$$

(4.18)

The resulting Lagrangian function is

$$\mathcal{L} = J(x) - \lambda^{\mathrm{T}} c(x) - \mu^{\mathrm{T}} [r - z + h(x)]$$

(4.19)

and the associated first-order optimality conditions

$$
\begin{aligned}
\partial \mathcal{L}(x)/\partial x = 0 &\Rightarrow C^{\mathrm{T}} \lambda + H^{\mathrm{T}} \mu = 0 \\
\partial \mathcal{L}(x)/\partial \lambda = 0 &\Rightarrow c(x) = 0 \\
\partial \mathcal{L}(x)/\partial r = 0 &\Rightarrow W r - \mu = 0 \\
\partial \mathcal{L}(x)/\partial \mu = 0 &\Rightarrow r - z + h(x) = 0
\end{aligned}
$$

(4.20)

The third equation above allows r to be eliminated ($r = R\mu$), leading the remaining three equations to an iterative process in which the following linear system is solved:

$$
\begin{bmatrix}
0 & 0 & C \\
0 & R & H \\
C^{\mathrm{T}} & H^{\mathrm{T}} & 0
\end{bmatrix}
\begin{bmatrix}
\lambda \\
\mu \\
\Delta x
\end{bmatrix}
=
\begin{bmatrix}
-c(x^k) \\
\Delta z^k \\
0
\end{bmatrix}
$$

(4.21)

where the coefficient matrix is termed the Hachtel's augmented matrix.

Obviously, if the multiplier vector μ is eliminated, the system Equation 4.16 containing the product $H^{\mathrm{T}} W H$ is obtained. Even though the LU factorization of both coefficient matrices is of similar complexity, the condition number of the Hachtel's matrix is quite smaller. The fact that this matrix is of larger dimension does not necessarily mean more computations if sparsity is preserved, as it contains many more null elements.

Furthermore, it is possible to implement the factorization of Hachtel's matrix without losing its initial symmetry. For this purpose, symmetrical permutations need to be performed, either beforehand or on the fly, so that whenever a very small or null diagonal pivot appears, it can be grouped with the next diagonal element to form a 2×2 invertible matrix. This way, the LU factorization will resort to either scalar arithmetic, if the current pivot is not very small, or 2×2 block arithmetic otherwise.

EXAMPLE 4.7

Let us again consider the case of Example 4.6, where the null injections at bus 2 will be modeled as equality constraints. Normalizing the elements W_{ii} so that the largest one is unity, the coefficient matrix in Equation 4.16 for the first iteration is

$$
F = \begin{bmatrix}
275.0275 & & 0.0000 & 0.0000 & & 45.3795 & -4.5380 \\
& 99.0099 & 0.0000 & & 0.0000 & -12.3762 & 1.2376 \\
0.0000 & 0.0000 & 375.0374 & -275.0275 & -99.0099 & 3.3003 & -33.0033 \\
0.0000 & & -275.0275 & 276.0275 & & 4.5380 & 45.3795 \\
& 0.0000 & -99.0099 & & 99.0099 & -1.2376 & -12.3762 \\
\hline
45.3795 & -12.3762 & -3.3003 & 4.5380 & -1.2376 & & \\
-4.5380 & 1.2376 & -33.0033 & 45.3795 & -12.3762 & &
\end{bmatrix}
$$

its condition number being 932. On the other hand, the condition number of the Hachtel's matrix for the same case would be just 165. Note that the gain matrix of Example 4.6 is much more ill-conditioned than both matrices in this example.

In this case, normalizing the coefficients W_{ii} with the largest one is helpful to reduce the condition number of the coefficient matrices involved in the solution, as the scaling of the columns of $H^T W H$ is improved with respect to those of C^T. Of course, this scaling affects also, in the same proportion, the objective function and the Lagrange multipliers, but not the estimate.

The estimate is very close to the one obtained when virtual measurements are treated as very accurate measurements, but the powers injected at bus 2 are closer to zero.

4.8 BAD DATA DETECTION AND IDENTIFICATION

The accuracy and reliability of the state estimator depends on the measurements received at the control center. Measurements are taken by various types of meters and then sent via telecommunication channels to a central location for further processing. Meters may be analog or digital and may contain metering errors due to their accuracy limits. Measurement signals may also become noisy due to imperfect telecommunication channels. Occasionally, following a substation maintenance or equipment repair, meter connections may be inadvertently changed and not reported to the control center. While not very common, all such incidents do occur during the daily operation of power systems and result in corruption of the measurement set by bad data. Unless these bad measurements are detected and identified, the estimated state of the system based on such measurement sets corrupted with bad data will yield incorrect results.

All state estimators have a preprocessor that checks obvious inconsistencies and errors in measurements. These may be voltage and current magnitude measurements whose values cannot be negative, or real power injection measurements at a load bus whose values must be negative, and so on. However, this rough filtering cannot pick up all errors particularly in weakly measured substations.

The WLS estimator uses postestimation procedures to detect and identify bad data in the measurement set. Other types of estimators that will be discussed briefly in Section 4.9 attempt to eliminate bad data as part of the estimation procedure. When using the WLS estimation method, bad data detection and identification are done after the estimation process, by processing the measurement residuals [9,22–24]. Methods of detecting and identifying bad data take advantage of the statistical properties of the residuals. These properties will first be reviewed below as a precursor to the discussion of these methods.

4.8.1 PROPERTIES OF MEASUREMENT RESIDUALS

Consider the state estimate in the linearized measurement model

$$
\Delta \hat{x} = (H^T R^{-1} H)^{-1} H^T R^{-1} \Delta z
$$

and the estimated value of Δz

$$
\Delta \hat{z} = H \Delta \hat{x} = K \Delta z
$$

where $K = H(H^T R^{-1} H)^{-1} H^T R^{-1}$.

Measurement residuals can be expressed in terms of measurement errors as follows:

$$
\begin{aligned}
r &= \Delta z - \Delta \hat{z} \\
&= (I - K)\Delta z \\
&= (I - K)(H\Delta x + e) \quad [\text{Note that } KH = H] \\
&= (I - K)e \\
&= Se
\end{aligned}
$$

Here, the matrix S is called the *residual sensitivity matrix*.

Assuming that the covariance matrix for the measurement errors R is known, then the residual covariance matrix Ω can be written as

$$
\begin{aligned}
E[rr^T] = \Omega &= S \cdot E[ee^T] \cdot S^T \\
&= SRS^T \\
&= SR
\end{aligned}
$$

Measurement error distribution is commonly assumed to be normal (Gaussian) with zero mean and a diagonal covariance matrix R and denoted by

$$
e \sim N(0, R)
$$

Thus, the distribution of the measurement residuals will also be normal (Gaussian) with following parameters:

$$
r \sim N(0, \Omega) \tag{4.22}
$$

4.8.2 Classification of Measurements

Bad data detection and identification is possible due to the redundancy in the measurement set. When such redundancy is lost, then detection and identification will not be possible irrespective of the method used. Measurements can belong to either one of two mutually exclusive sets, *critical* and *redundant* (or *noncritical*), according to the following properties:

- A measurement belongs to the critical set, if its elimination from the measurement set will result in an unobservable system. If a measurement is not critical, it will belong to the redundant set.
- The residuals of critical measurements will always be zero, and therefore errors in critical measurements can be neither detected nor identified.

4.8.3 Detecting Bad Data

Residuals can be used to decide whether or not there is sufficient reason to suspect bad data in the measurement set. We will review two tests that are commonly used for detecting bad data—the χ^2 test and the *largest normalized residual (r^N) test*. While the former is a test strictly for detection, the latter can be used for both detection and subsequent identification of bad data.

4.8.3.1 χ^2-Test for Detecting Bad Data

In statistics, the sum of squares of independent random variables, each distributed according to the standard normal distribution is known to have a χ^2-distribution. Consider the objective function of the WLS SE problem given by Equation 4.3. Since the measurement residuals are assumed to have a normal distribution with above derived parameters, the objective function can be shown to have a χ^2-distribution with $m - n$ degrees of freedom, n and m being the number of state variables and measurements, respectively.

Bad data detection by the χ^2-test will then proceed as follows:

- Solve the WLS estimation problem and compute the objective function

$$J(\hat{x}) = \sum_{i=1}^{m} \frac{[z_i - h_i(\hat{x})]^2}{\sigma_i^2}$$

where $\hat{x}$ is the estimated state vector of dimension n, $h_i(\hat{x})$ is the estimated measurement i, z_i is the measured value of the measurement i, σ_i^2 is the variance of the measurement i, and m is the number of measurements.

- Look up the value corresponding to p (e.g., 95%) probability and $(m - n)$ degrees of freedom, from the χ^2 distribution table. Let this value be $\chi^2_{(m-n),p}$, where $p = \Pr[J(\hat{x}) \leq \chi^2_{(m-n),p}]$.
- Test if $J(\hat{x}) \geq \chi^2_{(m-n),p}$. If yes, then bad data are detected. Else, the measurements are not suspected to contain bad data.

EXAMPLE 4.8

Consider the 3-bus system shown in Figure 4.5. The number of state variables is five, slack bus phase angle being excluded from the state list. Assume that there are altogether ten measurements, that is, two voltage magnitude measurements, two pairs of real/reactive flows and two pairs of real/reactive injections. Then the degrees of freedom will be

$$m - n = 10 - 5 = 5$$

Using .95 probability, the value looked up from the χ^2 table will be $\chi^2_{5,.95} = 11.07$.

Therefore, if the objective function $J(\hat{x})$ evaluated at the WLS estimate $\hat{x}$ is larger than 11.07, bad data will be suspected in the measurement set.

4.8.4 IDENTIFYING BAD DATA

Bad data identification refers to the selection of the erroneous measurements from the measurement set once bad data is suspected using a bad data detection method. Two methods of bad data identification that will be reviewed here are as follows:

- Normalized residual r^N test, which is easy to implement and performs satisfactorily unless there are multiple conforming bad data in the measurement set.
- Hypothesis testing identification (HTI) method, which is more complicated to implement, but is more reliable compared with the r^N test.

These two methods are discussed in detail in the following sections.

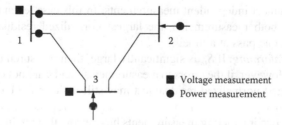

1   2   3

■ Voltage measurement
● Power measurement

FIGURE 4.5 Three-bus system to illustrate the χ^2-test.

4.8.4.1 Largest Normalized Residual (r^N) Test

Measurement residuals are assumed to have a normal distribution as derived earlier in Equation 4.22. Normalizing the residual for measurement i will yield

$$r_i^N = \frac{r_i}{\sqrt{\Omega_{ii}}} = \frac{r_i}{\sqrt{R_{ii}S_{ii}}}$$

Normalized residuals r_i^N will have a standard normal distribution with mean 0 and variance 1. It can be shown that if there is a single bad data in the measurement set (provided that it is not a critical measurement), the largest normalized residual will correspond to bad data. This also applies to multiple bad data cases as long as none of the bad measurements interact. The following are the steps of the largest normalized residual test for detection and identification of single and noninteracting multiple bad data:

1. Solve the WLS estimation and obtain the elements of the measurement residual vector:

$$r_i = z_i - h_i(\hat{x}) \quad i = 1, \ldots, m$$

2. Compute the normalized residuals:

$$r_i^N = \frac{|r_i|}{\sqrt{\Omega_{ii}}}, \quad i = 1, \ldots, m$$

3. Find k such that r_k^N is the largest among all r_i^N, $i = 1, \ldots, m$.
4. If $r_k^N > c$, then the kth measurement will be suspected as bad data. Else, stop, no bad data will be suspected. Here, c is an appropriately chosen identification threshold, for example, 3.0.
5. Eliminate the kth measurement from the measurement set and go to step 1.

4.8.4.2 Types of Bad Data

Bad data may appear in single or multiple measurements. In the case of multiple bad data, there may be three possible situations for each of which the r^N test performs differently. These cases are briefly discussed below.

4.8.4.2.1 Single Bad Data

When there is single bad data, the largest normalized residual will correspond to the bad measurement, provided that it is not critical. Therefore, the r^N test will successfully detect and identify bad data.

4.8.4.2.2 Multiple Bad Data

Multiple bad data may appear in three different ways:

- *Noninteracting*: If $S_{ik} \approx 0$, then measurements i and k are said to be noninteracting. This implies that there is little correlation between the two measurement errors and they can be considered as almost independent measurements. In this case, even if bad data appears simultaneously in both measurements, the largest normalized residual test can identify them sequentially, one pass at a time.
- *Interacting, nonconforming*: If S_{ik} is significantly large, then measurements i and k are said to be interacting. However, if the errors in measurements i and k are not consistent with each other, then the largest normalized residual test may still indicate the bad data correctly.
- *Interacting, conforming*: As in the above case, in this case too the S_{ik} is significantly large. Furthermore if the two interacting measurements have errors that are in agreement, then the largest normalized residual test may fail to identify either one.

EXAMPLE 4.9

In the 3-bus system of Figure 4.6, all branches have identical reactances, $x_i = j\,0.1$, and all measurements have the same error variance $\sigma_i = 0.01$. Single, multiple interacting nonconforming and multiple interacting conforming bad data are introduced into the measurements and the results are given in the following table. Note that all measurements are assumed to be zero except for the bad data. The largest normalized residual for each case is typed in bold in the following table. The largest r^N test successfully identifies the bad data in the single and multiple but nonconforming bad data cases. However, it fails and incorrectly selects a good measurement (flow 2–3) as the bad data in the last case where measurement errors are conforming.

Measurement Type	Single		Multiple Interacting			
			Nonconforming		Conforming	
	z_i	r_i^N	z_i	r_i^N	z_i	r_i^N
Flow 1–3	0	7.1	0	17.6	0	31.7
Flow 2–1	0	18.7	0	11.2	0	26.2
Flow 3–2	1	**88.0**	−1	**120.9**	1	55.0
Flow 2–3	0	25.6	0	7.3	0	**58.6**
Inj. 1	0	15.5	0	41.5	0	10.4
Inj. 3	0	41.7	1	111.3	1	27.9

4.8.4.3 Hypothesis Testing Identification (HTI)

This method differs from the largest r^N test in that it computes the estimates of measurement errors directly. Multiple bad data may be identified by using this method even when they are interacting and conforming, because, unlike the residuals, the estimated measurement errors are independent of one another. This method's effectiveness however depends upon the choice of an initial suspect measurement set, which should include all bad data. This choice, being based on the normalized residuals, constitutes the main shortcoming of this method. The method's essential features are summarized below.

A set of linearly independent and noncritical measurements are designated as the suspect set. The number of such measurements to be included in the suspect set is up to the user. The rest of the measurements are assumed to be error free. Then the sensitivity matrix S and the error covariance matrix R are partitioned according to the suspect and true measurements

$$r_s = S_{ss}e_s + S_{st}e_t$$
$$r_t = S_{ts}e_s + S_{tt}e_t$$

$$R = \begin{bmatrix} R_s & 0 \\ 0 & R_t \end{bmatrix}$$

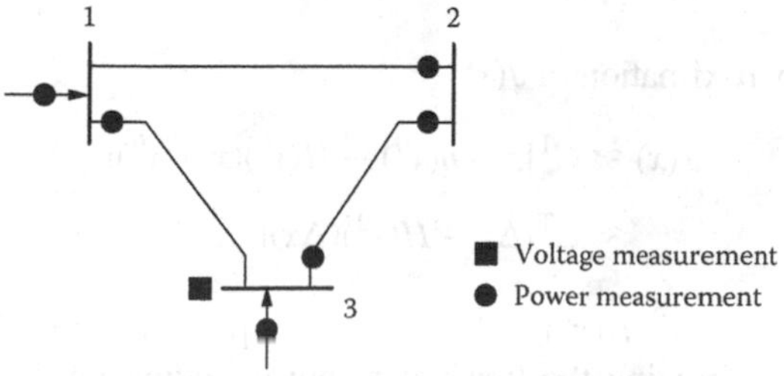

FIGURE 4.6 Three-bus system to illustrate the normalized residual test.

Treating the last term of the first equation as noise, an estimate for e_s can be obtained as

$$\hat{e}_s = S_{ss}^{-1} r_s$$

The HTI method essentially starts with a rather sizable suspect set of measurements and progressively eliminates measurements from this suspect set based on their estimated errors $\hat{e}_{s,i}$. Details of how the identification thresholds are adjusted for the decision of eliminating measurements from the initial suspect set are explained in [25] and [26]. The method's main advantage is that it can identify interacting conforming multiple bad data provided that bad data are initially placed in the suspect set. This initial placement however is still based on normalized or weighted residuals and therefore is vulnerable to effective masking of bad data due to strongly interacting measurements.

4.9   NONQUADRATIC STATE ESTIMATORS

The SE is commonly formulated as an unconstrained optimization problem as given in Equation 4.4. If the objective function $J(x)$ is chosen as in Equation 4.3, then the solution of the optimization problem will yield the WLS estimate for the system state. This estimate has several statistically desirable properties provided that the measurement errors have an approximately normal (Gaussian) distribution. However, once this assumption is violated, as in the case of a bad measurement, topology error, or the like, then the WLS estimate may become significantly biased. This was recognized early on by Schweppe et al. and alternative formulations that use nonquadratic objective functions have been proposed. In general the objective function can be written as a function of the residuals as follows:

$$J(x) = \rho(r) \tag{4.23}$$

where $r = z - \hat{z}$ is the residual vector and $\rho(r)$ is a real valued function. In case of WLS estimation, this function is defined as

$$\rho(r) = r^T W r \tag{4.24}$$

Choice of the function ρ affects the statistical robustness of the resulting estimator. Here, robustness is defined as the degree of insensitivity of the estimator toward randomly occurring gross errors in the measurements. Formal treatment of robustness based on the influence functions can be found in [27] and [28]. One of the commonly used alternative objective functions is given by

$$\rho(r) = C^T |r| \tag{4.25}$$

where C is a vector of weights assigned to each measurement residual, r_i. This function represents the weighted sum of the absolute values for measurement residuals. Using this as the objective function, the optimization problem will become

$$\text{Minimize } J(x) = C^T |z - h(\hat{x})| \tag{4.26}$$

The solution $\hat{x}$ of the above problem is referred to as the weighted least absolute value (WLAV) state estimate.

Using the first-order approximation of $J(x)$

$$J(x) \approx C^T |z - h(x^0) - H(x^0)(x - x^0)| \tag{4.27}$$
$$\approx C^T |\Delta z - H(x^0)(\Delta x)|$$

where $\Delta z = z - h(x^0)$, $\Delta x = x - x^0$, $H(x^0) = \partial h(x^0)/\partial x$, the optimization problem (Equation 4.26) can be solved by iteratively minimizing the first-order approximation until the changes in x become insignificant.

Minimization of the approximate $J(x)$ in Equation 4.27 can be shown to be equivalent to solving the following linear programing (LP) problem [29–31]:

$$
\begin{aligned}
\text{Minimize} \quad & C^T \cdot (r_+ + r_-) \\
\text{Subject to} \quad & H(x^0) \cdot \Delta x_+ - H(x^0) \cdot \Delta x_- + r_+ - r_- = \Delta z \\
& \Delta x_+, \Delta x_-, r_+, r_- \qquad \geq 0
\end{aligned}
$$

where

$$
\Delta x_+(i) = \begin{cases} \Delta x(i) & \text{if } \Delta x(i) \geq 0 \\ 0 & \text{otherwise} \end{cases} \qquad \Delta x_-(i) = \begin{cases} \Delta x(i) & \text{if } \Delta x(i) \leq 0 \\ 0 & \text{otherwise} \end{cases}
$$

$$
C = [C_1, C_2, \ldots, C_m]^T,
$$

r_+, r_- are $m \times 1$ non-negative slack vectors, such that
$$\min[r_+(i), r_-(i)] = 0 \text{ and}$$
$$\max[r_+(i), r_-(i)] = |r_i|.$$

The solution of this LP problem interpolates n out of the m available measurements. Therefore, at the solution, exactly n of the measurement equations will thus be satisfied, with zero residuals. This property allows the WLAV estimator to automatically eliminate bad measurements, provided there is enough redundancy in the measurement set.

There are different methods for solving the optimization problem (Equation 4.26). Some of them are based on LP [31–35], while others make use of the interior point-based algorithms [36,37]. WLAV estimation is known to fail when there are leverage measurements in the system. Leverage measurements are those measurements that have disproportionately high influence on the state estimate due to their location and types [28,38]. Their influence can be neutralized by other nonquadratic estimation methods such as in [39] and [40].

EXAMPLE 4.10

Consider the 3-bus system and its measurement configuration given in Figure 4.7. All measurements are assigned equal weights for simplicity. There are 13 measurements, four pairs of flows, two pairs of injections, and one voltage magnitude. Gaussian errors are added to perfect measurements generated by a power flow program when generating the measurement set for this example. These values are shown in the second column of the table below, corresponding to each measurement.

Initially, the WLAV estimate for the system state is obtained using the generated measurement set as described above. Then, a gross error is introduced in the real power measurement 11 by changing its sign and making it appear as 0.4 instead of −0.4 pu. The WLAV estimator is run again and the resulting state estimate and measurement residuals are given in the table below.

Note that, the WLAV estimator always interpolates through five measurements out of the existing 13. In the first case, measurements 2, 7, 9, 12, and 13 are chosen. In the second case, a different set of measurements, namely measurements 1, 5, 8, 10, and 13, is picked by the estimator. The estimator rejected the bad measurement 11 in the second case, with a very large residual, which actually reflects the difference between the measured (bad) and estimated (optimal) values. Also, in both cases, the state estimates are very close to one another, demonstrating the ability of WLAV estimator to remain robust against bad data in this example.

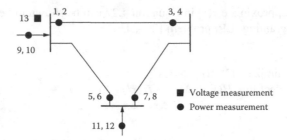

FIGURE 4.7 Three-bus system and its measurement configuration.

Measurement	Measured	Residuals	
No.	Value	No Errors	With Bad Data
1	0.98422	0.00035	0
2	0.69696	0	0.00003
3	−0.98422	0.00052	0.00075
4	−0.55152	0.00160	0.00141
5	−0.71580	0.00064	0
6	0.02565	0.00064	0.00100
7	0.31578	0	−0.00039
8	0.65038	0.00050	0
9	1.69989	0	0.00094
10	0.72262	0.00002	0
11	−0.40000	−0.00001	0.79859
12	0.67603	0	−0.00009
13	1.00000	0	0
Bus		**Estimated Voltage**	
No.		No Errors	With Bad Data
1		$1.0008\angle 0$	$1.00077\angle 0$
2		$0.9363\angle -6.03$	$0.9362\angle -6.03$
3		$1.0008\angle -4.1$	$1.0008\angle -4.09$

4.10 INCLUSION OF SYNCHRONIZED PHASOR MEASUREMENTS IN SE

In the previous sections the set of available measurements is assumed to contain the so-called SCADA measurements only. SCADA measurements are sampled every few seconds and communicated to the control centers by a front-end computer, sometimes referred to as the remote terminal unit (RTU). These measurements are not time synchronized and typically carry a small but nonzero time skew. Unlike SCADA measurements, those measurements that are provided by the recently introduced phasor measurement units (PMU) are time synchronized and are sampled at much faster rates in the order of 30–60 times a second. Incorporating these new types of measurements into SE presents some challenges as well as advantages which will be discussed in the next sections.

The main distinguishing feature of phasor measurements is the fact that they are synchronized with respect to the time reference provided by the global positioning system (GPS) satellites. Using this accurate global time reference, voltage and current samples that are concurrently recorded at various substations are assigned identical time stamps which are then aligned to create a coherent picture of the system state at a given point in time. It should also be noted that such time synchronization also eliminates the need to use an arbitrarily selected bus as the reference for the voltage phase angles as done when using SCADA measurements.

4.10.1 Synchronized Phasor Measurements

The measuring devices that can compute synchronized voltage and current phasors from time-stamped samples were initially developed by Phadke et al. [41,42]. The first implementation of GPS-synchronized phase-angle measurements in an industrial power system SE was presented in [43]. A PMU is a digital device providing synchronized voltage and current phasor measurements, referred to as synchrophasors [44]. Initial PMUs were designed as stand-alone units whose primary and only function was computation of synchrophasors. Gradually various types of intelligent electronic devices (IEDs) such as digital relays were retrofitted to function as PMUs in addition to their primary functions. Primary three-phase voltages and currents are converted to appropriate analog inputs by instrument transformers and anti-aliasing filtering. Each analog signal is digitized by an A/D converter with a sampling rate typically between 12 to 128 samples per cycle of the nominal power frequency. The sampling clock is phase-locked with the GPS clock pulse which provides the Universal Time Coordinated (UTC) time reference used to time-tag the outputs. Each PMU may have multiple input terminals that are used to connect three-phase voltage signal at the substation bus and three-phase current signals on one or more incident lines at the substation.

Phasors corresponding to the phase voltages and currents are computed from sampled data by PMU's microprocessor using one of several alternative signal processing techniques. The most commonly used technique relies on the Discrete Fourier Transform (DFT) [45]. The method uses a moving data window of length equal to one or more fundamental cycles. As the window moves forward in time, the oldest sample is dropped while the most recent sample is added yielding a computationally efficient recursive DFT algorithm. While the sampling clocks are commonly kept constant at a multiple of the fundamental frequency f_0, as frequency deviates by a small amount away from its nominal value, the resulting leakage errors introduced in phasor estimates can be readily compensated with high accuracy by a post-processing filter. It can be also shown that the computed phasor rotates in the complex plane with an angular velocity equal to the difference between f_0 and the actual frequency so that the frequency and the rate of change of frequency estimates can be obtained by the first and the second derivatives of the phasor angle respectively.

Among other alternatives, there are methods which replace the DFT by a wavelet transform [46], as well as others based on different nonlinear estimation techniques (nonlinear WLS, Kalman filtering, neural networks, etc.) [47–49]. These latter ones formulate the problem as a parameter estimation problem where the phase signal is considered to be a nonlinear function of the relevant variables namely the amplitude, phase angle, frequency, and rate of change of frequency to be estimated from the waveform samples. Once the phasors are obtained for individual phases, they are converted into symmetrical components, that is, positive, negative and zero sequence phasors. In addition, frequency and rate of change of frequency of the signals are also estimated. A digital communication network is used to transmit these phasor values to a coordination facility at rates of 10 to 60 frames per second. It should be mentioned that several PMUs also have capability to locally store the computed phasor values for other applications. However, SE function is executed at a central location and therefore phasor measurements need to be appropriately transmitted. This is accomplished via the help of Phasor Data Concentrators (PDCs) which collect phasor measurements from individual PMUs and align them in time based on their time stamps. As a result, measurements are streamed to the state estimator in a synchronized fashion. The actual architecture of the communication system may vary widely and may involve one or more levels of hierarchy where local PDCs report to higher level super PDCs, and so on.

Figure 4.8 shows the basic relation between a sinusoidal waveform with nominal frequency f_0 and its polar and rectangular coordinate phasor representation. The phase angle is defined as the angular difference between the time where the waveform peaks and the reference time $t = 0$. In the case of

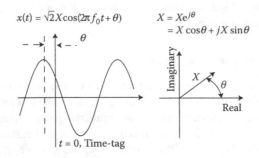

FIGURE 4.8 Sinusoidal waveform and its phasor representation.

synchrophasors this reference time corresponds to the time-tag assigned according to the GPS clock. If the recorded waveform deviates from a pure sinusoidal function, the computed phasor will represent its fundamental frequency component.

While PMU data are typically time tagged with precision better than 1 microsecond and magnitude accuracy that is better than 0.1%, the accuracy of PMU measurements often fail to reach such accuracy levels due to various errors in the instrumentation channels as well as system imbalances. It is crucial for PMU manufacturers to define and choose various PMU device specifications such as the window length, sampling rate and type of phasor estimation algorithm, the phasor estimate reporting rate, the communication protocol in order to achieve the required measurement accuracy. Furthermore, given the increasing number of manufacturers of this device, interoperability is a key concern. IEEE has done a lot of work towards developing a common standard for PMUs which is documented in IEEE C37.118-2011 standard publication [50]. It provides definitions of synchrophasors, frequency, and rate of change of frequency measurement under all operating conditions. It also specifies measurement evaluation techniques, time-tag and synchronization requirements and requirements for compliance with the standard under not only steady-state but also dynamic conditions. The standard defines a PMU either as a stand-alone physical unit or as a functional unit within another physical unit. It should be noted that this standard does not specify a technique or algorithm to compute phasors, frequency, or rate of change of frequency. Most recent amendment to this standard was published in 2014 [51].

4.10.2 Problem Formulation and Practical Issues

Utilization of PMU measurements will impact SE in different ways. On the one hand, since the number of PMU devices installed in existing power systems is not yet sufficient to carry out SE exclusively based on PMU measurements, SE formulation and solution remains nonlinear and iterative respectively. It is also shown that PMU measurements can be incorporated into SE via a linear recursive estimation step following the solution based on only SCADA measurements [52]. In the absence of phase-angle measurements, SE problem is formulated by choosing an arbitrary bus as the phase reference which is commonly assumed to be zero. This practice can be abandoned if phase-angle measurements are made available. Furthermore, keeping an arbitrarily chosen angle reference will create inconsistencies. Therefore, it is best to eliminate the reference bus [53] altogether and formulate the SE problem without a reference bus. This way, errors in any phasor measurement can be detected and identified provided there is enough measurement redundancy.

The PMUs not only provide synchronized voltage phasors but also current phasors as well. While having these current phasors adds to the measurement redundancy and consequently improves estimation variance, they also create certain numerical problems in particular if they are expressed in polar coordinates. SE solution algorithm initialization may become difficult when current phasor measurements are present. Entries of the measurement Jacobian corresponding to certain current phasors may become undefined when they are evaluated at flat start. This issue can be addressed in a variety of ways. A simple alternative is to exclude current phasor measurements in the first iteration and incorporate them back after the second iteration. However, this approach will fail if one of the current phasor measurements is critical, that is, excluding it will make the system unobservable. A slightly revised formulation using rectangular coordinates will allow the use of phasor voltage and current measurements without any numerical difficulties. It will also allow detection and identification of errors in these phasor measurements. It is earlier noted that the formulation of SE problem when using PMU measurements does not require a reference bus. Thus, the measurement Jacobian is formed using N columns for the phase angles and N columns for the voltage magnitude for an N-bus system.

Let us consider the standard measurement model (Equation 4.1), repeated here for convenience:

$$z = h(x) + e \tag{4.28}$$

Its first-order Taylor approximation can be written as:

$$\Delta z = H \Delta z + e \tag{4.29}$$

where
$\Delta x = x_k - x$
$\Delta z = z - h(x_k)$
x_k is the system state vector evaluated at the k-th iteration

$$H = \frac{\partial h(x)}{\partial x} \tag{4.30}$$

If there exists at least one phasor measurement, the Jacobian H will have $2N$ columns. This will allow estimation of all phase angles corresponding to all bus voltages in the system. If there are no phasor measurements, then an artificial phase-angle measurement of zero will be introduced in order to make the column rank of H full. Hence, this formulation will work with or without phasor measurements.

As mentioned above, when using polar coordinates, the expressions for current phasor measurements may lead to numerical difficulties during initialization of iterative SE solution. The expressions used to evaluate the entries in H corresponding to the derivatives of current phasor measurements will have apparent power flows in their denominators which will be identically equal to zero at flat start for lines with small or no line charging susceptances. This will lead to undefined values in H. Hence, the alternative formulation based on rectangular coordinates as described below is preferred.

Let

$$\mathcal{V}_k = V_{rk} + j V_{xk} \quad \mathcal{I}_{ij} = I_{rij} + j I_{xij}$$

denote the voltage phasor at bus k and current phasor along branch $i - j$, both in rectangular coordinates. Disregarding measurement errors, phasor measurements can be linearly related to the system states as:

$$\mathcal{V}_k^m = V_{rk}^m + jV_{xk}^m$$
$$V_{rk}^m = V_{rk} \qquad (4.31)$$
$$V_{xk}^m = V_{xk}$$

$$\mathcal{I}_{ij}^m = I_{rij}^m + jI_{xij}^m$$
$$I_{rij}^m = G_{ij}(V_{ri} - V_{rj}) - (B_{ij} + B_{ii})V_{xi} + B_{ij}V_{xj} \qquad (4.32)$$
$$I_{xij}^m = G_{ij}(V_{xi} - V_{xj}) + (B_{ij} + B_{ii})V_{ri} - B_{ij}V_{rj}$$

where
   $G_{ij} + jB_{ij}$: Series admittance of branch $i - j$
   B_{ii}: Shunt admittance at bus i

Consider a system with N buses and m phasor measurements. The corresponding measurement equations will take the following linear form [42]:

$$z = Hz + e \qquad (4.33)$$

where
   H is a constant matrix of dimension $(2m \times 2N)$, that is, it is not a function of system state
   $z^T = [z_{r1}, z_{r2}, \dots z_{rm} | z_{x1}, z_{x2}, \dots z_{xm}]$ is the measurement vector
   $x^T = [V_{r1}, V_{r2}, \dots V_{rm} | V_{x1}, V_{x2}, \dots V_{xm}]$ is the state vector
   e is the $(2m \times 1)$ measurement error vector

Then, the WLS estimator $\hat{X}$ for the system state will be given by:

$$\hat{x} = (H^T R^{-1} H)^{-1} H^T R^{-1} z \qquad (4.34)$$

where $R = E(e \cdot e^T)$ is the diagonal covariance matrix of measurement errors.

4.10.3   Phasor Only Linear WLS Estimator

As indicated earlier, when using only voltage and current phasor measurements, WLS SE problem formulation becomes linear. However, there may be additional computational benefits if the measurement set contains branch current and bus voltage phasors as the only types of measurements. Consider a power system which is fully measured, that is, every bus has a voltage phasor measurement and every branch is measured from one end by a current phasor measurement. The measurement vector for such a system will be given by:

$$z = \begin{bmatrix} \mathcal{V}^m \\ \mathcal{I}^m \end{bmatrix} = \begin{bmatrix} U \\ Y_b \cdot A \end{bmatrix} [\mathcal{V}] + e \qquad (4.35)$$

where
   U is the identity matrix
   $Y_b = [g + jb]$ is the complex network branch admittance matrix
   $\mathcal{V}$ is the complex vector of bus voltage phasors
   A is the branch to bus incidence matrix

In the above formulation, all shunt connections at system buses are intentionally ignored, but this assumption will be subsequently relaxed. Equation 4.35 involves complex numbers and can be easily converted into real form by substituting expanded real expressions in rectangular coordinates as follows:

$$
z = \begin{bmatrix} V_r^m \\ V_x^m \\ I_r^m \\ I_x^m \end{bmatrix} = \begin{bmatrix} U & 0 \\ 0 & U \\ g \cdot A & -b \cdot A \\ b \cdot A & g \cdot A \end{bmatrix} \begin{bmatrix} V_r \\ V_x \end{bmatrix} + e
$$

$$
= H_0 x_0 + e
$$

(4.36)

where
$x_0^T = [V_r, \ V_x]$
H_0 is the Jacobian which is built by neglecting shunt connections

Then, if weighting coefficients are ignored for convenience, the corresponding gain matrix $G_0 = H_0^T \cdot H_0$ will have the following block diagonal structure:

$$
G_0 = H_0^T H_0 = \begin{bmatrix} U + A^T(g^T g + b^T b)A & 0 \\ 0 & U + A^T(g^T g + b^T b)A \end{bmatrix}
$$

(4.37)

The following observations are made regarding Equation 4.37:

- Gain matrix G_0 is block diagonal.
- It has identical diagonal blocks.
- Diagonal blocks are constant, that is, independent of the system state.

The above decoupled and constant structure of G_0 allow computational shortcuts. This however will only be true if the shunts can be ignored which may not be possible in general. In order to ensure accuracy for systems containing branches with non-negligible shunt admittances, the following procedure can be used. Writing H as a sum of two matrices,

$$
H = H_0 + H_{shunt}
$$

(4.38)

where H_{shunt} is the Jacobian formed only using shunt connections, and substituting (Equation 4.38) into (Equation 4.36):

$$
z = [H_0 + H_{shunt}]x_0 + e
$$

(4.39)

$$
= [H_0]x_0 + \beta
$$

(4.40)

where

$$
\beta = H_{shunt} \cdot x_0 + e
$$

(4.41)

$$
E(\beta) = H_{shunt} \cdot E(x_0) = H_{shunt}\hat{x}_0; \quad E(e) = 0
$$

(4.42)

then, the corrected state estimate can be obtained by removing the expected value of β from the measurement vector:

$$
x^{corr} = G_0^{-1}H_0^T R^{-1}(z - H_{shunt}\hat{x}_0)
$$

(4.43)

$$
= \hat{x}_0 - G_0^{-1}\kappa\hat{x}_0
$$

(4.44)

where

$G_0 = H_0^T R^{-1} H_0$ is the decoupled gain matrix formed by neglecting shunts, including the diagonal weighting matrix R^{-1}

$\kappa = H_0^T R^{-1} H_{shunt}$ is a super sparse matrix

$E(\cdot)$ is the expected value of a variable

$\hat{x}_0 = E(x_0) = G_0^{-1} H_0^T R^{-1} z$ is the initially estimated state by neglecting all shunts

It should also be noted that the product $G_0^{-1} \kappa \hat{x}_0$ does not involve matrix inversion. Instead, the vector $\kappa \hat{x}_0$ is computed as a very sparse product of two arrays, followed by fast forward/back substitutions to solve $G_0 \delta x = \kappa \hat{x}_0$ for the correction term δx.

4.10.4 HYBRID SE

While exclusive use of PMU measurements has several benefits as seen in the previous section, due to the relatively large number of required PMUs to make the system fully observable, most power grids will still continue to be monitored by a mixture of PMU and SCADA measurements. Hence, state estimators should be properly modified to handle such sets of hybrid measurements. In addition to the differences in the types of measured quantities, there is also a significant difference in the scan rates at which SCADA and PMU measurements are received. Hence, it will be assumed that a set of SCADA measurements will be received at time instants t_n, t_{n+1}, etc. and in between the two consecutive scans there will be k scans of PMU measurements at time instants $t_1, t_2, \ldots, t_k$ such that $t_n < t_1 < \cdots < t_k < t_{n+1}$. Furthermore, it will be assumed that the system is fully observable by SCADA measurements and PMU measurements are used to improve redundancy and subsequently the variance of the estimated state. Since the SCADA measurements are typically expressed in polar coordinates, yet PMU measurements are written in rectangular coordinates (among other reasons to avoid computational issues at flat start) a nonlinear transformation needs to be used to switch between polar and rectangular coordinates [52,54]. Three alternative implementations of hybrid SE will now be described. All three implementations assume a full SCADA-based estimation at time t_n, followed by *improvements* introduced by the PMU measurements received at $t_1, t_2, \ldots, t_k$. The process is repeated starting at time t_{n+1} using the newly received SCADA measurements.

Let z_s be the vector of SCADA measurements scanned at $t = t_n$. Use the conventional WLS estimator to estimate the system state $\hat{x}_s$ based solely on z_s. The measurement equations can be compactly expressed as:

$$z_s = h_s(x_s) + e_s \qquad (4.45)$$

and the state estimate will be iteratively solved using:

$$x_s^{i+1} = x_s^i + G_s^{-1} H_s^T R_s^{-1} (z_s - h_s(x_s^i)) \qquad (4.46)$$

where

i is the iteration counter

$H_s = \partial h_s(x_s)/\partial x_s$, x_s is the state vector expressed in polar coordinates

$G_s = H_s^T G_s^{-1} H_s$ is the gain matrix

$\Omega_s = \text{cov}(\hat{x}_s) = G_s^{-1}$ is the covariance of estimated state

Note that this is the conventional WLS solution in polar coordinates using the SCADA measurements. The system is assumed to be fully observable by the SCADA measurements and inverse of the gain matrix evaluated at the last iteration provides the covariance matrix for the estimated states.

Following the estimation of x_s, a nonlinear transformation is applied to convert it to rectangular coordinates:

$$V_s = f(\hat{x}_s)$$

where $f(x_s)$ is a vector function having a pair of entries for each bus k as follows:

$$V_{rk} = V_k \cos \theta_k$$
$$V_{xk} = V_k \sin \theta_k$$

Then, the covariance of $\hat{V}_s$ can be obtained as:

$$\Omega_s = F \cdot \mathrm{cov}(\hat{x}_s) \cdot F^{\mathrm{T}} \tag{4.47}$$

$$= F \cdot G_s^{-1} \cdot F^{\mathrm{T}} \tag{4.48}$$

where $F = \partial f(x_s)/\partial x_s$ is the Jacobian of $f(\cdot)$ computed for x_s. Note that this Jacobian is a 2×2-block diagonal square matrix.

In the final step, the phasor measurements provided by PMUs, in rectangular coordinates, along with the estimate $\hat{x}_s$, are used to carry out the *improved* solution using the following linear measurement model (in more compact complex form):

$$\hat{V}_s = V + e_x \tag{4.49}$$

$$V^m = KV + e_V \tag{4.50}$$

$$\mathcal{I}^m = Y_r V + e_I \tag{4.51}$$

where the covariances of e_V and e_I are known diagonal matrices R_V and R_I, respectively, and that of e_x is Ω_s, given by Equation 4.47. Accordingly, the final estimate $\hat{V}$ will be given by the solution of the following Normal equations in complex form:

$$
\begin{bmatrix} U \\ K \\ Y_r \end{bmatrix}^{\mathrm{T}}
\begin{bmatrix} \Omega_s & & \\ & R_V & \\ & & R_I \end{bmatrix}^{-1}
\begin{bmatrix} U \\ K \\ Y_r \end{bmatrix} \hat{V} =
$$
$$
\begin{bmatrix} U \\ K \\ Y_r \end{bmatrix}^{\mathrm{T}}
\begin{bmatrix} \Omega_s & & \\ & R_V & \\ & & R_I \end{bmatrix}^{-1}
\begin{bmatrix} \hat{V}_s \\ V^m \\ \mathcal{I}^m \end{bmatrix} \tag{4.52}
$$

where U is the identity matrix, K is a binary matrix used to map measured voltage phasors to the state variable vector V and Y_r is the adequate subset of the bus admittance matrix. Note that R_V and R_I are customarily considered diagonal matrices, which disregards any correlation between the measurements provided by the same PMU at a substation. On the other hand, the inverse of the matrix Ω_s can be efficiently computed since the inverse of F is trivially obtained from the 2×2 inverse of its diagonal blocks. This matrix will no longer be diagonal but it will be very sparse.

The above solution procedure has the advantage that a conventional SE constitutes the first step, hence eliminating the need to develop any new code for the first stage. Also, the third stage involves a linear estimator, which is nondivergent by design even under bad data. It is noted that even though the final solution is not optimal, it is sufficiently accurate for all practical purposes.

The above procedure can be revised slightly in order to improve accuracy of the PMU-based estimation stages. This can be explained by using Figure 4.9, which shows a magnified view of the period between two consecutive SCADA scans, for example, at times t and $t + 1$. It is assumed that k PMU scans are received between these two instants. Typically, power systems will not have sufficient number of installed PMUs to make the entire system observable. The system will thus be unobservable at instants when only PMU measurements are received. An updated version of the most recently scanned SCADA measurements will then be used along with the most recent PMU measurements

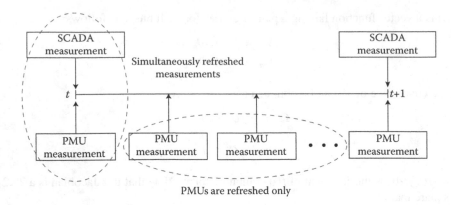

FIGURE 4.9   Timeline of SCADA and PMU scans.

in order to recover observability. Most recent SCADA measurements are replaced by their updated values which are calculated using the most recent state estimates based on the most recent PMU measurements. Given the relatively short time interval between PMU scans (typically PMU measurements are received 30–60 times per second) SCADA measurement updates can be approximated by using the linear first-order Taylor approximation for the measurement equations. Hence the PMU measurement-based estimation can be carried out using a linear estimator such as the WLS [55]. Another option is to use a more robust alternative such as the least absolute value (LAV) estimator [56] to minimize the influence of approximation errors from SCADA measurements and also automatically reject possible bad data from PMU measurements. This "hybrid state estimator" switches between a conventional WLS estimator and a robust weighted least absolute value (WLAV) estimator depending on the available set of measurements at that instant. As shown in Figure 4.9 at times t and $t + 1$, a conventional WLS state estimator that is based on both SCADA and PMU measurements is used to estimate the system state expressed in polar coordinates. Implementation of such an estimator can be easily accomplished by augmenting the SCADA measurement set with the additional PMU measurements [57]. At instances in between SCADA scans, a WLAV state estimator will be used for updating the estimated states. The WLAV estimator is formulated in rectangular coordinates and changes the weights assigned to SCADA measurements in order to account for the fact that their validity is progressively reduced in time during the period between two consecutive SCADA scans. Weights are reset at the time of each SCADA scan.

4.10.5   PLACEMENT OF PMUs

It was realized early on Reference 58 that since each PMU could measure not only the bus voltage but also the currents along all the lines incident to the bus, it would be possible to make the system observable by cleverly placing PMUs at a selected subset of buses. The strategy is rather straight forward: proceed to place PMUs at buses by excluding the neighbors of buses where a PMU is already placed. It was then shown that this problem can be formulated and solved using graph theoretic observability analysis and a linear integer optimization method [59]. This formulation allows easy analysis of network observability for mixed measurement sets, which may also include equality constraints in form of known (e.g., zero) injections. A simple revision to the basic formulation accomplishes that as shown in Reference 60. Determination of optimal locations of PMUs with or without existing PMUs can also be accomplished by effective removal of corresponding integer variables. Furthermore, optimal placement of PMUs to make a selected group of interconnected branches observable (a geographical region) can be similarly accomplished. Since the publication of Reference 58, a large volume of papers appeared in the literature on the general topic of PMU placement. While most of them have the objective of establishing an observable system, some also address robustness against loss of PMUs or topological changes. The types of PMUs being considered for placement may also be different. Some assume unlimited number of inputs, others consider PMUs that can monitor a single line only. A comprehensive list of published work on this topic can be found in Reference 61.

4.10.5.1 Problem Formulation

The objective of the PMU placement problem is to make the system observable by using a minimum number of PMUs. Problem formulation is dependent upon the assumed type of PMUs available for placement. PMUs can be broadly categorized as bus-type or branch-type. Bus-type PMUs are assumed to have either unlimited or limited number of inputs for measuring voltage and current phasors at a given bus and on its incident branches. Branch-type PMUs monitor individual branches by measuring the voltage and current phasors at the sending-end of the corresponding branch. In this section, formulation and solution of the PMU placement problem will be described for bus-type PMUs with unlimited inputs. It is possible to modify the basic formulation and apply the described approach to cases of bus-type PMUs with limited inputs or branch-type PMUs [62,63].

In order to formulate the bus-type PMU placement problem, consider an N-bus system and define a binary vector X with binary entries x_i depending on whether a PMU exists at bus i ($x_i = 1$) or not ($x_i = 0$). Then, the following linear integer programing problem, whose solution will yield the optimal placement for PMUs, can be formulated:

$$\text{Minimize} \quad \sum_{i=1}^{N} \omega_i \cdot x_i \tag{4.53}$$

$$\text{Subject to} \quad [A][X] \geq \hat{1} \tag{4.54}$$

where
ω_i is the installation cost for the PMU at bus i
$[A]$ is the binary connectivity matrix defined as
$$A(i, j) = \begin{cases} 1 & \text{if bus } i \text{ and } j \text{ are connected} \\ 0 & \text{otherwise,} \end{cases}$$
and $\hat{1}$ is a vector whose entries are all ones.

The general formulation will now be explained using the *IEEE* 14-bus test system whose diagram is given in Figure 4.10. The same example will be used to explain the implementation of zero injection constraints.

The only matrix that needs to be formed is the binary connectivity matrix A which is formed using the above given definition:

$$A = \begin{bmatrix}
1 & 1 & 0 & 0 & 1 & 0 & 0 & 0 & 0 & 0 & 0 & 0 & 0 & 0 \\
1 & 1 & 1 & 1 & 1 & 0 & 0 & 0 & 0 & 0 & 0 & 0 & 0 & 0 \\
0 & 1 & 1 & 1 & 0 & 0 & 0 & 0 & 0 & 0 & 0 & 0 & 0 & 0 \\
0 & 1 & 1 & 1 & 1 & 0 & 1 & 0 & 1 & 0 & 0 & 0 & 0 & 0 \\
1 & 1 & 0 & 1 & 1 & 1 & 0 & 0 & 0 & 0 & 0 & 0 & 0 & 0 \\
0 & 0 & 0 & 0 & 1 & 1 & 0 & 0 & 0 & 0 & 1 & 1 & 1 & 0 \\
0 & 0 & 0 & 1 & 0 & 0 & 1 & 1 & 1 & 0 & 0 & 0 & 0 & 0 \\
0 & 0 & 0 & 0 & 0 & 0 & 1 & 1 & 0 & 0 & 0 & 0 & 0 & 0 \\
0 & 0 & 0 & 1 & 0 & 0 & 1 & 0 & 1 & 1 & 0 & 0 & 0 & 1 \\
0 & 0 & 0 & 0 & 0 & 0 & 0 & 0 & 1 & 1 & 1 & 0 & 0 & 0 \\
0 & 0 & 0 & 0 & 0 & 1 & 0 & 0 & 0 & 1 & 1 & 0 & 0 & 0 \\
0 & 0 & 0 & 0 & 0 & 1 & 0 & 0 & 0 & 0 & 0 & 1 & 1 & 0 \\
0 & 0 & 0 & 0 & 0 & 1 & 0 & 0 & 0 & 0 & 0 & 1 & 1 & 1 \\
0 & 0 & 0 & 0 & 0 & 0 & 0 & 0 & 1 & 0 & 0 & 0 & 1 & 1
\end{bmatrix}$$

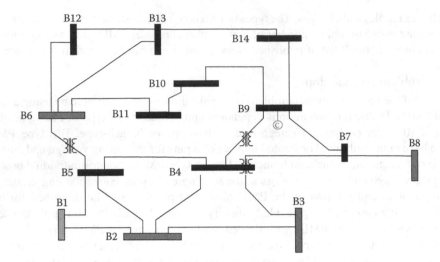

FIGURE 4.10   IEEE 14-bus system one-line diagram.

Substituting A in the linear binary programing problem of Equation 4.53 yields the binary solution vector X that will satisfy the observability condition imposed by the constraint (Equation 4.54), that is.

$$X^{\mathrm{T}} = \begin{bmatrix} 0 & 1 & 0 & 0 & 0 & 1 & 1 & 0 & 1 & 0 & 0 & 0 & 0 & 0 \end{bmatrix}$$

Hence, placing only 4 PMUs at buses 2, 6, 7, and 9 will be sufficient to observe the entire system state. Note that the type of PMUs considered in this placement problem are those with unlimited number of inputs, allowing measurement of all current phasors along branches that are incident to the bus where the PMU is placed.

4.10.5.2   Incorporating Existing Injections

Injection measurements whether they are actual measurements or virtual zero injections are treated the same way. Consider again the 14-bus system shown in Figure 4.10, where bus 7 is known to be a zero injection bus. In this case, it is easy to see that if the phasor voltages at any three out of the four buses 4, 7, 8, and 9 are known, then the fourth one can be calculated using the Kirchhoff's Current Law applied at bus 7 where the net injected current is known. Hence, the constraints associated with these buses will have to be modified accordingly as shown below:

$$x_2 + x_3 + x_4 + x_5 + x_7 + x_9 \geq s_4$$
$$x_4 + x_7 + x_8 + x_9 \geq s_7$$
$$x_7 + x_8 \geq s_8$$
$$x_4 + x_7 + x_9 + x_10 + x_14 \geq s_9$$
$$s_4 + s_7 + s_8 + s_9 \geq d_7$$

where
   s_j represents the *unknown* slack variable associated with bus i, such that the sum of these variables for the buses belonging to the set T_k is at least equal to the total number of neighbors of the zero injection bus k (in this example bus 7)
   T_k is the set of buses including the zero injection bus k and all its neighbors
   d_k is the number of directly connected buses to bus k, sometimes referred to as the degree of bus k (in this example $d_7 = 3$)

Note that this revision not only adds additional constraints (the last inequality) but also additional unknowns (s_4, s_7, s_8, s_9) to the optimization problem given by Equation 4.54 above. Solution of the modified linear binary optimization problem will yield

$$X^T = \begin{bmatrix} 0 & 1 & 0 & 0 & 0 & 1 & 0 & 0 & 1 & 0 & 0 & 0 & 0 & 0 \end{bmatrix}$$

$$s_4 = 1$$
$$s_7 = 1$$
$$s_8 = 0$$
$$s_9 = 1$$

As evident from the solution, having the zero injection at bus 7 reduces the number of PMUs required to make the system observable from 4 to 3, that is, it now becomes sufficient to place 3 PMUs at buses 2, 6, and 9 for full network observability.

4.11 TOPOLOGY AND PARAMETER ERRORS

Whenever an adequate redundancy level exists, it is possible to develop more sophisticated estimator models aimed at

- Detecting and, if possible, identifying switching devices' status errors
- Estimating certain parameter values (impedances, tap changers settings) in an attempt to improve the contents of the database

The first type of errors usually give rise to large inconsistencies in the estimated measurements and hence can be detected (not necessarily identified) easily. On the contrary, errors in network parameters or tap changers are less visible and may remain undetected for very long periods.

4.11.1 PARAMETER ESTIMATION

Although, in theory, it is possible to estimate the impedance of all lines when the measurement redundancy is sufficient, in practice, only a reduced subset of parameters is estimated. The suspected set of line parameters to be estimated can be selected manually, but an automated procedure based on the measurement residuals is preferable [64].

Let p_0 be the assumed parameter value, p the exact unknown value (in the following developments p can be either a scalar or a column vector comprising several suspected parameters), and z_s the set of measurements (branch power flows and terminal bus injections), which are explicit functions of such parameter(s). The nonlinear model relating those magnitudes can be written as

$$z_s = h_s(x, p) + e = h_s(x, p_0) + [h_s(x, p) - h_s(x, p_0)] + e \tag{4.55}$$

The term within brackets in this equation can be interpreted as an additional "noise," originated by the parameter error $e_p = p - p_0$, that affects the set of measurements z_s. For small parameter errors, the following linearized expression applies

$$h_s(x, p) - h_s(x, p_0) \approx \left[\frac{\partial h_s}{\partial p} \right] e_p \tag{4.56}$$

If e_p is sufficiently large, the normalized residuals corresponding to those measurements will be higher than expected and they can be even classified as bad data. Therefore, those lines or transformers with associated measurements carrying large residuals could be included in the suspect set.

Note that a necessary condition for a parameter error to be detectable is that the associated measurements are not critical. Otherwise, as explained in Section 4.8, the respective residuals will be null irrespective of the parameter errors.

It is also worth mentioning that the parameter estimation process can be carried out locally, as the influence of a parameter error on nearby measurements gets drastically reduced with electrical distance. Reciprocally, only sufficiently close measurements will have a significant effect on the estimated parameter value. In practice, working with a subnetwork composed of first, and perhaps second neighbors to the terminal nodes of the suspected branch will suffice.

Traditional parameter estimation methods are broadly classified into two groups [65]:

1. Methods based on the residual sensitivity matrix, applied at the local level when the estimation process finishes [66].
2. Methods augmenting the state vector with suspect parameters [67–69]. In turn, these methods can be based on either the normal equations (possibly using several measurement snapshots), or the Kalman filter.

Both techniques are summarized briefly in the following sections. In Section 4.11.4, a more comprehensive approach is presented which can be indistinctly applied to the identification of both topology and parameter errors.

4.11.1.1 Sensitivity-Based Parameter Estimation

As explained in Section 4.8, the residuals of estimated measurements are related with the measurement errors through the residual sensitivity matrix,

$$r = Se$$

Keeping in mind Equation 4.56, the following expression is obtained [70]:

$$r_s = \left(S_{ss}\frac{\partial h_s}{\partial p}\right)e_p + \bar{r}_s \tag{4.57}$$

where S_{ss} is the $s \times s$ submatrix of S corresponding to the s measurements directly related with the suspected parameters, and $\bar{r}_s$ is the residual vector that would be obtained in absence of parameter errors.

The above expression relates the residuals r_s with parameter errors in the presence of the "noise" $\bar{r}_s$. Therefore, determining e_p can be interpreted as solving a small scale, linear least-squares problem. The optimal value $\hat{e}_p$ is then obtained from

$$\hat{e}_p = \left[\left(\frac{\partial h_s}{\partial p}\right)^{\mathrm{T}} W_s S_{ss} \left(\frac{\partial h_s}{\partial p}\right)\right]^{-1} \left(\frac{\partial h_s}{\partial p}\right)^{\mathrm{T}} W_s r_s \tag{4.58}$$

leading to the improved parameter values

$$\hat{p} = p_0 + \hat{e}_p$$

Eventually, using the updated parameter values, the error estimation process can be repeated until negligible changes are obtained.

Note that the above development is based on the assumption that the measurements surrounding the suspected parameters do not contain bad data, which implies that measurement residuals are exclusively due to parameter errors. If this is not the case, the updated parameter values may be ironically less accurate than the original ones.

4.11.1.2 State-Augmentation Parameter Estimation

The state vector can be enlarged to include the set of suspected parameters p. Note that, in order not to compromise the network observability, the number of variables added to the state vector should be kept as small as possible.

Consider for instance the situation in which the unitary parameters defining the π model of a line are precisely known but the total length is suspected. In this case, there is no need to add the three line parameters as independent unknowns to the state vector, but only the line length. For a line between nodes i and j, the following admittances should be used in the power flow expressions:

$$\text{Series:} \quad (g_{ij} + jb_{ij})/L$$

$$\text{Shunt:} \quad jb_{ij}^{sh}L \tag{4.59}$$

where L is the actual line length normalized with the length contained in the database.

The new variable requires an extra column to be added to the Jacobian, whose elements are all null except for the measurements corresponding to its own power flows and power injections at terminal buses. The non-null derivatives can be easily found from the expressions above, much in the same way as the Jacobian elements corresponding to series regulating devices are computed in the load flow problem (see Chapter 3). For the power measurements related with bus i, such derivatives are

$$\frac{\partial P_{ij}}{\partial L} = \frac{\partial P_i}{\partial L} = \frac{-P_{ij}}{L^2}$$

$$\frac{\partial Q_{ij}}{\partial L} = \frac{\partial Q_i}{\partial L} = \frac{-[Q_{ij} + V_i^2 b_{ij}^{sh}(1 + L^2)]}{L^2}$$

where P_{ij} and Q_{ij} are the power flows computed with $L = 1$ (exchanging subindices provides the expressions for bus j).

Note that the above derivatives are very small or null at flat voltage profile, leading to severe ill-conditioning or even singularity during the first iteration. For this reason, the parameter L is only added at subsequent iterations.

Any other network parameter can be handled essentially in the same manner, the only difference arising in the derivative values.

Some authors propose to add the known value of p as a pseudomeasurement, to preserve the observability and to elude the ill-conditioning problem at flat voltage profile. However, the weight adopted for such pseudomeasurement has a noticeable influence on the estimate, which is not a desirable feature.

Irrespective of the pseudomeasurement being added or not, if the existing parameter value p_0 is accurate enough and the surrounding measurements are not, the estimate $\hat{p}$ will likely be worse than p_0. As there is no way to know whether this is the case or not, it is advisable to estimate only those parameters whose expected errors are larger than the average measurement errors. Note that, adding a new state variable while keeping intact the measurement set will lead to smaller residuals and lower objective function, which does not necessarily mean that the estimate is better (in fact, the residuals are null if $n = m$).

A simple way of increasing the redundancy, and hence the accuracy of the estimate, is by resorting to several measurement snapshots, all of them comprising the area of interest [69].

EXAMPLE 4.11

Considering the same data of Example 4.2, the relative length of line 1–2 that best suits the available measurements will be estimated, for which the state vector is augmented as explained above.

For the same tolerance, the iterative process converges in four iterations to the state:

Node	$\hat{V}$ (pu)	$\hat{\theta}$ (°)
1	0.9998	0.0
2	0.9740	−1.2605
3	0.9440	−2.7466

which is slightly different to that obtained in Example 4.2. The extra iteration needed is a consequence of having introduced the new state variable L at the second iteration.

The relative length estimated is $L = 1.0119$, that is, about 1% larger than that stored in the database. The accuracy of this length will depend on the redundancy level and accuracy of existing measurements.

As expected, the residuals and the objective functions are slightly smaller than those of Example 4.2, owing to the lower redundancy ($m = 8$, $n = 6$).

The gain matrix after the fourth iteration is

$$
G^{(4)} = 10^7
\begin{bmatrix}
3.1437 & -0.4433 & -0.0689 & -0.0312 & 0.0038 & 0.0589 \\
-0.4433 & 0.5957 & -0.0451 & 0.0046 & 0.0000 & -0.0070 \\
-0.0689 & -0.0451 & 3.1415 & -2.8229 & -0.2204 & -0.0646 \\
-0.0312 & 0.0046 & -2.8229 & 3.2436 & -0.4717 & 0.0702 \\
0.0038 & 0.0000 & -0.2204 & -0.4717 & 0.6685 & -0.0088 \\
0.0589 & -0.0070 & -0.0646 & 0.0702 & -0.0088 & 0.0027
\end{bmatrix}
$$

Note the reduced values of the elements in the last row/column, being responsible for the larger condition number of this matrix to be 1.04×10^5. The weak coupling between the new state variable and the complex voltages of nonterminal buses (in this case only bus 3) is also apparent.

4.11.2 Transformer Tap Estimation

Transformer taps constitute a particular case of network parameters and hence most techniques and comments discussed previously can be applied in this case [71–73]. Following are the distinctive features of the transformer tap estimation problem:

- Unlike other network parameters, transformer taps must be estimated on line so as to be useful. Therefore, only a single snapshot should be used unless it can be safely assured that the tap changer has not modified its position through several consecutive snapshots.
- As the tap changer shifts between discrete steps, the estimate should be rounded at the end to the closest feasible value.
- The presence of voltage magnitude measurements at the transformer terminal buses and reactive power flow measurements is essential to obtain an accurate estimate, as those magnitudes are strongly influenced by the tap setting.

The model and power–voltage relationships of regulating transformers are presented in Chapter 2. When the tap is included in the state vector, those expressions allow the respective Jacobian elements to be obtained in terms of existing components, as explained in Chapter 3.

4.11.3 Topology Errors

As discussed in the introductory section, the topology processor provides an electrical bus-branch network model by analyzing the substations at the physical level. Sectionalizing switches associated with CBs allow for seasonal topology changes and temporary arrangements required for maintenance. However, owing to economic reasons, only CBs are remotely operated and their status monitored. A physical CB, along with the set of associated switches, can be considered as one or several "logical" CBs, depending on the number of different topological configurations they give rise to. These logical devices are the ones referred to below.

In a majority of cases, the status of CBs is transmitted correctly to the EMS, as there is no analog transducer involved in capturing this information. However, owing to several reasons (failure of contacts, signal wires or interface cards, loss of information at subsequent stages, misunderstanding with the maintenance teams, etc.), it sometimes happens that the status of certain isolated devices is uncertain or simply wrong.

Assuming that a CB is closed when it is actually open, or vice versa, leads to a topology error, which may be catastrophic for the quality of the information provided by the estimator, at least locally. It can also globally affect the convergence rate of the iterative process [6]. This explains the significant research effort devoted in the last decade to develop improved techniques aimed at circumventing the limitations associated with existing topology processors [74–77].

According to the role a CB plays, two types of topology errors can be distinguished, namely:

- Error in the status (connected/disconnected) of a network component (line, transformer, reactor, or capacitor bank). It is termed an *exclusion error* when the element is actually closed but reported as open, and *inclusion error* in the opposite situation. This kind of errors, affecting one or several CBs, always involve a non-null impedance.
- Error in the status of a CB characterizing the inner topology of a substation (coupling of two bus bar sections, ring configuration, etc.). This kind of errors, not involving any finite impedance, usually affect the number of assumed electrical nodes.

Suspecting the status of all CBs would be wasteful and impractical, owing to the enormous amount of information that should be handled. Therefore, a procedure is needed to select the subset of suspected CBs, which can be based on the following information:

- Information provided *a priori* by the topology processor, related with ambiguous or unknown statuses as well as resulting configurations that are in contradiction with the nearby measurements.
- Information provided by the measurement residuals after the first SE run. Much like in the parameter estimation case, if the residuals of an entire substation are abnormally large, the CBs contained in such substation should be declared as suspicious. The detection capability is better in this case since topology errors give rise typically to much larger residuals than parameter errors. This approach cannot be used, however, in those cases where the SE does not converge as a consequence of severe topology errors.

Available techniques to estimate the status of dubious CBs belong to two different categories, although only the second one faces the problem in an integral manner.

The first technique takes advantage of the fact that each CB status is a binary variable, and hence determining the most plausible status of a reduced set of CBs is a matter of checking all feasible status combinations and retaining the one minimizing the scalar $J(\hat{x})$ (i.e., the one showing more compatibility with local measurements). To reduce the number of combinations to be checked, it is convenient to resort to former measurement records, assuming that all of them share the same topology.

The second group of methods, by far the most successful, relies on the state-augmentation technique for which the status of suspected CBs has to be somehow modeled. Depending on whether the topology error can or cannot be associated with a non-null impedance, different CB models are possible.

First consider the problem of estimating the status of a suspected line. As the disconnected status implies null admittances, the following π model can be adopted for a line potentially connected between buses i and j:

$$\text{Series:} \quad (g_{ij} + jb_{ij})k \quad \text{Shunt:} \quad jb_{ij}^{\text{sh}}k \tag{4.60}$$

where $k = 1$ and $k = 0$ represent the "on" and "off" statuses, respectively. Therefore, a single state variable suffices to model the status of a line (this ignores the possibility of a line being connected just through one of its terminal buses). Note the similarity of the expression in Equation 4.59. The non-null elements of the extra Jacobian column corresponding to magnitudes at bus i are

$$\frac{\partial P_{ij}(k)}{\partial k} = \frac{\partial P_i(k)}{\partial k} = P_{ij}$$

$$\frac{\partial Q_{ij}(k)}{\partial k} = \frac{\partial Q_i(k)}{\partial k} = Q_{ij}$$

where P_{ij} and Q_{ij} are the respective power flows computed for $k = 1$. Similar expressions are obtained for the derivatives with respect to magnitudes related with bus j.

In practice, the estimated value $\hat{k}$ may significantly differ from 0 or 1, owing to related measurement errors. Even worse, in the presence of bad data, $\hat{k}$ can even approach 0.5, leaving open the question, as to whether the line is connected or not. In addition, the estimate will be less accurate if $\hat{k}$ does not exactly correspond with one of its possible integer values. Such potential risks can be prevented by adding the following equality, forcing $\hat{k}$ to converge to either of its feasible values:

$$k(1 - k) = 0 \qquad (4.61)$$

If the measurements related with the suspected line are not critical then k is observable and the above constraint constitutes a redundant information intended to refine the estimate. Otherwise, such constraint makes k nonuniquely observable, as both solutions are equally acceptable.

Like in the parameter estimation case, the state augmentation technique should be applied only after the first iteration to avoid the ill-conditioning of the Jacobian matrix arising at flat start.

EXAMPLE 4.12

Again, the data of Example 4.2 will be used to estimate the status of line 1–2 for which the new variable k multiplying all line admittances is considered. The table below shows the number of required iterations for different initial values of k:

Initial Value	Number of Iterations
1	4
0.5	6
0.001	7

In all cases, the estimate agrees with that of Example 4.11, with $\hat{k} = 0.9882$ (just the inverse of $\hat{L}$ in that example). Note that, even when assuming an open line initially ($k^0 \approx 0$), the process converges to the state minimizing the objective function ($k \approx 1$).

As a line cannot be 98.8% closed, a second run would be needed with $k = 1$ to refine the estimate, unless the quadratic constraint 4.34 was added to the model during the first run. When this is done, the estimator converges in six iterations to the same state as that of Example 4.2 for $k^0 \geq 0.5$ but diverges to the wrong state (line open) for $k^0 < 0.5$. Hence, as the true status of the line is unknown in advance, when adding the constraint 4.34 it is advisable to initialize k with the neutral value $k^0 = 0.5$.

Clearly, the above technique cannot be applied to estimate the status of certain CBs, like bus bar couplers, associated with null impedances. For these cases, a more generic approach [77] has been developed, which is based on adding the active and reactive power flows to the state vector through the suspected CBs, P_{ij} and Q_{ij}.

The new variables should be compensated whenever possible with adequate equality constraints, reflecting the assumed CB status. For a closed CB between physical buses i and j such constraints are

$$\theta_i - \theta_j = 0 \quad V_i - V_j = 0 \tag{4.62}$$

whereas for an open CB, the associated constraints are

$$P_{ij} = 0 \quad Q_{ij} = 0 \tag{4.63}$$

Explicitly modeling closed CBs gives rises to extra electrical nodes with respect to those identified by a conventional topology processor. Note that, as many of those new nodes are of the null injection type, the respective equality constraints should be also added to the model.

At first glance, it might seem useless to explicitly model CBs whose status is known. Indeed, we are adding two new state variables, which, at the same time, are set to zero if the CB is open. The reason to do so is that this gives the opportunity to subsequently check the correctness of the assumptions made about the CB status, which is not possible otherwise. In this regard, it can be shown that the Lagrange multipliers corresponding to the equality constraints, when normalized with their respective covariances, play the same role for detecting/identifying a topology error as the normalized residuals for bad data in ordinary measurements. Therefore, an abnormally large normalized multiplier will indicate a topological error that would be masked as several nearby bad data in a conventional estimator without capability of modeling CBs [7].

A way of reducing the huge size of the state vector when all CBs are explicitly modeled, while keeping intact the capability to perform topology error processing, is by systematically exploiting the well-known topological properties (i.e., Kirchhoff laws) of the substation graph. With the help of a carefully selected spanning tree, the resulting linear relationships can be used to remove from the state vector as many CB variables, yielding a reduced model in which only the truly independent variables remain [78].

The incorporation of the features described in this section (network parameter estimation, topology error identification), along with other enhancements have led to the emergence of the so-called *Generalized State Estimator* [6,79]. Under this paradigm, a conventional state estimator is first run, followed by a generalized state estimator in which only suspected substations and parameters are fully modeled as explained above, in an attempt to reduce the computational burden that would arise if the entire system was modeled at the physical level.

4.11.4 IMPLICIT MODELS

An alternative to the two-stage process involved in the Generalized State Estimator lies in the use of the so-called *implicit model* for handling equality constraints. The basic idea, first applied to topology errors [80] and afterwards to parameter errors [81], consists of adding the CB known statuses or assumed parameter errors, respectively, as trivial equality constraints to be enforced. Then, the entire set of linear constraints are removed from the equality-constrained SE formulation, along with an equal number of variables from the augmented state vector, leading to a model whose size is roughly the same as that of the conventional bus-branch model. Yet, the required Lagrange multipliers associated with the implicit equality constraints can be easily recovered by first identifying large enough residuals and then computing the required subsets of sparse sensitivity matrices relating both types of quantities (residuals and multipliers).

To begin with, consider the topology error identification problem. Let the linear expression $As = 0$ represent, in compact form, the set of constraints (Equations 4.62 and 4.63), each reflecting the assumed status of a CB, where vector s contains the extra variables that should be added to the state vector of the bus-branch model. Then, the following equality-constrained SE problem arises (the reader is referred to the general case, presented in Section 4.7.2),

$$\text{Minimize} \quad J(x, s) = \frac{1}{2}[z - h(x, s)]^T W[z - h(x, s)]$$

(4.64)

$$\text{Subject to} \quad As = 0$$

As matrix A is of full rank (in fact, it reduces to the identity matrix provided the variables in s are suitably defined [78,80]), it turns out that $s = 0$. Keeping in mind this constraint, more compact functions can be defined where the explicit dependence on s is formally eliminated,

$$J_i(x) = J(x, s)|_{s=0} \qquad h_i(x) = h(x, s)|_{s=0}$$

Hence, a reduced model, fully equivalent to Equation 4.64, can be considered in which the constraint $s = 0$ is embedded:

$$\text{Minimize} \quad J_i(x) = \frac{1}{2}[z - h_i(x)]^T W[z - h_i(x)]$$

Apparently, the above implicit model provides the same information as the conventional bus-branch model. In fact, the residuals of both models are identical. However, if the Lagrangian of the constrained minimization problem (Equation 4.64) is formed,

$$\mathcal{L} = J(x, s) - \lambda^T As$$

and the first-order optimality conditions are derived, then the following relationship is obtained between the residuals of the implicit (or full) model and the Lagrange multipliers of the full model:

$$H_s^T W r_i + A^T \lambda = 0$$

(4.65)

where,

$$H_s = \frac{\partial h(x, s)}{\partial s} \qquad r_i = z - h_i(x)$$

At this point, it is worth comparing Equation 4.65 with the first equation of the system (Equation 4.15). In the general case, it is not possible to obtain λ given a set of residuals, as the Jacobian C is a rectangular matrix for which no inverse exists. However, in this particular case, A is a trivially invertible matrix, allowing the multipliers to be expressed as linear combinations of the estimated residuals provided by the implicit model:

$$\lambda = T\hat{r}_i$$

(4.66)

where

$$T = -A^{-T} H_s^T W$$

is denoted the Topological Sensitivity matrix. Note that in practice there is no need to compute the whole matrix T, but only those columns corresponding to entries of r_i which are large enough. In the presence of a single topology error, large residuals will arise at one or several interconnected substations, depending on whether it is a busbar coupling or branch status error.

The largest normalized residual test, as explained in Section 4.8.4.1, can be easily generalized to account for the possibility of topology errors, as follows:

1. Solve the implicit SE model and obtain the respective residuals.
2. If the residuals of a set of measurements for one or several substations are large enough, then build the corresponding components of matrix T and obtain the Lagrange multipliers from Equation 4.66.
3. Normalize both the residuals and the computed multipliers and rank all normalized quantities in a single list.
4. If the first element in the list is a normalized residual then a bad datum is suspected. Otherwise, the status of the CB associated with the largest multiplier(s) is suspected.
5. According to the results of the previous step, remove/correct the bad data or switch the CB status, and go to Step 1.

The same implicit approach can be adopted to detect abnormally high parameter errors, as explained in Reference 81. In this case, the extra unknowns are given by the parameter errors, e_p, relating the data base values, p_0, with the true parameters as follows:

$$p = p_0 + e_p$$

The corresponding equality-constrained WLS problem can then be formulated as:

$$\text{Minimize } J(x, e_p) = \frac{1}{2}[z - h(x, e_p)]^{\mathrm{T}} W[z - h(x, e_p)]$$

$$\text{Subject to } e_p = 0$$

(4.67)

where the initial assumption is that all parameter errors are null ($e_p = 0$). This trivial constraint can be used to eliminate e_p from the problem, leading again to an implicit model in which the required Lagrange multipliers can still be computed *a posteriori* from the estimated residuals:

$$\lambda = -H_p^{\mathrm{T}} W \hat{r}_i$$

(4.68)

where

$$H_p = \partial h(x, e_p)/\partial e_p$$

Note that Equation 4.68 is a generalization of Equation 4.58, which is actually an approximated expression obtained under the assumption that a single parameter error has only a local influence. Another major difference lies in the computational cost, as only sparse matrices appear in Equation 4.68 while selected subsets of the dense matrix S are involved in Equation 4.58.

The largest normalized quantity (residual/multiplier) test summarized above can be trivially extended to include the Lagrange multipliers associated with parameter errors. If the largest value corresponds to the constraint of a parameter, then that parameter should be explicitly added to the state vector, and the constraint removed from the model, before the implicit SE is run again.

Unlike topology errors, which can be easily detected and identified as long as the branch with wrong status carries a minimum load, most parameter errors will remain unnoticed, unless they are sufficiently large to trigger the largest normalized quantity test. How big parameter errors should be for them to be detectable depends both on the local redundancy level and measurement accuracy. As discussed previously, a way of enhancing the capability to refine parameter values is by resorting to multiple snapshots [69].

Normalizing the Lagrange multipliers requires that the diagonal of the respective covariance matrices be computed, which can be a cumbersome and costly process. In Reference 82, a computationally efficient implementation is provided to obtain the required covariance elements.

REFERENCES

1. F. Schweppe and B. Douglas, Power system static-state estimation, *IEEE Transactions on Power Apparatus and Systems*, PAS-89, 1970, 120–135.
2. F. Schweppe and E. Handschin, Static state estimation in electric power systems, *Proceedings IEEE*, 62, July 1974, 972–983.
3. R. Larson, W. Tinney, L. Hajdu, and D. Piercy, State estimation in power systems. Part II: Implementation and applications, *IEEE Transactions on Power Apparatus and Systems*, PAS-89(3), March 1970, 353–362.
4. T. Dy Liacco, The role and implementation of state estimation in an energy management system, *Electrical Power and Energy Systems*, 12(2), April 1990, 75–79.
5. F. Wu, Power system state estimation, *Electrical Power and Energy Systems*, 12(2), April 1990, 80–87.
6. A. Monticelli, *State Estimation in Electric Power System: A Generalized Approach*, Kluwer Academic Publishers, 1999, 1–13.
7. A. Abur and A. Gómez-Expósito, *Power System State Estimation: Theory and Implementation*, Marcel Dekker, New York, 2004.
8. J. J. Allemong, L. Radu, and A. M. Sasson, A fast and reliable state estimation algorithm for AEP's new control center, *IEEE Transactions on Power Apparatus and Systems*, PAS-101, April 1982, 933–944.
9. A. García, A. Monticelli, and P. Abreu, Fast decoupled state estimation and bad data processing, *IEEE Transactions on Power Apparatus and Systems*, PAS-98, September 1979, 1645–1652
10. J. Wang and V. Quintana, A decoupled orthogonal row processing algorithm for power system state estimation, *IEEE Transactions on Power Apparatus and Systems*, PAS-103, August 1984, 2337–2344.
11. A. Monticelli and F. Wu, Network observability: Identification of observable islands and measurement placement, *IEEE Transactions on Power Apparatus and Systems*, PAS-104(5), May 1985, 1035–1041.
12. A. Monticelli and F. Wu, Network observability: Theory, *IEEE Transactions on PAS*, PAS-104(5), May 1985, 1042–1048.
13. G. R. Krumpholz, K. A. Clements, and P. W. Davis, Power system observability: A practical algorithm using network topology, *IEEE Transactions on Power Apparatus and Systems*, PAS 99(4), July/August 1980, 1534–1542.
14. K. A. Clements, G. R. Krumpholz, and P. W. Davis, Power system state estimation with measurement deficiency: An observability/measurement placement algorithm, *IEEE Transactions on Power Apparatus and Systems*, 102(7), 1983, 2012–2020.
15. K. A. Clements, G. R. Krumpholz, and P. W. Davis, Power system state estimation with measurement deficiency: An algorithm that determines the maximal observable subnetwork, *IEEE Transactions on Power Apparatus and Systems*, 101(7), 1982, 3044–3052.
16. L. Holten, A. Gjelsvik, S. Aam, F. Wu, and W. Liu, Comparison of different methods for state estimation, *IEEE Transactions on Power Systems*, 3, November 1988, 1798–1806.
17. A. Simões-Costa and V. Quintana, A robust numerical technique for power system state estimation, *IEEE Transactions on Power Apparatus and Systems*, PAS-100, February 1981, 691–698.
18. M. Vempati, I. Slutsker, and W. Tinney, Enhancements to givens rotations for power system state estimation, *IEEE Transactions on Power Systems*, 6(2), May 1991, 842–849.
19. A. Monticelli, C. Murari, and F. Wu, A hybrid state estimator: Solving normal equations by orthogonal transformations, *IEEE Transactions on Power Apparatus and Systems*, 2, December 1985, 3460–3468.
20. F. Aschmoneit, N. Peterson, and E. Adrian, State estimation with equality constraints 10-th Power Industry Computer Applications (PICA) Conference Proceedings, Toronto, May 1977, pp. 427–430.
21. A. Gjelsvik, S. Aam, and L. Holten, Hachtel's augmented matrix method – a rapid method improving numerical stability in power system static state estimation, *IEEE Transactions on Power Apparatus and Systems*, PAS-104, November 1985, 2987–2993.
22. H. M. Merrill and F. C. Schweppe, Bad data suppression in power system static estimation, *IEEE Transactions on Power Apparatus and Systems*, PAS-90, November/December 1971, 2718–2725.
23. E. Handschin, F. C. Schweppe, J. Kohlas, and A. Fiechter, Bad data analysis for power system state estimation, *IEEE Transactions on Power Apparatus and Systems*, PAS-94(2), 1975, 329–337.
24. A. Monticelli and A. García, Reliable bad data processing for real-time state estimation, *IEEE Transactions on Power Apparatus and Systems*, PAS-102(3), 1983, 1126–1139.
25. L. Mili, T. Van Cutsem, and M. Pavella, Hypothesis testing identification: A new method for bad data analysis in power system state estimation, *IEEE Transactions on Power Apparatus and Systems*, PAS-103(11), 1984, 3239–3252.

26. L. Mili and T. Van Cutsem, Implementation of HTI method in power system state estimation, *IEEE Transactions on Power Systems*, 3(3), August 1988, 887–893.
27. F. R. Hampel, E. M. Ronchetti, P. J. Rousseeuw, and W. A. Stahel, *Robust Statistics: The Approach Based on Influence Functions*, John Wiley & Sons, Hoboken, NJ, 1986, 223–225.
28. P. J. Rousseeuw and A. M. Leroy, *Robust Regression and Outlier Detection*, John Wiley & Sons, Hoboken, NJ, 1987, 9–17.
29. M. R. Irving, R. C. Owen, and M. J. Sterling, Power system state estimation using linear programming, *Proceedings of IEE*, 125, 1978, 879–885.
30. W. W. Kotiuga and M. Vidyasagar, Bad data rejection properties of weighted least absolute value techniques applied to static state estimation, *IEEE Transactions on Power Apparatus and Systems*, PAS-101 (4), April 1982, 844–851.
31. R. H. Bartels, A. R. Conn, and J. W. Sinclair, Minimization techniques for piecewise differentiable functions: The ℓ_1 solution to an overdetermined linear system, *SIAM Journal on Numerical Analysis*, 15(2), April 1978, 224–241.
32. H. M. Wagner, Linear programming techniques for regression analysis, *Journal of American Statistical Association*, 54, 1959, 206–212.
33. I. Barrodale and F. D. K. Roberts, An improved algorithm for discrete ℓ_1 linear approximation, *SIAM Journal on Numerical Analysis*, 10(5), October 1973, 839–848.
34. I. Barrodale and F. D. K. Roberts, Applications of mathematical programming to ℓ_p approximation, *Nonlinear Programming* J. B. Rosen, O. L. Mangasarian, and K. Ritter, Eds. Academic Press, New York, 1970, pp. 447–464.
35. A. Abur and M. Çelik, A fast algorithm for the weighted least absolute value state estimation, *IEEE Transactions on Power Systems*, 6(1), February 1991, 1–8.
36. N. K. Karmarkar, *A new polynomial time algorithm for linear programming*, Combinatorica, vol. 4, Springer, Berlin/Heidelberg, 1984, pp. 373–395.
37. A. Arbel, *Exploring Interior-Point Linear Programming Algorithms and Software*, The MIT Press, Cambridge, MA, 1993, pp. 103–124.
38. S. P. Ellis and S. Morgenthaler, Leverage and breakdown in L_1 regression, *Journal of the American Statistical Association*, 87(417), March 1992, 143–148.
39. L. Mili, M. G. Cheniae, N. S. Vichare, and P. J. Rousseeuw, Robust state estimation based on projection statistics, *IEEE Transaction on Power Systems*, 11(2), May 1996, 1118–1127.
40. L. Mili, V. Phaniraj, and P. J. Rousseeuw, Robust estimation theory for bad data diagnostics in electric power systems, *Advances in Control and Dynamic Systems*, C.T. Leondes (ed.), Academic Press, New York, NY, 37, 1990, 271–325.
41. A. G. Phadke, Synchronized phasor measurements in power systems, *IEEE Computer Applications in Power*, 6(2) April 1993, 10–15.
42. A. G. Phadke, J. S. Thorp, and K. J. Karimi, State estimation with phasor measurements, *IEEE Transactions on Power Systems*, 1(1), February 1986, 233–241.
43. I. W. Slutsker, S. Mokhtari, L. A. Jaques, J. M. G. Provost, M. B. Perez, J. B. Sierra, F. G. Gonzalez, and J. M. M. Figueroa, Implementation of Phasor Measurements in State Estimator at Sevillana de Electricidad, Proc. of the Power Industry Computer Application Conference, May 1995, 392–398.
44. A. G. Phadke and J. S. Thorpe, *Synchronized Phasor Measurements and Their Applications*, (Book) Springer, 2008.
45. A. G. Phadke, J. S. Thorpe, and M. G. Adamiak, A new measurement technique for tracking voltage phasors, local system frequency, and rate of change of frequency, *IEEE Transactions on Power Apparatus and Systems*, PAS-102(5), 1983, 1025–1038.
46. W. Chi-Kong, L. Ieng-Tak, W. Jing-Tao, and H. Ying-Duo, A novel algorithm for phasor calculation based on wavelet analysis, *IEEE Power Engineering Society Summer Meeting*, 3(July 2001), 1500–1503.
47. V. V. Terzija, M. B. Djuric, and B. D. Kovacevic, Voltage phasor and local system frequency estimation using Newton type algorithm, *IEEE Transactions on Power Delivery*, 9(3), 1991, 1000–1007.
48. A. A. Girgis, and R. G. Brown, Application of Kalman filtering in computer relaying, *IEEE Transactions on Power Apparatus and Systems*, PAS-100(7), 1981, 3387–3397.
49. P. K. Dash, S. K. Panda, B. Mishra, and D. P. Swain, Fast estimation of voltage and current phasors in power networks using an adaptive neural network, *IEEE Transactions on Power Systems*, 12(4), 1997, 1494–1499.
50. IEEE standard for synchrophasor measurements for power systems, *IEEE Std C37.118.1-2011 (Revision of IEEE Std C37.118-2005)*, Dec. 28, 2011, 1–61.

51. IEEE standard for synchrophasor measurements for power systems – amendment 1: Modification of selected performance requirements, *IEEE Std C37.118.1a-2014 (Amendment to IEEE Std C37.118.1-2011)*, April 30, 2014, 1–25.

52. M. Zhou, V. A. Centeno, J. S. Thorp, and A. G. Phadke, An alternative for including phasor measurements in state estimators, *IEEE Transactions on Power Systems*, 21(4), November 2006, 1930–1937.

53. J. Zhu and A. Abur, Effect of Phasor Measurements on the Choice of Reference Bus for State Estimation, Proceedings of 2007 PES General Meeting, June 24–28 2007, Tampa, Florida.

54. A. Gómez-Expósito, A. Abur, P. Rousseaux, A. de la Villa and C. Gómez-Quiles, *On the Use of PMU in Power Systems State Estimation*, 17th Power Systems Computation Conference (PSCC), Stockholm, August 2011.

55. M. Glavic and T. Van Cutsem, Reconstructing and tracking network state from a limited number of synchrophasor measurements, *IEEE Transactions on Power Systems*, 28(2), May 2013, 1921–1929.

56. M. Göl, and A. Abur, LAV Based robust state estimation for systems measured by PMUs, *IEEE Transactions on Smart Grid*, 5(4), July 2014, 1808–1814.

57. M. Göl and A. Abur, A hybrid state estimator for systems with limited number of PMUs, *IEEE Transactions on Power Systems*, 30(3), May 2015, 1511–1517.

58. T. L. Baldwin, L. Mili, M. B. Boisen, and R. Adapa, Power system observability with minimal phasor measurement placement, *IEEE Transactions on Power Systems*, 8(2), May 1993, 707–715.

59. X. Bei, Y. Yoon, and A. Abur, Optimal Placement and Utilization of Phasor Measurements for State Estimation, 15th Power Systems Computation Conference, Liége (Belgium), August 22–26, 2005.

60. D. Dua, S. Dambhare, R.K. Gajbhiye, and S.A. Soman, Optimal multistage scheduling of PMU placement, an ILP approach, *IEEE Trans. On Power Delivery*, 23(4), October 2008, 1812–1820.

61. N. M. Manousakis, G. N. Korres, and P. S. Georgilakis, Taxonomy of PMU placement methodologies, *IEEE Transactions on Power Systems*, 27(2), May 2012, 1070–1077.

62. M. Korkali and A. Abur, Placement of PMUs with Channel Limits, Proceedings of the IEEE Power Engineering Society General Meeting, Calgary, Canada, July 26–30, 2009.

63. R. Emami and A. Abur, Robust measurement design by placing synchronized phasor measurements on network branches, *IEEE Transactions on Power Systems*, 25(1), February 2010, 38–43

64. I. Slutsker and K. Clements, Real time recursive parameter estimation in energy management systems, *IEEE Transactions on Power Systems*, 11(3), August 1996, 1393–1399.

65. P. Zarco and A. Gómez Expósito, Power system parameter estimation: A survey, *IEEE Transactions on Power Systems*, 15(1), February 2000, 216–222.

66. W. Liu, F. Wu, and S. Lun, Estimation of parameter errors from measurement residuals in state estimation, *IEEE Transactions on Power Systems*, 7(1), February 1992, 81–89.

67. A. Debs, Estimation of steady-state power system model parameters, *IEEE Transactions on Power Apparatus and Systems*, PAS-93(5), 1974, 1260–1268.

68. W. Liu and S. Lim, Parameter error identification and estimation in power system state estimation, *IEEE Transactions on Power Systems*, 10(1), February 1995, 200–209.

69. P. Zarco and A. Gómez Expósito. Off-line determination of network parameters in state estimation, Proceedings 12th Power System Computation Conference, Dresde, Germany, August 1996, pp. 1207–1213.

70. T. Van Cutsem and V. Quintana, Network parameter estimation using online data with application to transformer tap position estimation, *IEE proceedings, part C*, 135(1), January 1988, 31–40.

71. D. Fletcher and W. Stadlin, Transformer tap position estimation, *IEEE Transactions on Power Apparatus and Systems*, PAS-102(11), November 1983, 3680–3686.

72. E. Handschin and E. Kliokys, Transformer tap position estimation and bad data detection using dynamic signal modelling, *IEEE Transactions on Power Systems*, 10(2), May 1995, 810–817.

73. P. Teixeira, S. Brammer, W. Rutz, W. Merritt, and J. Salmonsen, State estimation of voltage and phase-shift transformer tap settings, *IEEE Transactions on Power Systems*, 7(3), August 1992, 1386–1393.

74. N. Singh and H. Glavitsch, Detection and identification of topological errors in online power system analysis, *IEEE Transactions on Power Systems*, 6(1), February 1991, 324–331.

75. K. A. Clements and A. Simoes-Costa, Topology error identification using normalized lagrange multipliers, *IEEE Transactions on Power Systems*, 13(2), May 1998, 347–353.

76. A. Monticelli and A. García, Modeling zero impedance branches in power systems state estimation, *IEEE Transactions on Power Systems*, 6(4), November 1991, 1561–1570.

77. A. Monticelli, Modeling circuit breakers in weighted least squares state estimation, *IEEE Transactions on Power Systems*, 8(3), August 1993, 1143–1149.

78. A. Gómez Expósito and A. de la Villa, Reduced substation models for generalized state estimation, *IEEE Transactions on Power Systems*, 16(4), November 2001, 839–846.
79. O. Alsac, N. Vempati, B. Stott, and A. Monticelli, Generalized state estimation, *IEEE Transactions on Power Systems*, 13(3), August 1998, 1069–1075.
80. A. de la Villa and A. Gómez Expósito, Implicitly-constrained substation model for state estimation, *IEEE Transactions on Power Systems*, 17(3), August 2002, 850–856.
81. Z. Jun A. Abur, Identification of network parameter errors, *IEEE Transactions on Power Systems*, 21(2), May 2006, 586–592.
82. Y. Lin and A. Abur, Highly efficient implementation for parameter error identification method exploiting sparsity, *IEEE Transactions on Power Systems*, 32(1), January 2017, 734–742.

5 Economics of Electricity Generation

Francisco D. Galiana and Antonio J. Conejo

CONTENTS

5.1 INTRODUCTION

This chapter includes two parts. In the first part, comprising Sections 5.2 and 5.3, we consider that the electric energy system is managed by a central operator with full information on the technical and economic data of the generating units, the demands, and the network. This constitutes a centralized, noncompetitive operation.

In the second part, presented in Sections 5.4 through 5.7, we consider that the system is operated within a market environment in which producers compete with each other to supply power to the consumers, whose demand may exhibit an elastic behavior.

In both parts, we consider a short-term operation perspective with a time framework ranging from several minutes to one week. Such operation problems can usually be formulated as constrained optimization problems, whose solution can be conveniently found through computational environments, such as GAMS [1] or Matlab [2], which enable focusing on modeling, thus avoiding the technicalities associated with solution algorithms. Such computational environments offer the benefit of using state-of-the-art solvers such as CPLEX [3] for mixed-integer linear programing or MINOS [3] for nonlinear programing.

Appendix B provides some background on optimization theory and an introduction to the most common solution algorithms.

5.2   ECONOMIC DISPATCH

The economic dispatch (ED) problem consists of allocating the total demand among generating units so that the production cost is minimized. Generating units have different production costs depending on the prime energy source used to produce electricity (mainly coal, oil, natural gas, uranium, and water stored in reservoirs). And these costs vary significantly; for example, the marginal costs for nuclear, coal, and gas units may take on values ranging between $0.03 and $0.20 per kWh. Furthermore, for any given generating technology, an individual unit may have variable marginal costs that depend on its MW output.

To appreciate the advantages of dispatching a power system according to the solution of the ED problem, consider the case where a power plant supplies 10,000 MW during 1 h at an average cost of $0.05/kWh; if the consumers buy this energy at the rate of $0.06/kWh, this arrangement results in a net profit to the supplier of $100,000/h. In this case, an improvement in supply efficiency of just 1% through the use of ED would result in a profit increment of $5000/h or $43.8 million in one year. Also, note that this increment in profit does not necessarily have to end up in its entirety in the pockets of the producer, but, rather, could also be used to reduce the consumer price. Therefore, it is clear that there is a strong incentive for both producers and consumers to increase the efficiency of the generating units, both individually and as a system.

In addition to the continuous decisions on how to allocate the demand among generating units (ED), a decision that involves calculating the MW outputs of all units (a set of continuous variables), the economics of electricity generation also requires the calculation of an optimum time schedule for the start-up and shutdown of the generating units (a set of discrete or binary variables). Since the units' start-up or shutdown costs can be significant, on/off scheduling decisions must be optimally coordinated with the ED of the continuous generation outputs. For instance, if the fixed cost of a unit is high, it may be more economical to shutdown the unit than to operate at a low power output. However, the decision to shutdown a unit when the demand is low must be made considering the cost involved and the feasibility of restarting the unit when the demand increases. Starting up and shutting down units involve binary decisions, which can be modeled mathematically through binary variables. The resulting model is significantly more complex than the ED model with only continuous variables. The problem of determining how to allocate the load among the generating units over a multiperiod time horizon, including the possibility of starting up and shutting down units, in a way such that the total cost (operating, start-up, and shutdown) is minimized, is denominated unit commitment (UC).

The ED problem is analyzed in detail below, together with a brief introduction to the UC problem.

5.2.1   CLASSIC ED

Each generating unit is assigned a function, $C_i(P_{Gi})$, characterizing its generating cost in $/h in terms of the power produced in MW, P_{Gi}, during 1 h. This function is obtained by multiplying the heat rate curve, expressing the fuel consumed to produce 1 MW during 1 h, by the cost of the fuel consumed during that hour. Note that the heat rate is a measure of the energy efficiency of the generating unit [4].

The cost function is generally approximated by a convex quadratic or piecewise linear function, as illustrated in Figure 5.1.

Considering n generating units, the total production cost is

$$C(P_G) = \sum_{i=1}^{n} C_i(P_{Gi}) \tag{5.1}$$

where P_G is the column vector of the unit generation levels P_{Gi}.

If the system total demand is P_D^{total} and all generating units contribute to supply this demand, the generation must equal the total demand plus transmission losses, P_{loss}, that is,

$$\sum_{i=1}^{n} P_{Gi} = P_D^{\text{total}} + P_{\text{loss}} \tag{5.2}$$

The ED problem consists of minimizing the total generation cost (Equation (5.1)) with respect to the unit outputs, P_{Gi}, subject to the power balance, Equation (5.2), and to the unit operational limits,

$$P_{Gi}^{\text{min}} \leq P_{Gi} \leq P_{Gi}^{\text{max}} \tag{5.3}$$

where the superscripts "min" and "max" indicate minimum and maximum, respectively.

5.2.2 Basic ED

We first consider the basic ED without losses ($P_{\text{loss}} = 0$) and without generation limits, where the system demand is assumed to be inelastic. The notion of elastic demand arises when, in addition to considering generation costs, the ED also accounts for the benefit derived by the consumers by consuming electricity. The objective of ED then becomes one of maximizing the difference between the demand benefit and the generation cost, or so-called social welfare. Since the sole constraint considered here is the one requiring that the demand be supplied—the power balance—the Lagrangian function, therefore, becomes

$$\mathcal{L}(P_G, \lambda) = \sum_{i=1}^{n} C_i(P_{Gi}) - \lambda\left(\sum_{i=1}^{n} P_{Gi} - P_D^{\text{total}}\right) \tag{5.4}$$

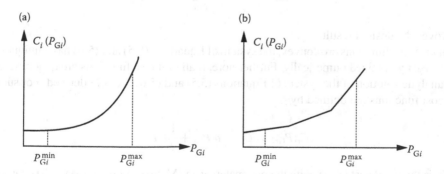

FIGURE 5.1 Examples of cost functions: (a) convex quadratic and (b) piecewise linear.

The first-order necessary optimality conditions are

$$\frac{\partial \mathcal{L}(\cdot)}{\partial P_{\mathrm{G}i}} = IC_i(P_{\mathrm{G}i}) - \lambda = 0; \quad i = 1, \ldots, n \tag{5.5}$$

$$\frac{\partial \mathcal{L}(\cdot)}{\partial \lambda} = -\sum_{i=1}^{n} P_{\mathrm{G}i} + P_{\mathrm{D}}^{\mathrm{total}} = 0 \tag{5.6}$$

where the function $IC_i(P_{\mathrm{G}i})$ is the incremental cost of unit i,

$$IC_i(P_{\mathrm{G}i}) = \frac{\mathrm{d}C_i(P_{\mathrm{G}i})}{\mathrm{d}P_{\mathrm{G}i}} \tag{5.7}$$

Equation (5.5) implies that under ED, all units must operate at identical incremental costs equal to the Lagrange multiplier λ. It is interesting to observe that the common incremental cost λ also coincides with the system marginal cost, that is, the sensitivity of the total cost with respect to the system demand,

$$\lambda = \frac{\mathrm{d}C(P_{\mathrm{G}})}{\mathrm{d}P_{\mathrm{D}}^{\mathrm{total}}}, \tag{5.8}$$

also referred to as the cost of the "last" MW added to the demand.

Equation (5.8) is derived as follows. Consider that there exists an optimal solution satisfying the necessary optimality conditions in Equations (5.5) and (5.6). If the system demand changes by a small amount, $\mathrm{d}P_{\mathrm{D}}^{\mathrm{total}}$, the corresponding optimum generation levels must also change so that

$$\sum_{i=1}^{n} \mathrm{d}P_{\mathrm{G}i} = \mathrm{d}P_{\mathrm{D}}^{\mathrm{total}} \tag{5.9}$$

Similarly, the total cost also changes according to

$$\begin{aligned}
\mathrm{d}C(P_{\mathrm{G}}) &= \sum_{i=1}^{n} \mathrm{d}C_i(P_{\mathrm{G}i}) \\
&= \sum_{i=1}^{n} IC_i(P_{\mathrm{G}i}) \mathrm{d}P_{\mathrm{G}i} \\
&= \sum_{i=1}^{n} \lambda \, \mathrm{d}P_{\mathrm{G}i} = \lambda \sum_{i=1}^{n} \mathrm{d}P_{\mathrm{G}i} = \lambda \, \mathrm{d}P_{\mathrm{D}}^{\mathrm{total}}
\end{aligned} \tag{5.10}$$

which proves the desired result.

If all n unit cost functions are convex, the system of Equations (5.5) and (5.6) has a unique solution that can be easily obtained numerically. Furthermore, if all unit cost functions are quadratic and convex, the analytic solution of the system of Equations (5.5) and (5.6) is easily derived. For such a case the unit cost functions are defined by

$$C_i(P_{\mathrm{G}i}) = C_{0i} + a_i P_{\mathrm{G}i} + \frac{1}{2} b_i P_{\mathrm{G}i}^2 \tag{5.11}$$

where C_{0i} is the fixed cost ($\$/\mathrm{h}$), with the parameters a_i ($\$/\mathrm{MWh}$) and b_i ($\$/(\mathrm{MW})^2\mathrm{h}$) characterizing the variable cost component depending on the generation level ($P_{\mathrm{G}i}$).

For notational convenience, the following vectors are defined:

$$C_0 = [C_{01}, \ldots, C_{0n}]^T$$

$$a = [a_1, \ldots, a_n]^T$$

$$b = [b_1, \ldots, b_n]^T \tag{5.12}$$

$$e = [1, \ldots, 1]^T$$

$$P_G = [P_{G1}, \ldots, P_{Gn}]^T$$

in addition to the diagonal matrix

$$B = \text{diag}(b) \tag{5.13}$$

The total cost can then be expressed as

$$C(P_G) = e^T C_0 + a^T P_G + \frac{1}{2} P_G^T B P_G \tag{5.14}$$

which is a quadratic function of the vector of generating levels.

The power balance equation can also be expressed in vector form as

$$e^T P_G = P_D^{\text{total}} \tag{5.15}$$

The first-order necessary conditions in Equations (5.5) and (5.6) then become

$$a + B P_G = \lambda e \tag{5.16}$$

$$e^T P_G = P_D^{\text{total}} \tag{5.17}$$

The analytic solution of Equations (5.16) and (5.17), in other words, the ED, can now be written as

$$P_G = \lambda B^{-1} e - B^{-1} a \tag{5.18}$$

where the system incremental cost is given by

$$\lambda = \frac{P_D^{\text{total}} + e^T B^{-1} a}{e^T B^{-1} e} \tag{5.19}$$

It is also useful to express the optimal generating levels in terms of the demand. Using Equation (5.19) to eliminate λ from Equation (5.18), we get

$$P_G = \alpha P_D^{\text{total}} + \beta \tag{5.20}$$

where α and β are defined by

$$\alpha = \frac{B^{-1} e}{e^T B^{-1} e} \tag{5.21}$$

and

$$\beta = \frac{B^{-1}e(e^T B^{-1}a)}{e^T B^{-1}e} - B^{-1}a \tag{5.22}$$

Observe that the vector α defines the load participation factors, namely, the increments in the generation outputs under ED following a system demand increment. Thus, if the system demand changes by dP_D^{total}, the generating outputs change by

$$dP_G = \alpha \, dP_D^{total} \tag{5.23}$$

Note that the participation factors are positive parameters that add up to 1, that is, $e^T \alpha = 1$, a property that arises from the power balance requirement.

EXAMPLE 5.1 BASIC ED

Consider two generating units supplying the system demand P_D^{total}. The quadratic unit cost functions are characterized by the parameters provided in the table as follows:

Unit	C_0 ($/h)	a ($/MWh)	b ($/(MW)^2$h)	P_G^{min} (MW)	P_G^{max} (MW)
1	100	20	0.05	0	400
2	200	25	0.10	0	300

The ED produces the following results:

$$\lambda = \frac{650 + P_D^{total}}{30} \ \$/\text{MWh} \tag{5.24}$$

$$P_G = \begin{bmatrix} \frac{2}{3}P_D^{total} + \frac{100}{3} \\ \frac{1}{3}P_D^{total} - \frac{100}{3} \end{bmatrix} \text{MW} \tag{5.25}$$

For the specific system demand levels of 40, 250, 300, and 600 MW, we obtain the corresponding ED generation levels, system incremental costs, and costs as indicated in the table below:

Case	P_D^{total} (MW)	P_{G1} (MW)	P_{G2} (MW)	λ ($/MWh)	C ($/h)
A-Ex.5.1	40	60	−20	23	1110
B-Ex.5.1	250	200	50	30	6675
C-Ex.5.1	300	233.3	66.7	31.67	8217
D-Ex.5.1	600	433.3	166.7	41.67	19,217

The following observations are in order:

1. Equation (5.24) implies that the incremental cost, λ, increases linearly with the demand.
2. Equation (5.25) results in asymmetrical allocations of the system demand among the generating units, as a consequence of the participation factors being 2/3 for unit 1 and 1/3 for unit 2. This is a reasonable result as the cost of unit 2 increases with its production at a higher rate than the cost of unit 1, that is, $b_1 = 0.05 < b_2 = 0.1$. In this sense, we can say that unit 1 is incrementally more efficient than unit 2.
3. The system demand is covered by both units, not only by the most efficient one, namely, unit 1. However, the most efficient unit does get a higher share of the load (2/3).

4. The optimal ED solution 5.25 satisfies the demand balance equation. However, if the demand is low enough ($P_D^{\text{total}} < 100\,\text{MW}$), as in case A-Ex.5.1, the generation level P_{G2} becomes negative, which violates the minimum power output of unit 2. On the other hand, if the demand is large enough ($P_D^{\text{total}} > 550\,\text{MW}$), as in case D-Ex.5.1, the generation level P_{G1} becomes greater than the capacity of unit 1, which is 400 MW. These two observations confirm that since generation limits have not been taken into account in solving the basic ED problem, some of the solutions are infeasible.

5.2.3 ED WITH ELASTIC DEMAND

We now extend the basic ED without transmission losses ($P_{\text{loss}} = 0$) and without generating limits to include elastic demands.

An elastic demand is characterized by its utility function, $U_j(P_{Dj})$ in \$/h, expressing the utility or benefit obtained by the demand by consuming P_{Dj} MW during 1 h.

Considering m loads, the total system utility is

$$U(P_D) = \sum_{j=1}^{m} U_j(P_{Dj}) \tag{5.26}$$

where P_D is a column vector containing the individual demands P_{Dj}.

The social welfare is then defined as

$$SW(P_G, P_D) = \sum_{j=1}^{m} U_j(P_{Dj}) - \sum_{i=1}^{n} C_i(P_{Gi}) \tag{5.27}$$

Without transmission losses, the system production should equal the system demand,

$$\sum_{i=1}^{n} P_{Gi} = \sum_{j=1}^{m} P_{Dj} \tag{5.28}$$

If the demands are elastic, the ED consists of maximizing the social welfare (Equation (5.27)) subject to the power balance equilibrium between production and demand (Equation (5.28)), including the operating limits on the generating units and the demands.

If such operating limits are ignored, the sole constraint considered is again the power balance (Equation (5.28)). The Lagrangian function then becomes

$$\mathcal{L}(P_D, P_G, \lambda) = \sum_{j=1}^{m} U_j(P_{Dj}) - \sum_{i=1}^{n} C_i(P_{Gi}) - \lambda\left(\sum_{j=1}^{m} P_{Dj} - \sum_{i=1}^{n} P_{Gi}\right) \tag{5.29}$$

The first-order necessary optimality conditions are

$$\frac{\partial \mathcal{L}(\cdot)}{\partial P_{Dj}} = IU_j(P_{Dj}) - \lambda = 0; \quad j = 1, \dots, m \tag{5.30}$$

$$\frac{\partial \mathcal{L}(\cdot)}{\partial P_{Gi}} = -IC_i(P_{Gi}) + \lambda = 0; \quad i = 1, \dots, n \tag{5.31}$$

$$\frac{\partial \mathcal{L}(\cdot)}{\partial \lambda} = \sum_{i=1}^{n} P_{Gi} - \sum_{j=1}^{m} P_{Dj} = 0 \tag{5.32}$$

where the functions $IU_j(P_{Dj})$ are the incremental utility or benefit functions given by

$$IU_j(P_{Dj}) = \frac{dU_j(P_{Dj})}{dP_{Dj}} \tag{5.33}$$

The above result again implies that all generating units must operate at identical incremental costs and that all demands operate at identical marginal utilities (conditions in Equations (5.30) and (5.31)). Moreover, the single incremental cost must be equal to the single incremental utility.

EXAMPLE 5.2 ED WITH ELASTIC DEMAND

The generating units of Example 5.1 as well as the elastic demands characterized below are considered in this example. Just like the generation costs, the demand utility is assumed to be quadratic. Note, however, that the b-coefficient is negative since the utility function is concave while the cost function is convex.

Load	a^D ($/MWh)	b^D [$/(MW)^{2}h]	P_D^{min} (MW)	P_D^{max} (MW)
1	55	−0.2	0	300
2	50	−0.1	0	350

Superscript and subscript "D" in the table above indicates "demand."
The ED with elastic demand produces the results shown below.

Case	P_D^{total} (MW)	P_{G1} (MW)	P_{G2} (MW)	P_{D1} (MW)	P_{D2} (MW)	λ ($/MWh)	SW ($/h)
A-Ex.5.2	180	153.3	26.7	68.3	111.7	27.7	2504.2

Note that the total generation is equal to the total demand (180 MW) and that the incremental cost is equal to the incremental utility $27.7/MWh. In general, elastic loads have the tendency to lower both consumption as well as the system incremental cost.

The ED with elastic demand is revisited in the second part of this chapter in the study of market-clearing procedures (Section 5.5.1).

5.2.4   ED WITH GENERATION LIMITS

If generating limits are considered and the demand is inelastic and no transmission losses are taken into account, the Lagrangian function becomes

$$\mathcal{L}(P_G, \lambda) = \sum_{i=1}^{n} C_i(P_{Gi}) - \lambda \left(\sum_{i=1}^{n} P_{Gi} - P_D^{total} \right)$$

$$- \sum_{i=1}^{n} \mu_i^{max}(P_{Gi} - P_{Gi}^{max}) \tag{5.34}$$

$$- \sum_{i=1}^{n} \mu_i^{min}(P_{Gi} - P_{Gi}^{min})$$

where new multipliers are incorporated, corresponding to the minimum and maximum power outputs of each generating unit (Equation (5.3)). Multiplier $\mu_i^{\max}$ is associated with the capacity or maximum power output of the unit, while multiplier $\mu_i^{\min}$ is associated with the minimum power output.

The first-order necessary optimality conditions become

$$\frac{\partial \mathcal{L}(\cdot)}{\partial P_{Gi}} = IC_i(P_{Gi}) - \lambda - \mu_i^{\max} - \mu_i^{\min} = 0; \quad i = 1, \ldots, n \tag{5.35}$$

$$\frac{\partial \mathcal{L}(\cdot)}{\partial \lambda} = -\sum_{i=1}^{n} P_{Gi} + P_D^{\text{total}} = 0 \tag{5.36}$$

including the complementarity slackness conditions [5],

$$\mu_i^{\max} \leq 0 \quad \text{if } P_{Gi} = P_{Gi}^{\max} \tag{5.37}$$

$$\mu_i^{\max} = 0 \quad \text{if } P_{Gi} < P_{Gi}^{\max} \tag{5.38}$$

$$\mu_i^{\min} \geq 0 \quad \text{if } P_{Gi} = P_{Gi}^{\min} \tag{5.39}$$

$$\mu_i^{\min} = 0 \quad \text{if } P_{Gi} > P_{Gi}^{\min} \tag{5.40}$$

If, as in the case here, generation limits are imposed, the equal marginal cost condition of the basic ED is no longer valid, replaced instead by the criterion below, derived from Equations (5.35) through (5.40),

$$IC_i(P_{Gi}) = \lambda + \mu_i^{\min} \geq \lambda \quad \text{if } P_{Gi} = P_{Gi}^{\min} \tag{5.41}$$

$$IC_i(P_{Gi}) = \lambda \quad \text{if } P_{Gi}^{\min} < P_{Gi} < P_{Gi}^{\max} \tag{5.42}$$

$$IC_i(P_{Gi}) = \lambda + \mu_i^{\max} \leq \lambda \quad \text{if } P_{Gi} = P_{Gi}^{\max} \tag{5.43}$$

It is interesting, however, that the Lagrange multiplier λ can still be interpreted as the system marginal cost, in other words, the sensitivity of the system cost with respect to the system demand. This result can be proved in a similar manner as in the case without generation limits, a derivation that the reader is encouraged to carry out.

Conditions in Equations (5.41) through (5.43) are interpreted as follows. Units operating within its upper and lower limits exhibit identical incremental costs equal to λ. Units operating at their maximum capacity have incremental costs lower than or equal to λ, while units operating at their minimum power output have incremental costs greater than or equal to λ. If the two units of Example 5.1 with null minimum power outputs are considered, Figure 5.2 illustrates the three possible dispatches.

For the case with $\lambda = \lambda_A$, unit 1 operates at its maximum capacity and meets the condition $\lambda > IC_1(P_{G1}^{\max})$, while unit 2 operates within its bounds meeting the condition $\lambda = IC_2(P_{G2})$. For $\lambda = \lambda_B$, both units operate within their respective bounds. Finally, for $\lambda = \lambda_C$, unit 2 operates at its minimum power output and meets the condition $\lambda < IC_2(P_{G2}^{\min})$, while unit 1 operates within its limits and satisfies the requirement that $\lambda = IC_1(P_{G1})$.

Example 5.3 below illustrates the solution of the ED with generation limits for four demand levels.

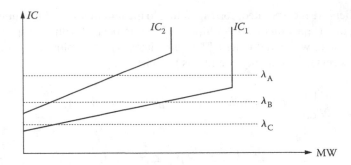

FIGURE 5.2   Three examples of ED with generation limits.

EXAMPLE 5.3 ED WITH GENERATION LIMITS

Considering the units with the cost functions of Example 5.1, the ED results in the table as follows:

Case	P_D^{total} (MW)	P_{G1} (MW)	P_{G2} (MW)	IC_1 ($/MWh)	IC_2 ($/MWh)	λ ($/MWh)	C ($/h)
A-Ex.5.3	40	40	0 (min)	22	25	22	1140
B-Ex.5.3	250	200	50	30	30	30	6675
C-Ex.5.3	300	233.3	66.7	31.67	31.67	31.67	8217
D-Ex.5.3	600	400 (max)	200	40	45	45	19,300

The following observations are in order:

1. For case A-Ex.5.3, the solution obtained implies that unit 2 operates at its minimum power output of 0 MW. Thus, the incremental cost of unit 2 ($IC_2 = \$25/MWh$) is greater than the marginal cost ($\lambda = \$22/MWh$).
2. For cases B-Ex.5.3 and C-Ex.5.3, both units operate inside their bounds, so that $\lambda = IC_1 = IC_2$.
3. For case D-Ex.5.3, unit 1 operates at its maximum capacity, with an incremental cost of $IC_1 = \$40/MWh$ meeting the requirement of being smaller than the marginal cost $\lambda = \$45/MWh$.
4. As expected, the system incremental cost grows piecewise linearly with the demand, while the total cost grows quadratically with the demand.

If the cost functions are convex, the solution of the ED with generation limits is unique and easy to compute numerically. However, an analytic solution is not easily obtained because it is necessary to consider all possible combinations of units either operating at their respective limits or not, a difficult combinatorial problem in general. Nevertheless, if the units operating at their limits are known, the remaining units must operate at equal incremental cost and meet the residual demand resulting from subtracting the binding generation limits from the original demand.

As an alternative to using mathematical programing solvers, such as CPLEX or MINOS [3], specific procedures can be used to efficiently solve the generation-constrained ED problem, particularly the algorithm known as λ-iteration, which works as follows:

1. The multiplier λ is approximated by $\lambda^{(v)}$.

2. The unit generation levels are computed so that the optimality conditions of Equations (5.41) through (5.43) are satisfied, that is,

$$\text{if} \quad IC_i(P_{Gi}^{\min}) \geq \lambda^{(\nu)}, \quad \text{then } P_{Gi} = P_{Gi}^{\min} \tag{5.44}$$

$$\text{else} \quad \text{if } IC_i(P_{Gi}^{\max}) \leq \lambda^{(\nu)}, \quad \text{then } P_{Gi} = P_{Gi}^{\max} \tag{5.45}$$

$$\text{otherwise, compute } P_{Gi} \text{ so that } IC_i(P_{Gi}) = \lambda^{(\nu)} \tag{5.46}$$

The total generation level is calculated adding the generation levels of all units, and the balance of generation and demand is checked. If the balance is satisfied within a given tolerance, the procedure concludes. Otherwise, it continues with step 3.

3. To update lambda, a bisection rule is used, $\lambda^{(\nu+1)} = [\lambda^{(\nu)} + \lambda^{(\nu-1)}]/2$, with the previous values of λ corresponding to a generation surplus and to a generation deficit, respectively. The procedure then continues with step 2.

5.2.5 ED with Losses

In this section, we analyze the ED accounting for transmission network losses, but excluding generating limits. The consideration of transmission losses in the ED problem produces two significant changes with respect to the basic ED. One is to alter the generation dispatch, and the second is to render the marginal cost of supplying the local demand nonunique across the transmission network, varying from bus to bus.

Losses can be incorporated into the ED through the following modified power balance equation:

$$\sum_{i=1}^{n} P_{Gi} - P_D^{\text{total}} - P_{\text{loss}}(P_G, P_D) = 0 \tag{5.47}$$

From Equation (5.47), we observe that the transmission losses modify the power balance equation in two respects, first by slightly increasing the net demand by the value of the losses (typically of the order of 3%–5% of the system load), and second by introducing a nonlinearity representing the functional relation between the losses P_{loss} and the vectors of generations and demands, P_G and P_D, respectively.

Considering the power balance equation with losses (Equation (5.47)), the Lagrangian function becomes

$$\mathcal{L}(P_G, \lambda) = \sum_{i=1}^{n} C_i(P_{Gi}) - \lambda \left[\sum_{i=1}^{n} P_{Gi} - P_D^{\text{total}} - P_{\text{loss}}(P_G, P_D) \right] \tag{5.48}$$

The first-order necessary conditions are

$$\frac{\partial \mathcal{L}(\cdot)}{\partial P_{Gi}} = IC_i(P_{Gi}) - \lambda \left(1 - \left. \frac{\partial P_{\text{loss}}}{\partial P_{Gi}} \right|_s \right) = 0, \quad i = 1, \ldots, n \tag{5.49}$$

$$\frac{\partial \mathcal{L}(\cdot)}{\partial \lambda} = -\sum_{i=1}^{n} P_{Gi} + P_D^{\text{total}} + P_{\text{loss}}(P_G, P_D) = 0 \tag{5.50}$$

We observe from Equation (5.49) that, as indicated earlier, the generating units do not operate at equal incremental costs (as is the case without losses), since the marginal costs of the units now depend on the sensitivity of the losses with respect to their generation levels. Note that subscript s in the sensitivity coefficients identifies the (arbitrarily selected) slack or swing bus s. The topic of the slack bus is discussed in more detail later in this section.

The system of Equations (5.49) and (5.50) can only be solved numerically, since both the loss function and the sensitivity coefficients are nonlinear functions of the generation levels, P_G. In general, it is not possible to describe these nonlinear functions in an explicit form, however, in the classical ED solution, losses were often represented by explicit loss formulae in P_G [4]. The more accurate and common approach today is to compute the losses and their sensitivities from the load flow equations, which define these quantities implicitly. We, therefore, now show how the load flow-based approach works.

For the sake of simplicity, we henceforth consider constant voltage magnitudes throughout the network. Then, load flow equations become

$$P_G - P_D = P(\delta) \tag{5.51}$$

The n-dimensional vectors P_G, P_D, and $P(\delta)$ are generations, demand, and injections, respectively, where the vector of injections is a nonlinear function of the $(n-1)$-dimensional voltage angle vector, δ (the reference voltage angle is set to zero without loss of generality).

The load flow equations are illustrated in the example below.

EXAMPLE 5.4 LOAD FLOW EQUATIONS

Consider the three-bus, three-line network (shunt susceptances neglected) described in the table as follows:

From Bus	To Bus	Conductance (pu)	Susceptance (pu)
1	2	1	−10
1	3	1	−10
2	3	1	−10

Assuming that each bus might include a generating unit and a fixed load and that bus 3 is the reference one, the load flow equations are

$$P_{G1} - P_{D1} = \cos(\delta_1 - \delta_2) + \cos(\delta_1 - 0) + 10\sin(\delta_1 - \delta_2) + 10\sin(\delta_1 - 0)$$
$$P_{G2} - P_{D2} = \cos(\delta_2 - \delta_1) + \cos(\delta_2 - 0) + 10\sin(\delta_2 - \delta_1) + 10\sin(\delta_2 - 0) \tag{5.52}$$
$$P_{G3} - P_{D3} = \cos(0 - \delta_1) + \cos(0 - \delta_2) + 10\sin(0 - \delta_1) + 10\sin(0 - \delta_2)$$

The system above includes three equations and two δ variables. Thus, two out of three generation levels can be specified to compute the unknown vector δ. The third generation is computed through Equations (5.52) once the unknown vector δ is determined.

To calculate the transmission losses, the corresponding load flow has to be solved numerically to compute the voltage angle vector. To do so, one equation has to be eliminated from Equation (5.51), namely, the equation corresponding to an arbitrarily selected slack bus. The remaining $n-1$ equations are then solved to determine the voltage angle vector δ. Then, the losses can be calculated using

$$P_{\text{loss}} = e^T P(\delta) \tag{5.53}$$

where the unitary vector e is defined in Equation (5.12).

The incremental loss around a given operating condition δ_0 is then

$$dP_{\text{loss}} = e^{\text{T}} \left[\frac{\partial P(\delta_0)}{\partial \delta} \right] d\delta \tag{5.54}$$

Now to compute the sensitivity coefficients of the losses with respect to the generation levels, $\partial P_{\text{loss}}/\partial P_{\text{G}}$, the load flow equations are first linearized around the same operating condition δ_0,

$$dP = dP_{\text{G}} - dP_{\text{D}} = \left[\frac{\partial P(\delta_0)}{\partial \delta} \right] d\delta \tag{5.55}$$

The $n \times (n-1)$-dimensional matrix $[\partial P(\delta_0)/\partial \delta]$ is computed taking derivatives of the load flow vector-Equation (5.51).

The next step is to express the vector of differential angles $d\delta$ as a function of the differential power injections using Equation (5.55). This requires eliminating one equation from vector-Equation (5.55) and leaving the corresponding injection at bus s, $dP_{\text{G}s} - dP_{\text{D}s}$, unspecified. Then, we define the $(n-1)$-vector of differential generations, $dP_{\text{G}}|_s$, obtained by suppressing the unspecified element $dP_{\text{G}s}$ from dP_{G}. Analogously, we define the vector $dP_{\text{D}}|_s$ and the $(n-1) \times (n-1)$-dimensional matrix $[\partial P(\delta_0)/\partial \delta|_s]$ obtained from $[\partial P(\delta_0)/\partial \delta]$ by eliminating row s. Note that this matrix is now square and usually invertible. Then,

$$dP_{\text{G}}|_s - dP_{\text{D}}|_s = \left[\frac{\partial P(\delta_0)}{\partial \delta} \bigg|_s \right] d\delta \tag{5.56}$$

Combining Equations (5.54) and (5.56), we get

$$dP_{\text{loss}} = e^{\text{T}} \left[\frac{\partial P(\delta_0)}{\partial \delta} \right] \left[\frac{\partial P(\delta_0)}{\partial \delta} \bigg|_s \right]^{-1} (dP_{\text{G}}|_s - dP_{\text{D}}|_s) \tag{5.57}$$

or

$$dP_{\text{loss}} = e^{\text{T}} \left[\frac{\partial P(\delta_0)}{\partial \delta} \right] \left[\frac{\partial P(\delta_0)}{\partial \delta} \bigg|_s \right]^{-1} dP|_s \tag{5.58}$$

From Equation (5.58), we obtained the sensitivity coefficient of the losses with respect to the power injections,

$$\frac{\partial P_{\text{loss}}}{\partial P|_s} = \left(\left[\frac{\partial P(\delta_0)}{\partial \delta} \bigg|_s \right]^{\text{T}} \right)^{-1} \left[\frac{\partial P(\delta_0)}{\partial \delta} \right]^{\text{T}} e \tag{5.59}$$

Note that the above sensitivities are with respect to power injections P, that is, with respect to $P_{\text{G}} - P_{\text{D}}$. This is a general result in the sense that it embodies sensitivities with respect to generations and demands. If demands are fixed (the most common situation), sensitivities with respect to injections and generations are identical, that is,

$$\frac{\partial P_{\text{loss}}}{\partial P_{\text{G}i}} \bigg|_s = \frac{\partial P_{\text{loss}}}{\partial P_i} \bigg|_s \tag{5.60}$$

We point out that the sensitivity of the losses with respect to the slack bus s is not defined in Equation (5.59). This is so because the generation $P_{\text{G}s}$ is a dependent variable. With a certain degree of arbitrariness, but consistently with the mathematics involved, we can say that $\partial P_{\text{loss}}/\partial P_{Gs}|_s$ is zero. Then, sensitivities can be defined for all buses, including the slack, and differential losses can be expressed as a function of all the n power injections,

$$dP_{\text{loss}} = \sum_{i=1}^{n} \left.\frac{\partial P_{\text{loss}}}{\partial P_i}\right|_s dP_i = \sum_{i=1}^{n} \left.\frac{\partial P_{\text{loss}}}{\partial P_i}\right|_s (dP_{Gi} - dP_{Di}) \tag{5.61}$$

Recalling that

$$\sum_{i=1}^{n} dP_i = dP_{\text{loss}} \tag{5.62}$$

the differential power balance Equation (5.61) can be conveniently expressed as

$$\sum_{i=1}^{n} \left(1 - \left.\frac{\partial P_{\text{loss}}}{\partial P_i}\right|_s \right) dP_i = 0 \tag{5.63}$$

In practice, loss sensitivities can be positive or negative, but they are usually small in absolute value. The impact of losses on the total cost of the ED is generally also small, but losses do influence individual generations as well as bus incremental or marginal costs throughout the network. The marginal cost at a given bus, traditionally called bus incremental cost, but also known as locational marginal price (LMP), is defined as the sensitivity of the total cost with respect to the bus demand, $\lambda_i = \partial C/\partial P_{Di}$. As shown in the second part of this chapter, LMPs are important in the operation of electricity markets.

The LMPs can be derived as follows. Assuming that the operating conditions meet the first-order optimality conditions of Equations (5.49) and (5.50), let the demand vector be marginally perturbed by dP_D. This implies that the generation vector changes by dP_G so that the optimality conditions remain satisfied. The total cost changes according to

$$
\begin{aligned}
dC &= \sum_{i=1}^{n} IC_i dP_{Gi} = \sum_{i=1}^{n} \lambda \left(1 - \left.\frac{\partial P_{\text{loss}}}{\partial P_i}\right|_s \right) dP_{Gi} \\
&= \lambda \left(\sum_{i=1}^{n} dP_{Gi} - \sum_{i=1}^{n} \left.\frac{\partial P_{\text{loss}}}{\partial P_i}\right|_s dP_{Gi} \right)
\end{aligned}
\tag{5.64}
$$

Considering now the incremental power balance

$$\sum_{i=1}^{n} (dP_{Gi} - dP_{Di}) = dP_{\text{loss}} \tag{5.65}$$

and using Equation (5.64), we get

$$
\begin{aligned}
dC &= \lambda \left(\sum_{i=1}^{n} dP_{Di} + dP_{\text{loss}} \right) - \lambda \left(\sum_{i=1}^{n} \left.\frac{\partial P_{\text{loss}}}{\partial P_i}\right|_s dP_{Gi} \right) \\
&= \lambda \left(\sum_{i=1}^{n} dP_{Di} + \sum_{i=1}^{n} \left.\frac{\partial P_{\text{loss}}}{\partial P_i}\right|_s (dP_{Gi} - dP_{Di}) \right) - \lambda \left(\sum_{i=1}^{n} \left.\frac{\partial P_{\text{loss}}}{\partial P_i}\right|_s dP_{Gi} \right) \\
&= \lambda \sum_{i=1}^{n} \left(1 - \left.\frac{\partial P_{\text{loss}}}{\partial P_i}\right|_s \right) dP_{Di}
\end{aligned}
\tag{5.66}
$$

The LMP at any arbitrary bus i, λ_i, is then obtained from

$$\lambda_i = \frac{\partial C}{\partial P_{Di}} = \lambda\left(1 - \frac{\partial P_{loss}}{\partial P_i}\bigg|_s\right) \tag{5.67}$$

Since the solution of the ED with losses must be independent of the arbitrary selection of the slack bus, a relationship must exist among the sets of sensitivities derived with different slack bus choices. This relationship is derived next. Consider the differential power balance equation for two different slack bus selections, s and r

$$\sum_{i=1}^{n}\left(1 - \frac{\partial P_{loss}}{\partial P_i}\bigg|_s\right)dP_i = 0 \tag{5.68}$$

and

$$\sum_{i=1}^{n}\left(1 - \frac{\partial P_{loss}}{\partial P_i}\bigg|_r\right)dP_i = 0 \tag{5.69}$$

Equation (5.68) can be obtained from Equation (5.69), if the loss sensitivity coefficients with slack bus r are related to the sensitivity coefficients with slack bus s by

$$1 - \frac{\partial P_{loss}}{\partial P_i}\bigg|_r = \frac{1 - \dfrac{\partial P_{loss}}{\partial P_i}\bigg|_s}{1 - \dfrac{\partial P_{loss}}{\partial P_r}\bigg|_s} \tag{5.70}$$

Thus, irrespective of which slack bus is chosen, the incremental power balance equations are equivalent.

Finally, we note that the ED (including losses) is most conveniently solved through nonlinear programing algorithms such as those implemented in solvers MINOS, CONOPT, or SNOPT [3]. Nonetheless, the analytic results derived in the sections above are important as they provide useful insight into the nature of the optimal solution.

Example 5.5 below illustrates the ED with losses.

EXAMPLE 5.5 ED WITH LOSSES

Consider the three-bus, three-line network characterized in the table below and depicted in Figure 5.3 (shut susceptances neglected).

From Bus	To Bus	Resistance (pu)	Reactance (pu)
1	2	0.02	0.1
1	3	0.02	0.1
2	3	0.02	0.1

Buses 1 and 2 include the two units characterized in Example 5.1. The only demand is located in bus 3 and changes with the cases considered below. Voltage magnitudes are constant and equal to 1. A base of 100 kV and 200 MVA is considered.

The table below provides results for the ED with and without losses.

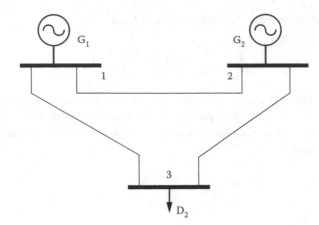

FIGURE 5.3 Three-bus network for Example 5.5.

Case	P_D (MW)	P_{G1} (MW)	P_{G2} (MW)	λ_1 ($/MWh)	λ_2 ($/MWh)	λ_3 ($/MWh)	C ($/h)
			No Loss Cases				
B-Ex.5.3	250	200.0	50.0	30.00	30.00	30.00	6675
C-Ex.5.3	300	233.3	66.7	31.67	31.67	31.67	8217
			Cases with Loss				
B-Ex.5.5	250	200.4	53.3	30	30.02	30.98	6786
C-Ex.5.5	300	234.4	70.8	31.72	32.08	32.92	8383

The following observations are in order:

1. For the lossless cases, the LMPs are identical. For the lossy cases, the LMPs vary through-out the network, these differences becoming more significant as the demand increases.
2. The system losses are below 2% of the demand.
3. The total cost increases for the lossy cases by about 2% due to losses.
4. The losses are allocated to the generating units unevenly. It is relevant and somewhat anti-intuitive that the most expensive generating unit (unit 2) gets the highest share of losses. Why is this so? The reasoning behind this result is left to the reader.

5.2.6 ED: NETWORK IMPACT

The power flow through a transmission line is usually limited by either thermal or stability consider-ations. Thermal limits are imposed by the ability of the line conductors to dissipate the heat created by I^2R losses. Stability limits are imposed on the line power flows by the ability of the system to recover synchronism following a major fault such as a short circuit on the line. Although networks are gen-erally operated and planned so that no transmission capacity limits are active, it is nonetheless pos-sible for line capacity limits to be reached under unusual loads or contingencies. In such cases, the ED must be solved, subject to line flow constraints.

To present an illustrative example that preserves simplicity, we now consider an ED with no losses and no generating limits but including transmission capacity limits on just one of the lines of the sys-tem. We also assume that the power flow through the line whose capacity is limited can be expressed linearly in terms of the power injections, that is,

$$P_F = \sum_{i=1}^{n} \beta_i(P_{Gi} - P_{Di}) = \beta^T(P_G - P_D) \tag{5.71}$$

The resulting ED problem consists in minimizing the generating cost (Equation (5.1)), subject to the power balance condition

$$\sum_{i=1}^{n} (P_{Gi} - P_{Di}) = 0 \qquad\qquad (5.72)$$

and the flow limits imposed on the single line with limited capacity,

$$-P_F^{max} \leq \beta^T(P_G - P_D) \leq P_F^{max} \qquad\qquad (5.73)$$

Observe that only one of the two constraints above may become binding.

The first-order optimality conditions are derived following the same Lagrangian approach as in the previous ED cases. Here, however, there are two equality constraints instead of one, namely, the power balance and one of the extreme line flow limits from the inequalities (Equation (5.73)). These conditions then state that

$$IC_i(P_{Gi}) = \lambda + \gamma\beta_i, \quad \forall i \qquad\qquad (5.74)$$

where γ is the Lagrange multiplier associated with the active line flow limit in one of the two constraints of Equation (5.73).

If the line becomes congested, γ is different from zero, and the condition of Equation (5.74) works similarly as the condition of Equation (5.50) (derived for the lossy case); in other words, forcing the generating units to operate at different incremental costs. The example below, solved using GAMS-MINOS [3], illustrates this ED case.

EXAMPLE 5.6 ED INCLUDING A LINE WITH LIMITED TRANSMISSION CAPACITY

We consider the data of Example 5.5, but impose a transmission limit of 140 MW on line 1–3. The ED results are as follows:

Case	P_D (MW)	P_{G1} (MW)	P_{G2} (MW)	λ_1 ($/MWh)	λ_2 ($/MWh)	λ_3 ($/MWh)	C ($/h)
Results without Losses and without Transmission Limit on Line 1–3							
B-Ex.5.3	250	200	50	30	30	30	6675
C-Ex.5.3	300	233.3	66.7	31.67	31.67	31.67	8217
Results without Losses but with Transmission Limit on Line 1–3							
B-Ex.5.6	250	170.06	79.94	28.50	32.99	37.49	6742
C-Ex.5.6	300	119.94	180.06	26.00	43.01	60.07	9181

The following observations are in order:

1. If line 1–3 is not congested, the LMPs are identical throughout the network. If line 1–3 is congested, the LMPs vary throughout the network and the relative differences increase with load. For case C-Ex.5.6, the LMP of bus 3 is more than twice the LMP of bus 1, indicating that it is comparatively very expensive to supply the demand at bus 3 under congestion.
2. As expected, the total cost increases with congestion. The most significative increment takes place in case C-Ex.5.6.
3. Both generating units share the change in generation symmetrically, one increasing its generating level and the other one reducing it. Since only two generating units are avail able and losses are neglected, no other allocation is possible.
4. The generation levels change much more significantly with congestion than with losses.

5.2.7 NETWORK-CONSTRAINED ED

In this section, we consider the ED with generation limits, losses, and transmission capacity limits. This represents a simplified instance of the optimal power flow (OPF) model, in which voltage magnitudes and reactive power injections are also decision variables. The OPF is treated in detail in Chapter 6.

This particular instance of OPF, denoted henceforth as network-constrained ED, or NC-ED, consists of minimizing the total cost (Equation (5.75)), subject to the generation limits (Equation (5.76)), the load flow equality constraints (Equation (5.77)), and the transmission capacity limits for all lines, that is,

$$\text{minimize}_{P_G, \delta} \quad \sum_{i=1}^{n} C_i(P_{Gi}) \tag{5.75}$$

subject to

$$P_G^{\min} \le P_G \le P_G^{\max} \tag{5.76}$$

$$P_G - P_D = P(\delta) \tag{5.77}$$

$$|P_F(\delta)| \le P_F^{\max} \tag{5.78}$$

It should be noted that all previous cases of ED are particular instances of the above formulation.

The first-order optimality conditions of problem (5.75–5.78) show that the LMPs are characterized by

$$\lambda_i = \frac{\partial C}{\partial P_{Di}}; \quad i = 1, \dots, n \tag{5.79}$$

where the LMPs λ_i are the Lagrangian multipliers associated with the power flow Equation (5.77). The proof of the statement above is suggested as an exercise to the interested reader. The example below illustrates the ED solution obtained for the above form of an NC-ED.

EXAMPLE 5.7 NETWORK-CONSTRAINED ED

We consider the same data as in Example 5.6. Results are provided in the table as follows:

Case	P_D (MW)	P_{G1} (MW)	P_{G2} (MW)	λ_1 ($/MWh)	λ_2 ($/MWh)	λ_3 ($/MWh)	C ($/h)
Cases with No Losses and No Transmission Capacity Limits but with Generation Limits							
A-Ex.5.3	40	40	0 (min)	22	22	22	1140
B-Ex.5.3	250	200	50	30	30	30	6675
C-Ex.5.3	300	233.3	66.7	31.67	31.67	31.67	8217
D-Ex.5.3	600	400 (max)	200	45	45	45	19,300
Cases with No Losses but with Generation and Transmission Capacity Limits							
B-Ex.5.6	250	170.06	79.94	28.50	32.99	37.49	6742
C-Ex.5.6	300	119.94	180.06	26.00	43.01	60.07	9181
Cases with Losses and Generation and Transmission Capacity Limits							
A-Ex.5.7	40	40.11	0 (min)	22.01	22.07	22.13	1142
B-Ex.5.7	250	166.4	87.0	28.32	33.70	39.78	6874
C-Ex.5.7	300	115.55	189.31	25.78	43.93	64.17	9469
D-Ex.5.7	600				Infeasible		

The following observations are in order:

1. For case A-Ex.5.7 (low demand), no transmission congestion occurs. This results in relatively small differences (due to losses) among the LMPs.
2. Cases A-Ex.5.7 and A-Ex.5.3 are similar, differing only in how losses affect the solution.
3. Cases B-Ex.5.7 and C-Ex.5.7, with line 1–3 congested, are similar to cases B-Ex.5.6 and C-Ex.5.6. Any difference again can be attributed to the effect of losses.
4. Case D-Ex.5.7 is infeasible since the demand surpasses the capacity of the transmission network. In the real world, a similar situation would result in a shortage of power to certain consumers.
5. The inclusion of both losses and congestion reduces the generation level of the cheapest unit. Why? The interested reader is encouraged to provide an answer.

Finally, it should be noted that firm bilateral contracts are easily incorporated into the NC-ED. Each firm bilateral agreement results in a constraint, forcing a unit to produce a prespecified quantity and a demand to consume such quantity, irrespective of the cost of the unit engaged in the firm bilateral contract. The example below illustrates the effect of bilateral contracts.

EXAMPLE 5.8 ED WITH BILATERAL CONTRACTS

Considering the units with the cost functions of Example 5.1, suppose that generating unit 2 and the load engage in a bilateral contract of 50 MW at $31.5/MWh. We assume that this contract is physical in nature, which means that the generator is obliged to generate the contracted amount. In contrast, a financial contract would be one in which the generator is only obliged to guarantee the bilateral price but need not generate the contracted power that is then generated by other sources. With this physical contract, unit 2 is obliged to generate 50 MW irrespective of the system's marginal cost.

The profit for generating unit 1, $Profit_1$, is computed as the system marginal cost times its production, minus its production cost. The profit for generating unit 2, $Profit_2$, is computed as the system marginal cost times its production reduced by 50 MW plus the contract price ($31.5/MWh) times 50 MW, minus its production cost.

Case	P_D (MW)	P_{G1} (MW)	P_{G2} (MW)	λ ($/MWh)	$Profit_1$ ($)	$Profit_2$ ($)
			Results with Bilateral Contracts			
A-Ex.5.8	50	0.00	50.00	20.00	−100.0	0.00
B-Ex.5.8	250	200.00	50.00	30.00	900.0	0.00
C-Ex.5.8	295	230.00	65.00	31.50	1222.5	11.25
D-Ex.5.8	400	300.00	100.00	35.00	2150.0	125.00
E-Ex.5.8	600	433.33	167.67	41.67	4594.4	680.56

In case A-Ex.5.8, we see that since the system marginal cost is $20/MWh, unit 2 is making money on the bilateral contract by selling its contracted amount of 50 MW at $31.5/MWh relative to what it would earn by selling at the system incremental cost. On the other hand, the load loses money from this contract since it is receiving power at a rate higher than what it would pay if it bought the power at the system incremental cost. The situation is reversed in case E-Ex.5.8, the selling generator 2 losing money by selling power below the system marginal cost, while the load wins by buying power below the system marginal cost. In case C-Ex.5.8, the bilateral contract is immaterial as the system incremental cost and the contract price are identical.

5.3   UNIT COMMITMENT

5.3.1   BASIC UC

In formulating the ED, we consider that all generating units are online and ready to produce. However, units can be either online or off-line. The extension of the ED problem to units that can be started up and shutdown over time is denoted as UC. The simple example below characterizes the UC problem.

EXAMPLE 5.9 UNIT COMMITMENT

We consider the data of Example 5.1 with neither losses nor transmission capacity limits, but with generation limits. We analyze all different combinations of units online and off-line, that is, $(1,0)$, $(0,1)$, and $(1,1)$, which yield the ED results shown in the table as follows:

Case	P_D (MW)	P_{G1} (MW)	P_{G2} (MW)	λ ($/MWh)	C ($/h)
$(1,0)$	40	40.00	0.00	22.00	940.0
$(0,1)$	40	0.00	40.00	29.00	1280.0
$(1,1)$	40	40.00	0.00	22.00	1140.0
$(1,0)$	250	250.00	0.00	32.50	6662.5
$(0,1)$	250	0.00	250.00	50.00	9575.0
$(1,1)$	250	200.00	50.00	30.00	6675.0
$(1,0)$	300	300.00	0.00	35.00	8350.0
$(0,1)$	300	0.00	300.00	55.00	12,200.0
$(1,1)$	300	233.33	66.67	31.67	8216.7

The following observations are in order:

1. For low demand, the ED with the unit 2 online is more expensive than the ED with just unit 1 online. If unit 2 is started up, its production is zero because its marginal cost ($22/MWh) is below the incremental cost of unit 2 ($25/MWh). The optimal commitment is, therefore, $(1,0)$.
2. For a demand of 250 MW, the optimal commitment remains $(1,0)$ with a production cost of $6662.5/h, slightly below the cost with commitment $(1,1)$ of $6675/h.
3. For the demand of 300 MW, the optimal commitment is $(1,1)$.
4. The number of combinations to be considered per demand level is 2^n (n being the number of generating units), a number that grows exponentially and reaches seemingly intractable values for real-world problems. For instance, if 100 units and 24 demand levels have to be considered, the number of possible combinations is $2^{100 \times 24}$. Nonetheless, as will be seen, such problems must be and are solved by power utilities.

The particular instance of UC analyzed above is denominated static UC since a single time period is considered. Using the OPF-ED formulation (Equations (5.75) through (5.78)), the static UC problem is easily formulated. To this end, the generating limits of unit i, P_{Gi}^{min} and P_{Gi}^{max}, are multiplied by a binary variable u_i. The operating status of a generating unit is then expressed by the couple (P_{Gi}, u_i), and the static UC takes the form

$$\text{minimize}_{u, P_G, \delta} \quad \sum_{i=1}^{n} C_i(u_i, P_{Gi}) \tag{5.80}$$

subject to

$$u_i P_{Gi}^{\min} \leq P_{Gi} \leq u_i P_{Gi}^{\max}; \quad i = 1, \ldots, n \tag{5.81}$$

$$P_G - P_D = P(\delta) \tag{5.82}$$

$$|P_F(\delta)| \leq P_F^{\max} \tag{5.83}$$

In this case, the unit cost is expressed as

$$C_i(u_i, P_{Gi}) = u_i C_{0i} + a_i P_{Gi} + \frac{1}{2} b_i P_{Gi}^2 \tag{5.84}$$

The following observations are pertinent:

1. The optimization variables consist of the triplet of vectors (u, P_G, δ), which includes continuous and binary variables.
2. Problem (5.80)–(5.83) is, in general, a mixed-integer nonlinear programing problem, which is very difficult to solve. However, if the functions are linearized, the resulting mixed-integer linear problem generally becomes much more tractable.
3. If a generating unit is off-line ($u_i = 0$), its generating limits are zero (as imposed by Equation 5.81). This implies that $P_{Gi} = 0$ and $C_i(u_i, P_{Gi}) = 0$.
4. The cost of a generating unit can be expressed as $C_i(u_i, P_{Gi}) = u_i C_{0i} + C_{Vi}(P_{Gi})$ where $C_{Vi}(P_{Gi})$ is a convex function depending solely on P_{Gi}.

More generally, the UC problem is a multiperiod scheduling problem that must include ramping constraints limiting the generating level of any unit between two consecutive time periods. The UC may also include minimum up and down times, which enforce the fact that thermal units if started up must remain online a certain number of time periods, and, conversely, if shutdown must remain off-line a given number of time periods. The process of starting up a generating unit implies an extra cost associated with the fuel consumed to drive the boiler to its working temperature and pressure.

The multiperiod UC formulation described below is sufficiently general to reflect the most significant features of the problem, yet is simple to understand

$$\text{Minimize}_{P_{Gjt}, \forall j, \forall t; \, u_{jt}, \forall j, \forall t}$$

$$\sum_{t=1}^{T} \sum_{j=1}^{n} C_{jt}(u_{jt}, P_{Gjt}) + C_{jt}^{SU} \tag{5.85}$$

subject to

$$C_{jt}^{SU} \geq C_j^{SU}(u_{jt} - u_{j,t-1}) \quad \forall j, \forall t \tag{5.86}$$

$$C_{jt}^{SU} \geq 0 \quad \forall j, \forall t \tag{5.87}$$

$$u_{jt} P_{Gj}^{\min} \leq P_{Gjt} \leq u_{jt} P_{Gj}^{\max} \quad \forall j, \forall t \tag{5.88}$$

$$P_{Gj,t-1} - P_{Gjt} \leq R_{Gj}^{\text{down}} \quad \forall j, \forall t \tag{5.89}$$

$$P_{Gjt} - P_{Gj,t-1} \leq R_{Gj}^{\text{up}} \quad \forall j, \forall t \tag{5.90}$$

$$\sum_{j=1}^{n} P_{Gjt} = P_{Dt} \qquad \forall t \tag{5.91}$$

$$u_{it} \in \{0, 1\} \qquad \forall j, \forall t \tag{5.92}$$

where $C_{jt}(u_{jt}, P_{Gjt})$ is the generating cost function of unit j (known), C_{jt}^{SU} is the start-up cost incurred by unit j at the beginning of period t (variable), C_j^{SU} is the constant start-up cost of unit j (known), P_{Gjt} is the generating level of unit j during period t (variable), P_{Gj}^{min} is the minimum power output of unit j (known), P_{Gj}^{max} is the capacity of unit j (known), P_{Dt} is the demand during time period t (known), R_{Gj}^{down} is the ramping down limit of unit j (known), R_{Gj}^{up} is the ramping up limit of unit j (known), u_{jt} is a binary variable that is equal to 1 if unit j is on line during time period t and zero otherwise (variable), n is the number of generating units, and P_{Gj0} and u_{j0} are the values of the generation outputs and on/off variables at the start of the UC optimization interval (known).

In the formulation above, we assume that $R_{Gj}^{up} \geq P_{Gj}^{min}$ and $R_{Gj}^{down} \geq P_{Gj}^{min}$ for any generating unit j. This implies that when a unit is turned on, it has enough ramping capability to be able to reach the minimum generation level within a single time period.

The objective function (5.85) includes operating and start-up costs throughout the planning horizon. Constraints (5.86) and (5.87) are needed to model start-up costs. Constraints (5.88) ensure that all units operate within their operating limits. Constraints (5.89) and (5.90) are ramping limits, while constraints (5.91) enforce the power balances for all time periods in the optimization horizon. Finally, constraints (5.92) state that the on/off generation variables are binary.

For the sake of simplicity, in the UC formulation above, no network constraint is imposed. Moreover, no minimum up or down times are enforced on generating units. Additionally, no shutdown cost is considered in the objective function.

Reserve constraints that guarantee a specific security level can be incorporated in the UC problem as

$$\sum_{j=1}^{n} u_{jt} P_{Gj}^{max} \geq P_{Dt} + P_{Rt} \qquad \forall t \tag{5.93}$$

where P_{Rt} is a specified reserve level. This inequality ensures that at any time t there is sufficient scheduled generation capacity to be able to balance the load after any random generation outage or load variation. The bigger the specified reserve P_{Rt}, the more secure the system. On the other side of the coin, if too much reserve is specified, the resulting cost may be excessive or the problem may become infeasible.

The above constraint usually leads to a larger number of online units with reduced generating levels. This circumstance increases the available spinning reserve ready to replace a unit that unexpectedly trips or to compensate for a sudden increase or decrease in demand. The example below illustrates the multiperiod UC problem.

EXAMPLE 5.10 MULTIPERIOD UC

The units of Example 5.1 including an additional unit are considered in this example. The reserve imposed, P_{Rt}, should be higher than or equal to 20% of the demand. Additional data pertaining to these generation units are provided in the following table.

Unit	C_0 ($/h)	a ($/MWh)	b [$/(MW)^2 h]	P_G^{min} (MW)	P_G^{max} (MW)
1	100	20	0.05	0	400
2	200	25	0.10	0	300
3	300	40	0.20	0	250

Unit	Start-Up Cost ($/h)	Ramp-Up Limit (MW/h)	Ramp-Down Limit (MW/h)	Initial Power Output (MW)	Initial Status (On/Off)
1	300	160	160	0	Off
2	400	150	150	0	Off
3	500	100	100	0	Off

The solutions of different instances of the UC problem are provided in the table as follows:

CASE A-EX.5.10
UC without Including Start-Up Costs and Ramping Limits

Period	P_D^{total} (MW)	P_{G1} (MW)	P_{G2} (MW)	P_{G3} (MW)	Cost ($)
1	40	40.0	0.0	0	940.0
2	250	250.0	0.0	0	6662.5
3	300	233.3	66.7	0	8216.7
4	600	400.0	200	0	19,300.0
Total	1190	923.3	266.7	0	35,119.2

CASE B-EX.5.10
UC Including Start-Up Costs but No Ramping Limits

Period	P_D^{total} (MW)	P_{G1} (MW)	P_{G2} (MW)	P_{G3} (MW)	Cost ($)
1	40	40.0	0.0	0	1240.0
2	250	250.0	0.0	0	6662.5
3	300	233.3	66.7	0	8616.7
4	600	400.0	200	0	19,300.0
Total	1190	923.3	266.7	0	35,819.2

CASE C-EX.5.10
UC Including Start-Up Costs and Ramping Limits

Period	P_D^{total} (MW)	P_{G1} (MW)	P_{G2} (MW)	P_{G3} (MW)	Cost ($)
1	40	40.0	0.0	0	1240.0
2	250	200.0	50.0	0	7075.0
3	300	240.0	60.0	0	8220.0
4	600	400.0	200.0	0	19,300.0
Total	1190	880.0	310.0	0	35,835.0

CASE D-EX.5.10
UC Including Start-Up Costs, Ramping Limits, and Reserve Constraints

Period	P_D^{total} (MW)	P_{G1} (MW)	P_{G2} (MW)	P_{G3} (MW)	Cost ($)
1	40	40.0	0.0	0	1240.0
2	250	200.0	50.0	0	7075.0
3	300	240.0	60.0	0	8220.0
4	600	400.0	183.3	16.7	20,058.3
Total	1190	880.0	293.3	16.7	36,593.3

The following observations are in order:

1. If neither start-up cost nor ramping limits are included (case A-Ex.5.10), the solution obtained is similar to the solution of Example 5.9 (sequence of static UCs).
2. Including start-up costs (case B-Ex.5.10) results in a higher total cost but no changes in the generating levels of the units. The cost increments during periods 1 and 3 are due to the start-up of some units in those periods.
3. Including ramping limits (case C-Ex.5.10) leads to changes in the generating levels of units 1 and 2 as well as to a higher total cost.
4. Including reserve constraints (case D-Ex.5.10) results in a higher cost and in changes to the UC. Note that in this case, unit 3 is committed on during hour 4 for reserve purposes.

The hydrothermal coordination problem is similar to a UC problem that also includes hydroelectric units generally coupled through water balance equations, reflecting the continuity of water flow through river systems. In this case, in addition to constraints enforcing water balances, flow and volume limits must also be considered.

5.3.2 UC under Production Uncertainty

We consider below a UC problem with production uncertainty. Specifically, besides thermal units, we consider a single stochastic unit, for example, a wind unit with zero production cost. We first assume that we have perfect knowledge of the hourly production of this stochastic unit, that is, P_{Wt}, $\forall t$. For simplicity we also assume that ramping constraints of thermal units are not binding and thus can be neglected, and that the operating costs of these units are linear, that is, $C_{jt}(u_{jt}, P_{Gjt}) = C_j P_{Gjt}$, where C_j is the production cost in \$/MWh of thermal unit j. Thus, the UC problem 5.85–5.92 becomes:

$$\text{Minimize}_{P_{Gjt}, \forall j, \forall t; \, u_{jt}, \forall j, \forall t}$$

$$\sum_{t=1}^{T} \sum_{j=1}^{n} C_{jt}^{SU} + C_j P_{Gjt} \tag{5.94}$$

subject to

$$C_{jt}^{SU} \geq C_j^{SU}(u_{jt} - u_{j,t-1}) \quad \forall j, \, \forall t \tag{5.95}$$

$$C_{jt}^{SU} \geq 0 \quad \forall j, \, \forall t \tag{5.96}$$

$$u_{jt} P_{Gj}^{\min} \leq P_{Gjt} \leq u_{jt} P_{Gj}^{\max} \quad \forall j, \, \forall t \tag{5.97}$$

$$\sum_{j=1}^{n} P_{Gjt} + P_{Wt} = P_{Dt} \quad \forall t \tag{5.98}$$

$$u_{it} \in \{0, 1\} \quad \forall j, \, \forall t. \tag{5.99}$$

Note that the production of the wind unit at time t, P_{Wt}, appears in Equation (5.98).

Next we consider that the production of the wind unit is uncertain, but can be characterized by s plausible production levels (that we call *scenarios*), each with an occurrence probability. That is:

$$P_{Wt} = \begin{cases} P_{Wt}^1 & \text{with probability } \pi_1 \\ \vdots & \vdots \\ P_{Wt}^\omega & \text{with probability } \pi_\omega \quad \forall t, \\ \vdots & \vdots \\ P_{Wt}^s & \text{with probability } \pi_s \end{cases} \tag{5.100}$$

and $\sum_{\omega=1}^{s} \pi_\omega = 1$. These probability levels may also vary with time t.

To accommodate wind fluctuations, we assume that the thermal units are flexible and provide *reserves* that can be deployed to increase ($0 \leq R_{Gjt}^{U\omega} \leq R_{Gj}^{U\max}$) or decrease ($0 \leq R_{Gjt}^{D\omega} \leq R_{Gj}^{D\max}$) their power outputs during operation. Specifically, $R_{Gjt}^{U\omega}$ and $R_{Gjt}^{U\omega}$ are the increase and decrease, respectively, in the power output of thermal unit j at time t to compensate for the random wind production changes.

We then consider a scheduling stage (that takes place well in advance of power delivery), identified by the superscript S, accounting for the s random operating conditions, corresponding to operating with each of the plausible wind production levels (scenarios). Considering this, the *stochastic* UC problem can be formulated as:

$$\text{Minimize}_{u_{jt}, \forall j, \forall t; P_{Gjt}^S, \forall j, \forall t; R_{Gjt}^{U\omega}, R_{Gjt}^{D\omega}, \forall j, \forall t, \forall \omega; P_{Wt}^S, \forall t}$$

$$\sum_{t=1}^{T} \sum_{j=1}^{n} C_j^{SU} + C_j P_{Gjt}^S + \sum_{\omega=1}^{s} \sum_{t=1}^{T} \sum_{j=1}^{n} \pi_\omega C_j (R_{Gjt}^{U\omega} - R_{Gjt}^{D\omega}) \tag{5.101}$$

subject to

$$C_{jt}^{SU} \geq C_j^{SU}(u_{jt} - u_{j,t-1}) \quad \forall j, \ \forall t \tag{5.102}$$

$$C_{jt}^{SU} \geq 0 \quad \forall j, \ \forall t \tag{5.103}$$

$$u_{jt} P_{Gj}^{\min} \leq P_{Gjt}^S \leq u_{jt} P_{Gj}^{\max} \quad \forall j, \ \forall t \tag{5.104}$$

$$\sum_{j=1}^{n} P_{Gjt}^S + P_{Wt}^S = P_{Dt} \quad \forall t \tag{5.105}$$

$$u_{jt} P_{Gj}^{\min} \leq (P_{Gjt}^S + R_{Gjt}^{U\omega} - R_{Gjt}^{D\omega}) \leq u_{jt} P_{Gj}^{\max} \quad \forall j, \ \forall t, \ \forall \omega \tag{5.106}$$

$$\sum_{j=1}^{n} (P_{Gjt}^S + R_{Gjt}^{U\omega} - R_{Gjt}^{D\omega}) + P_{Wt}^\omega = P_{Dt} \quad \forall t, \ \forall \omega \tag{5.107}$$

$$0 \leq R_{Gjt}^{U\omega} \leq R_{Gj}^{U\max} \quad \forall j, \ \forall t, \ \forall \omega \tag{5.108}$$

$$0 \leq R_{Gjt}^{D\omega} \leq R_{Gj}^{D\max} \quad \forall j, \ \forall t, \ \forall \omega \tag{5.109}$$

$$u_{it} \in \{0, 1\} \quad \forall j, \ \forall t. \tag{5.110}$$

The variables of this stochastic UC problem include u_{jt}, $\forall j$, $\forall t$, the binary commitment variables at the scheduling stage; P^S_{Gjt}, $\forall j$, $\forall t$, the scheduled production of each thermal unit; $R^{U\omega}_{Gjt}$, $R^{D\omega}_{Gjt}$, $\forall j$, $\forall t$, $\forall \omega$, the up/down reserve deployment (production changes) of each thermal unit to accommodate wind variability; and P^S_{Wt}, $\forall t$, the *production* of the wind unit at the scheduling stage.

In addition to the constants previously defined, $R^{U\,max}_{Gj}$, $\forall j$, $\forall t$, and $R^{D\,max}_{Gj}$, $\forall j$, $\forall t$, are the maximum up and down reserve deployments of the thermal units, respectively.

The objective function (5.101) represents the expected value of the cost based on the probabilities of the various wind scenarios. It includes start-up costs, operation costs at the scheduling stage and expected costs due to production changes in all the plausible operating conditions. Constraints (5.102)–(5.103) model the start-up cost at the scheduling state. Constraints (5.104) and (5.105) establish production limits and energy balance, respectively, at the scheduling stage. On the other hand, constraints (5.106) and (5.107) establish production limits and energy balance, respectively, for each operating condition. Constraints (5.108) and (5.109) enforce up and down reserve deployment limits. Constraints (5.110) are binary variable declarations.

The simple example below illustrates this stochastic UC formulation.

EXAMPLE 5.11 UC UNDER PRODUCTION UNCERTAINTY

Consider a 4-period scheduling horizon with 2 thermal unit and 1 wind unit.
  Demand data are provided in the table below:

Period	P^{Total}_D (MW)
1	40
2	250
3	300
4	600

Data for the two thermal units are given in the table below:

Production Unit	Initial Status (on/off)	Variable Cost ($/MWh)	Start-Up Cost ($)	P^{min}_{Gj} (MW)	P^{max}_{Gj} (MW)	R^{Umax}_{Gj} (MW)	R^{Dmax}_{Gj} (MW)
1	Off	20	300	0	400	400	400
2	Off	25	400	0	300	300	300

Data for the wind unit are provided in the table below:

	Wind production (MW)	
	0.2	0.8
Probability Period	Scenario 1 (low wind)	Scenario 2 (high wind)
1	10	30
2	40	100
3	30	70
4	20	60

Results at the scheduling stage are provided in the table below:

Period	P_D^{Total} (MW)	P_{G1}^S (MW)	P_{G2}^S (MW)	P_W^S (MW)
1	40	10	0	30
2	250	150	0	100
3	300	230	0	70
4	600	400	140	60

Scheduling results in the table above correspond to the "pre-positioning" of the generation system to face the two possible operating conditions: low wind and high wind.

Results for scenario 1 are provided in the table below:

Period	P_D^{Total} (MW)	P_{G1}^1 (MW)	P_{G2}^1 (MW)	P_W^1 (MW)	Cost ($)
1	40	30	0	10	900
2	250	210	0	40	4200
3	300	270	0	30	5400
4	600	400	180	20	12,900
Total	–	–	–	–	23,400

Finally, results for scenario 2 are provided in the table below:

Period	P_D^{Total} (MW)	P_{G1}^2 (MW)	P_{G2}^2 (MW)	P_W^2 (MW)	Cost ($)
1	40	10	0	30	500
2	250	150	0	100	3000
3	300	230	0	70	4600
4	600	400	140	60	11,900
Total	–	–	–	–	20,000

The following observations are in order. Since wind production is low in scenario 1, thermal unit 1 deploys up reserve in periods 1, 2 and 3 (20,60 and 40 MW, respectively) while thermal unit 2 deploys it in period 4 (40 MW). Since wind production is high in scenario 2, the total cost is lower in scenario 2 ($20,000) than in scenario 1 ($23,400).

Further details on UC with stochastic sources can be found in [6].

5.4 ELECTRICITY MARKET OPERATIONS

5.4.1 FUNDAMENTALS

An electricity market includes different trading mechanisms, primarily the pool, bilateral contracting, and forward contracting through the futures market. The pool is considered below.

Generating companies (GENCOS) submit generating offers to the pool, while loads submit consumption bids. The market operator (PX) then uses a market-clearing procedure to determine the accepted offers and bids, and the market-clearing prices. The market operator manages the day-ahead market, which is cleared one day in advance and spans typically 24 h. The market operator also manages shorter horizon markets for deviations and adjustments. These short-term balancing markets, which are cleared every few hours, generally have a comparatively small economic impact.

Considering this short-term economic viewpoint, this chapter, therefore, focuses particularly on the pool operation.

Bilateral contracting among producers and consumers is usually allowed, provided appropriate information is sent to the market operator in sufficient time. Forward contracts spanning longer time periods (one week to one year) allow both producers and consumers to hedge against the volatility of their respective profits or costs.

Retailers buy energy in the pool and through bilateral and forward contracting to supply their own consumers. As such, retailers face uncertainty from both directions: if buying, from the pool price volatility, and if selling, from demand uncertainty and in the longer term from the fact that customers might switch to rival retailers.

The market operator collaborates with the independent system operator (ISO) to manage ancillary service markets such as AGC, reserve, reactive power, and voltage control. However, these markets are outside the scope of this chapter.

It should be noted that in the medium and in the long term, both producers and consumers should resolve the trade off between bilateral/forward contracting and pool involvement. For producers/consumers, contracting (both bilateral and forward) generally leads to stable but low/high average selling/buying prices, while pool involvement generally results in uncertain but comparatively high/low selling/buying average prices.

In the above framework, short-term refers to spans from seconds to one week, medium term to spans ranging from one week to several months, and long term to spans from several months to several years.

The market is overseen by a regulator whose functions and capabilities differ from one market to another. Note also that since electricity transactions are implemented through transmission and distribution networks, such transactions should be approved by the system operator.

Relevant manuals on electricity markets include those by Ilic et al. [7], Chao and Huntington [8], Sheblé [9], Shahidehpour and Li [10], and Kirschen and Strbac [11]. In the following sections, market-clearing procedures are addressed first, followed by an analysis of the producer and consumer market strategies.

5.4.2   CENTRALIZED DISPATCH VERSUS MARKET OPERATION

In the basic ED (Section 5.2.2) analyzed in the first part of this chapter, a central operator solves the scheduling and dispatching problems with full knowledge of the economic and technical data of the generating units and demand.

Within a market environment, this basic ED problem can be solved alternatively as follows [12]:

1. A market operator broadcasts trial market-clearing prices.
2. Considering these prices, the producers determine their optimal productions (as in Section 5.6) and submit them to the market operator.
3. For every hour of the market horizon, the market operator computes the power mismatch. If mismatches are small enough, the procedure concludes. If not, the market operator modifies trial hourly prices proportionally to the respective mismatches, and the procedure continues from 1.

Note that in the above procedure, no information pertaining to producers or consumers is transferred to the market operator. Therefore, in contrast to the ED approach, this procedure is decentralized.

It can be proved that both procedures (ED and competitive market operation) lead to the same solution in terms of productions, consumptions, and prices. And this is also true if elastic demand is considered. However, both procedures might differ (although not significantly if the number of generating units is large enough) if commitment (on/off) decisions of generating units are considered.

The following sections of this chapter provide a market viewpoint to electricity supply and consumption.

5.5 MARKET-CLEARING PROCEDURES

For a given set of offers by producers and bids by consumers, a market-clearing procedure is an algorithm used by the market operator to determine (i) accepted offers, (ii) accepted bids, and (iii) clearing prices. These algorithms are auctions of diverse complexity.

Offers by producers are generally considered to be monotonically increasing, while bids by consumers are considered monotonically decreasing. Three auctions are considered below, namely, single-period auction, multiperiod auction [13], and network-constrained multiperiod auction [14].

5.5.1 SINGLE-PERIOD AUCTION

In the single-period auction, a single time period is considered, therefore, neglecting (or oversimplifying) intertemporal coupling. This assumption may result in operating infeasibilities for both producers and consumers.

The market operator collects generating offers (increasing in price) by producers and load bids (decreasing in price) by consumers and clears the market by maximizing the social welfare [15], or declared social welfare if producers/consumers do not submit actual costs/utilities.

A single-period auction can be formulated as

$$\text{maximize}_{P_{Djk}, \forall j,k; P_{Gib}, \forall i,b} \quad SW^S = \sum_{j=1}^{N_D} \sum_{k=1}^{N_{Dj}} \lambda_{Djk} P_{Djk} - \sum_{i=1}^{N_G} \sum_{b=1}^{N_{Gi}} \lambda_{Gib} P_{Gib} \tag{5.111}$$

subject to

$$0 \leq P_{Djk} \leq P_{Djk}^{\max} \quad \forall j, k \tag{5.112}$$

$$0 \leq P_{Gib} \leq P_{Gib}^{\max} \quad \forall i, b \tag{5.113}$$

$$\sum_{j=1}^{N_D} \sum_{k=1}^{N_{Dj}} P_{Djk} = \sum_{i=1}^{N_G} \sum_{b=1}^{N_{Gi}} P_{Gib} \tag{5.114}$$

where SW^S is the single-period social welfare or declared social welfare (objective function), P_{Djk} is the power block k bid by demand j (variable), P_{Gib} is the power block b offered by generating unit i (variable), $P_{Djk}^{\max}$ is the MW size of block k bid by demand j (constant), $P_{Gib}^{\max}$ is the MW size of block b offered by generating unit i (constant), λ_{Djk} is the price ($/MWh) of block k bid by demand j (constant), λ_{Gib} is the price ($/MWh) of block b offered by generating unit i (constant), N_{Dj} is the number of blocks bid by demand j, N_{Gi} is the number of blocks offered by generating unit i, N_D is the number of demands, and N_G is the number of generating units.

Objective function (5.111) is composed of the difference between two terms. The first one is the sum of the demand blocks times their respective bid prices. The second is the sum of the generating blocks times their respective offer prices. Their difference, SW^S, defines the social welfare for the single period considered.

Constraints (5.112) establish bounds and declare nonnegativity of the demand power block bids, while constraints (5.113) establish bounds and declare the nonnegativity of the generating units power block offers. Constraint (5.114) ensures that the market is cleared, that is, that the total accepted demand equal the total accepted generation.

Note that the optimization variables are the demand and the generation power blocks, P_{Djk}, $\forall j$, k and P_{Dib}, $\forall i$, b.

The marginal market-clearing price (MCP) is the dual variable of constraint (5.114). It coincides either with the price of the most expensive generation block that has been accepted or the price of the cheapest demand block that has been accepted. The problem above is a linear programing problem of moderate size, whose solution can be easily obtained [3].

If ramping up and ramping down limits for generating units are considered, the constraints below should be added to problem (5.111)–(5.114):

$$P_{Gi}^0 - \sum_{b=1}^{N_{Gi}} P_{Gib} \le R_{Gi}^{\text{down}} \quad \forall i \tag{5.115}$$

$$\sum_{b=1}^{N_{Gi}} P_{Gib} - P_{Gi}^0 \le R_{Gi}^{\text{up}} \quad \forall i \tag{5.116}$$

where P_{Gi}^0 is the generating level of unit i just before the auction period, and R_{Gi}^{down} and R_{Gi}^{up} are its ramping down and ramping up limits, respectively.

If minimum and maximum generating levels are imposed on the units, the constraints indicated below should also be added to problem (5.111)–(5.114). Note that in this case a binary variable, u_i, is required per generating unit to force the optimization scheme to consider all possible combinations of active power generation limits. These constraints are defined by inequalities (5.117) below:

$$u_i P_{Gi}^{\text{min}} \le \sum_{b=1}^{N_{Gi}} P_{Gib} \le u_i P_{Gi}^{\text{max}} \quad \forall i \tag{5.117}$$

Similarly a minimum demand level can be imposed on any load, that is,

$$P_{Dj}^{\text{min}} \le \sum_{k=1}^{N_{Dj}} P_{Djk} \quad \forall j \tag{5.118}$$

As minimum demands generally need to be supplied, no binary variables are used in constraints (5.118) above. In other words, the demand will always be above its specified minimum value since no binary variable allows the demand to become zero.

If the noncontinuous constraints (5.117) were considered, marginal prices could not be directly computed. In this case, it would be necessary to solve the mixed-integer problem (5.111–5.118) first, then fix the binary variables to their optimal values, and finally solve the resulting continuous problem. The dual variables corresponding to constraints (5.114) of this continuous problem become the marginal MCPs.

Example 5.11 below illustrates different instances of the single-period auction problem (5.111–5.118).

EXAMPLE 5.12 SINGLE-PERIOD AUCTION

We consider three generating units and two demands. Each unit offers three blocks, while each demand bids four blocks. The technical characteristics of the generating units are given in the table as follows:

Unit Data	Unit 1	Unit 2	Unit 3
Capacity (MW)	30	25	25
Minimum power output (MW)	5	8	10
Ramp-up/down limit (MW/h)	5	10	10
Initial status (on/off)	on	on	on
Initial power output (MW)	10	15	10

Generator offers and demand bids are provided in the tables as follow:

Offers		Unit 1			Unit 2			Unit 3	
Block	1	2	3	1	2	3	1	2	3
Power (MW)	5	12	13	8	8	9	10	10	5
Price ($/MWh)	1	3	3.5	4.5	5	6	8	9	10

Bids		Demand 1				Demand 2		
Block	1	2	3	4	1	2	3	4
Power (MW)	8	5	5	3	7	4	4	3
Price ($/MWh)	20	15	7	4	18	16	11	3

The solution of different instances of the single-period auction are provided in the tables as follows:

Figure 5.4a and b illustrates cases B-Ex.5.11 and C-Ex.5.11, respectively. Case A-Ex.5.11 is illustrated in Figure 5.4a. Thin lines represent the case with no constraint (ramping limit or minimum power output), while thick lines represent the cases with constraints (either ramping limits, Figure 5.4a, or minimum power outputs, Figure 5.4b).

CASE A-EX.5.11
No Ramping Limits and No Minimum Power Output

Accepted Offers (MW)					Accepted Bids (MW)					
Block	1	2	3	Total	Block	1	2	3	4	Total
Unit 1	5	12	13	30	Demand 1	8	5	5	0	18
Unit 2	3	0	0	3	Demand 2	7	4	4	0	15
Unit 3	0	0	0	0						
Total	—	—	—	33	Total	—	—	—	—	33

Social welfare $404

Market-clearing price $4.5/MWh

CASE B-EX.5.11
Ramping Limits Enforced but No Minimum Power Output

	Accepted Offers (MW)					Accepted Bids (MW)				
Block	1	2	3	Total	Block	1	2	3	4	Total
Unit 1	5	10	0	15	Demand 1	8	5	5	0	18
Unit 2	8	8	2	18	Demand 2	7	4	4	0	15
Unit 3	0	0	0	0						
Total	—	—	—	33	Total	—	—	—	—	33

Social welfare	$381
Market-clearing price	$6/MWh

CASE C-EX.5.11
Minimum Power Output Enforced but No Ramping Limits

	Accepted Offers (MW)					Accepted Bids (MW)				
Block	1	2	3	Total	Block	1	2	3	4	Total
Unit 1	5	12	13	30	Demand 1	8	5	2	0	15
Unit 2	0	0	0	0	Demand 2	7	4	4	0	15
Unit 3	0	0	0	0						
Total	—	—	—	30	Total	—	—	—	—	30

Social welfare	$396.5
Market-clearing price	$7.0/MWh

Observe that ramping limits lead to a price increase in the aggregated offer curve (Figure 5.4a). On the other hand, enforcing minimum power output yields a solution to the left of where the offer and bid curves cross (Figure 5.4b).

The following observations are in order:

1. The MCP for case A-Ex.5.11 corresponds to the most expensive generating block that has been accepted ($4.5/MWh).
2. Ramping constraints (case B-Ex.5.11) limit the production level of the cheapest unit, which results in smaller social welfare ($381 versus $404). The resulting MCP increases ($6 versus $4.5/MWh).

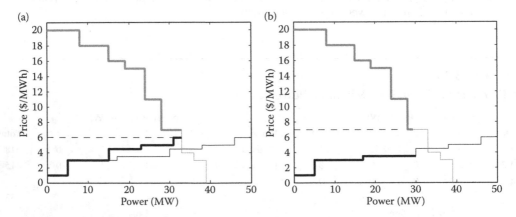

FIGURE 5.4 Single-period auction (Example 5.11): (a) ramping limits enforced but no minimum power output and (b) minimum power output enforced but no ramping limits.

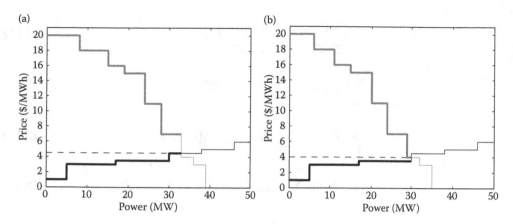

FIGURE 5.5 Single-period auction including no constraints (Example 5.11): (a) a generating block at margin and (b) a demand block at margin.

3. Minimum power outputs (case C-Ex.5.11) lead to unit 2 being scheduled out of the market, which results in a reduction in the demand supplied (30 versus 33 MWh), reduced social welfare ($396.5 versus $404), and a higher MCP ($7.0 versus $4.5/MWh).
4. The single-period auction is formally identical to the ED with elastic demand (Section 5.2.3) analyzed in the first part of this chapter.

If demand bids are modified as indicated in the table below, the resulting MCP in the case of no ramping limits and no minimum power output is $4.0/MWh, while the social welfare becomes $345.5. Observe that the crossing of the supply and demand blocks (a demand block is at margin) is the opposite of the crossing in case A-Ex.5.11 (a generating block is at margin). Figure 5.5 illustrates case A-Ex.5.11 (Figure 5.5a) and the opposite crossing case (Figure 5.5b).

Bids	Demand 1				Demand 2			
Block	1	2	3	4	1	2	3	4
Energy (MWh)	6	5	5	3	5	4	4	3
Price ($/MWh)	20	15	7	4	18	16	11	3

5.5.2 MULTIPERIOD AUCTION

In a multiperiod auction several time periods (typically 24) are considered simultaneously, and so, intertemporal constraints of the generating units must be taken into account. The objective is to maximize the social welfare of the multiperiod market horizon.

The multiperiod auction is then formulated as,

$$\text{maximize}_{P_{Djkt},\forall j,k,t; P_{Gibt},\forall i,b,t} \quad SW^M = \sum_{t=1}^{T}\left[\sum_{j=1}^{N_D}\sum_{k=1}^{N_{Djt}}\lambda_{Djkt}P_{Djkt} - \sum_{i=1}^{N_G}\sum_{b=1}^{N_{Git}}\lambda_{Gibt}P_{Gibt}\right] \quad (5.119)$$

subject to

$$0 < P_{Djkt} < P_{Djkt}^{max} \quad \forall j, k, t \quad (5.120)$$

$$0 \le P_{Gibt} \le P_{Gibt}^{max} \quad \forall i, b, t \quad (5.121)$$

$$\sum_{j=1}^{N_D} \sum_{k=1}^{N_{Djt}} P_{Djkt} = \sum_{i=1}^{N_G} \sum_{b=1}^{N_{Git}} P_{Gibt} \quad \forall t \tag{5.122}$$

$$\sum_{b=1}^{N_{Gi,t-1}} P_{Gib,t-1} - \sum_{b=1}^{N_{Git}} P_{Gibt} \leq R_{Gi}^{\text{down}} \quad \forall i, t \tag{5.123}$$

$$\sum_{b=1}^{N_{Git}} P_{Gibt} - \sum_{b=1}^{N_{Gi,t-1}} P_{Gib,t-1} \leq R_{Gi}^{\text{up}} \quad \forall i, t \tag{5.124}$$

$$u_{it} P_{Gi}^{\text{min}} \leq \sum_{b=1}^{N_{Git}} P_{Gibt} \leq u_{it} P_{Gi}^{\text{max}} \quad \forall i, t \tag{5.125}$$

$$P_{Djt}^{\text{min}} \leq \sum_{k=1}^{N_{Djt}} P_{Djkt} \quad \forall j, t \tag{5.126}$$

where SW^M is the multiperiod social welfare (objective function), P_{Djkt} is the power block k bid by demand j at time t (variable), P_{Gibt} is the power block b offered by generating unit i at time t (variable), P_{Djkt}^{max} is the MW size of block k bid by demand j at time t (constant), P_{Gibt}^{max} is the MW size of block b offered by generating unit i at time t (constant), λ_{Djkt} is the price (\$/MWh) of block k bid by demand j at time t (constant), λ_{Gibt} is the price (\$/MWh) of block b offered by generating unit i at time t (constant), N_{Djt} is the number of blocks bid by demand j at time t, N_{Git} is the number of blocks offered by generating unit i at time t, u_{it} is the status binary variable of unit i at time t (1 if on and 0 if off), P_{Djt}^{min} is the minimum load of demand j at time t, and T is the number of time period of the market horizon. Note that $\sum_{b=1}^{N_{Gi0}} P_{Gib0}$ is equal to the initial power output of unit i.

The objective function (5.119) is similar to the static case objective function in Equation (5.111) but includes an additional summation over time. It represents the social welfare (or declared social welfare) over the multiperiod market horizon, SW^M.

Constraints (5.120) and (5.121) establish bounds and declare the nonnegativity of the power blocks bid by the demands and offered by the generating units, respectively. Constraint (5.122) ensures that the market is cleared at every time period, that is, the accepted demand equals the accepted generation at any time period. Constraints (5.123) and (5.124) enforce down and up ramping limits for all generating units across all time periods. Constraints (5.125) and (5.126) impose minimum and maximum power outputs on the generating units and minimum load requirements on the demands across all time periods.

The decision variables are P_{Djkt}, $\forall j, k, t$; P_{Gibt}, $\forall i, b, t$; and u_{it}, $\forall i, t$; that is, the demand power blocks, the generating power blocks, and the unit status variables.

The MCPs are the dual variables of the power balance equations (5.122) once the binary variables have been fixed to their optimal values and the resulting continuous linear programing problem is solved.

The problem (5.119)–(5.126) is a moderate sized mixed-integer linear programing problem that can be easily solved [3].

Example 5.12 below illustrates different instances of the multiperiod auction considered in this section.

EXAMPLE 5.13 MULTIPERIOD AUCTION

We consider three generating units, two demands, and a market horizon spanning 2 h. The data of Example 5.11 are used in this example. Additional data are provided below. The generating units offer identically in both periods, while the demands bid in the first period as in Example 5.11 and in the second period as indicated in the table as follows:

Bids	Demand 1				Demand 2			
Block	1	2	3	4	1	2	3	4
Energy (MWh)	17	14	8	5	13	11	4	2
Price ($/MWh)	20	15	7	4	18	16	11	3

The solutions of different instances of the multiperiod auction problem are provided in the table as follows:

1. No ramping limits and no minimum power outputs, case A-Ex.5.12.

Accepted Offers (MW)										
Hour	1					2				
Block	1	2	3	4	Total	1	2	3	4	Total
Unit 1	5	12	13	—	30	5	12	13	—	30
Unit 2	3	0	0	—	3	8	8	9	—	25
Unit 3	0	0	0	—	0	4	0	0	—	4
Total	—	—	—	—	33	—	—	—	—	59

Accepted Bids (MW)										
Hour	1					2				
Block	1	2	3	4	Total	1	2	3	4	Total
Demand 1	8	5	5	0	18	17	14	0	0	31
Demand 2	7	4	4	0	15	13	11	4	0	28
Total	—	—	—	—	33	—	—	—	—	59

Prices, Revenues, Payments, and Social Welfare			
Hour	1	2	Total
Price ($/MWh)	4.5	8.0	—
Revenue unit 1 ($)	135	240	375
Revenue unit 2 ($)	13.5	200	213.5
Revenue unit 3 ($)	0	32	32
Total revenue ($)	148.5	472	620.5
Payment demand 1 ($)	81	248	329
Payment demand 2 ($)	67.5	224	291.5
Total payment ($)	148.5	472	620.5
Social welfare ($)		1159.5	
Producer surplus ($)		272.0	
Consumer surplus ($)		887.5	

Note that the consumer surplus is the difference between what consumers are willing to pay and what they actually pay. Similarly, the producer surplus is the difference between what producers actually receive and what they asked. The social welfare equals the sum of the consumer and the producer surpluses.

2. Ramping limits enforced but no minimum power outputs, case B-Ex.5.12.

			Accepted Offers (MW)							
Hour			1					2		
Block	1	2	3	4	Total	1	2	3	4	Total
Unit 1	5	10	0	—	15	5	12	3	—	20
Unit 2	8	8	2	—	18	8	8	9	—	25
Unit 3	0	0	0	—	0	10	0	0	—	10
Total	—	—	—	—	33	—	—	—	—	55
			Accepted Bids (MW)							
Hour			1					2		
Block	1	2	3	4	Total	1	2	3	4	Total
Demand 1	8	5	5	0	18	17	14	0	0	31
Demand 2	7	4	4	0	15	13	11	0	0	24
Total	—	—	—	—	33	—	—	—	—	55

Prices, Revenues, Payments, and Social Welfare			
Hour	1	2	Total
Price ($/MWh)	6.0	11.0	—
Revenue unit 1 ($)	90	220	310
Revenue unit 2 ($)	108	275	383
Revenue unit 3 ($)	0	110	110
Total revenue ($)	198	605	803
Payment demand 1 ($)	108	341	449
Payment demand 2 ($)	90	264	354
Total payment ($)	198	605	803
Social welfare ($)		1079.5	
Producer surplus ($)		418.5	
Consumer surplus ($)		661.0	

3. Minimum power output enforced but no ramping limits, case C-Ex.5.12.

			Accepted Offers (MW)							
Hour			1					2		
Block	1	2	3	4	Total	1	2	3	4	Total
Unit 1	5	12	13	—	30	5	12	13	—	30
Unit 2	0	0	0	—	0	8	8	9	—	25
Unit 3	0	0	0	—	0	0	0	0	—	0
Total	—	—	—	—	30	—	—	—	—	55
			Accepted Bids (MW)							
Hour			1					2		
Block	1	2	3	4	Total	1	2	3	4	Total
Demand 1	8	5	2	0	15	17	14	0	0	31
Demand 2	7	4	4	0	15	13	11	0	0	24
Total	—	—	—	—	30	—	—	—	—	55

Prices, Revenues, Payments, and Social Welfare			
Hour	1	2	Total
Price ($/MWh)	7.0	11.0	—
Revenue unit 1 ($)	210	330	540
Revenue unit 2 ($)	0	275	275
Revenue unit 3 ($)	0	0	0
Total revenue ($)	210	605	815
Payment demand 1 ($)	105	341	446
Payment demand 2 ($)	105	264	369
Total payment ($)	210	605	815
Social welfare ($)		1140	
Producer surplus ($)		512	
Consumer surplus ($)		628	

4. Minimum outputs and ramping limits enforced, case D-Ex.5.12. This case is identical to case B-Ex.5.12.

Figure 5.6a and b illustrates case B-Ex.5.12 or D-Ex.5.12 of Example 5.12. Figure 5.6a refers to hour 1, while Figure 5.6b refers to hour 2. In Figure 5.6a, the rise in the offer curve is clear while in Figure 5.6b, the solution is attained at the left of the crossing of the offer and bid curves.
The following observations are in order:

1. As in the single-period case (Example 5.11), the operating constraints (ramping and minimum production level constraints) always leads to a smaller social welfare and eventually to a smaller load supplied. Prices remain stable or increase with respect to those in the case of no operating constraints. Note that operating constraints reflect physical operating limitations.
2. Enforcing only ramping limits (case B-Ex.5.12) results in a smaller production of unit 1 in period 1 (compensated by a larger production of unit 2) and a reduction of the demand supplied in period 2. The price increases in both periods.
3. Enforcing only minimum power outputs (case C-Ex.5.12) results in a smaller demand supplied and a higher price in both periods. Unit 2 is scheduled out of the market in period 1 and unit 3 in period 2.

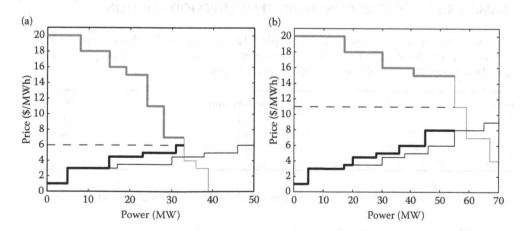

FIGURE 5.6 Multiperiod auction (Example 5.12). Minimum outputs and ramping limits enforced: (a) hour 1 and (b) hour 2.

4. Enforcing both minimum power outputs and ramping limits (case D-Ex.5.12) results in a production pattern equal to the case with only ramping limits enforced (case B-Ex.5.12).

5.5.3 Transmission-Constrained Multiperiod Auction

In the more general case, the "single bus" market equilibrium Equation (5.105) is substituted by the set of nodal balance equations:

$$\sum_{i\in\Omega_{Gr}}\sum_{b=1}^{N_{Git}}P_{Gibt}-\sum_{j\in\Omega_{Dr}}\sum_{k=1}^{N_{Djt}}P_{Djkt}=\sum_{s\in\Omega_r}B_{rs}(\delta_{rt}-\delta_{st})\quad\forall r,t \tag{5.127}$$

with transmission capacity limits incorporated as

$$-P_{Frs}^{\max}\le B_{rs}(\delta_{rt}-\delta_{st})\le P_{Frs}^{\max}\quad\forall\text{ line }rs,\forall t \tag{5.128}$$

where Ω_{Dr} is the set of indices of demands connected to bus r (data), Ω_{Gr} is the set of indices of generating units connected to bus r (data), Ω_r is the set of buses adjacent to bus r (data), B_{rs} is the susceptance of line rs (constant), δ_{rt} is the voltage angle of bus r at time t (variable), and $P_{Frs}^{\max}$ is the maximum power transmission capacity of line rs (constant).

The optimization problem, including Equations (5.119) through (5.121), Equations (5.123) through (5.126), and Equations (5.127) and (5.128), is a network-constrained multiperiod auction. Again, this is formulated as a mixed-integer linear program—albeit with more variables and constraints—but readily solved by a commercially available package.

Once binary variables have been fixed to their optimal values, the dual variables of Equation (5.110) provide the LMPs throughout the network for all time periods, that is,

$$\lambda_{rt}=\frac{\partial SW^M}{\partial P_{Drt}}\quad\forall r,t \tag{5.129}$$

Example 5.13 below illustrates two instances of a network-constrained multiperiod auction.

EXAMPLE 5.14 NETWORK-CONSTRAINED MULTIPERIOD AUCTION

The generating unit and demand data of Example 5.12 are considered in this example. Additionally, the network depicted in Figure 5.7 is considered. Data for this network are provided in the table below. Resistances and shunt susceptances are neglected.

From	To	Reactance (pu)	Capacity (MW)
1	2	0.1	9
1	3	0.1	9
2	3	0.1	9

The tables below provide the solution of this network-constrained multiperiod auction.

1. Case A-Ex.5.13. Transmission capacity limits are enforced but minimum power outputs and ramping limits are not. Line 1–3 is congested during hour 1 while during hour 2 both lines 1–2 and 1–3 are congested.

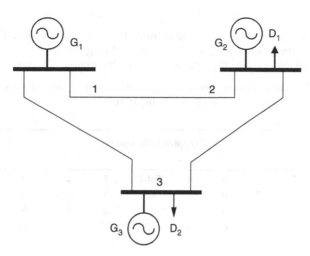

FIGURE 5.7 Network for the network-constrained multiperiod auction (Example 5.13).

	Accepted Offers (MW)									
Hour	**1**					**2**				
Block	**1**	**2**	**3**	**4**	**Total**	**1**	**2**	**3**	**4**	**Total**
Unit 1	5	9.5	0	—	14.5	5	12	1	—	18
Unit 2	8	8	0	—	16	8	8	9	—	25
Unit 3	2.5	0	0	—	2.5	10	9	0	—	19
Total	—	—	—	—	33	—	—	—	—	62
	Accepted Bids (MW)									
Hour	**1**					**2**				
Block	**1**	**2**	**3**	**4**	**Total**	**1**	**2**	**3**	**4**	**Total**
Demand 1	8	5	5	0	18	17	14	3	0	34
Demand 2	7	4	4	0	15	13	11	4	0	28
Total	—	—	—	—	33	—	—	—	—	62

	Prices, Revenues, Payments, and Social Welfare						**Total**
Hour	**1**			**2**			**Total**
Bus #	**1**	**2**	**3**	**1**	**2**	**3**	—
Price ($/MWh)	3.0	5.5	8.0	3.5	7.0	9.0	—
Revenue unit 1 ($)		43.5			63		106.5
Revenue unit 2 ($)		88			175		263
Revenue unit 3 ($)		20			171		191
Total revenue ($)		151.5			409		560.5
Payment demand 1 ($)		99			238		337
Payment demand 2 ($)		120			252		372
Total payment ($)		219			490		709
Social welfare ($)			1064				
Producer surplus ($)			95.5				
Consumer surplus ($)			020.0				
Merchandizing surplus ($)			148.5				

As LMPs vary across the buses, what consumers pay and producers receive is not the same, the difference being the so-called merchandizing surplus. The social welfare is then the sum of the consumer, producer, and merchandizing surpluses.

2. Case D-Ex.5.13. Minimum outputs, ramping limits, and transmission capacity limits are enforced. During both hours, line 1–3 is congested.

					Accepted Offers (MW)					
Hour			1					2		
Block	1	2	3	4	Total	1	2	3	4	Total
Unit 1	5	7	0	—	12	5	11.5	0	—	16.5
Unit 2	8	8	5	—	21	8	8	9	—	25
Unit 3	0	0	0	—	0	10	0	0	—	10
Total	—	—	—	—	33	—	—	—	—	51.5

					Accepted Bids (MW)					
Hour			1					2		
Block	1	2	3	4	Total	1	2	3	4	Total
Demand 1	8	5	5	0	18	17	14	0	0	31
Demand 2	7	4	4	0	15	13	7.5	0	0	20.5
Total	—	—	—	—	33	—	—	—	—	51.5

	Prices, Revenues, Payments, and Social Welfare						
Hour		1			2		Total
Bus #	1	2	3	1	2	3	—
Price ($/MWh)	3.0	6.0	9.0	3.0	9.5	16.0	—
Revenue unit 1 ($)		36			49.5		85.5
Revenue unit 2 ($)		126			237.5		363.5
Revenue unit 3 ($)		0			160		160
Total revenue ($)		162			447		609
Payment demand 1 ($)		108			294.5		402.5
Payment demand 2 ($)		135			328		463
Total payment ($)		243			622.5		865.5
Social welfare ($)				1026.5			
Producer surplus ($)				227.5			
Consumer surplus ($)				542.5			
Merchandizing surplus ($)				256.5			

The following observations are in order:

1. Enforcing just transmission capacity limits (case A-Ex.5.13) results in significant changes in production patterns if these limits are binding. Observe that the expensive unit 3 produces significantly more than that would be expected, as the cheap unit 1 is straddled by the constraints and cannot produce as much as would be desirable. The prices vary across the buses and the social welfare decreases dramatically.
2. Enforcing transmission capacity and ramping limits as well as minimum power outputs (case D-Ex.5.13) results in a reduction in the demand supplied during period 2. The

social welfare then decreases compared with its level in case A-Ex.5.13. Likewise, the bus prices vary from bus to bus and the price differences increase significantly.

3. The producer surplus is higher in case D-Ex.5.13 than in case A-Ex.5.13 (i.e., itself higher than in case A-Ex.5.12). These results indicate that the higher the number of constraints enforced, the higher the surplus of the producers and the lower the surplus of the consumers. The same trend is observed for the merchandizing surplus, while the consumer surplus exhibits the opposite trend.

Figure 5.8a and b illustrates case A-Ex.5.13. Observe that in both hours (Figure 5.8a and b) the solution occurs at the right of the crossing of the supply and demand curves. This implies that at the optimal solution it is possible to have a block of negative social welfare. The LMPs are not shown in these plots because they vary across buses.

5.5.4 PRICING

Two main pricing criteria are generally used: marginal and pay-as-bid pricing.

Under marginal pricing, generating units and demands, respectively, are paid and pay the marginal price. More generally, if network constraints are considered in the market-clearing process, generating units and demands are paid and pay the LMPs. This marginal pricing criterion has been used throughout the previous sections of this chapter. In comparison, under pay-as-bid pricing, accepted offers by generators are paid at the offered price and accepted bids by demand pay the bid price.

It is commonly accepted that marginal pricing provides a more "appropriate" economical signal than pay-as-bid pricing. A further discussion on pricing is outside the scope of this book. The interested reader is referred to [15] and [16].

5.5.5 DEMAND FLEXIBILITY

As described in the previous sections, demands exercise flexibility adjusting their price and quantity bids, and by explicitly declaring hourly consumption ranges, not levels. It is important to note that some consumers are flexible (e.g., a factory with capability of rescheduling tasks), but others are not (e.g., a hospital).

Demand flexibility is desirable because such flexibility generally decreases the actual load during peak periods and increases it during valley periods, which may result in a reduction in peak prices and usually not a large increase in valley prices.

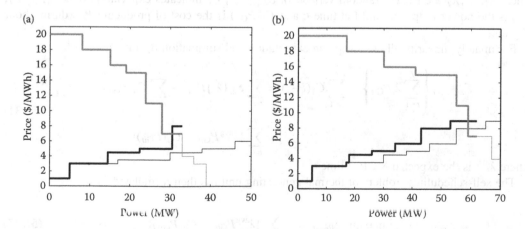

FIGURE 5.8 Network-constrained multiperiod auction (Example 5.13): (a) hour 1 and (b) hour 2.

Flexibility results in a better utilization of production units as they increase their load factors, which benefits electricity producers. Generally, flexibility also results in higher social welfare.

Flexibility helps integrating increasing amounts of stochastic production, since, through flexibility, actual load can be shifted to periods of high stochastic production and comparatively low demand.

Demand flexibility also helps in reducing the undesirable effect of rapid changes in load at the beginning of the day, when load increases rapidly, and in the late afternoon, when load decreases rapidly. In these cases, demand flexibility allows reducing the speed of demand increase or decrease in critical periods, by rapidly decreasing or increasing, respectively, the load levels of flexible demands.

Further details on demand flexibility can be found in [6].

5.6 PRODUCER SELF-SCHEDULING AND OFFER STRATEGIES

A producer should determine its optimal self-production. Then, it should devise an offer strategy to ensure that the optimal production is accepted by the pool. Alternatively, the producer might derive its offer curve directly. For clarity, in this chapter, we pursue the first approach.

First, we consider a producer with no capability of altering the MCPs (Section 5.6.1), that is, a price-taker producer, and derive its optimal self-schedule. Then, we consider a producer with the capability of altering the MCPs (Section 5.6.2), that is, a price-maker producer, and derive its optimal self-schedule.

We conclude this section with an overview of bidding strategies for both price-taker and price-maker producers (Section 5.6.3).

5.6.1 PRICE-TAKER PRODUCER

Since a (thermal) price-taker producer has no capability of altering the MCPs and its units are generally not physically interrelated, its profit maximization problem decomposes by generator. Then, the objective of a given generating unit is to maximize its expected profit over the market horizon. The expected total profit can be expressed as

$$\mathcal{E}_{\lambda_1,\ldots,\lambda_T}\left\{\sum_{t=1}^{T}\lambda_t P_{Git}\right\} - \sum_{t=1}^{T}C_{it}(P_{Git}) \tag{5.130}$$

where $\lambda_1,\ldots,\lambda_T$ are MCPs (random variables); $\mathcal{E}_{\lambda_1,\ldots,\lambda_T}\{\cdot\}$ indicates expectation over $\lambda_1,\ldots,\lambda_T$; P_{Git} is the power output of unit i at time t; and $C_{it}(P_{Git})$ is the cost of producing P_{Git} during hour t by unit i.

Fortunately, linearity allows swapping expectation and summation operators,

$$\mathcal{E}_{\lambda_1,\ldots,\lambda_T}\left\{\sum_{t=1}^{T}\lambda_t P_{Git}\right\} - \sum_{t=1}^{T}C_{it}(P_{Git}) = \sum_{t=1}^{T}\mathcal{E}_{\lambda_t}\{\lambda_t\}P_{Git} - \sum_{t=1}^{T}C_{it}(P_{Git})$$
$$= \sum_{t=1}^{T}[\lambda_t^{\mathrm{avg}}P_{Git} - C_{it}(P_{Git})] \tag{5.131}$$

where λ_t^{avg} is the expected MCP at time t.

The self-scheduling problem of thermal generating unit i is then formulated as

$$\mathrm{maximize}_{P_{Git},u_t;\forall t} \quad \sum_{t=1}^{T}[\lambda_t^{\mathrm{avg}}P_{Git} - C_{it}(P_{Git})] \tag{5.132}$$

subject to

$$u_{it}\,P_{Gi}^{min} \le P_{Git} \le u_{it}P_{Gi}^{max} \quad \forall t \tag{5.133}$$

$$P_{Gi,t-1} - P_{Git} \le R_{Gi}^{down}u_{it} + R_{Gi}^{sd}(u_{i,t-1} - u_{it}) + P_{Gi}^{max}(1 - u_{i,t-1}) \quad \forall t \tag{5.134}$$

$$P_{Git} - P_{Gi,t-1} \le R_{Gi}^{up}u_{i,t-1} + R_{Gi}^{su}(u_{it} - u_{i,t-1}) + P_{Gi}^{max}(1 - u_{i,t}) \quad \forall t \tag{5.135}$$

where P_{Gi0} is the production of unit i at the beginning of the market horizon, R_{Gi}^{sd} the shutdown ramping limit and R_{Gi}^{su} the start-up ramping limit.

Note that two new ramping limits are introduced in this section, namely, the start-up and shutdown ramping limits. These limits allow a more detailed description of the start-up and shutdown procedures. The start-up ramping limit sets the maximum production level of the unit in the start-up period, while the shutdown limit sets the maximum production level of the unit in the period preceding its shutdown.

The objective function (5.132) is the expected profit of generating unit i throughout the market horizon. Constraint (5.133) enforces that each unit works within its operating limits. Constraints (5.134) and (5.135) enforce ramping limits. Other constraints, such as minimum down and up times, can also be easily incorporated [16–18].

The solution of problem (5.132)–(5.135) provides the optimal self-scheduling of generating unit i, that is, its optimal production level at every time period so that the expected value of its profit is maximized.

Problem (5.132)–(5.135) is a mixed-integer linear programing problem of small size. It can be readily solved using available commercial software [3].

Example 5.14 below illustrates the self-scheduling problem faced by a price-taker thermal generating unit.

EXAMPLE 5.15 PRICE-TAKER PRODUCER

We consider a generating unit characterized by the data in the table as follows:

Capacity/minimum power output (MW)	13	2
Ramping up/down limits (MW/h)	5	5
Start-up/shutdown ramping limits (MW/h)	4	4
Linear cost ($/MWh)	2	
Initial status	Off	

The considered scheduling horizon spans 6 h. The MCP forecasts are provided in the table as follows:

Hour	1	2	3	4	5	6
Price forecast ($/MWh)	2	4	6	2	2	1

The solution of the optimal self-scheduling problem is given in the table as follows:

				Productions, Revenues, Costs, and Profits			
Hour	1	2	3	4	5	6	Total
Price forecast ($/MWh)	2	4	6	2	2	1	—
Hourly production (MWh)	4	9	13	8	4	0	38
Revenue ($)	8	36	78	16	8	0	146
Cost ($)	8	18	26	16	8	0	76
Profit ($)	0	18	52	0	0	0	70

The following observations are in order:

1. The unit starts up in period 1, therefore, the start-up ramping constraint is binding in that period. The ramping-up constraint is binding in period 2, whereas the ramping-down constraint is binding in period 4 and the shutdown ramping constraint in period 5. The unit produces at maximum power during period 3 (higher price period) and is shut down during period 6.
2. Even though no profit is achieved in hours 1, 4, and 5, the generating unit is not shut down during those hours due to ramping limitations.

Figure 5.9 illustrates the price-taker optimal self-schedule. Figure 5.9a depicts prices and optimal productions, while Figure 5.9b shows revenues, costs, and profits as well as prices.

Next, we consider a hydroelectric producer. The different hydroelectric units of this producer are physically coupled through the river system. Moreover, water scarcity results in temporal coupling among hydroelectric units. Therefore, all units need to be considered simultaneously.

The resulting profit maximization problem has the form:

$$\text{maximize}_{P_{Ght}, V_{ht}, Q_{ht}; \forall h, t} \quad \sum_{t=1}^{T} \left(\lambda_t^{avg} \sum_{h=1}^{N_H} P_{Ght} \right) \tag{5.136}$$

subject to

$$P_{Ght} = \rho_h Q_{Ght} \quad \forall h, t \tag{5.137}$$

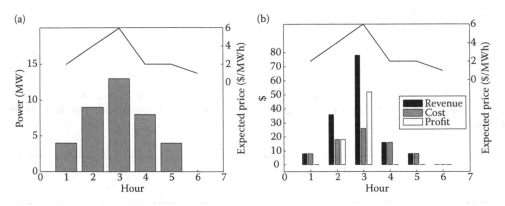

FIGURE 5.9 Price-taker optimal self-schedule: (a) productions and prices and (b) revenues, costs, profits, and prices.

$$V_{ht} = V_{h,t-1} - Q_{ht} + W_{ht} + \sum_{k \in \Omega_{Hh}} Q_{kt} \qquad (5.138)$$

$$V_{ht}^{\min} \leq V_{ht} \leq V_{ht}^{\max} \qquad (5.139)$$

$$Q_{ht}^{\min} \leq Q_{ht} \leq Q_{ht}^{\max} \qquad (5.140)$$

where P_{Ght} is the power produced by hydroelectric unit h at hour t (variable), V_{ht} is the volume of reservoir h at the end of period t (variable), Q_{ht} is the volume of water discharged by unit h during period t (variable), W_{ht} is the lateral inflow (volume) reaching reservoir h during period t (constant), N_H is the number of hydroelectric units of the considered hydro system (data), ρ_h is the MWh/Hm^3 conversion factor of hydroelectric unit h (constant), and Ω_{Hh} is the set of indices of reservoirs upstream and adjacent to reservoir h (data).

Objective function (5.136) is the expected profit (revenue) of the hydroelectric units in the considered river system. Constraints (5.137) establish the relationship between water volume released and electric energy produced. Nonlinear expressions considering the effect of reservoir contents and losses can also be derived and linearized [19]. However, for the sake of simplicity, these technical details are not considered in this chapter. Constraints (5.138) through (5.140) are water balances and variable limits.

It should be noted that in the above formulation the water discharges traveling times are considered negligible. If they are not, delays can be easily incorporated into problem 5.119–5.123 [19].

Problem (5.136)–(5.140) is a linear programing problem of moderate size, easy to solve using commercial software [3].

Example 5.15 below illustrates the self-scheduling problem faced by a hydroelectric producer.

EXAMPLE 5.16 PRICE-TAKER HYDROELECTRIC PRODUCER

The considered market horizon spans 6 h. The market-clearing price forecasts are provided in the table as follows:

Hour	1	2	3	4	5	6
Price forecast ($/MWh)	3	4	6	4	1	2

The hydroelectric producer manages a river system including two cascaded reservoirs, each one with a generating unit. Data are provided in the table as follows:

	Hydroelectric System Data	
	Upper	Lower
Reservoir	Unit 1	Unit 2
Initial volume (Hm^3)	2	3
Maximum volume (Hm^3)	10	10
Minimum volume (Hm^3)	2	2
Hourly lateral inflow (Hm^3)	2	1
Maximum discharge (Hm^3)	4	4
Energy volume conversion factor (MWh/Hm^3)	4	3

The solution of this problem is provided below. The optimal production of each unit is given in the following table.

	Optimal Production (MWh)						
Hour	1	2	3	4	5	6	Total
Hydroelectric unit 1	0	8	16	8	0	16	48
Hydroelectric unit 2	3	12	12	12	3	12	54
Total	3	20	28	20	3	28	102

The reservoir contents are provided in the table as follows:

	Reservoir Contents (Hm^3)					
Hour	1	2	3	4	5	6
Reservoir 1	4	4	2	2	4	2
Reservoir 2	3	2	3	2	2	3
Total	7	6	5	4	6	5

The revenues are given in the table as follows:

	Revenue ($)						
Hour	1	2	3	4	5	6	Total
Hydroelectric unit 1	0	32	96	32	0	32	192
Hydroelectric unit 2	9	48	72	48	3	24	204
Total	9	80	168	80	3	56	396

The following observations are in order:

1. Units do not work at maximum power during all periods due to the constraints imposed by the river system. Unit 1 works at full power during periods 3 and 6, whereas unit 2 works at full power during periods 2–4 and 6.
2. Unit 1 does not work during periods 1 and 5, so that water is stored to be released later during higher price periods.
3. The reservoir contents stay close to their respective lower limits. Reservoir 1 is at its minimum level during the last period but reservoir 2 is not.

5.6.2   PRICE-MAKER PRODUCER

A producer with the capability of altering the MCP coordinates the production of its units in such a way that the MCP is modified to the benefit of the producer, a strategy denoted as gaming. Rival producers also benefit as they obtain a price rise without gaming. Moreover, the revenue rise might be higher for rival producers than for the gaming producer.

For the sake of simplicity, we consider a single thermal producer. Note, however, that a hydrothermal producer can be treated similarly.

The profit maximization problem faced by a price-maker producer is formulated as

$$\text{maximize}_{P_{Git}, \forall i, t; \, P_{Gt}, \forall t} \quad \sum_{t=1}^{T} \left[\lambda_t^{\text{avg}}(P_{Gt}) P_{Gt} - \sum_{i=1}^{N_G} C_{it}(P_{Git}) \right] \tag{5.141}$$

subject to

$$P_{Gt} = \sum_{i=1}^{N_G} P_{Git} \quad \forall t \tag{5.142}$$

$$u_{it} P_{Gi}^{\min} \le P_{Git} \le u_{it} P_{Gi}^{\max} \quad \forall i, t \tag{5.143}$$

$$P_{Gi,t-1} - P_{Git} \le R_{Gi}^{\text{down}} \quad \forall i, t \tag{5.144}$$

$$P_{Git} - P_{Gi,t-1} \le R_{Gi}^{\text{up}} \quad \forall i, t \tag{5.145}$$

where P_{Gt} is the total production of the producer in hour t (variable), $\lambda_t^{\text{avg}}(P_{Gt})$ is the expected MCP as a function of the total production of the producer in hour t (data), and N_G is the number of units belonging to the producer considered (data). It should be noted that function $\lambda_t^{\text{avg}}(P_{Gt})$ is decreasing with the power output of the price-maker producer.

The objective function (5.141) is the profit of the producer over the whole market horizon. It is computed by subtracting costs from revenues. Note that revenues depend on prices that, in turn, are influenced by the levels of production of the producer.

Constraints (5.142) determine the total production of the price-maker producer. The remaining constraints are similar to the corresponding constraints in the case of a price-taker producer.

The solution of this problem provides the optimal production of the price-maker producer at each hour and subsequently the resulting hourly MCP.

Problem (5.141)–(5.145) is a mixed-integer nonlinear programing problem of moderate size. In general, this problem is not easy to solve.

Example 5.16 below illustrates the self-scheduling problem of a price-maker power producer.

EXAMPLE 5.17 PRICE-MAKER PRODUCER

The expected price ($/MWh) functions for the considered price-maker producer and for the six time periods under study are

$$\lambda_1 = 3 - \frac{1}{10}P_{G1}, \quad \lambda_2 = 5 - \frac{1}{10}P_{G2}$$

$$\lambda_3 = 5 - \frac{1}{4}P_{G3}, \quad \lambda_4 = 5 - \frac{1}{10}P_{G4}$$

$$\lambda_5 = 5 - \frac{1}{8}P_{G5}, \quad \lambda_6 = 4 - \frac{1}{10}P_{G6}$$

Unit data are given in the table below. For simplicity, no ramping constraints and no minimum power outputs are considered.

Data	Unit 1	Unit 2
Capacity (MW)	8	10
Minimum power output (MW)	0	0
Linear cost ($/MWh)	2	2.5

The solution is provided in the table as follows:

Productions, Prices, Revenues, Costs, and Profits							
Hour	1	2	3	4	5	6	Total
Production unit 1 (MW)	5	8	6	8	8	8	43
Production unit 2 (MW)	0	4.5	0	4.5	2	0	11
Total production (MW)	5	12.5	6	12.5	10	8	54
Clearing price ($/MWh)	2.50	3.75	3.50	3.75	3.75	3.20	—
Revenue ($)	12.50	46.87	21.00	46.87	37.5	25.60	190.35
Cost ($)	10.00	27.25	12.00	27.25	21.00	16.00	113.50
Profit ($)	2.50	19.62	9.00	19.62	16.50	9.60	76.85

The following observations are in order:

1. The producer does not produce at maximum capacity to keep prices high enough, thereby maximizing its profit. This is particularly apparent in periods 1 and 3.
2. In period 3, the production is low to keep the price high enough, but the resulting price is rather low due to the high (in absolute value) slope of change of the price in that period.
3. Since the price slope during period 1 is comparatively small (in absolute value), the resulting price is low due to the low intercept during that period.

An additional observation is that the rise in prices forced by the price-maker producer might benefit rival nongaming generating companies more than it benefits the price-maker.

5.6.3 Offer Strategies

A bidding strategy is a set of rules that enables a producer to offer in such a manner that its optimal self-scheduling is accepted during the market-clearing process [20].

If the producer is a price-taker, an appropriate rule is to offer at zero price during every hour for the blocks that should be accepted, and at infinite price for the remaining blocks [21].

If the producer is a price-maker, more elaborated rules are needed. However, the analysis of such rules is outside the scope of this book.

5.7 CONSUMER AND RETAILER VIEWPOINTS

A retailer buys energy in the pool (or through bilateral or forward contracts in medium- or long-term horizons) to sell to its clients at fixed prices [22,23]. A consumer buys energy for its own consumption. The problems these agents face are thus similar. Considering additionally that the producer/retailer owns a generating facility of limited size, the consumer/retailer problems can be formulated as

$$\text{maximize}_{P_{Gt}, P_{Bt}, u_t; \forall t} \quad \sum_{t=1}^{T} [U_t(P_{Dt}) - C_t(P_{Gt}) - \lambda_t^{\text{avg}} P_{Bt}] \tag{5.146}$$

subject to

$$P_{Gt} + P_{Bt} = P_{Dt} \quad \forall t \tag{5.147}$$

$$u_t P_G^{\text{min}} \leq P_{Gt} \leq u_t P^{\text{maxG}} \quad \forall t \tag{5.148}$$

$$P_{Gt-1} - P_{Gt} \leq R_G^{\text{down}} \quad \forall t \tag{5.149}$$

$$P_{Gt} - P_{G,t-1} \leq R_G^{\text{up}} \quad \forall t \tag{5.150}$$

where $U_t(P_{Dt})$ is the revenue/utility that is obtained by the retailer/consumer from selling/consuming P_{Dt}, P_{Gt} is the self-produced power in hour t, P_{Bt} is the power bought from the pool in hour t, P_{Dt} is the demand to be supplied in hour t, and $C_t(P_{Gt})$ is the self-production cost to produce P_{Gt} in hour t.

The objective function (5.146) represents the profit of the retailer/consumer. It includes three terms, the revenue/utility of the retailer/consumer, the cost of self-producing, and the cost of buying from the pool.

Constraints (5.147) ensure that the demand is met at all hours of the market horizon. Constraints (5.148) through (5.150) enforce the operating feasibility of the self-production facility. The solution of this problem provides the hourly optimal self-productions and the hourly energy to be bought from the pool.

Problem (5.146)–(5.150) is a moderate sized mixed-integer linear programing problem, which can be readily solved using commercially available software [3].

Example 5.17 below illustrates the optimal procurement problem faced by a retailer.

EXAMPLE 5.18 RETAILER

The retailer considered owns a self-production unit with capacity and minimum power output of 30 and 10 MW, respectively, with production cost of $2/MWh. The selling price to clients is $4/MWh.

A 6-h market horizon is considered. Market price forecasts and hourly demands are given in the table as follows:

Hour	1	2	3	4	5	6
Price forecast ($/MWh)	2	4	3	1	4	2
Demand (MW)	20	30	50	20	5	10

The solution is provided in the table as follows:

	Self-Productions, Purchases, Revenues, Costs, and Profits						
Hour	1	2	3	4	5	6	Total
Price ($/MWh)	2	4	3	1	4	2	—
Self-production (MWh)	0	30	30	0	30	0	90
Purchase (MWh)	20	0	20	20	−25	10	45
Demand (MWh)	20	30	50	20	5	10	135
Revenue ($)	80	120	200	80	20	40	540
Cost ($)	40	60	120	20	−40	20	220
Profit ($)	40	60	80	60	60	20	320

The following observations are in order:

1. Self-production takes place (at maximum capacity) during periods 2, 3, and 5 when the pool prices are comparatively high.
2. During period 2, the energy self-produced is entirely used to supply the demand of the retailer, while during period 3 the energy self-produced is only partly used to supply the demand of the retailer.
3. During period 5, the energy self-produced is mostly sold in the pool (negative purchase), 25 out of 30 MWh, but also used to supply the demand of the retailer, 5 MWh. This results in a net profit of $60 in period 5.

4. During periods 1, 4, and 6, the demand to be supplied by the retailer is entirely bought in the pool.
5. In this particular example no loss is incurred during any time period.

5.8 SUMMARY

This chapter consists of two parts. The first part reviews generation operating issues from a centralized viewpoint. Both the ED and the UC problems are described and analyzed through examples. The second part focuses on electricity markets. The perspectives of the market operator, the producers, the consumers, and the retailers are considered. Several examples clarify and illustrate the theoretical issues.

As many electric energy systems throughout the world evolve from a centralized environment to a market one, we consider it appropriate to provide the reader with both perspectives. Moreover, these perspectives are deeply interrelated.

REFERENCES

1. A. Brooke, D. Kendrick, A. Meeraus, and R. Raman, *GAMS: A User's Guide*, GAMS Development Corporation, Washington, 2008 (http://www.gams.com/).
2. The MathWorks, Inc. *Matlab 7.3*, Natick, MA, 2008 (http://www.mathworks.com/products/matlab/).
3. GAMS Development Corporation. *GAMS—The Solver Manuals*, GAMS Development Corporation, Washington, 2008 (http://www.gams.com/).
4. L. K. Kirchmayer, *Economic Operation of Power Systems*, John Wiley & Sons, Inc., New York, 1958.
5. M. S. Bazaraa, H. D. Sherali, and C. M. Shetty, *Nonlinear Programming: Theory and Algorithms*, Second Edition, John Wiley & Sons, Inc., New York, 1993.
6. J. M. Morales, A. J. Conejo, H. Madsen, P. Pinson, and M. Zugno, *Integrating Renewables in Electricity Markets*, Springer, New York. 2014.
7. M. D. Ilić, F. D. Galiana, and L. H. Fink, *Power System Restructuring: Engineering and Economics*, Kluwer Academic Publishers, Norwell, MA, 1998.
8. H. Chao and H. G. Huntington, *Designing Competitive Electricity Markets, Fred Hillier's International Series in Operations Research & Management Science*, Kluwer Academic Publishers, Boston, MA, 1998.
9. G. B. Sheblé, *Computational Auction Mechanisms for Restructured Power Industry Operation*, Kluwer Academic Publishers, Boston, MA, 1999.
10. M. Shahidehpour and Z. Li, *Electricity Market Economics*, John Wiley & Sons, New York, 2005.
11. D. S. Kirschen and G. Strbac, *Fundamentals of Power System Economics*, John Wiley & Sons, Ltd., Chichester, 2004.
12. E. Castillo, A. J. Conejo, R. Mínguez, and C. Castillo, A closed formula for local sensitivity analysis in mathematical programming, *Engineering Optimization*, 38(1), 2006, 93–112.
13. J. M. Arroyo and A. J. Conejo, Multi-period auction for a pool-based electricity market, *IEEE Transactions on Power Systems*, 17(4), 2002, 1225–1231.
14. A. L. Motto, F. D. Galiana, A. J. Conejo, and J. M. Arroyo, Network-constrained multiperiod auction for a pool-based electricity market, *IEEE Transactions on Power Systems*, 17(3), 2002, 646–653.
15. A. Mas-Colell, M. D. Whinston, and J. R. Green, *Microeconomic Theory*, Oxford University Press, New York, 1995.
16. J. M. Arroyo and A. J. Conejo, Optimal response of a thermal unit to an electricity spot market, *IEEE Transactions on Power Systems*, 15(3), 2000,1098–1104.
17. M. Carrión and J. M. Arroyo, A computationally efficient mixed-integer linear formulation for the thermal unit commitment problem, *IEEE Transactions on Power Systems*, 21(3), 2006, 1371–1378.
18. J. M. Arroyo and A. J. Conejo, Modeling of start-up and shut-down power trajectories of thermal units, *IEEE Transactions on Power Systems*, 19(3), 2004, 1562–1568.
19. A. J. Conejo, J. M. Arroyo, J. Contreras, and F. A. Villamor, Self-scheduling of a hydro producer in a pool-based electricity market, *IEEE Transactions on Power Systems*, 17(4), 2002, 1265–1272.

20. M. A. Plazas, A. J. Conejo, and F. J. Prieto, Multi-market optimal bidding for a power produce, *IEEE Transactions on Power Systems*, 20(4), 2005, 2041–2050.

21. A. J. Conejo, F. J. Nogales, and J. M. Arroyo, Price-taker bidding strategy under price uncertainty, *IEEE Transactions on Power Systems*, 17(4), 2002, 1081–1088.

22. A. J. Conejo, J. J. Fernández-González, and N. Alguacil, Energy procurement for large consumers in electricity markets, *IEE Proceedings Generation, Transmission and Distribution*, 152(3), 2005, 357–364.

23. A. J. Conejo and M. Carrión, Risk-constrained electricity procurement for a large consumer, *IEE Proceedings Generation, Transmission and Distribution*, 153(4), 2006, 407–415.

6 Optimal and Secure Operation of Transmission Systems

José L. Martínez-Ramos and Antonio J. Conejo

CONTENTS

6.1 INTRODUCTION

In the initial stages of electrical engineering, power systems were composed of isolated generators supplying energy to a certain number of local loads. Those configurations, relatively easy to control and supervise, have given place to current power systems, composed of multiple generators and loads, connected by a high-voltage transmission network. Consequently, due to their meshed topologies and the presence of distributed generation, the planning and operation of power systems have become very complex.

The growing complexity of power systems, with a clear tendency to increase the interconnection with adjacent systems, is mainly due to the importance of decreasing the electricity costs and improving the reliability of the electricity supply. In addition, the progressive liberalization of the electricity transactions enhances participation in different activities to facilitate the competition in the electricity market; this adds even more complexity to the operation of power systems.

As a result, it is mandatory nowadays to design adequate energy management systems (EMS) where the available information is collected, and the tasks such as supervision and control are carried out. For this reason, qualified staff for *operation* and *planning* are also needed to ensure the supply of electrical energy.

This chapter presents inherent concepts, activities, and tools for power system operation within the context of modern EMS. There will be special emphasis on tools for correcting transmission network problems and its optimization.

First, a classification of the power system state as a function of its degree of security is presented to identify the different activities involved in the system operation. Second, *contingency analysis* techniques for determining the security degree of the power system are presented constituting the base of many real-time operation or planning studies.

Then, a brief introduction to the *optimal power flow* (OPF) problem is presented, with different examples of the operation problems in transmission system operation, formulated as constrained optimization problems.

Finally, the application of OPF problems to the short-term secure operation of transmission networks is presented and discussed.

6.2 POWER SYSTEM STATES

Focusing on the operation of power systems, the objective of real-time control is basically to maintain the electrical magnitudes between predetermined limits. These magnitudes are mainly the system frequency, bus voltages, and power flows. The process involves correcting the effects of the demand evolution and the consequences of possible, nonpredictable events. As a result, for a responsible system operator, the "security" of the system can be quantified in terms of its capability of remaining in a feasible state, without violating any of the imposed operational limits. In other words, the ability to maintain the desired state against predictable changes (demand and generation evolution) and unpredictable events called *contingencies* [1,2].

The correct comprehension of the role played by the different activities involved in the system operation implies classifying the possible system states as a function of the security degree. This classification is based on the one proposed by DyLiacco in Reference 3. Using it as a starting point, we define the system states in Figure 6.1.

The power system is in a *normal state* when the demand and all the operational constraints are satisfied, that is, when generators as well as the rest of the equipment work within design limits.

If there is no limit violation but the accorded safety criteria are not satisfied, then the network is said to be in an *alert state*.

If the system enters an *emergency state*, defined as a state with variables beyond operational limits due to an unexpected demand evolution or a contingency, *corrective actions* will be required to eliminate all the violations and to bring the system back to a normal state (*corrective control*). In certain circumstances, the actions of protection devices or even the operator intervention to avoid major problems (*load shedding*) lead to service interruption to customers. In this new state, operators will have to manage the restoration of the interrupted service restoration [4].

The objective that guides the operator's actions depends on the state of the system. For example, if the whole system has suffered a *blackout*, the operator will try to restore the interrupted service as soon as possible. This objective affects not only the available control actions but also the design phase of the generation, transmission, and distribution systems to ensure that the recovery of the operating conditions can be achieved in a reasonable time.

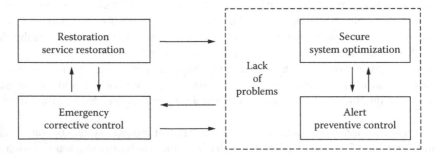

FIGURE 6.1   Operating states of a power system.

The aim of the corrective control is the restoration of the system to the normal state, either safe or unsafe, with absolute priority, as in emergency state the economic considerations are secondary. Once the variables are between limits again, the objective is basically economic: minimization of the production cost and allocation of the total generation among the most economical units. If system security is key, *preventive control* enters the scene to drive the system to a *secure state*. The decision of performing preventive control actions is always tied to the trade-off between economy and security, objectives that seldom go in the same direction.

Therefore, priorities will guide the control actions on the system. Mainly imposed by the system state, the priorities could be contradictory, as it happens when preventive control actions are required to ensure system security, but move the system far from the optimal state in terms of production cost.

Finally, the fundamental importance of control and supervisory systems (SCADA) must be emphasized, together with tools like *state estimators*, to follow the evolution of the different electrical magnitudes.

EXAMPLE 6.1 POWER SYSTEM STATES

Consider two generators supplying a 200 MW load, as shown in Figure 6.2. Both generators are in *economic dispatch* and the total cost is $2800/h.

The system is in *unsafe state* as the outage of line 2–3 would overload the line 1–3, also resulting in a voltage problem at the load bus, as shown in Figure 6.2b. If the outage occurs, *corrective actions* are required: to increase the power generation of node 1 in 50 MW, and consequently, to decrease the power generation of node 3 in the same amount; additionally, the voltage at bus 3 ought to be raised.

If the corrective actions were considered infeasible for some reason, it would be necessary to adopt *preventive actions* on the *base case* (Figure 6.2a), leading the system to the *safe state*

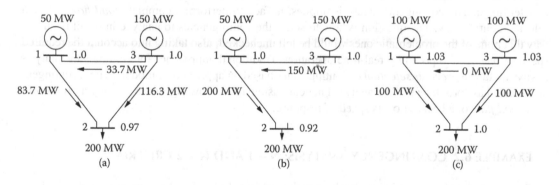

Generator	Cost	P_G^{max}	P_G^{min}		Q_G^{max}	Q_G^{min}
(Bus)	$/h	MW	MW	V_G^{sp} pu	Mvar	Mvar
1	$100 + 20\,P_G$	200	50	1.0	150	−150
3	$200 + 10\,P_G$	200	50	1.0	150	−150

	Reactance	P_f^{max}
Line	pu ($S_{Base} = 100$ MVA)	MW
1–2	0.1	200
2–3	0.1	200
1–3	0.1	100
Voltages	$0.95 \leq V \leq 1.05$	

FIGURE 6.2 3-Bus network for Example 6.1.

shown in Figure 6.2c. Clearly, the adoption of corrective actions (increasing P_{G_1} in 50 MW and decreasing P_{G_3} in the same amount, and rising the scheduled voltages to 1.03 pu) implies an increase in the production cost ($3300/h) with respect to the *economic dispatch* ($2800/h).

6.3   SECURITY ASSESSMENT: CONTINGENCY ANALYSIS

The evaluation of the *security degree* of a power system is a crucial problem, both in planning and in daily operation. Without considering dynamic issues, power system security must be interpreted as security against a series of previously defined contingencies. In this sense, a common criterion is to consider the following contingencies:

- A single outage of any system element (generator, transmission line, transformer, or capacitor/reactor bank), known as the $N - 1$ security criterion.
- Simultaneous outages of double-circuit lines that share towers in a significant part of the line path.
- In special situations, the outage of the largest generator in an area and any of the interconnection lines with the rest of the system.

In planning studies, which are more demanding with respect to security than the operation itself, simultaneous outages of any two elements in the systems are considered ($N - 2$ criterion). Besides, the *reliability analysis* of the system, commonly used in planning studies, is based on a detailed analysis considering both single and multiple simultaneous outages, making use of the associated probabilities of failure and repair times [5].

Consequently, the *security assessment*, better known as *contingency analysis* [1,2], basically consists of multiple studies in which the state of the network after the outage of one or multiple elements is determined. The contingency analysis implies, in fact, to perform a complete *load flow* for each selected contingency. The question is how to select the contingencies to analyze in detail in such a way that none of the problematic ones would be left unchecked, also taking into account the required speed of response imposed by real-time operation. Traditional approaches for contingency analysis always include a *previous selection* of contingencies based on approximate models. Then, contingencies labeled as "problematic" are analyzed in detail using a load flow algorithm, usually a *fast decoupled load flow* one because of its speed of response.

EXAMPLE 6.2   CONTINGENCY ANALYSIS: N − 1 AND N − 2 CRITERIA

The $N - 1$ and $N - 2$ contingency analyses of the system of Example 6.1 comprise a detailed study of the following states:

- $N - 1$ *analysis*: Single outage of five elements (two generators and three lines), as shown in Figure 6.3. These are cases (1)–(5).
- $N - 2$ *analysis*: Simultaneous outage of two arbitrary elements from the n elements, that is, $n \cdot (n - 1)/2 = 10$ double contingency cases (Figure 6.4). These are cases (6)–(15).

The existence of several critical cases can be observed:

- *Each of the states 6, 7, 11, and 13 implies a blackout*: In cases 6 and 13, it is due to insufficient generation and transmission resources, respectively. In cases 7 and 11, the blackout is caused by insufficient reactive power resources.
- *Cases 1, 2, 8, 9, 10, 12, 14, and 15 present low-voltage problems*: In case 2, this problem adds to an overload in line 1–3.

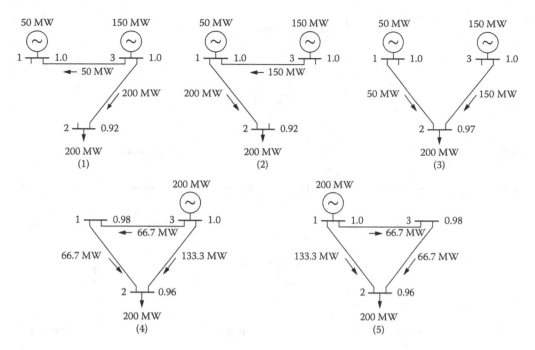

FIGURE 6.3 $N-1$ analysis in Example 6.2.

Earlier algorithms [2] for contingency selection are based on a classification of the contingencies following a descendent order of severity. This classification is performed using *ranking indexes* that intend to evaluate approximately the loading level for transmission lines and transformers after a given contingency [6]. These techniques make use of *distribution factors*, that is, linear factors that represent the per unit change of power flow in each transmission line or transformer as a result of the outage of a generator or a branch element. To calculate approximately the active power flow of each branch after a given contingency, it is sufficient to multiply the corresponding distribution factor by the power of the lost generator or the flow of the lost branch before the contingency.

Some authors [7] have proposed the use of distribution factors for detecting abnormal voltage problems. However, the strongly nonlinear characteristic of the problem that relates voltages and reactive powers, leads to question the results obtained by using this approach. This fact, in addition to masking problems inherent to the use of ranking indexes, justifies the development of a second group of techniques for detecting problematic contingencies, known as *contingency screening* [8]. Contingency screening techniques obtain the approximate state of the power system after a contingency by means of only one or two iterations of a load flow algorithm starting from the precontingency state, followed by a checking for branch overloads and out-of-limits voltages. Should a problem be detected, the exact postcontingency state is obtained for further checking. A comparison of the most used contingency screening techniques can be found in Reference 8.

6.3.1 CONTINGENCY ANALYSIS BASED ON DISTRIBUTION FACTORS

In transmission networks, it is possible to use an approximate linear model that only considers active power flows, the *DC load flow* presented in Chapter 3. The DC load flow provides a linear relation between active power injections and phase angles of nodal voltages,

$$P_i = \sum_j P_{ij} = \sum_j \frac{V_i V_j}{x_{ij}} \sin \theta_{ij} \sim \sum_j \frac{\theta_i - \theta_j}{x_{ij}} \tag{6.1}$$

where x_{ij} is the reactance of the branch between nodes i and j.

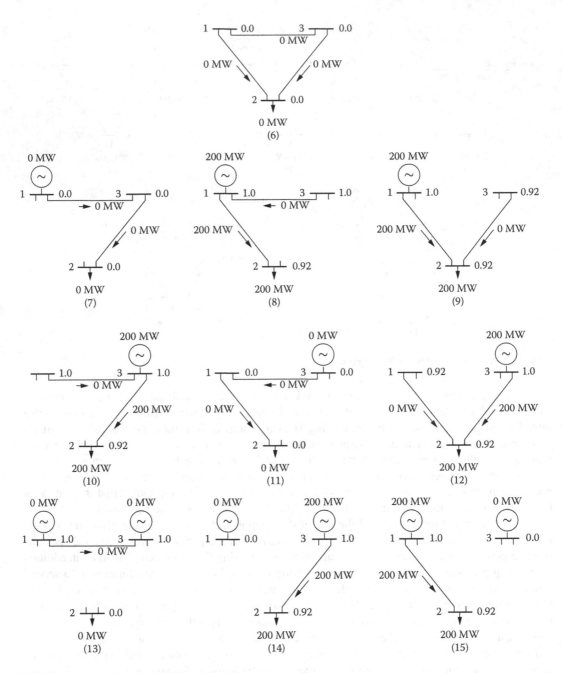

FIGURE 6.4 $N-2$ analysis in Example 6.2.

The above equation can be written in matrix form as $P = B\,\theta$. Then, phase angles can be eliminated to obtain a linear relationship between power flows, P_f, and nodal injected powers, P,

$$\left.\begin{array}{l} A^T\theta = X P_f \\ P_f = [X^{-1}A^T]\theta \\ P = A P_f \end{array}\right\} \Rightarrow P_f = [X^{-1}A^T B^{-1}]\,P = S_f\,P \qquad (6.2)$$

where A is the reduced branch-to-node incidence matrix (omitting the slack bus), X is a diagonal reactance matrix of branch elements, and S_f is the matrix of *sensitivities* between branch power flows and injected powers. Consequently, power flows after a change on the injected powers can be computed as:

$$P_f = S_f [P + \Delta P] = P_f^0 + S_f \Delta P \Rightarrow \Delta P_f = S_f \Delta P \tag{6.3}$$

Injection distribution factors are defined as the flow increase in a given element (line or transformer) that connects nodes m and n after a unitary increase in the power injected in bus i:

$$\rho_{mn}^i = \frac{\Delta P_{mn}^i}{\Delta P_i} = \frac{\Delta \theta_m - \Delta \theta_n}{x_{mn}} = S_{mn,i} \tag{6.4}$$

Note that distribution factors depend only on the network topology, and, consequently, they can be computed off-line using sparse matrix solution techniques.

The flow increase in a branch element, ΔP_{mn}, after the outage of a generator in node i, will be obtained as follows:

- If the lost generation is assumed by the slack bus

$$\Delta P_{mn} = \rho_{mn}^i \Delta P_i$$

where $\Delta P_i = -P_{G_i}^0$, that is, the active power generation before the outage.
- If the outaged generation is shared among the remaining generators to model the response of the automatic generation control (AGC), "sharing" factors γ_j^i must be used

$$\Delta P_{mn} = \rho_{mn}^i \Delta P_i - \sum_{j \neq i} \rho_{mn}^j \gamma_j^i \Delta P_i = \Delta P_i \left(\rho_{mn}^i - \sum_{j \neq i} \rho_{mn}^j \gamma_j^i \right)$$

with $\sum_{j \neq i} \gamma_j^i = 1$.

EXAMPLE 6.3 USING DISTRIBUTION FACTORS TO STUDY GENERATOR OUTAGES

Consider that the system in Figure 6.5 suffers the outage of the generator located at bus 3.

Matrices A, X, and B are as follows, assuming the slack bus is bus 5:

$$A = \begin{bmatrix} 1 & 1 & 1 & 0 & 0 & 0 \\ -1 & 0 & 0 & 1 & 0 & 0 \\ 0 & -1 & 0 & 0 & 1 & 0 \\ 0 & 0 & -1 & 0 & -1 & 1 \end{bmatrix}$$

$$X = \begin{bmatrix} 0.01 & 0 & 0 & 0 & 0 & 0 \\ 0 & 0.02 & 0 & 0 & 0 & 0 \\ 0 & 0 & 0.01 & 0 & 0 & 0 \\ 0 & 0 & 0 & 0.02 & 0 & 0 \\ 0 & 0 & 0 & 0 & 0.02 & 0 \\ 0 & 0 & 0 & 0 & 0 & 0.02 \end{bmatrix}$$

$$B = \begin{bmatrix} 250 & -100 & -50 & -100 \\ -100 & 150 & 0 & 0 \\ -50 & 0 & 100 & -50 \\ -100 & 0 & -50 & 200 \end{bmatrix}$$

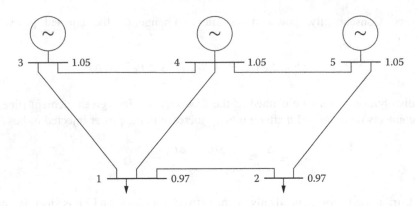

Node	P_L MW	Q_L Mvar	P_G MW	P_G^{max} MW	P_G^{min} MW	V^{sp} pu	Q_G^{max} Mvar	Q_G^{min} Mvar
1	1500	750	–	–	–	–	–	–
2	500	250	–	–	–	–	–	–
3	0	0	1000	1500	250	1.05	750	−750
4	0	0	750	1500	250	1.05	750	−750
5	0	0	309	1000	250	1.05	500	−500
Voltages				$0.95 \leq V \leq 1.05$				

Line i, j	Resistance	Reactance	Shunt Susceptance	P_f^{max} MW	P_f $i \rightarrow j$	$j \rightarrow i$
	pu ($P_{base} = 100$ MVA)					
L1 1–2	0.002	0.01	0.002	1000	96	−96
L2 1–3	0.004	0.02	0.004	1000	−699	721
L3 1–4	0.002	0.01	0.002	1000	−897	920
L4 2–5	0.004	0.02	0.004	1000	−404	414
L5 3–4	0.004	0.02	0.004	1000	279	−276
L6 4–5	0.004	0.02	0.004	1000	106	−105

FIGURE 6.5 5-Bus network for Example 6.3.

and the sensitivity matrix S_f:

$$
S_f = X^{-1}A^T B^{-1} = \begin{bmatrix}
0.4828 & -0.3448 & 0.4138 & 0.3448 \\
0.1034 & 0.0689 & -0.4828 & -0.0689 \\
0.4138 & 0.2759 & 0.0689 & -0.2759 \\
0.4828 & 0.6552 & 0.4138 & 0.3448 \\
0.1034 & 0.0689 & 0.5172 & -0.0689 \\
0.5172 & 0.3448 & 0.5862 & 0.6552
\end{bmatrix}
$$

If the slack bus fully compensates for the outaged generation,

$$\Delta P_i^T = [0 \quad 0 \quad -1000 \quad 0]$$

yielding

$$\Delta P_f^T = [-414 \quad 483 \quad -69 \quad -414 \quad -517 \quad -586]$$

Postcontingency power flows are shown in Table 6.1 together with the actual values obtained using a load flow program. It can be seen that the errors are acceptable, at least for detecting the existence of problems.

TABLE 6.1
Power Flows in Example 6.3

| | Base Case | | Imbalance Assumed by Generator 5 | | | | | Imbalance Assumed by Generators 4 and 5 | | | | |
| | | | Load Flow | | | | DF | Load Flow | | | | DF |
Line	P_{ij}	P_{ji}	P_{ij}	P_{ji}	ΔP_{ij}	ΔP_{ji}	ΔP_{ij}	P_{ij}	P_{ji}	ΔP_{ij}	ΔP_{ji}	ΔP_{ij}
1–2	96	−96	−294	296	−390	392	−414	−104	104	−200	201	−207
1–3	−699	721	−241	245	458	−476	483	−279	284	420	−438	441
1–4	−897	920	−965	994	−68	74	−69	−1117	1152	−219	231	−234
2–5	−404	414	−796	830	−392	416	−414	−604	626	−201	212	−207
3–4	279	−276	−245	248	−524	524	−517	−284	288	−562	564	−559
4–5	106	−106	−492	502	−598	608	−586	−89	90	−195	196	−193

In practice, in case of a generator outage, several generators respond to the power imbalance due to the action of the *automatic generation control*. In this example, the slack generator is not able to supply 1250 MW after the outage of generator 3. Then, we assume that the power imbalance is shared by the remaining generators in proportion to P_G^{max}. Thus, 600 MW will be produced by generator 4 ($P_{G_4}^{max} = 1500$ MW) and 400 MW by generator 5 ($P_{G_5}^{max} = 1000$ MW). This yields

$$\Delta P_i^T = [0 \quad 0 \quad -1000 \quad 600]$$

and consequently,

$$\Delta P_f^T = [-207 \quad 441 \quad -234 \quad -207 \quad -559 \quad -193]$$

Postcontingency power flows using sharing factors $\gamma_4^3 = 0.6$ and $\gamma_5^3 = 0.4$ are also shown in Table 6.1. Note the overload of line 1–4 that cannot be detected if the model does not consider the effect of the AGC on the power imbalance.

In case of contingencies due to a line or transformer outage, postcontingency flows can be obtained using the "compensation theorem" for linear systems. In this way, flow changes due to the outage of the branch between nodes i and j, carrying an active power flow P_{ij}^0 before the outage, are obtained as

$$P_f = P_f^0 + S_f' \Delta P_i \tag{6.5}$$

where matrix S_f' is obtained by modifying the original model to eliminate the outaged branch ij,

$$S_f' = (X')^{-1}(A')^T(B')^{-1}$$

and ΔP_i only contains P_{ij}^0 at node i and $-P_{ij}^0$ at node j. Consequently, the distribution factor corresponding to the branch element that ties nodes m and n is obtained as

$$\rho_{mn}^{ij} = \frac{\Delta P_{mn}^{ij}}{P_{ij}^0} = S_{mn,i}' - S_{mn,j}' \tag{6.6}$$

An alternative to the use of the compensation theorem to obtain the distribution factors is based on modeling the outage of the branch using two fictitious injections at both ends. This approach allows avoiding refactorizing matrix B to obtain S_f'. The fictitious injections must coincide with the power flow after the outage:

$$P_{ij} = P_{ij}^0 + S_{ij,i} \Delta P_i + S_{ij,j} \Delta P_j = P_{ij}^0 + (S_{ij,i} - S_{ij,j}) P_{ij} \tag{6.7}$$

The above equation yields

$$\Delta P_i = -\Delta P_j = \frac{P_{ij}^0}{1 - S_{ij,i} + S_{ij,j}} \tag{6.8}$$

Then, the flow through branch mn after branch ij outage is obtained as

$$P_{mn} = P_{mn}^0 + S_{mn,i}\,\Delta P_i + S_{mn,j}\,\Delta P_j = P_{mn}^0 + (S_{mn,i} - S_{mn,j})\,\Delta P_i \tag{6.9}$$

and the corresponding distribution factor, ρ_{mn}^{ij},

$$\rho_{mn}^{ij} = \frac{\Delta P_{mn}^{ij}}{P_{ij}^0} = \frac{S_{mn,i} - S_{mn,j}}{1 - S_{ij,i} + S_{ij,j}} \tag{6.10}$$

Multiple outage distribution factors can be obtained in a similar way by solving a linear system of equations to compute the fictitious injections at both ends of outaged branches, taking into account the interactions between the outages.

Note that both methods are equivalent and provide identical distribution factors.

EXAMPLE 6.4 USING DISTRIBUTION FACTORS TO STUDY THE OUTAGE OF A TRANSMISSION LINE

In the system of Example 6.3, suppose that line 1–3 suddenly trips. Modifying matrices A, X, and B to consider the line outage, the new sensitivity matrix is obtained as

$$S_f' = (X')^{-1}(A')^{\mathsf{T}}(B')^{-1} = \begin{bmatrix} 0.5 & -0.3333 & 0.3333 & 0.3333 \\ 0.5 & 0.3333 & -0.3333 & -0.3333 \\ 0.5 & 0.6667 & 0.3333 & 0.3333 \\ 0 & 0 & 1 & 0 \\ 0.5 & 0.3333 & 0.6667 & 0.6667 \end{bmatrix}$$

The new distribution factors are

$$\rho^{1,3} = S_f'\,[1 \quad 0 \quad -1 \quad 0]^{\mathsf{T}} = [0.1667 \quad 0.8333 \quad 0.1667 \quad -1 \quad -0.1667]^{\mathsf{T}}$$

Then, multiplying the distribution factors by the flow before the line fault $P_{L2} = -699$, measured at node 1, the postcontingency power flow changes are obtained (Table 6.2).

TABLE 6.2
Power Flows in Example 6.4

Line	Base Case P_{mn}	Base Case P_{nm}	Load Flow P_{mn}	Load Flow P_{nm}	Load Flow ΔP_{mn}	Load Flow ΔP_{nm}	DF ΔP_{mn}
1–2	96	−96	−53	53	−149	149	−109
1–3	−699	721	0	0	699	−721	
1–4	−897	920	−1447	1506	−550	586	−545
2–5	−404	414	−553	576	−149	162	−109
3–4	279	−276	1000	−964	721	−688	655
4–5	106	−106	207	−204	102	−98	109

Alternatively, instead of modifying all the matrices to consider the line outage, distribution factors can be obtained using two fictitious power injections at nodes 1 and 3,

$$\Delta P_1 = -\Delta P_3 = \frac{1}{1 - S_{L2,1} + S_{L2,3}} = 2.417$$

with $P_{L2}^0 = 1$. Then, the distribution factors are obtained from the original sensitivity matrix, S_f, as

$$\rho^{L2} = S_f \, [2.417 \quad 0 \quad -2.417 \quad 0]^T = [0.1667 \quad 1.417 \quad 0.8333 \quad 0.1667 \quad -1 \quad -0.1667]^T$$

Note that the outaged branch is "in service" when using this approach.

A commonly used method to detect critical contingencies is based on determining a contingency ranking in descendent order of severity. A *ranking index* that quantifies the loading level of the system after a given outage is typically used,

$$\text{RI} = \frac{1}{b} \sum_{k=1}^{b} \left(\frac{|P_{f_k}|}{P_{f_k}^{\max}} \right) \tag{6.11}$$

where P_{f_k} is the branch flow of element k, from a total of b lines and transformers, obtained approximately using distribution factors. In this case, the ranking index is just the average rate of lines and transformers.

In practice, several ranking indexes have been proposed, including weighting factors to give higher importance to relevant elements.

Once the ranking indexes have been obtained for all possible contingencies, they are arranged in descendent order. In this way, the analysis begins with the *a priori* most critical contingency, and goes down in the list until a nonproblematic contingency is analyzed.

The main drawbacks of methods based on ranking indexes are the errors inherent to distribution factors and the possibility of "masking" a problematic contingency as a result of "condensing" the loading rates of all branch elements in just one index, that is, the ranking approach might give priority to a contingency that results in slight overloads on several elements instead of a critical contingency in terms of highly overloading a particular branch.

EXAMPLE 6.5 CONTINGENCY ANALYSIS BASED ON A RANKING INDEX

Suppose that the contingency ranking index given by Equation 6.11 is adopted for the network of Example 6.3. The RI of the base case is 0.413.

Let us consider all outages except that of the slack generator, as it is responsible for supplying any generation imbalance. Then, using P_{ij}^0 as the base-case flows and obtaining the postcontingency flows by means of the corresponding distribution factors for each one of the eight $N-1$ contingencies, the following ranking indexes are obtained:

Contingency	G3	G4	L1	L2	L3	L4	L5	L6
RI	0.506	0.480	0.417	0.537	0.504	0.499	0.367	0.410

In consequence, postcontingency states can be classified in descendent RI order: L2, G3, L3, L4, G4, L1, L6, and L5.

A detailed analysis of postcontingency states using a load flow provides the following results:

Case	L1		L2		L3		L4		L5		L6		Overloads
	1	2	1	3	1	4	2	5	3	4	4	5	
L2	−53	53	0	0	−1447	1506	−553	576	1000	−963	207	204	L3, (L5)
G3	−294	296	−241	245	−965	994	−796	830	−245	248	−492	502	−
L3	−352	359	−1148	1240	0	0	−859	918	−240	243	507	−497	L2
L4	508	−500	−782	818	−1226	1271	0	0	182	−180	−341	345	L3
G4	−153	153	−655	681	−692	705	−653	677	319	−314	−392	399	−
L1	0	0	−680	701	−820	841	−500	513	299	−296	205	−203	−
L6	198	−197	−720	743	−979	1004	−302	310	257	−254	0	0	(L3)
L5	138	−137	−960	1000	−677	695	−362	372	0	0	55	−55	−

Note that the ranking analysis would have stopped after detecting a not overloaded case (G3), without paying attention to two critical contingencies: that is, L3 and L4 outages. This problem is usually known as *masking effect* and is inherent to ranking methods.

6.3.2 CONTINGENCY ANALYSIS BASED ON LOAD FLOWS

As discussed above, the use of distribution factors provides reasonable accuracy regarding over-loading after a contingency in transmission networks. However, distribution factors cannot be used for the analysis of abnormal voltages that could occur after a given contingency due to the strong nonlinearity of the voltage problem.

Due to its speed of response, a common approach to detect postcontingency voltage problems is to use a fast decoupled load flow program to compute approximately the postcontingency state. Preoutage complex voltages are used to initialize the load flow and only one complete iteration is carried out [8].

The method is based on checking for overloads and voltage problems on the approximate state obtained after the first iteration of the load flow. If problems are detected, the iterative algorithm continues until convergence, verifying the existence of overloads or out-of-limit voltages in the exact postcontingency state.

The main advantage of screening techniques with respect to ranking index techniques is the fact that all contingencies are analyzed, although approximately, without making any previous selection. However, identification errors may arise since flows and voltages are approximate values obtained after only one iteration of the load flow algorithm.

EXAMPLE 6.6 CONTINGENCY ANALYSIS BASED ON A FAST DECOUPLED LOAD FLOW

In this example, all $N - 1$ contingencies for Example 6.3 are analyzed by means of a fast-decoupled load flow, with the exception of the slack generator outage. After the first complete iteration, and always starting from the preoutage state, a first checking of problems is done using the approximated flow values. The iterative process continues if problems are detected.

To assess the accuracy of the results obtained after only one load flow iteration, Table 6.3 shows the flows and voltages after the first iteration and after the final convergence with a tolerance of 0.1 MVA. Outages of L2 and generator G3 are presented, and three iterations were required in both cases.

Note that the first iteration provides a good estimation of the final values, which allows identifying the overloads of lines L3 and L5 when line L2 trips, and possible problems in line L3 due to generator G3 outage. Regarding voltages, notice that only one iteration is not enough to correctly detect voltage problems at nodes 1 and 2 due to generator G3 outage. This fact suggests the convenience of performing two iterations to detect possible voltage problems, especially if some generators reach reactive power limits after an outage.

TABLE 6.3
Postcontingency State after the First Iteration and after Convergence

	Outage of L2				Outage of G3			
	One Iteration		Converged		One Iteration		Converged	
Line Flow	Origin	End	Origin	End	Origin	End	Origin	End
L1	−52	53	−52	53	−257	258	−293	295
L2	0	0	0	0	−214	217	−239	243
L3	−1376	1428	−1447	1506	−971	1000	−962	991
L4	−543	564	−553	576	−782	811	−797	831
L5	1020	−982	1000	−963	−278	282	−245	249
L6	171	−169	207	−203	−481	490	−495	505
Voltages								
1		0.916		0.909		0.960		0.920
2		0.935		0.929		0.962		0.934
3		1.050		1.050		1.003		0.961
4		1.016		1.014		1.050		1.004
5		1.050		1.050		1.050		1.050

The complete contingency analysis provides the information shown in Table 6.4. It can be seen how a preliminary checking of limit violations, after the first iteration, allows identifying problems for most contingencies, with the exception of some voltage problems (G3) and those pertaining to reactive power limits of generators (L3 and L4).

We conclude indicating that several techniques have been proposed on the basis of computing only a subset of the variables for each outage [7], determining the most influenced area

TABLE 6.4
Information Provided by the Contingency Analysis of Example 6.6

Outaged Element	First Iteration	After Convergence	No. of Iterations
G3	L3 overloaded	Low voltages in buses 1 and 2	3
G4	Low voltage in bus 1	Low voltage in bus 1	2
L1	–	–	1
L2	L3 and L5 overload	Overloads in L3 and L5	3
	Low voltage in buses 1 and 2	Low voltage in buses 1 and 2	
	G4 in reactive limit	G4 in reactive limit	
L3	L2 overloaded	L2 overloaded	5
	Critical voltages in buses 1 and 2	Critical voltages in buses 1 and 2	
		G3 in reactive limit	
L4	L3 overloaded	L3 overloaded	4
	Low voltages in buses 1 and 2	Very low voltages in buses 1 and 2	
		G4 in reactive limit	
L5	–	L2 close to rating	2
L6	L3 on rating	L3 on rating	1

and solving iteratively the load flow equations in that area. *Bounding* techniques can be incorporated not only for contingency ranking but also in load flow algorithms to analyze the critical contingencies [9].

6.4 OPTIMAL POWER FLOW

Historically, the first efforts on optimization in power systems arose in the generation dispatch, the classic ED, that is, on the problem of to allocate the total demand among generating units to minimize the cost of supplying this demand. Soon, clearly appeared the need to include the transmission losses in the economic objective, as well as the constraints on the power flows (see Chapter 5).

The introduction of sparse matrix techniques and new optimization algorithms have made it possible to solve increasingly complex power system optimization problems. This evolution resulted in formulating and solving the OPF problem, nowadays a commonly used tool in energy management centers, a tool which allows to optimize different objectives functions subject to the load flow equality constraints and operational limits imposed on electrical magnitudes.

The OPF has become an essential tool for the operation and planning of power systems. In operation, an OPF allows to determine optimal control actions considering all the operational constraints. In planning, the OPF is used to determine optimal system configurations in the future evolution of the power systems.

The first solution algorithms of the OPF were developed in the beginning of 1960s as a consequence of the introduction of new nonlinear programing techniques [10,11]. Since then, the new algorithms for the OPF have followed the evolution of the mathematical programing techniques, with many variants because of the peculiarities of the OPF problem. The great variety of applied optimization techniques (Reference 12 constitutes an useful survey of OPF algorithms) includes the generalized reduced gradient [10], Newton's method [13], sequential quadratic programing [14], sequential linear programing [15], and, more recently, interior point methods [16]. The majority of the currently commercial OPF tools use one of the last four techniques.

A summary of the key issues associated with OPFs is presented below. The interested reader may find specific references on the topic in References 17–19.

The OPF problem can be formulated as

$$
\begin{aligned}
&\text{Minimize}_{u,x} && f(x, u) \\
&\text{Subject to} && h(x, u) = 0 \\
& && g(x, u) \geq 0
\end{aligned}
\tag{6.12}
$$

where u is the set of decision variables, x is the set of dependent variables, f is the scalar objective function, h is the set of network equations, and g is the set of operational constrains.

In practice, the above formulation is not easy to implement as multiple objectives can be considered desirable attributes of the optimal state of a power system. Clearly, the first objective of an appropriate power system operation is to eliminate possible limit violations. Besides, the power system should remain in normal state even if certain contingencies occur. However, while some power systems might be operated in the most economic way, imposing only operational constraints on the actual state of the network (the base case), some others are operated in a preventive manner and the corresponding OPF must consequently include operational constraints imposed on the base case and on selected postcontingency scenarios, turning into a more complex problem known as the security-constrained optimal power flow (SCOPF) [16]. The most recent advances in optimization techniques allow the use of OPF for contingency analysis, including possible corrective control actions in the evaluation of the different postcontingency scenarios, and even considering the uncertainty associated with distributed generation and demand [20].

Besides, the following observations are pertinent:

- *Objective function*: Since it is required to formulate the objective of the power system optimization, a common approach is to vary the objective function according to the state of the power system:
  - In emergency state, the objective is to determine the minimum number of control actions to return the system to a normal state, either secure or insecure, and as quickly as possible. If the system cannot be driven to a normal state, then the objective should be changed to minimizing the size of the operational limit violations.
  - In normal state, if the system is insecure (alert state), the objective would be to drive the system to a safe state with the minimum possible increase in production cost. If the system is secure, the objective would be minimizing the production cost by dispatching the demand among the generators optimally; alternatively, if the generation dispatch has been previously established, it is possible to act on reactive power resources to minimize transmission losses.
- *Network equations*: As stated in Chapter 3, the state of a power system of n buses is determined by $2n$ nodal equations:

$$P_i = V_i \sum_{j=1}^{n} V_j \left(G_{ij} \cos \theta_{ij} + B_{ij} \sin \theta_{ij}\right) \quad i = 1, \ldots, n \tag{6.13}$$

$$Q_i = V_i \sum_{j=1}^{n} V_j \left(G_{ij} \sin \theta_{ij} - B_{ij} \cos \theta_{ij}\right) \quad i = 1, \ldots, n \tag{6.14}$$

Each node is characterized by four variables: P_i, Q_i, V_i, and θ_i. If two of the four variables are known, Equations 6.12 and 6.13 allow computing the other two variables. Besides, a phase reference must be established ($\theta_r = 0$ for the reference bus) and one or more active power injections must be released since the power losses are unknown *a priori*.

It is also possible to use a DC load flow model in active power optimization studies. The DC network model introduces errors into the OPF solution, but greatly simplifies the OPF calculations. The DC network model is determined by the following n nodal equations:

$$P_i = \sum_{j=1}^{n} \frac{\theta_{ij}}{x_{ij}} \quad i = 1, \ldots, n \tag{6.15}$$

where x_{ij} is the branch reactance between nodes i and j.
- *Inequality constraints*:
  - Limits on control variables: Generator active and reactive power, transformer tap changers, capacitor and/or reactor banks, *FACTS*, etc.
  - Operational constraints: Limits on bus voltages and power flows.
  Note that there is an essential difference between these two types of limits: while the limits imposed on the control variables must be considered "rigid," operational constraints might be considered "flexible" since they can be exceeded, for a short period of time in exceptional circumstances.
- *Additional modeling issues*:
  - Automatic controls. The response of automatic controls acting in the power system must be appropriately modeled. This is the case of devices (transformers, capacitors, reactors, FACTS) controlling voltages or reactive power flows, and power interchanges among different areas controlled by dedicated generators.
  - Components working coordinately, for example, parallel transformers and generators in the same power plant.

- "Operational preferences," such as preference for using certain control devices, priority for mitigating certain constraint limits over the rest, restrictions regarding the number of control actions on a device during a given period of time, and so on.

  The above-mentioned modeling issues ought to be properly included in the OPF formulation for the solution attained to be useful for the operator [14].

The example below, solved using GAMS [21], illustrate the complexity of the OPF problem.

EXAMPLE 6.7 USING AN OPF TO MINIMIZE THE PRODUCTION COST

In this example, the optimal state in terms of generation cost of the system presented in Figure 6.6 is obtained.

The costs of generating units are the following:

$$G4:100 + 10.5P_{G_4} \quad \$/h \quad G5:100 + 10.0P_{G_5} \quad \$/h$$

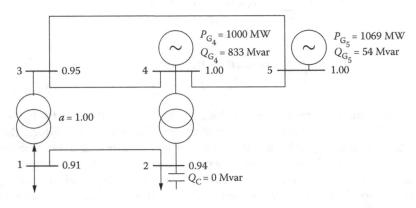

Node	P_D MW	Q_D Mvar	P_G MW	P_G^{max} MW	P_G^{min} MW	V^{sp} pu	Q_G Mvar	Q_G^{max} Mvar	Q_G^{min} Mvar
1	1000	250	–	–	–	–	–	–	–
2	1000	250	–	–	–	–	0	200	0
3	0	0	–	–	–	–	–	–	–
4	0	0	1000	1500	500	1.00	833	1000	−1000
5	0	0	1069	1500	500	1.00	54	1000	−1000
Voltages				$0.95 \leq V \leq 1.05$					

Line	Resistance	Reactance	Susceptance	S_f^{max}
	pu ($S_{base} = 100$ MVA)			MVA
L1 1–2	0.010	0.010	0.020	500
L2 1–3	0.001	0.015	0.000	2000
L3 2–4	0.001	0.010	0.000	2000
L4 3–4	0.005	0.020	0.020	1000
L5 3–5	0.005	0.010	0.020	1000
L6 4–5	0.005	0.020	0.020	1000

FIGURE 6.6   5-Bus network for Example 6.7.

Besides, the system contains voltage control elements: Generator controllers, a 200 Mvar capacitor bank with 20 steps of 10 Mvar, and transformer 1–3 that includes 21 regulation taps to change the turn ratio $\pm 10\%$ in increments of 1%. The capacitor bank is initially disconnected and the transformer is set to its middle tap.

Matrices G and B are

$$G = \begin{bmatrix} 50 + 4.4248/a^2 & -50.0 & -4.4248/a & 0.0 & 0.0 \\ -50.0 & 59.9010 & 0.0 & -9.9010 & 0.0 \\ -4.4248/a & 0.0 & 56.1895 & -11.7647 & -40.0 \\ 0.0 & -9.9010 & -11.7647 & 33.4304 & -11.7647 \\ 0.0 & 0.0 & -40.0 & -11.7647 & 51.7647 \end{bmatrix}$$

$$B = \begin{bmatrix} -49.99 - 66.3717/a^2 & 50.0 & 66.3717/a & 0.0 & 0.0 \\ 50.0 & -148.9999 & 0.0 & 99.0099 & 0.0 \\ 66.3717/a & 0.0 & -193.4105 & 47.0588 & 80.0 \\ 0.0 & 99.0099 & 47.0588 & -193.1076 & 47.0588 \\ 0.0 & 0.0 & 80.0 & 47.0588 & -127.0388 \end{bmatrix}$$

where a is the transformer pu turn ratio.

Note that matrix B should also depend on the susceptance of the capacitor bank. However, in this example, the capacitor is modeled as a reactive power injection in node 2.

The OPF objective is to minimize the production cost (powers in pu),

$$C = C_1 + C_2 = 200 + 1050 P_{G_4} + 1000 P_{G_5}$$

and the constraints are the following:

- *Network equations:*

$$P_i = P_{G_i} - P_{D_i} = \sum_{j=1}^{5} V_i V_j [G_{ij} \cos(\theta_i - \theta_j) + B_{ij} \sin(\theta_i - \theta_j)] \quad i = 1, \dots, 5$$

$$Q_i = Q_{G_i} - Q_{D_i} = \sum_{j=1}^{5} V_i V_j [G_{ij} \sin(\theta_i - \theta_j) - B_{ij} \cos(\theta_i - \theta_j)] \quad i = 1, \dots, 5$$

where the decision variables are P_{G_4}, P_{G_5}, Q_{G_4}, Q_{G_5}, as well as the reactive power injected by the capacitor bank, Q_{G_2} and the transformer turn ratio a included in matrices G and B. On the other hand, node 5 is the reference bus, $\theta_5 = 0$.

- *Control limits:*
  - Generator active and reactive power limits:

$$5 \le P_{G_4} \le 15 \quad -10 \le Q_{G_4} \le 10$$
$$5 \le P_{G_5} \le 15 \quad -10 \le Q_{G_5} \le 10$$

- Capacitor bank: $0 \le Q_{G_2} \le 2.0$, in discrete steps of 10 Mvar (0.1 in pu).
- Transformer turn ratio: $0.9 \le a \le 1.1$, in 0.01 pu increments.

The discrete nature of transformer taps and capacitor bank, considering the small magnitude of the steps in both cases, is taken into account by rounding the variables to the nearest discrete value and updating the system state by carrying out a load flow.

- *Operational limits:*

$$0.95 \le V_i \le 1.05 \quad i = 1, \dots, 5$$

$$S_{ij} = \sqrt{P_{ij}^2 + Q_{ij}^2} \leq S_{ij}^{max} \quad i,j = 1,\ldots,5$$

where P_{ij}, Q_{ij}, and S_{ij} are, respectively, the active power, reactive power, and apparent power flows in branch elements.

The solution to the above OPF problem leads to the state shown in Figure 6.7. Note the following observations on the optimal solution:

- The production cost is reduced to \$21,289/h, with a saving of \$101/h as compared to the base case.
- The losses have also been reduced from 69.2 to 65.2 MW. Note that generator 4 is favored due to its advantageous location in the network. In fact, an ED exclusively based on costs would drive that generator down to its minimum power output.
- All variables are optimized to minimize the cost, subject to all constraints. Note that the voltage problems that appear in the base case are corrected.
- The transformer turn ratio has been rounded off to 1.03 pu, and capacitors are all connected (200 Mvar rating) and inject 206 Mvar at 1.014 pu voltage.
- The Lagrange multipliers corresponding to active power balance equations provide the *bus incremental costs*, that is, the sensitivity of the production cost with respect to the bus active power demand,

$$\lambda^T = [11.08 \quad 10.70 \quad 10.76 \quad 10.50 \quad 10.00] \quad \$/MWh$$

Note that similar incremental costs can be obtained for reactive power demands.

Clearly, the application of an optimization algorithm to a power system problem is not straightforward in terms of implementation, since identifying the relevant constraints and the objective might be complex. Besides, the number of variables and equations increase with the number of buses. Consequently, its use in real time and the importance of ensuring the quality and continuity of the electrical service requires that an OPF algorithm has the following characteristics: (i) reliability in terms of its capability to give an acceptable solution in the most cases; (ii) fast solution time; and (iii) flexibility to adapt to new models and approaches. Besides, the need of including complex constraints and heuristic operational requirements, requires an iterative use of the OPF, evaluating the suitability of the obtained OPF's solution and updating the formulation if necessary.

Additional details on OPF algorithms successfully applied in practice, as well as issues being still researched are provided, for example, in [17–20,22].

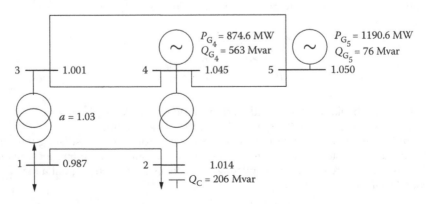

FIGURE 6.7   Optimal state in terms of generation costs of 5-bus network.

6.5   TRANSMISSION SYSTEM OPERATION

The objectives and possible control actions regarding transmission system operations are clearly conditioned to the network state. Consequently, appropriate tools for system supervision (SCADA, state estimator) and security assessment (contingency analysis) are essential for power system operations.

In this section, different examples of the application of an OPF as a decision tool in power system operations are presented. These examples can be easily solved using computational environments such as GAMS [21].

6.5.1   EMERGENCY STATE

A power system may enter an *emergency state* as a result of an unexpected demand increase or a contingency occurrence. In this case, the operator has to bring the system back to a normal state as quickly as possible, disregarding economical considerations.

The urgency of such situations dictates the reduction of the number of control actions to the minimum necessary for correcting the problem. Mathematically, this problem can be formulated as an OPF with the objective of minimizing control actions to bring all the variables within limits. This objective is usually realized by incorporating a quadratic or linear penalty function of the control actions in the objective function.

Taking advantage of the decoupling between the active and the reactive power equations, separate studies of overloads and out-of-limit voltages can be made. Clearly, some specific cases might require a complete, coupled OPF formulation. This is the case if it is necessary to correct voltages by rescheduling active power generations.

6.5.1.1   Overload Correction

To correct a branch overload, operators generally resort to rescheduling the active power generations. Alternative control elements for active power flows are phase shifters and FACTS devices, if available. Besides, if the available decision variables are not sufficient to correct the overload in a reasonable time, load shedding might be required.

Using a DC model, this OPF instance takes the form

$$
\begin{aligned}
\text{Minimize} \quad & c_u^T \Delta P^+ + c_d^T \Delta P^- \\
\text{Subject to:} \quad & P + \Delta P^+ - \Delta P^- = B\theta \\
& P^{\min} \leq P + \Delta P^+ - \Delta P^- \leq P^{\max} \\
& -P_f^{\max} \leq P_f = X^{-1} A^T \theta \leq P_f^{\max} \\
& \Delta P^+ \geq 0 \quad \Delta P^- \geq 0
\end{aligned}
\tag{6.16}
$$

where c_u and c_d are cost vectors associated with the decision variables, ΔP^+ and ΔP^-.

An alternative to the above formulation is using the sensitivity matrix, S_f, to obtaining a more compact model, yielding

$$
\begin{aligned}
\text{Minimize} \quad & c_u^T \Delta P^+ + c_d^T \Delta P^- \\
\text{Subject to:} \quad & -P_f^{\max} \leq P_f = P_f^0 + S_f(\Delta P^+ - \Delta P^-) \leq P_f^{\max} \\
& \Delta P^+ \leq P^{\max} - P \\
& \Delta P^- \leq P - P^{\min} \\
& \Delta P^+ \geq 0 \quad \Delta P^- \geq 0
\end{aligned}
\tag{6.17}
$$

EXAMPLE 6.8 USING AN OPF TO CORRECT OVERLOADS

The power system of Example 6.3 is in emergency state as presented in Figure 6.8, where line 3 is overloaded.

Power flows can be easily obtained using the DC load flow model as in Example 6.3:

$$
P_f =
\begin{bmatrix}
0.4828 & -0.3448 & 0.4138 & 0.3448 \\
0.1034 & 0.0689 & -0.4828 & -0.0689 \\
0.4138 & 0.2759 & 0.0689 & -0.2759 \\
0.4828 & 0.6552 & 0.4138 & 0.3448 \\
0.1034 & 0.0689 & 0.5172 & -0.0689 \\
0.5172 & 0.3448 & 0.5862 & 0.6552
\end{bmatrix}
\begin{bmatrix}
-12.5 \\
-12.5 \\
11.5 \\
11.0
\end{bmatrix}
=
\begin{bmatrix}
6.82 \\
-8.47 \\
-10.86 \\
-5.67 \\
3.03 \\
3.17
\end{bmatrix}
$$

Note that line L3 overload is 86 MW ($P_f^{max} = 1000$ MW and $S_{base} = 100$ MVA).

Adopting the linear model of Equation 6.15, the OPF problem can be formulated as follows:

- *Objective function*:

$$
c_u^T \Delta P^+ + c_d^T \Delta P^- = [1000 \quad 1000 \quad 2000]
\begin{bmatrix}
\Delta P_{G_3}^+ \\
\Delta P_{G_4}^+ \\
\Delta P_{G_5}^+
\end{bmatrix}
+ [1000 \quad 1000 \quad 2000]
\begin{bmatrix}
\Delta P_{G_3}^- \\
\Delta P_{G_4}^- \\
\Delta P_{G_5}^-
\end{bmatrix}
$$

$$
\Delta P^+ \geq 0 \quad \Delta P^- \geq 0
$$

Note that the rescheduling penalty costs are \$10/MW for generators 3 and 4, and \$20/MW for generator 5.

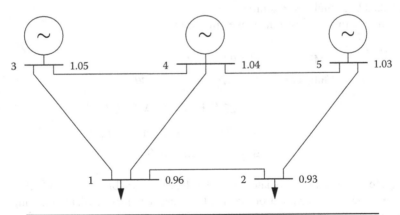

Node	P_G MW	Q_D Mvar	P_G MW	Q_G Mvar	V pu
1	1250	500	–	–	0.96
2	1250	500	–	–	0.93
3	0	0	1150	366	1.05
4	0	0	1100	682	1.04
5	0	0	351	456	1.03

FIGURE 6.8 5-Bus network for Example 6.8.

- *Network equations:* $P + \Delta P^+ - \Delta P^- = B\theta$

$$
\begin{bmatrix} -12.5 \\ -12.5 \\ 11.5 + \Delta P^+_{G_3} - \Delta P^-_{G_3} \\ 11.0 + \Delta P^+_{G_4} - \Delta P^-_{G_4} \end{bmatrix} = \begin{bmatrix} 250 & -100 & -50 & -100 \\ -100 & 150 & 0 & 0 \\ -50 & 0 & 100 & -50 \\ -100 & 0 & -50 & 200 \end{bmatrix} \begin{bmatrix} \theta_1 \\ \theta_2 \\ \theta_3 \\ \theta_4 \end{bmatrix}
$$

As the above model does not include the nodal equation of the slack bus, a further equation is required:

$$
\left(11.5 + \Delta P^+_{G_3} - \Delta P^-_{G_3}\right) + \left(11.0 + \Delta P^+_{G_4} - \Delta P^-_{G_4}\right) + \left(2.5 + \Delta P^+_{G_5} - \Delta P^-_{G_5}\right) = 25.0
$$

where 250 MW is the initial generation of generator 5, neglecting losses. The above equation is not required if the full $n \times (n-1)$ susceptance matrix $\hat{B}$ is used instead of B.

- *Limits:*

$$
\begin{bmatrix} 15.0 \\ 15.0 \\ 10.0 \end{bmatrix} \geq \begin{bmatrix} 11.5 + \Delta P^+_{G_3} - \Delta P^-_{G_3} \\ 11.5 + \Delta P^+_{G_4} - \Delta P^-_{G_4} \\ 2.0 + \Delta P^+_{G_5} - \Delta P^-_{G_5} \end{bmatrix} \geq \begin{bmatrix} 2.5 \\ 2.5 \\ 2.5 \end{bmatrix}
$$

$$
\begin{bmatrix} -10.0 \\ -10.0 \\ -10.0 \\ -10.0 \\ -10.0 \\ -10.0 \end{bmatrix} \leq \begin{bmatrix} 100 & -100 & 0 & 0 \\ 50 & 0 & -50 & 0 \\ 100 & 0 & 0 & -100 \\ 0 & 50 & 0 & 0 \\ 0 & 0 & 50 & -50 \\ 0 & 0 & 0 & 50 \end{bmatrix} \begin{bmatrix} \theta_1 \\ \theta_2 \\ \theta_3 \\ \theta_4 \end{bmatrix} \leq \begin{bmatrix} 10.0 \\ 10.0 \\ 10.0 \\ 10.0 \\ 10.0 \\ 10.0 \end{bmatrix}
$$

The solution to this linear programing problem provides the following rescheduling and power flow values:

$$
\Delta P^+ = \begin{bmatrix} 2.5 & 0 & 0 \end{bmatrix}^T \qquad \Delta P^- = \begin{bmatrix} 0 & 2.5 & 0 \end{bmatrix}^T
$$

$$
P_f = \begin{bmatrix} 7.0 & -9.5 & -10.0 & -5.5 & 4.5 & 3.0 \end{bmatrix}^T
$$

with a total cost of $50.

An alternative way to proceed is to formulate the OPF problem in a more compact form by eliminating voltage phases:

$$
P_f = S_f(P + \Delta P^+ - \Delta P^-) = P_f^0 + S_f(\Delta P^+ - \Delta P^-)
$$

$$
= \begin{bmatrix} 6.82 \\ -8.47 \\ -10.86 \\ -5.67 \\ 3.03 \\ 3.17 \end{bmatrix} + \begin{bmatrix} 0.4828 & -0.3448 & 0.4138 & 0.3448 \\ 0.1034 & 0.0689 & -0.4828 & -0.0689 \\ 0.4138 & 0.2759 & 0.0689 & -0.2759 \\ 0.4828 & 0.6552 & 0.4138 & 0.3448 \\ 0.1034 & 0.0689 & 0.5172 & -0.0689 \\ 0.5172 & 0.3448 & 0.5862 & 0.6552 \end{bmatrix} \begin{bmatrix} 0 \\ 0 \\ \Delta P^+_{G_3} - \Delta P^-_{G_3} \\ \Delta P^+_{G_4} - \Delta P^-_{G_4} \end{bmatrix}
$$

Clearly, both OPF problems provide identical solutions.

Additionally, it is possible to use an alternative, heuristic technique, using the distribution factors of overloaded line L3,

$$
\Delta P_{L3} = P_{L3} - P^0_{L3} = -10.0 - (-10.86) = 0.86 = \rho^i_{L3} \Delta P_i
$$

$$
\rho^{G3}_{L3} = 0.0689 \Rightarrow \Delta P_3 = 12.48 \qquad \rho^{G4}_{L3} = -0.2759 \Rightarrow \Delta P_4 = -3.12
$$

which implies increasing the generator 3 active power output by 1248 MW, or, alternatively, decreasing generator 4 power output 312 MW. In both cases, the imbalance is covered by the slack generator.

Economical considerations can also be included in the above procedure. For example, consider that an increase in the slack generator should be avoided due to high cost. In this case, the power not provided by generator 4 should be supplied by generator 3:

$$\Delta P_{L3}^{i} = 0.86 = \rho_{L3}^{G3}\,\Delta P_3 + \rho_{L3}^{G4}\,\Delta P_4 \Rightarrow \Delta P_3 = -\Delta P_4 = 2.49$$

Note that this heuristic solution coincides with the OPF solution.

In the above example, the use of an OPF instance based on linear programing to determine corrective actions for overloads is presented, along with the alternative use of a heuristic technique based on distribution factors. This last strategy is especially useful for obtaining efficiently corrective actions in real time.

It is important to note that a feasible solution based on rescheduling generation might not be obtained in some critical cases. In such situation, operators need to implement the necessary actions to mitigate the overload, and, if the overload cannot be reduced to an acceptable level, load shedding is required.

EXAMPLE 6.9 USING AN OPF TO CORRECT OVERLOADS INCLUDING LOAD SHEDDING

Consider that the 5-bus system of the previous example is in the state presented in Figure 6.9.

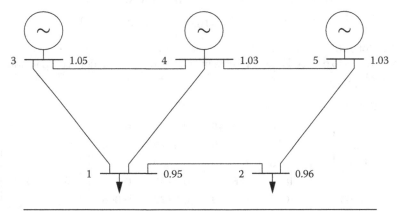

Node	P_D MW	Q_D Mvar	P_G MW	Q_G Mvar	V pu
1	2500	750	–	–	0.95
2	300	100	–	–	0.96
3	0	0	1300	479	1.05
4	0	0	1250	597	1.03
5	0	0	356	299	1.03

FIGURE 6.9   5-Bus system of Example 6.9.

The power flows for this state, computed using a DC model, are the following:

$$P_f = \begin{bmatrix} 0.4828 & -0.3448 & 0.4138 & 0.3448 \\ 0.1034 & 0.0689 & -0.4828 & -0.0689 \\ 0.4138 & 0.2759 & 0.0689 & -0.2759 \\ 0.4828 & 0.6552 & 0.4138 & 0.3448 \\ 0.1034 & 0.0689 & 0.5172 & -0.0689 \\ 0.5172 & 0.3448 & 0.5862 & 0.6552 \end{bmatrix} \begin{bmatrix} -25.0 \\ -3.0 \\ 13.0 \\ 12.5 \end{bmatrix} = \begin{bmatrix} -1.35 \\ -9.93 \\ -13.72 \\ -4.35 \\ 3.07 \\ 1.85 \end{bmatrix}$$

Notice the heavy overload affecting line L3, and that line L2 is practically at its limit (1000 MW). In this case, load shedding ΔP^d is required to eliminate the overload. Clearly, load shedding should be the last resort to alleviate the overload, and, consequently, a much higher "cost" than the highest generator redispatch cost must be assigned to unserved energy. Then, the OPF problem can be formulated as follows:

- *Objective function*:

$$c_u^T \Delta P^+ + c_d^T \Delta P^- + c_{ls}^T \Delta P^d = \begin{bmatrix} 1000 & 1000 & 2000 \end{bmatrix} \begin{bmatrix} \Delta P_{G_3}^+ \\ \Delta P_{G_4}^+ \\ \Delta P_{G_5}^+ \end{bmatrix} + \begin{bmatrix} 1000 & 1000 & 2000 \end{bmatrix} \begin{bmatrix} \Delta P_{G_3}^- \\ \Delta P_{G_4}^- \\ \Delta P_{G_5}^- \end{bmatrix}$$

$$+ \begin{bmatrix} 10,000 & 10,000 \end{bmatrix} \begin{bmatrix} \Delta P_{D_1} \\ \Delta P_{D_2} \end{bmatrix}$$

$$\Delta P^+ \geq 0 \quad \Delta P^- \geq 0 \quad \Delta P_D \geq 0$$

- *Network equations*:

$$P + \Delta P^+ - \Delta P^- + \Delta P^d = B\theta$$

$$\begin{bmatrix} -25.0 + \Delta P_{D_1} \\ -3.0 + \Delta P_{D_2} \\ 13.0 + \Delta P_{G_3}^+ - \Delta P_{G_3}^- \\ 12.5 + \Delta P_{G_4}^+ - \Delta P_{G_4}^- \end{bmatrix} = \begin{bmatrix} 250 & -100 & -50 & -100 \\ 100 & 150 & 0 & 0 \\ -50 & 0 & 100 & -50 \\ -100 & 0 & -50 & 200 \end{bmatrix} \begin{bmatrix} \theta_1 \\ \theta_2 \\ \theta_3 \\ \theta_4 \end{bmatrix}$$

and the additional equation for the slack bus,

$$(13 + \Delta P_{G_3}^+ - \Delta P_{G_3}^-) + (12.5 + \Delta P_{G_4}^+ - \Delta P_{G_4}^-) + (2.5 + \Delta P_{G_5}^+ - \Delta P_{G_5}^-) = 28 - \Delta P_{D_1} - \Delta P_{D_2}$$

- *Limits*:

$$\begin{bmatrix} 15.0 \\ 15.0 \\ 10.0 \end{bmatrix} \geq \begin{bmatrix} 13.0 + \Delta P_{G_3}^+ - \Delta P_{G_3}^- \\ 12.5 + \Delta P_{G_4}^+ - \Delta P_{G_4}^- \\ 2.5 + \Delta P_{G_5}^+ - \Delta P_{G_5}^- \end{bmatrix} \geq \begin{bmatrix} 2.5 \\ 2.5 \\ 2.5 \end{bmatrix}$$

$$\begin{bmatrix} -10.0 \\ -10.0 \\ -10.0 \\ -10.0 \\ -10.0 \\ -10.0 \end{bmatrix} \leq \begin{bmatrix} 100 & -100 & 0 & 0 \\ 50 & 0 & -50 & 0 \\ 100 & 0 & 0 & -100 \\ 0 & 50 & 0 & 0 \\ 0 & 0 & 50 & -50 \\ 0 & 0 & 0 & 50 \end{bmatrix} \begin{bmatrix} \theta_1 \\ \theta_2 \\ \theta_3 \\ \theta_4 \end{bmatrix} \leq \begin{bmatrix} 10.0 \\ 10.0 \\ 10.0 \\ 10.0 \\ 10.0 \\ 10.0 \end{bmatrix}$$

The solution to the above linear programing problem provides the following corrective action and power flow values:

$$\Delta P^+ - \begin{bmatrix} 2.00 & 0 & 6.00 \end{bmatrix}^T \quad \Delta P^- - \begin{bmatrix} 0 & 10.00 & 0 \end{bmatrix}^T \quad \Delta P^d - \begin{bmatrix} 2.0 & 0 \end{bmatrix}^T$$

$$P_f = \begin{bmatrix} -3.0 & -10.0 & -10.0 & -6.0 & 5.0 & -2.5 \end{bmatrix}^T$$

Generators G3 and G5 increase their active power outputs while generator G4 decreases its power output. A 200 MW load shedding is also required at node 1.

Note that different load-shedding costs can be assigned to different loads, introducing differentiation among customers.

6.5.1.2 Voltage Correction

When some voltages violate their operational limits, operators can make use of their knowledge and experience to determine appropriate corrective actions. In fact, heuristic rules such as "an excessively low voltage is more effectively corrected with a local reactive power injection or modifying a nearby transformer tap" are used. If the problem is not highly complex, the use of basic rules permits quickly obtaining a simple solution. However, the complexity of voltage problems due to nonlinearities calls for the use of appropriate tools to deal with such problems [23].

As previously noticed, several issues inherent to current OPF tools limit their use in control centers [18,19]. A clear example of this is that OPF tools provide solutions generally involving a large number of controls, making their implementation difficult. Besides, current power systems are equipped with a large variety of devices that directly affect the voltages, making the voltage control problem different from the active power one.

EXAMPLE 6.10 USING AN OPF TO CORRECT VOLTAGE PROBLEMS

The system of Example 6.7 is in the state shown in Figure 6.6. Note that excessively low voltages at nodes 1 and 2 (0.91 and 0.94 in pu, respectively) should be corrected.

To avoid large deviations on the decision variables with respect to their initial values, a quadratic penalty function of the control actions is used as the objective function:

$$\mathcal{F} = (\Delta V_4)^2 + (\Delta V_5)^2 + (\Delta a)^2 + (\Delta Q_C)^2$$

where $V_4 = 1.0 + \Delta V_4$, $V_5 = 1.0 + \Delta V_5$, $a = 1.0 + \Delta a$, and $Q_C = 0.0 + \Delta Q_C$.

Constraints are the same as those of Example 6.7 with two exceptions: $P_{G_4} = 10.0$, that is, P_{G_4} is fixed to 1000 MW, and P_{G_5} (the slack bus) compensates any power imbalance.

The solution to this optimization problem, rounding off *ex post* the variables corresponding to discrete controllers, yields the state shown in Figure 6.10.

Note the following observations on this OPF solution:

- The objective function is a combination of very different physical magnitudes (generator voltages, the turn ratio of a transformer, and the reactive power of a capacitor bank) with different scales and margins, but expressed in pu. In spite of being a local

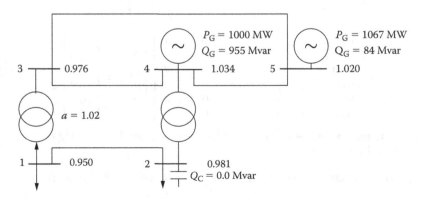

FIGURE 6.10   5-Bus network of Example 6.10 after correcting out-of-limit voltages using an OPF.

reactive power resource located in a low-voltage node, note that the capacitor bank is not used. This is the result of the scaling of the control variables in the objective function.
- The use of quadratic penalty functions of the control actions as the objective function does not prevent a solution involving a high number of control actions. In fact, all available controllers are used except the capacitor bank.
- The corrective measures imply a cost reduction due to the lower losses resulting from higher voltages. The new production cost is \$21,370/h, representing a saving of \$20/h as compared with the initial cost.

An alternative to using an OPF tool is using heuristic rules, including numeric algorithms for calculating voltage sensitivities with respect to control variables [23,24]. The example below illustrates the use of a heuristic approach based on sensitivities to correct voltage issues.

EXAMPLE 6.11 CORRECTIVE ACTIONS FOR VOLTAGE ISSUES USING SENSITIVITIES

In this example, the voltage issues of Example 6.7 (Figure 6.6), corrected in Example 6.10 using an OPF, is corrected in this example using sensitivities. First, the sensitivities of problematic voltages with respect to the available control actions are obtained using a linear approximation of the reactive power nodal equations,

$$I_{r_i} = \frac{Q_i}{V_i} = \sum_j V_j [G_{ij} \sin \theta_{ij} - B_{ij} \cos \theta_{ij}] \simeq \sum_j -B_{ij} V_j$$

where I_{r_i} is the reactive current injected in node i. The above expression can be written in matrix form, separating demand and generation nodes

$$\begin{bmatrix} \Delta I_{r_D} \\ \Delta I_{r_G} \end{bmatrix} = \begin{bmatrix} B_{D,D} & B_{D,G} \\ \hline B_{D,G}^T & B_{G,G} \end{bmatrix} \begin{bmatrix} \Delta V_D \\ \Delta V_G \end{bmatrix}$$

$$\begin{bmatrix} \Delta I_{r_1} \\ \Delta I_{r_2} \\ \Delta I_{r_3} \\ \Delta I_{r_4} \\ \Delta I_{r_5} \end{bmatrix} = \begin{bmatrix} 116.3616 & -50.0 & -66.3716 & 0 & 0 \\ -50.0 & 148.9999 & 0 & -99.0099 & 0 \\ -66.3717 & 0 & 193.4105 & -47.0588 & -80.0 \\ \hline 0 & -99.0099 & -47.0588 & 193.1075 & -47.0588 \\ 0 & 0 & -80.0 & -47.0588 & 127.0388 \end{bmatrix} \begin{bmatrix} \Delta V_1 \\ \Delta V_2 \\ \Delta V_3 \\ \Delta V_4 \\ \Delta V_5 \end{bmatrix}$$

Next, the effect of each control variable on problematic voltages is analyzed.

- *Capacitor bank*: The connection of the capacitor bank implies the injection of a reactive current ΔI_{r_2}, while generator voltages remain constant ($\Delta V_4 = \Delta V_5 = 0$). Consequently,

$$\Delta V_D = B_{D,D}^{-1} \Delta I_{r_D}; \quad \Delta I_{r_G} = B_{D,G}^T B_{D,D}^{-1} \Delta I_{r_D}$$

Note that 200 Mvar (reactive power at rated voltage) are available at node 2. The terms of interest are,

$$\begin{bmatrix} \Delta V_1 \\ \Delta V_2 \\ \Delta V_3 \end{bmatrix} = \begin{bmatrix} 0.00437 \\ 0.00818 \\ 0.00150 \end{bmatrix} \Delta I_{r_2}; \quad \begin{bmatrix} \Delta I_{r_4} \\ \Delta I_{r_5} \end{bmatrix} = \begin{bmatrix} -0.88021 \\ -0.11994 \end{bmatrix} \Delta I_{r_2}$$

The reactive current provided by the capacitor bank depends on the bus voltage:

$$\Delta I_{r_2} = \Delta Q_2/V_2 = 2.0\, V_2^2/V_2 = 2.0\, V_2$$

Considering that prior to connecting the capacitor bank, $V_2 = 0.942$,

$$\Delta V_2 = 0.00818 \times 2.0 \times (0.942 + \Delta V_2)$$

yielding $\Delta V_2 = 0.0157$ and $\Delta I_{r_2} = 1.9154$. Then

$$[\Delta V_1 \quad \Delta V_2 \quad \Delta V_3] = [0.0084 \quad 0.0157 \quad 0.0029]$$
$$[\Delta I_{r_4} \quad \Delta I_{r_5}] = [-1.686 \quad -0.23]$$

Note that the connection of 200 Mvar in node 2 would correct voltage V_2 ($V_2 = 0.958$), and also improve V_1 ($V_1 = 0.917$). Besides, generators would inject less reactive power.

- *Generators in buses 4 and 5*: Considering $\Delta I_{r_C} = 0$,

$$\Delta V_C = -B_{C,C}^{-1} B_{C,G} \Delta V_G; \qquad \Delta I_{r_G} = -B_{C,G}^T B_{C,C}^{-1} B_{C,G} \Delta V_G$$

yielding

$$\begin{bmatrix} \Delta V_1 \\ \Delta V_2 \\ \Delta V_3 \end{bmatrix} = \begin{bmatrix} 0.64283 & 0.35743 \\ 0.88021 & 0.11994 \\ 0.46391 & 0.53629 \end{bmatrix} \begin{bmatrix} \Delta V_4 \\ \Delta V_5 \end{bmatrix}$$

$$\begin{bmatrix} \Delta I_{r_4} \\ \Delta I_{r_5} \end{bmatrix} = \begin{bmatrix} 84.12692 & -84.17143 \\ -84.17143 & 84.13593 \end{bmatrix} \begin{bmatrix} \Delta V_4 \\ \Delta V_5 \end{bmatrix}$$

Note that to drive V_1 within limits ($\Delta V_1 = 0.041$) acting on V_4, $\Delta V_4 = 0.064$ and the generator would become overloaded ($\Delta I_{r_4} = 5.366$). Alternatively, $\Delta V_5 = 0.115$ would be required, and V_5 would go beyond its upper limit. Thus, a combined action on both generators is required.

- *Transformer turn ratio*: The action involving transformer turn ratio can be modeled as two reactive power injections in both ends of the transformer:

$$\frac{\partial Q_{ij}}{\partial a} = -\frac{V_i V_j}{a}(G_{ij}\sin\theta_{ij} - B_{ij}\cos\theta_{ij}) - \frac{2B_{ij}V_i^2}{a}$$

$$\frac{\partial Q_{ji}}{\partial a} = -\frac{V_j V_i}{a}(G_{ji}\sin\theta_{ji} - B_{ji}\cos\theta_{ji})$$

Using the usual approximations, the above equations become

$$\Delta I_{r_i} = -\left\{ \frac{B_{ij} V_j}{a} - \frac{2V_i B_{ij}}{a} \right\}\Delta a; \qquad \Delta I_{r_j} = -\frac{B_{ji} V_i}{a}\Delta a$$

yielding

$$\begin{bmatrix} \Delta V_1 \\ \Delta V_2 \\ \Delta V_3 \end{bmatrix} = \begin{bmatrix} 0.01302 & 0.00437 & 0.00447 \\ 0.00437 & 0.00818 & 0.00150 \\ 0.00447 & 0.00150 & 0.00670 \end{bmatrix} \begin{bmatrix} 57.4115 \\ 0 \\ -60.3319 \end{bmatrix} \Delta a = \begin{bmatrix} 0.4779 \\ 0.1604 \\ -0.1479 \end{bmatrix} \Delta a$$

In consequence, an increment $\Delta a = 0.086 \simeq 0.09$ would be required to correct V_1. However, such an action would reduce V_3 in $\Delta V_3 = -0.013$, in the opposite transformer end, creating a new issue in that node.

It is relevant to note that none of the possible corrective actions discussed above is able to correct the voltage of bus 1 without violating other operational limits. Thus, several combined actions

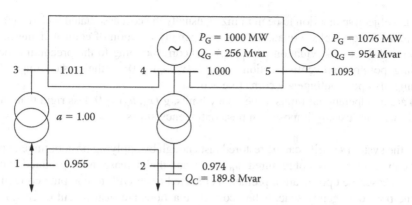

FIGURE 6.11   5-Bus network of Example 6.11. State after correcting voltage problems using sensitivities.

are needed. In this example, it seems reasonable to connect the whole capacitor bank, since V_2 is rather low. After this action, voltages become

$$[V_1 \quad V_2 \quad V_3] = [0.917 \quad 0.958 \quad 0.956]$$

Next, it is necessary to choose between the actions related to the two generators and the transformer. As generator G5 has a larger reactive power margin, its voltage is increased, $\Delta V_5 = (0.95 - 0.917)/0.35743 \simeq 0.093$, yielding

$$[V_1 \quad V_2 \quad V_3] = [0.950 \quad 0.969 \quad 1.001]$$

Figure 6.11 shows the final system state after both control actions, obtained using a load flow program. Note that linearization errors are acceptable and that a larger value of reactive power is provided by the capacitor bank due to a further increase of voltage V_2.

6.5.2 Alert State

The system is in alert state if all variables are within operational limits and operators are aware that one or more contingencies might drive the system to an unacceptable state.

Once the critical contingencies have been identified (those driving the system to an unacceptable state), operators must decide whether to implement preventive actions or to set corrective plans to be used if an outage occurs. Special attention is required in case a contingency might result in a blackout or if abnormal circumstances such as severe weather or terrorist alerts increase the risk of contingencies.

Mathematically, to determine preventive actions for a given contingency, an OPF including security constraints [16,17, 20] can be used, that is, an optimization problem known as SCOPF:

$$
\begin{aligned}
\text{Minimize}_{u_c, u_p, x_c, x_p} \quad & f(u_0, u_p, u_c) \\
\text{Subject to} \quad & h(x_p, u_p) = 0 \\
& h_c(x_c, u_c) = 0 \\
& g(x_p, u_p) \geq 0 \\
& g_c(x_c, u_c) \geq 0
\end{aligned}
$$

where

- u is the vector of decision variables, being u_0 the vector corresponding to the initial state, u_p is the vector of precontingency preventive actions, and u_c of postcontingency corrective actions.
- x is the vector of dependent variables, being x_p the vector of precontingency values, and x_c of postcontingency values.

- $f(u)$ is the objective function to be minimized, usually the cost associated with different control actions, for example, a quadratic or linear penalty function of control changes.
- $h(x, u)$ are the network equations, $h(x_p, u_p) = 0$ corresponding to the precontingency state including possible preventive actions, and $h_c(x_c, u_c) = 0$ to the postcontingency state including also postcontingency corrective actions.
- $g(x, u)$ are the operational limits. Emergency limits, $g_c(x_p, u_p) \geq 0$, less rigid than the operational ones, are usually imposed on postcontingency states.

Clearly, if the system security can be restored fast enough by applying postcontingency corrective actions, preventive actions are not required ($u_p = u_0$), and the increase in the production cost can be avoided. In this case, the optimization problem is reduced to an OPF to compute corrective control actions for the postcontingency state. When corrective actions are insufficient or cannot be implemented fast enough, preventive actions should be adopted, trying to minimize the unavoidable increase in the operation cost.

Traditionally, voltage problems have been ignored due to the complexity of the complete AC model, as well as the erroneous belief that there is always enough time to implement postcontingency corrective actions in case of voltage problems. As a consequence, the SCOPF algorithms used in practice are generally based on linear models of the transmission network. However, the need to resort to preventive actions to avoid catastrophic voltage events in critical cases, as the voltage collapse problem, should not be forgotten.

EXAMPLE 6.12 SECURITY-CONSTRAINED OPTIMAL POWER FLOW

The system of Example 6.3 is vulnerable to the outage of line L2 connecting buses 1 and 3. In this example, postcontingency corrective actions, as well as preventive measures if required, are computed to allow the system to survive this contingency.

Control actions are the redispatch of generation in the precontingency state ($\Delta P_{p_i}^+$ and $\Delta P_{p_i}^-$), the redispatch of generation after the contingency ($\Delta P_{c_i}^+$ and $\Delta P_{c_i}^-$) and postcontingency load shedding (ΔP_{d_i}).

Using a DC model for the network equations, a SCOPF can be formulated as follows:

- *Objective function*: A linear penalty function of generation changes ($\$10/MW$ for generators 3 and 4, and $\$20/MW$ for generator 5). Postcontingency generation reschedulings are not penalized, and a large penalty value is assigned to postcontingency load shedding ($\$1000/MW$). Note that power injections are in pu in the objective function.

$$
[1000 \quad 1000 \quad 2000] \begin{bmatrix} \Delta P_{p_3}^+ \\ \Delta P_{p_4}^+ \\ \Delta P_{p_5}^+ \end{bmatrix} + [1000 \quad 1000 \quad 2000] \begin{bmatrix} \Delta P_{p_3}^- \\ \Delta P_{p_4}^- \\ \Delta P_{p_5}^- \end{bmatrix} + [100{,}000 \quad 100{,}000] \begin{bmatrix} \Delta P_{d_1} \\ \Delta P_{d_2} \end{bmatrix}
$$

$$
\Delta P_p^+ \geq 0 \quad \Delta P_p^- \geq 0 \quad \Delta P_d \geq 0
$$

- *Constraints*:
  - Network equations prior to the contingency: $P + \Delta P_p^+ - \Delta P_p^- = B\,\theta_p$

$$
\begin{bmatrix} -15.0 \\ -5.0 \\ 10.0 + \Delta P_{p_3}^+ - \Delta P_{p_3}^- \\ 7.5 + \Delta P_{p_4}^+ - \Delta P_{p_4}^- \end{bmatrix} = \begin{bmatrix} 250 & -100 & -50 & -100 \\ -100 & 150 & 0 & 0 \\ -50 & 0 & 100 & -50 \\ -100 & 0 & -50 & 200 \end{bmatrix} \begin{bmatrix} \theta_{p_1} \\ \theta_{p_2} \\ \theta_{p_3} \\ \theta_{p_4} \end{bmatrix}
$$

and the additional equation to include the slack generator,

$$\left(10.0 + \Delta P_{p_3}^+ - \Delta P_{p_3}^-\right) + \left(7.5 + \Delta P_{p_4}^+ - \Delta P_{p_4}^-\right) + \left(2.5 + \Delta P_{p_5}^+ - \Delta P_{p_5}^-\right) = 20.0$$

- Network equations after the contingency:

$$P + \Delta P_p^+ - \Delta P_p^- + \Delta P_c^+ - \Delta P_c^- + \Delta P_d = B_c\,\theta_c$$

$$\begin{bmatrix} -15.0 + \Delta P_{d_1} \\ -5.0 + \Delta P_{d_2} \\ 10.0 + \Delta P_{p_3}^+ - \Delta P_{p_3}^- + \Delta P_{c_3}^+ - \Delta P_{c_3}^- \\ 7.5 + \Delta P_{p_4}^+ - \Delta P_{p_4}^- + \Delta P_{c_4}^+ - \Delta P_{c_4}^- \end{bmatrix} = \begin{bmatrix} 200 & -100 & 0 & -100 \\ -100 & 150 & 0 & 0 \\ 0 & 0 & 50 & -50 \\ -100 & 0 & -50 & 200 \end{bmatrix} \begin{bmatrix} \theta_{c_1} \\ \theta_{c_2} \\ \theta_{c_3} \\ \theta_{c_4} \end{bmatrix}$$

and

$$\left(10 + \Delta P_{p_3}^+ - \Delta P_{p_3}^- + \Delta P_{c_3}^+ - \Delta P_{c_3}^-\right) + \left(7.5 + \Delta P_{p_4}^+ - \Delta P_{p_4}^- + \Delta P_{c_4}^+ - \Delta P_{c_4}^-\right)$$
$$+ \left(2.5 + \Delta P_{p_5}^+ - \Delta P_{p_5}^- + \Delta P_{c_5}^+ - \Delta P_{c_5}^-\right) = 20 - \Delta P_{d_1} - \Delta P_{d_2}$$

- Generation limits:

$$\begin{bmatrix} 15.0 \\ 15.0 \\ 10.0 \end{bmatrix} \geq \begin{bmatrix} 10.0 + \Delta P_3^+ - \Delta P_3^- \\ 7.5 + \Delta P_4^+ - \Delta P_4^- \\ 2.5 + \Delta P_5^+ - \Delta P_5^- \end{bmatrix} \geq \begin{bmatrix} 2.5 \\ 2.5 \\ 2.5 \end{bmatrix}$$

$$\begin{bmatrix} 15.0 \\ 15.0 \\ 10.0 \end{bmatrix} \geq \begin{bmatrix} 10.0 + \Delta P_{p_3}^+ - \Delta P_{p_3}^- + \Delta P_{c_3}^+ - \Delta P_{c_3}^- \\ 7.5 + \Delta P_{p_4}^+ - \Delta P_{p_4}^- + \Delta P_{c_4}^+ - \Delta P_{c_4}^- \\ 2.5 + \Delta P_{p_5}^+ - \Delta P_{p_5}^- + \Delta P_{c_5}^+ - \Delta P_{c_5}^- \end{bmatrix} \geq \begin{bmatrix} 2.5 \\ 2.5 \\ 2.5 \end{bmatrix}$$

- Power flow limits, $-P_f^{max} \leq P_f = X^{-1}A^T\theta \leq P_f^{max}$:

$$\begin{bmatrix} -10.0 \\ -10.0 \\ -10.0 \\ -10.0 \\ -10.0 \\ -10.0 \end{bmatrix} \leq \begin{bmatrix} 100 & -100 & 0 & 0 \\ 50 & 0 & -50 & 0 \\ 100 & 0 & 0 & -100 \\ 0 & 50 & 0 & 0 \\ 0 & 0 & 50 & -50 \\ 0 & 0 & 0 & 50 \end{bmatrix} \begin{bmatrix} \theta_{p_1} \\ \theta_{p_2} \\ \theta_{p_3} \\ \theta_{p_4} \end{bmatrix} \leq \begin{bmatrix} 10.0 \\ 10.0 \\ 10.0 \\ 10.0 \\ 10.0 \\ 10.0 \end{bmatrix}$$

$$\begin{bmatrix} -10.0 \\ -10.0 \\ -10.0 \\ -10.0 \\ -10.0 \end{bmatrix} \leq \begin{bmatrix} 100 & -100 & 0 & 0 \\ 50 & 0 & -50 & 0 \\ 0 & 50 & 0 & 0 \\ 0 & 0 & 50 & -50 \\ 0 & 0 & 0 & 50 \end{bmatrix} \begin{bmatrix} \theta_{c_1} \\ \theta_{c_2} \\ \theta_{c_3} \\ \theta_{c_4} \end{bmatrix} \leq \begin{bmatrix} 10.0 \\ 10.0 \\ 10.0 \\ 10.0 \\ 10.0 \end{bmatrix}$$

The solution to the above SCOPF provides the following "emergency plan":

$$\Delta P_{c_3}^- = 550 \text{ MW} \quad \Delta P_{c_4}^- = 500 \text{ MW} \quad \Delta P_{c_5}^+ = 750 \text{ MW} \quad \Delta P_{d_1} = 300 \text{ MW}$$

Note that no preventive actions are required.

Active power flows after the emergency actions are shown in the following table:

	DC Load Flow	AC Load Flow	
Line	P_f	P_{ij}	P_{ji}
L1 1–2	−200	−202.1	203.0
L2 1–3	–	–	–
L3 1–4	−1000	−997.9	1027.6
L4 2–5	−700	−703.0	728.0
L5 3–4	450	450.0	−442.7
L6 4–5	−300	−334.9	339.1

The resulting postcontingency emergency actions could be considered infeasible since they require large generation reschedulings that are difficult or even impossible to be carried out fast enough. Then, the problem should be resolved imposing additional constraints on the postcontingency actions. Suppose that it is impossible to implement generation changes larger than 200 MW in an acceptable time. The SCOPF solution, including the additional constraints ($\Delta P_{c_i} \leq 2.0$ pu), leads to the following measures:

- *Preventive actions*: $\Delta P_{p_5}^+ = 650$ MW, $\Delta P_{p_4}^- = 350$ MW, and $\Delta P_{p_3}^- = 300$ MW.
- *Postcontingency corrective actions*: $\Delta P_{c_5}^+ = 100$ MW, $\Delta P_{c_4}^- = 200$ MW, $\Delta P_{c_3}^- = 200$ MW, and $\Delta P_{d_1} = 300$ MW

An alternative to the use of a SCOPF, which is quite useful for determining preventive actions in the precontingency network state, is to use "outage compensated" distribution factors. This distribution factor is the power flow increment in branch mn due to an increment in the active power injected at node g, if the branch ij is out of service, and can be obtained as

$$\Delta P_{mn}^g|_{ij=\text{out}} = \rho_{mn}^g \, \Delta P_g + \rho_{mn}^{ij} \rho_{ij}^g \, \Delta P_g \tag{6.18}$$

$$\rho_{mn}^g|_{ij=\text{out}} = \frac{\Delta P_{mn}}{\Delta P_g}\bigg|_{ij=\text{out}} = \rho_{mn}^g + \rho_{mn}^{ij} \rho_{ij}^g \tag{6.19}$$

The above distribution factors allows formulating the optimization problem using the state before the contingency, including preventive actions, and imposing constraints on the postcontingency power flows.

EXAMPLE 6.13 PREVENTIVE ACTIONS USING DISTRIBUTION FACTORS

The 5-bus system shown in Figure 6.5 is vulnerable to the outage of line L3 that connect nodes 1 and 4. The objective of this example is to obtain preventive actions using distribution factors.

The possible preventive actions are generation reschedulings (ΔP_p^+ and ΔP_p^-). Distribution factors are used to impose constraints in the optimization problem:

- *Precontingency power flows*: $\Delta P_f = S_f \, \Delta P_g$

$$\begin{bmatrix} \Delta P_{1,2} \\ \Delta P_{1,3} \\ \Delta P_{1,4} \\ \Delta P_{2,5} \\ \Delta P_{3,4} \\ \Delta P_{4,5} \end{bmatrix} = \begin{bmatrix} 0.4138 & 0.3448 \\ -0.4828 & -0.0690 \\ 0.0690 & -0.2759 \\ 0.4138 & 0.3448 \\ 0.5172 & -0.0690 \\ 0.5862 & 0.6552 \end{bmatrix} \begin{bmatrix} \Delta P_3^+ - \Delta P_3^- \\ \Delta P_4^+ - \Delta P_4^- \end{bmatrix}$$

- *Postcontingency power flows*: $\Delta P'_f = S'_f \, \Delta P_g$

$$
\begin{bmatrix} \Delta P'_{1,2} \\ \Delta P'_{1,3} \\ \Delta P'_{1,4} \\ \Delta P'_{2,5} \\ \Delta P'_{3,4} \\ \Delta P'_{4,5} \end{bmatrix}
=
\begin{bmatrix}
0.4444 & 0.2222 \\
-0.4444 & -0.2222 \\
0.0 & 0.0 \\
0.4444 & 0.2222 \\
0.5556 & -0.2222 \\
0.5556 & 0.7778
\end{bmatrix}
\begin{bmatrix} \Delta P_3^+ - \Delta P_3^- \\ \Delta P_4^+ - \Delta P_4^- \end{bmatrix}
$$

Thus, the SCOPF problem is formulated as follows:

- *Objective function*: Linear penalty function of generation changes ($10/MW for generators 3 and 4, and $20/MW for generator 5).

$$
[1000 \quad 1000 \quad 2000]
\begin{bmatrix} \Delta P_{p_3}^+ \\ \Delta P_{p_4}^+ \\ \Delta P_{p_5}^+ \end{bmatrix}
+ [1000 \quad 1000 \quad 2000]
\begin{bmatrix} \Delta P_{p_3}^- \\ \Delta P_{p_4}^- \\ \Delta P_{p_5}^- \end{bmatrix}
$$

$$
\Delta P_p^+ \geq 0 \quad \Delta P_p^- \geq 0
$$

- *Constraints*:
  - On the power flows of the precontingency state: $-P_f^{max} \leq P_f \leq P_f^{max}$, where

$$
P_f = P_{f_0} + \Delta P_f = P_{f_0} + S_f \, \Delta P_g
$$

The initial power flows P_{f_0} obtained using a DC load flow are

$$
P_{f_0} = [120.7 \quad -724.1 \quad -896.6 \quad -379.3 \quad 275.9 \quad 129.3]^T \quad \text{MW}
$$

  - On the power flows of the postcontingency state: $-P_f^{max} \leq P'_f \leq P_f^{max}$, where

$$
P'_f = P'_{f_0} + \Delta P'_f = P_{f_0} + \rho^{L3} P_{f_0}^{L3} + S'_f \, \Delta P_g
$$

P'_{f_0} are the power flows after the contingency without generation changes, and ρ^{L3} are the distribution factors for the outage of line L3,

$$
\rho^{L3} = [0.444 \quad 0.556 \quad -0.444 \quad 0.556 \quad -0.444]^T
$$

- Network power balance neglecting losses, required to include the slack generator,

$$
(10 + \Delta P_{p_3}^+ - \Delta P_{p_3}^-) + (7.5 + \Delta P_{p_4}^+ - \Delta P_{p_4}^-) + (2.5 + \Delta P_{p_5}^+ - \Delta P_{p_5}^-) = 20
$$

- Generation limits:

$$
\begin{bmatrix} 15.0 \\ 15.0 \\ 10.0 \end{bmatrix}
\geq
\begin{bmatrix} 10.0 + \Delta P_3^+ - \Delta P_3^- \\ 7.5 + \Delta P_4^+ - \Delta P_4^- \\ 2.5 + \Delta P_5^+ - \Delta P_5^- \end{bmatrix}
\geq
\begin{bmatrix} 2.5 \\ 2.5 \\ 2.5 \end{bmatrix}
$$

The above SCOPF problem yields the following preventive reschedulings: $\Delta P_3^- = 500$ MW and $\Delta P_4^+ = 500$ MW, with a total operation overcost of $15,000 per hour.

The active power flows in the base case, after implementing the preventive measures and after the contingency, obtained using a DC load flow (P_f) and an AC load flow (P_{ij} and P_{ji}) in all cases, are provided in Figure 6.12.

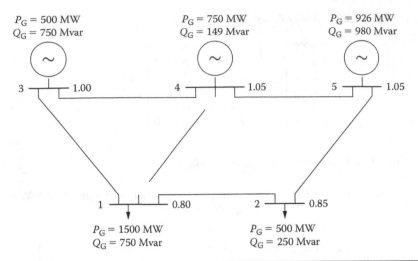

	Base Case			Preventive Actions			Postcontingency State		
Line	P_f	P_{ij}	P_{ji}	P_f	P_{ij}	P_{ji}	P_f	P_{ij}	P_{ji}
L1 1–2	120	96	−96	−86	−92	92	−500	−530	541
L2 1–3	−724	−699	721	−482	−476	488	−1000	−969	1043
L3 1–4	−896	−897	920	−931	−931	955	0	0	0
L4 2–5	−379	−403	414	−586	−592	609	−1000	−1041	1118
L5 3–4	275	278	−275	17	11	−11	−501	−543	556
L6 4–5	129	105	−105	−164	−193	195	249	194	−192

FIGURE 6.12   5-Bus network after the outage of line L3.

Note the following observations on the SCOPF solution:

- The solution based on a DC model is an adequate approximation for practical purposes. However, the effect of the simplifications should be analyzed. Note, for instance, that the postcontingency state presents overloads in lines L2 and L4, since the corridor they constitute, with a total capacity of 2000 MW, transfers 2000 MW plus the losses of both lines and L1.
- The DC model only considers active powers and no losses, and, consequently, the feasibility of the different states should be tested in terms of reactive powers and voltages. Note that the postcontingency state (Figure 6.12) presents very low voltages, questioning the feasibility of this state.

6.5.3 Secure State

When the system is in a normal, secure state, all operational limits are fulfilled and the considered security criteria is satisfied. Consequently, the objective of the system operation focuses on further reducing the operation cost.

The problem of determining the optimal control actions in secure state is a particular case of the network constrained ED using a detailed network model (Chapter 5). In this type of OPF problems (see Example 6.7), demand and grid losses are dispatched among generators in an optimal way as a function of the generator costs. Besides, the outcome of this OPF problem provides the optimal set-points of voltage control variables to minimize transmission losses.

There are two main trends regarding short-term operation of the electrical systems. A common approach is to perform first the generation scheduling, either by a central operator with full knowledge of generator economic and technical data, or by a market operator. Then, an OPF is used to correct the generation scheduling to prevent possible overloads and voltage problems (Section 6.5.1). A SCOPF may be used to fulfill security criteria, usually the $N-1$ criteria (Section 6.5.2). Finally, once the definitive generation scheduling has been obtained, the optimal voltage profile is determined to satisfy voltage limits and further reduce transmission losses. On the other hand, it is possible to use an OPF to schedule the generation taking into account the operational constraints of the network, possibly including contingency security requirements (SCOPF). This alternative is addressed in the following section.

The example below illustrates the use of an OPF to obtain an optimal voltage profile.

EXAMPLE 6.14 OPF TO DETERMINE THE OPTIMAL VOLTAGE PROFILE

In this example, the optimal voltage profile to minimize the transmission losses of the system presented in Figure 6.6 is computed. It is assumed that the generation scheduling has been previously decided ($P_{G_4} = P_{G_5} = 1000$ MW). Besides, generator G5 covers any power imbalance, and the initial production cost is $21,390/h.

The decision variables are the generator voltage set points, the 200 Mvar capacitor bank (20 steps of 10 Mvar), and the transformer turn ratio (21 taps with increments of 0.01 pu).

The objective function is the sum of the individual losses in all lines and transformers:

$$P_{losses} = \sum_{i,j} -G_{ij}(V_i^2 + V_j^2 - 2V_iV_j \cos\theta_{ij})$$

which is equivalent in this case to the power generated by the slack generator, considering that $P_{G_5} = 1000 + P_{losses}$.

The solution to the OPF problem, subject to the same constraints than those in Example 6.7, yields the state shown in Figure 6.13.

The following observations on the OPF solution are pertinent:

- Discrete control variables have been rounded off *ex post*, updating the final state using an AC load flow model.
- Losses are decreased by 9.3 MW, from 69.2 to 59.5 MW, representing a saving of $95/h as compared with the base case. Note that the optimum in terms of generation costs (Example 6.7) implies a cost reduction of $101 per hour.

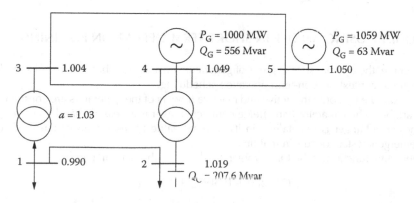

FIGURE 6.13 Optimal voltage profile of the 5-bus, two-generator system.

- Lagrange multipliers corresponding to active power network equations provide the incremental loss coefficients,

$$\lambda^T = [1.103 \quad 1.062 \quad 1.071 \quad 1.041 \quad 1.000]$$

Note that the slack bus incremental loss coefficient, the one that covers the losses, equals one.

In parallel to the problem of identifying *ex ante* the optimal setpoints of all voltage control variables, some control actions can be implemented to further decrease losses in real-time operation. A useful alternative to a full OPF to reduce losses is to use sensitivities of losses to voltage control variables, taking into account the actual state of the system. Then, based on the sensitivities and considering the margin of each control, control actions can be easily determined to reduce losses [23].

6.6 SHORT-TERM SECURE OPERATION

We consider below the viewpoint of the ISO minutes prior to power delivery, when operation security needs to be guaranteed.

Considering problem (6.12), an OPF problem for the "current" operating condition identified by the superscript "0" can be formulated as

$$
\begin{aligned}
&\text{Minimize}_{u^0, x^0} \quad && f(x^0, u^0) \\
&\text{Subject to} \quad && h^0(x^0, u^0) = 0 \\
& && g^0(x^0, u^0) \geq 0,
\end{aligned}
\tag{6.20}
$$

where x^0 is the vector of state variables (voltage magnitudes in all nodes and voltage angles in all nodes but the reference one) and other optimization variables, u^0 is the vector of control variables (active and reactive powers in all generating units and tap-changing positions of regulating transformers), $h^0(x^0, u^0) = 0$ are the active and reactive power flow equations, $g^0(x^0, u^0) \geq 0$ are operation limits of generating units, transmission lines and state variables, and $f(x^0, u^0)$ is the objective function representing operating cost, active power losses, or other objective.

The output of this OPF problem guarantees a secure operation solely for the "current" operating condition, but not necessarily if a contingency occurs.

To guarantee a secure operation under contingencies, OPF (Equation 6.20) needs to be modified as explained in the next subsection.

Example 6.15 illustrates an OPF that solely considers the "current" operating condition.

EXAMPLE 6.15 USING AN OPF IN SHORT-TERM OPERATION PLANNING

In this example, the optimal state in terms of generation cost of the 5-bus system of Example 6.7 is obtained in the context of short-term operation scheduling.

Unlike Example 6.7, only the active and reactive powers of the generators are considered to be control variables, the capacitor bank being connected, as it is a peak-demand scenario, and the transformer tap changer is maintained in its nominal value to reserve all its range of action in case an emergency state occurs in real time.

The objective function of the OPF problem, $f(x^0, u^0)$, is the operating cost,

$$C = 200 + 1050P_{G_4} + 1000P_{G_5}$$

and the problem is subject to the following constraints:

- Active and reactive power flow equations, $h^0(x^0, u^0) = 0$,

$$P_i = P_{G_i} - P_{D_i} = \sum_{j=1}^{5} V_i V_j [G_{ij} \cos(\theta_i - \theta_j) + B_{ij} \sin(\theta_i - \theta_j)] \quad i = 1, \ldots, 5$$

$$Q_i = Q_{G_i} - Q_{D_i} = \sum_{j=1}^{5} V_i V_j [G_{ij} \sin(\theta_i - \theta_j) - B_{ij} \cos(\theta_i - \theta_j)] \quad i = 1, \ldots, 5$$

where the decision variables, u^0, are P_{G_4}, P_{G_5}, Q_{G_4}, Q_{G_5}, and the state variables, x^0, are voltage magnitudes in the five nodes and voltage angles in all nodes but the reference one (bus 5).
- Operation limits, $g^0(x^0, u^0) \geq 0$,

$$5 \leq P_{G_4} \leq 15, \quad -10 \leq Q_{G_4} \leq 10$$

$$5 \leq P_{G_5} \leq 15, \quad -10 \leq Q_{G_5} \leq 10$$

$$0.95 \leq V_i \leq 1.05 \quad i = 1, \ldots, 5$$

$$S_{ij}^2 = P_{ij}^2 + Q_{ij}^2 \leq \left(S_{ij}^{max}\right)^2 \quad i, j = 1, \ldots, 5$$

where S_{ij}, P_{ij}, and Q_{ij} are the apparent power, active power and reactive power flow in transmission elements (lines and transformers).

The solution to the above OPF problem leads to the following operational scheduling of the generators:

$$P_{G_4} = 878.3 \text{ MW} \quad Q_{G_4} = 552.5 \text{ Mvar} \quad V_4 = 1.039 \text{ pu}$$

$$P_{G_5} = 1187.6 \text{ MW} \quad Q_{G_5} = 89.8 \text{ Mvar} \quad V_5 = 1.05 \text{ pu}$$

Note the following observations on this optimal operational scheduling:

- Production cost is $21,298 per hour.
- Transmission losses are 65.7 MW.
- Bus incremental costs (Lagrange multipliers corresponding to active power network equations) are

$$\lambda^T = [11.08 \quad 10.70 \quad 10.76 \quad 10.50 \quad 10.00] \quad \$/MWh$$

- None of the apparent power flow limits is reached at the optimal state. The following table lists the power flows in the programed operating state:

Line	S_{ij}	S_{ji}	S_f^{max}
		MVA	
L1 1–2	236	243	500
L2 1–3	811	840	2000
L3 2–4	1218	1263	2000
L4 3–4	175	179	1000
L5 3 5	774	010	1000
L6 4–5	382	386	1000

It is suggested that the reader compares this solution with that of Example 6.7, where more control variables are optimized.

6.6.1 SECURITY-CONSTRAINED OPTIMAL POWER FLOW

If, in addition to the "current" operating condition "0", the ISO needs to consider in advance alternative operating conditions due to contingencies, and to impose, in a corrective manner, that the system safely and rapidly transitions from the "current" operating condition to any of the postcontingency ones, the OPF problem (6.20) becomes a SCOPF, and can be formulated as follows:

$$
\begin{aligned}
&\text{Minimize}_{x^0, u^0; x^1, u^1, \ldots, x^c, u^c} \quad f(x^0, u^0; x^1, u^1, \ldots, x^c, u^c) \\
&\text{Subject to} \quad h^0(x^0, u^0) = 0 \\
&\qquad\qquad g^0(x^0, u^0) \geq 0 \\
&\qquad\qquad h^\omega(x^\omega, u^\omega) = 0 \quad \omega = 1, \ldots, c \\
&\qquad\qquad g^\omega(x^\omega, u^\omega) \geq 0 \quad \omega = 1, \ldots, c \\
&\qquad\qquad l^\omega(u^0, u^\omega) \leq 0 \quad \omega = 1, \ldots, c,
\end{aligned}
\tag{6.21}
$$

where, in addition to the symbols already defined (previous subsection), x^ω is the vector of state variables and other optimization variables under contingency ω, u^ω is the vector of control variables under contingency ω, $h^\omega(x^\omega, u^\omega) = 0$ are the active and reactive power flow equations under contingency ω, $g^\omega(x^\omega, u^\omega) \geq 0$ are operation limits on generating units, transmission lines and state variables under contingency ω, $f(u^0, x^0; u^1, x^1, , u^c, x^c)$ is the objective function representing operating cost, active power losses, or other objective, under the "current" operating condition and all contingency conditions, $l^\omega(u^0, u^\omega) \leq 0$ are linking constraints relating control variables of the "current" operating condition and control variables of any of the contingency conditions, and c is the number of considered contingency conditions.

The example below illustrates an OPF that considers the "current" operating condition and a number of contingencies in a corrective manner.

EXAMPLE 6.16 USING AN SCOPF IN SHORT-TERM OPERATION PLANNING

The optimal operation scenario computed in Example 6.15 is insecure with respect to $N-1$ contingencies of lines L5 and L6. In consequence, a SCOPF will be used to compute a $N-1$ secure scheduling of generators.

As in Example 6.15, the active and reactive powers of the generators are considered to be control variables, both in the "current" operating scenario (P_{G_4}, P_{G_5}, Q_{G_4}, Q_{G_5}), and in contingency scenarios L5 and L6 ($P_{G_4}^{L5}$, $P_{G_5}^{L5}$, $Q_{G_4}^{L5}$, $Q_{G_5}^{L5}$, $P_{G_4}^{L6}$, $P_{G_5}^{L6}$, $Q_{G_4}^{L6}$, $Q_{G_5}^{L6}$). Besides, to ensure the feasibility of postcontingency scenarios, emergency load shedding is considered in load buses 1 and 2 ($\Delta P_{D_1}^{L5}$, $\Delta P_{D_2}^{L5}$, $\Delta P_{D_1}^{L6}$, $P_{D_2}^{L6}$) at constant power factor,

$$
\Delta Q_{D_i}^\omega = \frac{Q_{D_i}^0}{P_{D_i}^0} \cdot \Delta P_{D_i}^\omega \quad i = 1, 2 \quad \omega = L5, L6
$$

In consequence, the objective function of the SCOPF problem becomes

$$
C = 200 + 1050 \cdot P_{G_4} + 1000 \cdot P_{G_5} + 10{,}000 \cdot \left(\Delta P_{D_1}^{L5} + \Delta P_{D_2}^{L5} + \Delta P_{D_1}^{L6} + \Delta P_{D_2}^{L6} \right)
$$

Note that an unserved energy cost has been introduced for emergency load shedding, and that no penalization is introduced for the emergency rescheduling of generators.

The problem is subject to the active and reactive power flow equations in the current operating scenario and contingency scenarios, and voltage limits are relaxed to emergency limits in postcontingency scenarios,

$$0.90 \leq V_i^\omega \leq 1.1 \quad i = 1, \ldots, 5 \quad \omega = L5, L6$$

Finally, in order to guarantee the feasibility of postcontingency rescheduling of generators' active power, a limit is imposed on the emergency reschedulings,

$$-1\,\text{MW} \leq P_{G_i} - P_{G_i}^\omega \leq 1\,\text{MW} \quad i = 1, \ldots, 5 \quad \omega = L5, L6$$

The solution to the SCOPF problem leads to the following scheduling:

Scenario	P_{G_4}	P_{G_5}	V_4	V_5
	MW		pu	
Base case	1006	1054	1.042	1.05
Contingency L5	1106	1000	1.075	1.10
Contingency L6	1068	1000	1.10	1.10

The production cost increases to \$29,902/h, an increment of 40.4% with respect to Example 6.15, but transmission losses decrease to 59.9 MW.

The following table lists the power flows in the programed operating state, $N-1$ secure with respect to the outage of lines L5 and L6:

	Base case		Contingency L5		Contingency L6		
	S_{ij}	S_{ji}	S_{ij}	S_{ji}	S_{ij}	S_{ji}	S_f^{max}
Line	MVA		MVA		MVA		MVA
L1 1–2	249	257	482	511	232	226	500
L2 1–3	793	822	549	566	864	884	2000
L3 2–4	1236	1281	1503	1561	1163	1202	2000
L4 3–4	180	184	566	604	261	269	1000
L5 3–5	718	750	–	–	962	1000	1000
L6 4–5	312	314	977	1000	–	–	1000

Note that there is a slight overload on line L1 in the scenario corresponding to the L5 contingency. This overload is perfectly acceptable for a transient period in case of contingency.

6.6.2 Security-Constrained Economic Dispatch

Solving SCOPF (6.21) is often challenging, particularly for a large enough number of contingencies. In some cases, it might be acceptable to use a linearized version of SCOPF (6.21) based on the DC power flow equations, which we refer to as security-constrained economic dispatch, SCED for short. Needless to say, potential voltage issues are not represented in the SCEC formulation since a DC model is used. The linearized version of problem (6.21) has the form:

$$
\begin{aligned}
\text{Minimize}_{x^0, u^0; x^1, u^1, \ldots, x^c, u^c} \quad & f^T[x^0, u^0; x^1, u^1, \ldots, x^c, u^c] \\
\text{Subject to} \quad & H^0[x^0, u^0] = 0 \\
& G^0[x^0, u^0] \geq 0 \\
& H^\omega[x^\omega, u^\omega] = 0 \quad \omega = 1, \ldots, c \\
& G^\omega[x^\omega, u^\omega] \geq 0 \quad \omega = 1, \ldots, c \\
& L^\omega[u^0, u^\omega] \leq 0 \quad \omega = 1, \ldots, c
\end{aligned}
\tag{6.22}
$$

where state variable vectors $x^0, x^1, , x^c$ include voltage angles for all nodes other than the reference one (and other optimization variables) for the "current" operating condition and contingency conditions, and control variable vectors $u^0, u^1, , u^c$ include active powers in all generating units for the "current" operating condition and the contingency conditions. Matrices H^0, H^1, , H^c pertain to DC power flow equations, G^0, G^1, , G^c to constraints on power productions from generating units and to power flows through lines, and L^1, , L^c to linking conditions regarding power productions between the "current" operating condition and any contingency condition.

The example below illustrates an ED that considers the "current" operating condition and a number of contingencies.

EXAMPLE 6.17 USING AN SCED IN SHORT-TERM OPERATION PLANNING

The SCOPF problem of Example 6.16 is addressed below using a SCED. The objective function is the same, but the load flow equations in the different scenarios become

- Base case:

$$
\begin{pmatrix} P_{L1} \\ P_{L2} \\ P_{L3} \\ P_{L4} \\ P_{L5} \\ P_{L6} \end{pmatrix} = \begin{pmatrix} 0.4043 & -0.3830 & 0.0851 & -0.1702 \\ 0.5957 & 0.3830 & -0.0851 & 0.1702 \\ 0.4043 & 0.6170 & 0.0851 & -0.1702 \\ -0.0426 & -0.1702 & 0.1489 & -0.2979 \\ 0.6383 & 0.5532 & 0.7660 & 0.4681 \\ 0.3617 & 0.4468 & 0.2340 & 0.5319 \end{pmatrix} \cdot \begin{pmatrix} -P_{D_1} \\ -P_{D_2} \\ 0 \\ P_{G_4} \end{pmatrix}
$$

$$ P_{G_4} + P_{G_5} = P_{D_1} + P_{D_2} $$

- Contingency scenario L5:

$$
\begin{pmatrix} P_{L1}^{L5} \\ P_{L2}^{L5} \\ P_{L3}^{L5} \\ P_{L4}^{L5} \\ P_{L5}^{L5} \\ P_{L6}^{L5} \end{pmatrix} = \begin{pmatrix} 0.6364 & -0.1818 & 0.3636 & 0.0 \\ 0.3636 & 0.1818 & -0.3636 & 0.0 \\ 0.6364 & 0.8182 & 0.3636 & 0.0 \\ 0.3636 & 0.1818 & 0.6364 & 0.0 \\ 0.0 & 0.0 & 0.0 & 0.0 \\ 1.0 & 1.0 & 1.0 & 1.0 \end{pmatrix} \cdot \begin{pmatrix} -P_{D_1} + \Delta P_{D_1}^{L5} \\ -P_{D_2} + \Delta P_{D_2}^{L5} \\ 0 \\ P_{G_4}^{L5} \end{pmatrix}
$$

$$ P_{G_4}^{L5} + P_{G_5}^{L5} = P_{D_1} - \Delta P_{D_1}^{L5} + P_{D_2} - \Delta P_{D_2}^{L5} $$

- Contingency scenario L6:

$$
\begin{pmatrix} P_{L1}^{L6} \\ P_{L2}^{L6} \\ P_{L3}^{L6} \\ P_{L4}^{L6} \\ P_{L5}^{L6} \\ P_{L6}^{L6} \end{pmatrix} = \begin{pmatrix} 0.2727 & -0.5455 & 0.0 & -0.3636 \\ 0.7273 & 0.5455 & 0.0 & 0.3636 \\ 0.2727 & 0.4545 & 0.0 & -0.3636 \\ -0.2727 & -0.4545 & 0.0 & -0.6364 \\ 1.0 & 1.0 & 1.0 & 1.0 \\ 0.0 & 0.0 & 0.0 & 0.0 \end{pmatrix} \cdot \begin{pmatrix} -P_{D_1} + \Delta P_{D_1}^{L6} \\ -P_{D_2} + \Delta P_{D_2}^{L6} \\ 0 \\ P_{G_4}^{L6} \end{pmatrix}
$$

$$ P_{G_4}^{L6} + P_{G_5}^{L6} = P_{D_1} - \Delta P_{D_1}^{L6} + P_{D_2} - \Delta P_{D_2}^{L6} $$

Note that the compact formulation of the DCLF based on distribution factors is used in the equations above, and that a balance equation between generation and demand is included since the slack generator is not represented in these DCLF equations.

The operational limits of the generators are only related with the active power,

$$ 5 \leq P_{G_4} \leq 15 \quad 5 \leq P_{G_5} \leq 15 $$

$$5 \leq P_{G_4}^{\omega} \leq 15 \quad 5 \leq P_{G_5}^{\omega} \leq 15 \quad \omega = L5, L6$$

and the operational limits on the power flows on lines and transformers become

$$-P_{ij}^{\max} \leq P_{ij} \leq P_{ij}^{\max} \quad i, j = 1, \ldots, 5$$

$$-P_{ij}^{\max} \leq P_{ij}^{\omega} \leq P_{ij}^{\max} \quad i, j = 1, \ldots, 5 \quad \omega = L5, L6$$

Finally, the limits imposed on the emergency reschedulings,

$$-1\,\text{MW} \leq P_{G_i} - P_{G_i}^{\omega} \leq 1\,\text{MW} \quad \omega = L5, L6$$

The solution to the SCED problem leads to the following scheduling:

Scenario	P_{G_4}	P_{G_5}
	MW	
Base case	900	1100
Contingency L5	1000	1000
Contingency L6	1000	1000

The production cost is now \$29,902/h, and branch flows provided by the SCED (active power flows) and obtained by an ACLF using the scheduled generations (with $V_4 = V_5 = 1.05$ in all scenarios), are the following:

	Base case			Contingency L5			Contingency L6			
	P_{ij}	S_{ij}	S_{ji}	P_{ij}	S_{ij}	S_{ji}	P_{ij}	S_{ij}	S_{ji}	$S_f^{\max}$
Line	MW	MVA		MW	MVA		MW	MVA		MVA
L1 1–2	−175	250	258	−455	482	514	−91	230	236	500
L2 1–3	−826	802	828	−545	549	569	−909	884	908	2000
L3 2–4	−1175	1227	1273	−1455	1508	1575	−1091	1150	1197	2000
L4 3–4	−55	203	209	−545	569	611	91	304	316	1000
L5 3–5	−770	758	789	–	–	–	−1000	1036	1085	1000
L6 4–5	−330	387	387	−1000	1118	1133	–	–	–	1000

Comparing the results of the SCED with those of the SCOPF solved in Example 6.16 highlight drawbacks such as ignoring the effect of the losses in the dispatch of the generators in the base-case scenario.

An alternative is to use the ACOPF formulated for the base-case operating scenario in Example 6.15, along with the linear constraints based on the distribution factors for the postcontingency scenarios.

6.6.3 OPF with Stability Constraints

In this subsection, we consider the "current" operating condition and d contingencies that may involve stability issues. These contingencies are identified with the superscript s, which stands for stability. The OPF below ensures security over these stability-related contingencies.

$$
\begin{aligned}
&\text{Minimize}_{x^0, u^0; x^1, u^1, \ldots, x^d, u^d} \quad f(u^0, x^0; u^1, x^1, \ldots, u^d, x^d) \\
&\text{Subject to} \quad h^0(x^0, u^0) = 0 \\
&\qquad\qquad\quad g^0(x^0, u^0) \geq 0 \\
&\qquad\qquad\quad h^s(x^s, u^s) = 0 \quad s = 1, \ldots, d \\
&\qquad\qquad\quad g^s(x^s, u^s) \geq 0 \quad s = 1, \ldots, d \\
&\qquad\qquad\quad l^s(u^0, u^s) \leq 0 \quad s = 1, \ldots, d
\end{aligned}
\tag{6.23}
$$

where x^s is the vector of state variables and other optimization variables under stability-related contingency s, u^s is the vector of control variables under stability-related contingency s, $h^s(x^s, u^s) = 0$ are the active and reactive power flow equations and other stability-related equations under stability-related contingency s, $g^s(x^s, u^s) \geq 0$ are operation limits on generating units, transmission lines and state variables under stability-related contingency s, $f(u^0, x^0; u^1, x^1, , u^c, x^c)$ is the objective function representing operating cost, active power losses, or other objective, under the "current" operating condition and all stability-related contingency conditions, $l^s(u^0, u^s) \leq 0$ are linking constraints relating control variables of the "current" operating condition and control variables of any of the stability-related contingency conditions, and d is the number of considered contingency conditions.

Additional details pertaining to the stability-constrained OPF are provided, among others, in Reference 25.

REFERENCES

1. Balu N., Bertram T., Bose A., Brandwajn V. et al., On-Line power system security analysis, *Proceedings of the IEEE*, 80(2), 262–282, February 1992.
2. Wood A. J. and Wollenberg B. F., *Power Generation, Operation and Control*, John Wiley & Sons, EEUU, New York, NY, February 1996.
3. DyLiacco T. E., Real-time computer control of power systems, *Proceedings of the IEEE*, 62, 884–891, July 1974.
4. Adibi M. M. ed., *Power System Restoration: Methodologies & Implementation Strategies*, IEEE Press Series in Power Engineering, IEEE, Piscataway, NJ, 2000.
5. Allan R.N. and Billinton R., *Reliability Evaluation of Power Systems*, Springer US, New York, 1996.
6. Ejebe G. C. and Wollemberg B. F., Automatic contingency selection, *IEEE Trans. on Power Apparatus & Systems*, 98, 92–104, January/February 1979.
7. Brandwajn V. and Lauby M. G., Complete bounding for AC contingency analysis, *IEEE Trans. on Power Systems*, 4, 724–729, May 1990.
8. Sakis Meliopoulos A. P., Cheng C. S., and Xia F., Performance evaluation of static security analysis methods, *IEEE Trans. on Power Systems*, 9, 1441–1449, August 1994.
9. Bacher R. and Tinney W. F., Faster local power flow solutions: The zero mismatch approach, *IEEE Trans. on Power Systems*, 4, 1345–1354, November 1989.
10. Dommel H. W. and Tinney W. F., Optimal power flow solutions, *IEEE Trans. on Power Apparatus & Systems*, PAS-87(10), 1866–1876, October 1968.
11. Peschon J., Piercy D. S., Tinney W. F., Tveit O. J., and Cuénod M., Optimum control of reactive power flow, *IEEE Trans. on Power Apparatus & Systems*, PAS-87(1), 40–48, January 1968.
12. Huneault M. and Galiana F. D., A survey of the optimal power flow literature, *IEEE Trans. on Power Systems*, PWRS-6(2), 762–770, May 1991.
13. Sun D. I., Ashley B., Brewer B., Hughes A., and Tinney W. F., Optimal power flow by Newton approach, *IEEE Trans. on Power Apparatus & Systems*, PAS-103(10), 2864–2880, 1984.
14. Quintana V. H. and Santos-Nieto M., Reactive-power dispatch by successive quadratic programming, *IEEE Trans. on Power Systems*, PWRS-4(3), 425–435, September 1989.
15. Alsaç O., Bright J., Prais M., and Stott B., Further developments in LP-based optimal power flow, *IEEE Trans. on Power Systems*, PWRS-5(3), 697–711, August 1990.
16. Vargas L. S., Quintana V. H., and Vanelli A., A tutorial description of an interior point method and its application to security-constrained economic dispatch, *IEEE Trans. on Power Systems*, 8(3), 1315–1323, August 1993.
17. Frauendorfer K., Glavitsh H., and Bacher R., *Optimization in Planning and Operation of Electric Power Systems*, Physica-Verlag, Germany, October 1993.
18. Tinney W. F., Bright J. M., Demaree K. D., and Hughes B. A., Some deficiencies in optimal power flow, *IEEE Trans. on Power Systems*, 3(2), 676–683, May 1988.
19. Capitanescu F., Critical review of recent advances and further developments needed in AC optimal power flow, *Electric Power Systems Research*, 136, 57–68, 2016.
20. Panciatici P., Hassaine Y., Fliscounakis S., Platbrood L., Ortega-Vazquez M., Martinez-Ramos J.L., Capitanescu F., and Wehenkel L., Security management under uncertainty: From day-ahead planning to intraday operation, Bulk Power System Dynamics and Control, August 1–6, 2010, Rio de Janeiro, Brasil. DOI 10.1109/IREP.2010.5563278

21. GAMS Development Corporation, *GAMS-The Solver Manuals*, Washington, 2007 (http://www.gams.com).
22. Capitanescu F., Martinez-Ramos J.L., Panciatici P., and Kirschen D. et al., State-of-the-art, challenges, and future trends in security constrained optimal power flow, *Electric Power Systems Research*, 81(8), 1731–1741, 2011.
23. Martínez J. L., Gómez A., Cortés J., Méndez E., and Cuéllar Y., A hybrid tool to assist the operator in reactive power/voltage control and optimization, *IEEE Trans. on Power Systems*, 10, 760–768, May 1995.
24. Gómez A., Martínez J. L., Ruiz J. L., and Cuéllar Y., Sensitivity-based reactive power control for voltage profile improvement, *IEEE Trans. on Power Systems*, 8(3), 937–945, August 1993.
25. Zarate-Minano R., Van Cutsem T., Milano F., and Conejo A. J., Securing transient stability using time-domain simulations within an optimal power flow, *IEEE Trans. on Power Systems*, 25(1), 243–253, February 2010.

7 Three-Phase Linear and Nonlinear Models of Power System Components

Enrique Acha and Julio Usaola

CONTENTS

7.1   INTRODUCTION

An electrical system is considered to be in steady state if the magnitude variations in the system's operational parameters are slow in comparison with its electrical time constants. If such a condition is not fulfilled then the system will be in a transient state. Under ideal steady-state conditions, the electricity supply would exhibit a perfectly sinusoidal waveform with constant amplitude and fixed frequency. In a three-phase supply system, the three sinusoidal waveforms should have equal amplitudes and their phase angles be phase-shifted by 120 electrical degrees from each other; a condition that is amenable to a perfect quality of supply.

To a large extent, such operating conditions prevail in transmission networks, and use of the power system models described in Chapter 2 is justified when conducting application studies of the network. There are instances, however, where more detailed studies, and use of complex models, are required because ideal operating conditions cannot be assumed. For instance, when an electromagnetic disturbance affects power quality in the network, the use of complex models is justified. Poor power quality introduces different adverse effects on the grid components and end users and should be ameliorated as much as practicable.

A common power quality disturbance is the steady-state waveform distortion. For the study of the network, the distorted waveform may be decomposed, using, for instance, Fourier analysis, into fundamental and harmonic frequency components. This disturbance is caused by the presence of nonlinear plant components in the network. Distorted waveforms may result in malfunctioning of equipment and controllers, additional power system losses, premature aging of insulation equipment, and telephone interference.

Voltage fluctuations are another form of adverse power quality effect and refer to changes in the amplitude of the waveform, $< 10\%$, and with frequency changes in the range of 0.1–30 Hz. The source of the disturbance is fluctuating loads like electric arc furnaces and motors with frequent stop-starts. The dominant effect is the flicker produced by incandescent lamps fed by such fluctuating voltages.

Voltage dips are sudden reductions of the voltage at a point in the electrical system, followed by voltage recovery after a short period of time, from half a cycle to several cycles (a few seconds). A short supply interruption represents the loss of voltage supply for a period of up to 1 min. These disturbances may be caused by switching operations involving heavy currents or by the operation of protective devices in response to short-circuit faults. Their effect is the maloperation or damage of electronic controllers and disconnection of equipment, such as motors.

Voltage unbalance is a condition in which one or more phase voltages differ in amplitude, or are displaced from their intended 120° phase relationship or both. This is a phenomenon caused mainly by single-phase loads being fed from the three-phase supply. Depending on the severity of the unbalance, it may cause malfunctioning and heating in induction machines and in certain kinds of power electronic converters.

The study of electrical systems involving some of these issues requires representation and knowledge of power plant components' behavior at frequencies other than the fundamental, that is, 50 or 60 Hz. At high frequencies, network behavior is markedly affected by structural imbalances in power plant components and by supply imbalances. The aim of this chapter is to detail, in a coherent and systematic way, suitable models of power plant components with which to carry out these kinds of studies.

The power network components presented in this chapter may be classified into linear and nonlinear plant components. The former are those whose behavior can be modeled by means of linear algebraic or differential equations. The latter can only be modeled by means of nonlinear equations. Linear plant components respond to a sinusoidal input (voltage or current) with a sinusoidal output (current or voltage), while the response of a nonlinear plant component is a distorted waveform. The solution tools for the analysis of circuits involving nonlinear components are given in Chapter 11, together with the theoretical background.

The models of power plant components presented in this chapter are necessarily of various kinds. Some of them are linear but their impedances vary with frequency in a nonlinear fashion, such as transmission lines, while some others are decidedly nonlinear, such as power converters. Other plant components may behave only as a mild nonlinearity, such as power transformers. The chapter begins by presenting a popular tool used in the study of unbalanced systems—the long—enduring symmetrical components. Here only a short review of these concepts will be given. A thorough approach may be found in Reference 1.

7.2   UNBALANCED NETWORKS: SYMMETRICAL COMPONENTS

A symmetrical or balanced set of voltages in a three-phase system has the following expression for each of the three phases a, b, and c, where phase a is taken to be the reference and has a phase angle of 0 rad:

$$\mathcal{U}_{a1} = V\angle 0$$
$$\mathcal{U}_{b1} = V\angle -2\pi/3 = a^2 \cdot \mathcal{U}_{a1} \qquad (7.1)$$
$$\mathcal{U}_{c1} = V\angle 2\pi/3 = a \cdot \mathcal{U}_{a1}$$

where $a = 1\angle 2\pi/3$ is a complex number used to simplify the notation. An expression similar to (7.1) could be written for the three-phase currents of a balanced system. Such a sequence of the three phases is called the *positive sequence*.

However, the voltages of a balanced system could also be expressed as

$$\mathcal{U}_{a2} = V\angle 0$$
$$\mathcal{U}_{b2} = V\angle 2\pi/3 = a \cdot \mathcal{U}_{a2} \qquad (7.2)$$
$$\mathcal{U}_{c2} = V\angle -2\pi/3 = a^2 \cdot \mathcal{U}_{a2}$$

This sequence of the three phases is called the *negative sequence*. It is also possible to have a balanced systems whose voltages are

$$\mathcal{U}_{a0} = V\angle 0$$
$$\mathcal{U}_{b0} = V\angle 0 = \mathcal{U}_{a0} \qquad (7.3)$$
$$\mathcal{U}_{c0} = V\angle 0 = \mathcal{U}_{a0}$$

This is called the *homopolar sequence*.

Most grids are balanced, or can be assumed to be balanced, and the phases are chosen to be of positive sequence. Sometimes it is necessary to deal with unbalanced networks. The analysis of an unbalanced grid is usually made by decomposing the unbalanced system into three balanced systems exhibiting positive, negative, and homopolar sequence. This is possible because the three-phase voltages of an unbalanced voltage system may be written as the sum of three balanced systems of positive, negative, and homopolar sequences voltages. Incidentally, this is called the Fortescue decomposition [1].

This decomposition may be expressed mathematically, using Equations 7.1 through 7.3, as follows:

$$\begin{pmatrix} \mathcal{U}_a \\ \mathcal{U}_b \\ \mathcal{U}_c \end{pmatrix} = \begin{pmatrix} \mathcal{U}_{a0} \\ \mathcal{U}_{b0} \\ \mathcal{U}_{c0} \end{pmatrix} + \begin{pmatrix} \mathcal{U}_{a1} \\ \mathcal{U}_{b1} \\ \mathcal{U}_{c1} \end{pmatrix} + \begin{pmatrix} \mathcal{U}_{a2} \\ \mathcal{U}_{b2} \\ \mathcal{U}_{c2} \end{pmatrix} = \begin{pmatrix} 1 & 1 & 1 \\ 1 & a^2 & a \\ 1 & a & a^2 \end{pmatrix} \begin{pmatrix} \mathcal{U}_0 \\ \mathcal{U}_1 \\ \mathcal{U}_2 \end{pmatrix} \qquad (7.4)$$

where $\mathcal{U}_0$, $\mathcal{U}_1$, and $\mathcal{U}_2$ stand for $\mathcal{U}_{a0}$, $\mathcal{U}_{a1}$, and $\mathcal{U}_{a2}$. In a more compact form, Equation 7.4 may be written as

$$\mathcal{U}_{abc} = A \cdot \mathcal{U}_{012} \tag{7.5}$$

where A is a regular matrix whose inverse is

$$A^{-1} = \frac{1}{3} \begin{pmatrix} 1 & 1 & 1 \\ 1 & a & a^2 \\ 1 & a^2 & a \end{pmatrix} \tag{7.6}$$

The fact that matrix A is a regular (nonsingular) matrix means that for a given unbalanced three-phase voltage system, the Fortescue decomposition is unique. The sequence components of a unbalanced voltage system have a unique value and may be found as

$$\mathcal{U}_{012} = A^{-1} \cdot \mathcal{U}_{abc} \tag{7.7}$$

Of course, the same decomposition could be applied to a system of unbalanced currents.

7.2.1 SEQUENCE NETWORKS

Use of the Fortescue decomposition enables a three-phase network to be represented as three single-phase-like networks, which are easier to handle than the one three-phase network. Hence, three networks one for each sequence should be formed. These are called the *sequence networks* and must be formed by suitably linking the grid elements together.

For example, consider a general grid element, whose relationship between terminal voltages and currents is given by

$$\begin{pmatrix} \mathcal{U}_a \\ \mathcal{U}_b \\ \mathcal{U}_c \end{pmatrix} = \begin{pmatrix} \mathcal{E}_a \\ \mathcal{E}_b \\ \mathcal{E}_c \end{pmatrix} + \begin{pmatrix} \mathcal{Z}_s & \mathcal{Z}_m & \mathcal{Z}_m \\ \mathcal{Z}_m & \mathcal{Z}_s & \mathcal{Z}_m \\ \mathcal{Z}_m & \mathcal{Z}_m & \mathcal{Z}_s \end{pmatrix} \begin{pmatrix} \mathcal{I}_a \\ \mathcal{I}_b \\ \mathcal{I}_c \end{pmatrix} \tag{7.8}$$

or, more compactly

$$\mathcal{U}_{abc} = \mathcal{E}_{abc} + \mathcal{Z}_{abc} \cdot \mathcal{I}_{abc} \tag{7.9}$$

where $\mathcal{E}_{abc}$ is an internal voltage vector, and $\mathcal{Z}_{abc}$ is a symmetrical matrix that represents the coupling between the impedances of the three phases.

Equation 7.9 may be transformed into sequence magnitudes, arriving at

$$\mathcal{U}_{012} = \mathcal{E}_{012} + \mathcal{Z}_{012} \cdot \mathcal{I}_{012} \tag{7.10}$$

where $\mathcal{U}_{012}$, $\mathcal{E}_{012}$, and $\mathcal{I}_{012}$ result from applying Fortescue decompositions to the individual terms of Equation 7.9, where

$$\mathcal{Z}_{012} = A^{-1} \cdot \mathcal{Z}_{abc} \cdot A \tag{7.11}$$

For a given $\mathcal{Z}_{abc}$, $\mathcal{Z}_{012}$ is diagonal, with elements $\mathcal{Z}_0$, $\mathcal{Z}_1$, and $\mathcal{Z}_2$, which are the homopolar, positive, and negative sequence impedances of the grid element. These impedances represent the behavior of this element when fed with homopolar, positive, or negative sequence supplies, respectively.

The fact of $\mathcal{Z}_{012}$ being a diagonal matrix means that in this element the three sequences are fully decoupled, which means, for instance, that a supply of positive sequence does not produce magnitudes of other sequences. It should be said that this situation only presents itself when the matrix $\mathcal{Z}_{abc}$, as shown in Equation 7.8, represents a structurally balanced power system element. Any unbalance in the three-phase impedances produces nondiagonal elements in matrix $\mathcal{Z}_{012}$, and therefore, the sequences become coupled and cannot be studied separately with Fortescue decomposition losing its simplifying appeal.

For each element in the grid, it is necessary to obtain its sequence network to form the different sequence systems of the whole network.

7.2.2 COMPLEX POWER IN SEQUENCE COMPONENTS

The general expression of complex power in a three-phase system is

$$S = \mathcal{U}_a\mathcal{I}_a^* + \mathcal{U}_b\mathcal{I}_b^* + \mathcal{U}_c\mathcal{I}_c^* = (\mathcal{U}_a \quad \mathcal{U}_b \quad \mathcal{U}_c)\begin{pmatrix} \mathcal{I}_a \\ \mathcal{I}_b \\ \mathcal{I}_c \end{pmatrix}^* = \mathcal{U}_{abc}^t\mathcal{I}_{abc}^* \qquad (7.12)$$

where t stands for transposition and * for complex conjugate. If Equation 7.12 is transformed into its sequence components, it yields

$$S = \mathcal{U}_{012}^t A^t A^* \mathcal{I}_{012}^* = 3\mathcal{U}_{012}^t\mathcal{I}_{012}^* = 3\mathcal{U}_0\mathcal{I}_0 + 3\mathcal{U}_1\mathcal{I}_1 + 3\mathcal{U}_2\mathcal{I}_2 \qquad (7.13)$$

EXAMPLE 7.1 SEQUENCE NETWORKS OF A GRID

In the grid shown in Figure 7.1, the data of the different elements of the grid are

G1:	$U_n = 20\,\text{kV}$	$S_n = 20\,\text{MVA}$	$x' = x_2 = 0.12\,\text{pu}$	$x_0 = 0.04\,\text{pu}$	$X_n = 0.4\,\Omega$
G2:	$U_n = 23\,\text{kV}$	$S_n = 25\,\text{MVA}$	$x' = x_2 = 0.10\,\text{pu}$	$x_0 = 0.034\,\text{pu}$	
T1:	$132/20\,\text{kV}$	$S_n = 25\,\text{MVA}$	$z_{cc} = 0.12\,\text{pu}$	$\cos\varphi_{cc} = 0.1\,\text{Dyn11}$	
T2:	$132/23\,\text{kV}$	$S_n = 30\,\text{MVA}$	$z_{cc} = 0.12\,\text{pu}$	$\cos\varphi_{cc} = 0.1\,\text{YNd1}$	
LA:	$\mathcal{Z}' = 0.31 + j0.37\,\Omega/\text{km}$	$\mathcal{Z}_0' = 0.31 + j0.52\,\Omega/\text{km}$	$l = 100\,\text{km}$		
LB:	$\mathcal{Z}' = 0.3 + j0.35\,\Omega/\text{km}$	$\mathcal{Z}_0' = 0.31 + j0.5\,\Omega/\text{km}$	$l = 100\,\text{km}$		
Grid:	$S_{cc} = 500\,\text{MVA}$	$\cos\varphi_{cc} = 0, 2$			

The positive, negative, and homopolar sequence networks must be obtained.
Notes:

- The values in pu are referred to the base powers of each element.
- The grid must be modeled as a balanced positive sequence voltage source. It is solidly grounded, and the impedances of the three sequences are equal.

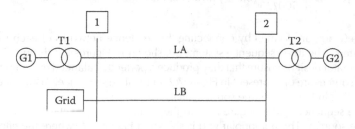

FIGURE 7.1　Grid of Example 7.1.

- The neutrals of both transformers are solidly grounded.
- The voltage in all nodes is the rated voltage.
- Phase shifts introduced by the delta–star connected transformer is not considered.
- Base power $S_b = 25$ MVA.

The models of the different elements are given in subsequent sections. Here, they are assumed as known.

The various sequence impedances referred to the system base power are

- Generator G1:

$$x_1 = x_2 = 0.12\frac{25}{20} = 0.15 \, \text{pu}$$

$$x_0 = 0.04\frac{25}{20} = 0.05 \, \text{pu}$$

$$Z_b = \frac{20^2}{25} = 16 \, \Omega$$

$$x_n = \frac{0.4}{16} = 0.0025 \, \text{pu}$$

- Transformer T1:

$$\mathcal{Z}_{cc} = 0.12(0.1 + j\sin\arccos 0.1) = 0.012 + j0.1194 \, \text{pu}$$

- Transformer T2:

$$\mathcal{Z}_{cc} = 0.12\frac{25}{30}(0.1 + j\sin\arccos 0.1) = 0.01 + j0.0995 \, \text{pu}$$

- Line LA:

$$Z_b = \frac{132^2}{25} = 697 \, \Omega$$

$$\mathcal{Z}_1 = \mathcal{Z}_2 = \frac{100(0.31 + j0.37)}{697} = 0.0444 + j0.0531 \, \text{pu}$$

$$\mathcal{Z}_0 = \frac{100(0.31 + j0.52)}{697} = 0.0444 + j0.0746 \, \text{pu}$$

- Line LB:

$$\mathcal{Z}_1 = \mathcal{Z}_2 = \frac{100(0.3 + j0.35)}{697} = 0.043 + j0.0502 \, \text{pu}$$

$$\mathcal{Z}_0 = \frac{100(0.3 + j0.5)}{697} = 0.043 + j0.0717 \, \text{pu}$$

- Grid:

$$\mathcal{Z} = \frac{1}{500/25}(0.2 + j\sin\arccos 0.2) = 0.01 + j0.049 \, \text{pu}$$

Sequence networks are formed by connecting the sequence networks of each individual element in the system. The three sequence systems are shown in Figures 7.2 through 7.4.

Modern generator designs ensure that they produce a perfectly balanced excitation voltage and, hence, voltage sources are only present in Figure 7.2 for positive sequence. Also, since there are no currents circulating in the system, the source voltages are 1 pu.

The negative sequence grid in Figure 7.3 has the same topology as the positive sequence but with no voltage sources. The homopolar grid is shown in Figure 7.4, where the effect of the two transformer connections shows up.

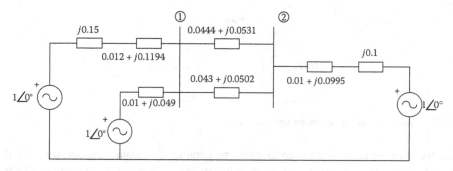

FIGURE 7.2 Positive sequence grid of Example 7.1.

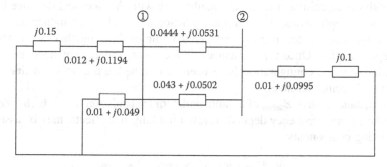

FIGURE 7.3 Negative sequence grid of Example 7.1.

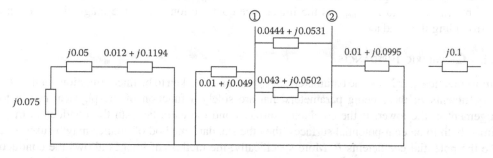

FIGURE 7.4 Homopolar sequence grid of Example 7.1.

7.3 TRANSMISSION LINES

Overhead transmission lines intended for high and extra high-voltage transmission use a group of phase conductors (bundle) to transmit the electrical energy. Moreover, one or two ground wires are used to shield the phase conductors against lightning strokes. It is not uncommon for a double-circuit three-phase transmission line operating at, say, 400 kV to contain up to 26 individual conductors in close proximity; four bundle conductor per phase and two ground wires; with a strong electromagnetic coupling existing between them. In practical applications, all conductors in power network transmission lines are located at a finite distance from the earth's surface and may use the ground as a return path. Accordingly, it becomes necessary to take all these issues into consideration when calculating transmission line parameters.

In power system studies, the overall transmission line model is well represented by either a nominal π-circuit or an equivalent π circuit, depending on the application study. The inductive and resistive effects of the multiconductor transmission line are modeled as a series impedance matrix, and the capacitive effects as a shunt admittance matrix, as illustrated in Figure 7.5.

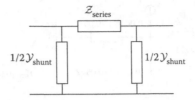

FIGURE 7.5　Transmission line π-circuit representation.

The computation of three-phase transmission line parameters becomes cumbersome by the existence of inductive and capacitive couplings between conductors, and between conductors and ground. Moreover, resistances and self- and mutual inductances vary nonlinearly with frequency and, together with the capacitive effects, vary nonlinearly with the electrical distance of the line.

In power systems applications, it is a normal practice to calculate the inductive and capacitive effects of transmission lines independently and then to combine them together to give the final transmission line representation. Once the resistances, inductances, and capacitances associated with a particular transmission line configuration have been determined, a transmission line model in the form of a π-circuit becomes feasible.

The series impedance matrix $\mathcal{Z}_{\text{series}}$ of a multiconductor transmission line, which takes account of geometric imbalances and frequency dependency, but not long-line effects, may be assumed to consist of the following components:

$$\mathcal{Z}_{\text{series}} = \mathcal{Z}_{\text{internal}} + \mathcal{Z}_{\text{geometric}} + \mathcal{Z}_{\text{ground}} \tag{7.14}$$

where $\mathcal{Z}_{\text{internal}}$ is the impedance inside the conductors, $\mathcal{Z}_{\text{ground}}$ is the impedance contribution of the ground return path, and $\mathcal{Z}_{\text{geometric}}$ is the impedance contribution due to the magnetic fluxes in the air surrounding the conductors.

7.3.1　Geometric Impedances

For most practical purposes, the parameters $\mathcal{Z}_{\text{geometric}}$ may be taken to be linear functions of the potential coefficients P; these being parameters that are solely a function of the physical conductor's arrangement in the tower. If the conductor surfaces and the earth beneath the conductors may be assumed both to be equi-potential surfaces, then the standard method of images may be used to calculate the potential coefficients P. More specifically, the method of images allows the conducting plane to be replaced by a fictitious conductor located at the mirror image of the actual conductors. Figure 7.6 shows the case when phase conductors a, b, and c above the ground have been replaced by three equivalent conductors and their images.

The self-potential coefficient of an overhead conductor is solely a function of the height of the conductor above ground, say h, and the conductor's external radius, say r_{ext}. On the other hand, the mutual-potential coefficient between two conductors is a function of the separation, d, between the two conductors, and the separation, D, between one conductor and the image of the second conductor. For the three conductors in the transmission line shown in Figure 7.6, the matrix of potential coefficients is

$$P = \begin{pmatrix} \log\dfrac{2h_a}{r_{\text{exta}}} & \log\dfrac{D_{ab}}{d_{ab}} & \log\dfrac{D_{ac}}{d_{ac}} \\[2ex] \log\dfrac{D_{ba}}{d_{ba}} & \log\dfrac{2h_b}{r_{\text{extb}}} & \log\dfrac{D_{bc}}{d_{bc}} \\[2ex] \log\dfrac{D_{ca}}{d_{ca}} & \log\dfrac{D_{cb}}{d_{cb}} & \log\dfrac{2h_a}{r_{\text{exta}}} \end{pmatrix} \tag{7.15}$$

It should be noted that potential coefficients are dimensionless and reciprocal.

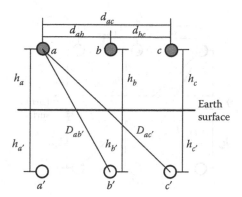

FIGURE 7.6   Line geometry and its image.

The geometric impedance matrix for the circuit of Figure 7.6 is

$$\mathcal{Z}_{\text{geometric}} = j\frac{\omega\mu_0}{4\pi}P \ \Omega/\text{km} \tag{7.16}$$

where $\mathcal{Z}_{\text{geometric}}$ varies linearly with the base frequency f, $\omega = 2\pi f$ and the permeability of the free space is $\mu_0 = 4\pi \times 10^{-4}\,\text{H/km}$.

7.3.2   SHUNT ADMITTANCES

Shunt admittance parameters vary linearly with frequency and are completely defined by the inverse potential coefficients. The matrix of shunt admittance parameters for the circuit of Figure 7.6 is

$$\mathcal{Y}_{\text{shunt}} = j\omega 2\pi\varepsilon_0 P^{-1} \ \text{S/km} \tag{7.17}$$

where the permittivity of the free space is $\varepsilon_0 = 8.85 \times 10^{-9}\,\text{F/km}$.

7.3.3   GROUND RETURN IMPEDANCES

The impedance of the ground return path varies nonlinearly with frequency and exhibits an effect similar to that of the skin effect in conductors, where the effective area available for the current to flow reduces with frequency.

The problem of current carrying wires above a flat earth of homogeneous conductivity, and the related issue of transmission line parameters calculation, was satisfactorily solved in 1926 by Carson [10]. More recent formulations use the concept of a complex mirroring surface beneath the ground. Rigorous mathematical analyses have shown these formulations to be good physical and mathematical approximations to Carson's solution. The most popular equations in power systems applications are those attributed to Dubanton [14]. The reason behind this is their simplicity and good accuracy for the whole span of frequencies for which Carson's equations are valid.

With reference to Figure 7.7, the equations for calculating the self-impedance of conductor l, and the mutual impedance between conductors l and m take the following form:

$$\mathcal{Z}_{ll} = j\frac{\omega\mu_0}{2\pi} \cdot \log\frac{2(h_l + p)}{r_{\text{ext}\,l}} \ \Omega/\text{km} \tag{7.18}$$

$$\mathcal{Z}_{lm} = j\frac{\omega\mu_0}{2\pi} \cdot \log\frac{\sqrt{(h_l + h_m + 2p)^2 + d_{lm}^2}}{\sqrt{(h_l - h_m)^2 + d_{lm}^2}} \ \Omega/\text{km} \tag{7.19}$$

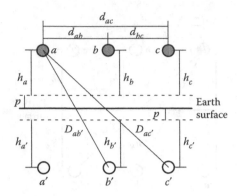

FIGURE 7.7 Line geometry showing the complex depth.

where $p = 1/\sqrt{j\omega\mu_0\sigma_g}$ is the complex depth beneath the ground at which the mirroring surface is located.

It should be noted that the use of Equations 7.18 and 7.19 yields combined information of $Z_{\text{geometric}} + Z_{\text{ground}}$.

7.3.4 INTERNAL IMPEDANCES

It has long been recognized that both the internal resistance and the inductance of conductors vary with frequency in a nonlinear manner. The reason for this effect is attributed mainly to the nonuniform distribution of current flow over the full area available, with current tending to flow on the surface. This trend increases with frequency and is termed skin effect. At power frequencies, the skin effect is negligibly small, and there is little error in calculating the internal impedance of conductors by assuming that the magnetic field inside the conductor is confined to an area lying between the conductor's external radius (r_{ext}) and the conductor's geometric mean radius (gmr), as illustrated in Figure 7.8.

The gmr is normally measured and made available by the manufacturer. An approximated, frequency-independent relationship is $\text{gmr} = e^{-1/4}r_{\text{ext}}$.

If the frequency of interest is low enough for the skin effect to be of no consequence, then the concept of potential coefficients can be applied to calculate the internal impedance of conductors. For instance, the internal impedance of a generic conductor l is

$$P_{\text{internal }l} = \log\frac{r_{\text{ext }l}}{\text{gmr}_l} \tag{7.20}$$

and the internal impedance of conductor l is

$$Z_{\text{internal }l} = R_{\text{ac }l} + j\frac{\omega\mu_0}{4\pi}P_{\text{internal }l}\ \Omega/\text{km} \tag{7.21}$$

where $R_{\text{ac }l}$ is the AC power-frequency resistance (50 or 60 Hz) of conductor l. It should be noted that there are no mutual impedances due to internal effects.

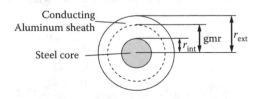

FIGURE 7.8 Cross section of a power conductor.

In higher frequency power systems applications, skin effects become significant, and more sophisticated formulations are used to calculate the internal impedances of transmission line conductors. The classical procedure for evaluating the impedance of an annular conductor at a given frequency uses the Bessel functions of zero order—first kind, second kind, and their derivatives—which are solved, within specified accuracy, using their associated infinite series [4]. Alternatively, a closed-form solution, which applies the concept of complex penetration, uses the following set of equations to calculate the internal impedance of conductor l [5]:

$$Z_{\text{internal } l} = \sqrt{R_0^2 + Z_\infty^2}$$

$$R_0 = \frac{1}{\pi \left(r_{\text{ext}}^2 - r_{\text{int}}^2\right)\sigma_c}$$

$$Z_\infty = \frac{1}{2\pi r_{\text{ext}}\sigma_c p_c}$$

(7.22)

where $p_c = 1/\sqrt{j\omega\mu_0\sigma_c}$ is the parameter of complex penetration beneath the conductor's surface, R_0 is the conductor's resistance at zero frequency, Z_∞ is the conductor's complex impedance at infinite frequency, and σ_c is the conductor's conductivity in siemens per meter.

7.3.5 Ground Wires Reduction

Overhead transmission lines use one or two ground wires to shield the phase conductors against lightning strokes. However, in power systems studies, the interest is normally limited to the phase conductors, with the effect of ground wires only accounted for in an implicit manner. In the transmission line parameters calculation stage, this requires a mathematical elimination of all the impedance entries associated with ground wires by means of a matrix reduction procedure known as Kron's reduction. This is exemplified below for the case of a three-phase transmission line with phase conductors a, b, and c; and two ground wires, w and v, having the following voltage drop equation:

$$\begin{pmatrix} U_a \\ U_b \\ U_c \\ U_w \\ U_v \end{pmatrix} = \begin{pmatrix} Z_{aa-g} & Z_{ab-g} & Z_{ac-g} & Z_{aw-g} & Z_{av-g} \\ Z_{ba-g} & Z_{bb-g} & Z_{bc-g} & Z_{bw-g} & Z_{bv-g} \\ Z_{ca-g} & Z_{cb-g} & Z_{cc-g} & Z_{cw-g} & Z_{cv-g} \\ Z_{wa-g} & Z_{wb-g} & Z_{wc-g} & Z_{ww-g} & Z_{wv-g} \\ Z_{va-g} & Z_{vb-g} & Z_{vc-g} & Z_{vw-g} & Z_{vv-g} \end{pmatrix} \begin{pmatrix} I_a \\ I_b \\ I_c \\ I_w \\ I_v \end{pmatrix} + \begin{pmatrix} U'_a \\ U'_b \\ U'_c \\ U'_w \\ U'_v \end{pmatrix}$$

(7.23)

It is assumed that the individual impedance elements are calculated using Equations 7.18, 7.19, and 7.22. In compact notation, we have

$$\Delta U_{abc} = A I_{abc} + B I_{wv}$$
$$\Delta U_{wv} = C I_{abc} + D I_{wv}$$

(7.24)

where

$$A = \begin{pmatrix} Z_{aa-g} & Z_{ab-g} & Z_{ac-g} \\ Z_{ba-g} & Z_{bb-g} & Z_{bc-g} \\ Z_{ca-g} & Z_{cb-g} & Z_{cc-g} \end{pmatrix} \quad B = \begin{pmatrix} Z_{aw-g} & Z_{av-g} \\ Z_{bw-g} & Z_{bv-g} \\ Z_{cw-g} & Z_{cv-g} \end{pmatrix}$$

$$C = \begin{pmatrix} Z_{wa-g} & Z_{wb-g} & Z_{wc-g} \\ Z_{va-g} & Z_{vb-g} & Z_{vc-g} \end{pmatrix} \quad D = \begin{pmatrix} Z_{ww-g} & Z_{wv-g} \\ Z_{vw-g} & Z_{vv-g} \end{pmatrix}$$

$$\Delta \mathcal{U}_{abc} = (\mathcal{U}_a - \mathcal{U}'_a \quad \mathcal{U}_b - \mathcal{U}'_b \quad \mathcal{U}_c - \mathcal{U}'_c)^{\mathrm{T}}$$

$$\Delta \mathcal{U}_{wv} = (\mathcal{U}_w - \mathcal{U}'_w \quad \mathcal{U}_v - \mathcal{U}'_v)^{\mathrm{T}}$$

$$\mathcal{I}_{abc} = (\mathcal{I}_a \quad \mathcal{I}_b \quad \mathcal{I}_c)^{\mathrm{T}}$$

$$\mathcal{I}_{wv} = (\mathcal{I}_w \quad \mathcal{I}_v)^{\mathrm{T}}$$

In transmission line modeling, it is assumed that $\Delta \mathcal{U}_{wv} = 0$, since it is normal practice to connect ground wires to earth at both ends of every transmission span. Solving Equation 7.24 for $\mathcal{I}_{wv}$, we arrive at

$$\mathcal{I}_{wv} = D^{-1} C \mathcal{I}_{abc} \tag{7.25}$$

and substitution of Equation 7.25 into Equation 7.24 yields

$$\Delta \mathcal{U}_{abc} = [A - BD^{-1}C]\mathcal{I}_{abc} = \mathcal{Z}_{abc-vw-g}\mathcal{I}_{abc} \tag{7.26}$$

where

$$\mathcal{Z}_{abc-vw-g} = A - BD^{-1}C$$

Equation 7.26 can be written in expanded form as

$$\begin{pmatrix} \Delta \mathcal{U}_a \\ \Delta \mathcal{U}_b \\ \Delta \mathcal{U}_c \end{pmatrix} = \begin{pmatrix} \mathcal{Z}_{aa-wv-g} & \mathcal{Z}_{ab-wv-g} & \mathcal{Z}_{ac-wv-g} \\ \mathcal{Z}_{ba-wv-g} & \mathcal{Z}_{bb-wv-g} & \mathcal{Z}_{bc-wv-g} \\ \mathcal{Z}_{ca-wv-g} & \mathcal{Z}_{cb-wv-g} & \mathcal{Z}_{cc-wv-g} \end{pmatrix} \begin{pmatrix} \Delta \mathcal{I}_a \\ \Delta \mathcal{I}_b \\ \Delta \mathcal{I}_c \end{pmatrix} \tag{7.27}$$

The reduced equivalent matrix Equation 7.27 is fully equivalent to matrix Equation 7.23, where the ground wires have been mathematically eliminated. For most system analysis purposes, Equation 7.27 provides a suitable representation for transmission lines with ground wires. Symmetrical components can be applied to Equation 7.27, and it is therefore preferred over Equation 7.23.

7.3.6 BUNDLE CONDUCTORS REDUCTION

The use of more than one conductor per phase, that is, bundle conductors, reduces the equivalent transmission line impedance and allows for an increase in power transmission. It also allows for a reduction in corona loss and radio interference due to a reduction in conductor-surface voltage gradients. For cases of 400 kV transmission lines and above, it is standard practice to have four bundle conductors per phase, while for 230 kV lines, only three- or two-bundle conductors per phase are required.

In power systems studies, the interest is rarely on individual conductors but, rather, on the individual phases. Hence, steps are taken to find reduced equivalents that involve only one conductor per phase. The equivalent conductors correctly account for the original configuration but keep essential information only.

The bundle conductor reduction may be achieved in a number of ways. Using the concept of equivalent gmr is one of them, and although it is a frequency independent method, it yields reasonable accurate solutions, particularly at power frequencies. A more rigorous approach, which includes frequency dependency for the reduction of the series impedance matrix, involves matrix reduction techniques. Using this method, all conductor impedances are calculated explicitly and after a suitable manipulation of terms in the impedance matrix, the mathematical elimination of bundle conductors is carried out. The actual elimination uses Kron's reduction procedure.

To illustrate the elimination procedure used when bundle conductors are present, take the case of a three-phase transmission line (a, b, c) with two conductors per phase (1,2) and no ground wires. The matrix of series impedance parameters representing such a transmission line would be

$$
\begin{pmatrix} \Delta \mathcal{U}_{a1} \\ \Delta \mathcal{U}_{b1} \\ \Delta \mathcal{U}_{c1} \\ \Delta \mathcal{U}_{a2} \\ \Delta \mathcal{U}_{b2} \\ \Delta \mathcal{U}_{c2} \end{pmatrix} = \begin{pmatrix} \mathcal{Z}_{a1a1-g} & \mathcal{Z}_{a1b1-g} & \mathcal{Z}_{a1c1-g} & \mathcal{Z}_{a1a2-g} & \mathcal{Z}_{a1b2-g} & \mathcal{Z}_{a2c2-g} \\ \mathcal{Z}_{a1b1-g} & \mathcal{Z}_{b1b1-g} & \mathcal{Z}_{b1c1-g} & \mathcal{Z}_{b1a2-g} & \mathcal{Z}_{b1b2-g} & \mathcal{Z}_{b1c2-g} \\ \mathcal{Z}_{c1a1-g} & \mathcal{Z}_{c1b1-g} & \mathcal{Z}_{c1c1-g} & \mathcal{Z}_{c1a2-g} & \mathcal{Z}_{c1b2-g} & \mathcal{Z}_{c1c2-g} \\ \mathcal{Z}_{a2a1-g} & \mathcal{Z}_{a2b1-g} & \mathcal{Z}_{a2c1-g} & \mathcal{Z}_{a2a2-g} & \mathcal{Z}_{a2b2-g} & \mathcal{Z}_{a2c2-g} \\ \mathcal{Z}_{b2a1-g} & \mathcal{Z}_{b2b1-g} & \mathcal{Z}_{b2c1-g} & \mathcal{Z}_{b2a2-g} & \mathcal{Z}_{b2b2-g} & \mathcal{Z}_{b2c2-g} \\ \mathcal{Z}_{c2a1-g} & \mathcal{Z}_{c2b1-g} & \mathcal{Z}_{c2c1-g} & \mathcal{Z}_{c2a2-g} & \mathcal{Z}_{c2b2-g} & \mathcal{Z}_{c2c2-g} \end{pmatrix} \begin{pmatrix} \mathcal{I}_{a1} \\ \mathcal{I}_{b1} \\ \mathcal{I}_{c1} \\ \mathcal{I}_{a2} \\ \mathcal{I}_{b2} \\ \mathcal{I}_{c2} \end{pmatrix} \tag{7.28}
$$

Each phase consists of two parallel conductors, with the current split shown in Figure 7.9.

The individual elements of Equation 7.28 are calculated using Equations 7.18, 7.19, and 7.22. In compact notation, we have

$$
\begin{pmatrix} \Delta \mathcal{U}_{abc1} \\ \Delta \mathcal{U}_{abc2} \end{pmatrix} = \begin{pmatrix} \mathcal{Z}_{abc11} & \mathcal{Z}_{abc12} \\ \mathcal{Z}_{abc21} & \mathcal{Z}_{abc22} \end{pmatrix} \begin{pmatrix} \mathcal{I}_{abc1} \\ \mathcal{I}_{abc2} \end{pmatrix} \tag{7.29}
$$

If it is assumed that the two conductors in the bundle are at equal potential, then the row and column corresponding to one of the conductors in the bundle, say the second, is mathematically eliminated. There are three main steps involved in the elimination process:

1. The following voltage equality constraint $\Delta \mathcal{U}_{abc2} - \Delta \mathcal{U}_{abc2} = 0$ is incorporated into Equation 7.29:

$$
\begin{pmatrix} \Delta \mathcal{U}_{abc1} \\ \Delta \mathcal{U}_{abc2} - \Delta \mathcal{U}_{abc2} \end{pmatrix} = \begin{pmatrix} \mathcal{Z}_{abc11} & \mathcal{Z}_{abc12} \\ \mathcal{Z}_{abc21} - \mathcal{Z}_{abc11} & \mathcal{Z}_{abc22} - \mathcal{Z}_{abc12} \end{pmatrix} \tag{7.30}
$$

2. Matrix symmetry is restored. This is achieved by adding and subtracting the terms $\mathcal{Z}_{abc11}\mathcal{I}_{abc2}$ and $(\mathcal{Z}_{abc21} - \mathcal{Z}_{abc11})\mathcal{I}_{abc2}$ in rows 1 and 2, respectively:

$$
\begin{pmatrix} \Delta \mathcal{U}_{abc1} \\ 0 \end{pmatrix} = \begin{pmatrix} \mathcal{Z}_{abc11} & \mathcal{Z}_{abc12} - \mathcal{Z}_{abc11} \\ \mathcal{Z}_{abc21} - \mathcal{Z}_{abc11} & (\mathcal{Z}_{abc22} + \mathcal{Z}_{abc11} - \mathcal{Z}_{abc12} - \mathcal{Z}_{abc21}) \end{pmatrix} \tag{7.31}
$$

3. A Kron's reduction is carried out:

$$
\Delta \mathcal{U}_{abc-b} = \Delta \mathcal{Z}_{abc-b-g} \mathcal{I}_{abc-b} \tag{7.32}
$$

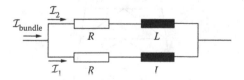

FIGURE 7.9   Current in a two-bundle conductor.

where

$$\begin{aligned}
\mathcal{Z}_{abc-b-g} &= A - BD^{-1}C \\
A &= \mathcal{Z}_{abc11} \\
B &= \mathcal{Z}_{abc12} - \mathcal{Z}_{abc11} \\
C &= \mathcal{Z}_{abc21} - \mathcal{Z}_{abc11} \\
D &= \mathcal{Z}_{abc22} + \mathcal{Z}_{abc11} - \mathcal{Z}_{abc12} - \mathcal{Z}_{abc21} \\
\Delta\mathcal{U}_{abc-b} &= \Delta\mathcal{U}_{abc1} \\
\mathcal{I}_{abc-b} &= \mathcal{I}_{abc1} + \mathcal{I}_{abc2}
\end{aligned} \tag{7.33}$$

As illustrated in Figure 7.5, the current $\mathcal{I}_{abc-b}$ may be interpreted as the phase current just before it splits into the individual currents $\mathcal{I}_{abc1}$ and $\mathcal{I}_{abc2}$ in the bundle.

The reduced equivalent matrix Equation 7.32 can be written in expanded form as

$$\begin{pmatrix} \Delta\mathcal{U}_{a-b} \\ \Delta\mathcal{U}_{b-b} \\ \Delta\mathcal{U}_{c-b} \end{pmatrix} = \begin{pmatrix} \mathcal{Z}_{aa-b} & \mathcal{Z}_{ab-b} & \mathcal{Z}_{ac-b} \\ \mathcal{Z}_{ba-b} & \mathcal{Z}_{bb-b} & \mathcal{Z}_{bc-b} \\ \mathcal{Z}_{ca-b} & \mathcal{Z}_{cb-b} & \mathcal{Z}_{cc-b} \end{pmatrix} \begin{pmatrix} \Delta\mathcal{I}_{a-b} \\ \Delta\mathcal{I}_{b-b} \\ \Delta\mathcal{I}_{c-b} \end{pmatrix} \tag{7.34}$$

7.3.7 DOUBLE-CIRCUIT TRANSMISSION LINES

Often two or more three-phase transmission lines are operated in parallel. A common arrangement is to place two three-phase circuits in the same tower as shown in Figure 7.10.

In this case, the magnetic interaction between the phase conductors of both the three-phase circuits can be represented by the following impedance matrix equation:

$$\begin{pmatrix} \Delta\mathcal{U}_a \\ \Delta\mathcal{U}_b \\ \Delta\mathcal{U}_c \\ \Delta\mathcal{U}_A \\ \Delta\mathcal{U}_B \\ \Delta\mathcal{U}_C \end{pmatrix} = \begin{pmatrix} \mathcal{Z}_{aa-g} & \mathcal{Z}_{ab-g} & \mathcal{Z}_{ac-g} & \mathcal{Z}_{aA-g} & \mathcal{Z}_{aB-g} & \mathcal{Z}_{aC-g} \\ \mathcal{Z}_{ba-g} & \mathcal{Z}_{bb-g} & \mathcal{Z}_{bc-g} & \mathcal{Z}_{bA-g} & \mathcal{Z}_{bB-g} & \mathcal{Z}_{bC-g} \\ \mathcal{Z}_{ca-g} & \mathcal{Z}_{cb-g} & \mathcal{Z}_{cc-g} & \mathcal{Z}_{cA-g} & \mathcal{Z}_{cB-g} & \mathcal{Z}_{cC-g} \\ \mathcal{Z}_{Aa-g} & \mathcal{Z}_{Ab-g} & \mathcal{Z}_{Ac-g} & \mathcal{Z}_{AA-g} & \mathcal{Z}_{AB-g} & \mathcal{Z}_{AC-g} \\ \mathcal{Z}_{Ba-g} & \mathcal{Z}_{Bb-g} & \mathcal{Z}_{Bc-g} & \mathcal{Z}_{BA-g} & \mathcal{Z}_{BB-g} & \mathcal{Z}_{BC-g} \\ \mathcal{Z}_{Ca-g} & \mathcal{Z}_{Cb-g} & \mathcal{Z}_{Cc-g} & \mathcal{Z}_{CA-g} & \mathcal{Z}_{CB-g} & \mathcal{Z}_{CC-g} \end{pmatrix} \begin{pmatrix} \mathcal{I}_a \\ \mathcal{I}_b \\ \mathcal{I}_c \\ \mathcal{I}_A \\ \mathcal{I}_B \\ \mathcal{I}_C \end{pmatrix} \tag{7.35}$$

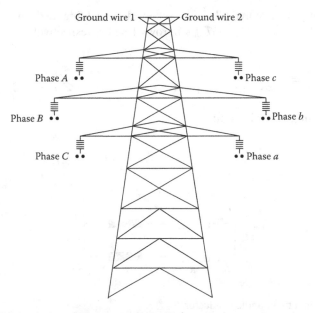

FIGURE 7.10 Double-circuit three-phase transmission line.

It is assumed here that neither ground wires nor bundle conductors are present in the double-circuit transmission line, or that they have been mathematically eliminated using the methods discussed previously. It is also assumed that the individual elements of Equation 7.35 were calculated using Equations 7.18, 7.19, and 7.22.

7.3.8 LONG-LINE EFFECTS

The transmission line models required for long-distance transmission applications are more involved than those covered above, which are only suitable to represent short to medium distance transmission lines. In actual applications, however, it is not common practice to see transmission lines of more than 300 km without series compensation, which fall within the category of medium distance transmission lines for the purpose of fundamental frequency operation. However, at frequency applications higher than the fundamental, it is certainly mandatory to incorporate long-line effects since the electrical distance increases rapidly with frequency. Even transmission lines of only a few tens of kilometers may be seen as a very long line at 1 kHz.

Calculation of multiconductor transmission line parameters, including long-line effects, requires the use of formulations derived from the wave propagation equation. This introduces a degree of extra complexity since these formulations invariably involve square roots and circular and hyperbolic functions of matrices. Several options are available to carry out such nonconventional matrix operations, but perhaps the best well-known method is to simply apply suitable eigenvector techniques to the relevant transmission line parameter matrices. This enables all calculations to be performed in the frame-of-reference of the modes and then referred back to the frame-of-reference of the phases.

Arguably, the best well-known formulation derived from the wave propagation equation is the *ABCD* parameters formulation:

$$\begin{pmatrix} \mathcal{U}_S \\ \mathcal{I}_S \end{pmatrix} = \begin{pmatrix} A & B \\ C & D \end{pmatrix} \begin{pmatrix} \mathcal{U}_R \\ -\mathcal{I}_R \end{pmatrix} \tag{7.36}$$

where $\mathcal{U}_S$ and $\mathcal{I}_S$ are the phase voltage and currents of the three-phase line in its sending end, $\mathcal{U}_R$ and $\mathcal{I}_R$ are the phase voltage and currents of the three-phase line in its receiving end, and

$$A = T_v \cdot \text{diag}(\cosh \gamma_m l) T_i^{-1}$$

$$B = T_v \cdot \text{diag}(z_m \cosh \gamma_m l) T_i^{-1}$$

$$C = T_i \cdot \text{diag}(y_m \cosh \gamma_m l) T_v^{-1} \tag{7.37}$$

$$D = T_i \cdot \text{diag}(\cosh \gamma_m l) T_i^{-1}$$

In Equation 7.37, diag is a diagonal matrix; m is the subscript for modes 0, α, and β; l is the length of the line; T_v and T_i are transformation matrices made up of the eigenvectors of the matrix products $\mathcal{Z}\mathcal{Y}$ and $\mathcal{Y}\mathcal{Z}$, respectively, and $\mathcal{Z}$ and $\mathcal{Y}$ are lumped transmission line parameters.

The modal parameters γ_m, z_m, and y_m in Equation 7.37 are calculated by first making $\mathcal{Z}$ and $\mathcal{Y}$ diagonal

$$\mathcal{Z}_m = T_v^{-1} \mathcal{Z} T_i$$

$$\mathcal{Y}_m = T_i^{-1} \mathcal{Z} T_v \tag{7.38}$$

and then performing the following operations:

$$\gamma_m = \begin{pmatrix} \sqrt{\mathcal{Z}_0 \mathcal{Y}_0} & 0 & 0 \\ 0 & \sqrt{\mathcal{Z}_\alpha \mathcal{Y}_\alpha} & 0 \\ 0 & 0 & \sqrt{\mathcal{Z}_\beta \mathcal{Y}_\beta} \end{pmatrix} \tag{7.39}$$

$$z_m = \frac{1}{y_m} = \begin{pmatrix} \sqrt{\mathcal{Z}_0 \mathcal{Y}_0^{-1}} & 0 & 0 \\ 0 & \sqrt{\mathcal{Z}_\alpha \mathcal{Y}_\alpha^{-1}} & 0 \\ 0 & 0 & \sqrt{\mathcal{Z}_\beta \mathcal{Y}_\beta^{-1}} \end{pmatrix} \tag{7.40}$$

where $\mathcal{Z}_0$, $\mathcal{Z}_\alpha$, $\mathcal{Z}_\beta$ and $\mathcal{Y}_0$, $\mathcal{Y}_\alpha$, $\mathcal{Y}_\beta$ are the diagonal terms of matrices $\mathcal{Z}_m$ and $\mathcal{Y}_m$, respectively.

Alternative formulations, derived from the wave propagation equation, are available, which may present advantages in certain applications. The two obvious ones are the impedance and the admittance representations:

$$\begin{pmatrix} \mathcal{U}_S \\ \mathcal{U}_R \end{pmatrix} = \begin{pmatrix} Z' & Z'' \\ Z'' & Z' \end{pmatrix} \begin{pmatrix} \mathcal{I}_S \\ \mathcal{I}_R \end{pmatrix} \tag{7.41}$$

$$\begin{pmatrix} \mathcal{I}_S \\ \mathcal{I}_R \end{pmatrix} = \begin{pmatrix} Y' & Y'' \\ Y'' & Y' \end{pmatrix} \begin{pmatrix} \mathcal{V}_S \\ \mathcal{V}_R \end{pmatrix} \tag{7.42}$$

where

$$\begin{aligned} Z' &= T_v \operatorname{diag}(\mathcal{Z}_m \coth \gamma_m l) T_i^{-1} \\ Z'' &= T_v \operatorname{diag}(\mathcal{Z}_m \operatorname{csch} \gamma_m l) T_i^{-1} \\ Y' &= T_i \operatorname{diag}(\mathcal{Y}_m \coth \gamma_m l) T_v^{-1} \\ Y'' &= T_i \operatorname{diag}(\mathcal{Y}_m \operatorname{csch} \gamma_m l) T_v^{-1} \end{aligned} \tag{7.43}$$

7.3.9 SYMMETRICAL COMPONENTS AND SEQUENCE DOMAIN PARAMETERS

If Equations 7.27 and 7.34 can be assumed to be perfectly balanced then they can be replaced by the following impedance matrix equation:

$$\begin{pmatrix} \Delta \mathcal{U}_a \\ \Delta \mathcal{U}_b \\ \Delta \mathcal{U}_c \end{pmatrix} = \begin{pmatrix} \mathcal{Z}_s & \mathcal{Z}_m & \mathcal{Z}_m \\ \mathcal{Z}_m & \mathcal{Z}_s & \mathcal{Z}_m \\ \mathcal{Z}_m & \mathcal{Z}_m & \mathcal{Z}_s \end{pmatrix} \begin{pmatrix} \Delta \mathcal{I}_a \\ \Delta \mathcal{I}_b \\ \Delta \mathcal{I}_c \end{pmatrix} \tag{7.44}$$

Equation 7.44, written in compact notation, is subjected to the following treatment:

$$A^{-1} \Delta \mathcal{U}_{abc} = A^{-1} \mathcal{Z}_{abc} A A^{-1} \mathcal{I}_{abc}$$

where A is defined as in Equation 7.5. This yields the sequence domain representation

$$\Delta \mathcal{U}_{012} = \mathcal{Z}_{012} \mathcal{I}_{012} \tag{7.45}$$

The subscripts 0, 1, and 2 stand for homopolar, positive, and negative sequence components, respectively. Furthermore,

$$\begin{aligned} \Delta \mathcal{U}_{012} &= A^{-1} \Delta \mathcal{U}_{abc} \\ \mathcal{I}_{012} &= A^{-1} \mathcal{I}_{abc} \\ \mathcal{Z}_{012} &= A^{-1} \mathcal{Z}_{abc} A \end{aligned} \tag{7.46}$$

Equation 7.45 in expanded form is

$$
\begin{pmatrix} \Delta \mathcal{U}_0 \\ \Delta \mathcal{U}_1 \\ \Delta \mathcal{U}_2 \end{pmatrix} = \begin{pmatrix} \mathcal{Z}_0 & 0 & 0 \\ 0 & \mathcal{Z}_1 & 0 \\ 0 & 0 & \mathcal{Z}_2 \end{pmatrix} \begin{pmatrix} \Delta \mathcal{I}_0 \\ \Delta \mathcal{I}_1 \\ \Delta \mathcal{I}_2 \end{pmatrix} \tag{7.47}
$$

where

$$
\begin{aligned}
\mathcal{Z}_0 &= \mathcal{Z}_s + 2\mathcal{Z}_m \\
\mathcal{Z}_1 &= \mathcal{Z}_s - \mathcal{Z}_m \\
\mathcal{Z}_2 &= \mathcal{Z}_s - \mathcal{Z}_m
\end{aligned} \tag{7.48}
$$

This is a useful result that enables the calculation of the homopolar, positive, and negative sequence impedances from known self- and mutual impedances.

The reverse problem, where the self- and mutual impedances of a perfectly balanced transmission line are to be determined from known sequence impedances, is of great practical interest. Suitable equations can be derived from

$$
\begin{aligned}
\mathcal{Z}_s &= \tfrac{1}{3}(\mathcal{Z}_0 + 2\mathcal{Z}_1) \\
\mathcal{Z}_m &= \tfrac{1}{3}(\mathcal{Z}_0 - \mathcal{Z}_1)
\end{aligned} \tag{7.49}
$$

However, it should be marked that if Equation 7.27 or 7.34 cannot be assumed to be perfectly balanced then the use of symmetrical components does not yield a decoupled matrix equation and the use of symmetrical components is of limited value.

To some extent this problem arises when the perfectly balanced counterpart of matrix Equation 7.23 is represented in the sequence domain. If Equation 7.23 can be assumed to be perfectly balanced then it is replaced by the following matrix equation:

$$
\begin{pmatrix} \Delta \mathcal{U}_a \\ \Delta \mathcal{U}_b \\ \Delta \mathcal{U}_c \\ \Delta \mathcal{U}_A \\ \Delta \mathcal{U}_B \\ \Delta \mathcal{U}_C \end{pmatrix} = \begin{pmatrix} \mathcal{Z}_s & \mathcal{Z}_m & \mathcal{Z}_m & \mathcal{Z}_m & \mathcal{Z}_m & \mathcal{Z}_m \\ \mathcal{Z}_m & \mathcal{Z}_s & \mathcal{Z}_m & \mathcal{Z}_m & \mathcal{Z}_m & \mathcal{Z}_m \\ \mathcal{Z}_m & \mathcal{Z}_m & \mathcal{Z}_s & \mathcal{Z}_m & \mathcal{Z}_m & \mathcal{Z}_m \\ \mathcal{Z}_m & \mathcal{Z}_m & \mathcal{Z}_m & \mathcal{Z}_s & \mathcal{Z}_m & \mathcal{Z}_m \\ \mathcal{Z}_m & \mathcal{Z}_m & \mathcal{Z}_m & \mathcal{Z}_m & \mathcal{Z}_s & \mathcal{Z}_m \\ \mathcal{Z}_m & \mathcal{Z}_m & \mathcal{Z}_m & \mathcal{Z}_m & \mathcal{Z}_m & \mathcal{Z}_s \end{pmatrix} \begin{pmatrix} \Delta \mathcal{I}_a \\ \Delta \mathcal{I}_b \\ \Delta \mathcal{I}_c \\ \Delta \mathcal{I}_A \\ \Delta \mathcal{I}_B \\ \Delta \mathcal{I}_C \end{pmatrix} \tag{7.50}
$$

Using compact notation to represent Equation 7.50 and applying symmetrical components

$$
\begin{pmatrix} A^{-1} & 0 \\ 0 & A^{-1} \end{pmatrix} \begin{pmatrix} \Delta \mathcal{U}_{abc} \\ \Delta \mathcal{U}_{ABC} \end{pmatrix} = \begin{pmatrix} A^{-1} & 0 \\ 0 & A^{-1} \end{pmatrix} \begin{pmatrix} \mathcal{Z}_s & \mathcal{Z}_m \\ \mathcal{Z}_m & \mathcal{Z}_s \end{pmatrix} \begin{pmatrix} A & 0 \\ 0 & A \end{pmatrix} \begin{pmatrix} A^{-1} & 0 \\ 0 & A^{-1} \end{pmatrix} \begin{pmatrix} \Delta \mathcal{I}_{abc} \\ \Delta \mathcal{I}_{ABC} \end{pmatrix} \tag{7.51}
$$

leads to the following result:

$$
\begin{pmatrix} \Delta \mathcal{U}_{012} \\ \Delta \mathcal{U}'_{012} \end{pmatrix} = \begin{pmatrix} \mathcal{Z}_{s012} & \mathcal{Z}_{m012} \\ \mathcal{Z}_{m012} & \mathcal{Z}_{s012} \end{pmatrix} \begin{pmatrix} \Delta \mathcal{I}_{012} \\ \Delta \mathcal{I}'_{012} \end{pmatrix} \tag{7.52}
$$

Equation 7.52 in expanded form is

$$
\begin{pmatrix} \Delta\mathcal{U}_0 \\ \Delta\mathcal{U}_1 \\ \Delta\mathcal{U}_2 \\ \Delta\mathcal{U}_0 \\ \Delta\mathcal{U}'_1 \\ \Delta\mathcal{U}'_2 \end{pmatrix} =
\begin{pmatrix}
\mathcal{Z}_s + 2\mathcal{Z}_m & 0 & 0 & 3\mathcal{Z}_m & 0 & 0 \\
0 & \mathcal{Z}_s - \mathcal{Z}_m & 0 & 0 & 0 & 0 \\
0 & 0 & \mathcal{Z}_s - \mathcal{Z}_m & 0 & 0 & 0 \\
3\mathcal{Z}_m & 0 & 0 & \mathcal{Z}_s + 2\mathcal{Z}_m & 0 & 0 \\
0 & 0 & 0 & 0 & \mathcal{Z}_s - \mathcal{Z}_m & 0 \\
0 & 0 & 0 & 0 & 0 & \mathcal{Z}_s - \mathcal{Z}_m
\end{pmatrix}
\begin{pmatrix} \Delta\mathcal{I}_0 \\ \Delta\mathcal{I}_1 \\ \Delta\mathcal{I}_2 \\ \Delta\mathcal{I}'_0 \\ \Delta\mathcal{I}'_1 \\ \Delta\mathcal{I}'_2 \end{pmatrix}
$$

$$\text{(7.53)}$$

where the sequence domain voltages and currents corresponding to circuit two are primed to differentiate them from those in circuit one. Also, note the impedance coupling between the two homopolar sequence circuits.

Examples of the computation of some of these parameters and models are shown in Section 12.7.2, where this particular issue is discussed in some detail in the context of the use of transmission line parameters and models in fast electromagnetic transient studies.

7.4 TRANSFORMERS

7.4.1 THREE-PHASE MODELING OF TWO-WINDING TRANSFORMERS

When unbalanced networks are to be considered in power system analysis, the transformer modeling of Chapter 2 needs expanding. For a three-phase two-winding transformer shown in Figure 7.11, the relationships between currents and voltages for the six nonconnected windings are given by Equation 7.54; voltage and current references are shown in the figure. At this point in the discussion, the magnetizing branch is neglected, but it will be incorporated at a later stage into the model:

$$
\begin{pmatrix} \mathcal{I}_I \\ \mathcal{I}_i \\ \\ \mathcal{I}_{II} \\ \mathcal{I}_{ii} \\ \\ \mathcal{I}_{III} \\ \mathcal{I}_{iii} \end{pmatrix} =
\begin{pmatrix}
\mathcal{Y}_{sc} & -\mathcal{Y}_{sc} & \mathcal{Y}'_{m} & \mathcal{Y}''_{m} & \mathcal{Y}'_{m} & \mathcal{Y}''_{m} \\
-\mathcal{Y}_{sc} & \mathcal{Y}_{sc} & \mathcal{Y}''_{m} & \mathcal{Y}'''_{m} & \mathcal{Y}''_{m} & \mathcal{Y}'''_{m} \\
\mathcal{Y}'_{m} & \mathcal{Y}''_{m} & \mathcal{Y}_{sc} & -\mathcal{Y}_{sc} & \mathcal{Y}'_{m} & \mathcal{Y}''_{m} \\
\mathcal{Y}''_{m} & \mathcal{Y}'''_{m} & -\mathcal{Y}_{sc} & \mathcal{Y}_{sc} & \mathcal{Y}''_{m} & \mathcal{Y}'''_{m} \\
\mathcal{Y}'_{m} & \mathcal{Y}''_{m} & \mathcal{Y}'_{m} & \mathcal{Y}''_{m} & \mathcal{Y}_{sc} & -\mathcal{Y}_{sc} \\
\mathcal{Y}''_{m} & \mathcal{Y}'''_{m} & \mathcal{Y}''_{m} & \mathcal{Y}'''_{m} & -\mathcal{Y}_{sc} & \mathcal{Y}_{sc}
\end{pmatrix}
\begin{pmatrix} \mathcal{U}_I \\ \mathcal{U}_i \\ \\ \mathcal{U}_{II} \\ \mathcal{U}_{ii} \\ \\ \mathcal{U}_{III} \\ \mathcal{U}_{iii} \end{pmatrix}
$$

$$\text{(7.54)}$$

In Equation 7.54, $\mathcal{I}_I$ stands for the current through winding I. Capital letters in the subscripts are used for the high-voltage windings, while lowercase letters refer to low-voltage windings. The admittance $\mathcal{Y}_{sc}$ is the short-circuit admittance between high-voltage and low-voltage windings, referred to

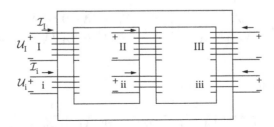

FIGURE 7.11 References for the three-legged transformer.

the high-voltage side. The admittance $\mathcal{Y}'_m$ is the mutual admittance between high-voltage windings, $\mathcal{Y}''_m$ is the mutual admittance between high-voltage and low-voltage windings of different transformer legs, and $\mathcal{Y}'''_m$ is the mutual admittance between low-voltage windings. The primed admittance terms have low values compared with $\mathcal{Y}_{sc}$ (they are zero for three-phase transformer banks). In subsequent equations these terms are dropped.

It should be noted that if a tertiary winding is present, the primitive network consists of nine, as opposed to six coupled coils, and its mathematical model would be a 9×9 admittance matrix. Equation 7.54 may be written in a more compact form as

$$\mathcal{I}_b = \mathcal{Y}_b \mathcal{U}_b \tag{7.55}$$

where the subscript b stands for *branch* quantities.

For studies at frequencies different from the fundamental frequency ($\omega_1 = 50$ or $60\,\text{Hz}$), the dependance of $\mathcal{Y}_{sc}$ with frequency is given by [6]:

$$Y_{sc}(\omega) = \cfrac{1}{R_p\sqrt{\dfrac{\omega}{\omega_1}} + jX_{p1}\left(\dfrac{\omega}{\omega_1}\right)} \tag{7.56}$$

where R_p and X_{p1} are the parallel resistance and reactance at ω_1, respectively.

Equation 7.54 does not incorporate the transformer connection. To take this into account, a connection matrix C, which relates the branch and nodal parameters is set up. A different connection matrix exists for each winding connection. For instance, for a grounded wye–delta connection such as that shown in Figure 7.12, matrix C takes the following form:

$$C = \begin{pmatrix} 1 & 0 & 0 & 0 & 0 & 0 \\ 0 & 0 & 0 & 1 & -1 & 0 \\ 0 & 1 & 0 & 0 & 0 & 0 \\ 0 & 0 & 0 & 0 & 1 & -1 \\ 0 & 0 & 1 & 0 & 0 & 0 \\ 0 & 0 & 0 & -1 & 0 & 1 \end{pmatrix} \tag{7.57}$$

With this matrix, relation between branch and node magnitudes may be written as follows:

$$\begin{aligned} \mathcal{U}_b &= C\mathcal{U}_n \\ \mathcal{I}_n &= C^T\mathcal{I}_b \end{aligned} \tag{7.58}$$

where $\mathcal{U}_n = (\mathcal{U}_A \ \ \mathcal{U}_B \ \ \mathcal{U}_C \ \ \mathcal{U}_a \ \ \mathcal{U}_b \ \ \mathcal{U}_c)^T$ and $\mathcal{I}_n = (\mathcal{I}_A \ \ \mathcal{I}_B \ \ \mathcal{I}_C \ \ \mathcal{I}_a \ \ \mathcal{I}_b \ \ \mathcal{I}_c)^T$ are vectors of nodal voltages and currents, respectively.

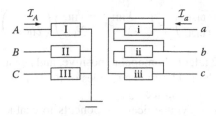

FIGURE 7.12 Connections in a wye–delta transformer.

Combining Equations 7.55 and 7.58 yields

$$\mathcal{I}_n = \mathcal{Y}_n \mathcal{U}_n \tag{7.59}$$

where $\mathcal{Y}_n = C^{\mathrm{T}} \mathcal{Y}_n C$. Once the transformer representation is in nodal form, it is straightforward to incorporate the transformer magnetizing branch as a shunt admittance. For the grounded wye–delta connection, the nodal equation between voltages and currents, in per unit, including the shunt admittance $\mathcal{Y}_p$ connected to the high-voltage side is given by Equation 7.60. The same discussion applies to the transformer's tap-changing ratio. Writing it in per unit implies that the terms $\sqrt{3}$ and 3 divide some entries in the following equation:

$$\begin{pmatrix} \mathcal{I}_A \\ \mathcal{I}_B \\ \mathcal{I}_C \\ \\ \mathcal{I}_a \\ \mathcal{I}_b \\ \mathcal{I}_c \end{pmatrix} = \begin{pmatrix} \mathcal{Y}_{\mathrm{sc}} + \mathcal{Y}_{\mathrm{p}} & 0 & 0 & -t\mathcal{Y}_{\mathrm{sc}}/\sqrt{3} & t\mathcal{Y}_{\mathrm{sc}}/\sqrt{3} & 0 \\ 0 & \mathcal{Y}_{\mathrm{sc}} + \mathcal{Y}_{\mathrm{p}} & 0 & 0 & -t\mathcal{Y}_{\mathrm{sc}}/\sqrt{3} & t\mathcal{Y}_{\mathrm{sc}}/\sqrt{3} \\ 0 & 0 & \mathcal{Y}_{\mathrm{sc}} + \mathcal{Y}_{\mathrm{p}} & t\mathcal{Y}_{\mathrm{sc}}/\sqrt{3} & 0 & -t\mathcal{Y}_{\mathrm{sc}}/\sqrt{3} \\ \\ -t\mathcal{Y}_{\mathrm{sc}}/\sqrt{3} & 0 & t\mathcal{Y}_{\mathrm{sc}}/\sqrt{3} & 2t^2\mathcal{Y}_{\mathrm{sc}}/3 & -t^2\mathcal{Y}_{\mathrm{sc}}/3 & -t^2\mathcal{Y}_{\mathrm{sc}}/3 \\ t\mathcal{Y}_{\mathrm{sc}}/\sqrt{3} & -t\mathcal{Y}_{\mathrm{sc}}/\sqrt{3} & 0 & -t^2\mathcal{Y}_{\mathrm{sc}}/3 & 2t^2\mathcal{Y}_{\mathrm{sc}}/3 & -t^2\mathcal{Y}_{\mathrm{sc}}/3 \\ 0 & t\mathcal{Y}_{\mathrm{sc}}/\sqrt{3} & t\mathcal{Y}_{\mathrm{sc}}/\sqrt{3} & -t^2\mathcal{Y}_{\mathrm{sc}}/3 & -t^2\mathcal{Y}_{\mathrm{sc}}/3 & 2t^2\mathcal{Y}_{\mathrm{sc}}/3 \end{pmatrix} \begin{pmatrix} \mathcal{U}_A \\ \mathcal{U}_B \\ \mathcal{U}_C \\ \\ \mathcal{U}_a \\ \mathcal{U}_b \\ \mathcal{U}_c \end{pmatrix} \tag{7.60}$$

A generic, compact representation of Equation 7.60, corresponding to the star–delta connection, may be expressed as

$$\begin{pmatrix} \mathcal{I}_{ABC} \\ \mathcal{I}_{abc} \end{pmatrix} = \begin{pmatrix} \mathcal{Y}_{NN} & -\mathcal{Y}_{Nn} \\ -\mathcal{Y}_{nN} & \mathcal{Y}_{nn} \end{pmatrix} \begin{pmatrix} \mathcal{U}_{ABC} \\ \mathcal{U}_{abc} \end{pmatrix} \tag{7.61}$$

7.4.2 SEQUENCE GRIDS OF TRANSFORMERS

Transformer parameters are also amenable to representation in the frame-of-reference of the sequences. The matrix of symmetrical components and its inverse are used to generate such effect. This requires that the order of all matrices involved in the exercise to be a multiple of three. This characteristic is met by matrices representing the star–star connected transformer where both star points are solidly grounded, and the star–delta transformer with the star point is solidly grounded. It should be noted that the symmetrical components transformed cannot directly be applied to cases of star-connected windings, where one or two star points are either not grounded at all or grounded through earthing impedances. In such cases, Kron's reductions are applied first to find out reduced equivalent representations, which is a function of phase terminals only.

The sequence domain representation of a transformer, in compact form, is given by

$$\begin{pmatrix} \mathcal{I}_{012H} \\ \mathcal{I}_{012L} \end{pmatrix} = \begin{pmatrix} \mathcal{Y}_{HH} & -\mathcal{Y}_{HL} \\ -\mathcal{Y}_{LH} & \mathcal{Y}_{LL} \end{pmatrix} \begin{pmatrix} \mathcal{U}_{012H} \\ \mathcal{U}_{012L} \end{pmatrix} \tag{7.62}$$

where the subscripts 0, 1, and 2 refer to homopolar, positive, and negative sequence quantities. Also, subscripts H and L refer to high- and low-voltage sides of the transformer.

In Equation 7.61, the order of matrices $\mathcal{Y}_{NN}$, $\mathcal{Y}_{Nn}$, $\mathcal{Y}_{nN}$, and $\mathcal{Y}_{nn}$ is 3×3; and suitable for direct treatment by the matrix of symmetrical components to enable representation in the frame-of-reference of the sequences. This is achieved by applying the following symmetrical

TABLE 7.1

Transformer Sequence Domain Admittances

Matrix	Star–Star	Delta–Delta	Star–Delta
$\mathcal{Y}_{HH}$	$\begin{pmatrix} \mathcal{Y}_{sc} & 0 & 0 \\ 0 & \mathcal{Y}_{sc} & 0 \\ 0 & 0 & \mathcal{Y}_{sc} \end{pmatrix}$	$\begin{pmatrix} 0 & 0 & 0 \\ 0 & \mathcal{Y}_{sc} & 0 \\ 0 & 0 & \mathcal{Y}_{sc} \end{pmatrix}$	$\begin{pmatrix} \mathcal{Y}_{sc} & 0 & 0 \\ 0 & \mathcal{Y}_{sc} & 0 \\ 0 & 0 & \mathcal{Y}_{sc} \end{pmatrix}$
$\mathcal{Y}_{HL}$	$\begin{pmatrix} t\mathcal{Y}_{sc} & 0 & 0 \\ 0 & t\mathcal{Y}_{sc} & 0 \\ 0 & 0 & t\mathcal{Y}_{sc} \end{pmatrix}$	$\begin{pmatrix} 0 & 0 & 0 \\ 0 & t\mathcal{Y}_{sc} & 0 \\ 0 & 0 & t\mathcal{Y}_{sc} \end{pmatrix}$	$\begin{pmatrix} 0 & 0 & 0 \\ 0 & t\mathcal{Y}_{sc}\angle{-\pi/6} & 0 \\ 0 & 0 & t\mathcal{Y}_{sc}\angle{\pi/6} \end{pmatrix}$
$\mathcal{Y}_{LH}$	$\begin{pmatrix} t\mathcal{Y}_{sc} & 0 & 0 \\ 0 & t\mathcal{Y}_{sc} & 0 \\ 0 & 0 & t\mathcal{Y}_{sc} \end{pmatrix}$	$\begin{pmatrix} 0 & 0 & 0 \\ 0 & t\mathcal{Y}_{sc} & 0 \\ 0 & 0 & t\mathcal{Y}_{sc} \end{pmatrix}$	$\begin{pmatrix} 0 & 0 & 0 \\ 0 & t\mathcal{Y}_{sc}\angle{\pi/6} & 0 \\ 0 & 0 & t\mathcal{Y}_{sc}\angle{-\pi/6} \end{pmatrix}$
$\mathcal{Y}_{LL}$	$\begin{pmatrix} t^2\mathcal{Y}_{sc} & 0 & 0 \\ 0 & t^2\mathcal{Y}_{sc} & 0 \\ 0 & 0 & t^2\mathcal{Y}_{sc} \end{pmatrix}$	$\begin{pmatrix} 0 & 0 & 0 \\ 0 & t^2\mathcal{Y}_{sc} & 0 \\ 0 & 0 & t^2\mathcal{Y}_{sc} \end{pmatrix}$	$\begin{pmatrix} 0 & 0 & 0 \\ 0 & t^2\mathcal{Y}_{sc} & 0 \\ 0 & 0 & t^2\mathcal{Y}_{sc} \end{pmatrix}$

components operations (see Equation 7.11):

$$\mathcal{Y}_{HH} = A^{-1}\mathcal{Y}_{NN}A$$

$$\mathcal{Y}_{HL} = A^{-1}\mathcal{Y}_{Nn}A$$

$$\mathcal{Y}_{LH} = A^{-1}\mathcal{Y}_{nN}A \tag{7.63}$$

$$\mathcal{Y}_{LL} = A^{-1}\mathcal{Y}_{nn}A$$

Table 7.1 shows matrices $\mathcal{Y}_{HH}$, $\mathcal{Y}_{HL}$, $\mathcal{Y}_{LH}$, and $\mathcal{Y}_{LL}$ in explicit form, for the star–star, delta–delta, and star–delta transformer connections. A set of these matrices for different transformer connections can be found in Reference 2.

Careful examination of the sequence domain parameters indicates that three independent transfer admittance matrix equations, leading to three independent circuits, are generated for a three-phase transformer.

The star–star and star–delta connections share the same positive and negative sequence equivalent circuits given in Figure 7.13.

On the other hand, the homopolar sequence equivalent circuits for the three connections differ from one another. The equivalent circuits are shown in Figure 7.14a–c for the star–star, delta–delta, and star–delta connections, respectively.

It should be noted that for the star–delta transformer connection the primary and secondary terminals of the homopolar sequence equivalent circuit are not electrically connected. However, the primary terminal contains an admittance $\mathcal{Y}_{sc}$ connected between this terminal and reference. It is also interesting to note that the positive and negative transfer admittances contain an asymmetrical phase shift of 30° between the primary and secondary terminals, giving rise to nonreciprocal equivalent

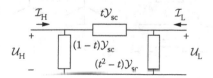

FIGURE 7.13 Positive and negative sequence equivalent circuits for the star–star and star–delta connections.

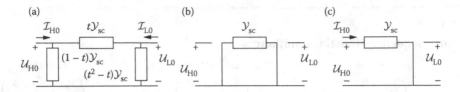

FIGURE 7.14 Positive and negative sequence equivalent circuits for the (a) star–star, (b) delta–delta, and (c) star–delta connections.

circuits. It is common practice in application studies, such as positive sequence power flow and sequence domain-based fault levels, to ignore the phase shift during the calculations and then to account for it during the analysis of results.

Cases of star–star and star–delta connections where star points are grounded through earthing impedances can also be expressed in symmetrical components form. This requires a Kron's reduction to eliminate mathematically the star points. In the case of a star–star connection where admittances $\mathcal{Y}_N$ and $\mathcal{Y}_n$ are used to ground star points N and n, the admittance $\mathcal{Y}_{sc}$ in the homopolar sequence equivalent circuit in Figure 7.14 is replaced by the compound admittance $\mathcal{Y}_0$:

$$\mathcal{Y}_0 = \frac{3\mathcal{Y}_{sc}\mathcal{Y}_N + 3\mathcal{Y}_{sc}\mathcal{Y}_n + \mathcal{Y}_N\mathcal{Y}_n}{\mathcal{Y}_{sc}\mathcal{Y}_N\mathcal{Y}_n}$$

which when expressed in the form of impedances yields a simpler expression

$$\mathcal{Z}_0 = \mathcal{Z}_{sc} + 3\mathcal{Z}_N + 3\mathcal{Z}_n$$

where $\mathcal{Z}_{sc} = 1/\mathcal{Y}_{sc}$, $\mathcal{Z}_N = 1/\mathcal{Y}_N$, and $\mathcal{Z}_n = 1/\mathcal{Y}_n$, respectively. If only one star point is grounded, say N through $\mathcal{Z}_N$, then the value of $\mathcal{Y}_0$ is

$$\mathcal{Y}_0 = \frac{3\mathcal{Y}_{sc} + \mathcal{Y}_N}{\mathcal{Y}_{sc}\mathcal{Y}_N}$$

and the value of the impedance is

$$\mathcal{Z}_0 = \mathcal{Z}_{sc} + 3\mathcal{Z}_N$$

Obviously, when the neutral of either the high or the low voltage of the transformer is ungrounded, the homopolar impedance is infinite, that is, the impedances of Figure 7.14 are an open circuit. A summary of the homopolar sequence grid for the different transformer connections is given in Figure 7.15, with solid grounding of the neutral $\mathcal{Z}_0 = \mathcal{Z}_{sc}$.

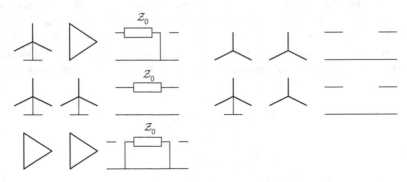

FIGURE 7.15 Summary of homopolar sequence grids for different transformer connections.

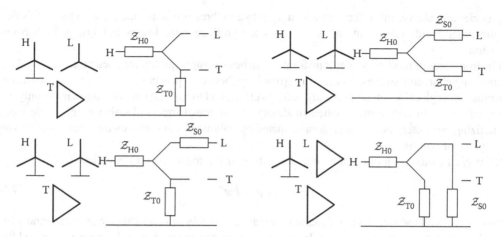

FIGURE 7.16 Summary of homopolar sequence grids for different three windings transformer connections.

It should be noted that earthing impedance schemes do not affect the transformer's positive and negative sequence equivalent circuits, hence, the equivalent circuit in Figure 7.13 remains unaltered.

Figure 7.16 shows the circuit connections of three-winding transformers. Letters H, L, and T are for the high voltage, low voltage, and tertiary voltage, respectively. A detailed analysis of the transformer sequence impedances may be found in References 2,11.

7.4.3 TRANSFORMER NONLINEARITIES

The transformer model of the previous sections comprises the copper windings, but the magnetic core was assumed to be linear. In practice, however, all ferromagnetic cores introduce three distinct effects, which are said to be nonlinear in nature, namely saturation, hysteresis, and eddy currents. The open circuit *B–H* characteristic of a ferromagnetic core obtained with a 50 or 60 Hz excitation will normally show the three effects. This is a looped, nonlinear curve whose width is dependent on the rate at which the contour is traversed. The loop is frequency-dependent and includes the effect of hysteresis and eddy currents. An example of the hysteresis loop is given in Figure 7.17.

The hysteresis phenomenon consists of two parts, one that is frequency-independent and is attributable to domain wall movements, nonmagnetic inclusions, and imperfections; and the other that is frequency-dependent and is referred to as the anomalous losses. Losses associated with hysteresis are expected to be small in modern power transformers. Eddy currents, on the other hand, are not present when very slow (essentially zero frequency) traverses of the loop occur but any increase beyond zero frequency will give rise to eddy currents, which increase the loss per cycle. At power

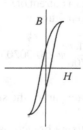

FIGURE 7.17 Hysteresis loop in a nonlinear inductance.

frequencies and above, this effect can account for two to three times as much loss as hysteresis does. On the other hand, saturation in itself does not introduce core losses but brings in harmonic distortion.

The iron cores of modern power transformers are built of high-quality magnetic materials and their magnetizing characteristics exhibit only a narrow loop, hence, in power systems harmonic studies it is customary to neglect hysteresis and eddy currents effect, and to assume that saturation is the only non-linear effect worth representing when modeling power transformers. Furthermore, single-valued magnetizing characteristics are well approximated by polynomial and exponential equations of varying degrees of complexity.

By way of example, a fitted polynomial equation of the form given in equation

$$i = a\psi + b\psi^n \tag{7.64}$$

approximates well the single-valued (saturation) characteristics of practical power transformers for a class of harmonic distortion problems. In this expression, the parameters a, b, and n are derived from specific points of the experimental saturation characteristic under treatment. The information required are (i) knee coordinates, (ii) coordinates of a point above the knee, and (iii) slope of linear part of the curve.

EXAMPLE 7.2 POLYNOMIAL FITTING OF A NONLINEAR INDUCTOR

The saturation characteristic for one leg of a three-phase 25 MVA, 110/44/4 kV, Y/Y/D transformer is fitted to a polynomial expression. Hysteresis is not considered. Data of the saturation are

Knee point:	$\psi_{nom} = 1$ pu	$i_{nom} = 0.008$ pu
Higher point:	$\psi_{max} = 1.2$ pu	$i_{nom} = 0.008$ pu
Slope:	$M = 1700$ pu	

Substituting these parameters in Equation 7.64 yields

$$0.026 = \frac{1.2}{1700} + b \cdot 1.2^n \Rightarrow b = \frac{0.025294}{1.2^n}$$

So that for odd values of n (since an odd function is required).

for $n = 3$	$b = 0.014638$
for $n = 5$	$b = 0.010165$
for $n = 7$	$b = 0.007057$

The experimental characteristic requires a current of $i = 0.008$ pu when $\psi = 1$ pu, and therefore, $n = 7$ is the best approximation. The chosen fitting is

$$i = \frac{1}{1700}\psi + 0.007057\psi^7$$

The nonlinearities of the magnetic core are part of the shunt inductance, and therefore they affect the term $\mathcal{Y}_p$ of Equation 7.59. When the application study involves transformer magnetizing harmonics, a suitable representation of the magnetizing branch, which is voltage-dependent, is established and connected to the terminal nodes of the transformer [6,7].

7.5 SYNCHRONOUS MACHINES

In general, synchronous machines are grouped into two main types, according to their rotor structure—cylindrical rotor and salient-pole machines. Steam turbine-driven generators (turbo-generators) work at high speed and have cylindrical rotors. Hydro units work at low speed and have salient-pole rotors, a feature known by power system engineer as saliency. The rotor carries a DC-excited field winding.

For simulation purposes, a three-phase synchronous machine may be assumed to have three stator windings and three rotor windings. This is illustrated in Figure 7.18, where all six windings are magnetically coupled. The relative position of the rotor with respect to the stator is given by the angle, θ, between the rotor's direct axis and the stator's phase a-axis, termed d-axis and a-axis, respectively. In the stator, the axes of phases a, b, and c are displaced from each other by 120 electrical degrees. In the rotor, the d-axis is magnetically centered in the north pole of the machine. A second axis, located 90 electrical degrees behind the d-axis is termed quadrature axis or q-axis.

Three main control systems directly affect the turbine-generator set, namely the boiler's firing control, the governor control, and the excitation system control. The excitation system consists of the exciter and the automatic voltage regulator (AVR). The latter regulates the generator terminal voltage by controlling the amount of current supplied to the field winding by the exciter. For the purpose of steady-state analysis, it is assumed that the three control systems act in an idealized manner, enabling the synchronous generator to produce constant power output, to run at synchronous speed, and to regulate voltage magnitude at the generator's terminal with no delay and up to its reactive power design limits.

The solution of some power system studies requires a model of the synchronous generator where the stator three-phase voltages and the currents are available to enable the machine model to be connected to a given three-phase bus of an unbalanced power system representation.

A simple three-phase model may be established by referring to the terminal voltage of the generator in sequence quantities:

$$\mathcal{E}_{012} = \mathcal{U}_{012} + \mathcal{Z}_{012}\mathcal{I}_{012} \tag{7.65}$$

$$\mathcal{Z}_{012} = \begin{pmatrix} \mathcal{Z}_0 & 0 & 0 \\ 0 & \mathcal{Z}_1 & 0 \\ 0 & 0 & \mathcal{Z}_2 \end{pmatrix} \tag{7.66}$$

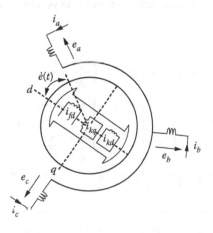

FIGURE 7.18 Schematic representation of a three-phase synchronous machine.

where $\mathcal{Z}_{012}$ represents the machine's homopolar, positive, and negative sequence impedances ($\mathcal{Z}_0$, $\mathcal{Z}_1$, $\mathcal{Z}_2$), respectively; $\mathcal{E}_{012}$ represents the homopolar, positive, and negative sequence internal excitation voltages, respectively; $\mathcal{U}_{012}$ represents the homopolar, positive, and negative sequence terminal voltages, respectively; and $\mathcal{I}_{012}$ represents the homopolar, positive, and negative sequence currents, respectively. In any practical synchronous machine $\mathcal{E}_0 = 0$ and the sequence impedances are obtained from measurements.

Use of the symmetrical components transform converts Equation 7.65 into an expression suitable for connecting the machine model to a given three-phase bus of an unbalanced power system representation:

$$\mathcal{E}_{abc} = \mathcal{U}_{abc} + \mathcal{Z}_{abc}\mathcal{I}_{abc} \tag{7.67}$$

$$\mathcal{Z}_{abc} = A\mathcal{Z}_{012}A^{-1} = \frac{1}{3}\begin{pmatrix} \mathcal{Z}_0 + \mathcal{Z}_1 + \mathcal{Z}_2 & \mathcal{Z}_0 + a\mathcal{Z}_1 + a^2\mathcal{Z}_2 & \mathcal{Z}_0 + a^2\mathcal{Z}_1 + a\mathcal{Z}_2 \\ \mathcal{Z}_0 + a^2\mathcal{Z}_1 + a\mathcal{Z}_2 & \mathcal{Z}_0 + \mathcal{Z}_1 + \mathcal{Z}_2 & \mathcal{Z}_0 + a\mathcal{Z}_1 + a^2\mathcal{Z}_2 \\ \mathcal{Z}_0 + a\mathcal{Z}_1 + a^2\mathcal{Z}_2 & \mathcal{Z}_0 + a^2\mathcal{Z}_1 + a\mathcal{Z}_2 & \mathcal{Z}_0 + \mathcal{Z}_1 + \mathcal{Z}_2 \end{pmatrix} \tag{7.68}$$

where $\mathcal{Z}_{abc}$ represents the machine's three-phase impedances; $\mathcal{E}_{abc}$ represents the three-phase internal excitation voltages; $\mathcal{U}_{abc}$ represents the three-phase terminal voltages; and $\mathcal{I}_{abc}$ represents the three-phase currents.

7.6   POWER SYSTEM LOADS AND FILTERS

In general, power system loads can be classified into rotating and static loads. A third category corresponds to power electronics-based loads. Rotating loads consist mainly of induction and synchronous motors, and their steady-state operation is affected by voltage and frequency variations in the supply. Power electronics-based loads are also affected by voltage and frequency variations in the supply. There is general agreement that such loads are more difficult to operate because in addition to being susceptible to supply variations, they inject harmonic current distortion back into the supply point.

Detailed representation of a synchronous motor load in a three-phase power flow study requires use of Equation 7.67 (given in the previous section), with changed signs to reflect the motoring action. An expression of comparable detail can be derived for the induction motor load. However, owing to the large number and diversity of loads that exist in power networks, it is often preferable to group loads and to treat them as bulk load points. Only the very important loads are singled out for detailed representation. It is interesting to note that a group of rotating loads operating at constant torque may be adequately represented as a static load that exhibits the characteristic of a constant current sink.

In steady-state applications, most system loads are adequately represented by a three-phase power sink, which may be connected either in star or in delta, depending on requirements. Figure 7.19a

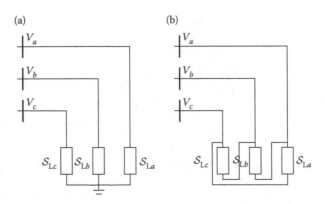

FIGURE 7.19   System load representation.

shows the schematic representation of a star-connected load with the star point solidly grounded, whereas Figure 7.19b shows that corresponding to a delta-connected load.

In three-phase power flow studies, it is normal to represent bulk power load points as complex powers per phase, on a per unit basis, as Equation 7.69

$$
\begin{aligned}
S_{La} &= P_{La} + jQ_{La} \\
S_{Lb} &= P_{Lb} + jQ_{Lb} \\
S_{Lc} &= P_{Lc} + jQ_{Lc}
\end{aligned}
\tag{7.69}
$$

Refinements can be applied to the above equations to make the power characteristic to be more responsive to voltage performance,

$$
S'_{La} = P_{La}\left(\frac{V_a}{V_{a0}}\right)^{\alpha} + jQ_{La}\left(\frac{V_a}{V_{a0}}\right)^{\beta}
$$

$$
S'_{Lb} = P_{Lb}\left(\frac{V_b}{V_{b0}}\right)^{\alpha} + jQ_{Lb}\left(\frac{V_b}{V_{b0}}\right)^{\beta}
\tag{7.70}
$$

$$
S'_{Lc} = P_{Lc}\left(\frac{V_c}{V_{c0}}\right)^{\alpha} + jQ_{Lc}\left(\frac{V_c}{V_{c0}}\right)^{\beta}
$$

In Equation 7.70, α and β take values in the range 0–2, and V_a, V_b, V_c are the per unit three-phase nodal voltage magnitudes at the load point. Note that when $\alpha = \beta = 0$, the complex power expressions in Equation 7.70 coincide with those in Equation 7.69. However, if $\alpha = \beta = 1$, expressions (7.70) resemble complex current characteristics more than complex power characteristics. Also, if $\alpha = \beta = 2$, the complex powers in Equation 7.69 would behave like complex admittances.

The admittance-like characteristic in Equation 7.69 may be expressed in matrix form for both kinds of load connections, star and delta, respectively, as follows:

$$
\begin{pmatrix}
S_{La}/V_a^2 & 0 & 0 \\
0 & S_{Lb}/V_b^2 & 0 \\
0 & 0 & S_{Lc}/V_c^2
\end{pmatrix}
\tag{7.71}
$$

$$
\frac{1}{3}\begin{pmatrix}
S_{La}/V_a^2 + S_{Lb}/V_b^2 & -S_{Lb}/V_b^2 & -S_{La}/V_a^2 \\
-S_{Lb}/V_b^2 & S_{Lb}/V_b^2 + S_{Lc}/V_c^2 & -S_{Lc}/V_c^2 \\
-S_{La}/V_a^2 & -S_{Lc}/V_c^2 & S_{Lc}/V_c^2 + S_{La}/V_a^2
\end{pmatrix}
\tag{7.72}
$$

Moreover, if it is assumed that the load powers and voltage magnitudes are taken to be balanced, $S_{La} = S_{Lb} = S_{Lc} = S_L$ and $V_a = V_b = V_c = V$, then application of the following symmetrical components operation, $\mathcal{Y}_{012} = A^{-1}\mathcal{Y}_{abc}A$, leads to the load model representation for homopolar, positive, and negative (0, 1, 2) sequences of equations:

$$
\begin{pmatrix}
S_L/V^2 & 0 & 0 \\
0 & S_L/V^2 & 0 \\
0 & 0 & S_L/V^2
\end{pmatrix}
\tag{7.73}
$$

$$
\begin{pmatrix}
0 & 0 & 0 \\
0 & S_L/V^2 & 0 \\
0 & 0 & S_L/V^2
\end{pmatrix}
\tag{7.74}
$$

Note that no homopolar sequence loads exist for the case of a three-phase delta-connected load, but only positive and negative sequences exist.

As an extension of the above result, the positive, negative, and homopolar sequence expression of a star-connected load with its star point solidly grounded may be expressed as

$$S'_{L(1)} = S'_{L(2)} = S'_{L(0)} = P_L\left(\frac{1}{V}\right)^{\alpha} + jQ_L\left(\frac{1}{V}\right)^{\beta} \tag{7.75}$$

whereas for the case of a delta-connected load, we have

$$S'_{L(1)} = S'_{L(2)} = P_L\left(\frac{1}{V}\right)^{\alpha} + jQ_L\left(\frac{1}{V}\right)^{\beta} \tag{7.76}$$

and $S'_{L(0)} = 0$.

It should be remarked that the exponents α and β are not confined to take only integer values and that a wide range of load characteristics can be achieved by judicious selection of α and β, depending on the group of loads present in the study.

Also, a three-phase delta-connected load can always be transformed into an equivalent star circuit by using a delta–star transformation. However, notice that the transformation will generate an extra bus in the form of the star point, which yields no physical meaning.

For studies at frequencies different from the fundamental ω_1 (50 or 60 Hz), the dependance of this model with frequency has to be considered. Owing to the great number of elements comprised in the aggregated load, there is no generally acceptable load model. However, some approximations have been proposed in literature.

When load is formed mostly by motors or electronic devices, the detailed modeling of Sections 7.5 and 7.7 may be necessary. When the loads are passive, they are modeled as admittances, and the configuration of these admittance is important.

From Equation 7.73 or 7.74, the equivalent admittance of the load $\mathcal{Y}_L$ is defined as follows:

$$\mathcal{Y}_L = \frac{P_L}{V^2} + j\frac{Q_L}{V^2} = \frac{1}{R_{Lp}} - j\frac{1}{X_{Lp}} \tag{7.77}$$

The equivalent circuit of this model is a parallel connected resistance and reactance, whose value is given by the active and reactive power consumption of the load. The admittance of this association for frequencies different from the fundamental is given by

$$\mathcal{Y}_L(\omega) = \frac{1}{R_{Lp}} - j\frac{1}{\dfrac{\omega}{\omega_1}X_{Lp}} \tag{7.78}$$

This means that the admittance of the load is more resistive at higher frequencies, and that at these frequencies, the admittance module is smaller.

The parallel model is used when aggregated load is considered. A single passive load behaves usually in a different way, and it is represented by a series admittance. The value of $\mathcal{Y}_L$ may be written as (for fundamental frequency ω_1)

$$\mathcal{Z}_L = \frac{1}{\mathcal{Y}_L} = \frac{R_p X_p^2}{R_p^2 + X_p^2} + j\frac{R_p^2 X_p}{R_p^2 + X_p^2} = R_s + jX_s \tag{7.79}$$

The dependance of R_s and X_s with frequency has been widely discussed in the literature. In Reference 6, the following expression is proposed:

$$\mathcal{Z}_L(\omega) = R_s\sqrt{\frac{\omega}{\omega_1}} + jX_s\frac{\omega}{\omega_1} \tag{7.80}$$

But in Reference 15 a more complex model is proposed. The series impedance of Equation 7.80 increases with frequency, in a different way than the parallel model. This difference may have important effects.

Sometimes, a combination of series and parallel models is used, trying to fit better than the actual behavior of loads.

7.6.1 Harmonic Filters

Passive shunt filters is a very popular resort used to control the propagation of harmonics currents. They are normally designed as a series combination of reactors and capacitors. Passive filters are also referred to as sinks because they absorb harmonics currents. A passive filter presents a low impedance path to the harmonic to which it is tuned. Power harmonic filters are installed at the AC terminals of rectifiers, motor drives, uninterruptible power supplies, and other nonlinear loads, to reduce voltage and current distortion to acceptable limits at the points of connection. Filter design is normally carried out assuming static operating conditions. Although, in practice, filters may be operated dynamically, switching them on and off by sections, according to system operations. Examples of applications where switched filters are used are elevators drives, adjustable speed pump drives, and reactive power compensators. More information about filters can found in References 9,12.

7.6.1.1 Active Filters

Active filters are the new trend in harmonics filtering technology. They make use of state-of-art power electronic switches and advanced control techniques. The basic principle of operation of an active filter is to inject a suitable nonsinusoidal voltage and current into the system to achieve a clean voltage and current waveform at the point of filtering.

The most promising active filter technology uses the pulse width modulation (PWM) voltage-source inverter as its core component. In theory, the main difference between the PWM STATCOM and the active filter lies on the control blocks used by both applications. The STATCOM's primary concern is to provide adaptive filtering action. Nevertheless, the great progress made in the development of advanced digital signal processors and sophisticated algorithms is enabling both applications to convergence. At the research level at least, prototypes are being developed where a PWM STATCOM has the ability to provide reactive power support, adaptive voltage regulation, and harmonic filtering.

7.6.1.2 Single-Tuned First-Order Filter

These filters are mostly used at low harmonic frequencies. They are probably the most common shunt filters in use today and comprise a series RLC circuit, as shown in Figure 7.20, and are tuned to a particular harmonic frequency.

Some characteristics of first-order filters are as follows: (i) They act as a very low impedance path at the frequency at which they are tuned; (ii) When the source impedance is inductive, there is a resonant peak, which always occurs at a frequency lower than the frequency at which the filter

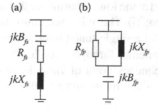

FIGURE 7.20 Common passive filter configuration: (a) single-tuned first-order filter and (b) high-pass second-order filter.

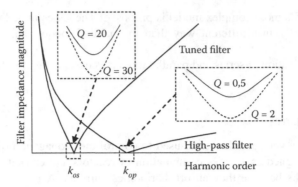

FIGURE 7.21 Typical frequency response of filters.

is tuned; (iii) There is a sharp increase in the impedance below the tuned frequency due to the proximity of the resonant frequency and (iv) The impedance rises with frequency for frequencies above the tuning frequency. The expression for the impedance is given by

$$\mathcal{Z}_{fs,k} = R_{fs} + j\left(kX_{fs} - \frac{1}{kB_{fs}} \right) \tag{7.81}$$

The tuning frequency $\omega_{os} = k_{os}\omega_1$ of the first-order single-tuned filter is given by

$$k_{os} = \frac{1}{\sqrt{X_{fs}B_{fs}}} \tag{7.82}$$

The tuning characteristic of the filter is described by its quality factor Q given by

$$Q = \frac{kX_{fs}}{R} \tag{7.83}$$

The parameters of these formulae are given in Figure 7.20. Typically, the value of R consists of only the resistance of the inductor, resulting in a large value of Q and a very sharp filtering characteristic. The value of the resistance may be obtained by selecting an appropriate value of quality factor $20 < Q < 30$. Figure 7.21 shows a typical response for this kind of filter for two quality factors.

7.6.1.3 High-Pass Second-Order Filters

In cases with high order harmonic currents, like 11th, 13th and higher, high-pass filters are normally used. High-pass filters receive their name from the low impedance characteristic that they observe above a corner frequency. Frequently, only one high-pass filter is used to eliminate a range of harmonics, whose corner frequency is located at the lowest harmonic that it is meant to eliminate. Two factors may discourage such an application: (i) The minimum impedance of the high-pass filter never achieves a value comparable to that of a single-tuned filter at tuned frequency and (ii) The shunting of a percentage of all the system harmonics through a single-filter may require the filter to be vastly over-rated from the fundamental frequency point of view. In contrast to the single-tuned filter, the quality factor of a high-pass filter is given by its inverse relation.

$$Q = \frac{R}{kX_{fp}} \tag{7.84}$$

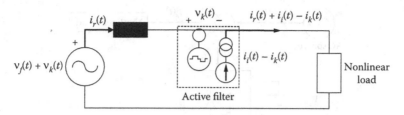

FIGURE 7.22 Basic concept of active filters. Voltages and currents are shown as time-dependent functions.

The impedance of the high-pass filter is given by

$$\mathcal{Z}_{fp,k} = \frac{jkX_{fp}R_{fp}}{R_{fp} + jkX_{fp}} - j\frac{1}{kB_{fp}} \tag{7.85}$$

and the frequency at which this filter has minimum impedance is $\omega_{op} = k_{op}\omega_1$, where

$$k_{op} = \frac{1}{\sqrt{B_{fp}X_{fp} - X_{fp}^2/R_{fp}^2}} \tag{7.86}$$

7.6.1.4 Basic Principles of Operation of an Active Filter

Figure 7.22 illustrates the case when a general nonsinusoidal voltage source composed by the fundamental and harmonics frequencies, that is, $\mathcal{V}_1 + \mathcal{V}_k$, supplies a nonlinear load. The load current may be represented as the vector sum of the active part of the current at fundamental frequency $\mathcal{I}_r$, the reactive part of the current at fundamental frequency $\mathcal{I}_i$, and the harmonic currents $\mathcal{I}_k$. An ideal active filter would remove all harmonics from the line current and from the voltage supply, that is, $\mathcal{I}_k$ and $\mathcal{V}_k$, respectively. Also, an ideal active filter would supply reactive current compensation $\mathcal{I}_i$, without affecting the flow of the active load current $\mathcal{I}_r$.

In principle, such a filter could be realized by harmonic currents and voltage generators and by an ideal reactive power compensator. Under such a condition, the nonlinear load is supplied from a perfect sinusoidal voltage V_t and draws a perfect sinusoidal current I_r, from the voltage source.

7.6.1.5 Shunt Active Filter

Provided a comprehensive control function is included in the active filter, the device may be used to replace the function of a conventional voltage and Var compensator, but with the added advantage that it can perform harmonic current compensation. Shunt active filters are becoming quite popular owing to their light construction and above all their expandable design. They are controlled with ease and are mostly used in low and medium power applications.

Figure 7.23 shows the schematic representation of a shunt active filter and the external network, where the filter is connected between an equivalent voltage source and a nonlinear load. At this high level of representation, it is not possible to distinguish between an active filter and a PWM STATCOM—the difference between the two applications may lie in the control scheme.

7.7 POWER ELECTRONIC CONVERTERS AND CONTROLLERS

This section includes the fundamentals of power electronic converters. A more complete insight into these devices may be found in Chapter 11, where its nonlinear behavior is examined thoroughly.

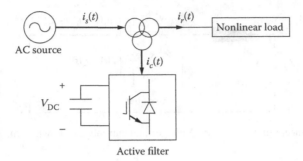

FIGURE 7.23 Shunt active filters. Voltages and currents are shown as time-dependent functions.

7.7.1 CONVENTIONAL CONVERTERS

The six-pulse rectifier is the most common three-phase converter used in the power range of 2–1000 kVA and in the voltage range of 220 V–13.8 kV. As the power level rises, it becomes increasingly attractive to use 12-pulse rectifiers, with one six-pulse unit connected to the secondary side of a star–star transformer. The second six-pulse unit is connected to the secondary side of a delta–star transformer. The two rectifiers are connected in series on the DC side to achieve the high voltages required in DC transmission. On the primary side, the two rectifiers are connected in parallel to achieve cancelation of mainly the fifth and the seventh harmonic currents. However, this is only possible if the two star connections in both secondary windings are shifted by 30° that naturally exists in the delta–star connection of the second unit.

The classic model of this converter is described in detail in Reference 8. This analysis is made under the following assumptions:

- Balanced and sinusoidal power supply.
- The transformer is modeled just as a reactance (the commutation reactance).
- The current in the DC side of the convertor is perfectly smooth.

Although these assumptions are rather restricted and are rarely found in real practice, the equations obtained with them are a good approach and useful for fundamental frequency studies. A more general model of convertors, especially suitable for harmonic analysis, may be found in Chapter 11.

Under the classic assumptions, the main magnitudes in the convertor are given in Table 7.2. In this table, E is the AC line voltage, α is the firing angle in the convertor, L_c is the inductance of the commutating reactance, I_d is the DC current, and $\delta = \alpha + u$, where u is the overlap angle,[*] and K_1 has the following expression:

$$K_1 = \frac{\sqrt{(\cos 2\alpha - \cos 2\delta)^2 + (2u + \sin 2\alpha - \sin 2\delta)^2}}{4(\cos \alpha - \cos \delta)}$$

The waveforms of these magnitudes are given in Figures 7.24 through 7.26.

When the rectifier is working under nonideal conditions such as unbalanced AC excitation, unbalanced firing angle control, unbalanced AC impedance, as well as preexisting harmonics distortion in the AC supply, a wide range of noncharacteristic harmonics will be generated on both the AC side and the DC side of the rectifier. These noncharacteristic harmonics cannot be calculated with the classic

---

[*] Time in radians that the AC phase current takes in becoming zero from their maximum value.

TABLE 7.2

Main Values in the Three-Phase Controlled Convertor

Magnitude	Expression
DC voltage	$U_d = \dfrac{3\sqrt{2}}{\pi}\cos\alpha - \dfrac{3}{\pi}\omega_1 L_c I_d$
AC phase current (fundamental)	$I_1 = K_1 \dfrac{\sqrt{6}}{\pi} I_d$
Power factor (approximate)	$\dfrac{\cos\alpha + \cos\delta}{2}$

equations, since the key assumptions are no longer fulfilled. These assumptions mean that at the end of the commutation period the current waveform will be perfectly flat-topped, as shown, for instance, in Figure 7.25. This is an assumption that quite often does not conform to realistic converter design and operation.

7.7.2 PWM CONVERTERS

There are several voltage-sourced converters (VSC) topologies currently in use in actual power systems operation [3]. Common aims of these topologies are (i) to minimize the switching losses of the semiconductors inside the VSC and (ii) to produce a high-quality sinusoidal voltage waveform with minimum or no filtering requirements. By way of example, the topology of a conventional two-level VSC using IGBT switches is illustrated in Figure 7.27.

The VSC shown above comprises six IGBTs, with two IGBTs placed on each leg. Moreover, each IGBT is provided with a diode connected in antiparallel to make provisions for possible voltage

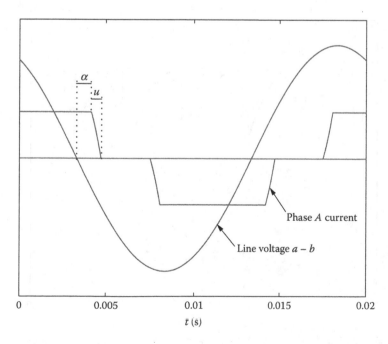

FIGURE 7.24 Voltage and current in a six-pulse converter with a Yy transformer. $\alpha = 15$ and $u = 10$.

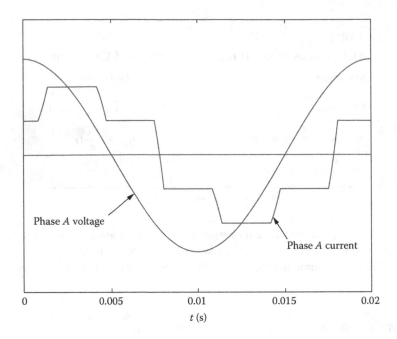

FIGURE 7.25 Voltage and current in a six-pulse converter with a Dy transformer.

reversals due to external circuit conditions. Two equally sized capacitors are placed on the DC side to provide a source of reactive power.

Although not shown in the VSC circuit, the switching control module is an integral component of the VSC. Its task is to control the switching sequence of the various semiconductor devices in the VSC, aiming at producing an output voltage waveform, which is as near to a sinusoidal waveform

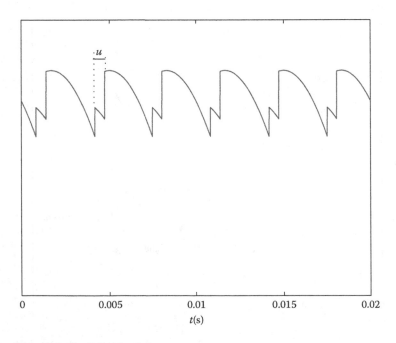

FIGURE 7.26 Voltage in the DC side of the converter.

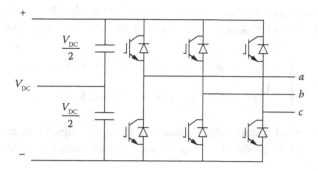

FIGURE 7.27   Voltage-sourced converter.

as possible, with high-power controllability and minimum switching loss. Current VSC switching strategies aimed at utility applications may be classified into two main categories:

1. *Fundamental frequency switching:* The switching of each semiconductor device is limited to one turn-on and one turn-off per power cycle.
2. *PWM:* This control technique enables the switches to be turned on and off at a rate considerably higher than the fundamental frequency.

The basic VSC topology, with fundamental frequency switching, yields a quasi-square-wave output, which has an unacceptable high harmonic content. It is normal to use several six-pulse VSCs, arranged to form a multipulse structure, to achieve better waveform quality and higher power ratings.

The output waveform is chopped with PWM control, and the width of the resulting pulses is modulated. Undesirable harmonics in the output waveform are shifted to the higher frequencies, and filtering requirements are much reduced.

From the viewpoint of utility applications, both the switching techniques are far from perfect. The fundamental frequency switching technique requires complex transformer arrangements to achieve an acceptable level of waveform distortion. Such a drawback is offset by its high semiconductor switch utilization and low switching losses; and it is, at present, the switching technique used in high-voltage, high-power applications.

The PWM technique incurs high switching loss, but it is envisaged that future semiconductor devices would reduce this by a significant margin, making PWM the universally preferred switching technique, even at high and extra high-voltage transmission applications.

7.7.2.1   PWM Control

This is a case of linear voltage control, where $m_a < 1$, but this is not the only possibility. Two other forms of voltage control exist, namely overmodulation and square-wave. The former takes place in the region $1 < m_a < 3.24$, and the latter applies when $m_a > 3.24$ [13].

To determine the magnitude and frequency of the resulting fundamental and harmonic terms, it is useful to use the concept of amplitude modulation ratio m_a, and frequency modulation ratio, m_f.

$$m_a = \frac{\hat{V}_{control}}{\hat{V}_{tri}}$$

$$m_f = \frac{f_s}{f_1}$$

With reference to one leg of the three-phase converter, the switches T_{a+} and T_{a-} are controlled by straightforward comparison of $\hat{V}_{control}$ and $\hat{V}_{tri}$, resulting in the following output voltages:

$$v_{Ao} = \begin{cases} \dfrac{1}{2}U_d & \text{when } T_{a+} \text{ is on in response to } \hat{V}_{control} > \hat{V}_{tri} \\ -\dfrac{1}{2}U_d & \text{when } T_{a-} \text{ is on in response to } \hat{V}_{control} < \hat{V}_{tri} \end{cases}$$

In the basic PWM method, a sinusoidal, fundamental frequency signal is compared against a high-frequency triangular signal, producing a square-wave signal, which controls the firing of the converter valves.

The sinusoidal and triangular signals, and their associated frequencies, are termed reference and carrier signals and frequencies. By varying the amplitude of the sinusoidal signal against the fixed amplitude of the carrier signal, which is normally kept at 1 pu, the amplitude of the fundamental component of the resulting control signal varies linearly.

The case of linear voltage control ($m_a < 1$) is of utmost interest. The peak amplitude of the fundamental frequency component is m_a times $U_d/2$; and the harmonics appear as sidebands, centered around the switching frequency and its multiples, following a well-defined pattern given by

$$f_k = (\beta m_f \pm \kappa)f_1$$

Harmonic terms exist only for odd values of β with even values of κ. Conversely, even values of β combine with odd values of κ. Moreover, the harmonic m_f should be an odd integer to prevent the appearance of even harmonic terms in v_{Ao}.

The fundamental frequency component is shown in the Figure 7.28 for the case of $m_f = 9$ and $m_a = 0.8$. The corresponding harmonic voltage spectrum in normalized form is shown in Figure 7.29.

7.7.2.2 Principles of Operation of the VSC

The interaction between the VSC and the power system may be explained in simple terms by considering a VSC connected to the AC mains through a loss-less reactor, as illustrated in the single-line diagram shown in Figure 7.30.

The premise is that the amplitude and the phase angle of the voltage drop, ΔV_x, across the reactor X_l, can be controlled; defining the amount and direction of active and reactive power flows through X_l. The voltage at the supply bus is taken to be sinusoidal, of value $V_s\angle 0$, and the fundamental frequency component of the VSC's AC voltage is taken to be $V_{vR}\angle\delta_{vR}$.

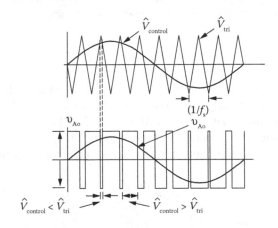

FIGURE 7.28 Operation of a PWM converter. f_s is nine times f_1.

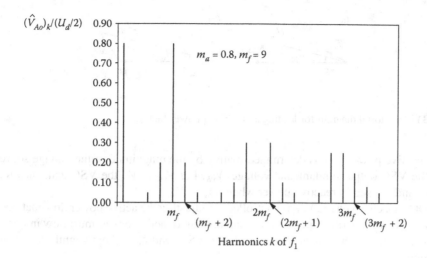

FIGURE 7.29   Operation of a PWM converter. f_s is nine times f_1.

The positive sequence, fundamental frequency vector representation is shown in Figure 7.31 for leading (a) and lagging (b) VAR compensation, respectively.

For leading and lagging VAR, the active and reactive powers can be expressed as

$$P = \frac{V_s V_{vR}}{X_l} \sin \delta_{vR}$$

$$Q = \frac{V_s^2}{X_l} - \frac{V_s V_{sR}}{X_l} \cos \delta_{vR}$$

With reference to the previous figures and equations, the following observations are derived:

- The VSC output voltage V_{vR} leads the AC voltage source V_s by an angle δ_{vR}, and the input current either leads or lags the voltage drop across the reactor ΔV_x by 90°.
- The active power flow between the AC source and the VSC is controlled by the phase angle δ_{vR}. Active power flows into the VSC from the AC source at lagging δ_{vR} ($\delta_{vR} > 0$) and vice versa for leading δ_{vR} ($\delta_{vR} < 0$).

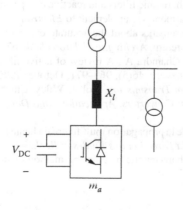

FIGURE 7.30   VSC connected to a single bus.

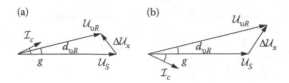

FIGURE 7.31   Vectorial diagram for leading and lagging power factor.

- The reactive power flow is determined mainly by the magnitude of the voltage source, V_s, and the VSC output fundamental voltage, V_{vR}. For $V_{vR} > V_s$, the VSC generates reactive power and consumes reactive power when $V_{vR} < V_s$.
- The DC capacitor voltage V_{DC} is controlled by adjusting the active power flow that goes into the VSC. During normal operation, a small amount of active power must flow into the VSC to compensate for the power losses inside the VSC, and δ_{vR} is kept slightly larger than $0°$ (lagging).
- For leading and lagging VAR, the active and reactive powers can be expressed as

$$P = \frac{V_s V_{vR}}{X_l} \sin \delta_{vR}$$

$$Q = \frac{V_s^2}{X_l} - \frac{V_s V_{sR}}{X_l} \cos \delta_{vR}$$

REFERENCES

1. Fortescue C. L., Method of symmetrical coordinates applied to solution of polyphase networks, *Transactions of the AIEE*, 37, 1027–1040, 1918.
2. Wakileh H., *Power Systems Harmonics*, Springer, Berlin, 2001.
3. Acha E., Agelidis V. G., Anaya-Lara O., and Miller T. J. E., *Power Electronic Control in Electrical Systems*, Elsevier, France, Paris, 2001.
4. Densem T. J., Three-phase power system harmonic penetration *PhD Thesis*, University of Canterbury, Christchurch, New Zealand, 1983.
5. Semlyen A., and Deri A., Time domain modelling of frequency dependent three-phase transmission line impedance, *IEEE Trans. on Power Apparatus and Systems*, PAS-104, 1549–1555, 1985.
6. Arrillaga J., and Watson N., *Power Systems Harmonics*, 2nd ed, Wiley, Chichester, 2003.
7. Acha E., and Madrigal M., *Power Systems Harmonics, Computer Modelling and Analysis*, Wiley, Chichester, 2001.
8. Anderson P., *Analysis of Faulted Power Systems*, Iowa State University Press, Ames, Iowa, 1991(1973).
9. CIGRE WG-14 03, 1989, AC harmonic filters and reactive compensation for HVDC with particular reference to non-characteristic harmonics, Supplement to *Electra* no. 63.
10. Tevan G., and Deri A., Some remarks about the accurate evaluation of the carson integral for mutual impedance of lines with earth return, *Archiv für Elektrotechnik*, 67, 83–90, 1984.
11. Singh R., Al-Haddad K., and Chamdra A., A review of active filters for power quality improvement, *IEEE Trans. on Industry Electronics*, 46(5), 960–971, October 1999.
12. Kimbark E. W., *Direct Current Transmission*. Vol. 1, Wiley, Chichester, 1971.
13. Mohan N., *Power Electronics: Converters, Applications and Design*, John Wiley & Sons, Chichester, 1995.
14. Gary C., Approche complète de la propagation multifilaire en haute fréquence par utilisation des matrices complexes, *EDF Bulletin de la Direction des Études et Recherches, Séries B*, 4(3), 5–20, 1976.
15. Pesonenn J. A., Harmonics, characteristic parameters, method of study, estimates of existing values in the network, *Electra*, 77, 35–56, 1981.

8 Fault Analysis and Protection Systems

José Cidrás, José F. Miñambres, and Fernando L. Alvarado

CONTENTS

8.1   PURPOSE AND ASSUMPTIONS OF FAULT ANALYSIS

In an electric network a short circuit occurs when two or more points, which are at a different voltage under normal operation conditions, accidentally come into contact with each other through a small or zero impedance. This generally happens when the insulation fails owing to different causes: loss of environmental insulation properties (overheating, contamination, etc.), overvoltages (external and internal origin), or diverse mechanical effects (breakages, deformations, displacements, etc.).

Short circuits should be studied both during the planning and the operation of an electric network. Short-circuit studies are particularly important during selection and design of the system conductors, system structures, and the relaying and protection system.

The damaging effects of the faults are numerous and are mainly related to the high currents which can appear in the system. Usually, these current values can be several times the value of the normal operational current and can occasionally produce overvoltage phenomena. The main effects of the faults can be summarized as follows:

- Conductor heating owing to the Joule effect, which depending on the value and duration of the short circuit can provoke irreversible damage.
- Electromechanical forces, which can create breakages and sudden conductor displacements possibly resulting in new faults.
- Voltage variations, which are voltage drops in the failed phases and eventual voltage increases in the others.

The analysis of a short-circuited electric network is made more or less difficult depending on the objective required, the complexity of the network being studied, and the degree of precision necessary. If the objective is the calculation of a LV and MV industrial installation, the maximum short circuit current must be obtained (normally a balanced short circuit) to determine the short-circuit breaking capacity of the circuit breakers, the thermal and dynamic limits of the conductors, and apparatus of the installation. On the other hand, if the objective is to adjust the protections of a HV-meshed electric network, it is necessary to calculate short-circuit currents at any point and for any type of short circuit (both balanced and unbalanced faults).

Short circuits are generally classified as balanced or unbalanced. In the first case, the three phases are short circuited at the same point and time, and therefore the resulting circuit is also balanced. For this reason, the study can be carried out using a single-phase equivalent circuit. In the second case, if the electric system resulting from the short circuit is unbalanced (not all phases are equally implicated), it is necessary to use three-phase models which, after using adequate mathematical techniques (symmetrical components), can be reduced to three coupled single-phase circuits.

8.1.1 RELATIONS BETWEEN STEADY-STATE AND TRANSIENT SHORT-CIRCUIT CONDITIONS

Usual techniques for modeling and studying electric networks can be classified into three different operational time frames: steady state, electro-mechanic transient, and electromagnetic transient. Fault analysis is an electromagnetic process, where the network under steady-state operation conditions suffers an abrupt topology modification. Thus, the network has to be modeled by resorting to the differential equations characterizing its components to accurately analyze any fault. Although detailed differential equation modeling of the fault process can be studied for certain cases, it is not practical while dealing with networks of hundreds of nodes and branches. For this reason, common practice for short-circuit calculation uses a quasi-steady-state model of the electric network as an adequate approximation.

To establish the equations for the transient behavior of a network in a fault situation, a simple case of an RL branch in series with a voltage source feeding a load will be studied, as shown in Figure 8.1, which is short circuited like in Figure 8.1. A more realistic analysis should consider the transient behavior of real voltage sources, where the magnetic inertia of rotating machines becomes crucial (see Appendix C).

The following differential equation represents the behavior of the previous circuit during a fault:

$$i(t) = Ri + L\frac{di}{dt} \tag{8.1}$$

where $R = R_g + R_L$ and $L = L_g + L_L$.

The solution to Equation 8.1 is as follows:

$$i(t) = Ke^{-(R/L)t} + i_{st}(t)$$

where $i_{st}(t)$ is the steady-state current of the faulted circuit, and K is a constant obtained from the currents at the initial instant, $K = i(0) - i_{st}(0)$. So, if the voltage source has the following expression $e(t) = \sqrt{2}E\sin(\omega t + \theta)$, the current $i(t)$ can be calculated as follows:

$$i(t) = (i_0 - \sqrt{2}I_{cc}\sin(\theta - \varphi))e^{-(R/L)t} + \sqrt{2}I_{cc}\sin(\omega t + \theta - \varphi) \tag{8.2}$$

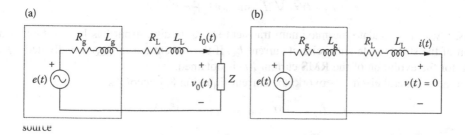

(a)   (b)

source

FIGURE 8.1 Circuits before the fault and during the fault: (a) initial circuit and (b) fault circuit.

where

$$I_{cc} = \frac{E}{\sqrt{R^2 + (L\omega)^2}}$$

$$\mathrm{tg}(\varphi) = \frac{L\omega}{R} = \frac{X}{R}$$

$$i(0) = \sqrt{2}I_0 \sin(\theta - \varphi_0)$$

$$I_0 = \frac{E}{\sqrt{(R + R_c)^2 + (L\omega + X_c)^2}}$$

$$\mathrm{tg}(\varphi_0) = \frac{L\omega + X_c}{R + R_c}$$

$$R_c + jX_c = \frac{1}{Y} = Z$$

The above Equation 8.2 contains two terms: a decreasing exponential and a sinusoidal component. The first term only exists during the so-called transient period, and disappears with time. It becomes almost equal to zero for $t > 5L/R$ s (beginning of the steady state). This unidirectional or asymmetrical component corresponds to the circuit's natural response (in mathematical terminology, it is the solution of the homogeneous equation). The second term, the sinusoidal component (also called the symmetric or periodic component) represents the steady state or permanent response of the circuit (in mathematical terminology, the particular solution of the differential equation).

Owing to the comparatively small value of the series elements R and L, when compared to the corresponding values of the load to which systems are typically connected, the rated current I_0 is often neglected relative to the much larger short-circuit current I_{cc}. That is, the total fault current $i(t)$ is not affected by the prefault initial current at $t = 0$ [$i(0)$]. Under this assumption, Equation 8.2 can be expressed as follows:

$$i(t) = -\sqrt{2}I_{cc} \sin(\theta - \varphi)e^{-(R/L)t} + \sqrt{2}I_{cc} \sin(\omega t + \theta - \varphi)$$

The sinusoidal behavior of $i(t)$ components results in a theoretical maximum value of $2\sqrt{2}I_{cc}$. This value occurs when $R = 0$ and $\theta - \varphi = -90$. Assuming the dominance of inductance over resistance in the electric networks (i.e., $L\omega > R$), the maximum values of $i(t)$ occur when $\theta = 0$ and $\omega t = \pi, 3\pi, 5\pi$.... Consequently, if the maximum instantaneous current values are obtained when $\theta = 0$, the evolution of the peak current values in subsequent oscillatory cycles, [$i_{max}(t)$], can be approximately expressed as follows:

$$i_{max}(t) = i(0)e^{-(R/L)t} + \sqrt{2}I_{cc} \sin(\varphi)(1 + e^{-(R/L)t}) \tag{8.3}$$

and, if the initial current $i(0)$ is neglected

$$i_{max}(t) = \sqrt{2}I_{cc} \sin(\varphi)(1 + e^{-(R/X)\omega t}) \tag{8.4}$$

Thus, a way to calculate the maximum transient short-circuit current has been established as a function of the steady-state short-circuit current I_{cc}. If the current curve $i_{max}(t)$ is divided by $\sqrt{2}$, a formula for the evolution of the RMS current $I(t)$ is obtained.

For a more general electric network, the current Equation 8.4 becomes

$$i_{max}(t) = \sqrt{2}I_{cc} \sin(\varphi_{eq})(1 + e^{-(R_{eq}/X_{eq})\omega t})$$

where R_{eq}, X_{eq}, and φ_{eq} are the equivalent resistance, reactance, and angle, respectively. These values are calculated at the faulted node and are obtained using the nodal impedance matrix (Thevenin impedance at the faulted node). International standards recommend the calculation of the equivalent impedance not taking into account shunt elements (such as loads, line capacitances, and regulating transformers).

The first maximum of the current $i_{max}(t)$, Equation 8.3, occurs approximately when $\omega t = \pi$, $t = 10$ ms (1/2 cycle). This maximum current value is called the peak current. The largest possible value for this peak current for the circuit shown in Figure 8.1b is as follows:

$$I_{peak} = \sqrt{2}I_{cc} \sin(\varphi)(1 + e^{-(R/X)\pi})$$

In the same way, the minimum short-circuit current value occurs when $\theta - \varphi = 0$ and $\omega t = \pi$, such that

$$I_{min} = \sqrt{2}I_{cc}$$

The IEC 909 standard recommends the following three methods to calculate the peak current:

- Method A is the least exact and calculates the peak current as

$$I_{peak} \approx \sqrt{2}I_{cc}(1.02 + 0.98e^{-(R/X)3})$$

  where R/X is the minimum existing value in all the network branches.
- In Method B, the short-circuit peak current is calculated differently for LV, MV, and HV as follows:
  - Low Voltage (LV):

$$I_{peak} \approx 1.8\sqrt{2}I_{cc}$$

- Medium Voltage (MV):

$$I_{peak} \approx 1.15\sqrt{2}I_{cc}(1.02 + 0.98e^{-(R/X)3})$$

  - High Voltage (HV):

$$I_{peak} \approx 2\sqrt{2}I_{cc}$$

  where R and X are the real and reactive parts of the nodal impedance at the faulted point.
- Method C is more exact. It defines the peak current as

$$I_{peak} \approx \sqrt{2}I_{cc}\left(1.02 + 0.98e^{-(R/X)3}\right)$$

  where the ratio R/X is calculated as $R/X = 20R_c/50X_c$, R_c and X_c being the real and reactive parts of the nodal impedance at the short-circuit point, respectively, both defined using a network steady-state frequency of 20 Hz, when the rated network frequency is 50 Hz (typical in Europe). In the case of a rated network frequency of 60 Hz, the calculation frequency should be 24 Hz.

The most evident consequence of a short circuit is the resulting high intensity current, where the values change from amperes under rated conditions (I_0) to thousands of amperes under faulted conditions (I_{cc}). This creates severe thermal and mechanical stresses in the electric installation elements and equipment. Thus, the fault should be cleared as soon as possible, isolating the short circuit from the generator using system protection equipment (relays and circuit breakers). The relays are used to detect the existence of a fault and send an order to the circuit breakers to carry out the fault clearing, and in this way interrupt the current flowing to the fault. The main objectives of system protection are to minimize the duration of a fault, to reduce its thermal and mechanical effects, and to minimize the number of other devices affected by the fault [1] (Figure 8.2).

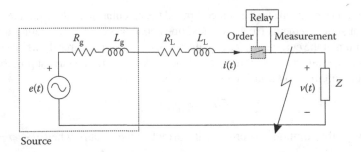

FIGURE 8.2 Short-circuit protection system in an electric network.

In general, the maximum fault current values are useful to determine the required interrupting rating of fuses and circuit breakers. The minimum fault values are useful to determine and coordinate the operation of the various switching devices (relay operation).

The switching device behavior is evaluated under the worst conditions (maximum current situations). Hence, if in the circuit of Figure 8.1a there is a circuit breaker between the generation and the short circuit (Figure 8.1b), it should be designed and calculated to withstand the following conditions [2]:

- When the breaker is closed (nonactive), it has to permanently withstand the steady-state rated current I_n (sinusoidal periodic).
- When the breaker is closed (nonactive), it has to withstand the following current from the onset of the fault ($t = 0$) until it clears at $t = t_c$.

$$i_{max} = \sqrt{2}I_{cc} \sin(\varphi)(1 + e^{-(R/L)t})$$

- When the breaker is closed, it should be in a position to break the maximum current in the worst conditions to clear the fault. For the clearing (or breaking) time (t_c)—normally between 0.06 and 0.08 s (three and four cycles)—the maximum current will be:

$$I_c = \sqrt{2}I_{cc} \sin(\varphi)(1 + e^{-(R/L)t_c})$$

Owing to the difference between the breaking of a periodic and DC current, a DC asymmetrical short-circuit current component I_a is defined as

$$I_a = \sqrt{2}I_{cc}e^{-(R/L)t_c}$$

and, to characterize all the breaking elements, another current called RMS total breaking current I is defined as follows:

$$I = \sqrt{I_a^2 + I_{cc}^2}$$

The breaking capacity is defined as $S_b = \sqrt{3}U_nI$, where U_n is the RMS-rated voltage and I is the RMS breaking current. Another commonly used term in electric networks is the short-circuit capacity (SCC), defined as $SCC = \sqrt{3}U_nI_{cc}$, where U_n is the RMS-rated voltage and I_{cc} is the RMS steady-state fault current.

As mentioned before, the order to open the breaking element is delivered by the so-called protection relay after it detects the presence of the short circuit, and the worst detection condition is when the short-circuit current has the value $I_{min} = \sqrt{2}I_{cc}$.

To summarize, the following currents are of interest during short-circuit calculations:

- From the switching device perspective: the maximum instantaneous value of the current $[i_{max}(t)]$, the peak current (I_{peak}), the maximum current at the breaking instant (I_c), the asymmetrical breaking current (I_a), the RMS total breaking current (I), and the steady-state short-circuit current I_{cc}. In general, the maximum fault current is calculated based on the assumption of all generators in service and solid (zero impedance) fault.
- From the relay perspective: all of the above, but in addition also the minimum short-circuit current (I_{min}). The minimum fault current is calculated based on the minimum likely number of generators and nonsolid faults.

8.2 MODELING OF COMPONENTS

This section describes the models (or equivalent circuits) of electrical system components suitable for short-circuit analysis. The components include transmission lines, three-phase transformers, generators, and loads. The component models of an electric network were presented in Chapter 7 for the case of balanced parameters to which nonbalanced electric variables are applied. The components have the same parameters for the three phases but they may not present a balanced set of electric variables (voltage and current). The models are based on symmetrical component theory. The application of symmetrical component theory to the different electric network elements (lines, transformers, generators, and loads) produces the so-called symmetrical component models, also known as sequence network. These sequence networks are useful for the analysis of power systems where a portion can be isolated for study (the fault location), with the rest of the system remaining balanced [3,4].

8.2.1 MODELING OF LINES

The model of three-phase balanced lines with unbalanced voltages and currents is reduced to three single-phase lines, one for each sequence (positive, negative, and zero). These π-sequence circuits are shown in Figure 8.3. In the analysis of short circuits, the parallel admittance in the positive and negative sequence networks are neglected. However, in the zero sequence this approximation is unacceptable with single line to ground fault in isolated-neutral systems.

The following equations represent the nodal analysis model for positive–negative circuits and zero-sequence circuit, where the matrix is the primitive sequence admittance matrix

$$\begin{bmatrix} y_{s,1}+y_{p,1} & -y_{s,1} \\ -y_{s,1} & y_{s,1}+y_{p,1} \end{bmatrix}\begin{bmatrix} V_i \\ V_j \end{bmatrix}=\begin{bmatrix} I_i \\ I_j \end{bmatrix} \quad \text{and} \quad \begin{bmatrix} y_{s,0}+y_{p,0} & -y_{s,0} \\ -y_{s,0} & y_{s,0}+y_{p,0} \end{bmatrix}\begin{bmatrix} V_i \\ V_j \end{bmatrix}=\begin{bmatrix} I_i \\ I_j \end{bmatrix}$$

8.2.2 MODELING OF TRANSFORMERS

The sequence models of three-phase power transformers are shown in Figure 8.4 where the shunt parameters (magnetization branch) are neglected.

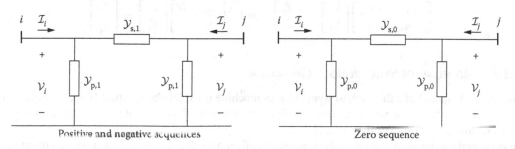

Positive and negative sequences   Zero sequence

FIGURE 8.3 Symmetrical sequences of lines.

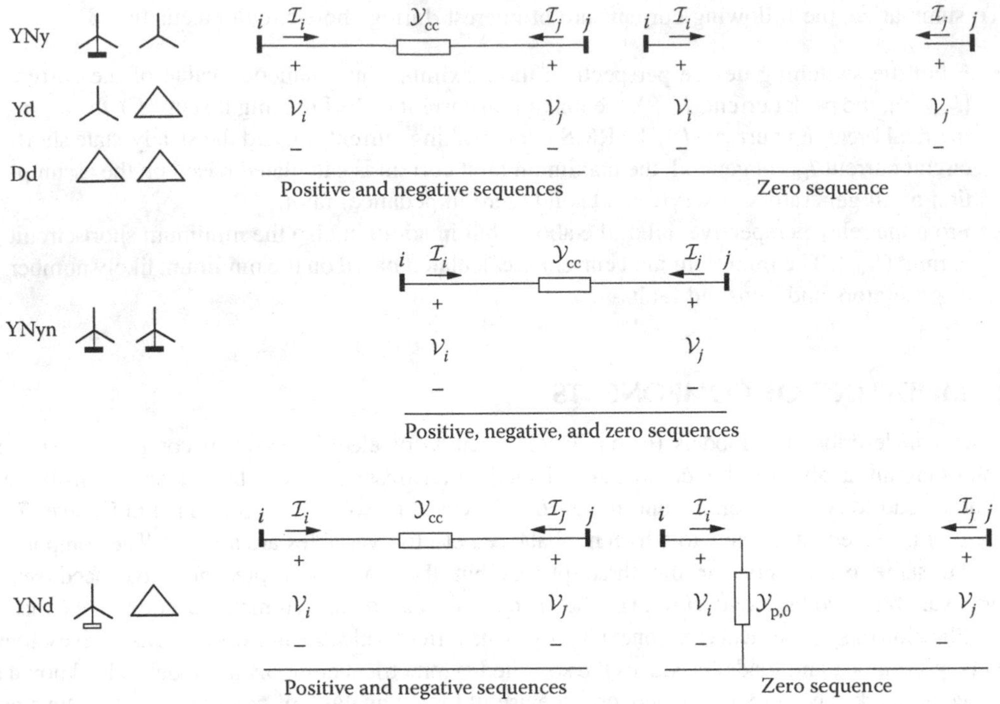

FIGURE 8.4 Symmetrical sequences of transformers.

The nodal analysis models for YNy, Yd, and Dd transformers for the sequence domain equivalents shown in Figure 8.4 are as follows:

$$\begin{bmatrix} y_{cc} & -y_{cc} \\ -y_{cc} & y_{cc} \end{bmatrix} \begin{bmatrix} \mathcal{V}_i \\ \mathcal{V}_j \end{bmatrix} = \begin{bmatrix} \mathcal{I}_i \\ \mathcal{I}_j \end{bmatrix} \quad \text{and} \quad \begin{bmatrix} 0 & 0 \\ 0 & 0 \end{bmatrix} \begin{bmatrix} \mathcal{V}_i \\ \mathcal{V}_j \end{bmatrix} = \begin{bmatrix} \mathcal{I}_i \\ \mathcal{I}_j \end{bmatrix}$$

For the YNy transformer, the zero-sequence exact model have a shunt admittance parameter $y_{p,0}/y_{cc}$. Likewise for a YNd transformer.

The following equations represent the sequence domain models for YNyn transformers illustrated in Figure 8.4:

$$\begin{bmatrix} y_{cc} & -y_{cc} \\ -y_{cc} & y_{cc} \end{bmatrix} \begin{bmatrix} \mathcal{V}_i \\ \mathcal{V}_j \end{bmatrix} = \begin{bmatrix} \mathcal{I}_i \\ \mathcal{I}_j \end{bmatrix}$$

The nodal analysis of transformers YNd for sequence models is as follows:

$$\begin{bmatrix} y_{cc} & -y_{cc} \\ -y_{cc} & y_{cc} \end{bmatrix} \begin{bmatrix} \mathcal{V}_i \\ \mathcal{V}_j \end{bmatrix} = \begin{bmatrix} \mathcal{I}_i \\ \mathcal{I}_j \end{bmatrix} \quad \text{and} \quad \begin{bmatrix} y_{p,0} & 0 \\ 0 & 0 \end{bmatrix} \begin{bmatrix} \mathcal{V}_i \\ \mathcal{V}_j \end{bmatrix} = \begin{bmatrix} \mathcal{I}_i \\ \mathcal{I}_j \end{bmatrix}$$

8.2.3 MODELING OF SYNCHRONOUS GENERATORS

The dynamic model of a three-phase synchronous machine is established from differential equations which relate voltages, fluxes, and currents. However, these models are not very useful for the purposes of this chapter due to their complexity. According to Appendix C, the synchronous machine models used for the short-circuit analysis are balanced circuits defined for the fault periods of interest (subtransient, transient, and eventually, steady state).

The following equations represent the nodal analysis of generators YN for the sequence models in Figure 8.5:

$$[y_1][\mathcal{V}_i] = [\mathcal{I}_i] + [\mathcal{E}_i y_1], \quad [y_2][\mathcal{V}_i] = [\mathcal{I}_i], \quad \text{and} \quad [y_0][\mathcal{V}_i] = [\mathcal{I}_i]$$

The following equations represent the nodal analysis of generators Y and D for the sequence models in Figure 8.5:

$$[y_1][\mathcal{V}_i] = [\mathcal{I}_i] + [\mathcal{E}_i y_1], \quad [y_2][\mathcal{V}_i] = [\mathcal{I}_i], \quad \text{and} \quad [0][\mathcal{V}_i] = [\mathcal{I}_i]$$

8.2.4 MODELING OF ASYNCHRONOUS MACHINES

The asynchronous machine models used for the short-circuit analysis, like those of synchronous machines, are balanced steady-state circuits defined for the transient behavior of balanced fault. The transient model presented is adequate to analyze the currents with 100 ms duration.

The following equations represent the nodal analysis of asynchronous machines YN for the sequence models in Figure 8.6:

$$[y_{cc}][\mathcal{V}_i] = [\mathcal{I}_i] + [\mathcal{E}_i y_{cc}], \quad [y_{cc}][\mathcal{V}_i] = [\mathcal{I}_i], \quad \text{and} \quad [y_0][\mathcal{V}_i] = [\mathcal{I}_i]$$

The following equations represent the nodal analysis of asynchronous machines Y and D for the sequence models in Figure 8.6:

$$[y_{cc}][\mathcal{V}_i] = [\mathcal{I}_i] + [\mathcal{E}_i y_{cc}], \quad [y_{cc}][\mathcal{V}_i] = [\mathcal{I}_i], \quad \text{and} \quad [0][\mathcal{V}_i] = [\mathcal{I}_i]$$

8.3 THE SEQUENCE MATRICES

A precondition to efficient and systematic fault calculations is the availability of the appropriate sequence domain nodal admittance matrices (sometimes called the Y-bus matrices). These matrices are constructed based on the sequence models for the various components. For the zero-sequence matrix, attention must be closely paid to the grounding patterns of the transformers and generators. For lines, one must be aware that the zero-sequence impedance of the line is often much greater than its positive and negative sequence impedance.

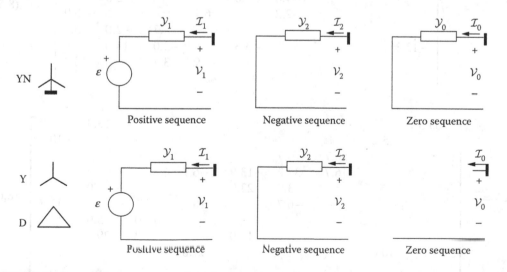

FIGURE 8.5 Symmetrical sequences of synchronous generators.

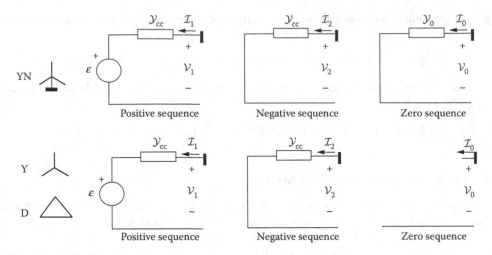

FIGURE 8.6 Symmetrical sequences of asynchronous machines.

The fault calculation techniques that follow are best illustrated by considering the example in Figure 8.7. Before building the sequence nodal admittance matrices, we construct the equivalent circuits for the sequence networks. The sequence network equivalent circuits for this example are illustrated in Figures 8.8 through 8.10. The topology (nonzero pattern) of the zero-sequence network is slightly different from the one for the positive and negative sequence network as a result of the delta–wye transformers (three-phase core with three legs).

The (primitive) impedances z for all the components in our test network are listed in Figure 8.7. The sequence nodal admittance matrices for the example are as follows:

$$
\mathbf{Y}^0 = -j
\begin{bmatrix}
58.8 & & & & & & & & -13.3 \\
& 30.3 & & & & & & & \\
& & 117.5 & & & & & & \\
& & & 6.9 & -4.6 & -2.2 & & & \\
& & & -4.6 & 8.0 & & & -3.3 & \\
& & & -2.2 & & 6.1 & & & -3.9 \\
& & & & & & 5.7 & -2.0 & -3.7 \\
-13.3 & & & & -3.3 & & -2.0 & 18.6 & \\
& & & & & -3.9 & -3.7 & & 18.2
\end{bmatrix}
\tag{8.5}
$$

$$
\mathbf{Y}^1 = -j
\begin{bmatrix}
63.3 & & & & & & & -13.3 & \\
& 25.2 & & & & & & & -10.0 \\
& & 56.7 & -16.7 & & & & & \\
& & -16.7 & 37.2 & -13.9 & -6.7 & & & \\
& & & -13.9 & 23.9 & & & -10.0 & \\
& & & -6.7 & & 18.4 & & & -11.8 \\
& & & & & & 17.0 & -5.9 & -11.1 \\
-13.3 & & & & -10.0 & & -5.9 & 29.2 & \\
& -10.0 & & & & -11.8 & -11.1 & & 32.9
\end{bmatrix}
\tag{8.6}
$$

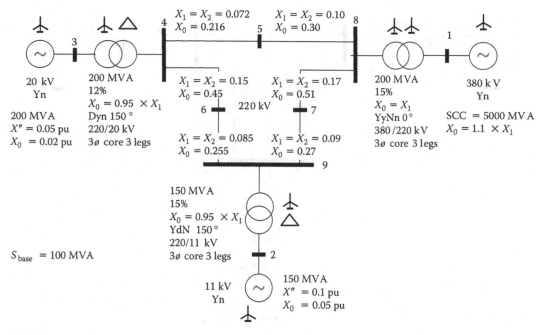

FIGURE 8.7　One-line diagram for the 9-node system.

The diagonal entries of each of these matrices correspond to the sum of all the primitive *admittances* connected to each node (including node to ground admittances) in the corresponding sequence network. The off-diagonal entries correspond to reduce the respective primitive *admittance* connecting the corresponding two nodes. Whenever there is no direct connection between two nodes, the corresponding entry is zero (or better yet, not present). This makes the matrices sparse.

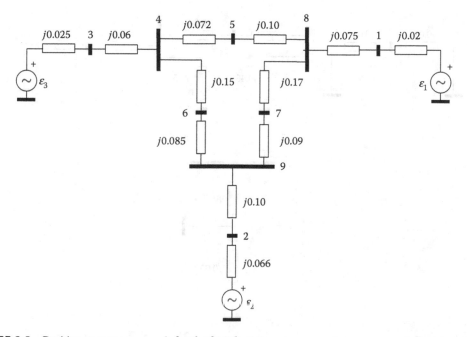

FIGURE 8.8　Positive sequence network for the 9-node system.

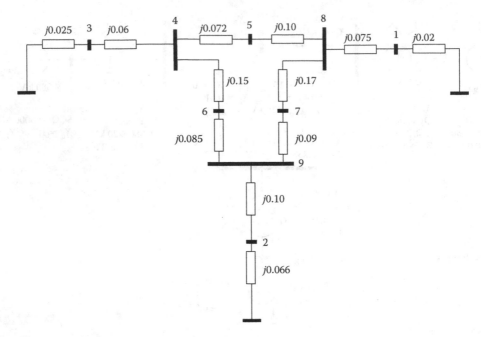

FIGURE 8.9   Negative sequence network for the 9-node system.

Because of our assumptions, the positive and negative sequence nodal admittance matrices are identical. That is, $\mathbf{Y}^2 = \mathbf{Y}^1$. Also, as indicated elsewhere, the construction of these matrices relies on certain assumptions, such as the notion of a quasi-steady-state solution and the notion of synchronous, transient, or subtransient periods.

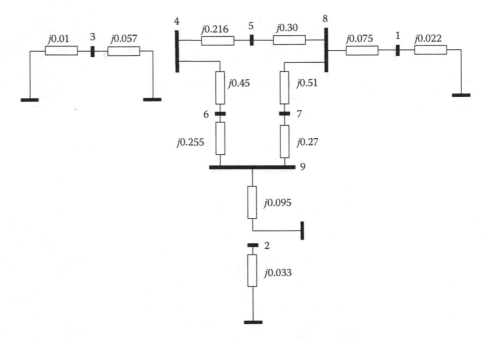

FIGURE 8.10   Zero-sequence network for the 9-node system.

8.3.1 CONSTRUCTION OF SEQUENCE NODAL ADMITTANCE MATRICES USING THE PRIMITIVE Y-NODE CONCEPT

This section describes a method for the efficient construction of nodal admittance matrices, also known as **Y** matrices. The methods for construction of these matrices are similar for the zero, positive, and negative sequence matrices. They also apply to the construction of phase-domain matrices. The method for construction of these matrices is a slight generalization of the "by inspection" method that many students learn at first. The procedure is based on the notion of "primitive admittance matrices" defined by applying nodal analysis to each network component (see Section 8.2). That is, every component in the system is assumed to have a known primitive admittance matrix representation in each the zero, positive, and negative sequence domains. The method illustrates how to combine these individual component primitive admittance matrices **y** into an overall complete **Y** for each domain [5].

This section illustrates the most important component primitive admittance matrices, and then illustrates how to aggregate these component models by using matrix overlapping. This is done using the same example illustrated in Figure 8.7. The example consists of six lines, three transformers, and three generators.

For each of the lines, the primitive admittance matrix in the phase domain is a 2×2 matrix. If the π-equivalent of the lines are known (these can be measured or calculated based on standard formulae), the primitive admittance matrices can be constructed by inspection (Table 8.1).

Likewise, the primitive admittance matrices for the transformers are illustrated in Table 8.2 and the primitive admittance matrices for the generators are illustrated in Table 8.3. The structure of the transformers differs between the zero and the positive and negative sequences as a result of the ground connectivity, but each consists of a 2×2 matrix. For the generators, the matrices are 1×1 (scalars) and the grounding pattern matters (for an ungrounded generator, the zero-sequence value would be zero).

For each positive, negative, and zero sequence, the assembly process involves overlapping these matrices by simply adding their values after expanding each matrix to the full 9×9 matrix dimension (the total number of nodes in the network). The complete zero and positive sequences matrices are presented in Equations 8.5 and 8.6. The process of overlapping applies equally well in the phase domain [6–8].

TABLE 8.1
Primitive Sequence Admittance Matrices for All the Lines in the Example

From	To	Zero	Positive	Negative
4	5	$\begin{bmatrix} -j4.6 & j4.6 \\ j4.6 & -j4.6 \end{bmatrix}$	$\begin{bmatrix} -j13.9 & j13.9 \\ j13.9 & -j13.9 \end{bmatrix}$	$\begin{bmatrix} -j13.9 & j13.9 \\ j13.9 & -j13.9 \end{bmatrix}$
5	8	$\begin{bmatrix} -j3.3 & j3.3 \\ j3.3 & -j3.3 \end{bmatrix}$	$\begin{bmatrix} -j10 & j10 \\ j10 & -j10 \end{bmatrix}$	$\begin{bmatrix} -j10 & j10 \\ j10 & -j10 \end{bmatrix}$
8	7	$\begin{bmatrix} -j2 & j2 \\ j2 & -j2 \end{bmatrix}$	$\begin{bmatrix} -j5.9 & j5.9 \\ j5.9 & -j5.9 \end{bmatrix}$	$\begin{bmatrix} -j5.9 & j5.9 \\ j5.9 & -j5.9 \end{bmatrix}$
6	9	$\begin{bmatrix} -j3.9 & j3.9 \\ j3.9 & -j3.9 \end{bmatrix}$	$\begin{bmatrix} -j11.8 & j11.8 \\ j11.8 & -j11.8 \end{bmatrix}$	$\begin{bmatrix} -j11.8 & j11.8 \\ j11.8 & -j11.8 \end{bmatrix}$
7	9	$\begin{bmatrix} -j3.7 & j3.7 \\ j3.7 & -j3.7 \end{bmatrix}$	$\begin{bmatrix} -j11.1 & j11.1 \\ j11.1 & -j11.1 \end{bmatrix}$	$\begin{bmatrix} -j11.1 & j11.1 \\ j11.1 & -j11.1 \end{bmatrix}$
4	6	$\begin{bmatrix} -j2.2 & j2.2 \\ j2.2 & -j2.2 \end{bmatrix}$	$\begin{bmatrix} -j6.7 & j6.7 \\ j6.7 & -j6.7 \end{bmatrix}$	$\begin{bmatrix} -j6.7 & j6.7 \\ j6.7 & -j6.7 \end{bmatrix}$

TABLE 8.2

Primitive Sequence Admittance Matrices for All the Transformers in the Example

From	To	Zero	Positive	Negative
3	4	$\begin{bmatrix} -j17.5 & 0 \\ 0 & 0 \end{bmatrix}$	$\begin{bmatrix} -j16.7 & j16.7 \\ j16.7 & -j16.7 \end{bmatrix}$	$\begin{bmatrix} -j16.7 & j16.7 \\ j16.7 & -j16.7 \end{bmatrix}$
8	1	$\begin{bmatrix} -j13.3 & j13.3 \\ j13.3 & -j13.3 \end{bmatrix}$	$\begin{bmatrix} -j13.3 & j13.3 \\ j13.3 & -j13.3 \end{bmatrix}$	$\begin{bmatrix} -j13.3 & j13.3 \\ j13.3 & -j13.3 \end{bmatrix}$
9	2	$\begin{bmatrix} -j10.5 & 0 \\ 0 & 0 \end{bmatrix}$	$\begin{bmatrix} -j10 & j10 \\ j10 & -j10 \end{bmatrix}$	$\begin{bmatrix} -j10 & j10 \\ j10 & -j10 \end{bmatrix}$

8.4 SOLID FAULTS

Once the sequence domain nodal admittance matrices are available, we can perform fault calculations. This section presents a "formula based" approach to fault calculations. The steps are:

1. Determine the total fault current(s) $\mathcal{I}^{f,s}$ in the sequence domain. These are the currents flowing through the fault itself. Generally, these quantities are not of direct interest, unless one is
   interested in the current in the arc. However, knowing the total fault currents makes subsequent steps simpler.
2. Determine the voltages at all points of interest (this could be every node in the system or just
   the nodes in the vicinity of the fault). The voltages represent the *change* in voltage resulting
   from the application of the fault (what we will call the *faulting* voltages ΔV). Sometimes
   voltages are of direct interest, but more commonly, voltages are a useful second step.
3. Determine the sequence fault current flows in all desired lines, transformers, breakers, and
   any other network component.
4. Convert sequence voltages or currents to the phase domain.
5. Superimpose any preexisting flows or voltages on the results obtained. Adjust any currents
   or voltages for unidirectional transient components, as described in Section 8.1.

The determination of the total fault current in the sequence domain depends on the type of fault. In
this section, we consider the following "classic four" faults of Figure 8.11:

- A *solid three-phase fault*: Although not very common, this fault is often the most severe one
  resulting in the largest fault currents.
- A *solid line to ground fault*: This is generally the most common type of fault. It also can,
  sometimes, be the one that results in the largest fault currents.
- A *line to line fault*: Often this fault results in some of the lowest fault currents. Knowledge of
  the lowest fault currents is often of direct interest in relay-setting applications.
- A double line to ground fault.

TABLE 8.3

Primitive Sequence Admittance Matrices for All the Generators in the Example

Bus	Zero	Positive	Negative
3	$[-j100]$	$[-j40]$	$[-j40]$
1	$[-j45.5]$	$[-j50]$	$[-j50]$
2	$[-j30.3]$	$[-j15.2]$	$[-j15.2]$

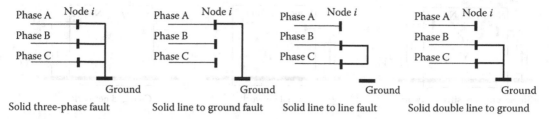

Solid three-phase fault   Solid line to ground fault   Solid line to line fault   Solid double line to ground

FIGURE 8.11 "Classic four" faults.

8.4.1 FAULT ANALYSIS METHOD BY USING NODAL ANALYSIS

The model of a three-phase power system with a fault at node i is illustrated in Figure 8.12, where the power system is balanced except for the affected node (fault condition). Using the symmetrical component theory, the three-phase power system is decomposed in three decoupled networks: positive, negative, and zero-sequence networks. The fault conditions are simulated by three unknown current sources or faulting currents[*]: $\mathcal{I}_i^{f,1}, \mathcal{I}_i^{f,2}, \mathcal{I}_i^{f,0}$. So, the short-circuit analysis can be approached by the nodal admittance method of electric circuits:

$$\mathbf{Y}^1 \begin{bmatrix} \mathbf{V}_*^1 \\ \mathcal{V}_i^1 \end{bmatrix} = \begin{bmatrix} \mathbf{I}_g \\ \mathcal{I}_i^{f,1} \end{bmatrix} \quad \mathbf{Y}^2 \begin{bmatrix} \mathbf{V}_*^2 \\ \mathcal{V}_i^2 \end{bmatrix} = \begin{bmatrix} \mathbf{0} \\ \mathcal{I}_i^{f,2} \end{bmatrix} \quad \mathbf{Y}^0 \begin{bmatrix} \mathbf{V}_*^0 \\ \mathcal{V}_i^0 \end{bmatrix} = \begin{bmatrix} \mathbf{0} \\ \mathcal{I}_i^{f,0} \end{bmatrix}$$

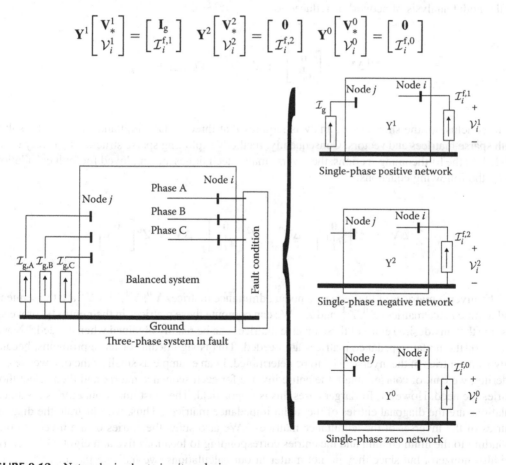

FIGURE 8.12 Networks in short-circuit analysis.

---

[*] Usually, the value of "fault current" is expressed from the faulted node to ground, so the "fault current" is opposite to "faulting current."

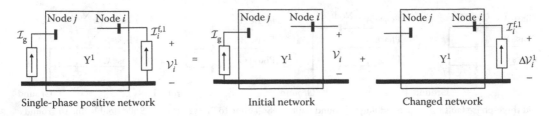

FIGURE 8.13 Networks of positive sequence.

where $\mathbf{V}_*^1$, $\mathbf{V}_*^2$, and $\mathbf{V}_*^0$ are the nodal voltage vectors, except the nodal voltage $\mathcal{V}_i^1$, $\mathcal{V}_i^2$, and $\mathcal{V}_i^0$, respectively.

The nodal admittance analysis of the positive sequence circuit of Figure 8.12, with generator components $\mathcal{I}_g$, can be simplified using the superposition theorem. The superposition theorem allows the positive network to be decomposed in: the "initial network" (prefault network without "fault source" $\mathcal{I}^{f,1}$) and the "changed network" without generator sources $\mathcal{I}_g$, as shown in Figure 8.13. The "initial network" is the balanced three-phase system where the electric variables are known, and so the nodal analysis of network is reduced to

$$\mathbf{Y}^1 \Delta \mathbf{V}^1 = \begin{bmatrix} \mathbf{0} \\ \mathcal{I}_i^{f,1} \end{bmatrix} \quad \text{and} \quad \mathcal{V}_k^1 = \mathcal{V}_{k,\text{initial}} + \Delta \mathcal{V}_k^1$$

In conclusion, the short-circuit analysis requires that three nodal admittance systems be solved with sparse matrices and vectors. Consequently, methods exploiting sparse structures are very useful [9–13]. In the forthcoming sections, the inverse matrix techniques are employed for fault calculations using the following equations:

$$\Delta \mathbf{V}^1 = \mathbf{Z}^1 \begin{bmatrix} \mathbf{0} \\ \mathcal{I}_i^{f,1} \end{bmatrix} \quad \Delta \mathbf{V}^2 = \mathbf{Z}^2 \begin{bmatrix} \mathbf{0} \\ \mathcal{I}_i^{f,2} \end{bmatrix} \quad \Delta \mathbf{V}^0 = \mathbf{Z}^0 \begin{bmatrix} \mathbf{0} \\ \mathcal{I}_i^{f,0} \end{bmatrix}$$

The inverse matrices of the sequence nodal admittance matrices $\mathbf{Y}^0$, $\mathbf{Y}^1$, and $\mathbf{Y}^2$ are the sequence nodal impedance matrices $\mathbf{Z}^0$, $\mathbf{Z}^1$, and $\mathbf{Z}^2$. We can compute these matrices in their entirety, or we can assume that any desired entry of these inverse matrices can be readily obtained when needed.* Not all entries of the nodal impedance matrices are needed. This is significant for large problems, because only some entries of the inverse need to be determined. In an example as small as the one we are considering, we can, of course, obtain the entire inverse for each sequence matrix and then select those entries we need. However, for larger cases this is impractical. The most important entries for all calculations are the diagonal entries of the nodal impedance matrices. Thus, we illustrate the diagonal entries of the inverses of the admittance matrices.† We also show the entries of the inverses corresponding to the ninth column and the entries corresponding to locations five and eight. Other entries are also nonzero, but since they do not matter in our calculations, we indicate them as "*". Those

---

* The nodal impedance matrices tend to be full, that is, all their entries are nonzero.
† There are efficient methods that calculate the diagonal entries of the inverse without having to find every inverse entry, and also to calculate selected entries of any column of the inverse.

entries that we actually use in later calculations are shown in boldface:

$$
Z^1 = j \begin{bmatrix}
0.0184 & * & * & * & * & * & * & * & \mathbf{0.0047} \\
* & 0.0530 & * & * & * & * & * & * & \mathbf{0.0332} \\
* & * & 0.0224 & * & * & * & * & * & 0.0065 \\
* & * & * & 0.0555 & * & * & * & * & 0.0222 \\
* & * & * & * & \mathbf{0.0809} & * & * & \mathbf{0.0367} & 0.0223 \\
* & * & * & * & * & 0.1058 & * & * & \mathbf{0.0614} \\
* & * & * & * & * & * & 0.1118 & * & \mathbf{0.0624} \\
* & * & * & * & \mathbf{0.0367} & * & * & \mathbf{0.0596} & 0.0223 \\
* & * & * & * & * & * & * & * & \mathbf{0.0836}
\end{bmatrix}
$$

$$(8.7)$$

$$
Z^0 = j \begin{bmatrix}
0.0213 & * & & * & & * & & * & 0.0031 \\
* & 0.0330 & * & * & * & & * & & \mathbf{0} \\
* & * & 0.0085 & * & * & * & & * & 0 \\
* & * & * & 0.3469 & * & * & * & & 0.0424 \\
* & * & * & * & \mathbf{0.2835} & * & * & \mathbf{0.0659} & 0.0304 \\
* & * & * & * & * & 0.2609 & * & * & \mathbf{0.0674} \\
* & * & * & * & * & * & 0.2276 & * & \mathbf{0.0581} \\
* & * & * & * & \mathbf{0.0659} & * & & \mathbf{0.0829} & 0.0138 \\
* & * & * & * & * & * & & * & \mathbf{0.0815}
\end{bmatrix}
$$

$$(8.8)$$

As stated before, the positive and negative sequence matrices are, for this example, identical.

8.4.2 Solid Three-Phase Faults (3φ)

Once the diagonal entries of the positive sequence matrix are known, the total positive sequence fault current $\mathcal{I}_i^{3\phi,1}$ for a solid three-phase fault at node i (assuming that the prefault voltages were nominal, or 1 per unit, voltages) can be obtained from the following formula:

$$
\mathcal{I}_i^{3\phi,1} = \frac{-1}{\mathcal{Z}_{ii}^1} \tag{8.9}
$$

The negative of the inverse of the faulted node diagonal entry of the nodal impedance positive sequence matrix gives the total fault current (the current flowing through the fault from ground to the network).[*] The zero and negative sequence fault currents are zero.

The second step is the determination of the postfault nodal voltages. The vector of *changes* in nodal voltages ΔV as a result of the fault can be determined by multiplying the total fault current times the column i of $\mathbf{Z}$, the column corresponding to the fault location:

$$
\Delta V^{3\phi,1} = \mathcal{Z}_{*i}^1 \mathcal{I}_i^{3\phi,1} \tag{8.10}
$$

The * subscript denotes all values of the index. Thus, $\mathcal{Z}_{*i}^1$ denotes the ith column of $\mathbf{Z}^1$. The faulting voltages $\Delta V^{3\phi,0}$ and $\Delta V^{3\phi,2}$ are zero for three-phase faults.

Once the (sequence) nodal faulting voltages $\Delta V^{3\phi}$ have been determined, and knowing the sequence primitive impedance z_{ik} connecting nodes j and k, the (in this case positive) sequence current

---

[*] Total fault currents are denoted by single subscripts.

flow on any line, transformer, or generator can be determined (flows have double subscripts):

$$\mathcal{I}_{jk}^1 = \frac{\Delta \mathcal{V}_j^{3\phi,1} - \Delta \mathcal{V}_k^{3\phi,1}}{z_{jk}^1} \tag{8.11}$$

Any sequence faulting voltages $\Delta \mathcal{V}_k^0$, $\Delta \mathcal{V}_k^1$, and $\Delta \mathcal{V}_k^2$ or sequence flows I_{jk}^0, I_{jk}^1, and I_{jk}^2 on any line can be converted back to phase-domain values by using the inverse symmetrical component transformation. For the 3ϕ fault, the phase a voltages are the same as the positive sequence voltages. Thus,[*]

$$\Delta \mathcal{V}_j^a = \Delta \mathcal{V}_j^1 \quad \Delta \mathcal{V}_j^b = a^2 \Delta \mathcal{V}_j^1 \quad \Delta \mathcal{V}_j^c = a \Delta \mathcal{V}_j^1 \tag{8.12}$$

$$\mathcal{I}_{jk}^a = \mathcal{I}_{jk}^{3\phi,1} \quad \mathcal{I}_{jk}^b = a^2 \mathcal{I}_{jk}^{3\phi,1} \quad \mathcal{I}_{jk}^c = a \mathcal{I}_{jk}^{3\phi,1} \tag{8.13}$$

EXAMPLE 8.1

We illustrate these formula using our example. Assume that we want to study a fault at node 9 ($i = 9$), and that we are interested in the fault currents on all the lines and transformers directly feeding the fault (i.e., the currents feeding the fault from nodes 2, 6, and 7). Using Equation 8.9, we obtain the total fault current[†]

$$\mathcal{I}_9^{3\phi,1} = -j11.966 \tag{8.14}$$

From Equation 8.10, we obtain the faulting voltages of interest

$$\Delta \mathcal{V}_2^{3\phi,1} = -0.3976 \quad \Delta \mathcal{V}_6^{3\phi,1} = -0.7345 \quad \Delta \mathcal{V}_7^{3\phi,1} = -0.7463 \quad \Delta \mathcal{V}_9^{3\phi,1} = -1$$

From Equation 8.11, we obtain the flows on the lines of interest

$$\mathcal{I}_{6-9}^1 = -j3.1233 \quad \mathcal{I}_{7-9}^1 = -j2.8186 \quad \mathcal{I}_{2-9}^1 = -j6.0241$$

Because there are no negative or zero-sequence voltages or currents to worry about, the phase a voltages and currents are the same as the positive sequence voltages and currents. You can verify that the sum of the fault current contributions from all adjacent nodes add up to the total fault current. (The j multiplier denotes that the fault currents are purely reactive.)

8.4.3 SINGLE LINE TO GROUND FAULTS

The formula for the total fault currents in the sequence domain for a solid single line to ground faults (SLG) is

$$\mathcal{I}_i^{SLG,0} = \mathcal{I}_i^{SLG,1} = \mathcal{I}_i^{SLG,2} = \frac{-1}{\mathcal{Z}_{ii}^0 + \mathcal{Z}_{ii}^1 + \mathcal{Z}_{ii}^2} \tag{8.15}$$

The faulting voltages for each sequence are determined as in the previous case, by multiplying the corresponding total fault current times the corresponding entries in the ith column of the impedance matrix. Although all three sequence fault currents are equal, the differing values of the sequence impedances give rise to different sequence faulting voltages:

$$\Delta \mathcal{V}^{SLG,0} = \mathcal{Z}_{*i}^0 \mathcal{I}_i^{SLG,0} \quad \Delta \mathcal{V}^{SLG,1} = \mathcal{Z}_{*i}^1 \mathcal{I}_i^{SLG,1} \quad \Delta \mathcal{V}^{SLG,2} = \mathcal{Z}_{*i}^2 \mathcal{I}_i^{SLG,2} \tag{8.16}$$

---

[*] $a = 1\angle 120°$ is the operator that rotates a phasor by $120°$.
[†] "Faulting current"$= j11.966$.

The sequence fault current flows on any component jk can be obtained from

$$I_{jk}^0 = \frac{\Delta V_j^0 - \Delta V_k^0}{z_{jk}^0} \quad I_{jk}^1 = \frac{\Delta V_j^1 - \Delta V_k^1}{z_{jk}^1} \quad I_{jk}^2 = \frac{\Delta V_j^2 - \Delta V_k^2}{z_{jk}^2} \tag{8.17}$$

The phase voltages and flows can be found using the symmetrical component transformations:

$$\begin{bmatrix} \Delta V_j^a \\ \Delta V_j^b \\ \Delta V_j^c \end{bmatrix} = T \begin{bmatrix} \Delta V_j^0 \\ \Delta V_j^1 \\ \Delta V_j^2 \end{bmatrix} \quad \text{and} \quad \begin{bmatrix} I_{jk}^a \\ I_{jk}^b \\ I_{jk}^c \end{bmatrix} = T \begin{bmatrix} I_{jk}^0 \\ I_{jk}^1 \\ I_{jk}^2 \end{bmatrix} \tag{8.18}$$

EXAMPLE 8.2

For our example, the calculations for the total fault current for a single line to ground fault are, from Equation 8.15,

$$I_9^{SLG,0} = I_9^{SLG,1} = I_9^{SLG,2} = -j4.022$$

If these sequence total fault currents are converted back to the phase domain using the symmetrical component transformation, that is, Equation 8.18, we obtain

$$I_9^{SLG,a} = -j12.066 \quad I_9^{SLG,b} = 0 \quad I_9^{SLG,c} = 0 \tag{8.19}$$

In this example, the total phase a fault current is greater for the SLG fault than for the 3ϕ fault. The sequence faulting voltages at the nodes of interest are, from Equation 8.16,

$$\begin{array}{lll} \Delta V_2^0 = 0 & \Delta V_2^1 = -0.1336 & \Delta V_2^2 = -0.1336 \\ \Delta V_6^0 = -0.2709 & \Delta V_6^1 = -0.2469 & \Delta V_6^2 = -0.2469 \\ \Delta V_7^0 = -0.2335 & \Delta V_7^1 = -0.2509 & \Delta V_7^2 = -0.2509 \\ \Delta V_9^0 = -0.3278 & \Delta V_9^1 = -0.3361 & \Delta V_9^2 = -0.3361 \end{array}$$

From these voltages, we can determine the desired sequence flows using Equation 8.17:

$$\begin{array}{lll} I_{7-9}^0 = -j0.3491 & I_{7-9}^1 = -j0.9474 & I_{7-9}^2 = -j0.9474 \\ I_{6-9}^0 = -j0.2230 & I_{6-9}^1 = -j1.0498 & I_{6-9}^2 = -j1.0498 \\ I_{2-9}^0 = 0 & I_{2-9}^1 = -j2.0348 & I_{2-9}^2 = -j2.0248 \\ I_{0-9}^0 = -j3.4500 & I_{0-9}^1 = 0 & I_{0-9}^2 = 0 \end{array}$$

There is no zero-sequence flow into bus 9 from bus 2 because of the transformer connections. Instead, there is a zero-sequence flow from ground to bus 9 that significantly contributes to the fault current. For each sequence, the sum of the individual fault contributions into bus 9 adds up to the total corresponding fault current.

The sequence flows and voltages can be converted to phase-domain quantities using the symmetrical component transformation.

$$\mathcal{I}^a_{7-9} = -j2.2439 \quad \mathcal{I}^b_{7-9} = -j0.5984 \quad \mathcal{I}^c_{7-9} = -j0.5984$$
$$\mathcal{I}^a_{6-9} = -j2.3226 \quad \mathcal{I}^b_{6-9} = -j0.8268 \quad \mathcal{I}^c_{6-9} = -j0.8268$$
$$\mathcal{I}^a_{2-9} = -j7.4997 \quad \mathcal{I}^b_{2-9} = -j1.4252 \quad \mathcal{I}^c_{2-9} = -j1.4252$$

$$\Delta\mathcal{V}^a_2 = -0.2673 \quad \Delta\mathcal{V}^b_2 = 0.1336 \quad \Delta\mathcal{V}^c_2 = 0.1336$$
$$\Delta\mathcal{V}^a_6 = -0.7647 \quad \Delta\mathcal{V}^b_6 = -0.0240 \quad \Delta\mathcal{V}^c_6 = -0.0240$$
$$\Delta\mathcal{V}^a_7 = -0.7352 \quad \Delta\mathcal{V}^b_7 = 0.0173 \quad \Delta\mathcal{V}^c_7 = 0.0173$$
$$\Delta\mathcal{V}^a_9 = -1.0000 \quad \Delta\mathcal{V}^b_9 = 0.0084 \quad \Delta\mathcal{V}^c_9 = 0.0084$$

These voltages are, for the case of the SLG fault, all real (some positive, some negative). These values must be superimposed to the preexisting phase voltages prior to the fault. For phases b and c, these values must be added to a complex number. The resulting magnitudes of the *true* voltages as a result of the fault are

$$|\mathcal{V}^a_2| = 0.7327 \quad |\mathcal{V}^b_2| = 0.9403 \quad |\mathcal{V}^c_2| = 0.9403$$
$$|\mathcal{V}^a_6| = 0.2353 \quad |\mathcal{V}^b_6| = 1.0122 \quad |\mathcal{V}^c_6| = 1.0122$$
$$|\mathcal{V}^a_7| = 0.2648 \quad |\mathcal{V}^b_7| = 0.9914 \quad |\mathcal{V}^c_7| = 0.9914$$
$$|\mathcal{V}^a_9| = 0.0000 \quad |\mathcal{V}^b_9| = 0.9958 \quad |\mathcal{V}^c_9| = 0.9958$$

As expected, the voltage in phase a at the fault location is zero. Voltages can be of direct interest when the voltage in the unfaulted phases, b and c increases as a result of a fault (in our example, phases b and c at node 6 experience a slight overvoltage). Under some conditions, these overvoltages can be quite substantial.

8.4.4 LINE TO LINE AND DOUBLE LINE TO GROUND FAULTS

Line to line (LL) faults do not involve zero sequence flows (which require ground flows). Thus, the formula for the total fault current for the LL fault does not require zero-sequence impedances. The formulae for the determination of the sequence domain total fault currents for a solid LL fault at node i are

$$\mathcal{I}^{LL,1}_i = -\mathcal{I}^{LL,2}_i = \frac{-1}{\mathcal{Z}^1_{ii} + \mathcal{Z}^2_{ii}} \quad \mathcal{I}^{LL,0}_i = 0$$

EXAMPLE 8.3

For our example, the resulting total sequence fault currents for the LL fault are

$$\mathcal{I}^{LL,0}_9 = 0 \quad \mathcal{I}^{LL,1}_9 = -j5.9830 \quad \mathcal{I}^{LL,2}_9 = j5.9830$$

A double line to ground solid fault (2LG), on the other hand, involves zero-sequence flows. The "formulae" for the determination of the total sequence fault currents for the 2LG are easier to state if we first define some admittances $\hat{y}$ and $\tilde{y}$

$$\hat{y}^0 = \frac{1}{\mathcal{Z}^0_{ii}} \quad \hat{y}^1 = \frac{1}{\mathcal{Z}^1_{ii}} \quad \hat{y}^2 = \frac{1}{\mathcal{Z}^2_{ii}} \quad \tilde{y} = \hat{y}^0 + \hat{y}^1 + \hat{y}^2$$

The formulae for the total 2LG fault currents in the sequence domain are

$$\mathcal{I}^{2LG,0}_i = \frac{-\hat{y}^0\hat{y}^1}{\tilde{y}} \quad \mathcal{I}^{2LG,1}_i = \frac{-\hat{y}^1\hat{y}^1}{\tilde{y}} + \hat{y}^1 \quad \mathcal{I}^{2LG,2}_i = \frac{-\hat{y}^2\hat{y}^1}{\tilde{y}}$$

EXAMPLE 8.4

For the 2LG fault, the resulting sequence total fault currents at node 9 are

$$\mathcal{I}_9^{2LG,0} = -j4.0560 \quad \mathcal{I}_9^{2LG,1} = j8.011 \quad \mathcal{I}_9^{2LG,2} = -j3.9550$$

After the total sequence fault currents are obtained, all other steps are identical to those of the previous case. One can obtain the sequence faulting voltages, the sequence flows, and go back to calculate the phase-domain voltages and flows using the same formulae and methods already outlined.

8.4.5 PREEXISTING CONDITIONS AND OTHER ADJUSTMENTS

The most common assumption during fault calculations is that, prior to the fault, the system was in its nominal state: every voltage was 1 per unit, all currents were zero. As we illustrated in the case of the SLG fault, the faulting voltages obtained in our calculations are not the true voltages but rather the changes relative to preexisting voltages. To calculate actual voltages, one must superimpose the calculated faulting voltages to the prefault voltages, which are assumed to be 1 per unit for the positive sequence at all locations and zero for the zero and negative sequences.

Given that fault currents tend to be overwhelmingly larger than normal flow currents, the assumption that all preexisting currents are zero is not a severe limitation. Unlike the voltages, the above calculations for currents do not generally need adjustment by superimposing preexisting conditions. However, for some systems and under some conditions, preexisting currents can be superimposed to those obtained for the fault. Simple addition of the preexisting flow is all that is required. Of course, if this type of accuracy is required, it is likely that one should not assume that the initial voltages were 1 per unit everywhere, and instead adjust the voltages applied to the fault current calculations to the actual values of the prefault voltages.

It is common practice to give the formulae above with the option of explicit inclusion of fault impedances. In most cases, the changes to the formulae are quite simple. We prefer to deal with anything other than solid faults by means of the much more general procedure outlined in the next section.

Another common adjustment in the fault calculation process is to adjust the calculated fault currents for transient unidirectional components. The assumptions used for fault calculations presume a quasi-steady-state system. However, depending on the exact instant of fault onset, there can be a significant offset on both the currents seen and the voltages observed, over and above those obtained from the transient or subtransient models. Fortunately, these unidirectional components die out within a few cycles. Thus, for many calculations these can be ignored. However, for some applications, it is necessary to adjust for these effects. This is commonly done by means of a "fudge factor" that depends on how long after fault onset is one interested in (see Section 8.1). This unidirectional flow correction factor can be as large as 1.67 (representing a 67% excess current).

8.5 GENERALIZED THEVENIN EQUIVALENTS

The calculations in the previous sections work because the entries of the nodal impedance matrix represent entries of a generalized Thevenin equivalent for each of the sequence networks. For more complex types of faults, faults through impedances, simultaneous faults, faults along a line, open conductor faults, and many other similar calculations it is best to be explicit about these equivalents. If the reader is not interested in faults other than the four basic types, this section may be skipped. However, understanding this section will result in a greater understanding of any type of fault or change in the system.

Simply stated, the diagonal entries of each of the sequence nodal impedance matrices represent the Thevenin impedance of the sequence network as seen from that specific location. A Thevenin impedance allows us to relate voltages and currents injected at a node to each other. A generalized Thevenin equivalent relates a vector of voltages to a vector of currents at more than one node.[*]

A generalized Thevenin equivalent can be converted to a generalized Norton equivalent. The generalized Norton admittances are obtained from $\hat{\mathbf{Y}} = (\hat{\mathbf{Z}})^{-1}$ and the generalized Norton current sources from $\hat{\mathbf{I}} = \hat{\mathbf{Y}}\hat{\mathbf{V}}$. Figures 8.14 and 8.15 illustrate the generalized Thevenin and Norton equivalents in the phase domain.

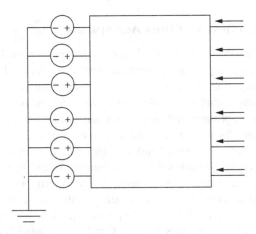

FIGURE 8.14   A generalized Thevenin equivalent in the phase domain as seen from two three-phase nodes.

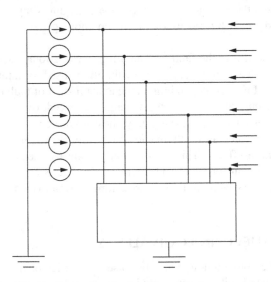

FIGURE 8.15   A generalized Norton equivalent in the phase domain as seen from two three-phase nodes.

---

[*] To distinguish Thevenin impedances (and admittances) from other admittances and impedances, we will use the hat "^" symbol modifier.

8.5.1 Phase-Domain Fault Analysis

A general procedure for fault analysis of any type of balanced or unbalanced fault using the generalized Thevenin or Norton equivalents is as follows:

1. Determine the selected elements of the sequence impedance matrices corresponding to all the locations that will be involved in the fault or network change.
2. Arrange these entries into square sequence domain Thevenin impedance matrices $\hat{\mathbf{Z}}^0$, $\hat{\mathbf{Z}}^1$, and $\hat{\mathbf{Z}}^2$. For the case of a fault at a single node, these Thevenin impedance "matrices" are just scalars; for a fault with two nodes j and k, the Thevenin impedance matrices have 2 × 2 dimension.
3. Invert a subset of the three $\hat{\mathbf{Z}}$ sequence matrices to obtain three Norton equivalent admittance matrices $\hat{\mathbf{Y}}$ for the nodes of interest in the sequence domain.
4. Convert the three $\hat{\mathbf{Y}}^0$, $\hat{\mathbf{Y}}^1$, and $\hat{\mathbf{Y}}^2$ sequence admittance matrices into a phase domain equivalent Norton admittance matrix $\hat{\mathbf{Y}}^\phi$ by using the inverse symmetrical component transformation on each entry.
5. Apply any network change or any desired fault to the Norton phase-domain model and obtain the desired phase-domain fault current injections.
6. Once the phase-domain total fault currents are obtained, the process can be unraveled: the fault currents converted back to the sequence domain, the fault injections applied to the individual sequence networks, voltages obtained, fault current flows calculated, and everything converted once more back to the phase domain if desired.

To convert the Norton sequence admittance matrices $\hat{\mathbf{Y}}^0$, $\hat{\mathbf{Y}}^1$, and $\hat{\mathbf{Y}}^2$ to the phase quantities, we use the inverse symmetrical component transformation on each of the respective matrix entries.* So, for a fault between the nodes j and k, the phase-domain equivalent Norton admittance 6 × 6 matrix $\hat{\mathbf{Y}}^\phi$ is

$$
\hat{\mathbf{Y}}^\phi = \begin{bmatrix} [\hat{\mathbf{Y}}^\phi]_{jj} & [\hat{\mathbf{Y}}^\phi]_{jk} \\ [\hat{\mathbf{Y}}^\phi]_{kj} & [\hat{\mathbf{Y}}^\phi]_{kk} \end{bmatrix} \tag{8.20}
$$

The four 3 × 3 sub-matrices $[\hat{\mathbf{Y}}^\phi]_{jj}$, $[\hat{\mathbf{Y}}^\phi]_{jk}$, $[\hat{\mathbf{Y}}^\phi]_{kj}$, and $[\hat{\mathbf{Y}}^\phi]_{kk}$ are defined using,

$$
[\hat{\mathbf{Y}}^\phi]_{jj} = \mathbf{T} \begin{bmatrix} \hat{Y}_{jj}^0 & 0 & 0 \\ 0 & \hat{Y}_{jj}^1 & 0 \\ 0 & 0 & \hat{Y}_{jj}^2 \end{bmatrix} \mathbf{T}^{-1} \tag{8.21}
$$

To convert the vector of Thevenin positive sequence nominal voltages to a vector of Norton sequence domain currents $\hat{\mathbf{I}}^1$, we use the expression (the zero and negative sequence components are zero)

$$
\hat{\mathbf{I}}^1 = \begin{bmatrix} \hat{I}_j^1 \\ \hat{I}_k^1 \end{bmatrix} = \hat{\mathbf{Y}}^1 \begin{bmatrix} 1 \\ 1 \end{bmatrix}
$$

---

* Where, as a reminder:

$$
\mathbf{T} = \begin{bmatrix} 1 & 1 & 1 \\ 1 & a & a^2 \\ 1 & a^2 & a \end{bmatrix} \qquad a = 1\angle 120°
$$

To convert the vector of sequence Norton currents to the phase domain $\hat{\mathbf{I}}^{\phi}$, we use the expressions

$$\hat{\mathbf{I}}^{\phi} = \begin{bmatrix} [\hat{\mathbf{I}}^{\phi}]_j \\ [\hat{\mathbf{I}}^{\phi}]_k \end{bmatrix} \quad [\hat{\mathbf{I}}^{\phi}]_j = \mathbf{T} \begin{bmatrix} 0 \\ \hat{i}_j^1 \\ 0 \end{bmatrix} \quad [\hat{\mathbf{I}}^{\phi}]_k = \mathbf{T} \begin{bmatrix} 0 \\ \hat{i}_k^1 \\ 0 \end{bmatrix}$$

The best way to illustrate the process is with application examples.

8.5.2 APPLICATION: FAULTS THROUGH IMPEDANCES

A powerful capability of the Thevenin–Norton approach to fault analysis is that very complex faults (or network unbalanced conditions) can be studied with ease. To illustrate the possibilities of such analysis, consider the application of a "fault" consisting of a large but unbalanced resistive load at a node.

Assume that we want to apply a shunt 0.1 per unit ohm resistive load to phase a of the system, and connect a resistance of 0.2 per unit between phases b and c. Assume we wish to do this at node 9. The "fault" and the connection of the fault network to the Norton equivalent circuit is illustrated in Figure 8.16.

The sequence Thevenin equivalents for this case are scalars:

$$\hat{\mathbf{Z}}^0 = j0.0815 \quad \hat{\mathbf{Z}}^1 = j0.0836 \quad \hat{\mathbf{Z}}^2 = j0.0836$$

The Norton admittances in the sequence domain are their simple inverses:

$$\hat{\mathbf{Y}}^0 = -j12.2717 \quad \hat{\mathbf{Y}}^1 = -j11.9660 \quad \hat{\mathbf{Y}}^2 = -j11.9660$$

These Norton equivalents are converted to the phase domain to obtain

$$\hat{\mathbf{Y}}^{\phi} = -j \begin{bmatrix} 12.0679 & 0.1019 & 0.1019 \\ 0.1019 & 12.0679 & 0.1019 \\ 0.1019 & 0.1019 & 12.0679 \end{bmatrix} \tag{8.22}$$

The positive sequence prefault network voltage at node 9 (the Thevenin voltage at node 9) can be converted to the positive sequence Norton current at node 9:

$$\hat{\mathbf{I}}^0 = 0 \quad \hat{\mathbf{I}}^1 = -j11.9660 \quad \hat{\mathbf{I}}^2 = 0$$

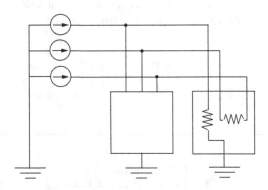

FIGURE 8.16 Application of a resistive network fault to the phase-domain Norton equivalent.

The vector of sequence Norton currents can be converted to the phase domain:

$$\hat{\mathbf{I}}^{\phi} = \begin{bmatrix} 11.9660\angle{-90°} \\ 11.9660\angle 150° \\ 11.9660\angle 30° \end{bmatrix} \tag{8.23}$$

At this point, we are prepared to do any kind of analysis. The Norton currents and Norton injections have the interpretation of a simple three-node network. In parallel with this network we connect a second network, the fault itself. To do so, we obtain the admittance matrix for the fault. On the basis of the description of the fault, the matrix that characterizes the fault has the following admittance matrix description in the phase domain:

$$\mathbf{Y}^{f} = \begin{bmatrix} 10 & 0 & 0 \\ 0 & 5 & -5 \\ 0 & -5 & 5 \end{bmatrix} \tag{8.24}$$

To connect this fault network to the original network, we add the admittance matrices keeping in mind that the fault admittance is real and the Norton admittance is purely imaginary:

$$\mathbf{Y}^{net} = \hat{\mathbf{Y}}^{\phi} + \mathbf{Y}^{f} = \begin{bmatrix} 10 - j12.0679 & -j0.1019 & -j0.1019 \\ -j0.1019 & 5 - j12.0679 & -5 - j0.1019 \\ -j0.1019 & -5 - j0.1019 & 5 - j12.0679 \end{bmatrix} \tag{8.25}$$

We apply this matrix to the Norton currents to obtain the resulting voltages:

$$\begin{bmatrix} \mathcal{V}^{a} \\ \mathcal{V}^{b} \\ \mathcal{V}^{c} \end{bmatrix} = \left[\mathbf{Y}^{net}\right]^{-1} \hat{\mathbf{I}}^{\phi} = \begin{bmatrix} 0.7699\angle{-39.66°} \\ 1.0523\angle{-151.28°} \\ 0.5188\angle 97.80° \end{bmatrix}$$

If we apply the admittance matrix for the fault $\mathbf{Y}^{f}$ to the phase voltages, we obtain the phase-domain total fault currents:

$$\begin{bmatrix} \mathcal{I}_{9}^{f,a} \\ \mathcal{I}_{9}^{f,b} \\ \mathcal{I}_{9}^{f,c} \end{bmatrix} = \mathbf{Y}^{f} \begin{bmatrix} \mathcal{V}^{a} \\ \mathcal{V}^{b} \\ \mathcal{V}^{c} \end{bmatrix} = \begin{bmatrix} 7.7013\angle{-39.66°} \\ 6.6466\angle{-129.90°} \\ 6.6466\angle 50.10° \end{bmatrix}$$

We can now calculate the sequence domain current injections. We can then apply these sequence fault currents to the sequence networks and obtain the sequence voltages at any desired node, determine the sequence flows, and so on. In the interest of brevity, we choose not to elaborate on the details.

In terms of steps, the process would be no more complicated if we wanted to apply a very complex simultaneous unbalanced impedance fault involving several different network nodes.[*]

8.5.3 APPLICATION: FAULT ALONG A LINE

To further illustrate the approach, consider a fault at any point along a line. To do this, first construct the Thevenin impedance including the line in question as seen from both ends of the line. Then remove the line from the equivalent. Finally, add the line as two segments connecting the two line ends to ground. The diagram corresponding to this procedure is illustrated in Figure 8.17.

---

[*] Faults across primary to secondary of transformers or faults explicitly involving a transformer neutral point that is not solidly grounded require considerable more care because the assumptions made during the per-unit model construction eliminate any phase shift that occurs between primary and secondary during the model construction phase. Study of these faults is beyond the scope of this book.

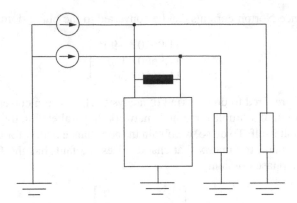

FIGURE 8.17   Removal of a line from the positive sequence Norton model and reinsertion as an impedance fault to ground.

We illustrate this only for the case of a three-phase fault, but a similar procedure would allow us to consider any other type of fault. We choose the line from bus 5 to bus 8. We calculate the total fault currents for a fault at 10%, 80%, and 95% of the distance from bus 5 to bus 8. We are interested in the total fault current, the voltages at both ends of the line during the fault and the current feeding the fault from either end.

The process starts with the construction of the sequence Thevenin matrices. These matrices are extracted from the appropriate entries of the respective sequence nodal impedance matrices, entries (5,5), (5,8), (8,5), and (8,8):

$$\hat{\mathbf{Z}} = j \begin{bmatrix} 0.0809 & 0.0367 \\ 0.0367 & 0.0596 \end{bmatrix}$$

The Thevenin voltages are all 1. The Norton admittance matrix (by inversion) is

$$\hat{\mathbf{Y}} = -j \begin{bmatrix} 17.1505 & -10.5621 \\ -10.5621 & 23.2783 \end{bmatrix}$$

The Norton currents are

$$\hat{\mathbf{I}} = -j \begin{bmatrix} 6.5883 \\ 12.7162 \end{bmatrix}$$

We now remove from $\hat{\mathbf{Y}}$ the effect of the line from bus 5 to bus 8. We do this by connecting in parallel with the Norton matrix, a matrix that models without the impedance of the line to be removed. This admittance "canceling matrix" is

$$\mathbf{Y}^{\text{line}} = -j \begin{bmatrix} 10.0000 & -10.0000 \\ -10.0000 & 10.0000 \end{bmatrix}$$

If we subtract this matrix from the Norton matrix, we get $\mathbf{Y}^{\text{wo}}$, a matrix without the line from bus 5 to bus 8.

$$\mathbf{Y}^{\text{wo}} = \hat{\mathbf{Y}} - \mathbf{Y}^{\text{line}} = -j \begin{bmatrix} 7.1505 & -0.5621 \\ -0.5621 & 13.2783 \end{bmatrix}$$

We reintroduce the line from bus 5 to bus 8 as two segments, one connecting node 5 to ground and the second connecting node 8 to ground, as illustrated in Figure 8.17. The result is a new admittance

TABLE 8.4

Voltages and Currents as a Result of a Mid-Line Three-Phase Fault

Distance from 5	V_5^a	V_8^a	$\mathcal{I}_5^{f,a}$	$\mathcal{I}_8^{f,a}$
10%	0.0642	0.5229	$-j6.4230$	$-j5.8096$
80%	0.3411	0.2040	$-j4.2639$	$-j10.1993$
95%	0.3746	0.0606	$-j3.9436$	$-j12.1220$

matrix $\mathbf{Y}^{net}$ including the effect of the fault:

$$\mathbf{Y}^{net} = -j\begin{bmatrix} 107.1505 & -0.5621 \\ -0.5621 & 24.3894 \end{bmatrix}$$

We obtain the voltages at the two ends of the line by applying $\hat{\mathbf{I}}$ to this new matrix:

$$\begin{bmatrix} V_5^1 \\ V_8^1 \end{bmatrix} = (\mathbf{Y}^{net})^{-1} \ \hat{\mathbf{I}} = \begin{bmatrix} 0.0642 \\ 0.5229 \end{bmatrix}$$

Knowing these voltages, we can calculate the flows between node 5 and the fault and node 8 and the fault. The result of this calculation, as well as the results of calculations for other fault locations, are summarized in Table 8.4 (phase *a* and currents from node 5 or 8 to fault point).

8.6 NEED FOR A PROTECTION SYSTEM

Continuity and quality of service [14] are two requirements closely linked to the smooth running of an electric power system.

Continuity means that the power system must guarantee that the energy produced in the generation centers will be supplied without interruptions to the consumption centers. This characteristic is particularly important because electric energy, unlike other types of energy, cannot be stored in significant amounts. Therefore, a break in the supply has a direct and immediate impact on the processes that are based on electric energy consumption.

The quality requirement implies that the energy must be supplied in specific conditions, to guarantee that the different devices connected to the electrical network operate as expected. The acceptable margins for each magnitude (wave parameters, frequency, phase balance, harmonic distortion, etc.) depend on the installation.

When a fault occurs, the power system associated magnitudes reach values outside their normal operating ranges and certain areas of the system may begin to operate under unbalanced conditions, with the resulting risk for the different elements that make it up. If measures are not adopted to limit the fault propagation, the impact of the fault would spread throughout the network. The quality of supply would be affected even in zones far away from the point where the fault occurred.

It is impossible to prevent faults from occurring for both technical and economic reasons. The design of an electrical system must contemplate the possibility that unexpected and random faults are going to occur. The system needs to be able to deal with them. The power system must include a protection system intended to minimize the effects arising from the different types of faults that can occur.

The performance of the protection system is aimed at maintaining the quality and the continuity of the service, with the purpose of both characteristics to be minimally impacted during a minimum time. The network therefore needs to be planned in such a way that operating alternatives can be offered to provide the suitable power supply for all consumption points even should faults occur that affect the generation, transmission, or distribution elements.

Even though a fault may occur in almost any of the system's elements, about 90% of all faults occur in overhead lines, with one-phase fault (SLG) being the most prevalent. The reason for this disproportionate number is that overhead lines cover large stretches of land and are exposed to external actions that are impossible to control, whereas other types of elements, such as generators, transformers, and so on often operate under much more controlled conditions.

Independently of the point where the fault occurs, the first reaction of the protection system must be to disconnect the faulty circuit, to insulate the circuit from spreading, and to reduce the time that the most directly affected devices are subject to extreme conditions. Disconnecting the faulty circuit using automatic circuit breakers causes a transient situation that, likewise, may imply a series of changes, such as overvoltages, imbalance between generation and consumption, and a consequent change in the frequency, and so on. When these fault consequences result in unacceptable conditions for certain elements, the protection system must operate to disconnect circuits and components that, even though the elements are not directly affected by the fault, they would be affected by its impact.

Once the fault and its impact have been neutralized, actions must be taken to return the system to normal operating conditions as quickly as possible.

8.7 FUNDAMENTALS OF FAULT PROTECTION SYSTEMS

8.7.1 FUNCTIONAL FEATURES OF A PROTECTION SYSTEM

A protection system as a whole and each of its protection relays must meet the following functional characteristics:

8.7.1.1 Sensitivity

The protection relay must unequivocally differentiate between fault conditions and normal operating conditions. To accomplish this objective consider the following:

- The necessary minimum magnitudes that allow fault conditions to be differentiated from normal operating conditions must be established for each type of protection relay.
- "Borderline conditions" that separate fault conditions from normal operating conditions must be identified.

Borderline conditions are a broader concept than "borderline values." On many occasions, just knowing the value of a magnitude is not enough to determine whether this value has been reached as a result of a fault or it is the result of a normal incident within the operating system.

Such is the case, for example, during the energization of a power transformer. This energization causes an inrush current which, if it is analyzed solely and exclusively from the point of view of its high value, may result in incorrect interpretations. A wider analysis, which includes studying the wave shape through its harmonic components, reveals whether the sudden increase of the current is due to the energizing of the transformer or has been caused by a fault situation.

8.7.1.2 Selectivity

Once a fault has been detected, selectivity is the capacity that the protection relay must have to discern if the fault has occurred within or outside its surveillance area and, therefore, give the order to the automatic circuit breakers to trip to remove the fault.

It is as important for a protection relay to operate when it must operate as to not operate when it must not operate. If a fault has occurred within the surveillance area of the relay, it must order the opening of the circuit breakers that insulate the faulty circuit. If, on the contrary, the fault has occurred outside its surveillance area, the protection relay must allow other protections to operate to remove it, as its tripping would leave a higher number of circuits nonoperational than the number that is strictly necessary to isolate the fault and, therefore, would result in an unnecessary weakening of the system.

There are various ways to provide the protection systems with selectivity. In some case, the very configuration of the protection system means that it is only sensitive to faults occurring in its protection area and, therefore, selectivity is an inherent quality to the very operating of the relay. In cases where the relays are sensitive to faults occurring outside their surveillance area, selectivity may be achieved, for instance, by means of the suitable setting of operating times and conditions of the various protection elements of the system.

8.7.1.3 Speed

After being detected, a fault must be removed as quickly as possible. The shorter the time it takes to isolate a fault, the smaller its impact will be and the lesser the damages to the different elements operating under the abnormal conditions until the fault is isolated. Fast fault isolation reduces the costs and recovery times of the system after a fault. Faster fault isolation allows greater and better use of the resources offered by the power system.

The rapidity with which a protection relay may act depends on the technology used for the protection system as well as the response speed of the control system and the response of the relevant automatic circuit breakers. However, it does not follow that all the relays that detect should always operate immediately. The protection relays are classified into

1. Instantaneous protection.

    They are those that act as quickly as possible due to the fault having occurred within their direct surveillance area. The usual removal time of an HV fault by means of instantaneous protection is currently around two or three cycles. If the removal time is lower, the relay is called a high speed relay.
2. Time delay protection.

    They are those where a time delay is intentionally introduced to delay the relay being tripped, in other words, that it delays the start of the maneuver to open the circuit breakers once the decision to trip them has been taken. The delay facilitates, for example, the coordination between relays so that only those that allow the fault to be isolated are tripped, which disconnect the minimum possible portion of the power system.

8.7.1.4 Reliability

Reliability means that the protection relay should always respond correctly. That means that the relay must securely and effectively respond to any situation that occurs.

The response of the relay must not be confused with its tripping or operating. The protection relay is continuously monitoring what happens in the system and, therefore, it is responding at each moment according to the conditions that occur there. Therefore, the response of the protection relay may be to operate or not to operate. Security means that no unnecessary tripping may occur or necessary tripping omitted.

On the other hand, when the relay must operate, all the stages of the process to remove the fault need to be effectively performed. The failure of any of them would mean that the tripping order given by the relay could not be duly implemented by the relevant automatic circuit breaker.

Therefore, the great importance that defining a suitable preventative maintenance program has for the protection system [15] must be highlighted. It has to be remembered that a protection relay is only tripped in fault conditions and that these conditions are rare and exceptional in any modern power system. Therefore, even though a relay throughout its useful life will only be tripped on very few occasions (possibly never), there has to be the assurance that it will operate correctly every time it may be needed.

8.7.1.5 Economy and Simplicity

Installing a protection system must be justified for technical and economic reasons. Protecting a line is important, but it is much more important to prevent that the effects of the fault reach the installations supplied by the line. The protection system is a key part of the power system as it allows

- The fault spreading through the system and reaching other devices and installations causing a deterioration in the quality and continuity of the service to be prevented.
- The costs of repairing the damage to be reduced.
- The time that devices and installations remain out of service to be reduced.

Therefore, the economic assessment must not only be restricted to the directly protected element, but rather must take into account the consequences that the fault or the abnormal operating of the aforementioned element would imply.

Finally, it must be pointed out that a relay or protection system must avoid unnecessary complexities, as they would be sources of risk that would affect the compliance of the properties that must characterize its behavior.

8.7.2 STRUCTURE OF A PROTECTION SYSTEM

The great importance of the function performed by the protection system means that it is recommendable to provide it with a structure that prevents the failure of any of its components leaving the power system unprotected and causing a series of undesired consequences.

A technical analysis would recommend covering the possible failure of the protection devices by means of backup devices. However, economic considerations make the use of backup devices infeasible in the cases where experience shows that the probability of a fault occurring is minimum. On the other hand, in cases such as that of protecting overhead lines that statistically withstand around 90% of the faults that occur in a power system, establishing backup protection systems is fundamental.

Thus, the network protection system is structured into two forms:

1. Primary relaying.
2. Backup relaying.

8.7.2.1 Primary Relaying

Primary relaying is responsible for initially removing the fault. They are defined to disconnect the minimum number of elements needed to insulate the fault.

To optimize its services, the power system is divided into primary relaying zones defined around each important element, as indicated in Figure 8.18. Each zone overlaps with the adjacent zones to

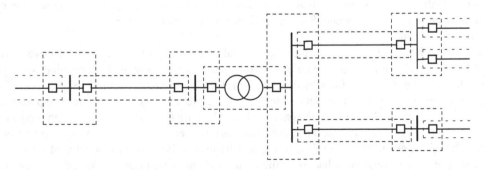

FIGURE 8.18 Primary protection zones.

ensure there are no dead zones not covered by primary relaying. The overlap between two zones is established around a communal circuit breaker for both areas that is used to separate the two adjoining elements.

When a fault occurs inside a zone, the relevant primary relays must trip its circuit breakers, but only those and no other must be tripped to remove the fault. Only in the case, which is highly unlikely but possible, that the fault occurs in the overlap zone, the primary relaying may disconnect a wide area than the one strictly necessary to insulate the fault.

8.7.2.2 Backup Relaying

The backup relays are those that are responsible for removing the fault as a second alternative, in other words, they must only operate at the relevant primary relays which have failed. It is, therefore, very important to make the causes of failure in the primary and backup relays independent from each other, so that nothing that may cause the primary relay to fail is also capable of causing the backup relay to fail. This is usually achieved by using different elements, control circuits, supply circuits, and so on in either type of relay.

The backup relays must be setting with a time delay with respect to the relevant primary relays. Once primary relays have been tripped, the backup relays must be reset to prevent unnecessary openings of the circuit breakers.

Local backup relaying [16] is the term used when it is located in the same substation as the relevant primary relaying. Duplicity of elements, such as the instrument transformers for protection, means that it is essential that the reasons for the failure in one or other type of relay must be made independent.

When the backup relay is installed in a substation adjoining the one that contains the primary relay, it is referred to as remote backup relaying. The remote backup relays offer the advantage of separating, as the result of their very installation philosophy, the failure causes with respect to the relevant primary relays. However, there is the drawback that every time they are tripped, an area of the network bigger than that strictly necessary to isolate the fault is always disconnected.

Finally, it should be pointed out that a single relay may act as a primary relay for a specific element and, at the same time, as backup relaying for another element. Likewise, when the primary relays are out of service for repairs or maintenance, the relevant backup relays are used as primary relays for any faults that may occur.

8.7.3 ELEMENTS OF A PROTECTION DEVICE

A protection device is not only the relay, as such, but also includes all those components that allow the fault to be detected, analyzed, and removed [17,18]. The main elements that make up a protection device are

- Battery.
- Instrument transformers for protection.
- Protection relay.
- Automatic circuit breaker.

8.7.3.1 Battery

The battery is the element that guarantees the continuity of the supply of energy needed to operate the protection device. The protection device cannot be supplied directly from the line. If that were the case, a fault that left a substation without power, or caused the energy supply to be defective, would also leave all the protection devices there out of service. This would have serious consequences due to the fact that it is precisely when there is a fault that the protection system must operate.

Therefore, a protection device has to have its own power supply that allows it to operate as an island, without depending on external sources, during sufficient time. Generally speaking, the direct current battery is permanently connected through a charger to the alternating current of the auxiliary services of the substation and, should the ac line fail, it has a battery life of around 10–12 h.

8.7.3.2 Instrument Transformers for Protection

The input data to the protection relay must reflect the status of the power system. Although several exceptions exist, the data that is normally used are those relating to current and voltage magnitudes. Logically, due to its high value, the existing current and voltages in the network cannot be used directly as input signals to the relay and elements that reduce them to a suitable level must be used. These elements are the instrument transformers for protection [19].

The instrument transformers reproduce on a reduced scale in their secondary terminals, the high value magnitude that supplies their primary terminal. For the information to correctly arrive at the protection relay, the secondary connections also need to be performed in respect to the directions marked by the relevant primary and secondary terminals, particularly if it is taken into account that some types of relays are sensitive to the polarity of the signal that reaches them.

The data provided by the instrument transformers is affected by a specific error. The accuracy class is a characteristic data of each instrument transformer that refers to the maximum error that may incorporate the information provided by the transformer when it functions within the conditions for which it is defined. The lower the value of the accuracy class, the lower the maximum error, and the greater the accuracy of the data obtained using the transformer.

Conventional instrument transformers provide reliable information when they are working in the range of values corresponding to the normal operation of the power system. However, it is under fault conditions when it is more important for the protection relays to receive reliable data. Therefore, the network data must be supplied to the relays by means of instrument transformers for protection, which are projected and constructed to guarantee precision in the extreme conditions that arise when a fault occurs.

Depending on the magnitude that they transform, the instrument transformers for protection may be

- Voltage transformers (VT).
- Current transformers (CT).

Voltage transformers have the same operating principle as power transformers. Usually, their secondary nominal voltage is 100 or 110 V in European countries and 120 V in the United States and Canada. They may be phase–phase, used only for voltages under 72.5 kV, or phase-earth. It is very common to use capacitor voltage transformers which, basically, consist of a capacitor divider that reduces the voltage applied to the primary of a conventional inductive voltage transformer. Depending on the voltage that is to be obtained, the voltage transformers may be connected according to various wiring diagrams. For example, to obtain homopolar voltage, the secondaries must be arranged in open delta.

The current transformers are connected in series with the conductor where the current circulates that is to be measured. Their secondary nominal current is usually 5 A, although the 1 A-one is also usually used. The greatest danger for their accuracy is that the major current that are produced as the result of a fault, cause it to saturate.

It is very common for current transformers to have various secondary ones with different characteristics, as each secondary one has its own nucleus and is independent from the others. If a current transformer has, for example, two secondaries, it is normal for it to use one to measure and the other to protect. Depending on the current that is to be obtained, the current transformers may be connected according to various wiring diagrams. For example, to obtain the homopolar current, the secondaries must be arranged in wye.

8.7.3.3 Protection Relay

The protection relay, which is usually known as the relay, is the most important element of the protection device. It can figuratively be said that it acts as the brain of the device, as it receives the information, processes it, takes the decisions, and orders the measures to be taken.

To do so, independently of the technology used to build it, a protection relay internally develops three fundamental stages:

1. Signal processing.
2. Protection functions application.
3. Tripping logic.

The relays require data that, in general, cannot be directly provided by the instrument transformers that supply them. Therefore, the first stage consists of adjusting the input signals to the format that the relay needs to work. The input data are usually instantaneous values of the phase magnitudes (voltage and current). These are used, according to the specific necessities of each relay, to determine RMS values, maximum values, aperiodic components, sequence components, upper order, or fundamental harmonics, and so on.

Once the protection relay has the data that it requires, it proceeds to apply the decision criteria that have been implemented. The decision criteria are constructed by means of basic protection functions that will be explained later. The element where each basic function is performed is called the measurement unit. Owing to the complexity and variety of factors that need to be taken into account, various basic functions are usually needed to be included for a relay to operate correctly. Therefore, a relay usually consists of various measurement units.

The results provided by the various functions that make up the relay are analyzed together by means of the tripping logic, which is responsible for deciding how the relay must act. This action is implemented by means of auxiliary control circuits of the circuit breakers associated to the operation of the relay. The order is transmitted through the contacts that energize the trip circuits of the circuit breakers that have been defined by the tripping logic as those that need to be opened to isolate the fault.

Likewise, the protection relay governs another series of auxiliary control circuits that are used, for example, to activate alarms, send information to the operating control center, and so on.

8.7.3.4 Automatic Circuit Breaker

The automatic circuit breaker [20] is the element that allows a powered circuit to be opened or closed, breaking or current circulation establishing. It operates under the control of the protection relay and its opening, coordinated with that of other circuit breakers, allows the point where the fault occurred to be isolated. It basically consists of

- Control circuit, which is controlled by the relevant protection relay.
- Main contacts, which as they separate or come together implies, respectively, the opening or closing of the circuit breaker.
- Auxiliary contacts, which reflect the status of the circuit breaker. They are used to supply the relay and other devices with the information about whether the circuit breaker is open or closed and, therefore, establish whether the circuit breaker has operated correctly following the order given by the protection relay.
- Arc extinction chamber, where a high dielectric rigidity environment is created; this environment encourages the extinction of the arc that is produced as a result of the separation

of the main contacts of the circuit breaker that are immersed in it. Oil and sulfur hexafluoride are currently among the most common dielectric means.

From the point of view of the relay, independently of the technology used for its construction, the two main characteristics that the circuit breaker must meet are

- Rapid separation of the main contacts, to minimize the time need to carry out the opening operation. When the relay gives the order to perform the opening to insulate the fault, the trip circuit is activated and, therefore, the contacts begin to separate. However, the initial separation of the contacts does not imply the immediate opening of the circuit, as an arc is initially established that maintains the current circulation between the two contacts. The circuit break occurs in the first cross zero current, but, if the separation of the contacts is not sufficient at that time, the voltage between them means that the arc is established again. The definitive circuit break, and therefore the opening of the circuit, occurs in the subsequent cross zero current, as the contacts have then had time to sufficiently separate to prevent the arc from linking up. The faster it takes for the contacts to separate, the less time will be needed to reach the distance that guarantees the opening of the circuit. Usually, the definitive circuit break in general occurs during the second or third cross zero current.
- Sufficient breaking capacity to guarantee the interruption of the maximum short-circuit current that may occur at the point where the circuit breaker is installed. The breaking capacity is closely linked to the capacity of the dielectric medium to also undertake the function of cooling medium, as it must be capable of channeling outside the energy released in the arc extinction process. It is common practice to use various arc extinction chambers in series, whose contacts must be operated in a synchronized manner, to increase the breaking capacity on HV lines. This fact does not introduce any modification from the relay point of view, as it is a single tripping order in any case and it is the circuit breaker that must incorporate the necessary mechanisms to ensure the synchronization.

In the case of overhead lines, it is very usual that the causes of a fault are temporary, in other words, that they disappear after having caused it and the fault been removed. The reclosure operation is therefore used in the protection of overhead lines [21]. After a prudential time from a fault having occurred and after having performed a three- or single-phase trip to remove it, the reclosure operation comprises closing the circuit again, by means of the relevant single- or three-phase closure. The waiting time from when the circuit breaker opens until its closure is attempted again is necessary to deionize the medium contained in the arc extinction chamber. If the causes were temporary, the reclosure will be successful and the system will continue to operate satisfactorily, with it being not available for a minimum time. If, on the other hand, the causes leading to the fault still persist, the relay will order that the circuit breakers are tripped again.

On occasions, above all in distribution networks, various reclosure attempts are programed, separated by increasing time intervals, to ensure that the causes resulting in the fault have not disappeared by themselves at the end of a certain time. To have an idea of the size of the generally programed times, it should be pointed out that, in distribution lines, the first reclosure attempt is performed after approximately 0.2 s and that the two or three successive reclosure attempts, should they be necessary, are performed between the following 10 and 150 s.

Therefore, the relay controls both the trip circuit and the circuit to close the automatic circuit breaker. When the importance of the installations or protected devices thus justifies it, the control circuits are installed in duplicate to ensure, for example, that even when a failure occurs that puts the main trip circuit out of service, the opening of the circuit breaker is guaranteed by the action of the reserve trip circuit.

8.7.4 PROTECTIVE FUNCTIONS

Even though the functions carried out by the relays are very varied and complex, an abstraction of them can be performed that allows them to be classified into the following basic four types [18]:

1. Level function of a single magnitude.
2. Quotient function of two magnitudes.
3. Phase comparison function.
4. Magnitude comparison function.

The particularization and combination of those basic functions result in the different functions that characterize the operation of the different types of existing relays.

Each of these basic functions are summarized below and the main types of protection relays used in electric power networks are explained in subsequent points.

8.7.4.1 Level Function of a Single Magnitude

This function takes as data the value of a single magnitude at a single point of the system and compares it with a threshold level that has been established to characterize the abnormal system operation that must be detected by the relay. Not all the functions belonging to this category take the same type of value as data. Thus, for example, it may be convenient to consider for some of them the RMS value and for others, more suitable to take into account the maximum or peak value.

The overcurrent function, together with the under-voltage function and the frequency function, are the most characteristic example of this type of functions.

8.7.4.2 Quotient Function of Two Magnitudes

This function uses quotient to establish the relation between the two input data magnitudes to the relay. The result of the quotient is the factor used to analyze the situation monitored by the protection relay and, based on its value, the decision to act in one or other way is taken.

The impedance function is the most important and most used of this type of functions. This function is used in distance protections and it uses the voltage and current magnitudes existing at the end of the line where the relay is installed as input data.

8.7.4.3 Phase Comparison Function

What this function basically does is a comparison between the time sequence of two magnitudes. The above function can be explained in terms of vectorial or phasorial representation by saying that this function compares the phase angles corresponding to two magnitudes, which gives the angle existing between both. This type of functions may be used to monitor, for example, the direction of the power flow and detect its inversion, and so on.

This group of functions includes the directional function, which uses voltage and current as input data, or two currents on occasions, and establishes the existing angle between both magnitudes. Using the angle obtained as a result, the relevant directional protection takes the decision to act in one or other way.

8.7.4.4 Magnitude Comparison Function

This type of functions are used to compare the result of the combination of the input magnitudes with the reference value established in each case, to differentiate the normal operating conditions from those that are not. Even though each specific function has its own characteristics, the magnitudes taken into account are usually currents.

This group of basic functions includes the differential function that, basically, uses the currents at two points as input data to determine if there is any current leak in the stretch of circuit included between both.

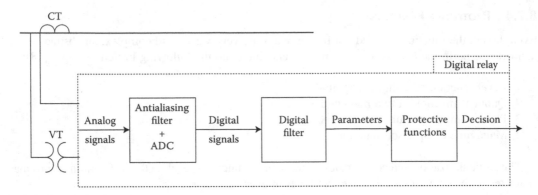

FIGURE 8.19   Components and stages of a digital relay.

8.7.5 DIGITAL RELAYS

The structure and elements exposed in this Section 8.7 can be implemented using different technologies, each with its own distinguishing features. Due to their intrinsic advantages (flexibility, adaptability, possibility of communication with remote systems, interoperability, etc.), digital protection systems have become the standard choice in transmission systems and will be progressively adopted in distribution systems.

A digital relay is a protection device based on digital technology. One of its main features is the possibility to implement multiple protective functions and other additional functions on a single unit. The major components and stages of a digital relay are shown in Figure 8.19. They are

1. Data acquisition and signal conditioning: The input data are analog signals from the CT and/ or VT. If it is necessary, they must be scaled down to match the input range according to the digital relay requirements. In order to avoid aliasing effect, the analog signals pass through an antialiasing (low pass) filter that eliminates the high frequency components. Then, an Analog to Digital Converter (ADC) provides the discrete samples that will be numerically processed in the following step.
2. Digital filtering: A Digital Signal Processor (DSP) carries out this process. A digital filter is implemented on the DSP in order to estimate the parameters that will be used by the protective functions. These parameters are usually the amplitude and angle of harmonic components (in most cases the fundamental component). The most popular algorithms implemented on DSP are based on Discrete Fourier Transform (DFT). Taking into account that DFT provides correct results during steady state, these algorithms modify the DFT in order to also guarantee correct results during transient periods.
3. Protective functions: The different protective functions are implemented through software on a microprocessor. They are a set of logical and mathematical criteria which, based on the output data of the digital filter, determine the correct decision to be made by the relay.

In a digital relay, hardware and software must be able to work in real time. In addition, the cutoff frequency of the antialiasing filter and the sampling rate of the ADC must be set taking into account the Nyquist-Shannon theorem and the input data required for the protective functions.

8.8 OVERCURRENT PROTECTION

The overcurrent protections are the simplest of all. Their operation is based on the overcurrent function that consists in comparing the value of the current used as input data to the relay with a reference value (pick-up current). This reference value is established according to the conditions that converge

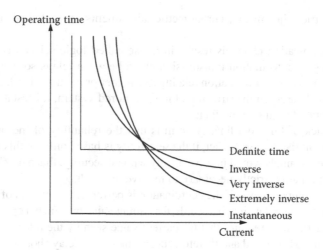

FIGURE 8.20 Operating characteristic curves of overcurrent protections.

at the point where the relay is installed. It must therefore be readjusted as required if the configuration of the system changes. The relay operates when the input current exceeds the value of the pick-up current. Therefore, overcurrent protection may only be used when the current that circulates through the point where it is installed complies with the condition that the maximum load current, corresponding to normal operation conditions of the system, is less than the minimum fault current.

Depending on the operating time, the overcurrent protection is classified into

1. Instantaneous overcurrent protection.
2. Time delay overcurrent protection.
   - Definite time.
   - Inverse time.

The instantaneous overcurrent protections are those that operate immediately; in other words, they do not introduce any intentional delay time in their operation from the moment that the input current exceeds the reference value.

The time delay overcurrent protections are those that include an intentional time delay in their operation. When this time is independent of the value of the input current, it is known as definite time overcurrent protection.

When the delay time is a function of the input current magnitude, it is called inverse time over-current protection. In those cases, the larger the value of the current the lower the introduced delay time and, therefore, the less time that the protection relay takes to operate.

Figure 8.20 shows that the characteristic curves correspond to the aforementioned different types of overcurrent protection. It shows how the characteristics corresponding to the inverse time protections may be, according to its slope, inverse, very inverse, or extremely inverse.

In each specific protection unit, the factors that define its characteristic curve may be adjusted within the range for which it was designed. Therefore, each relay has a family of characteristic curves. When the relay is installed, it must be set so that it operates with the characteristic curve that best responds to the specific characteristics of the point in which it is located and is most suitable to the coordination with other relays.

The main disadvantage of the overcurrent protections is that individually they are barely selective due to the fact that their response solely depends on the value of the current that they are seeing, irrespective of the underlying cause, the direction of its circulation, or the point where the fault occurred. Therefore, the selectivity of the overcurrent protections is defined overall, in other words, suitably coordinating all of their responses. This coordination may take place by means of communication channels, amperemetric adjustments, chronometric adjustments, or technical ones resulting from their combination.

The use of communication channels results in the so-called logic selectivity. This technique is based on establishing an information transmission chain between relays so that each one receives and transmits, when necessary, information relating to whether or not there is a fault current. According to the logic defined based on the structure of the monitored system, it is established which of the relays should be tripped to remove the fault.

The main drawback of this coordination form is that the reliability of the protection system is affected by a significant risk factor. In fact, if the selectivity is based only on this technique, a failure in the communication channels would involve, nearly with all security, that a much wider zone would be out of service than that is strictly necessary to remove the fault.

Selectivity is called amperemetric if the coordination is performed by means of adjusting the reference value of each protection relay, in other words, the current value on which it operates. This technique is seldom used in practice as it requires that the current value seen by the relay univocally defines the zone in which the fault has occurred and therefore, determines if the relay should or should not operate.

When the selectivity is achieved by duly adjusting the operating times of the relays, it is called chronometric selectivity. Logically, this type of selectivity is applied to the time delay overcurrent protections, as the instantaneous overcurrent protections do not allow their operating time to be set. A great drawback of this technique is that the majority of the cases cause a delay in the removal of the fault, which may result in serious consequences. Despite this, chronometric selectivity is the one that is most used among overcurrent protections.

The poor selectivity services and the drawbacks inherent to their coordination mean that overcurrent protections are fundamentally used in the sphere of distribution networks. They are used to a much lesser extent in HV transmission lines. This is due to the fact that the faults in distribution lines do not generally affect the stability of the electric power system, therefore economic cost is the main factor when establishing their protection system. On the contrary, in HV transmission lines, it is very important to remove the fault as quickly as possible so as to isolate it by disconnecting the smallest number of circuits. Otherwise, the stability of the electric power system would be endangered and weakened unnecessarily. Therefore, other more sophisticated and costly types of protection relay, such as distance protections, are used to protect HV transmission lines.

On the other hand, the overcurrent protection application philosophy greatly depends on the system topology being monitored and, therefore, it differs notably when radial networks or meshed networks have to be protected.

8.8.1 RADIAL NETWORKS

The power flow is always in the same direction for a distribution radial network without intermediate points of power injection.* In each line that makes up this radial network, power flow always runs from the end nearest to the generation and to the end nearest to the load. If a fault occurs in a radial network, it may be considered, after ruling out the transient currents potentially injected by motors, that the fault current is only contributed from the generation side. Therefore, in this type of networks, there is *a priori* additional information of the circulation direction of the current in each line. This data is fundamental at the time of applying overcurrent protections as, by facilitating their coordination, it allows the essential degree of selectivity required for any protection system.

---

* Nowadays, the network operating conditions are changing as distributed generation is incorporated to power systems. The possibility of a bidirectional power flow must be taken into account to guarantee a correct behavior of the protection system. In this case, the protection system must be modified and coordination techniques similar to the ones described in the following sections must be carried out.

The faults in radial networks must be removed, in each case, by opening the automatic circuit breaker which, as it is situated on the side of the generation, is closest to the point where the fault has occurred. Therefore, if a fault occurs at point F in the radial distribution network shown in Figure 8.21, the automatic circuit breaker that must remove it is B.

The diagram clearly shows how opening circuit breakers C or D do not eliminate the fault, as the fault current comes from the generation side through A and B. On the other hand, the opening of circuit breaker A allows the fault to be removed, but causes circuits Z1–Z5 to be disconnected. Only opening circuit breaker B removes the fault and, at the same time, leaves the minimum number of circuits out of service, which in this case are circuits Z4 and Z5, due to the fact that they are supplied by the line where the fault has occurred.

In general, this selectivity is achieved by means of the chronometric coordination of the overcurrent protections that govern the automatic circuit breakers. Therefore, they must be set in such a way that, referring to equivalent currents, their operating times will increase the closer they are located to the generation and further from the consumption.

Even though this time scalability can be achieved with definite time overcurrent protections, this is usually performed with inverse time units, due to the fact that they allow the times under fault conditions to be reduced as the value of the fault current increases and, therefore, they mitigate the severity of the consequences of its circulation.

Figure 8.22a shows the characteristic curves of the inverse time overcurrent protections that control the circuit breakers involved in the fault shown in Figure 8.21. It can be seen how, for the circulation currents caused by the fault at F, the relays at C and D, even though they are the quickest, will not trip as they do not see the fault current, which is only seen by the relays at A and B. In this case, the relay at B will trip as it is the quickest of the two and, therefore, performs primary protection functions in the line in which it is installed.

The relay at A will only remove the fault if the relay at B fails. In that case, the relay at A performs the backup protection functions of the faulty line.

Establishing a logic selectivity would be much more expensive as it would require communication channels between all the protection relays. In that case, a relay will only operate when it detects the fault current and, furthermore, receives information that the relays located immediately downstream do not detect fault current. In the analyzed case, the relay at B would open the circuit breaker as it sees the fault current and receives the information that C and D do not detect the fault current. The relay at A would not trip as even though it sees the fault current, it receives the information that B also detects

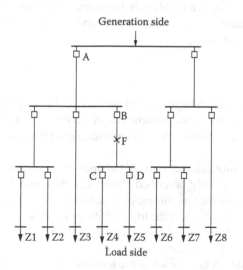

FIGURE 8.21   Radial network.

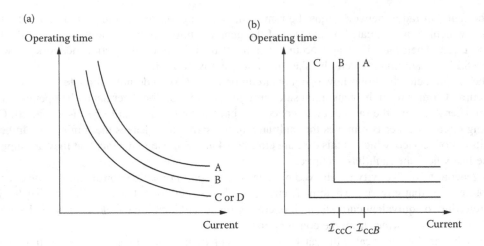

FIGURE 8.22 Overcurrent protections coordination.

the fault current. This technique, combined with the chronometric selectivity, would allow the times needed to remove the fault to be reduced, as it allows B to cancel its delay time as it receives data that confirms that it must remove the fault.

The values of the short-circuit currents, increasing as we get near to the side of the generation, allow amperemetric selectivity to be established in a radial network. Each relay must be set in such a way that it only operates if the current is greater than the short-circuit current for the position of the adjacent circuit breaker located toward the load side. Therefore, for the network shown in Figure 8.21, the reference value for the relay in B must be the short-circuit current in C, while the value of the short-circuit current in B will be taken as the reference for the protection in A.

One drawback of amperemetric selectivity is that to ensure that the faults that occur just at a circuit breaker input point are correctly removed, the relays need to be set for reference values that are a little lower than those indicated. However, this may lead to an erroneous selectivity between the relays. For example, if fault F occurs at a point that is very close to the output point of circuit breaker B, the relay at A may remove it, if it is quicker than that located at B, as both relays see fault currents within their operating range. Therefore, to ensure that the protection system reacts satisfactorily, the amperemetric selectivity should be completed with a suitable chronometric selectivity. Figure 8.22b shows the characteristic curves corresponding to joint amperemetric and chronometric selectivity, when definite time overcurrent protections are used.

8.8.2 Meshed Networks

It is not possible to define the selectivity in meshed networks only using overcurrent functions. Therefore, for this type of networks, directional overcurrent protections need to be used, whose name comes from the fact that they operate when the current surpasses the set reference value and, also, circulates in a specific direction.

A directional overcurrent protection consists of two measurement units inside. The first is an overcurrent unit, similar to those already described above, and the second is a directional unit, which allows the circulation direction of the monitored current to be known. The trip logic developed means that it only operates when both units meet the trip conditions for which they have been set. The directional function, belonging to the group of basic phase comparison functions, takes as data the angles corresponding to the phasors of the polarization and operation magnitudes with which it is supplied.

The polarization magnitude is the one set as the reference. In order for the directional function to provide satisfactory results, the polarization magnitude must meet the requirements of not being

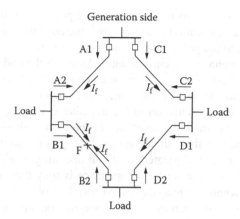

FIGURE 8.23 Overcurrent protections in meshed networks.

canceled when a fault occurs and to keep its direction invariant even when the direction of the operation magnitude changes. The operation magnitude is what is compared with the polarization magnitude and, in this case, agrees with the current monitored by the overcurrent function that is implemented together with the directional function.

Figure 8.23 shows a loop belonging to a meshed distribution network where a fault has occurred at point F that causes the fault currents in each line to circulate in the direction indicated there. In this case, the opening of a single circuit breaker is not sufficient to eliminate the fault as, due to the existence of the loop, this would remain through the other circuit breakers that remain closed. Even though all the relays see the fault current, to correctly remove it, only the two circuit breakers located at the end of the faulty line must be opened. Thus, the relays must be set to operate only when the fault current goes toward the interior of the line at whose end they are located, in other words, when the direction of the fault current is that shown in Figure 8.23 by means of an arrow located next to each circuit breaker.

Thus, in the analyzed case, the F fault must be removed by means of opening B1 and B2 circuit breakers. The B2 relay will operate as the fault current is seen in the direction in which it has been polarized, while the A2, C2, and D2 relays will not operate as, even though they see the fault current, its direction is leaving the line they are supervising. Finally, the B1 selectivity compared to A1, C1, and D1 will be achieved by means of other coordination methods, such as the chronometric or logic ones indicated above.

8.9 DISTANCE PROTECTION

The distance protections [22] are the ones that are most used to protect HV transmission lines. Even though the overcurrent protections are still used on some occasions, mainly for phase-earth fault protection or as backup to others, distance protections are more suitable due to the fact that, even though they are more complex, they offer better services and this fact prevails over the cost when the aim is to protect an element as important as the HV transmission network.

Some advantages of the distance protections compared to the overcurrent ones are:

* They offer better selectivity, due to the fact that they are easier to inter coordinate.
* Their coordination allows quicker responses.
* The configuration changes of the electric system have less impact on their adjustment values.
* The power oscillations have less impact.

Despite what may be indicated by its name, the distance protections do not calculate the distance from the fault, but rather determine if it is inside or outside the zone that they monitor. Therefore, they perform functions belonging to the group of basic quotient functions of two magnitudes that, in this case, are obtained using the voltage and current data relating to the end of line where the distance protection is located. On the basis of these, the impedance seen by the relay is obtained. Therefore, the distance protections are also known as impedance protections.

This conversion allows the representation of the impedance seen by the relay and its operating characteristic on Cartesian axes where the resistance or real part is shown on the abscissa axis and the reactance or imaginary part on the ordinate axis, resulting in the so-called R–X diagram [23].

The impedance seen by the relay is greater in normal operating conditions of the system than in faulty conditions, as in the latter case, such an impedance is only that corresponding to the circuit included between the point where the relay is located and the point where the fault is occurred. To apply this principle, the impedance seen by the relay and its operating characteristic is represented on the R–X diagram. The operating characteristic represents the limit value of the zone that is to be protected and defines an operating area on the R–X diagram. The protection relay only must operate if the point defined by the coordinates of the impedance seen by the relay is within the operating area, as to the contrary, either there is no fault or it is outside the zone that it must protect.

The way in which the operating area is defined results in different types of distance protection units. Even though other more sophisticated forms can be currently achieved, the existing types are basically impedance type, reactance type, and admittance type, commonly known as mho type. Figure 8.24 shows a characteristic operating curve for each of those types, where the operating point should lie within the shaded area for the relay to trigger. Two possible impedance values seen by the relay are shown on the R–X diagram. In the case corresponding to point P, the protection relay must operate as it is an internal point in the operation area. In the case corresponding to point Q, the protection relay must not operate as it is an external point in the operation area.

If we look at Figure 8.24, it can be seen that only the mho-type units are directional. The reactance and impedance units, as they can operate for points located in the four quadrants, do not have this characteristic. Therefore, to assure selectivity, the reactance and impedance units must be used together with other units that allow them to acquire the directionality characteristic. On the other hand, the reactance-type units take the operation decision based only on the value of the reactance seen by the relay and, therefore, are the most suitable for protecting short lines and those that best respond to highly resistive faults.

Setting the parameters that define the operation area allow the reach of the unit to be defined in each case. The reach of the unit is, in other words, the zone that is under its protection. In principle, it could be thought that to protect a line, the unit that is located at its end should be set to protect its whole length. The application of this methodology to the AB line of Figure 8.25 would involve setting the unit that is located at A in such a way that it covers the whole length AB. However, by applying this procedure, any cause that results in an error in the calculation of the impedance seen by the relay may lead to an erroneous selectivity, as the fault occurring in the initial stretch of the CD line

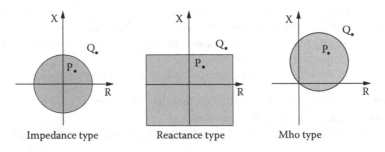

FIGURE 8.24 Operating characteristics of distance protections.

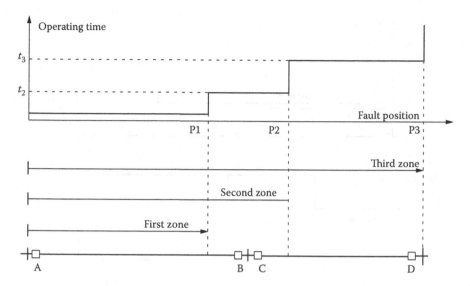

FIGURE 8.25 Reaches and operating times of a distance protection located at A.

may be seen by the relay in A within its operating zone. This fact implies the risk that circuit breaker A instead of C is opened, while C should open in that case to correctly remove the fault.

Therefore, to endow the distance protections with suitable selectivity, and facilitate their coordination, it is usual to define, from the end of the line where the distance protection is located, three protection zones with increasing operating times and reaches.

The first zone covers around 80%–90% of the length of the line, taken from the end where the relay is located. This protection relay is implemented by means of instantaneous units that, therefore, operate as quickly as their technology allow, due to the fact that no intentional time delay is introduced.

The second zone includes the whole line and, also, extend to 20% or 30% of the adjacent line. In this case, time delay units are used and are set to operate with a time delay of around 0.3–0.4 s.

The third zone covers the whole line and 100% of the adjacent line. This zone usually even extends rather more to guarantee that it includes the whole of the adjacent line. This protection relay is also performed by means of time delay units with operating times greater than those of the second zone and which, typically, are set in values around 0.8–1 s.

In practice, the real reach of each zone is influenced, as well as by the errors that may occur in the gathering and processing of data, by the value of the fault resistance.

The units that cover each of the zones, within a single distance protection, do not have to be all of the same type. A usual practice is to use reactance-type units for the primary and secondary zones and a mho-type unit for the third zone. Using mho-units for the third zone has the advantage that they may be used to endow directionality to the primary and secondary zone units that, as in the case in question, lack it.

Figure 8.25 shows the reach of the zones corresponding to a distance protection installed at end A of line AB. Likewise, the staggering corresponding to the operating times of the units of each zone is indicated on the schematic of the line.

If a fault occurs in the first zone, the first zone unit acts as instantaneous primary protection. Should this unit not operate, provided the causes for its failure do not affect the second zone unit, the latter will operate in a t_2 time as local backup protection for faults occurring in the first zone. Likewise, should these two units fail for reasons that do not imply the failure of the third zone unit, it would operate in a t_3 time. In this case, the third zone unit would act as local backup protection with regard to the other two units.

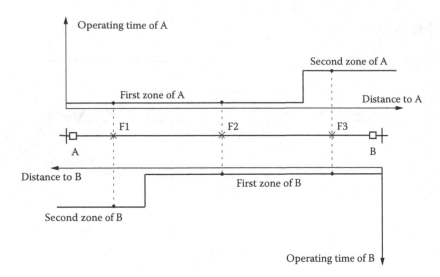

FIGURE 8.26   Distance protections coordination.

If the fault occurs between P1 and P2, but within the AB line, the second zone unit will act as primary time delay protection and the third zone unit can be considered to be playing the role of local backup protection.

If the fault occurs between P1 and P2, but within the adjacent CD line, the second and third zone units will play the role of remote backup protections with respect to the protection installed in C, which will remove the fault as it is the primary instantaneous protection with respect to the point where the fault has occurred.

Finally, if the fault occurs between P2 and P3, the third zone unit plays the role of remote backup protection with respect to the distance protection installed in C. This case clearly shows the importance of establishing the staggering of times, as to guarantee the selectivity, it is essential that the operating time in second zone of the distance protection in C is less than the operation time in the third zone of the distance protection in A.

Therefore, in summary, it can be said that the first and second zone units offer primary protection to the AB line and that the second and third zone units acts as remote backup protection for the CD line.

The above facts only refers to the behavior of the distance protection installed in A. However, it should be remembered that to remove a fault in lines with possibility to be fed by the two ends, as is generally the case of HV transmission lines, the circuit breakers of the two ends of the faulty line must be opened. Therefore, to protect the AB line, one distance protection needs to be installed at end A, with directionality that runs from A to B, and another distance protection at end B, with directionality from B to A.

Figure 8.26 shows how, for example, if a fault occurs in F1, near to end A, distance protection A would see the fault in the first zone and would operate instantaneously by opening circuit breaker A. The distance protection in B would see the fault in the second zone and would operate by opening circuit breaker B after the time delay set for the second zone operation. If the fault occurs in the central zone of the line, in a point such as F2, both distance protections would see the fault in the first zone and would operate by opening circuit breakers A and B without an intentional delay. Finally, if the fault occurs in a point such as F3, near to end B, the distance protection in A would operate in the second zone and distance protection in B would operate in the first zone. However, due to its great impact on the system, it should be remembered that quickly removing the fault is fundamental when protecting HV transmission lines. To avoid the delay resulting from the operation in the second zone, a communication may be established between the two distance protections in such a way that when one operates in first zone, it sends a signal to the other so that its circuit breaker opens without a time delay, as in that case, there is the certainty that the fault is internal to the line that they protect. On the other hand, should any failure occur in the operation in any of the units, the backup protections would respond as has been indicated beforehand.

Therefore, the selectivity between distance protections is achieved more easily than between overcurrent protections, as establishing the staggering of the indicated operating times and reaches is sufficient to achieve a selective response of the protection system. In addition, establishing communication between the distance protections helps to increase the speed of response as it eliminates the intentional time delays in their operation.

Special consideration should be given to the multiterminal lines, due to their specific characteristics. The intermediary branches of the line make contributions, in one way or another, to the currents seen by the distance protections located at its ends. This fact causes a distortion in the impedance seen by the distance protection that makes it difficult to establish a suitable selectivity. Therefore, it is usual to intercommunicate the distance protection located at each end in multiterminal lines to establish a logic selectivity that allows them to know whether or not they must operate.

EXAMPLE 8.5

In this example, the behavior of the distance protections to faults occurred in line 5–8 of Figure 8.7 is analyzed.

The faults occurred in line 5–8 must be removed by means of opening the circuit breakers located at the respective ends of the line. Therefore, the distance protections located in busbars 5 and 8 will act in this case as primary protections, operating in first or second zone, according to the specific point where the fault is located.

First, we are going to see how to calculate the impedance seen by each of the distance protections.

The distance protections measure the positive sequence impedance and, in order for its operation to be correct, the impedance seen by the distance protection ($\mathcal{Z}_V^1$) must be the impedance existing between the point where the distance protection is located and the fault point. Taking into account the existing relations between the sequence networks, the expressions to calculate this impedance are easily deducible. According to the fault type that has occurred, these expressions are:

- One-phase fault (phase a, ground)

$$\mathcal{Z}_V^1 = \frac{V^a}{\mathcal{I}^a + k \cdot \mathcal{I}^0}$$

where

$$k = \frac{\mathcal{Z}^0 - \mathcal{Z}^1}{\mathcal{Z}^1}$$

its absolute value ranges, normally, between 1.25 and 2.50 due to the existing relationship between the positive and zero-sequence impedances in overhead lines. In the case of the 5–8 line, it holds that

$$k = \frac{X_0 - X_1}{X_1} = \frac{0.30 - 0.10}{0.10} = 2$$

- Two-phase fault (phase a – phase b)

$$\mathcal{Z}_V^1 = \frac{V^a - V^b}{\mathcal{I}^a - \mathcal{I}^b}$$

- Three-phase fault

$$\mathcal{Z}_V^1 = \frac{V^a}{\mathcal{I}^a} = \frac{V^b}{\mathcal{I}^b} = \frac{V^c}{\mathcal{I}^c}$$

In these expressions, the phase voltages refer to the point where the distance protection is located and the phase currents are those corresponding to the input direction to the protected line.

Therefore, a distance protection must include for each zone the number of units needed to guarantee that, whatever the type of fault that occurs, the impedance that exists between the point where the protection is located and the point where the fault has occurred can be calculated. However, the number of units required is not as high as it might be thought at first, as, for example, in the case of a three-phase fault both the one- and two-phase fault provide the same impedance value, given that the homopolar current in the three-phase faults is zero and balance exists between the three phases.

The above development has been performed with the fault impedance being zero. Logically, if the units are set to operate according to the given expressions, the existence of a fault impedance distorts the value of the impedance seen and may result in the misoperation of distance protection.

The setting of the distance unit must be performed taking into account that the voltage and current signal reach the relay through the relevant voltage (VT) and current (CT) transformers. Therefore, the values used for the setting must be those for the seen impedance in the secondary system of those transformers:

$$\mathcal{Z}_{VS}^1 = \mathcal{Z}_V^1 \cdot \frac{r_C}{r_V}$$

where $\mathcal{Z}_{VS}^1$ is the impedance seen in the secondary and r_C, r_V are, respectively, the transformation ratios of the CT and the VT. Taking into account the base voltage value (220 kV between phases) and base power value (100 MVA), the base value of the impedance is

$$Z_{base} = \frac{(220\,kV)^2}{100\,MVA} = 484\,\Omega$$

Using this base value and the relevant per unit values, given in Figure 8.7, the positive sequence impedances of line 5–8 and adjacent are those indicated in Table 8.5.

If the transformation ratios of the VT and CT that supply the distance protections were those indicated in Table 8.6 and that the first and second zones of the distance protections located in busbars 5 and 8 were defined according to the following criteria:

- Mho-type characteristic.
- Angle of the pick-up impedance coinciding with the line impedance angle.

TABLE 8.5

Positive Sequence Impedances (Ω)

Line 5–8	Line 4–5	Line 8–7
$j48.4$	$j34.848$	$j82.28$

TABLE 8.6

Transformations Ratios of the VT and CT

Busbar	r_V	r_C
5	2000	400
8	2000	500

TABLE 8.7
Adjustment Values of First and Second Zone

Busbar	First Zone (mho type)	Second Zone (mho type)
5	$8.23\angle90\,\Omega$	$14.62\angle90\,\Omega$
8	$10.29\angle90\,\Omega$	$14.72\angle90\,\Omega$

- Module of the pick-up impedance equal to 85% of the line impedance for the first zone and equal to the line impedance plus 30% of the adjacent line for the second zone.

the pick-up values, referred to the secondary of the VT and CT, would be those given in Table 8.7.

The representation on the R–X diagram of the pick-up impedance defines the operating characteristic. Therefore, due to the simplified hypothesis adopted, the mho characteristics are in this case circles whose diameter coincides with the vertical axis, as indicated in Figure 8.27.

After examining the characteristics of the distance protections located at the line ends, their response to faults in line 5–8 is then analyzed. Therefore, the results of the three-phase zero impedance faults shown in Table 8.4 will be taken as starting data. The conversion of pu current and voltage values into absolute values, in V and A, is performed by multiplying them, respectively, by the base value of the simple voltage (127 kV) and by the base current value which, in this case, is

$$I_{base} = \frac{100\,\mathrm{MVA}}{\sqrt{3}\cdot220\,\mathrm{kV}} = 262.43\,\mathrm{A}$$

The cases relating to faults occurred at a distance of 0.10, 0.80, and 0.95 of the busbar 5 that, due to their position, are similar to faults F1, F2, and F3 indicated in Figure 8.26 if A is considered as busbar 5 and B as busbar 8. The results are those indicated, respectively, in Tables 8.8 through 8.10. The voltage is expressed in kV, the current in A and the impedance in Ω.

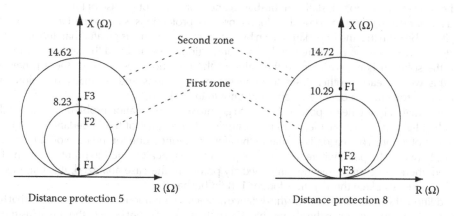

FIGURE 8.27 First and second zone operating characteristics.

TABLE 8.8
Three-Phase Fault at 0.10 from 5 in Line 5-8 (F1)

Busbar	V^a	I^a	Z^1_V	Z^1_{VS}	Operating
5	$8.15\angle0$	$1685.58\angle-90$	$4.84\angle90$	$0.968\angle90$	First zone
8	$66.40\angle0$	$1524.61\angle-90$	$43.56\angle90$	$10.89\angle90$	Second zone

TABLE 8.9

Three-Phase Fault at 0.80 from 5 in Line 5-8 (F2)

Busbar	$\mathcal{V}^a$	$\mathcal{I}^a$	$\mathcal{Z}_V^1$	$\mathcal{Z}_{VS}^1$	Operating
5	43.32∠0	1118.97∠−90	38.72∠90	7.744∠90	First zone
8	25.91∠0	2676.60∠−90	9.68∠90	2.419∠90	First zone

TABLE 8.10

Three-Phase Fault at 0.95 from 5 in Line 5-8 (F3)

Busbar	$\mathcal{V}^a$	$\mathcal{I}^a$	$\mathcal{Z}_V^1$	$\mathcal{Z}_{VS}^1$	Operating
5	47.57∠0	1034.92∠−90	45.98∠90	9.20∠90	Second zone
8	7.69∠0	3181.17∠−90	2.42∠90	0.60∠90	First zone

TABLE 8.11

Zones for the Distance Protections Located in 4 and 7

	Distance Protection 4	Distance Protection 7
F1 (0.10)	Second zone	Third zone
F2 (0.80)	Third zone	Second zone
F3 (0.95)	Third zone	Second zone

The operating times of each distance protection, in each of the case presented, are those indicated in Figure 8.26.

The distance protections located in busbars 4 and 7 must see these faults in the zones indicated in Table 8.11. Logically, these protections must not operate as they are providing backup protection and the primary protections installed in busbars 5 and 8 are, except in case of failure, more rapid.

If a similar study to the one described for the primary protections is performed for these protections, it can be seen that the impedance seen by distance protection 7 is greater than the one strictly included between the point where the distance protection is located and the fault point. In these cases, the solution is not to increase the value of the pick-up impedance of these protections since this would lead in other cases to unnecessary operations and, in many occasions, it is worse for a protection to operate when it must not than to do so.

Even though it is not the purpose of this text to go into the coordination problem in detail, it does aim to highlight the cause that leads to this distortion in the value of the impedance seen by the distance protection. The reason is similar to the aforementioned one for the case of multiterminal lines as, between the point where the distance protection is located and the point where the fault occurred, there are current injection intermediary points. In the case analyzed, this intermediate contribution takes place through transformer 1–8 (in busbar 8).

In addition, the impedance seen by the distance protection is influenced by other factors that have not been taken into account in the above analysis, such as the inherent error to the use of instrument transformers, the possible saturation of the current transformers, the existence of fault impedances, and so on. Therefore, each manufacturer includes their own techniques and technology to minimize the distortions in the impedance seen and improve the behavior of the distance protection.

8.10 DIFFERENTIAL PROTECTION

The differential protections are based on comparing the composition of two or more magnitudes with an established pick-up value that is fixed by means of the setting of the relevant protection. The most commonly used type is the current differential protection.

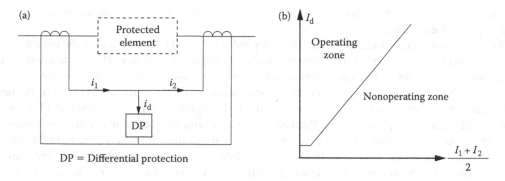

FIGURE 8.28 Differential protection.

Figure 8.28a shows the scheme for the differential protection of a generic element of the electric power system. The transformation ratios of the current transformers located at its ends must be chosen in such a way that, under normal operating conditions, the currents at either side of the protected element imply equal current circulation in the secondary circuits of the current transformers. The differential protection relay only operates when the differential current that circulates through it surpasses the value of the pick-up current, which is the reference value that has been set to be able to distinguish if an internal fault has occurred in the protected element.

If there is no internal fault in the protected element, the differential current is less than the pick-up current and the differential protection does not operate. An internal fault in the protected element implies a difference between the secondary currents of the current transformers that makes the differential current be higher than the pick-up current, which implies the operation of the differential protection that will trip the circuit breakers isolating the protected element.

However, selectivity based only on the value of the differential current may lead to misoperations of the relay. In fact, an external fault implies the circulation of high currents, both at the input and output of the protected element, that may lead to a different saturation of the current transformers. This fact would introduce different and important errors in the secondary currents that would make the differential current rise over the pick-up value and, consequently, cause the differential protection to operate when not needed.

To improve the selectivity of the differential protection, preventing erroneous tripping, the so-called percentage differential protection is used. This type of protection relay has an absolute minimum value of the pick-up current and the operating decision is taken based on the relative value of the differential current. Its operation characteristic is represented on Figure 8.28b and it shows that it only operates when the differential current exceeds a determined percentage of the restraining current, which is defined as the average of the input and output current of the protected element. It is also usual that the percentage is taken as the basis of the current at the output of the protected element, which is commonly known as the through current. This percentage value is, in general, constant and may be set in each specific differential protection.

8.10.1 Transformer Differential Protection

When the protected element is a power transformer, the transformation ratios of the current transformers must be in proportion to the power transformer ratio and the absolute minimum value of the pick-up current must be set to the no-load current of the power transformer, extended as far as it is considered suitable to prevent undue tripping of the differential protection.

There are two typical situations in which the differential current reaches high values, not resulting from internal faults in the transformer. These two situations are the energization and the overexcitation of the power transformer. In both cases, the differential protection tripping must be blocked.

Therefore, the differential protection must be capable to distinguish the origin of the differential current to guarantee a correct selectivity.

The waveform of transformer inrush current depends on many factors, including the instant when the switch is closed, the level of transformer saturation, and the residual flux. The greatest content of harmonics, considering the first cycle after switching on the transformer, will occur when the switch is closed at voltage zero crossing. In this case, the second harmonic can reach around 60% of the fundamental component of current. On the other hand, if the transformer is overexcited, the Fifth harmonic will reach the highest values. Because of this, the sum of Second and Fifth harmonics of differential current constitutes the so-called restraining current since it is used to restrain the differential protection tripping. Therefore, the Second and Fifth harmonics are used in transformer differential protection to discriminate the origin of the differential current and to improve the selectivity by subordinating the differential protection operation to the harmonic components of current [24].

8.10.2 BUSBAR DIFFERENTIAL PROTECTION

In the case of the busbar differential protection [25] the number of magnitudes that must be composed is equal to the number of lines that come together in the busbar. As indicated in Figure 8.29, the differential current is established by adding the secondary current corresponding to the input line currents to the busbar. In order for this sum to represent to scale the composition of the line currents, all the used current transformers need to have the same transformation ratio. Therefore, the differential current must be zero if there is no fault in the busbar.

The need to use a larger number of currents introduces a new problem that is an added risk factor that threatens the selectivity of the differential protection. If in the case of a busbar, as indicated in Figure 8.29, a fault occurs at point F outside it, all the fault current from the busbar flows through the line where the fault occurred. Owing to this, as the fault current is divided between the other lines that converge in the busbar, the current transformer located in the faulty line will reach a much greater saturation than the others. The errors that this implies in the secondary currents mean that their sum is not canceled and, therefore, they can result in an incorrect operation of the differential protection. To avoid this, there are various possibilities. One possibility is to increase the setting values of the differential protection, which may result in other cases in a poor selectivity by omission in the operation. Another possibility may be to introduce an intentional time delay so that the transient current is sufficiently reduced or other relays are tripped. This last possibility has the drawback of also delaying the operation in the presence of faults in the busbar that must be removed as quickly as possible by opening the circuit breakers of all the incident lines.

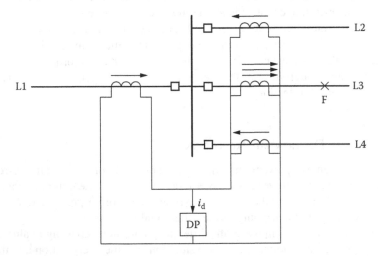

FIGURE 8.29 Busbar differential protection.

Given the disadvantages offered by the two possibilities exposed above, in practice other solutions are taken. For example, the use of air core current transformers eliminates the CT saturation problem and it is very simple. Other solution is the so-called high impedance differential protection. This last option is the most common as it offers the advantage of using conventional current transformers and operate correctly even when the transformers are saturated. It is fundamentally based on using a maximum voltage relay, that must be set to operate when the voltage exceeds a determined value, as the faults in the busbar produce greater voltages than the external faults.

8.10.3 Line Differential Protection

In addition to considerations regarding the saturation of CTs, the operation characteristic of a line differential protection [26] must be defined taking into account the charging current due to line capacitance. Under normal operating conditions this current represents the differential current and, consequently, the minimum pick-up threshold must be set to a higher value.

In line differential protection the communication channel is a key part of the protective relay. During many years, in spite of its evident advantages, the application of the differential technique to protection of the transmission lines has not been possible. This is due to the fact that the long length of the lines, compared to that of the other elements, such as transformers, generators, or busbars, implies a great distance between the current transformers located at its ends. Therefore, setting up a physical connection between its secondary circuits involves using very long cables. The longer the cables the more expensive the installation and, above all, the greater its accuracy is damaged. Therefore, the differential protection has been traditionally barely used in lines and its application has been limited to the sphere of short lines.

In the course of time, two important advances in technology have made it possible to extend its field of application to longer transmission lines. These two advances are the development of communication technologies and the development of digital relays.

In order to guarantee a correct behavior of the relay, by the reasons explained above, the communication between the substations must meet a set of conditions. Present communication technologies, such as optical fiber, make possible to have a reliable communication channel that fulfills this set of conditions. However, having the information is not enough. It is also necessary to process it adequately in order to synchronize data, apply the corresponding algorithms, and more. Nowadays, it is possible to carry out all these processes by means of digital relays.

Figure 8.30 illustrates the differential protection of a two-terminal line. As can be seen, one relay is located at each end of the line. Each relay receives directly the current data corresponding to the end where it is located and can operate in one of two modes: master or slave. Master mode implies that the relay:

- Receives the current information from all other relays on the line.
- Performs the corresponding differential algorithm.
- Makes the adequate decision.
- Transmits to the slave relays on the line the order to trip when an internal fault is detected.

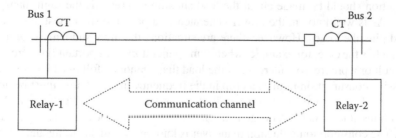

FIGURE 8.30 Line differential protection.

Slave mode implies that the relay:

- Transmits current data.
- Receives from master relay the order to trip when an internal fault is detected.

In the case of a two-terminal line at least one of the two relays must be a master relay.

The differential current technique can be also applied to protect multiterminal lines. In this case, the differential current is obtained by composing the currents corresponding to each end of the line. Consequently, one relay must be placed at each end of the line and at least one of them must be a master relay.

8.11   OTHER PROTECTION RELAYS

There is a great spectrum of causes and abnormal conditions that may change and threaten the correct operation of the electric power system. Even though some of them directly or indirectly arise from the types of faults outlined in this chapter, in other cases, they are the result of the evolution of the operating conditions of the system.

This is the case, for example, of the supplying of a motor with an abnormally low voltage, which can be caused by abnormally high loading or weakened topology. Irrespective of the origin of the low voltage, measures should be taken to ensure that the motor does not operate in critical conditions as, if the voltage is lowered sufficiently, the reduction may result in the axis being blocked, an increase in the losses, poorer cooling, increased temperature, and so on. Therefore, the motor is protected against under voltages by means of a relay that operates when the voltage is lower than the value for which it has been set, with a suitable time delay that prevents its operation if the decrease in voltage is temporary and does not result in the aforementioned effects.

This example shows that, to safeguard each of the elements that make up a system in any situation that may occur, a large variety of protection relays needs to be used and coordinated [18,27]. Even though the aim of this section is not to consider all of them in detail, some of them are set out that, together with the ones already discussed, are essential to guarantee the protection of the elements comprising the power system.

The generator protection must fundamentally include protection functions that operate in case of short circuits in its windings (both stator and rotor), nonremoved external faults, excitation loss, over excitation, abnormal frequencies, loss of synchronism, overloads, phase imbalance, or motorization (inversion of the power flow).

In the power transformers, in addition to the purely electrical protection relays (against, for example, internal faults or overexciting), the use of the Buchholz relay is typical. It detects gases that are inside the transformer tank and, by means of its analysis, enables the origin of abnormal conditions to be identified.

The motor protection basically involves relays that react to internal faults or external conditions that imply a deficient supply to the motor due to overvoltages, under voltages, phase imbalance, and so on.

Special mention should be made about the load shedding as regards the joint management of a power system. As is well known, the power generated in a power system must always be equal to that consumed plus the losses. However, there are situations that may suddenly or gradually break this balance. Such is the case, for example, when some generators or important lines are disconnected because of a fault or a progressive increase of the load that cannot be followed by the available generation. Excessive consumption implies a drop in the frequency that, initially, must be neutralized to be subsequently able to recover the frequency to its nominal value (see Chapters 9 and 10).

When it is impossible to increase the generation to reestablish the power balance, loads need to be disconnected. The consumption reduction maneuver is known as load shedding and must be considered as a last resource. It means interrupting the service to some consumption centers, but it is useful to avoid an interruption that affects a much larger area of the system which, in general, worsens the quality of the service.

Given that it is impossible to predict the numerous situations that may occur, as well as the specific load to be eliminated in each case, a load shedding program is established beforehand that sets the frequency from which load must begin to be reduced and defines, by steps in the frequency value, the load percentage that must be separated. Later on, before reconnecting the loads in a sequential manner, the frequency must be recovered to its nominal value and all the possible supply lines connected.

In the last years, electric power systems are experiencing great changes. Emergent technologies such as, among others, renewable energies, distributed generation or HVDC (High Voltage Direct Current) have increasing importance. All these changes are affecting the operating environment but not the final target. Continuity and quality of electrical energy supply remain the ultimate goals of any power system. Therefore, against all these changes, protection systems must continue to guarantee all the functional features exposed in the present chapter. The challenge is to adapt the protection systems to this new operating environment.

In the case of renewable energies, it is necessary to take into account the special features of each type of technology in order to define the structure and coordination of its protection system. The renewable energies require, besides particular elements (solar panels, inverters, wind turbines, etc.), other important electrical equipments such as transformers, generators or lines that must be protected according to the known protection principles of each one of them.

The incorporation of distributed generation has modified some aspects of electric power systems. The most important one is that in a traditional electric power system the energy generation and the consumption are clearly separated. In this environment, the power flows unidirectionally from generation side to load side through transmission and distribution lines. The distributed generation modifies this premise as it implies the possibility of bidirectional power flow through the lines. Consequently, in order to guarantee its correct behavior, the protection system must be modified and adequately coordinated using similar techniques to the ones exposed in this chapter.

In DC systems the use of DC instead of AC implies important differences. The objective of their protection system is the same as that corresponding to AC systems but the need of using specific devices results in a higher cost. This is the case, for example, of HVDC circuit breakers which must include some additional mechanism to force current to zero in order to allow reliable switching. The future development and spread of HVDC systems, including grids and point-to-point links, are largely subject to specific devices being commercially available. This will allow to design and implement control and protection systems technically reliable and economically viable.

Finally, it should be pointed out that there are other types of relays whose mission is not to protect the system when a fault has occurred, but rather to prevent the fault. Such is the case, for example, of the synchronism-check relay, whose goal is to grant the execution of the connection maneuver between two parts of a circuit. The relay therefore authorizes the connection when the two parts are synchronized and prevents it when that is not the case.

8.12 EVOLUTION OF PROTECTION RELAYS

The technology used to manufacture protection relays has undergone great changes over time [17,18, 27]. The two major steps of this evolution are, in chronological order, electromechanical relays and static relays.

Electromechanical relays are, fundamentally, electromagnetic relays or induction relays. In the first case, the energization of the operating or trip circuit takes place by means of the attraction force that is created between two contacts that close the circuit. In the second case, the circuit is closed thanks to a disk or cup being turned which closes the contacts when it reaches a specific position.

Static relays are divided into electronic relays and digital relays. These relays are known as static, since, unlike the electromechanical ones, they do not have any moving parts. This means that they respond more quickly than the electromechanical ones and, furthermore, they significantly modify some aspects of their behavior, such as the reset time. The reset time may be defined as that necessary to reset the relay, in other words, to return it to its initial status once it has operated or begun to operate. In the case of the electromechanical relays, this time may be appreciable due to the fact that it requires the displacement or turning of moving parts that, in addition, may involve a determined inertia. However, in the static relays, this time may be considered null, which implies a series of consequences such as, for example, lower times to be able to perform a reclosure.

The digital relays are very versatile thanks to the use of micro-processors where a large variety of functions may be implemented. Therefore, there is no point in talking about independent units for each function as regards this type of relays, which are integrated to make up the relay, as they are all implemented in the microprocessor. This means that the size of a digital relay is much lower than that of an electromechanical relay. In addition, the digital relay offers a much wider range of services that nowadays is usually included into it, as for example, a fault locator, a wave analyzer, a recorder, interface with communication systems, and so on.

In addition to the already mentioned ones, other advantages of digital relays are:

- Very fast speed of response.
- Very wide range of settings.
- Programable.
- Digital communication capability.
- Connectivity with SCADA.
- Self-resetting capability.
- Self-diagnosis capability.
- Almost no maintenance.
- Low burden.
- Low cost.

However, it is also necessary to take into account that the use of digital relays implies some drawbacks. These drawbacks are common to all applications based on digital technology and especially to those using communications. The most important are:

- Risk of hacking.
- Risk of interferences.

The numerical algorithms (implemented on the DSP or microprocessor of the digital relay) can be a source of another potential drawback. They can originate numerical oscillations which imply risk of relay misoperation.

As digital relays use numerical algorithms, they are also called numerical relays. Some authors distinguish between digital and numerical relays but their differences are so little that they can be considered only one type.

On the other hand, the communication technology between relays has also undergone great changes, evolving from the use of pilot wire, carrier wave, or microwaves to the use of optical fiber. This development has resulted in a new type of protection that some authors call teleprotection. Independently of how it is called, what is clear is that the better, more reliable and quicker the communication between relays, the easier it is to coordinate them and, therefore, the better the selectivity and behavior of the protection system.

The adaptive protections are, for the moment, the last field that is being developed. Their fundamental characteristic is that they use the possibilities offered by the present-day technologies to modify their settings and action guidelines according to the conditions in the system that they are protecting. This modification allows the behavior of the protection system to be optimized, as it is adapted in each situation to the needs of the monitored system.

REFERENCES

1. Gönen T., *Electric Power Distribution System Engineering*, McGraw-Hill, New York (USA), 1986.
2. Khalifa M., *High-Voltage Engineering*, Dekker, New York (USA), 1990.
3. Anderson P., *Analysis of Faulted Power Systems*, Iowa State University Press, Ames (Iowa-USA), 1973.
4. Gönen T., *Modern Power System Analysis*, Wiley, New York (USA), 1988.
5. Alvarado F. L., Formation of Y-node using the primitive Y-node concept, *Transactions on Power Apparatus and Systems*, PAS-101, December 1982, 4563–4572.
6. Alvarado F. L., Mong S. K., and Enns M. K., A fault program with macros, monitors, and direct compensation in mutual groups, *IEEE Transactions on Power Apparatus and Systems*, PAS-104(5), May 1985, 1109–1120.
7. Alvarado F. L., Tinney W. F., and Enns M. K., Sparsity in large-scale network computation, *Advances in Electric Power and Energy Conversion System Dynamics and Control*, (Leondes C. T., ed.), Control and Dynamic Systems, 41, Academic Press, London (UK), 1991, Part 1, 207–272.
8. Tinney W. F. and Walker J. W., Direct solutions of sparse network equations by optimally ordered triangular factorization, *Proceedings IEEE*, 55, 1967, 1801–1809.
9. Betancourt R., An efficient heuristic ordering algorithm for partial matrix refactorization, *IEEE Transaction on Power Systems*, 3(3), August 1988, 1181–1187.
10. Cidras J. and Ollero A., New algorithms for electrical power systems analysis using sparse vector methods, 4th IFAC/IFORS Symposium on Large Scale Systems: Theory and Applications Proceedings, Zurich, August 1986, 783–788.
11. Gomez A. and Franquelo L. G., Node ordering algorithms for sparse vector method improvement, *IEEE Transactions on Power Systems*, 3(1), February 1988, 73–79.
12. Takahashi K., Fagan J., and Chen M. S., Formulation of a sparse bus impedance matrix and its application to short circuit study, PICA Proceedings, May 1973, 63–69.
13. Tinney W. F., Brandwajn V., and Chan S. M., Sparse vector methods, *IEEE Transaction on Power Apparatus and Systems*, PAS-104(2), 1985, 295–301.
14. Blom J. H., Arriola F. J., Calvo J., and Pastor J. I. (eds), Planteamientos y Soluciones Tecnológicos que permiten un mejor Servicio Eléctrico a los Clientes, International Symposium on Seguridad y Calidad del Suministro Eléctrico, Iberdrola Instituto Tecnológico, Bilbao (Spain), March 1995, 1–377.
15. Bueno A., Carmena I., Marin S., Agrasar M., Miñambres J. F., Ruiz J., Amantegui J., and Criado R., Advanced protection models for power system operation and maintenance, C.I.G.R.E. 1992 Session, Paris, August 1992, 34–103.
16. Berdy J., *Local Back-up Protection for an Electric Power System*, GE Publication, Pennsylvania (USA), 1998.
17. Horowitz S. H. and Phadke A. G., *Power System Relaying*, Research Studies Press, Taunton (UK), 1993.
18. Iriondo A., *Protecciones de Sistemas de Potencia*, Editorial Service of the University of the Basque Country UPV/EHU, Bilbao, (Spain), 1997.
19. Berrosteguieta J., *Introduction to Instrument Transformers*, Arteche-EAHSA Publication, Spain, 2005.
20. Flurscheim C. H. (Ed.), *Power Circuit Breaker Theory and Design*, Peter Peregrinus Ltd.-IEE Power Engineering Series 1, London (UK), 1985.
21. Goff L. E. , *Automatic Reclosing of Distribution and Transmission Line Circuit Breaker*, GE Publication, Pennsylvania (USA), 1998.
22. Andrichak J. G. and Alexander G. E., *Distance Relay Fundamentals*, GE Publication, Pennsylvania (USA), 1998.
23. Switchgear Department, *The Use of the R-X Diagram in Relay Work*, GE Publication, Pennsylvania (USA), 1998.
24. Zorrozua M. A., Miñambres J. F., Sánchez M., Larrea B., and Infante P., Influence of a TCSC in the behaviour of a transformer differential protection, *Electric Power Systems Research*, 74, 2005, 139–145.
25. Andrichak J. G. and Cárdenas J., Bus differential protection, Twenty Second Annual Western Protective Relay Conference, Spokane-Washington (USA), 1995, 1–16.
26. Altuve-Ferrer H. J., Kasztenny B., and Fischer N., *Line Current Differential Protection: A Collection of Technical Papers Representing Modern Solutions*, Schweitzer Engineering Laboratories, Incorporated, Washington (USA), 2014.
27. Mason C. R., *The Art and Science of Protective Relaying*, GE Publication, Pennsylvania (USA), 1998.

9 Frequency and Voltage Control

Claudio A. Cañizares, Carlos Álvarez Bel, and Göran Andersson

CONTENTS

9.1   INTRODUCTION

When the first power systems were introduced more than a hundred years ago the controls used were very primitive and rudimentary. These systems were usually rather small, and the requirements regarding quality and stability were not as high as they are today, so the control could be performed manually in most cases. Fairly soon the end users started to put higher demands on the quality of supply, that is, on the voltage profile and the frequency, so automatic controllers and regulators had to be introduced to meet these requirements. Furthermore, the first systems very often had ample stability margins, so usually no controls were needed to guarantee stability in most operating conditions. When the systems grew in size and the loading increased, the stability margins decreased drastically and automatic controls had to be introduced to enhance stability. A consequence of the higher demands on quality in some cases called for the installation of high-gain controllers, which actually destabilized the system, and additional controllers had to be installed to restore the desired stability margins.

A main reason for frequency and voltage control is the variation of the load. It is obvious that the frequency cannot be maintained within desired limits without control, but also the voltage is influenced by the loading of the system. The daily load forecasting forms the basis for the unit commitment and dispatch of generators, in many systems today through one or more market clearings, but the faster variations must be balanced through automatic control. These control actions can also be supplied through a market for ancillary services, as discussed in Section 9.1.2.

Power system control can also be used to improve the economic performance of the power system. The unit commitment and economic dispatch are not only done to optimize economic criteria, in many systems today through markets, but also the voltage profile and the power flows could be controlled to minimize system losses and consequently enhance economic performance. Thus, there are three important reasons for power systems control, namely

- Quality
- Security
- Economy

The first controllers in power systems were local controllers. This means that the input to the controller is a signal measured locally, and the quantity controlled is also available at the same location. The voltage control and the primary frequency control of generators are examples of local controls. When the systems increased in size and tie-lines between different subsystems were built, local controllers were not enough for a secure and an economic operation of the system. The different local controllers had to be coordinated, and information had to be exchanged between control centers and local plants. Thus, a hierarchical control structure was introduced, as discussed in Section 9.2.

9.1.1   REQUIREMENTS IN DIFFERENT SYSTEMS

The requirements on the control performance of an interconnected system are normally prescribed and regulated by a coordinating "authority." This authority sets and regulates various performance criteria and responsibilities for the different system operators. These criteria and responsibilities are different in different systems; examples for the European and North American systems are discussed below.

9.1.1.1 Europe

Most of continental Europe is interconnected in a synchronous system referred to as the *Union for the Coordination of Transmission of Electricity* (UCTE) system. UCTE coordinates the operational activities of transmission system operators (TSO) of 22 European countries. Through the UCTE grids, 450 million people are supplied with electric energy, and the annual electricity consumption totals approximately 2100 TWh. The operation of the UCTE system is regulated through the UCTE Operation Handbook [34]. Some examples of these requirements are given in this section.

One part of the UCTE Operation Handbook prescribes how the frequency control should be performed by the participating countries. The frequency control is divided into the following tasks where the responsibilities of the participants are clearly stated:

1. Primary control
2. Secondary control
3. Tertiary control
4. Time control
5. Measures for emergency conditions

The frequency control, which is discussed in detail in Section 9.3, is organized according to the block diagram depicted in Figure 9.1. It is out of the scope of this book to discuss all the requirements given in the UCTE Operation Handbook here, but as an example, the primary frequency control requirements are given here:

1. *Calling up of primary control:* Primary control is activated if the frequency deviation exceeds ± 20 mHz.
2. *Maximum quasi-steady-state frequency deviation:* The quasi-steady-state deviation in the synchronous area must not exceed ± 180 mHz.
3. *Minimum instantaneous frequency:* The instantaneous frequency must not drop below 49.2 Hz.

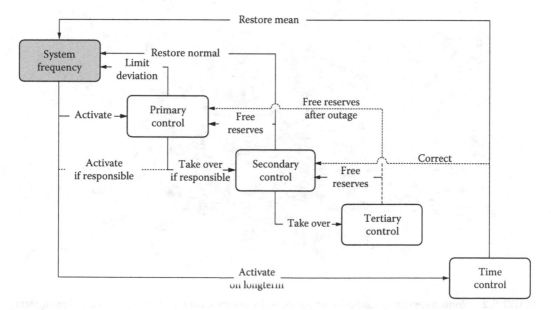

FIGURE 9.1 Block diagram describing the frequency control in the UCTE system [34].

4. *Load shedding frequency criterion:* Load shedding (automatic or manual, including the possibility to shed pumping units) starts at a system frequency of 49.0 Hz (or below).

5. *Maximum instantaneous frequency:* The instantaneous frequency must not exceed 50.8 Hz.

The procedures for the voltage control are also prescribed in the Operation Handbook. As with the frequency control, the voltage control is also divided into primary, secondary, and tertiary voltage control. In general, the voltages of the 380/400 kV network should be kept within the range 380–420 kV, and ranges for the 220/225 kV are also given. In most cases, a single TSO is responsible for the primary voltage control, whereas the other control modes might involve several TSOs. Load shedding can also be initiated when voltages have declined to abnormal levels.

9.1.1.2 North America

The North American Electric Reliability Corporation (NERC) is the entity nowadays in charge of setting reliability standards, which include basic frequency and voltage control requirements for the interconnected North American power grid. The system under NERC's purview, which provides approximately 4800 TWh/year of electricity to nearly 400 million people, is basically divided into eight regions, each one under the purview of a regional reliability coordinator, as shown in Figure 9.2. The figure shows that the system is divided into four asynchronous systems interconnected through High-Voltage DC (HVDC) links, namely:

- The western interconnection, which covers most of the western part of North America, from British Columbia and Alberta in the north to California and a small part of Baja California,

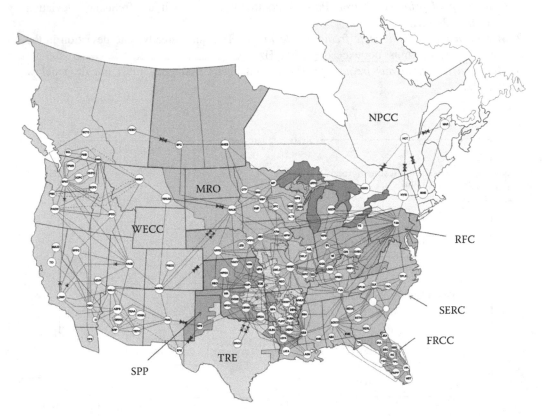

FIGURE 9.2   North American reliability coordinators, major interconnections among control areas and system operators, as of August 2007 [35].

Mexico, in the south, and from east of the Rocky Mountains in the east to the Pacific coast in the west.

- The eastern interconnection includes most of the rest of Canada and the United States, from Ontario and Quebec in the north to Florida and the Gulf of Mexico in the south, and from east of the Rocky Mountains in the west to the Atlantic coast in the east. The east and west systems are interconnected through relatively small back-to-back HVDC links.
- Most of Texas is served by a grid interconnected with both east and west systems through back-to-back HVDC links.
- The province of Quebec is connected to the rest of the eastern interconnection through major HVDC links.

NERC establishes the minimum frequency and voltage control requirements to guarantee the reliable operation of the full interconnected North American grid; regional coordinators and entities in charge of the various control areas may set more stringent requirements. NERC's requirements can be found at www.nerc.com. Thus, for frequency control, specific target, monitoring and compliance figures and rules are established for the entities in charge of the control areas (balancing authorities) for the following items:

1. Real-power balancing control performance
2. Disturbance control performance
3. Frequency response and bias
4. Time error correction
5. Automatic generation control
6. Inadvertent interchange
7. Underfrequency load shedding

These define various minimum performance requirements for frequency controls, especially for automatic generation control and area control errors, that is, for secondary frequency controls, the concepts which are explained in detail in Section 9.3.

The performance requirements for voltage controls are not dealt with in great detail by NERC, thereby providing somewhat vague performance measurements and targets in a couple of documents (Voltage and Reactive Control and Generator Operation for Maintaining Network Voltage Schedules), thus leaving it for the most part to the regional coordinating and operating entities to define these requirements more specifically. For example, the Western Electricity Coordinating Council (WECC) defines more detailed performance requirements for automatic voltage regulators (see www.wecc.biz).

9.1.2 CONTROL AS ANCILLARY SERVICE

One of the main results of the liberalization of electricity markets has been the unbundling or separation of the various services associated with the generation, transmission, and distribution of electrical energy, as discussed in detail in Chapter 5. In this context, ancillary services have been basically defined as the services needed to support the reliable transmission of electric power from suppliers to users [1]. On the basis of this definition, it is clear that frequency and voltage controls fall under the category of ancillary services and hence are directly associated with and affected by the particular rules and mechanisms used for management and pricing of these services in competitive electricity markets.

On the one hand, competitive markets for ancillary services have been or are being designed and developed, particularly for frequency control, given the fact that these services can be directly linked with active power/energy (MW/MWh) production, and hence can be readily priced. On the other hand, markets for voltage control services have not really evolved in good part because these services are associated with reactive power (Mvar), that is, "imaginary" power, management and nonspinning reserves for frequency control, from both generatogement and hence cannot be easily priced.

Although there is no unique market structure for the provision of frequency controls throughout most jurisdictions, frequency control markets, which are also referred to as balance service markets, are based on bidding and contract mechanisms designed for the purchase of spinning reserves [2]. For example, in Ontario, Canada, a bidding-based auction takes place to purchase 10-min spinning, 10-min nonspinning, and 30-min nonspinning reserve services.

Because voltage control is directly associated with reactive power control, the management and payment mechanisms for the provision of reactive power services have a direct effect on voltage control performance. In most jurisdictions, these services are defined by mandatory requirements for generators selling energy to the electricity market and through contracts between system operators and reactive power providers. These contracts are usually bilateral agreements based on system operator experience and traditional practices used for reactive power management, rather than through competition mechanisms [3]. For example, the Independent Electric System Operator (IESO) in Ontario, Canada, mandates generators to operate within a power factor range of 0.9 lag to 0.95 lead and within $\pm 5\%$ range of its terminal voltage, recognizing losses associated with reactive power generation. For the provision of services beyond these mandatory requirements, the IESO signs contracts for reactive power support and voltage control with generators, recognizing payments for the incremental cost of energy loss in the windings because of the increased reactive power generation. Generators are also paid if they are required to generate reactive power levels that affect their real-power dispatch, receiving an opportunity-cost payment at the energy market clearing price for the power not generated [4]. In most jurisdictions, particularly in North America, the provision of reactive power services from providers other than generators are either contracted or provided by transmission companies and/or utilities as part of their delivery services, as is the case of the SVCs in Ontario's transmission grid.

Another alternative for frequency control is consumers participating in the balancing of generation and loads through energy consumption changes when needed, which is an integral part of what is referred to as Demand Response (DR), as discussed in Section 9.6. There are traditional DR programs mostly oriented to solve capacity problems or emergencies. Thus, there is the possibility in many jurisdictions for large industrial consumers to participate in load shedding programs where load is interrupted if necessary, receiving incentives such a electricity cost rebates. Also, in North America, there have been for many years rebates or other incentives for residential customers who participate in DR programs, where mostly air conditioners and/or electric water heaters are directly controlled by utilities to reduced peak power demand, as for example, the Peaksaver Plus^{TM} program in Ontario, Canada [5]. However, the need to reduce consumption and a better integration of non-controllable, variable, and difficult to predict Renewable Energy (RE), especially wind and solar, generation motivated by environmental concerns, plus the recognition of the consumer as an important actor in the power grid in the context of smart grids and related technologies, have led to a paradigm change, in which consumers (individually or aggregated) participate in ancillary services via competitive auctions or contracts. The current status of demand participation in power grids and markets is depicted in Figure 9.3, where operating services are associated with some of the countries where demand resources may competitively participate [6,7].

9.1.3 RE GENERATION CONTROL

The penetration of RE sources in power systems has increased significantly in the past few years, with wind and solar power generation reaching energy penetration average levels of around 30% in some power systems [8]. Wind generation and solar PV systems have an impact on the grid, not only due to the special characteristics of the resource that is characterized by its hard-to-predict variability, but also because of the power-electronic technologies used to deliver power to the network. As opposed

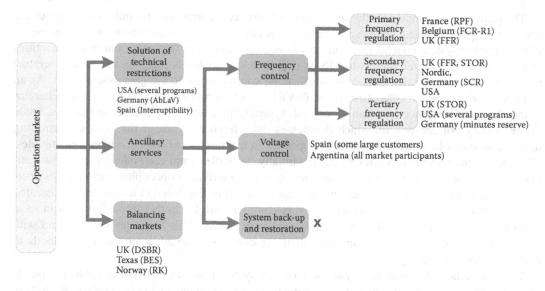

FIGURE 9.3 Demand response for ancillary services.

to synchronous generators, the dynamics of power-electronic converters are fast, and their interaction with the grid is mainly determined by their controls, which are mostly designed to extract maximum power from the source, with limited consideration for their impact on the system. As the share of these generators in the energy mix increases, either through distributed generators at the distribution system level or large-scale power plants connected at the transmission level, it is important to understand the impact of these systems, designing adequate control strategies and possible system upgrades to mitigate negative effects on the power grid.

There are several works that discuss the various effects of wind and solar generation in power networks. Thus, the integration of distributed generators is reviewed in Reference 9, and control strategies for distribution networks with a large penetration of these generators are discussed in Reference 10. At the transmission level, for instance, Munoz and Canizares [11] studies the impact of Doubly-Fed Induction Generators (DFIG) wind power plants in angle and voltage stability of a benchmark system for different controls, and Tamimi et al. [12] analyzes the effect of large and distributed solar PV generation in transient and voltage stability of a real-power network for different control strategies. More recently, the provision of synthetic inertia to ameliorate the impact of RE power generators on angle and frequency stability of the grid are discussed in, for example, Reference 13, and this concept is extended to the emulation of a synchronous machine, referred to as virtual synchronous generators, in Reference 14, applying it to a real-power network in Reference 15, where its practical advantages and limitations are demonstrated. In these studies, the relevance for and impact on power grids of converter controls of wind and solar generators is clearly demonstrated, concluding that the closer these controls emulate existing synchronous generator plants, the less the undesirable or unexpected effects of RE generators on the system. The various wind and solar generation control strategies and their impact on the power grid are discussed in some detail in Section 9.5, whereas their impact on system stability is discussed in Chapter 10, Section 10.5.

9.2   CONTROL IN POWER SYSTEMS

In this section, three very important concepts with regard to power system control, which were already mentioned in the Introduction of this chapter, will be further discussed and elaborated. Thus, the meaning of hierarchical control, local and remote control, and centralized and decentralized control will be clarified.

The power system comprises the subsystems: Electricity Generation, Transmission, Distribution, and Consumption (Loads); and the associated control system has a hierarchic structure. This means that the control system consists of a number of nested control loops that control different quantities in the system. In general, the control loops on lower system levels (e.g., locally in a generator) are characterized by smaller time constants than the control loops on a higher system level. As an example, the automatic voltage regulator (AVR), which regulates the voltage of the generator terminals to the reference (set) value, responds typically in a timescale of a second or less, while the secondary voltage control, which determines the reference values of the voltage controlling devices, among which are the generators, operates in a timescale of tens of seconds or minutes. That means that these two control loops are virtually decoupled from each other. This is also generally true for other controls in the system, resulting in a number of decoupled control loops operating in different time- scales. A schematic diagram showing the different timescales is presented in Figure 9.4. The overall control system is very complex, but due to the decoupling, it is in most cases possible to study the different control loops individually. This facilitates the task, and with appropriate simplifications, one can quite often use classical standard control theory methods to analyze these controllers.

The task of the different control systems of a power system is to keep the system within acceptable operating limits in such a way that security is maintained and that the quality of supply (e.g., voltage magnitudes and frequency) is within specified limits as exemplified in Section 9.1.1. In addition, the system should be operated in an economically efficient way. This has resulted in a hierarchical control system structure as shown in Figure 9.5. For a large power system, the System control level in Figure 9.5 is quite often hierarchically organized. A national control center then constitutes the highest level, and a number of regional control centers form the level below. Some decisions are taken in the national control center and communicated to the regional centers, and with these as basis, control inputs are sent to the local controllers. If the system is large and consists of many interconnected national systems (e.g., the UCTE system), a coordination between the national control centers is needed concerning certain control functions.

In a local controller, as already mentioned, the input and the output of the controller are at the same location. This has the advantages that no communication is needed, which is also favorable from the reliability point of view, and the response can be very fast. Consequently, local controllers are to be preferred when this is possible. However, there are control tasks that cannot be solved by local controllers, or the performance can be enhanced when nonlocal control inputs are used. An example is the secondary frequency control (see Section 9.3.3), where one of the objectives is to control the power flows on certain tie-lines. Because the power plants that should do this regulation most often are geographically located far away from the tie-lines, secondary frequency control involves remote signals. Owing to the need for communication, nonlocal controllers cannot usually act as fast as local controllers.

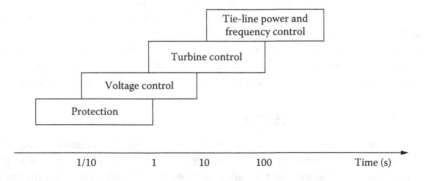

FIGURE 9.4 Schematic diagram of different timescales of power system controls.

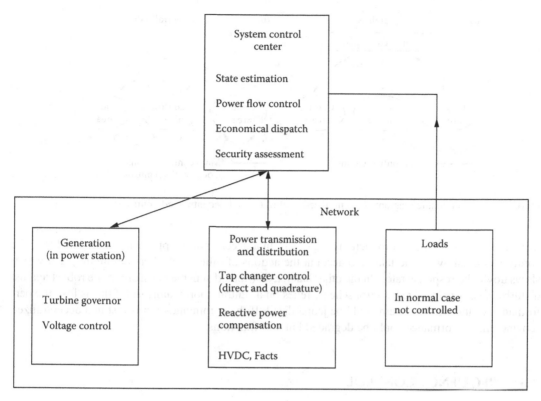

FIGURE 9.5 The structure of the hierarchical control systems of a power system.

In recent years, a new type of nonlocal controllers has been introduced in power systems. These are often used for emergency control and often referred to as remedial action schemes (RAS) or as wide area control systems (WACS). Particularly in the western part of North America, several such systems have been implemented. An example of such a controller is the automatic reduction of the output of a generator station when a critical line in the system is lost. With the relatively recent advent of GPS-based phasor measurement units (PMUs), such systems have become more prevalent, attracting much attention in the power industry and academia.

Some control functions can be implemented as centralized or decentralized. A typical example constitutes the control performed by control centers. In most countries, the control centers are organized so that a national control center coordinates the control actions between the regional control centers. This means that, for the performance of the control action, the regional control centers communicate with each other only indirectly through the national control center. This structure is shown in Figure 9.6a and is referred to as centralized control.

Another organization of the control is provided by the interaction between national control centers, as in the case of the UCTE, or area control centers, as in the case of NERC. For most control tasks, there is no pan-European of North American control center to coordinate control actions; instead the control centers exchange needed information with each other, and based on information from other control centers and the information available from its own control area, each control center determines the adequate control actions. The coordination is thus achieved through the exchange of appropriate information between control centers. This is an example of decentralized control and is illustrated in Figure 9.6b.

Sometimes a control function can be implemented both in a centralized and in a decentralized or distributed way. The advantage with a centralized approach is that the control response is usually faster. As the overall optimization is done by the central controller, the controllers on the next level

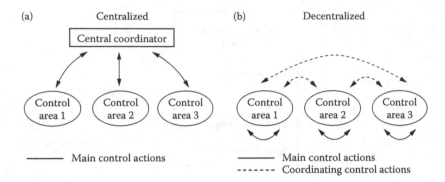

(a)   Centralized   (b)   Decentralized

FIGURE 9.6   Schematic diagrams of centralized control (a) and decentralized control (b).

just have to execute the control actions requested. In a decentralized approach, on the one hand, the controllers must await and take into account the actions of other controllers before they act, which slows down the response rate. On the other hand, a distributed scheme is usually more robust against disturbances. A centralized control scheme relies on a central coordinator, and if this is lost, the performance of the whole system could be jeopardized. If the communication is lost in a decentralized scheme, the performance could be degraded but still functions.

9.3   FREQUENCY CONTROL

9.3.1   INTRODUCTION

As pointed out before, maintaining the frequency constant in a power system is a basic operational requirement, as many loads (e.g., clocks and control equipment) may be very sensitive to frequency behavior and deviations. Furthermore, a tight frequency control is needed to keep a close balance between the real-power generated and consumed in a power system.

The interaction mechanisms between reactive power and voltage magnitude (QV), and real-power and voltage angles (Pδ) are discussed in other chapters. The global performance of power-angle (and frequency) interactions indicates that the imbalance in real powers is felt, after a few seconds, in the frequency across the system; this is analyzed in the next example.

EXAMPLE 9.1

A power system is supplying a load of 1000 MW, and the kinetic energy stored in all the rotating devices is 10,000 MJ. Assuming a sudden increment in the load of 10 MW, and assuming that the generation dispatch is not changed and that the frequency changes are spread uniformly in the system, find the initial frequency change rate.

The kinetic system energy can be related to the frequency in the following way:

$$W_K = \frac{1}{2} M \omega^2 = W_K^0 \left(\frac{\omega}{\omega^0}\right)^2 = W_K^0 \left(\frac{f}{f^0}\right)^2$$

Therefore, and provided that initially all the excess power is supplied from the kinetic energy:

$$10\,\text{MW} = -\frac{dW_K}{dt} = -W_K^0 \frac{d\left(\frac{f^0 + \Delta f}{f^0}\right)^2}{dt} \cong -\frac{2W_K^0}{f^0}\frac{d\Delta f}{dt}$$

This results in an initial frequency decrease of 25 mHz/s. It is important to note that this frequency deviation, if no corrected, will result in a permanent frequency deviation that is unacceptable. This can be corrected by modifying the primary torque in some generators as described below.

It is important to point out that changes in the QV mechanism (e.g., changes in the reactive power generated in a node of the system results in significant voltage changes in some nodes) may result, depending on the load voltage sensitivity, in significant changes in the real and reactive power balance, thus leading to needed adjustments in some of the control variables.

9.3.1.1 Control Elements and Structure

The main means for controlling the real-power balance and flows in a power system are the generator primary energy input in those generators that can be controlled (mainly steam and water in the driving turbines) and the load demand, which can be interrupted (load shedding). This last control action, when affecting loads that are not supposed to participate in secondary or tertiary frequency control actions associated with, for example, demand management or response programs, is only taken in emergency conditions to save the system. Another way of influencing the real-power balance in the system is by the redirection of the flows across the transmission lines by means of line switching actions, phase-angle transformers, and Flexible AC Transmission System (FACTS) controllers; this affects the system losses and, consequently, the real-power balance.

It is important to state that the frequency control that is discussed in this chapter refers to the control that is performed in normal operation, that is, when the frequency deviations are very small (<1% of the nominal frequency). Under these conditions, small disturbance analyses, which assume linear behavior and linear models of the system elements, can be performed.

As mentioned earlier, power system control is performed in a hierarchical scheme for both frequency and voltage and implemented in three steps: primary (local response of the generators), secondary (at the control area level), and tertiary (at the system level). There are clear interactions between these levels, with each having different objectives, time response, and geographical implications.

The primary response corresponds to the control performed locally at the generator, to stabilize the system frequency (generator speed) after a disturbance in the power balance. Typical reaction times for this primary control are in the order of a few seconds (2–20 s are typical). This control is not responsible for restoring the reference value of the system frequency.

The secondary control, which interacts with the dispatched generators in the control area, is designed to maintain the power balance in this area as well as help maintain the system frequency. This control acts in a time range of a few seconds to minutes (20 s–2 min are typical).

Finally, at the top of the hierarchy, the tertiary control level is responsible, in a time frame of minutes (15 min is a typical value), for modifying the active power set points in the generators to achieve a desired global power system operating strategy. It considers not only frequency and real-power controls but also voltage and reactive power controls.

The existence of the necessary means for controlling the real-power balance in the system has to be considered during the system planning process, where the availability of the proper means for a reliable and secure operation under all realistic operating conditions has to be foreseen. Thus, adequate generation reserves to account for abnormal conditions have to be planned for. These reserves can be classified into primary, secondary, and tertiary. The primary or spinning reserves are provided by power sources already connected to the network that can be available very rapidly (in a few minutes) in response to a major outage. Secondary reserves, also referred to as supplementary, are provided by generation units that can be easily switched on and whose full power can be available in a relatively short time (typically 15 min). Finally, the tertiary or replacement reserve is used to restore primary and secondary reserves to precontingency values within, typically, 30 min.

9.3.2 PRIMARY CONTROL

The generator is the main element for frequency and real-power balance in a power system. Therefore, this section is dedicated to analyze its quasi-stationary behavior, assuming that the power generated is being constantly adjusted through small reference value changes, which can be represented by a succession of steady states. Other generator control strategies to respond to sudden and large power imbalances are discussed in Chapter 10.

The generator is capable of modifying continuously and in a controlled way both the real and reactive power it is supplying to the network. The real power may be modified by regulating the primary energy input (water, steam, gas) to the driving turbine, while the reactive power may be changed by modifying the field current and therefore the induced primary voltage. These controls are depicted in Figure 9.7.

The meaningful variables in the operation of the synchronous generator are the real generated power (P_{elec}), which matches instantaneously with the electromagnetic power that is being transferred from the rotor, where the field winding is normally located, to the stator, where the armature windings are usually found. This power is balanced in steady state with the mechanical power (P_{m}) supplied by the driving turbine to the generator. These two powers are related to the electrical (T_{elec}) and mechanical (T_{mec}) torques through the angular speed (ω), which is proportional to the frequency f of the generated voltage. The current circulating in the field winding and the rotating speed induce a no-load electromotive force in the armature (E_{a}).

The coupling of the network and generator can be weak or strong, depending on the short-circuit level of the network at the connecting point. A weak coupling, which is nowadays the case for most generators committed to frequency control, does not yield large increments in the aforementioned generator variables (Δ increments with respect to the reference values). In this case, the generator behavior can be analyzed by means of linear models, with the associated equations and transfer functions being treated indistinctly in the time t or complex frequency s (after a Laplace transform) domains.

9.3.2.1 Control Objectives

The objective for the control applied to the primary energy input of the generator, which affects the mechanical torque on the generator shaft is twofold: first, maintain the rotating speed of the generator

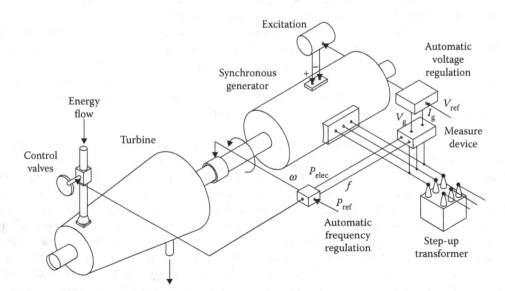

FIGURE 9.7 Generator controls.

as close as possible to the rated (synchronous) speed and, second, modify the generator real-power output. It is important to note that, because the generator speed is closely tied to the system frequency, the speed recovery after a disturbance has to be shared in a coordinated way by all generators in the system committed to frequency control (it is reasonable to assume that every generator should participate according to its rated power).

A controllable generator reacts to a change in the rotating speed by modifying its input power. A drop in the speed leads to the primary power–frequency (Pf) control system increasing the input power to the driving turbine. This has been traditionally implemented by sensing the speed deviation in a Watt centrifugal regulator within the mechanical control system. Novel control schemes based on more advanced digital technologies are in use today to perform this control more accurately.

Three different systems coexist in the generator environment which influence the performance of the primary Pf control: the mechanical turbine-generator system, or elements responsible for transforming the primary energy input into mechanical torque in the generator shaft; the electrical generator system, where a set of winding interactions are responsible for the conversion of mechanical into electrical power; and, finally, the control system, designed to react to changes in the generator speed or power reference settings.

9.3.2.2 Turbine-Generator System

This system is responsible for converting the primary power into mechanical power on the generator shaft. The description function for control purposes provides an increment in the mechanical power in the shaft (ΔP_m) resulting from an increment in the input steam or water because of a change in the input valve (ΔP_{val}). The system behavior is modeled through the transfer function $G_T(s)$ depicted in Figure 9.8, which depends on the turbine type.

Although the physical processes in the generator-turbine set are usually complex, it is possible to use, with no significant loss of accuracy, simpler models to analyze their dynamic response. It is clear that this dynamic behavior depends on the primary powering source, that is, water, steam, or gas. Moreover, it is also important how these are controlled through the variable P_{val}. For example, in steam thermal units, the steam flow entering the turbine can be controlled in two different ways. One option is to directly control the valves that regulate the steam input to the turbine (direct turbine control). Very fast response can be achieved by implementing this control, with time responses in the order of seconds for small variations in the output power (<5%), but slower for larger variations. Alternatively, the control may be implemented by directly controlling the fuel input to the boiler (direct boiler control), both for large and for small power variations. This control is slower than the previous one. Normally, these two means of control are implemented in a complementary way, depending on the desired control objectives.

For a single steam turbine with no reheating system, the following first-order transfer function can be assumed:

$$G_T(s) = \frac{1}{1 + sT_T} \tag{9.1}$$

where the time constant T_T may range between 100 and 500 ms (300 ms is a typical value).

FIGURE 9.8 Turbine-generator transfer function.

More complex multistage turbines require more complicate transfer functions with several time constants. Thus, usually, the maximum energy for power generation from steam can be obtained using several turbine stages, where the steam is used at different pressure and temperature conditions. In this case, different turbines (stages) are placed on the same shaft as the generator, with a high-pressure (HP) body being directly fed from the main boiler. The output steam of this stage, with degraded temperature and pressure, is then injected, after being previously reheated in another boiler or reheater, in a low pressure (LP) turbine. The joint dynamic response depends very much on the rated power relation between the two turbines. Because of the reheater, a new delay is introduced, which can be modeled by the following transfer function:

$$G_T(s) = \frac{1}{1 + sT_T} \frac{1 + s\alpha T_{RH}}{1 + sT_{RH}}$$ (9.2)

where α is the rated HP power divided by the total power, and T_{RH} is the reheater time constant, whose value is in the 4–11 s range. It is common also to find units with a third turbine body intermediate pressure (IP) between the HP and LP stages. This adds a new time constant (T_{IP}) with values in the 0.3–0.5 s range. The resulting transfer function $G_T(s)$ in this case is depicted in Figure 9.9, where α, β, and γ represent the pu contributions of the HP, LP, and IP stages, respectively, to the total power.

The response for hydro turbines is different depending on the type of turbine, that is, Pelton, Francis, or Kaplan. Nevertheless, all of these can be essentially represented using the following transfer function:

$$G_T(s) = \frac{1 - 2sT_\omega}{1 + sT_\omega}$$ (9.3)

where typical values for T_ω are in the 0.25–2.5 s range.

Both hydro and thermal generation units can participate in the real-power and frequency control in a power system. However, thermal units present limitations with respect to the speed at which the output power can be incremented or reduced. These limits, which are basically increment or decrement rate limits in MW/s, are due to the thermodynamic behavior of the boilers and their significant thermal inertias, which have to be accounted for. As discussed in the following section, these ramp limits may impose additional restrictions on how these units can participate in the primary and secondary Pf control. These limits are represented both in terms of the rate of change in power and in terms of the length of time this rates can be maintained.

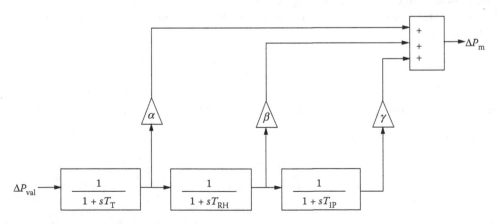

FIGURE 9.9   Block diagram of a three-stage turbine.

9.3.2.3 Hydraulic Amplifier

The admission valves for typical turbines are rather heavy elements; therefore, any changes to these valves are made through hydraulic amplifiers, where input changes are amplified as a result of a high-pressure fluid (usually oil). These changes are translated into valve positions, so that when the amplifier input ΔP_{ha} is positive (negative), there is an increment (decrement) in the primary fluid admission proportional to the input. Thus, the changes in the turbine valve position, that is, ΔP_{val}, are proportional to the integral of the input signal ΔP_{ha}. This process is depicted in Figure 9.10, which can be modeled through the following transfer function:

$$G_H(s) = \frac{\Delta P_{val}}{\Delta P_{ha}} = -\frac{K_H}{S} \tag{9.4}$$

9.3.2.4 Electrical Generator

This system is responsible for converting into electrical power the electromagnetic energy resulting from the electrical interactions between the stator and the rotor field winding. Therefore, the associated processes are very fast in comparison with the thermo-mechanical systems previously described, because the time constants in electromagnetic circuits are much faster. Consequently, for Pf control purposes, the response of this system is assumed to be instantaneous, that is, any increment in the mechanical power in the shaft of the generator is assumed to be immediately translated into an increment in the generated electrical power.

Because the electrical output of the generator is very much related to the angle of the internal-induced voltage, as discussed in detail in Chapter 10, the perturbations in the generator speed, internal angle, and frequency are given by

$$\Delta \delta = \int \Delta \omega \, dt = \frac{1}{2\pi} \int \Delta f \, dt \tag{9.5}$$

The power supplied by the generator is a function of this angle and the rest of the network, depending on whether the machine is in steady or transient state, and on the type of generator, that is, wound-rotor or salient poles.

9.3.2.5 Primary Regulator Models

The primary regulator is a control system designed to actuate upon the aforementioned hydraulic amplifier to modify the energy input to the turbine with two objectives: maintain the generator speed, and hence the frequency, as close to its synchronous value as possible; and modify, as desired, the power output of the generator. Generator speed variations are usual during operation, as loads

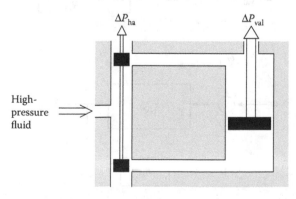

FIGURE 9.10 Hydraulic amplifier.

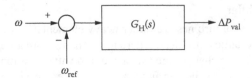

FIGURE 9.11   Speed control implementation.

continuously change, together with many other "perturbations" that take place under normal system operating conditions. Thus, a generator output is modified continuously according to a set of operating strategies, such as cost minimization or social benefit maximization, as discussed in Chapter 5.

The generator speed can be readily restored to its rated value using the hydraulic amplifier and implementing the integral control depicted in Figure 9.11. According to this control, on the one hand, any positive speed variation with respect to the synchronous speed will result, after some delay, in a reduction of the turbine input power. On the other hand, any speed reduction will lead to an increase in the primary power input to the turbine that will eventually drive the speed error to zero. This control is not suitable for multigenerator systems, because any controlled generator will compete with other generators to restore the frequency, exceeding its rated power in some cases. This would result in a highly unstable control that can damage the controlled generators. Hence, a speed droop R_ω and a power set point P_{ref}, to compare the actual output power with respect to a desired value, must be added to the control, as depicted in Figure 9.12, where the new error signals ΔP_{val} and $-\Delta P_{ref}$ are introduced. The resulting control, which allows steady-state speed (frequency) deviations and regulates the generated power, used to be realized through sophisticated mechanical gadgets and is nowadays accomplished by means of digital control systems.

The total transfer function for the controller in Figure 9.12 is

$$\Delta P_{val} = \frac{1}{1 + sT_R}\left[\Delta P_{ref} - \frac{1}{R_\omega}\Delta\omega\right] = G_R(s)\left[\Delta P_{ref} - \frac{1}{R_\omega}\Delta\omega\right] \tag{9.6}$$

where R_ω is the regulator constant, and $T_R = 1/(K_H R_\omega)$ is the time delay introduced by the regulator. This yields the block diagram shown in Figure 9.13.

In steady state, the increment in the generated power is

$$\Delta P_{val}^{SS} = \Delta P_m^{SS} = \Delta P_g^{SS} = \left[\Delta P_{ref} - \frac{1}{R_\omega}\Delta\omega\right] \tag{9.7}$$

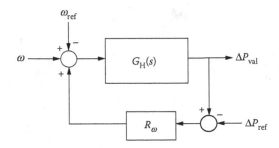

FIGURE 9.12   Primary frequency control block diagram.

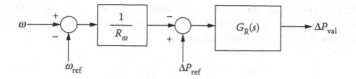

FIGURE 9.13 Frequency control simplified block diagram.

This equation can be restated in terms of the frequency instead of the angular speed

$$\Delta P_{val}^{SS} = \Delta P_m^{SS} = \Delta P_g^{SS} = \left[\Delta P_{ref} - \frac{1}{R_f} \Delta f \right] \tag{9.8}$$

where $R_f = R_\omega / 2\pi$. Observe from these formulae that the generator output will change to match the reference power set point, correcting the speed (frequency) by modifying its output according to the regulator constant R_ω (R_f). Notice that a small value of R_ω will result in a large participation in speed and frequency control, with large increments in the generated power for small deviations in the angular speed. This issue is analyzed in the following example.

EXAMPLE 9.2

A 250 MW rated power generator supplies 200 MW at a rated frequency of 50 Hz. The frequency regulation parameter R_f is such that a decrement of 10% in the frequency will yield an increment in the generated power of 100% of the rated power.

The steady state generated power P_g^{SS} with respect to the system frequency can be obtained by means of the incremental formula:

$$P_g^{SS} - 200 = -\frac{1}{R_f}(f - 50)$$

where

$$R_f = \frac{0.1 \cdot 50}{250} = 0.02 \, \text{Hz/MW}$$

This relation is depicted in Figure 9.14, which shows the given operating condition.

If this generator were to share the system frequency control with other system generators, it would be desirable that the share of each generator of the frequency correction was proportional to its rated power. Thus, if the 250 MW generator was supplying a load together with a 1000 MW

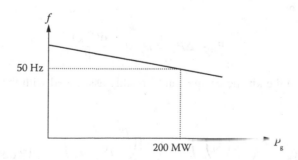

FIGURE 9.14 Static frequency power characteristic.

generator, a static increment or decrement in the system frequency should force a increment or decrement in the power four times larger in the second generator. Consequently, the frequency regulation parameter for this second unit should be

$$R_{f2} = \frac{0.002}{4} = 0.005 \, \text{Hz/MW}$$

This implies that both generators have the same pu regulation parameter R_f with respect to the generators' rated parameters; thus

$$R_{f1} = R_{f2} = (0.02 \, \text{Hz/MW}) \cdot 250 \, \text{MW} = (0.005 \, \text{Hz/MW}) \cdot 1000 \, \text{MW} = 5 \, \text{Hz/pu}$$

9.3.2.6 Power System Response

The previously described control systems refer to controllers associated with the synchronous generator's prime mover. In these discussions, it was assumed that the mechanical power transferred from the turbine to the generator shaft is directly transferred to the electrical network in the form of electric power to supply a set of electrical loads. An unbalance between the mechanical power given to the generator and the supplied electrical power will lead, with some delay, to a change in the generator's rotating speed and, therefore, a frequency perturbation. If the power reference P_{ref} is not modified, a surplus in the mechanical power will result, after some delay, in the rotor acceleration of the turbine-generator system. On the one hand, this will lead to an increment in the output frequency, until a new equilibrium is found at a higher frequency where the electrical power absorbed by the loads again matches the mechanical power supplied by the turbine. On the other hand, the frequency will drop with a shortfall in the mechanical power or a surplus in the electrical power absorbed by the loads.

The system response, for a specific operating point, is controlled by the kinetic energy stored at that moment in all mechanical systems, and by a damping parameter D that models the variation in the consumed electric power with respect to a speed (D_ω) or frequency (D_f) change. This change is assumed to be uniform for every load and generator in the system. A sudden change in the power balance leads to a smooth change in the system frequency because of the inertia of all rotating elements such as generator and motor loads in the system. Therefore, a sudden change in the shaft generator power, that is, a mechanical power ΔP_m different from the electrical power demand ΔP_D, is suddenly translated into a chance in the system kinetic energy according to

$$\Delta P_m - \Delta P_D = \frac{d(W_K)}{dt} \tag{9.9}$$

Furthermore, after the frequency change due to the power unbalance is felt by the loads, the absorbed power will change because of the damping parameter D by $D_f \Delta f$. Hence, the instantaneous power balance will be driven by

$$\Delta P_m - \Delta P_D = D_f \Delta f + \frac{d(W_K)}{dt} \tag{9.10}$$

The rate of change in the kinetic energy can be readily associated with the frequency variations as follows:

$$\frac{dW_K}{dt} = \frac{d\left(W_K^0 \left(\frac{f}{f^0}\right)^2\right)}{dt} = \frac{d\left(W_K^0 \left(\frac{f^0 + \Delta f}{f^0}\right)^2\right)}{dt} = \frac{2W_K^0}{f^0}\frac{d\Delta f}{dt} \tag{9.11}$$

Therefore, from the last two equations, the transfer function relating the frequency changes with respect to the power imbalance can be shown as

$$G(s) = \frac{\Delta f}{\Delta P_m - \Delta P_D} = \frac{K_S}{1 + sT_S} \tag{9.12}$$

where

$$T_S = \frac{2W_K^0}{f_0 D_f}; \quad K_S = \frac{1}{D_f}$$

The effect of a power imbalance in the system frequency (Pf mechanism) can now be investigated in an isolated system with uniform propagation of frequency variations, considering primary frequency controls. This is illustrated in the following example, where the frequency behavior is analyzed for a power system with a set of electric loads that are connected by low-impedance lines, that is, "electrically close" loads.

EXAMPLE 9.3

Consider an electrically uniform power system with a rated power demand of 1000 MW, where a sudden load increment of 50 MW takes place. The loads are characterized by an equivalent frequency sensitivity $D_f = 20$ MW/Hz, and by an initially stored kinetic energy of 10 GJ. Assume that these loads are fed by a set of identical generators with the same dynamics and pu regulation parameters, whose combined response can be characterized by an equivalent generator with the following parameters: $T_R = 80$ ms, $T_T = 300$ ms, and $R_f = 0.005$ Hz/MW.

For a 1 GW power base, the pu values for R_f and D_f are, respectively, 5 Hz/pu and 0.02 Hz/pu. The initial stored kinetic energy also needs to be referred to the base values as follows:

$$W_{K,pu}^0 = \frac{W_K^0}{P_b} = \frac{10\,\text{GWs}}{1\,\text{GW}} = 10\,\text{s}$$

This normalized kinetic energy is usually referred to as the Inertia Constant (H). K_S and T_S are, respectively,

$$T_S = \frac{2W_K^0}{f_0 D} = \frac{2 \times 10}{50 \times 0.02} = 20\,\text{s}$$

$$K_S = \frac{1}{D} = \frac{1}{0.02} = 50\,\text{Hz/pu}$$

The equivalent block diagram to study the evolution of the frequency after a change in the electric demand for the sample system is depicted in Figure 9.15. The resulting frequency response for a permanent 50 MW (0.05 pu) change in the demand is shown in Figure 9.16.

The steady-state error in the frequency after the 50 MW change in the load is as follows:

$$\Delta f_{ss} = -\frac{\Delta P_D}{D + \frac{1}{R}} = \frac{\Delta P_D}{\beta} = -0.2273\,\text{Hz}$$

This implies that, in steady state, the generator's output power will be as follows:

$$\Delta P_m = \frac{-0.2273}{5}\,\text{pu} = 0.04546\,\text{pu} = 45.46\,\text{MW}$$

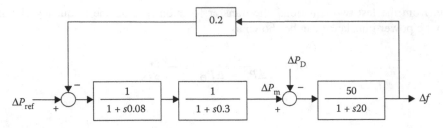

FIGURE 9.15 Block diagram for Example 9.3.

Hence, the sudden 50 MW load initial increase results in the following change in power absorbed by the load because of the frequency reduction:

$$\Delta P_D = D_f \Delta f = 20(-0.2273) = -4.54 \, \text{MW}$$

Figure 9.16 shows that the dynamic evolution of the frequency resulting from the increment in demand fits a first-order model, rather than the third order model associated with the block diagram of Figure 9.15. The reason for this is the quick response of the primary regulator and the turbine-generator system in comparison with the power system response. Thus, assuming an instantaneous response of these two subsystems, and after some basic algebra, the equivalent transfer function relating the increments in the frequency and the demand can be shown as equivalent to a first-order function with the following gain and time constant:

$$K_{EQ} = \frac{K_S R_f}{K_S + R_f} = 4.54 \quad T_{EQ} = \frac{R_f T_S}{R_f + K_S} = 1.82 \, \text{s}$$

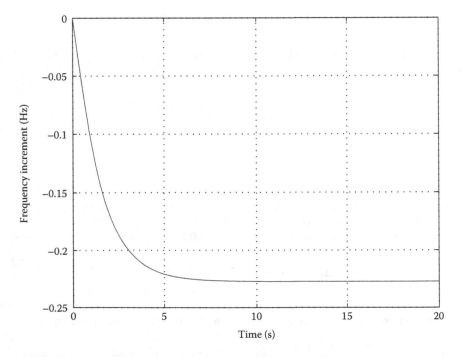

FIGURE 9.16 Frequency response after a load increment in Example 9.3.

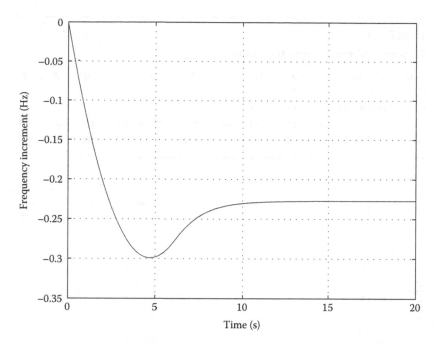

FIGURE 9.17 Frequency response with ramp limits in Example 9.3.

For a 20 MW (0.05 pu) increment in the demand, the time response according to this equivalent transfer function is as follows:

$$f(t) = -0.2273 \left(1 - e^{-t/1.82}\right)$$

By means of this equation, it can be shown that the variation of the stored kinetic energy at the initial time when the load is increased is as follows:

$$\left[\frac{d(W_K)}{dt}\right]_0 = \left[\frac{d\left(\frac{1}{2}I\omega^2\right)}{dt}\right]_0 = \frac{2W_K^0}{f_0}\left[\frac{df}{dt}\right]_0 = -0.05 \, \text{pu}$$

This matches the demand increase, since there is no initial response from the primary frequency control.

In this example, besides the uniformity in the frequency in all the system components, it is assumed that the demand is supplied by a single equivalent generator. However, very similar results can be obtained in cases where there are several generators with similar regulation parameter R values (with respect to each generator's rated power), and with similar dynamic response.

The frequency response in cases where limitations in the speed at which the power supplied by the generator is raised or lowered is degraded in comparison with the no ramp case. This can be observed in Figure 9.17, which depicts the results of a 50 MW demand increase with a ramp limit of 10 MW/s; observe can be observed that the frequency deviation is higher and with a slower response.

The relation between an increment in the demand and the frequency, provided that there is only primary frequency control, is given by the parameter $\beta = D + 1/R$ (MW/Hz). This is referred to as the "static frequency response" of the control area.

As mentioned earlier, generally, all the generators connected to a power system participate in the primary speed control. This situation is analyzed in the next example.

TABLE 9.1

Generators' Parameters for Example 9.4

Gen	Type	R	T_R	T_W	T_T	T_{RC}	T_{PI}	α	β	γ
G1	Hydro	30	0.2	0.5	—	—	—	—	—	—
G2	Thermal 1 stage	15	0.15	—	0.3	—	—	—	—	—
G3	Thermal 3 stages	10	0.12	—	0.3	9	0.4	0.3	0.3	0.4
G4	Thermal 3 stages	.52	0.15	—	0.35	10	0.45	0.3	0.3	0.4

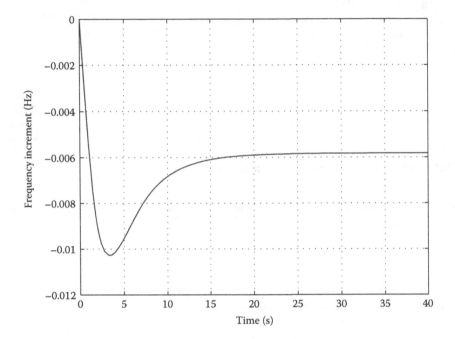

FIGURE 9.18 Frequency response for the various generators in Example 9.4.

EXAMPLE 9.4

Consider a 5 GW rated power system, where the frequency deviations are instantaneously perceived across all the system, and with the following four different types of generators (or sets of generators with the same regulation parameter R value and dynamic response) participating in the primary frequency control: one hydro unit (G1); one thermal unit with a single stage (G2); and two thermal units with high, medium, and low pressure stages (G3 and G4). The parameters of these units are given in the Table 9.1, where all parameter R values are referred to the same base power, the time constants are in seconds, and the α, β, and γ parameters are in pu. The power system response parameters are $K_S = 100\,\text{Hz/MW}$ pu and $T_S = 32$ s. The simulated frequency response for a 10 MW demand increment is shown in Figure 9.18.

9.3.3 Intertie and Frequency Secondary Control

From the primary generator Pf control discussed in the previous sections, it can be concluded that any change in the balance between power supply and demand will result in a change in power output for all generators, and a steady-state frequency different from the original (rated) one. This error in the frequency is generally unacceptable and should be corrected by implementing a secondary control level, slower than the primary control. This is usually referred to as automatic generation control (AGC), and its main objective is to return, in a stable fashion, the frequency error to zero after any imbalances in active power by modifying the power references of all the generators participating in this secondary control. These power reference changes have to account for the specific technical characteristics of the participating units, such as maximum and minimum powers as well as power-ramp limitations. Furthermore, the resulting generation patterns should consider the intertie power commitments agreed between neighboring utilities, and this is usually referred to as tie-line control. It should be mentioned that this is somehow changing as many new market participants get involved in energy transactions (e.g., generators, energy service companies, and transmission and distribution system operators); nevertheless, the same principles behind tie-line control can also be applied to a variety of transactions. Thus, the implementation of secondary frequency control in an area depends very much on the physical structure and characteristics of the controlled power system, as well as the general strategies in place to operate the interconnections with other areas, which imply that, in normal operating conditions, agreements among market participants should be taken into consideration, and that maximum support to solve emergency situations should be provided.

Besides the previously mentioned hierarchical structure of Pf controls, the specific spatial (or geographical) characteristics of the power system should also be considered in these types of controls. This spatial structure will result in several control areas in the power system, with each area being characterized by a uniform and instantaneous propagation of frequency perturbations throughout all area nodes. In other words, all the loads in an area will see the same frequency, and all generators will experience the same perturbations in their rotating speed, despite the different generator dynamic characteristics associated with the various generator and turbine types. In practice, this means that the electrical grid in the area is assumed to be rigid, that is, the transmission system impedances in the area are negligible, and that there exist proper controllers to damp spurious inter-machine dynamic oscillations among the generators within the area.

To implement AGC, the power input for every generator participating in the control should be modified. These generators are usually located in various control areas. Thus, ΔP_{ref} should be modified for each generator depicted in Figure 9.19 according to the signals coming from the AGC of the corresponding areas.

The AGC, as any controller of a complex physical system, must fulfill some requirements that guarantee that certain control objectives are achieved for any reasonable operating conditions. Thus, the tie-line and frequency deviation errors should be canceled as fast as possible, without excessively stressing the controls and ensuring the stability of the system. Additionally, it is usually required that the accumulated error, that is, the value of the integral of the error with respect to time, of the frequency and the intertie powers should be relatively small. The latter is because the accumulated errors in frequency (measured in Hz/s) produce a permanent bias in clocks and control systems fed by the system, and accumulated errors in the intertie powers between areas (measured in J or kWh) yields permanent errors in the programmed energy exchange. The way to fulfill these control requirements is by defining an area control error (ACE), which is directly introduced in an integral control that guarantees that the input power of the controlled generators is modified until the ACE is returned to zero in steady state. The ACE usually has two components: one to account for the frequency error in the area, and another for the tie-line power errors between neighboring areas. Thus, for each control area i, the corresponding ACE is defined as follows:

$$\mathrm{ACE}_i = B_i \Delta f_i + \sum_{j=1}^{n} \Delta P_{ij} \qquad (9.13)$$

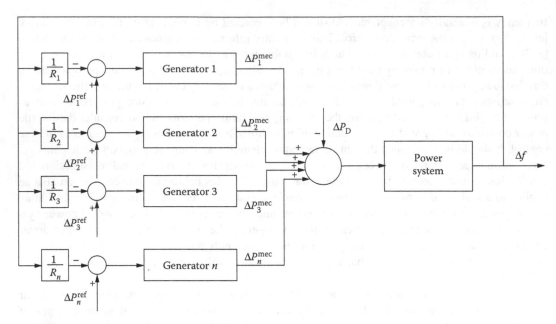

FIGURE 9.19 Block diagram of the primary frequency control in a control area.

where B_i is a chosen frequency bias in MW/Hz (the larger this value, the faster the action of the secondary frequency control), and ΔP_{ij} is the change in the power with respect to its scheduled value in the tie-line between areas i and j.

The concepts and characteristics of secondary frequency control and intertie control are explained with the help of the examples shown below, which are presented in order of complexity.

EXAMPLE 9.5

Consider the system of Example 9.3, where the permanent error in the frequency is to be returned to zero. This can be accomplished by making the ΔP_{ref} generator input proportional to the negative value of the integral of the frequency error.

$$\Delta P_{ref} = -K_I \int \Delta f(t)\, dt \tag{9.14}$$

or in the s domain

$$\Delta P_{ref}(s) = -\frac{K_I}{s}\Delta f(s) \tag{9.15}$$

where K_I (MW/cycle or MW/Hz s) indicates the speed at which the frequency error is reduced and basically defines the control behavior. Thus, large values of K_I will force a fast frequency recovery but may either result on an unstable control or undesirable oscillatory response, which indicate excessive burden on the control system. As an exercise, the reader should compute the critical value of K_I at which the control turns unstable for the data provided in Example 9.3. The frequency response for a 50 MW demand increase and $K_I = 0.3$ pu/cycle is illustrated in Figure 9.20.

The system considered here is formed by a single control area with a unique equivalent generator, which is not the usual case where various interconnected areas with different generators may coexist. Therefore, the ACE has a frequency deviation component only [ACE $= \Delta f(t)$].

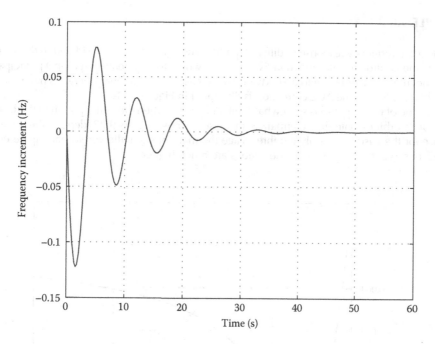

FIGURE 9.20 Frequency response with AGC.

AGC has traditionally relied on hydro-powered generators mainly because of the following two reasons: these units have the capability of responding quickly to changes in power reference set points, and they do not have restrictive ramp limitations, other than the ones imposed by the maximum generation capacity. Other units that are suitable for this control are turbine-controlled thermal units, although the ramp limitations in these units are more restrictive.

AGC used to be offered by suitable generation units as a service associated with the generation of electricity. With the liberalization of electricity markets, AGC has been unbundled from other services, and it is nowadays usually offered as an ancillary service, as discussed in Section 9.1.3.

9.3.3.1 Participation Strategies

The share of generating units participating in secondary frequency control, that is, the participation levels of thermal or hydro units in ACE correction, is a determinant factor in AGC behavior. This participation has been typically decided according to economic criteria, based on various dispatch and operation planning criteria and strategies. Thus, based on economic dispatch criteria and the cost functions of the participating generators, the AGC share of the generators can be set so that the changes in generated power are such that the total costs of responding to load changes are minimized. These participation factors can be readily determined once the participant generators and their cost characteristics are known. These factors should be recomputed as the load conditions in the system change (typically every 5–15 min). On the one hand, the participation levels of thermal units may be determined based on maintaining hydro production levels, while considering their ramp limitations, so that load changes are compensated with thermal units to maintain hydro reserves scheduled based on certain operation planning criteria and tools. In competitive electricity markets, on the other hand, participation strategies are based on the trading of reserve and control capacity in frequency regulation (reserve) markets, where generation units and loads compete to provide frequency control services through various auction and contract mechanisms, as discussed in Section 9.1.3.

EXAMPLE 9.6

The effect on frequency response of different participation strategies is considered in this example based on the control area described in Example 9.4, where four generators (groups) participate on AGC. The frequency response to a 10 MW load increase when all four generators have the same share, that is, 25%, in the ACE correction is illustrated in Figure 9.21.

The shape of the response is somewhat similar but faster when the participation of the fast units (hydro and turbine-controlled thermal) is increased. Thus, Figure 9.22 depicts the frequency response for the case when only the three-stage thermal units are allowed to participate equally in AGC, observe that the frequency deviations are much higher.

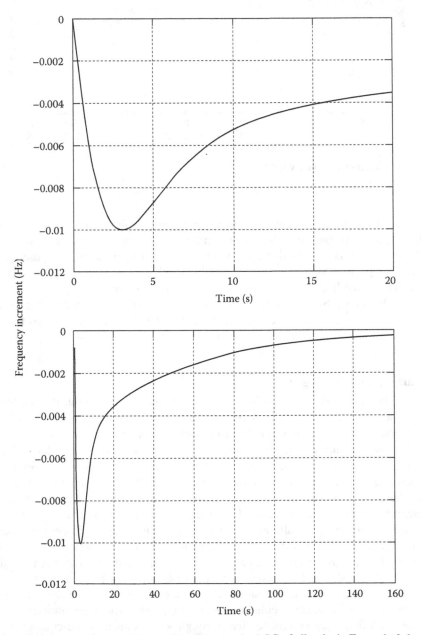

FIGURE 9.21   Frequency response with equal participation in AGC of all units in Example 9.6.

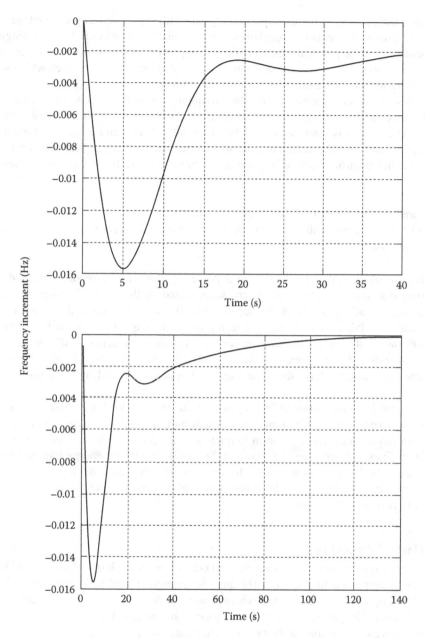

FIGURE 9.22 Frequency response with participation in AGC of only the three-stage thermal units in Example 9.6.

9.3.3.2 Multiarea Systems

So far, the frequency has been assumed to be uniform across a power system, resulting in a single control area. This is certainly not the case in practice, where various control areas can be identified, depending on the criteria, in a large interconnected power system. These control areas have been traditionally associated with the vertically integrated utilities in charge of well-defined geographical areas, which usually had their own secondary frequency controls. These utilities were responsible for the frequency and tie-line controls in their control areas, wherein the assumption of short electrical distances among the area nodes was reasonable. Alternatively, a single generation company may provide frequency regulation services in different and distant geographical areas that span various control areas. Furthermore, in highly populated regions with high-electricity demand, several companies may be providing frequency regulation services, that is, several utilities in a single control area. Independently of the physical characteristics of the power system, it is necessary for the various energy companies to establish strict energy interchanges at the transmission level, which sometimes involve transmission systems from other utilities (wheeling).

The design and implementation of secondary frequency controls or AGC should take into consideration the electric characteristics of the power system as well as the types and nature of the energy transactions. The former is associated with the electrical stiffness and dynamic characteristics of the power system, which may contain one or more control areas (geographical structure). The latter will determine the control structure and the area control errors. In general, secondary controls require that

1. Each area should contribute to its frequency control.
2. The interchange agreements between different areas have to be respected in normal (secure) operating conditions.

The existence of several control areas in a power system can be readily modeled based on the previously discussed concepts, which are summarized in the block diagram of Figure 9.12, and considering an additional block to account for the balance among the powers generated, demanded, and scheduled on tie-lines. The resulting block diagram is shown in Figure 9.23 for a system with two control areas. In this case, two new variables are defined: ΔP_{12} for the increment in the power transferred from Area 1 to Area 2, and ΔP_{21} for the power flowing from Area 2 to Area 1. These two variables have the same value and opposite sign when neglecting losses in the tie-lines.

To analyze multiarea AGC, two different possible situations should be considered. The first corresponds to two control areas interconnected through high-impedance (elastic) transmission lines, with each area having a different, uniform frequency. The second is the case of two areas that are linked by short lines and hence have a common frequency, but where energy transactions between these areas are to be included in the AGC. In the latter, the two areas belong to the same control area, as per physical considerations; however, it also allows to consider the common situation of energy trading between close agents.

9.3.3.3 Elastic Interconnection

Consider a power system with two interconnected control areas, each one characterized by a given frequency and system dynamic response. The areas are interconnected by lossless transmission lines with equivalent impedance X_L (Ω/phase), where power P_{12} flows from Area 1 to 2, and whose value is to be maintained. The generators in each area participate on their local AGC.

Different frequency variations in the two areas translate into increments in the exchanged power, according to the following approximated relation:

$$\Delta P_{12} = \left.\frac{V_1 V_2}{X_L}\cos(\delta_1 - \delta_2)\right|_0 (\Delta\delta_1 - \Delta\delta_2) = T_{12}^0\left(\int \Delta f_1\, dt - \int \Delta f_2\, dt\right) \tag{9.16}$$

where $V_1 \angle \delta_1$ and $V_2 \angle \delta_2$ are the phasor voltages at the connecting nodes in Areas 1 and 2, and

$$T_{12}^0 = \left.\frac{2\pi V_1 V_2}{X_L}\cos(\delta_1 - \delta_2)\right|_0 \tag{9.17}$$

is the electrical stiffness of the coupling at the initial operating point 0 that is to be perturbed.

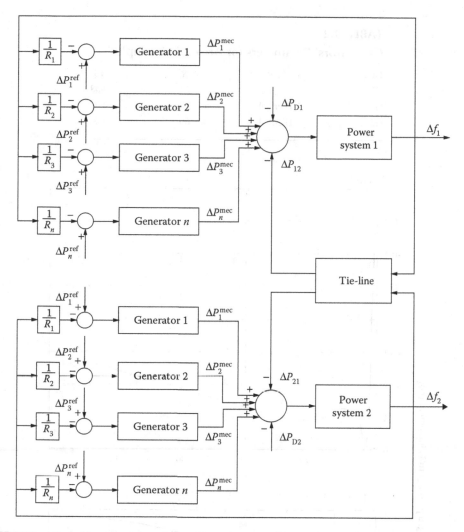

FIGURE 9.23 Block diagram of the primary frequency control in a two control area system.

To meet the secondary control specifications previously discussed, the ACE in each area is defined, as per Equation 9.13, as follows:

$$\text{ACE}_1 = B_1\,\Delta f_1 + \Delta P_{12} \tag{9.18}$$

$$\text{ACE}_2 = B_2\,\Delta f_2 - \Delta P_{12} \tag{9.19}$$

These errors are forced to zero in steady state, independently from the bias B values, so that the errors in frequency and power exchange vanish. In traditional implementations of AGC, the electric power system parameters have been used to define $B = \beta = D + 1/R$, with satisfactory results.

EXAMPLE 9.7

Consider two interconnected areas—one of them as defined in Examples 9.4 and 9.6, and the second one formed by one-stage thermal units with fast primary regulators. The turbine time constants and participation factors for Area 2 are given in Table 9.2. The parameters for the Area 2 electrical

TABLE 9.2

Generators' Parameters for Area 2 in Example 9.7

Gen	Type	R	T_R
G1	Thermal	24	0.08
G2	Thermal	40	0.07
G3	Thermal	60	0.06

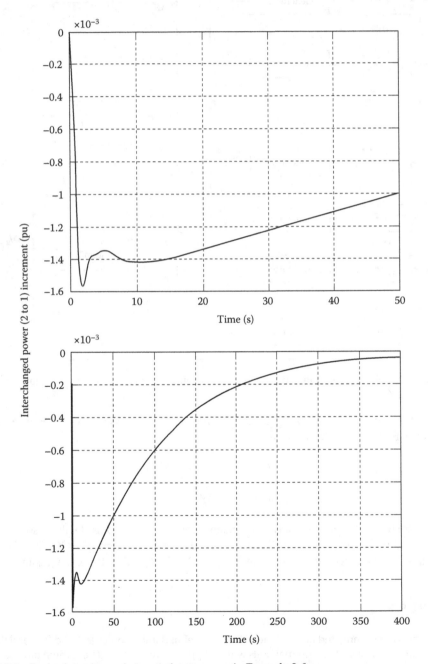

FIGURE 9.24 Power interchange between the two areas in Example 9.6.

system are $K_S = 500\,\text{Hz/pu}$, and $T_S = 20\,\text{s}$; and $K_I = 0.08533$ for the AGC, all referred to the same base values as in Example 9.6. The two areas are interconnected by a line with impedance $X_L = 13.26\,\text{pu/phase}$.

The variations in the interchanged power between the two areas, after a sudden 10 MW increment in the Area 2 power demand, are shown in Figure 9.24.

This elastic model could be extended to a larger number of interconnected areas by defining an ACE signal for each of the areas with secondary control, including the area frequency deviations and variations in the agreed exchanges with neighboring areas. However, such a decentralized control could easily become unstable when all the areas are fully interconnected through a mesh of tie-lines. This problem can readily be studied with the help of a power system example with three or more areas similar to the ones defined in the aforementioned examples. Besides these stability problems, the AGC requirements would have to be maintained using a proper control design, which could be based on a master control (e.g., pan-European or pan-North American) to generate some components of the AGC of the local areas.

9.3.3.4 Rigid Interconnection

In the rigid interconnection paradigm, the frequencies associated with the different areas are basically the same, that is, the frequency is uniform throughout the interconnected system. Hence, the power exchanges resulting from demand or generation changes cannot be associated with local frequency perturbations. It is necessary in this case to directly evaluate the interchanged power as the difference between the power supplied and demanded ($\Delta P_{Gi} - \Delta P_{Di}$) locally at each interconnected area. This interconnection model is illustrated in the next example.

EXAMPLE 9.8

Consider the rigid power system formed by two identical areas as described in Example 9.5, with a sudden demand increment of 50 MW taking place in Area 2. For the case where only frequency deviations are considered in the ACE signals, the resulting tie-line power is shown in Figure 9.25.

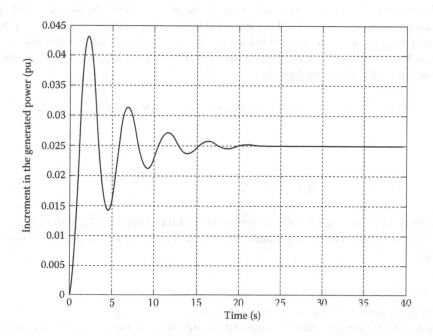

FIGURE 9.25 Power exchanged between two rigid areas in Example 9.8 with frequency bias only.

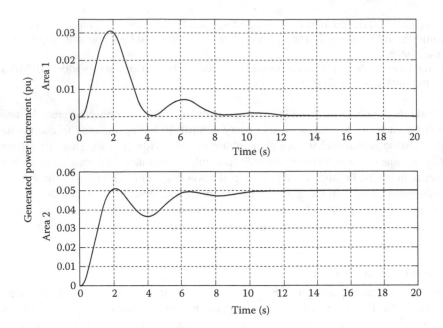

FIGURE 9.26 Power exchanged between two rigid areas in Example 9.8 considering frequency and area power differences.

Observe a likely undesirable oscillatory behavior, as well as an unacceptable permanent bias of half the value of the demand increment.

The permanent error in the interconnection can be corrected by including the difference between generated and load powers, $\Delta P_{Gi} - \Delta P_{Di}$, in each area. The resulting generation for each area is shown in Figure 9.26.

The aforementioned control for the rigid areas can be problematic if the areas are of significantly different sizes. Several strategies have been proposed in this case, where the interchange power component is weighted according to some physical parameters that account for the dynamic behavior of each system. For example, one approach for a two-area system is to weigh the power exchange increments according to the following formula:

$$\Delta P_{12} = -\Delta P_{21} = (\Delta P_{G1} - \Delta P_{D1})(1 - \alpha) - (\Delta P_{G2} - \Delta P_{D2})\alpha - \beta\,\Delta f \qquad (9.20)$$

where

$$\alpha = \frac{H_1 P_{b1}}{H_1 P_{b1} + H_2 P_{b2}}; \quad \beta = D_1 - \alpha(D_1 + D_2) \qquad (9.21)$$

It is left to the reader to evaluate the advantages of this control strategy for the system in Example 9.8, assuming that Area 1 is 10 times larger than Area 2, and to compare the results with respect to those obtained with $\alpha = 1$.

9.3.3.5 Multiarea Control

AGC in power systems where different areas coexist, both from the physical and from the administrative point of view, presents further complexities, especially in the case where these services are considered ancillary services in competitive electricity markets. The elastic interconnection models may turn unstable when the interconnection lines form loops, and the controls described for rigidly interconnected areas may result in excessive stresses on the control system.

AGC schemes are a major concern in current power systems, where the tendency of these systems is to expand geographically mostly because of economic and reliability reasons. Each control area has traditionally operated its power system according to its own criteria, with a centralized unique system operator (usually operating different control areas) or various independent and often electrically close system operators. Nowadays, however, pf control mechanisms are very much related to the way in which the associated electricity markets are organized either in each jurisdiction or, more commonly, in neighboring jurisdictions with agreements to open their electrical systems and markets. In this case, frequency and tie-line regulation services are being offered and traded as independent ancillary services, as discussed earlier on this chapter, considerably affecting how AGC is defined and implemented in each area.

9.4 VOLTAGE CONTROL

9.4.1 INTRODUCTION

9.4.1.1 Voltage and Reactive Power

In steady-state operation both active power balance must be maintained and reactive power balance must be kept in such a way that the voltages are within acceptable limits. If the active power balance is not kept, the frequency in the system will be influenced as described in Section 9.3, while an improper reactive power balance will result in the voltages in the system differing from the desired ones.[*]

Normally, the power system is operated so that the voltage drops along the lines are small. The node voltages of the system will then almost be equal (flat voltage profile). In this case, the transmission system is effectively used; primarily for transmission of active power, and not for transmission of reactive power. The voltage magnitudes can thus be controlled to desired values by the control of the reactive power. Increased production of reactive power gives higher voltages nearby the production source, while an increased consumption of reactive power gives lower voltages. While the active power is entirely produced by the generators in the system, there are several sources and sinks of reactive power. The reactive power, in contrast with the active power, cannot be transported over long distances in the system, because normally $X \gg R$ in a power system, and the reactive power can thus be regarded as a fairly local quantity.

Important generators of reactive power are

- Overexcited synchronous machines
- Capacitor banks
- The capacitance of overhead lines and cables
- FACTS controllers/devices (see Chapter 7)

Important consumers of reactive power are

- Inductive static loads
- Under-excited synchronous machines

---

[*] The frequency deviation is a consequence of an imbalance between power fed into system by the prime movers and the electric power consumed by the loads and losses. However, the reactive power generated and consumed is always equal, which is a consequence of Kirchhoff's laws, and the voltage magnitudes are always adjusted so that this balance is maintained. If the voltages settle at too low values, the reason is that the reactive generation is too small and one says that there is a lack of reactive power, and vice versa when reactive generation is too high. However, it does not mean that the generated and consumed reactive power are not equal as in the case of active power.

- Induction motors
- Shunt reactors
- The inductance of overhead lines and cables
- Transformer inductances
- FACTS controllers/devices

For some of these, the reactive power is easy to control, while for others it is practically impossible. The reactive power of the synchronous machines is controlled easily by means of the excitation system. Switching of shunt capacitors and reactors can also be used to control reactive power. FACTS devices also offer a possibility to control reactive power.

It is most effective to compensate the reactive power as close as possible to the reactive load. There are certain high-voltage tariffs to encourage large consumers (e.g., industrial loads) and electrical distribution companies to compensate their loads in an effective way. These tariffs are generally designed in such a manner that the reactive power is only allowed to reach a certain percentage of the active power. If this percentage is exceeded, the consumer has to pay for the reactive power. In this way, the high-voltage network is primarily used for transmission of active power.

The reactive losses of power lines and transformers depend on the size of the reactance. For overhead transmission lines, the reactance can be slightly reduced by the use of multiple conductors. The only possibility to radically reduce the total reactance of a transmission line is to connect a series capacitor or a series FACTS device.

9.4.1.2 Voltage Control Mechanisms

The following factors influence primarily the voltages in a power system:

- Terminal voltages of synchronous machines
- Impedances of lines
- Transmitted reactive and active power
- Turns ratio of transformers

A suitable use of these leads to the desired voltage profile.

Generators are often operated at constant voltage by using an AVR. The output from this AVR controls, via the electric field exciter, the excitation of the machine, so that the voltage is equal to the set value (see Figure 9.7). The voltage drop caused by the generator transformer is sometimes compensated totally or partially, and the voltage can consequently be kept constant on the high-voltage side of the transformer. Synchronous compensators can be installed for voltage control; these are synchronous machines without turbine or mechanical load, which can produce and consume reactive power by controlling the excitation. Nowadays, new installations of synchronous compensators are very rare, and power-electronics-based solutions are preferred if fast voltage control is needed.

The reactive power transmitted over a line has a great impact on the voltage profile. Large reactive power transmissions cause large voltage drops, and hence should be avoided. Instead, the production of reactive power should be as close as possible to the reactive loads. This can be accomplished through the excitation of the synchronous machines, as described previously. However, often there are no synchronous machines close to the load, so the most cost-effective way is to use shunt capacitors which are switched according to the load variations. A power-electronics-based device can be economically justified if fast response or accuracy in the regulation is required. Shunt reactors must sometimes be installed to limit the voltages to reasonable levels. In networks that contain large number of cables this is also necessary, because the reactive generation from these is much larger than those from overhead lines (C is much larger than L).

9.4.2 PRIMARY VOLTAGE CONTROL

The task of the primary voltage control is to control the reactive output from a device so that the voltage magnitude is kept at or close to the set value of the controller. Usually, the node of the controlled voltage is at the same or very close to the node of the reactive device, because reactive power is a fairly local quantity. The set values for the voltage controllers are selected so that the desired voltage profile of the system is obtained. The selection of set values is the task of the secondary voltage control, as discussed in Section 9.4.3.

The most important devices for reactive power and voltage control are described hereafter.

9.4.2.1 Synchronous Machine AVR

The reactive power output of synchronous machine can, for a given active power level, be adjusted within the limits of the capability curve by the excitation system. This offers a very efficient and fast way to control the terminal voltage of the machine, which is the most important way of voltage control in most power systems.

The main purpose of the excitation system of a synchronous machine is to feed the field winding of the synchronous machine with direct current, so that the main flux in the rotor is generated. Furthermore, the terminal voltage of the synchronous machine is controlled by the excitation system, which also performs a number of protection and control tasks. A schematic picture of a generator with its excitation system is depicted in Figure 9.27; and the functions of the different blocks depicted in this figure are, briefly:

- *The exciter* supplies the field winding with direct current and thus comprises the "power part" of the excitation system.
- *The controller* treats and amplifies the input signals to a level and form that is suitable for the control of the exciter. Input signals are pure control signals as well as functions for stabilizing the exciter system.
- *The voltage measurement and load compensation unit* measures the terminal voltage of the generator, and rectifies and filters it. Load compensation can be implemented if the voltage at a point away from the generator terminals, such as at a fictional point inside the generator's transformer, should be kept constant.

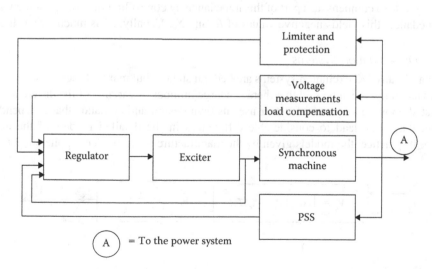

FIGURE 9.27 Schematic picture of a synchronous machine with excitation system with several control, protection, and supervisory functions [33].

- *The power system stabilizer (PSS)* gives a signal that increases the damping to the controller. Usual input signals for the PSS are deviations in rotor speed, accelerating power, or voltage frequency. A more detailed discussion of this controller can be found in Chapter 10.
- *The limiter and protection* can contain a large number of functions that ensure that different generator and exciter physical and thermal limits are not exceeded. Usual functions are current limiters, over-excitation protection, and under-excitation protection. Many of these ensure that the synchronous machine does not produce or absorb reactive power outside the limits it is designed for.

Different types of exciter systems are used. Three main types can be distinguished as follows:

- *DC excitation system*, where the exciter is a DC generator, often on the same axis as the rotor of the synchronous machine.
- *AC excitation system*, where the exciter is an AC machine with rectifier.
- *Static excitation system*, where the exciting current is fed from a controlled rectifier that gets its power either directly from the generator terminals or from the power plant's auxiliary power system, normally containing batteries. In the latter case, the synchronous machine can be started against an unenergized net (black start). The batteries are usually charged from the grid.

A more comprehensive treatment of some of the functions described above follows.

9.4.2.1.1 Compensation Equipment

Figure 9.28 shows the block diagram of a compensation circuit, consisting of a converter for measured values, a filter, and a comparator.

There are several reasons for the use of compensation in voltage control of synchronous machines. If two or more generators are connected to the same bus, the compensation equipment can be used to create an artificial impedance between the generators. This is necessary to distribute the reactive power in an appropriate way between the machines. The voltage is measured somewhat inside the generator, corresponding to positive values of R_c and X_c in Figure 9.28. If a machine is connected with a comparatively large impedance to the system, which usually is the case, since the generator's transformer normally has an impedance in the order of magnitude of 10% on the machine base values, it can be desirable to compensate a part of this impedance by controlling the voltage somewhat inside of that impedance; this yields negative values of R_c and X_c. Usually, X_c is much larger than R_c.

9.4.2.1.2 DC Excitation Systems

Presently hardly any DC excitation systems are being installed, but many of these systems are still in operation. Generally, it can be said that there is a large number of variants of the different excitation systems listed above. Every manufacturer uses its own design and demands that it depends on the application, that often lead to considerable differences in the detailed models of the devices in each group. In practice, the models given by the manufacturers and power suppliers must be used.

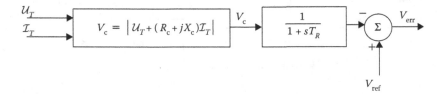

FIGURE 9.28   Block diagram of compensating circuit.

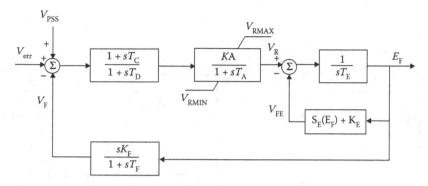

FIGURE 9.29 Model of DC exciter system (IEEE type DC1).

One example of a DC excitation system, the IEEE type DC1 system, is given in Figure 9.29. The input signal for the controller is the voltage error V_{err} from the compensation equipment (see Figure 9.28). The stabilizing feedback V_F is subtracted, and sometimes a signal from the PSS is added; both these signals vanish in steady state. The controller is mainly described by the dominating time constant T_A and the amplification K_A. The limits can represent saturation effects or limitations of the power supply. The time constants T_B and T_C can be used to model internal time constants in the controller; these are often small and hence can usually be neglected.

The output signal V_R from the voltage controller controls the exciter. The exciter consists of a DC machine that can be excited independently or shunt excited. For shunt excited machines, the parameter K_E models the setting of the field regulator. The term S_E represents the saturation of the exciter and is a function of the exciter's output voltage E_F. If saturation is neglected, that is, $S_E = 0$, the effective time constant of the exciter becomes T_E/K_E, and its effective amplification is $1/K_E$.

9.4.2.1.3 AC Excitation Systems

For AC excitation systems, the exciter consists of a smaller synchronous machine that feeds the exciter winding through a rectifier. The output voltage of the exciter in this case is influenced by the loading. To represent these effects, the exciter current is used as an input signal in the model. In Figure 9.30, the model of an ac exciter systems is shown (IEEE type AC1). The structure of the model is basically the same as for the DC excitation system, but with some added functions. The exciter's rectifier prevents (in most exciters) the exciter current from being negative. The feedback with the constant K_D represents the reduction of the flux caused by a rising field current I_F, that constant depends on the synchronous and transient reactances of the exciter. The voltage drop inside the rectifier is described by the constant K_C, and its characteristic is described by F_{EX}, which is a function of the load current I_H.

DC and ac excitation systems are sometimes called rotating exciters, since they contain rotating machines. This distinguishes them from static excitation systems, which are described hereafter.

9.4.2.1.4 Static Excitation Systems

In static excitation systems, the exciter winding is fed through a transformer and a controlled rectifier. By far, most exciter systems installed today are of that type, and a large number of variants exists. The primary voltage source can be a voltage transformer that is connected to the generator terminals, but even a combination of voltage and current transformers can be found. With the latter arrangement, an exciter current can be obtained even if the voltage at the generator terminals is low (e.g., during a ground fault at or near the power plant). Sometimes, it is possible to supplement these voltage sources by using the auxiliary power of the power plant as a voltage source, this makes it possible to start the generator in an unenergized grid.

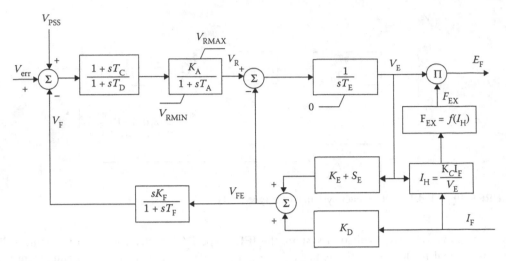

FIGURE 9.30   Model of an AC exciter system (IEEE type AC1).

Static excitation systems can often deliver negative field voltage and even negative field current. However, the maximum negative field current is usually considerably lower than the maximum positive field current.

An example of a model of a static exciter system is shown in Figure 9.31. The time constants are often so small that a stabilizing feedback is not needed; the constant K_F can then be set to zero. Since the exciter system is normally supplied directly from the generator bus, the maximum exciter voltage depends on the generator's output voltage (and possibly its current). This is modeled by the dependency of the limitations of the exciter output on the generator's output voltage. The constant K_C represents the relative voltage drop in the rectifier.

9.4.2.2 Reactive Shunt Devices

In many systems, the reactive control capabilities of the synchronous machines are not sufficient to keep the voltage magnitudes within prescribed limits at all loading conditions. The varied loading condition of the system during low and peak load situations implies that the reactive power needed to keep the desired voltage magnitudes vary significantly. It is impossible and also unwise to use the reactive power capabilities of synchronous machines to compensate for this, since many synchronous machines will be driven to their capability limits, and the fast and continuous reactive power control offered by the synchronous machine will not be available. Therefore, in most systems, breaker-switched shunt capacitor banks and shunt reactors are used for a coarse control of reactive power, so that the synchronous generators can be used for the fast and continuous control.

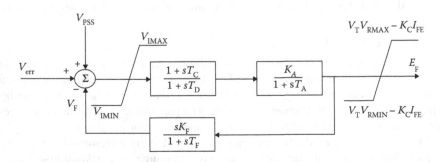

FIGURE 9.31   Model of a static exciter system.

Generally shunt capacitors are switched on during high-load conditions, when the reactive consumption from loads and line reactances is the highest. Since these load variations are rather slow and predictable, no fast control is needed and the capacitor banks can be breaker switched. In systems with long high-voltage lines, the reactive power generation of these lines can be very high during light load conditions. To keep the voltages at acceptable levels, shunt reactors might be needed.

The sizes of the reactive shunt elements determine how accurate the control can be. Capacitor banks, in particular, can often be switched in smaller units, while shunt reactors are most often installed in one single unit, because of high costs. A factor limiting the element size is the transient voltage change resulting from switching. The fundamental frequency voltage change in pu caused by switching a shunt element can be estimated as $\Delta V = Q_{shunt}/S_{sc}$, where Q_{shunt} is the size of the shunt element and S_{sc} is the short-circuit power at the node.

Reactive shunt elements are also used for reactive power and voltage control at HVDC terminal stations. Since typically a line commutated HVDC converter station consumes about 50% as much reactive power as active power transmitted, reactive compensation is needed. Part of the reactive compensation is usually provided by harmonic filters needed to limit the harmonic current injection into the ac networks. These filters are almost purely capacitive at fundamental frequency.

9.4.2.3 Transformer Tap-Changer Control

An important method for controlling the voltage in power systems is by changing the turns ratio of transformers. Certain transformers are equipped with a number of taps on one of the windings. Voltage control can be obtained by switching between these taps, as illustrated in Figure 9.32. Switching during operation by means of tap changers is very effective and useful for voltage control. Normally, the taps are placed on the high-voltage winding (the upper side), since then lower currents needs to be switched.

If N_1 is the number of turns on the high-voltage side and N_2 is the number of turns on the low-voltage side, the turns ratio of transformer is defined as

$$\tau = \frac{N_1}{N_2} \tag{9.22}$$

Then, the relation between the voltage phasors on the high-voltage side U_1 and on the low-voltage side U_2, at no load, is

$$U_2 = \frac{U_1}{\tau} \tag{9.23}$$

If the voltage decreases on the high-voltage side, the voltage on the lower side can be kept constant by decreasing τ, that is, by switching out a number of windings on the high-voltage side. When the transformer is loaded, Equation 9.23 is of course incorrect, since the load current yields a voltage drop over the leakage reactance of the transformer Z_k, but the same principle can still be applied for voltage control.

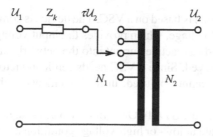

FIGURE 9.32 Transformer with variable turns ratio (tap changer).

Transformers with automatic tap changer control are often used for voltage control in distribution networks. The voltage at the consumer side can therefore be kept fairly constant even though voltage variations occur on the high-voltage network. Time constants in these regulators are typically in the order of tenths of seconds. In some transformers, the turns ratio cannot be changed during operation, but just manually when the transformer is unloaded. In this case, voltage variations in the network cannot be controlled, with voltage levels changing in large steps.

9.4.2.4 FACTS Controllers

As previously mentioned, power-electronics-based equipment, usually referred to as FACTS controllers or devices, can be used for fast voltage and reactive power control. The first devices introduced were based on thyristors as active elements, but more recently, devices using voltage source converters (VSCs) have been introduced (see Section 7.7.2); with the latter, a faster and more powerful control can be achieved. FACTS devices used for voltage control are connected in shunt. Two such devices are discussed here: the Static var compensator (SVC), and the STATic synchronous COMpensator (STATCOM). The SVC is based on thyristors, while the STATCOM is based on a VSC.

There are also series-connected FACTS devices that in principle can be used for voltage control, but the main reason for installing these devices is to control active power, which indirectly influences voltages. Among these devices, the unified power flow controller (UPFC) is equipped with a shunt part for voltage control, which basically operates as a STATCOM.

9.4.2.4.1 Static Var Compensator

The SVC may be composed of two different shunt elements, that is, a thyristor controlled reactor (TCR) and a thyristor switched capacitor (TSC) banks. If fast switching of the capacitor banks is not needed, one can also use breaker-switched capacitors. The TCR is depicted in Figure 9.33, by delaying the firing of the thyristors, a continuous control of the current through the reactor can be obtained, with the reactive power consumption varying between 0 and V^2/X, where X is the reactance of the reactor. By combining the TCR with a suitable number of capacitor banks, a continuous control of the reactive power can be achieved by a combination of capacitor bank switching and control of the reactor current. Usually, the TCR and TSC are connected to the high-voltage grid through a transformer, as shown in Figure 9.34.

The control system of the SVC controls the reactive output so that the voltage magnitude of the controlled node is kept constant. Usually, a certain slope is introduced in the control, as shown in Figure 9.35, which shows the reactive current as a function of the voltage. In the (almost) horizontal part of the curve, around the voltage set point, the SVC control is active. When the SVC has reached its maximum or minimum reactive output, the voltage cannot be controlled, and the device will behave as a pure reactor or pure capacitor; thus, in extreme voltage situations, the SVC behaves as a reactor or capacitor bank.

The control of the reactor current is based on thyristors, which limits the bandwidth of the voltage control. If a fast control is needed to compensate for flicker and voltage dips, one has to use technologies based on VSCs.

9.4.2.4.2 STATCOM

The STATCOM is a device which is based on a VSC. The device is connected in shunt and consists of a capacitor charged with a DC voltage, which provides the input voltage for a VSC, as illustrated in Figure 9.36. The converter feeds a reactive current into the network, and by controlling this reactive current, voltage control is achieved. Since a VSC needs semiconductor elements with current interrupting capabilities, thyristors cannot be used; instead, elements such as GTOs or IGBTs have to be used.

In comparison with the SVC, the STATCOM offers two advantages. First, the STATCOM's output reactive current is not limited at low- or high-voltage conditions; rather, the output current is only limited by the converter ratings and is not dependent on the system voltage. This means that the

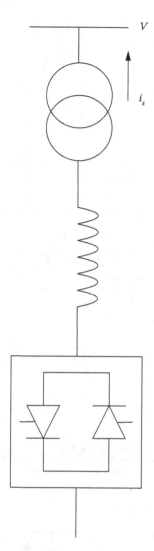

FIGURE 9.33 Thyristor controlled reactor.

reactive support during extreme voltage situations is much better with respect to the SVC, as shown in Figure 9.37. Second, the control response is much faster, since it is limited by the switching frequency of the VSC (usually around 1 kHz). The STATCOM can hence be used to reduce flicker and other fast voltage variations effectively.

The DC capacitor of the VSC constitutes an active power storage; hence, an active current can also be injected into the network. The STATCOM can thus also be used for active power control (e.g., to damp power oscillations). However, the energy stored in the DC capacitor is fairly small; therefore, to truly control the active power output, a battery must be installed on the DC side.

9.4.3 SECONDARY VOLTAGE CONTROL

Secondary voltage regulation (SVR), also referred to as secondary voltage control, has been developed and implemented in some European power grids, particularly in France [16] and Italy [17], with the objective of improving system voltage stability (VS) (see Chapter 10 for a detailed discussion on VS). The effects of SVR in the VS of power systems have been well studied and

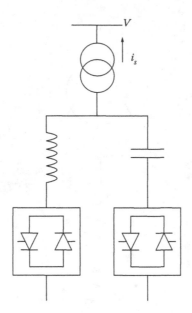

FIGURE 9.34   Static var compensator.

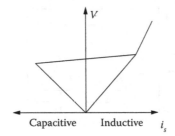

FIGURE 9.35   Static var compensator control.

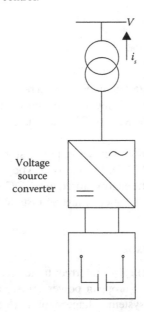

FIGURE 9.36   STATCOM.

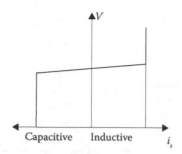

FIGURE 9.37 Voltage control and reactive capability of a STATCOM.

documented [18], demonstrating that systems operating under SVR schemes show an increase in transmission capacity directly associated with improvements in VS.

SVR is a hierarchical, centralized voltage control approach similar to the secondary frequency control approach discussed in Section 9.3.3. This type of control coordinates and supervises the primary voltage regulators or AVR of generators and other reactive power sources in a given geographical region to improve the VS characteristics of the interconnected grid. The main idea behind SVR is to coordinate the various regional reactive power suppliers, since reactive power is basically supplied locally as previously discussed, so that the reactive power reserves of each supplier are properly coordinated as the load changes. Thus, from the steady-state point of view, these controls are designed in such a manner that groups of reactive power sources, typically generators, control the voltage at a given "pilot" node (a remote, high-voltage bus) by moving together to deliver reactive power proportionally to their own capability, with all regional suppliers reaching their reactive power limits at the same time. It should be noted that, for generators, reactive power reserves vary depending on their loading conditions, and hence this should be taken into consideration.

The pilot nodes are chosen based on their short-circuit capacity, that is, those nodes with the largest short-circuit level in a given region are chosen as the pilot nodes. The generators that control each pilot node are then chosen through a sensitivity analysis based on the sensitivity $\partial V_p/\partial Q_g$, where V_p correspond to the pilot node voltages and Q_g are the generator reactive powers in the region. The largest inputs on this matrix define the generators that should be associated with the pilot nodes.

This hierarchical, centralized voltage control approach can be further enhanced using a tertiary voltage regulation (TVR) scheme, which is typically based on an overall system optimization approach that defines optimal voltage set points for the SVR pilot nodes to, for example, minimize losses. A typical generic TVR and SVR scheme is illustrated in Figure 9.38. A similar control structure has been implemented in the Italian power grid. In the depicted control scheme, the system control center determines the optimal voltage set points for the pilot buses based on a given optimization criterion applied to the system as a whole. These set points are then fed to regional voltage regulators (RVRs), which are basically integral controls that generate a pu signal q, where $-1 \leq q \leq 1$. This q signal is then used by local reactive power regulators (QRs) at reactive power sources, to control their reactive power output with respect to their own reserves by modulating the input of the corresponding primary voltage regulators (AVRs).

The actual control blocks used to model these various regulators in stability studies are depicted in Figure 9.39. Observe the hierarchy of the time constants, that is, the AVR time constant (typically about 0.5 s) is about 10 times less than the QR time constant ($T_{\mathrm{vsc}} = 5$ s), which is about 10 times less than RVR time constant ($T_{\mathrm{qsc}} = 50$ s). Notice that the limits on the QR control block restricts the AVR input to a 15%–20% range, which are somewhat typical voltage regulation limits in AVRs.

The basic control characteristics of the SVR system have been implemented into PSAT [19], which is a power system analysis tool described in more detail in Chapter 10. A power-flow-based

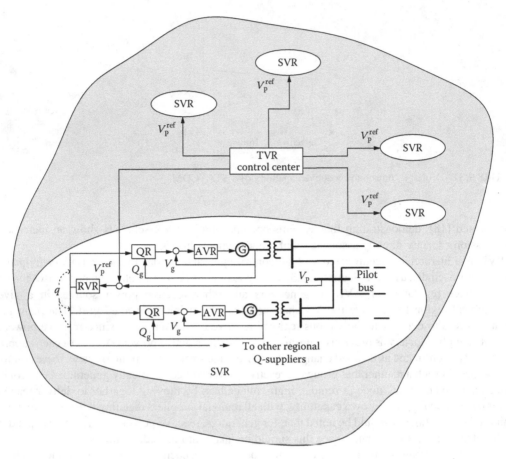

FIGURE 9.38 Secondary voltage regulation overall structure.

remote voltage control approach is used to model SVR, as proposed in Reference 32; thus, generator buses directly associated with a pilot bus are basically treated as PQ buses whose delivered reactive power changes in proportion to a K_g factor and the desired voltage magnitude at the pilot node. In standard remote voltage control, the K_g factor is a fixed value, whereas in the case of SVR, K_g should change according to the available capability of the device. Hence, the K_g for each control generator is computed based on fixed maximum or minimum reactive power limits, depending on whether the generator is over- or under-excited, respectively, as follows:

$$K_g = \frac{Q_{g_{M/m}}}{\sum_{G \in p} Q_{G_{M/m}}} \Rightarrow Q_g = K_g \sum_{G \in p} Q_G$$

where $Q_{g_{M/m}}$ stands for the maximum/minimum reactive power limits of the *gth* generator controlling a pilot node p (this factor changes during the solution process depending on whether the generator is over- or under-excited); and $G \in p$ represents the set of generators controlling the bus voltage at the pilot node. This is a first approximation of the actual SVR control, as the effect of generator loading levels on the corresponding reactive power limits is not modeled, that is, the generator reactive power limits are simply represented here using fixed values.

The effect of an SVR control approach on VS, which, as explained in more detail in Chapter 10, is directly associated with system loadability limits, is clearly illustrated in the following example:

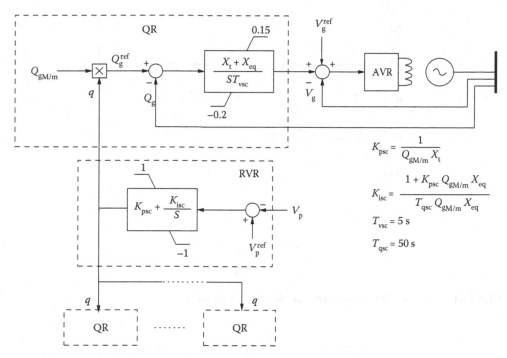

FIGURE 9.39 Generator SVR controller. The X_t and X_{eq} parameters stand for the generator's transformer reactance and the equivalent reactance from the generator terminals to the pilot bus p, respectively; Q_g represents the generator g reactive power, $Q_{g_m} \le Q_g \le Q_{g_M}$; V corresponds to the bus voltage magnitudes; and K_{psc} and K_{isc} are the PI gains of the RVR controller.

EXAMPLE 9.9

For the three-bus test system depicted in Figure 9.40, the PV curves are obtained using PSAT [19], assuming a constant power factor for the load.

Figure 9.41 depicts the PV curves for the system without SVR, and Figure 9.42 shows the PV curves for the system with SVR. From these figures, the maximum loadability of the system is shown to be about 230 MW for the system without SVR, and approximately 260 MW for the system with SVR, which clearly indicates that SVR enhances the system VS. Observe that the voltages with SVR are overall higher at all buses, which may lead to relatively high generator terminal voltages, as shown in Figure 9.42; this could be an issue. Notice as well that, under SVR, both generators

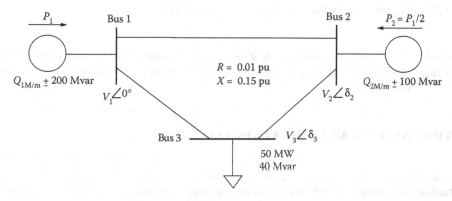

FIGURE 9.40 Secondary voltage regulation sample system.

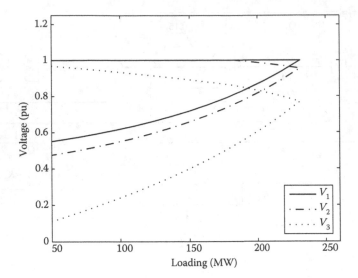

FIGURE 9.41   PV curves for the three-bus sample system without SVR.

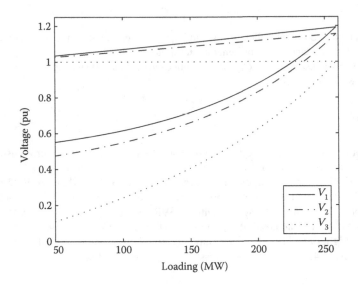

FIGURE 9.42   PV curves for the three-bus sample system with SVR.

reach their maximum reactive power limits at the same time, which is not the case without SVR, as depicted in Figure 9.41; this clearly illustrates the coordination of regional reactive power reserves on which SVR is based.

9.5   WIND AND SOLAR GENERATION CONTROLS

The requirements, development, and deployment of controls for RE, particularly wind and solar, power generators is closely tied to the technological development of wind power generation. Thus, starting with simple generators based on squirrel-cage induction machines, referred to as Type A wind generators, with mechanical blade pitch controls, these generators have evolved into full power-electronics interfaced systems, referred to as Type D and generally based on perma-nent-magnet synchronous machines and VSCs, with torque, power, voltage, and some frequency controls [20].

The first types of wind power generators to be developed and deployed were Type A generators, illustrated in Figure 9.43. These wind generators, which can be found nowadays mostly in rural and remote locations, are typically composed of fixed wind turbines with controllable blades through their pitch angle β, so that their mechanical torque T_m and/or power P_m output can be controlled as the wind speed v_w changes, based on the P_m–ω_m–β characteristics depicted in Figure 9.44. Usually, this control, which is slower than the power-electronic-based controls found in other types of wind power generators, is used to keep T_m and/or ω_m at rated values or within specified limits to avoid wind turbine mechanical issues, but this control can also be used to complement frequency controls as discussed in Section 9.5.3.

Since Type A wind power generators are based on induction machines, these generators require reactive power support from the network, thus requiring capacitive compensation to keep the machine reactive power demand relatively low. However, for faults near the generator, low terminal voltages may stall the induction machine due to its high reactive power demand, given the characteristics of the machine, leading to voltage stability issues, as shown in Figure 9.45 and discussed in Section 10.3.5. For this reason, utilities have required wind power generators to disconnect during faults, so that volt-ages are not hindered during fault recovery; this mandated disconnection has resulted in other prob-lems such as frequency stability issues due to the sudden loss of power generation. The lack of reactive power and terminal voltage controls of Type A generators has led to the development and deployment of better wind generator technologies capable of providing these types of controls, as dis-cussed in Section 9.5.2.

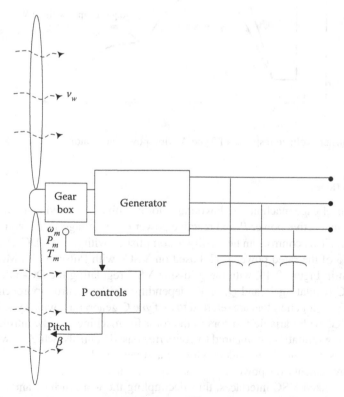

FIGURE 9.43 Induction machine, fixed speed Type A wind power generator.

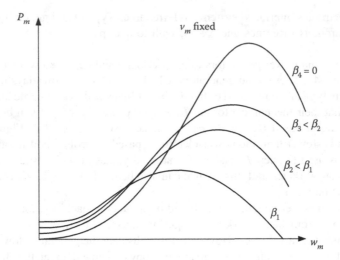

FIGURE 9.44 Wind turbine pitch angle β control.

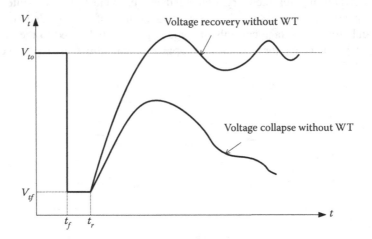

FIGURE 9.45 Transient voltage response of Type A wind power generators.

9.5.1 PQ CONTROL

Replacing the squirrel-cage machines with wound-rotor machines, referred to as DFIGs, has allowed the control of terminal active power P_t and reactive power Q_t or voltage V_t, through the regulation of the rotor currents I_R. This control can be readily accomplished with an AC–DC–AC converter interface, of about 25% of the rated power [21], based on VSCs with Pulse-Width-Modulation (PWM) controls, as shown in Figure 9.46, with the grid-side VSC regulating the DC-link voltage V_{dc}, and the rotor-side VSC regulating P_t and Q_t or V_t, depending on the control approach. These types of DFIG-based wind power generators are referred to as Type C generators, with Type B generators corresponding to DFIGs with variable resistors in the rotor for machine speed control, with its obvious control and operating limitations compared to converter-based controls, which is why these types of wind power systems have not been widely adopted and deployed.

The rapid improvements on power-electronic systems and switching devices has allowed the development of full-rated VSC interfaces, thus decoupling the wind turbine and its associated generator from the grid, providing more control flexibility. These types of generators, referred to as

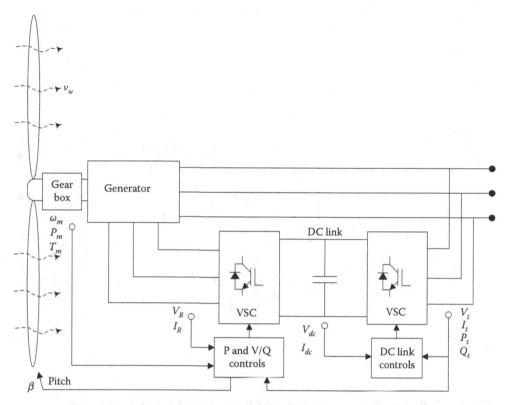

FIGURE 9.46 DFIG, variable speed Type C wind power generator.

Type D and illustrated in Figure 9.47, eliminate the need for a gear box, thus significantly reducing maintenance and wear out, and are hence the preferred wind power generation technology nowadays. These systems are typically based on Squirreled-Cage Induction Generators (SCIGs), with reactive power demands that impact the size the generator-side VSC, but relatively cheap in terms of equipment and maintenance costs, or utilize wound-rotor and Permanent-Magnet Synchronous Generators (PMSGs), which do not require reactive power support and are hence more efficient, but are more expensive, especially permanent-magnet machines due to the cost of magnets, or require more maintenance in the case of wound-rotor machines [21].

Type C and D wind power generators allow for the optimal control of output active power P_t through MPPT controls. This control technique is based on the P_m–ω_m–v_m characteristics of wind power generators depicted in Figure 9.48, where for a given value of β (regulated to keep T_m or ω_m within specified limits), the power varies with ω_m and v_w, presenting a maximum P_m value for given machine and wind speeds. Thus, the approach is to vary ω_m by controlling the current on the DFIG rotor I_R or the PMSG/SCIG stator I_G with the generator-side VSC, typically through perturb-and-observe or speed-power-curve (dashed bold line in Figure 9.48) controls, to yield maximum power output. Given the versatility of PWM-based VSCs, reactive power output can also be controlled through the DFIG I_R or PMSG/SCIG I_t currents, based on dq decomposition and associated current controls.

Solar photo-voltaic (SPV) generators with full-converter interfaces, as illustrated in Figure 9.49, are the prevalent solar power technology currently in use. These systems have the capacity of also controlling P_t with the solar-panel-side DC/DC converter, and Q_t or V_t with the grid-side VSC, and hence have similar control abilities as Type D generators. The output power is also controlled using an MPPT approach, based on the I_{PV}–V_{PV}–H characteristics of the PV panels depicted in Figure 9.50, which shows that for an irradiation level H, there is a maximum value of panel power

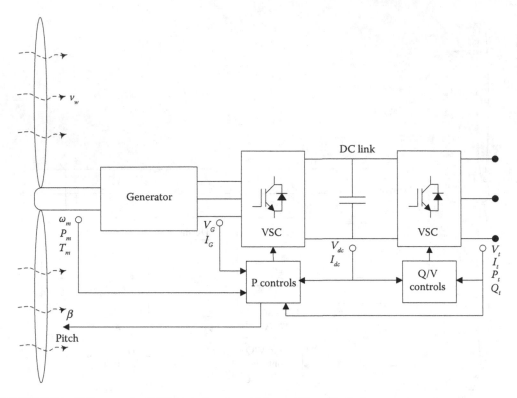

FIGURE 9.47 Full-converter (VSC) interfaced, variable seed Type D wind power generator.

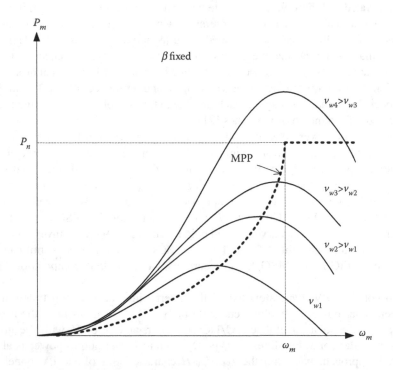

FIGURE 9.48 Wind generator Maximum Power Point Tracking (MPPT) control.

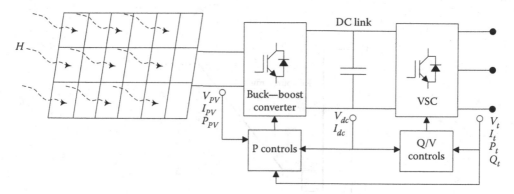

FIGURE 9.49 SPV generator.

P_{PV}, for certain values of panel voltages V_{PV} and currents I_{PV}. Hence, by controlling I_{PV} with the DC/DC converter, it is possible to obtain maximum power output P_t from the SPV generator.

Based on the operating characteristics of Type A generators, utilities initially required $Q_t = 0$ (unity power factor) from Types C and D and SPV generators, so that these RE generators did not represent a reactive power demand for the system. It was also required that these generators would trip during faults, so that they would not contribute negatively to the fault currents and voltages. However, given the capability to control Q_t or V_t of these RE sources, which allow them to provide reactive power or voltage control during normal operation and fault conditions, with all its advantages, as discussed next.

9.5.2 VOLTAGE CONTROL

The aforementioned need for RE generators to support fault recovery led utilities to define Low-Voltage Ride-Through (LVRT), also referred to as Fault Ride-Through (FRT), requirements for these types of generators. A typical LVRT or FRT response of a RE generator is depicted in Figure 9.51, mandating that the generator remains online during the fault, from its occurrence at t_f, clearance at t_{r1}, initial recovery t_{r2}, and final recovery at t_d, which is defined as the point at which the terminal voltage recovers to a minimum value V_{tmin} from a minimum fault value V_{tf}. The LVRT requirements and corresponding parameter values depend on the jurisdiction; thus, for example, the US FERC requirements for a wind power plant are: $t_f = t_{r1} = 0$, $t_{r2} = 0.625$ s, $t_d = 3$ s, $V_{tf} = V_{tr} = 0.15$ p.u.; and $V_{tmin} = 0.9$ p.u. [22]. LVRT/FTR controls are designed so that, during the fault and recovery process,

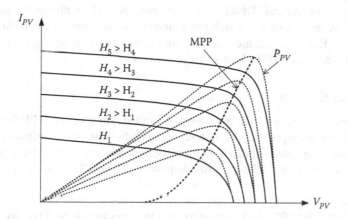

FIGURE 9.50 SPV generator MPPT control.

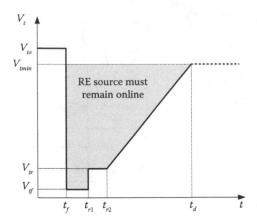

FIGURE 9.51 LVRT/FRT of RE generators.

the generator mainly provides reactive power support to reduce the fault voltage sag; thus, typically, Q_t output is linearly reduced from a maximum value at V_{tr} to a minimum value at V_{tmin}, while $P_t \approx 0$ to avoid excessive terminal currents that would contribute to the fault currents and/or trip the generator protections. These requirements certainly have an impact on the generator converter controls, and may even affect the converter ratings, especially if the LVRT/FTR time response parameters are long and/or the minimum required voltages are high. If the fault voltage profile is such that the time duration is too long and/or the voltages are too low, thus exceeding the LVRT/FRT requirements, as in the case of faults near the RE generation plant, its protections should operate.

Utilities and/or regulators are now requiring that RE generators provide voltage regulation and thus reactive power support not only during faults, as in the case, for example, of China, ISO-NE and ERCOT [23], as well as most European countries [24], based on the inherit capacity to control output voltage of today's RE power generators. This requirement increases the ratings of the generator and converters, since providing reactive power support increases the generator and terminal currents, during transients and normal operating conditions; however, it has no significant effect on the control systems, since reactive power controls are an integral part of RE generators nowadays.

The actual implementation and controls vary among manufactures, and are in fact considered trade secrets, being hence confidential; in fact, only black-box models are provided to customers for their simulation packages for evaluation and study purposes, without releasing the actual control details. Researchers have published a large number of control implementations, as for example in Figure 9.52 [25], where the details of "typical" DFIG controls are provided. Many phasor models for power system dynamic studies such as transient stability analyses can be found in the technical literature as well; for example, in Reference 12, the relatively simple and typical model illustrated in Figure 9.53 is used for stability studies of SPV plants in large networks.

9.5.3 FREQUENCY CONTROL

The frequency control capacity of RE generators is limited, since under MPPT control, P_t is controlled to yield maximum output. However, in several jurisdictions, such as China and United States [23], as well as most European countries [24], there are now requirements for frequency support from RE generators, which are being asked to provide power reserves. Thus, for example, in Denmark, these types of generators, in particular wind power plants, are required to provide power output in a 60%–100% range for up to 15 min in response to frequency changes in the 50–52 Hz range, and deliver power output in the 80%–100% range for 20 s to 3 min for 47–48.5 Hz frequency variations, and 90%–100% power output from 30 min to 5 h for frequency changes in the 48.5–50 Hz range [24].

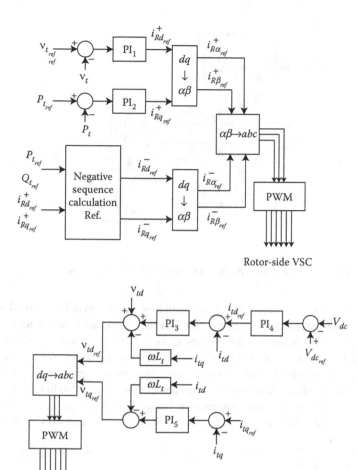

Rotor-side VSC

Grid-side VSC

FIGURE 9.52 DFIG detailed convert controls for Figure 9.46 [25], where L_t represents the grid-side inductance, and v and i correspond to time-domain voltages and currents, respectively, in the dq and $\alpha\beta$ domains.

Over-frequency response can be readily accomplished by reducing P_t by disabling MPPT control. However, underfrequency response requires that RE generators be derated, which means that the plant should maintain a power reserve with respect to its maximum possible output at any given wind speed or irradiation level. This control can be readily accomplished by introducing droop in the $P_{t_{ref}}$ regulation signal used to determine the dq-axis current references for the RE converter

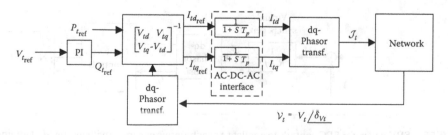

FIGURE 9.53 SPV plant phasor model for power system dynamic studies.

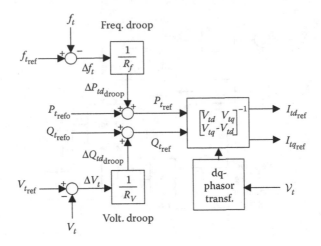

FIGURE 9.54   Frequency R_f and voltage R_V droop controls for RE generators.

controls, possibly also adding droop to the $Q_{t_{ref}}$ control loop, as shown in Figure 9.54 and discussed in Reference 15, where this approach is referred to as an Instantaneous Power Theory (IPT) control.

In addition to providing frequency control, researchers have proposed the introduction of virtual inertia in the RE generation control loop, so that inertial response is added to the generator for system frequency stability improvement. Thus, several approaches have been proposed in the literature to accomplish this; for example, in Reference 15, the control system depicted in Figure 9.55 is proposed for an SPV generator, which basically simulates the behavior of a synchronous machine through the converter interface, being hence referred to as a Synchronous Power Controller (SPC). The

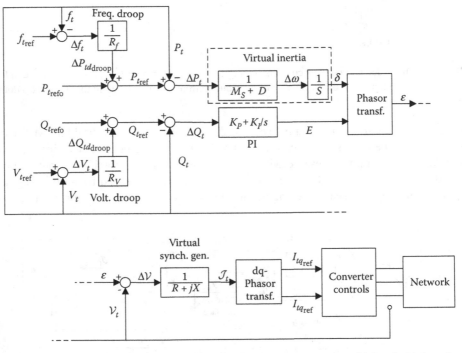

FIGURE 9.55   SPV generator SPC controls to emulate a synchronous generator with inertia M, damping D, and internal phasor voltage $\mathcal{E} = E\angle\delta$ behind impedance $R + jX$ [15].

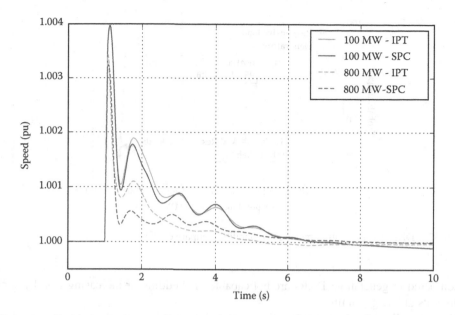

FIGURE 9.56 Frequency response of an important thermal generator for a simulated line contingency in Chile's northern system with an SPV plant with frequency controls and/or virtual inertia [15].

application of the SPC to the power grid of northern Chile with two different sized SPV plants is depicted in Figure 9.56, demonstrating the better frequency response of the plant for the SPC versus the inertia-less IPT control, especially for the larger SPV plant; this assumes that enough reserves are available in the plant, otherwise the plant would not able to provide frequency control due to its maximum power output.

9.6 DR FREQUENCY CONTROL

9.6.1 DR Resources

A significant part of customer loads (residential, commercial and industrial) have some degree of flexibility that could be used to provided services to utilities, if the benefits for the customers of providing this flexibility are larger than the incurred costs [26]. The consumer capability to interrupt, reduce, or increment their consumption while meeting technical and economical requirements for supplying energy balancing or reserve market products is referred to as Demand Response Resources (DRRs). DRRs have technical properties, such as size (kW or MW), duration (h), and notification time (min or h), each having an associated price, and may complement the resources provided by generators in ISO/RTO portfolios for balancing and ancillary services for system operation. DRRs participation in power system operation depends on the resources capability to react to meet the ISOs/RTOs requirements, and on DRRs providing in a more economic and efficient way to operate the system.

There are three major differences in the technical response of DRR with respect to generation resources [27]:

- DRRs may be relatively small in comparison with generator resources. For example, a large car manufacturer may drop 3 MW in ventilation with a very short notification, whereas a residential consumer may participate typically with a few kW. Hence, small DRRs require aggregation in order to represent significant resources to be useful for network operation; this aggregation may be performed directly by ISOs/RTOs or through a specialized agent known as an "Aggregator".

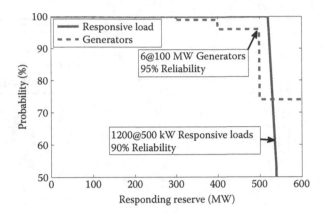

FIGURE 9.57 DRR and generation resources reliability comparison [29].

- Compared to generation, DRRs are not capable of shedding or increasing load by similar amounts at any given time.
- DRRs usually provide limited response at full power compared to generation, due to load requirements.

These DRR characteristics have to be taken into account in the design of ancillary service markets and contracts, where both generation and DRR compete. On the other hand, and because of the distributed nature of DRRs, the provision of ancillary services with DRRs is more reliable than the same service provided by generators. This is supported by a study carried out at the Oak Ridge National laboratory in the United States, where 500 MW of reserves were provided either by six generators of 100 MW or by 540 MW of DR, concluding that responsive loads provide higher global reliability in spite of being individually less reliable, as illustrated in Figure 9.57. Regarding the technical ability of the demand to follow the requirements imposed by network operators, Figure 9.58 shows the simulated response of a set of 250 residential AC loads to a test control signal from an RTO (PJM in this case), showing a better response compared to a similar response by generators, as discussed in Reference 28.

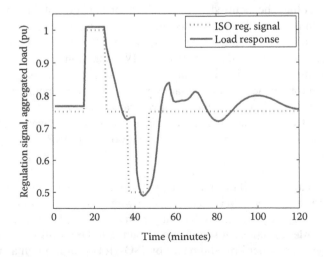

FIGURE 9.58 DRR response to RTO test signals.

9.6.2 DR PARTICIPATION

As it has been demonstrated by different experiences worldwide, the cost of using customer flexibility for ancillary services is competitive compared to the cost of similar services provided by generators. This fact, together with the technical suitability to provide these services, makes DRRs suitable for trade in the context of ancillary service markets or contracts. As an example of how this competitiveness between demand and generation may be implemented, consider an example from [30] depicted in Figure 9.59, where the offers for spinning reserves from generators correspond to the higher price trace, resulting in a market clearing price of 155/MW for a 10 MW reserve requirement. On the other hand, the offers including generation and demand, ordered together in increasing prices, correspond to the lower price trace, resulting in a 130/MW clearing price for the same reserve demand, thus decreasing operating costs and increasing market social welfare.

A good description of the state of demand participation in power system operation can be found in Reference 6. Thus, load participation in frequency control and restoration, strategic reserves, and balancing markets for some locations in Europe is illustrated in Table 9.3, where it can be observed that participation is at a pilot stage in some cases, but for the most part, there are good levels of demand involvement in the provision of reserve services. Demand participation in operating reserves in the United States has a long history, starting more than 40 years ago for reliability and capacity purposes. Table 9.4 provides the requirements for demand participation in the provision of spinning reserve for various US RTOs [27].

9.6.3 AGGREGATION

DRRs are supplied by customers that are flexible in their consumption, provided that they receive a payment that would, at least, cover the costs they may incur by varying their flexible demand. These costs include direct costs, which are equipment and facilities costs required to perform the required control actions, and indirect costs, which are derived from the impact of demand changes in the process (product, comfort, etc.) that is affected by the required service. DRRs are normally used by

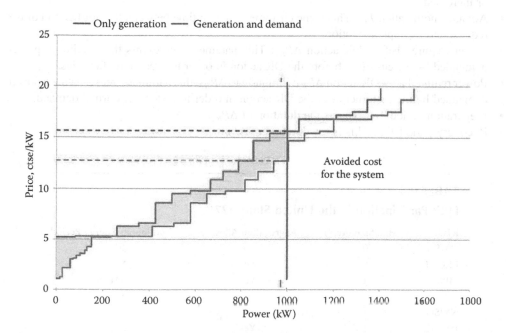

FIGURE 9.59 Ancillary services market with DRRs.

TABLE 9.3
DRR Participation in Europe [6]

ENTSO-E	TSO	Market Size	Load Access and Participation	Aggregated Load Accepted
FCR	Frequency controlled normal operation reserve (FCR-N)	140 MW	Pilots	Yes
	Frequency controlled disturbance reserve (FCR-D)	260 MW	70 MW	Yes
FRR-A	Automatic frequency restoration reserve (FRR-A)	70 MW	Pilots	Yes
FRR-M	Fast disturbance reserve (FRR-M)	1614 MW	385 MW	Yes
RR	Strategic Reserves	365 MW	40 MW	Yes
	Balancing Market (RPM)	300 MW	100–400 MW	Yes

network operators when the resources are large (MWs) and available over long periods of time (h); hence, smaller DRRs are useful for operators when pooled together by Aggregators.

DR can be decomposed in various parts, each associated with the flexibility of a single process, and is characterized by the following technical parameters, as illustrated in Figure 9.60:

- Flexible power ΔP_{R1}: This is the amount of power to be decreased or increased during the DR action. Depending on the service provided and the country, the minimum flexible power required to participate in DR services could be quite high (e.g., the minimum required capacity for tertiary reserves in Germany is 15 MW). However, small amounts of flexible power offered by customers can be pooled together by an Aggregator to meet or exceed minimum DR service requirements.
- DR duration T_D: This parameter represents the time during which the demand is decreased or increased.
- Advance notification T_{IA}: This is the customer reaction time between the notification of the request and its implementation.
- Power required before DR action ΔP_{R2}: This parameter represents the maximum power demanded by the customer before the DR action in order to prepare its facilities.
- Power required after DR action ΔP_{R3}: Similarly to ΔP_{R2}, this parameter represents the power demanded by the customer after the DR action in order to restore its normal demand.
- Preparation period T_{PR}: This is the duration of ΔP_{R2}.
- Recovery period T_{RC}: This is the duration of ΔP_{R3}.

TABLE 9.4
DRR Participation in the United States [27]

RTO	Min. Size (MW)	Aggregation Allowed	Continuous Energy Period
CAISO	0.5	No	30 min
ERCOT	0.1	No	NA
MISO	1	Yes	60 min
PJM	0.1	Yes	NA
NYISO	1	No	60 min
ISO-NE	1	Yes	NA

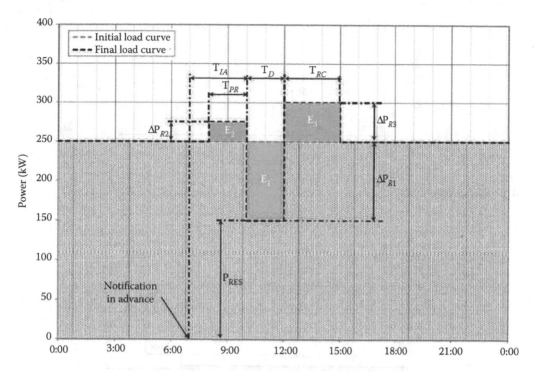

FIGURE 9.60 Load curve DRR components and parameters [28].

- Operating times: These parameters define the time frame during which the service could be made available and include:
  - The days in which the action can be performed.
  - Starting and final hours of availability during the days of availability.
- Minimum time between interruptions T_{MIN}: This parameter represents the time that must pass between the end of a DR action and the beginning of the next one, including T_{IA}.

These technical parameters are complemented by economic constraints indicating the minimum price for each DR action. The DRR parameters have to be identified for each suitable process, and should be included in the total DRR step-wise function a customer offers to the operator, directly or through an Aggregator.

The Aggregator role is very important to transform the customer DRR offers into valuable products for transmission and distribution system operators. These offers have to be evaluated in the short term, depending on the time frame of the service in which the customer is going to participate. When the DRR offers that suit a particular service are identified, these can be aggregated (for one or several consumers) in a step function as the one shown in Figure 9.61 [26], which represents the power that can be curtailed for a specified time (hourly, 30 min, or 15 min); the price assigned to each reduction depends on the associated costs. Figure 9.62 shows the DRR function for an office customer in Northern Europe, through an Aggregator, for a balancing market at 09:00 a.m. in a typical summer and winter day [30].

As discussed in this section, customers play an essential role in power grids nowadays, since they have in many cases the ability to manage their loads so as to provide operators such as TSOs and DSOs with resources similar to generators, which may significantly contribute to grid operation, especially in the context of proper integration of distributed generation technologies. Therefore, customers should be considered not as passive entities which just consume the energy supplied, but as active players and key elements in the efficient management and reliable operation of power grids.

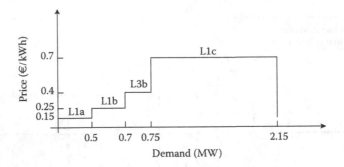

FIGURE 9.61 DRR offers.

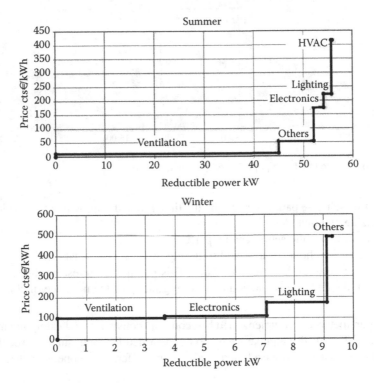

FIGURE 9.62 DRR offer of a Northern European office customer at 09:00 a.m. on a typical summer and winter day [30].

However, securing DRRs is a challenge, as customers are used to relatively cheap energy and operating services, but this is rapidly changing with the need for new and more sophisticated resources coming from customers, who will need extensive support and training in order for DRRs to be effective [31].

REFERENCES

1. U.S. Federal Energy Regulatory Commission (FERC), Promoting wholesale competition through open access nondiscriminatory transmission services by public utilities, Docket RM95-8-000, Washington, D.C., March 1995.

2. Bhattacharya K., Bollen M. H. J., and Daalder J. E., *Operation of Restructured Power Systems*, Kluwer Academic Publishers, 2001.

3. El-Samahy I., Bhattacharya K., Cañizares C. A., Anjos M., and Pan J., A procurement market model for reactive power services considering system security, *IEEE Transactions on Power Systems*, 23(1), 137–149, February 2008.

4. The Independent Electricity Market Operator (IESO), Market rules for the Ontario electricity market, December 2005.

5. Hydro One, Peaksaver PLUS frequently asked questions. Available at http://www.hydroone.com/ MyHome/SaveEnergy/Pages/peaksaverPLUS FAQs.aspx.

6. Smart Energy Demand Coalition (SEDC), Mapping demand response in Europe today, 2015.

7. Lawrence Berkeley National Lab. (LBNL), Demand response providing ancillary services: A comparison of opportunities and challenges in the USA wholesale markets, LBLN-5958E, Berkeley, CA, USA, 2012.

8. European Commission, A policy framework for climate and energy in the period from 2020 to 2030, 2014. Available at http://eur-lex.europa.eu/legal-content/EN/TXT/PDF/?uri=CELEX:52014DC0015&fr om=EN.

9. Adefarati T. and Bansal R. C., Integration of renewable distributed generators into the distribution system: A review, *IET Renewable Power Generation*, 10(7), 873–884, 2016.

10. Armendariz M., Babazadeh D., Brodén D., and Nordström L., Strategies to improve the voltage quality in active low-voltage distribution networks using DSOs assets, *IET Generation, Transmission & Distribution*, 11(1), 73–81, 2017.

11. Munoz J. C. and Canizares C. A., Comparative stability analysis of DFIG-based wind farms and conventional synchronous generators, IEEE-PES Power Systems Conference and Exposition, Phoenix, Arizona, USA, 1–7, March 2011.

12. Tamimi B., Canizares C. A., and Bhattacharya K., System stability impact of large-scale and distributed solar photovoltaic generation: The case of Ontario, Canada, *IEEE Transactions on Sustainable Energy*, 4(3), 680–688, 2013.

13. Miller N., Clark K., and Shao M., Frequency responsive wind plant controls: Impacts on grid performance, IEEE PES General Meeting, 1–8, July 2011.

14. Bevrani H., Ise T., and Miura Y. Virtual synchronous generators: A survey and new perspectives, *Electrical Power and Energy Systems*, 54, 244–254, 2014.

15. Remon D., Canizares C. A., and Rodriguez P., Impact of 100-MW-scale PV plants with synchronous power controllers on power system stability in northern Chile, to appear in, *IET Generation, Transmission & Distribution*, 11(11), 2958–2964, 2017.

16. Bourgin F., Testud G., Heilbronn B., and Verseille J., Present practices and trends on the French power system to prevent voltage control, *IEEE Transactions on Power Systems*, 8(3), 778–788, 1993.

17. Corsi S., Marannino P., Losignore N., Moreschini G., and Piccini G., Coordination between the reactive power scheduling function and the hierarchical voltage control of the EHV ENEL system, *IEEE Transactions on Power Systems*, 10(2), 686–694, 1995.

18. Berizzi A., Bresesti P., Marannino P., Granelli G. P., and Montagna M., System-area operating margin assessment and security enhancement against voltage collapse, *IEEE Transactions on Power Systems*, 11 (3), 1451–1462, 1996.

19. PSAT. Available at http://www.power.uwaterloo.ca/downloads.htm.

20. Anaya-Lara O., Jenkins N., Ekanayake J. B., Cartwright P., and Hughes M., *Wind Energy Generation: Modelling and Control*, E Wiley, USA, 2009.

21. Baroudi J. A., Dinavahi V., and Knight A. M., A review of power converter topologies for wind generators, *Renewable Energy*, 32, 2369–2385, 2007.

22. System Operator, Generator fault ride through (FRT) investigation, TRANSPOWER New Zealand Ltd., GEN FRT: S1, February 2009.

23. Gao D. W., Muljadi E., Tian T., Miller M., and Wang W., Comparison of standards and technical requirements of grid-connected wind power plants in China and the United States, NREL, NREL/TP-5D00-64225, September 2016.

24. Sourkounis C. and Tourou P., Grid Code Requirements for Wind Power Integration in Europe, Conference Papers in Energy, Hindawi Publishing Corp., 1–9, 2013.

25. Nasr E., Canizares C. A., and Bhattacharya K., Stability Analysis of Unbalanced Distribution Systems With Synchronous Machine Based Distributed Generators, *IEEE Transactions on Smart Grid*, 5(5), 2326–2338, 2014.

26. Alvarez C., Gabaldon A., and Molina A., Assessment and simulation of the responsive demand potential in end user facilities: Application to a university customer, *IEEE Transactions on Power Systems*, 9(2), 1123–1231, 2004.
27. MacDonald J., Kiliccote S., Boch J., Chen J., and Nawy R., Commercial building loads providing ancillary services in PJ&M, *ACEEE Summer Study on Energy Efficient Buildings*, 3, 192–206, 2014.
28. PJM State & Member Training Dept., Demand side response in the ancillary services markets, 2011.
29. Kirby B. J., Demand Response for Power System Reliability: FAQ, Oak Ridge National Laboratory, ORNL TM-2006/565, December 2006.
30. EU-DEEP European project, The birth of a European Distributed EnErgy Partnership (DEEP) that will help the large implementation of distributed energy resources in Europe, 2009. Available at http://www.eudeep.com.
31. Alvarez Bel C., Alczar Ortega M., Escriv Escriv G., and Gabaldn Marn A., Technical and economical tools to assess customer demand response in the commercial sector, *Energy Conversion and Management*, 2605–2612, 2009.
32. Canizares C. A., Cavallo C., Pozzi M., and Corsi S., Comparing secondary voltage regulation and shunt compensation for improving voltage stability and transfer capability in the Italian power system, *Electric Power Systems Research*, 73(1), 67–76, January 2005.
33. Kundur P., *Power System Stability and Control*, McGraw-Hill, USA, 1994.
34. European Network of Transmission System Operators for Electricity (ENTSOE), Continental Europe Operation Handbook, 2009. Available at https://www.entso.eu/publications/system-operations-reports/operation-handbook/pages/default.aspx.
35. FERC, NERC-Regions and Balancing Authorities, 2012. Available at https://www.ferc.gov/market-oversight/mkt-electric/nerc-balancing-authorities.pdf.

10 Angle, Voltage, and Frequency Stability

Claudio A. Cañizares, Luis Rouco, and Göran Andersson

CONTENTS

10.1 INTRODUCTION

Understanding stability issues in the power grid is rather important, since one of the main objectives of system operators, regardless of the basic structure of the electricity markets in which power systems are operated, is to guarantee that the system is securely operated, that is, the system remains stable even in the case of certain unexpected events or contingencies such as a major line or generator trip. Furthermore, in the context of competitive electricity markets, system security has a direct and significant effect on electricity prices. Therefore, this chapter concentrates on presenting and explaining the main concepts and tools associated with power system stability and security analysis, within the framework of competitive electricity markets.

Before defining the main concepts associated with power system stability, it is important to briefly review some basic and general stability concepts for nonlinear systems, of which power systems is a class, that will be used throughout this chapter. For that purpose, the rolling-ball example of Figure 10.1 is quite useful, as it allows to intuitively understand some important issues that otherwise may be difficult to explain and grasp. Thus, observe that the rolling-ball system has two types of equilibrium points or resting states, which correspond to system states at which the ball will not move if its speed v is zero, that is, if the ball is not pushed or perturbed; these are as follows:

- A stable equilibrium point (sep), which corresponds to the bottom of the valley and to which the ball of mass m will return after small and some large perturbations.
- Two unstable equilibrium points (uep's), which correspond to the tops of the hills marking the valley's edge and from which the ball will move away if it is perturbed slightly.

The stability of the ball is then defined as the ability of the ball to return to its permanent resting state or sep (normal operating point in power system jargon, as explained below) after small or large perturbations, that is, after pushing the ball so that $v \neq 0$ and $h \neq 0$. The system is stable if the ball returns to its sep after being pushed and is unstable if it leaves the valley through one of the uep's.

Power system stability can be defined in similar terms as in the case of the rolling-ball system. Hence, paraphrasing the IEEE-CIGRE stability definitions in Reference 1, power system stability is the ability of the grid to return to a *normal* operating condition after being subjected to a perturbation. Notice the emphasis on "normal," which is not typically found in stability definitions of nonlinear systems; the reason for this is that the operating condition that the system reaches after perturbation cannot be just any sep, but an equilibrium point at which the main system variables (angles, voltages, and frequency) are within "acceptable" ranges, as defined by system operators, and the topology of the system remains basically intact, that is, the action of protection devices and system controls activated by the perturbation does not lead to losing major parts of the system or a separation of the grid into islands, which are typical protective/control mechanisms to avoid total blackouts through the isolation of some parts of the system. The stability of power systems is

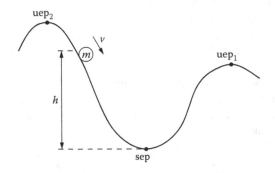

FIGURE 10.1   Stability of a rolling-ball.

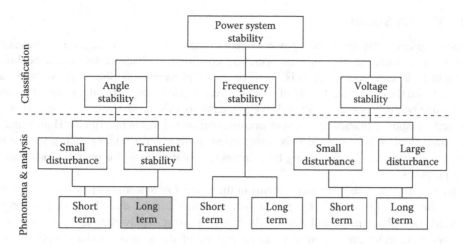

FIGURE 10.2 IEEE-CIGRE stability classification, with a small modification highlighted in gray.

classified as illustrated in Figure 10.2. This figure has a slight difference with respect to the one found in Reference 1, as explained in detail below. Observe that the power system stability is classified in terms of the main system variables: generator rotor angles, bus voltage magnitudes, and system frequency. The definitions and characterization of each one of these stability classes are given below, as per Reference 1.

10.1.1 ANGLE STABILITY

Rotor angle stability refers to the ability of the grid's synchronous machines to remain in synchronism after large or small disturbances and can be directly associated with maintaining or restoring the equilibrium between electromagnetic torque and mechanical torque at each synchronous machine in the system. In this case, stability problems become apparent mainly through power/frequency swings in some generators due to their loss of synchronism.

The further classification of angle stability shown in Figure 10.2 has to do with the characteristics of the physical phenomena as well as the techniques used for their analysis. Thus, small-disturbance (also referred to small signal or oscillatory) rotor angle stability is associated with the ability of generators to maintain synchronism after *small* disturbances, with "small" meaning that the perturbations should be of a magnitude such that the phenomenon can be studied through a linearization of system model equations. This problem is usually associated with the sudden appearance of low-damped or undamped oscillations in the system due to lack of sufficient damping torque, when contingencies occur. Even though these oscillations become apparent in the short term, the weakening of system damping is better understood using a long-term analysis approach, as more recently proposed by means of the application of Hopf bifurcation theory to the study of this phenomenon, as explained in more detail in Section 10.2; this is the reason for the proposed modification to the IEEE-CIGRE classification.

Large-disturbance rotor angle stability, more commonly referred to as transient stability, corresponds to the ability of generators to maintain synchronism when subjected to severe disturbances such as a short circuits or outages of major transmission lines. In this case, the system nonlinearities govern the system response; hence, equation linearizations do not work, requiring short-term analysis tools that fully account for the main nonlinear system characteristics. On the basis of nonlinear system theory, this analysis can be viewed as determining whether the system fault trajectory at the "clearance" point is outside or inside of the stability region associated with the postcontingency sep. For example, in the case of the rolling-ball system, the main objective of this analysis would be to determine whether the ball has left the sep valley, which can be directly linked with its stability region.

10.1.2 VOLTAGE STABILITY

This concept refers to the ability of a power system to maintain steady voltages at all buses after a small or large disturbance of a normal operating condition. Voltage stability can be linked to short-term or long-term problems, that is, "fast" or "slow" phenomenon. Short-term voltage stability is associated with the design and tuning of generator Automatic Voltage Regulators (AVR) and other system voltage controllers such as Static Var Compensators (SVC); fast voltage changes associated with converter response characteristics, such as commutation failures in inverters of High Voltage DC (HVDC) transmission links may also be categorized as short-term voltage stability phenomena. These issues are typically studied using time-domain simulation tools for operation and long-term planning purposes.

Voltage stability analysis of power systems in the context of system operation has concentrated mostly on the study of voltage collapse problems, which can be directly linked to long-term, small-perturbation phenomena, and thus can be analyzed using steady-state analysis tools, such as power-flow-based techniques, as well as linearizations of the system model equations; Section 10.3 mostly focuses on presenting and discussing these issues. It is important to highlight the fact that, as in the case of angle oscillation problem, even though small perturbations are assumed to study the problem, large perturbations such as major line or generator trips are actually the ones that typically trigger voltage stability problems.

More recently, fast voltage collapse problems have been identified associated with increased demand from motor loads, such as air-conditioning compressors, worsening sustained low voltages after severe system contingencies. This phenomenon is referred to as Fault-Induced Delayed Voltage Recovery (FIDVR) [2], and is discussed in detail at the end of Section 10.3.

10.1.3 FREQUENCY STABILITY

This type of stability issue is associated with the recovery of system frequency after large active power imbalances between generation and load due to system disturbances. Hence, frequency stability analysis concentrates on studying variations in system frequency due to large sudden changes in the generation–load balance, as discussed in detail in Section 10.4.

10.1.4 SYSTEM MODELING

For stability studies, it is typically assumed that the system is balanced and that system frequencies do not change significantly during transient events triggered by small or large perturbations. Hence, positive sequence, phasor domain models are used to model the grid in this case, from generators (see synchronous machine models in Chapter 7 and Appendix C), to the transmission system (see line and transformer models in Chapter 7), to loads (see Chapter 7 and induction motor models in Appendix C). Examples of typical models used for stability studies are presented and discussed throughout this chapter.

The typical power system model used in stability studies is a differential algebraic equation (DAE) model (e.g., Example 10.1 below), which can be generically defined as follows:

$$\dot{x} = f(x, y, p, \lambda)$$
$$0 = g(x, y, p, \lambda)$$

(10.1)

where x stands for the n state variables (e.g., generator internal angles and speeds); y represents m algebraic variables (e.g., static load voltages and angles); p stands for k controllable parameters (e.g., AVR set-points); λ represents l uncontrollable parameters (e.g., load levels); $f(\cdot)$ stands for n nonlinear differential equations (e.g., generator mechanical equations); and $g(\cdot)$ represents m nonlinear algebraic equations (e.g., power flow equations). If the Jacobian $D_y g = [\partial g_i / \partial y_j]_{m \times m}$ is nonsingular, that is, invertible, along the trajectory solutions of Equation 10.1, the system can be transformed into ordinary differential equations (ODE) based on the Implicit Function Theorem:

$$y = h(x, p, \lambda) \implies \dot{x} = f(x, h(x, p, \lambda), p, \lambda) = F(x, p, \lambda) \tag{10.2}$$

In practice, this is a theoretical exercise that is not carried out due to its complexity; the system model should be revised when $D_y g$ becomes singular, since these singularities can be typically removed by means of representing loads with dynamic models. Therefore, throughout this chapter stability principles associated with standard ODE models of nonlinear systems will be applied.

EXAMPLE 10.1

Figure 10.3 depicts a generator connected to a large system through a step-up transformer and a transmission line, and Figure 10.4 shows the corresponding positive sequence model. The generator is modeled using a simple second-order direct-axis transient model, that is, an internal transient voltage source $E' \angle \delta$ behind a direct-axis transient reactance X' with damping D and mechanical inertia H (see Appendix C), plus a simple integral control representing an AVR. The system is modeled as an infinite bus, that is, a constant voltage source $V_\infty \angle 0°$ of infinite inertia.

The DAE model resulting from the system model shown in Figure 10.4 is

$$\dot{\delta} = \Delta\omega = \omega - \omega_0 \tag{10.3}$$

$$\dot{\omega} = \frac{\omega_0}{2H}\left(P_m - \frac{E'V_\infty}{X_e}\sin\delta - D\Delta\omega\right) \tag{10.4}$$

$$\dot{E}' = K_v(V_{1o} - V_1) \tag{10.5}$$

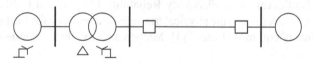

FIGURE 10.3 Single-line diagram of a generator connected to a large system through a step-up transformer and a transmission line.

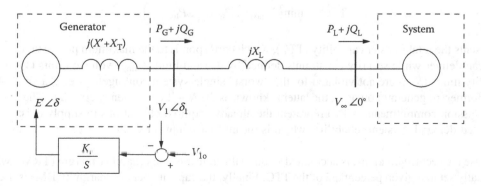

FIGURE 10.4 Generator-infinite bus with AVR example.

$$0 = \frac{V_\infty V_1}{X_L} \sin \delta_1 - \frac{E' V_\infty}{X_e} \sin \delta \tag{10.6}$$

$$0 = V_\infty^2 \left(\frac{1}{X_L} - \frac{1}{X_e} \right) + \frac{E' V_\infty}{X_e} \cos \delta - \frac{V_\infty V_1}{X_L} \cos \delta_1 \tag{10.7}$$

where $X_e = X' + X_T + X_L$. Hence, for this DAE model, $x = [\delta \; \omega \; E']^T$, $y = [V_1 \; \delta_1]^T$, $p = [V_{1o} \; V_\infty]^T$; and $\lambda = P_m = P_d$, which stands for the demand of the system, that is $P_L = P_d$ in this case. Thus,

$$f(x, y, p, \lambda) = \begin{bmatrix} \frac{\omega_0}{2H} \left(P_m - \frac{E' V_\infty}{X_e} \sin \delta - D \Delta \omega \right) \\ K_v (V_{1o} - V_1) \end{bmatrix}$$

$$g(x, y, p, \lambda) = \begin{bmatrix} \frac{V_\infty V_1}{X_L} \sin \delta_1 - \frac{E' V_\infty}{X_e} \sin \delta \\ V_\infty^2 \left(\frac{1}{X_L} - \frac{1}{X_e} \right) + \frac{E' V_\infty}{X_e} \cos \delta - \frac{V_\infty V_1}{X_L} \cos \delta_1 \end{bmatrix}$$

Equations 10.3 and 10.4 are usually referred to as the generator swing equations; Equation 10.5 represents a simple AVR integral control; and Equations 10.6 and 10.7 are basic active and reactive power flow equations, respectively, defining the algebraic variables V_1 and δ_1 in this model.

10.1.5 TRANSMISSION CONGESTION

System stability has a significant effect on competitive electricity markets, directly influencing electricity prices, since maintaining the system stability even in the case of certain major contingencies, that is, maintaining system security, takes precedence over economic issues during system operation. Thus, the most inexpensive generators may not necessarily be dispatched if dispatching them lead to violations of system security margins, with more expensive generators being dispatched to resolve these violations.

Security margins are defined based on the system's available transfer capability (ATC) concept, which is defined by North American Electricity Reliability Corporation (NERC) as the "measure of the transfer capability remaining in the physical transmission network for further commercial activity over and above already committed uses" [3]. Mathematically, this is defined as

$$\text{ATC} = \text{TTC} - \text{ETC} - \text{TRM} - \text{CBM} \tag{10.8}$$

where

$$\text{TTC} = \min \left(P_{\max_{I_{\lim}}}, P_{\max_{V_{\lim}}}, P_{\max_{S_{\lim}}} \right)$$

represents the total transfer capability (TTC), which corresponds to the maximum power that the system can deliver while security constraints defined by thermal limits ($I_{\lim}$), voltage limits ($V_{\lim}$), and stability limits ($S_{\lim}$) are not violated for the "worst" single-system contingency (e.g., a single-line, transformer, or generator outage); the latter is known as the $N-1$ contingency criterion. The existing transmission commitments (ETC) represent the already acquired obligations to supply existent and expected demand. System reliability, which is meant to account for further uncertainties in system operation not considered in the $N-1$ criterion (e.g., "credible" double contingencies) and thus increase the security margins, is accounted through the transmission reliability margin (TRM), which is usually set to a given percentage of the TTC. Finally, the capacity benefit margin (CBM) is used to define transmission capacity reserves needed to supply possible increases in demand, and is typically set to zero in practice, since expected demand increments are usually accounted for in the ETC value, as loads tend to be rather inelastic, that is, not very responsive to prices and hence relatively easy to forecast with reasonable accuracy.

From Equation 10.8, it is clear that system stability has a very direct effect on the value of ATC through the computation of the TTC. An example of the computation of the ATC is presented and discussed in detail in Section 10.3.3.

10.1.6 RENEWABLE ENERGY GENERATION

As discussed in Chapter 9, the rapid increase in penetration of generation from Renewable Energy (RE) sources in the grid, particularly wind- and solar-power generators, has had a significant impact on system stability, due to the unique characteristics of RE-based plants, particularly the unpredictability and variability of their power output, plus their converter interfaces and related fast controls. Depending on the types of RE generators and their associated controls, RE plants may either hinder or aid system stability; however, in general, RE generation tends to have a negative effect on grid stability when compared to traditional synchronous-generator power plants, due to limited voltage and power controls as well as lack of inertia. This is the reason why grid codes have been changing as RE penetration levels increase, demanding more voltage and power regulation services from RE plants [4,5], so that their possible negative impact on system stability can be minimized. This has been motivated and made possible by changes in RE generation controls and converters, with manufacturers' designing and developing systems that make these sources behave more like synchronous generators (see, e.g., [6]), thus providing somewhat similar voltage and power services as traditional generating plants.

The impact on voltage, angle, and frequency stability of RE generation, with its different types of technologies and associated controls presented in Section 9.5, are discussed in Section 10.5. Furthermore, given the particular fast switching and control characteristics of converter-based RE generators, especially full-converter-interfaced wind and Solar Photovoltaic (SPV) plants, this section also discusses particular stability issues directly associated with fast converter control loops, especially Phase-Locked Loop (PLL) synchronization issues and what is commonly known as "harmonic instability," which refers to fast stability problems related to resonant neighboring converter controls and harmonic filters.

10.2 ANGLE STABILITY

In this section, the main concepts and techniques used to analyze large- and small-disturbance angle stability problems are presented.

10.2.1 LARGE-DISTURBANCE STABILITY

The problem of large-disturbance stability analysis is discussed here in four distinct steps. The first step is the presentation of the equal area criterion (EAC), which allows to comprehensibly explain the large-disturbance stability of a single machine connected to an infinite bus. The second step explains the fundamentals of numerical integration of the differential equations associated with the single machine connected to an infinite-bus example. The third step addresses the numerical integration of ODE and DAE models of a multimachine system. The presentation is completed with a study on the effect of the machine modeling detail on its large-disturbance response.

10.2.1.1 Equal Area Criterion

Consider the case of a single generator connected to an infinite bus through a step-up transformer and a transmission line, as depicted in the single-line diagram of Figure 10.1. In this case, both transformer and transmission line can be modeled simply with inductances. One is interested in analyzing

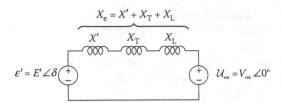

FIGURE 10.5 Equivalent circuit of a generator connected to an infinite bus through a step-up transformer and a transmission line.

the generator stability when a three-phase fault first occurs somewhere along the line, and it is then cleared by opening and reclosing the line (the fault is not permanent). The process has the following three distinct sequential stages: (i) a three-phase fault occurs in the line; (ii) the line is then quickly switched off at both ends, which results in the extinction of the fault; and (iii) finally, the line is switched back on.

The analysis of large-disturbance stability in this case consists of determining how long the fault can be on and the line kept opened, so that the generator does not lose synchronism once the line is reclosed. The generator loses synchronism because the energy gained during the rotor acceleration in the period when the fault is on and the line stays opened, which is associated with the electrical power applied to the rotor by the generator being zero while the mechanical power supplied by the prime mover remains constant, cannot be absorbed when the line is switched back on. The energy balance between acceleration and deceleration is evaluated through the so-called EAC.

If one assumes that the damping factor is zero*, the swing equation described in Appendix C becomes

$$\frac{2H}{\omega_0}\frac{d\omega}{dt} = P_m - P_e = P_d - P_e = P_{ace} \tag{10.9}$$

where P_m is the power applied to the prime mover, P_d is the demand of the system ($P_m = P_d$ to guarantee that $\omega = \omega_0$ in steady state, as discussed in Section 10.4), and P_{ace} is the acceleration power. In this case, according to the equivalent circuit of Figure 10.5, the electrical power supplied to the generator is given by the expression

$$P_e = \frac{E'V_\infty}{X_e}\sin\delta \tag{10.10}$$

where X_e is the equivalent reactance between the source behind the transient reactance and the infinite-bus voltage. It should be noted that the rotor angle δ is the angle of the phasor voltage behind the transient reactance $\mathcal{E}'$ with respect to the phasor voltage of the infinite bus $\mathcal{U}_\infty$.

If Equation 10.9 is multiplied by $d\delta$, it becomes

$$\frac{d\omega}{dt}d\delta = \frac{\omega_0}{2H}P_{ace}d\delta \tag{10.11}$$

This equation can also be written as

$$\frac{d\delta}{dt}d\omega = \frac{\omega_0}{2H}P_{ace}\,d\delta \tag{10.12}$$

---

* This is unrealistic, since synchronous machines are designed to yield significant electromechanical damping through their damper windings, as discussed at the end of Section 10.2.1, but simplifies the explanation of the energy balance issues.

or

$$(\omega - \omega_0)\, d\omega = \Delta\omega\, d\Delta\omega = \frac{\omega_0}{2H} P_{\text{ace}}\, d\delta \tag{10.13}$$

If Equation 10.13 is integrated between δ_0 and δ, which correspond to 0 and $\Delta\omega$, respectively, in the integral of $\Delta\omega$

$$\int_0^{\Delta\omega} \Delta\omega\, d\Delta\omega = \frac{\omega_0}{2H} \int_{\delta_0}^{\delta} P_{\text{ace}}\, d\delta \tag{10.14}$$

it yields

$$\frac{1}{2}\Delta\omega^2 = \frac{\omega_0}{2H} \int_{\delta_0}^{\delta} P_{\text{ace}}\, d\delta \tag{10.15}$$

The generator is stable if the integral of the speed deviation as time $t \to \infty$ is zero. In this case, Equation 10.15 can be written as

$$\int_{\delta_0}^{\delta_{\text{clr}}} P_{\text{ace}}\, d\delta = \int_{\delta_{\text{clr}}}^{\delta_{\text{max}}} P_{\text{dec}}\, d\delta \tag{10.16}$$

$$A_{\text{ace}} = A_{\text{dec}} \tag{10.17}$$

where δ_{clr} is the angle at the time of clearing; δ_{max}, the maximum rotor angle; P_{dec}, the decelerating power; A_{ace}, the acceleration area; and A_{dec}, the deceleration area.

Equation 10.16 indicates that the generator is stable if a clearing rotor angle exists such that the accelerating area is equal to the decelerating area. This is known as the EAC, which is illustrated in Figure 10.6. Figure 10.6a shows the application of this criterion to the stability analysis of a generator when the fault is cleared at the rotor angle δ_{clr}.

The "critical" value of the clearing angle δ_{cri} is obtained when the maximum angle δ_{max} is $\pi - \delta_0$. Hence,

$$\int_{\delta_0}^{\delta_{\text{cri}}} P_{\text{ace}}\, d\delta = \int_{\delta_{\text{cri}}}^{\pi - \delta_0} P_{\text{dec}}\, d\delta \tag{10.18}$$

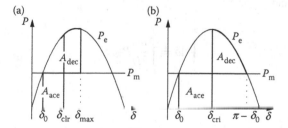

FIGURE 10.6 EAC for a generator-infinite-bus example.

Figure 10.6b shows how the critical clearing angle is determined through the application of the EAC. Thus, if the expressions of the accelerating and the decelerating power are substituted in Equation 10.18, the equation becomes

$$\int_{\delta_0}^{\delta_{cri}} P_m \, d\delta = \int_{\delta_{cri}}^{\pi-\delta_0} (P_e - P_m) \, d\delta \tag{10.19}$$

If Equation 10.10 of the electrical power expression is substituted into Equation 10.19 and the resulting equation is integrated, it results in (the mechanical power supplied by the prime mover is constant through the transient process)

$$P_m(\pi - 2\delta_0) = \frac{EV_\infty}{X_e}(\cos \delta_{cri} + \cos \delta_0) \tag{10.20}$$

Taking into account that the mechanical power is equal to the electrical power at the original equilibrium point, one has

$$P_m = P_d = \frac{E'V_\infty}{X_e} \sin \delta_0 \tag{10.21}$$

In this case, an analytical expression of the critical clearing angle can be obtained as follows:

$$\delta_{cri} = \arccos \left[\sin (\pi - 2\delta_0) - \cos \delta_0 \right] \tag{10.22}$$

Notice that in practice, when the damping is not zero, this critical angle has a larger value, as explained in more detail below.

The electric power is zero either when a three-phase fault is applied or when the line is opened. Hence, since it has also been assumed that the mechanical power supplied to the prime mover is constant, the acceleration is also constant. The time function of the angular speed variation can be easily obtained in this case integrating Equation 10.9 between 0 and t; thus,

$$\int_{\omega_0}^{\omega} d\omega = \int_0^t \frac{\omega_0}{2H} P_{ace} \, dt \tag{10.23}$$

which results in

$$\omega - \omega_0 = \frac{\omega_0}{2H} P_{ace} t \tag{10.24}$$

This equation shows the speed deviation varying linearly with time. In addition, if the speed deviation $\omega - \omega_0$ is substituted in Equation 10.24 by the derivative of the angle with respect to time $d\delta/dt$, and integrating between 0 and t, one has

$$\int_{\delta_0}^{\delta} d\delta = \int_0^t \frac{\omega_0}{2H} P_{ace} t \, dt \tag{10.25}$$

which becomes

$$\delta - \delta_0 = \frac{\omega_0}{4H} P_{ace} t^2 \tag{10.26}$$

This equation shows that the angle varies quadratically with time. Therefore, the clearing time t_{clr} of a fault can be obtained once the clearing angle δ_{clr} is known, as follows:

$$t_{clr} = \sqrt{\frac{4H}{\omega_0 P_{ace}}(\delta_{clr} - \delta_0)} \tag{10.27}$$

EXAMPLE 10.2

Consider a 100 MVA, 15 kV generator, whose inertia and transient reactance are, $H = 3$ s and $X' = 0.3$ pu on the generator MVA base, connected to an infinite bus through a 100 MVA, 220 kV/15 kV step-up transformer with reactance $X_T = 0.15$ pu, and a 220 kV transmission line with reactance $X_L = 0.1$ pu on a 100 MVA base. Assume that the generator is working at full load, 0.8 power factor lag, and rated terminal voltage. Determine the critical clearing time of a three-phase fault that occurs at the sending end of the line, which is cleared by switching it off and on.

The angle of the voltage source behind the transient reactance with respect to the infinite-bus voltage is determined first. Thus, from the terminal conditions of the generator, the initial conditions of the angles of the voltage behind the transient reactance and the infinite-bus voltage with respect to the terminal voltage are

$$\mathcal{E}' = \mathcal{U}_t + jX'\mathcal{I}_t = 1.0\angle 0° + j0.3 \times 1\angle -36.87° = 1.2042\angle 11.50°$$

$$\mathcal{U}_\infty = \mathcal{U}_t - j(X_T + X_L)\mathcal{I}_t = 1.0\angle 0° + j0.25 \times 1\angle -36.87° = 0.8732\angle -13.24°$$

Hence, the rotor angle, that is, the angle of the voltage behind the transient reactance with respect to the infinite-bus voltage is

$$\delta_0 = 11.50° + 13.24° = 24.74°$$

The critical angle is determined from the angle of the voltage behind the transient reactance with respect to the infinite bus applying Equation 10.22:

$$\delta_{cri} = 87.42°$$

In this example, the critical clearing angle is not greater than 90°. However, in general, this angle can be greater than 90°, which is the steady-state stability limit, as discussed in Section 10.2.2.

The critical clearing time is computed applying Equation 10.27:

$$t_{cri} = 0.228 \text{ s} = 228 \text{ ms}$$

A critical clearing time smaller than 100 ms might not be compatible with the time required by protections and circuit breakers to clear a fault. However, if one considers that the generator damping D value for this simple model can be significant, this critical value is larger than that computed with Equation 10.27, as shown below; hence, this is seldom a concern nowadays. Critical clearing times are used in practice as relative measures of stability regions; thus, the larger this value, the larger the stability region and hence the more stable the system.

The EAC is basically a Lyapunov stability criterion, which can be viewed as a general energy balance analysis technique. This type of stability analysis is typically referred to as a direct stability analysis method based on energy functions, as explained in more detail in References 7–9. Unfortunately, as with any Lyapunov-based stability technique, it is only a sufficient but not necessary test of stability, that is, it can be used to determine whether a perturbed system is stable, but it cannot be accurately used to determine whether the system is unstable.

Using the analogy of the rolling-ball in Figure 10.2, if the kinetic plus potential energy of the perturbed system is less than the potential energy of the "closest" uep, that is, the uep with the least potential energy value, the system is stable. However, if the system energy is greater than this value, the system may or may not be stable, depending on its trajectory and friction (damping), which in power systems can be significant, as previously mentioned. Nevertheless, if the perturbed system energy locally exceeds the maximum potential energy value along its trajectory, the system can be said to be unstable, since it has enough energy to leave the valley defined by the potential energy well around the sep.

In the case of the generator-infinite-bus example of Figure 10.3, a similar criterion can be applied, with the kinetic W_K and potential W_P energies being defined as

$$W_K = \frac{1}{2} M (\omega - \omega_0)^2$$

$$W_P = \int (T_e(\delta) - T_m) d\delta \approx \int (P_e(\delta) - P_m) d\delta \quad \text{for } \Delta\omega \text{ small}$$

$$\approx \int_{\delta_0}^{\delta} \left(\frac{E'V_\infty}{X_e} \sin \delta - P_d \right) d\delta$$

$$\approx -E'V_\infty B_e(\cos \delta - \cos \delta_0) - P_d(\delta - \delta_0)$$

where $M = 2H/\omega_0$. In this case, the potential-energy well around the sep can be readily depicted, as shown in Figure 10.7; note that this energy looks very much the same as the potential-energy well around the sep in the rolling-ball example. Therefore, the EAC can be equivalently restated based

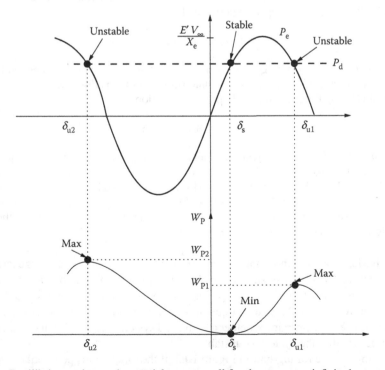

FIGURE 10.7 Equilibrium points and potential-energy well for the generator-infinite-bus example.

on the following transient energy function (TEF) definition:

$$\text{TEF} = \vartheta(x, x_0)$$

$$= \vartheta([\delta\ \omega]^{\text{T}}, [\delta_0\ 0]^{\text{T}})$$

$$= \frac{1}{2}M(\omega - \omega_0)^2 - E'V_\infty B_e(\cos\delta - \cos\delta_0) - P_d(\delta - \delta_0)$$

where $\vartheta(\cdot)$ is a Lyapunov function. Thus,

- If $\text{TEF} \le W_{\text{P1}}$, that is, $A_{\text{ace}} \le A_{\text{dec}}$ in the EAC, the system is stable.
- If $\text{TEF} > W_{\text{P1}}$, that is, $A_{\text{ace}} > A_{\text{dec}}$, the system may or may not be stable depending on the system damping.

To better understand the transient stability phenomenon, the more complex case of a single machine connected to an infinite bus through a step-up transformer and two parallel transmission lines, as illustrated in the single-line diagram of Figure 10.8, is now studied. The system is perturbed by applying a three-phase fault at an intermediate point of one of the parallel circuits, which is then cleared by opening the faulted line. In this case, the fault is permanent and can only be cleared by opening the line.

The transient can be separated into three periods: prefault, fault, and postfault. For each period, the electrical power delivered by the generator is a different function of the rotor angle, depending on the value of the transfer impedance between the voltage behind the transient reactance and the infinite bus:

1. For the prefault period

$$X_e^{\text{pref}} = X' + X_T + \frac{X_{L1}X_{L2}}{X_{L1} + X_{L2}} \tag{10.28}$$

2. For the fault period and assuming that the fault occurs in an intermediate point of the line defined by a factor f of the line reactance, Figure 10.9 shows the transformations of the equivalent circuit to determine the transfer impedance, yielding

$$X_e^{\text{f}} = X' + X_T + X_{L1} + \frac{(X' + X_T)X_{L1}}{fX_{L2}} \tag{10.29}$$

3. For the postfault period

$$X_e^{\text{posf}} = X' + X_T + X_{L1} \tag{10.30}$$

Figure 10.10 shows a plot of the electric power delivered by the generator as a function of the rotor angle in the three periods previously defined. This figure also shows the application of the EAC to this

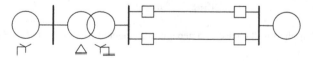

FIGURE 10.8 Single-line diagram of a generator connected to an infinite bus through a step-up transformer and two parallel lines.

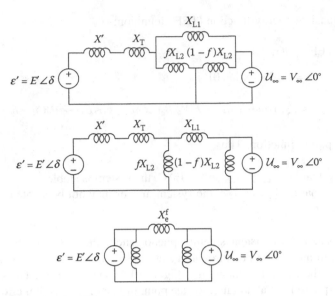

FIGURE 10.9 Equivalent circuit of Figure 10.8 during the fault.

problem.[*] During the prefault period, the generator is operating at the intersection of the prefault function $P_e^{pref}(\delta)$ and P_m. When the fault occurs, the generator operates on the fault function $P_e^f(\delta)$. Once the fault is cleared, the generator operates on the postfault function $P_e^{posf}(\delta)$. The critical clearing angle δ_{cri} can then be computed applying the EAC, that is,

$$A_{ace} = A_{dec}$$

$$\int_{\delta_{pref}}^{\delta_{cri}} \left(P_m - P_e^f\right) d\delta = \int_{\delta_{cri}}^{\pi-\delta_{posf}} \left(P_e^{posf} - P_m\right) d\delta \tag{10.31}$$

$$\int_{\delta_{pref}}^{\delta_{cri}} \left(P_m - \frac{E'V_\infty}{X_e^f} \sin \delta\right) d\delta = \int_{\delta_{cri}}^{\pi-\delta_{posf}} \left(\frac{E'V_\infty}{X_e^{posf}} \sin \delta - P_m\right) d\delta$$

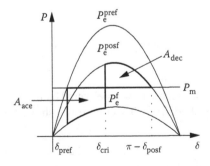

FIGURE 10.10 EAC applied to the case of a generator connected to an infinite bus through a step-up transformer and two parallel lines.

---

[*] It is assumed that the maximum power in the fault period $P_{e,max}^f$ is smaller than P_m.

Integrating Equation 10.31 leads to

$$
\begin{aligned}
0 = {} & \frac{1}{X_e^{\text{pref}}} \sin \delta_{\text{pref}} (\pi - \delta_{\text{pref}} - \delta_{\text{posf}}) \\
& + \frac{1}{X_e^{\text{f}}} (\cos \delta_{\text{cri}} - \cos \delta_{\text{pref}}) - \frac{1}{X_e^{\text{posf}}} (\cos \delta_{\text{posf}} + \cos \delta_{\text{cri}})
\end{aligned}
\tag{10.32}
$$

Equation 10.32 is a nonlinear function of the critical clearing angle and does not have an analytical solution; hence, it has to be solved numerically using, for instance, a Newton-based method. Furthermore, the critical clearing time cannot be determined from the critical clearing angle using Equation 10.9, since this equation was obtained assuming that the accelerating power P_{ace} is constant, whereas in this case, P_{ace} depends on the angle as follows:

$$
P_{\text{ace}}(\delta) = P_{\text{m}} - P_e^{\text{f}}(\delta) = P_{\text{m}} - \frac{E'V_\infty}{X_e^{\text{f}}} \sin \delta
\tag{10.33}
$$

The differential equations that describe the dynamic behavior of the generator during the fault period are nonlinear in this case, as opposed to the equations found in the case when the fault takes place at the sending end or receiving end of the line. Hence, to compute the critical clearing time from the critical clearing angle, the system's nonlinear equations have to be integrated numerically, even for the case of zero damping. The following section presents the analysis of large disturbances of a single generator connected to an infinite bus using numerical simulations, that is, numerically integrating the corresponding differential equations.

EXAMPLE 10.3

Consider the generator of Example 10.2 and assume that the generator is now connected to an infinite bus through a step-up transformer and two identical parallel lines, each with a reactance of $X_L = 0.2$ pu referred to a 100 MVA base. Determine the critical clearing angle assuming that a three-phase fault occurs at the midpoint of one of the lines and that the fault is cleared by opening the faulted line.

Since Equation 10.32 is expressed in terms of the transfer reactance during each transient period, as well as the steady-state value of rotor angle in the prefault and the postfault periods, these values need to be computed first. Thus, the transfer reactance during the prefault, fault, and postfault periods are, respectively,

$$
X_e^{\text{pref}} = 0.3 + 0.15 + \frac{0.1 \times 0.1}{0.1 + 0.1} = 0.55 \text{ pu}
$$

$$
X_e^{\text{f}} = 0.3 + 0.15 + 0.1 + \frac{(0.3 + 0.15) \times 0.1}{0.2 \times 0.5} = 1.55 \text{ pu}
$$

$$
X_e^{\text{posf}} = 0.3 + 0.15 + 0.2 = 0.65 \text{ pu}
$$

From the fault-period transfer reactance X_e^{f}, the maximum value of the electric power in such period is determined to be

$$
P_{e,\text{max}}^{\text{f}} = \frac{1.2042 \times 0.8732}{1.55} = 0.6784 \text{ pu}
$$

Since $P_{e,\text{max}}^{\text{f}} < P_e^{\text{pref}}$, the critical clearing angle can then be obtained solving Equation 10.32. The pre-fault angle has exactly the same value as the pre-fault angle in Example 10.2, since the generator presents the same terminal conditions and the equivalent impedance between the bus behind the transient reactance and the infinite bus is also the same; thus,

$$\delta_{pref} = 24.74°$$

The postfault angle is determined taking into consideration that the electric power supplied by the generator during the postfault conditions is equal to the pre-fault power, and that the voltage of the bus behind the transient reactance and the infinite bus does not change. Hence,

$$\delta_{posf} = \arcsin\frac{P_e^{pref} X_e^{posf}}{E'V_\infty} = 29.64°$$

The numerical solution of the nonlinear Equation 10.32 then yields the critical clearing angle value:

$$\delta_{cri} = 106.58°$$

Observe that this critical clearing angle is greater than the one obtained in Example 10.2, as the generator accelerates less during the fault period in this case, due to the fact that it supplies electric power during the fault.

So far, only three-phase faults have been considered. However, single-phase-to-ground, double-phase, and double-phase-to-ground faults can occur. Among them, single-phase-to-ground faults are the most frequent (in transmission systems, these types of faults amount to 98% of the total number of faults). Therefore, it is important to study how the large-disturbance stability of a single-generator-infinite-bus system is affected by a single-phase-to-ground fault. For this study, the fault is assumed to be cleared by opening the three phases of the line, which is the case in practice, and the fault is assumed to take place at the sending end of line.

The fault type affects the generator behavior mainly during the fault. For the study of this behavior one has to consider the positive, negative, and zero sequences equivalent circuits and their interconnection depending on the fault type, as discussed in Chapter 8. Thus, for a single-phase-to-ground fault, the sequence equivalent circuits are connected in series, as shown in Figure 10.11, considering

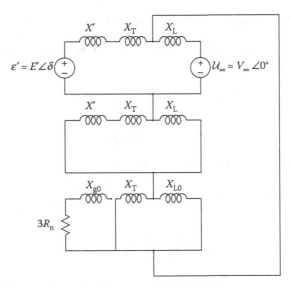

FIGURE 10.11 Connection of the positive, negative, and zero sequence circuits for a single-phase-to-ground fault for the generator-infinite-bus system.

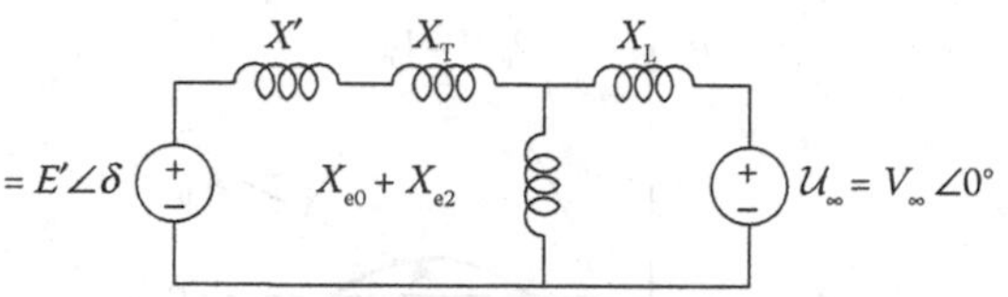

FIGURE 10.12 Equivalent circuit for the single-phase-to-ground fault illustrated in Figure 10.11.

the neutral connections of the generator and the transformer windings depicted in Figure 10.3, which mainly affect the zero sequence equivalent circuit. The negative sequence impedances of the generator, transformer, and line are assumed to be equal to the positive sequence ones.

Figure 10.12 shows the equivalent circuit that results from the reduction of the circuit of Figure 10.11, observe that this equivalent circuit is similar to the one in Figure 10.5. Hence, the electric power during the fault period in this case is determined by the equivalent reactance

$$X_e^f = X' + X_T + X_L \frac{(X' + X_T)X_L}{X_{e2} + X_{e0}}$$

$$= X' + X_T + X_L \frac{(X' + X_T)X_L}{\dfrac{(X' + X_T)X_L}{(X' + X_T) + X_L} + \dfrac{X_T X_{L0}}{X_T + X_{L0}}} \qquad (10.34)$$

where X_{L0} is the zero sequence reactance of the transmission line. It is interesting to note that the fault equivalent reactance X_e^f depends on neither the ground resistance of the generator R_n nor the zero sequence reactance of the generator X_{g0}. This is due to the connection of the low voltage windings of the step-up transformer.

The method just described for the large-disturbance stability analysis of a single-phase-to-ground fault can be readily extended to the cases of double-phase and double-phase-to-ground faults, considering the corresponding connections of the sequence equivalent circuits. However, observe that all these computations are just approximations, since the unbalances associated with these faults are neglected here. For accurate studies of unbalanced faults, the models and techniques described in Chapter 12 are needed.

EXAMPLE 10.4

Consider the generator of Example 10.2. The negative sequence reactances of the generator, transformer, and line are assumed to be identical to the corresponding negative sequence values, and the zero sequence reactance of the line is $X_{L0} = 0.3$ pu. Determine the critical clearing angle when a single-phase-to-ground fault takes place at the sending end of the line.

One has to first determine the transfer reactance during the fault period according to Equation 10.34:

$$X_e^f = 0.3 + 0.15 + 0.1 + \frac{(0.3 + 0.15) \times 0.1}{\dfrac{(0.3 + 0.15) \times 0.1}{0.3 + 0.15 + 0.1} + \dfrac{0.15 \times 0.3}{0.15 + 0.3}} = 0.7975 \text{ pu}$$

To apply the EAC, the maximum electric power during the fault-period needs to be computed; thus,

$$P_{e,\max}^f = \frac{1.2042 \times 0.8732}{0.7975} = 1.3185 \text{ pu}$$

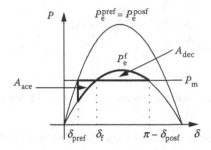

FIGURE 10.13 Application of the EAC in case of a single-phase-to-ground fault at the sending end of the line.

Since this maximum electric power is greater than the prefault generator power, the EAC is applied as illustrated in Figure 10.13. Hence, the accelerating and the decelerating areas are determined to be

$$
A_{\text{ace}} = \int_{\delta_{\text{pref}}}^{\delta_{\text{f}}} \left(P_{\text{m}} - P_{\text{e}}^{\text{f}} \right) d\delta
$$

$$
= P_{\text{m}}(\delta_{\text{f}} - \delta_{\text{pref}}) + \frac{E'V_{\infty}}{X_{\text{e}}^{\text{f}}} (\cos \delta_{\text{f}} - \cos \delta_{\text{pref}})
$$

$$
A_{\text{dec}} = \int_{\delta_{\text{f}}}^{\pi - \delta_{\text{f}}} \left(P_{\text{e}}^{\text{f}} - P_{\text{m}} \right) d\delta
$$

$$
= 2 \frac{E'V_{\infty}}{X_{\text{e}}^{\text{f}}} \cos \delta_{\text{f}} - P_{\text{m}}(\pi - 2\delta_{\text{f}})
$$

Since

$$
\delta_{\text{f}} = \arcsin \frac{P_{\text{e}}^{\text{pref}} X_{\text{e}}^{\text{f}}}{E'V_{\infty}} = 37.36°
$$

then

$$
A_{\text{ace}} = 0.0267 \text{ pu} \times \text{rad}
$$
$$
A_{\text{dec}} = 0.6260 \text{ pu} \times \text{rad}
$$

As the decelerating area is greater than the accelerating area, it can be concluded that, even if the single-phase-to-ground fault is permanent, the generator does not lose synchronism.

10.2.1.2 Time-Domain Analysis of a Single-Machine-Infinite-Bus System

The EAC for the large-disturbance stability analysis of a single machine connected to an infinite bus is based on the assumption that the generator damping factor D is zero, which is not very realistic, since the damping introduced by the damper windings of the generators can be quite significant. In practice, a relatively large value of D is used in the typical transient models of the generator used in the analyses presented so far, to somewhat model the damping effect of the damper windings. Hence, time-domain analyses are used instead of the EAC in practical applications, as this technique does not present any modeling constraints. However, direct stability analysis methods based on the Lyapunov stability theory, of which the EAC is an example, as previously discussed, have been developed for complex systems models [7–9]; for example, an "extended" EAC method has been proposed [10], and it is being used together with time-domain simulations for contingency screening in practical stability analysis applications.

Time-domain simulations consist of the numerical integration of the nonlinear differential equations that describe the generator dynamic behavior. This approach is used here for the computation of the critical clearing time of a fault, simulating the fault clearing process for several values of the clearing time until the generator loses synchronism. The generator loss of synchronism is detected by inspecting the time trajectories of the generator angle.

As previously discussed, the nonlinear equations that describe the dynamic behavior of a single generator connected to an infinite bus can be written in state-space as follows:

$$\frac{d\delta}{dt} = \omega - \omega_0 \tag{10.35}$$

$$\frac{d\omega}{dt} = \frac{\omega_0}{2H}\left[P_\mathrm{m} - \frac{EV_\infty}{X_\mathrm{e}}\sin\delta - \frac{D}{\omega_0}(\omega - \omega_0)\right] \tag{10.36}$$

or in a more compact form, as per Equation 10.2:

$$\dot{x} = F(x, p) \tag{10.37}$$

where $x = [\delta\ \omega]^\mathrm{T}$ and $p = [V_\infty\ E\ P_\mathrm{m}]^\mathrm{T}$, for a given set of parameters $p = p_0$:

$$\dot{x} = F(x, p_0) = F(x) \tag{10.38}$$

A numerical integration algorithm of the set of nonlinear differential Equation 10.38 determines the time evolution of the state variables x in discrete time instants separated by a time period Δt, that is, $(x_0, \ldots, x_k, \ldots)$. Such an algorithm must fulfill the following requirements: stability, accuracy, and simplicity [11]. Stability requires that the error between the exact and the approximate solution is bounded throughout the integration process; accuracy is associated with the magnitude of the integration error, and simplicity is concerned with how the state variables' values are determined at each time step. Of these properties, the method's stability is one of the most relevant in power systems, given the wide range of times constants' values of the system dynamic models (from fractions of a second for controls such as voltage regulators, to several seconds for the machines' inertias, and even minutes if certain controls such as transformer tap-changers are simulated); these types of "stiff" systems require Absolute-stable (A-stable) numerical integration methods. Accuracy and simplicity are opposite features of an integration method, since greater accuracy requires more complex algorithms.

One of the most simple integration algorithms is the Euler method, which is based on the following discrete formula:

$$x_{k+1} = x_k + \dot{x}_k\Delta t = x_k + F(x_k)\Delta t \tag{10.39}$$

This method basically approximates the function $x(t)$ at $k + 1$ by the first two terms of the Taylor-series expansion around $x(t_k)$.

The stability of the Euler method can be analyzed considering the simple linear differential equation $\dot{x} = \mu x$, where μ is a complex number with a real negative part. In this case, the analytical solution of the differential equation is $x = e^{\mu t}x_0$, which is stable given the choice of μ. On the other hand, the numerical solution at step k using Equation 10.39 yields the discrete equation $x_k = (1 + \mu\Delta t)^k x_0$. Observe that, if $|1 + \mu\Delta t| < 1$, the numerical solution will tend to zero as k increases, as expected from the analytical solution. Hence, assuming that $\mu\Delta t = \sigma + j\omega$ $(\sigma < 0)$, the aforementioned discrete equation is stable if $|(1 + \sigma) + j\omega| < 1$; in other words, the algorithm is stable if $\mu\Delta t$ is inside a circle of center $(-1, 0)$ and radius 1. However, note that there are values of $\mu\Delta t$ that violate this condition (for $\omega^2 \geq 1 - (1 + \sigma)^2$), yielding an unstable numerical solution even though the system is stable.

An alternative to the Euler method is the Euler Predictor–Corrector method. This method computes the values of the state variables at step $k + 1$ in two steps:

1. Predictor

$$x_{k+1,p} = x_k + \dot{x}_k \Delta t \qquad (10.40)$$

2. Corrector

$$x_{k+1,c} = x_k + (\dot{x}_k + \dot{x}_{k+1,p}) \frac{\Delta t}{2}$$

$$\qquad (10.41)$$

$$= x_k + [F(x_k) + F(x_{k+1,p})] \frac{\Delta t}{2}$$

The stability of this method can be analyzed following the same procedure used to analyze the stability of the Euler method. Thus, the algorithm is stable if $\left| (1 + \frac{|\mu|^2 \Delta t^2}{2} + \sigma) + j\omega \right| < 1$; in other words, the algorithm is stable if $\mu \Delta t$ is inside a circle of center $(-1 - \frac{|\mu|^2 \Delta t^2}{2}, 0)$ and radius $1 + \frac{|\mu|^2 \Delta t^2}{2}$. The stability region of the Euler Predictor–Corrector method is bigger than the stability region of the Euler method; however, there are values of μ that still make this method unstable, that is, the method is not A-stable.

The Euler and Euler Predictor–Corrector methods belong to the class of *explicit* numerical integration algorithms. These types of methods compute the values of the state variable at the current time step as a function of the values of the state variables and their derivatives at the preceding steps, that is, $x_{k+1} = \Gamma(x_k, \dot{x}_k) = \Gamma[x_k, F(x_k)]$. These methods are simple, since the solution of the method's discrete equation only requires function evaluations; however, their stability regions are small and are hence affected by the size of the integration step.

An alternative to explicit integration algorithms are the *implicit* methods. These algorithms obtain the values of the state variables not only from the values of the state variables and their derivatives at the preceding steps, but also from the state variable derivatives' values at the current step, that is, $x_{k+1} = \Gamma(x_k, \dot{x}_k, \dot{x}_{k+1}) = \Gamma[x_k, F(x_k), F(x_{k+1})]$. Of known implicit algorithms, the Trapezoidal Rule offers a good comprise between stability and simplicity. This method is based on the following discrete equation:

$$x_{k+1} = x_k + (\dot{x}_k + \dot{x}_{k+1}) \frac{\Delta t}{2}$$

$$\qquad (10.42)$$

$$= x_{k+1} = x_k + [F(x_k) + F(x_{k+1})] \frac{\Delta t}{2}$$

The stability of the Trapezoidal Rule can be analyzed following the same procedure used for the stability analysis of the Euler and Euler Predictor–Corrector methods; thus, the method can be readily shown to be stable if $\sigma < 0$, that is, the discrete Equation 10.42 is stable for a stable linear system (A-stable). The computation of the state variables' values at step $k + 1$ from Equation 10.42 requires the solution of a set of nonlinear equations using, for example, a Newton-based method.

EXAMPLE 10.5

Consider the generator of Example 10.2. Determine by time-domain simulations that the critical clearing time for a fault at the sending end of the line has the same value as the one determined using the EAC.

The critical clearing time is determined by repeatedly simulating numerically (using the Euler Predictor–Corrector method) the disturbance by increasing the clearing time 1 ms at a time, until the generator loses synchronism. Figure 10.14 depicts the rotor angle variations with time when the clearing time is 228 ms (solid line) and 229 ms (dashed line), respectively. The fault is applied at $t = 1$ s so that the prefault steady state can be seen. Observe that the generator is unstable (rotor

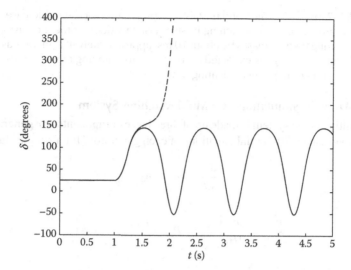

FIGURE 10.14 Time-domain simulation of a generator connected to an infinite bus after a three-phase fault at the sending end of the line for $D = 0$.

angle does not recover) when the clearing time is 229 ms, whereas it is stable when the clearing time is 228 ms. The rotor angle oscillates because the generator damping factor D is zero. The critical clearing time results agree exactly with those obtained in Example 10.2.

EXAMPLE 10.6

Consider the same generator of the previous example but assume now that $D = 2$ pu. Determine by time-domain simulations the critical clearing time for a three-phase fault at the sending end of the line.

The same approach and numerical technique as in the previous example are used here. Figure 10.15 illustrates the rotor angle variation with time for a 236 ms clearing time (solid line

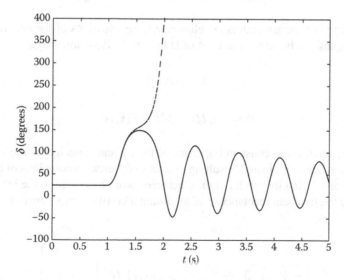

FIGURE 10.15 Time-domain simulation of a generator connected to an infinite bus after a three-phase fault at the sending end of the line for $D \neq 0$.

and stable) and 237 ms (dashed line and unstable). Observe that for the stable case, the generator angle oscillations are damped, approaching its steady-state value as time increases, since $D \neq 0$.

The critical clearing time increases by about 10 ms (approximately 4%) in this case with respect to the case with zero damping, as expected. Thus, the zero-damping model yields a pessimistic/conservative estimation of the critical clearing time.

10.2.1.3 Time-Domain Simulation of a Multi-Machine System

The model of a multimachine system is made up of three basic components: the generators, the electric network, and the loads. The differential equations of each generator i for a simple classical model are

$$\frac{d\delta_i}{dt} = \omega_i - \omega_0 \tag{10.43}$$

$$\frac{d\omega_i}{dt} = \frac{\omega_0}{2H_i}\left[P_{mi} - P_{ei} - \frac{D_i}{\omega_0}(\omega_i - \omega_0)\right] \tag{10.44}$$

$$P_{ei} = \Re\left\{\mathcal{E}'_i \mathcal{I}^*_{gi}\right\} \tag{10.45}$$

$$\mathcal{I}_{gi} = \frac{\mathcal{E}'_i - \mathcal{U}_{gi}}{jX'_i} \tag{10.46}$$

where $\mathcal{E}'$ is the internal transient phasor voltage (phasor voltage behind the transient reactance X'); $\mathcal{I}_g$ is the generator's output phasor current; and $\mathcal{U}_g$ is the generator's terminal phasor voltage.

The electric network is represented by its admittance matrix. The current injected by the generators depend on their state variables and the generator terminal bus voltages, whereas the currents injected by the loads $\mathcal{I}_l$ depend on the load bus voltages $\mathcal{U}_l$ (this dependency depends on the load model). Hence,

$$\begin{bmatrix} \mathcal{Y}_{gg} & \mathcal{Y}_{gl} \\ \mathcal{Y}_{lg} & \mathcal{Y}_{ll} \end{bmatrix}\begin{bmatrix} \mathcal{U}_g \\ \mathcal{U}_l \end{bmatrix} = \begin{bmatrix} \mathcal{I}_g(x, \mathcal{U}_g) \\ \mathcal{I}_l(\mathcal{U}_l) \end{bmatrix} \tag{10.47}$$

If the currents injected by the generators are eliminated, the equations of the generators, the electric network, and the loads can be written as a set of DAEs, as in Equation 10.1:

$$\dot{x} = f(x, \mathcal{U}) \tag{10.48}$$

$$0 = g(x, \mathcal{U}) = \mathcal{Y}\mathcal{U} - \mathcal{I}(x, \mathcal{U}) \tag{10.49}$$

If the loads were modeled as constant impedances, the currents injected by the loads and the load bus voltages can be readily eliminated, resulting in a set of differential equations of the form in Equation 10.38. Thus, considering the model of the electric network expanded to the internal buses of the generators (behind the transient reactance) and assuming a constant impedance model for the loads, one obtains

$$\begin{bmatrix} \mathcal{Y}' & -\mathcal{Y}' & 0 \\ -\mathcal{Y}' & \mathcal{Y}' + \mathcal{Y}_{gg} & \mathcal{Y}_{gl} \\ 0 & \mathcal{Y}_{lg} & \mathcal{Y}_{ll} + \mathcal{Y}_l \end{bmatrix}\begin{bmatrix} \mathcal{E}' \\ \mathcal{U}_g \\ \mathcal{U}_l \end{bmatrix} = \begin{bmatrix} \mathcal{I}_g \\ 0 \\ 0 \end{bmatrix} \tag{10.50}$$

where

$$\mathcal{Y}' = \text{diag}\left\{\frac{1}{jX'_i}\right\}$$

$$\mathcal{Y}_l = \text{diag}\left\{\frac{P_{li} - jQ_{li}}{\mathcal{U}_{li}^2}\right\}$$

Equation 10.50 can also be written as

$$\left[\begin{array}{c|c} \mathcal{Y}_{11} & \mathcal{Y}_{12} \\ \hline \mathcal{Y}_{21} & \mathcal{Y}_{22} \end{array}\right]\left[\begin{array}{c} \mathcal{E}' \\ \mathcal{U}_g \\ \mathcal{U}_l \end{array}\right] = \left[\begin{array}{c} \mathcal{I}_g \\ 0 \\ 0 \end{array}\right] \tag{10.51}$$

Hence, applying Kron's reduction to this equation yields

$$(\mathcal{Y}_{11} - \mathcal{Y}_{12}\mathcal{Y}_{22}^{-1}\mathcal{Y}_{21})\mathcal{E}' = \mathcal{Y}_R\mathcal{E}' = \mathcal{I}_g \tag{10.52}$$

On the basis of the reduced representation of the network Equation 10.52, the electric power supplied by each generator can be represented in terms of the state variables as follows:

$$P_{ei} = \Re\left\{\mathcal{E}'_i\mathcal{I}^*_{gi}\right\} = \Re\left\{\mathcal{E}'_i\sum_{j=1}^{N_g}\mathcal{Y}^*_{Rij}(\mathcal{E}'_j)^*\right\}$$
$$= (E'_i)^2 G_{Rii} + \sum_{\substack{j=1 \\ j\neq i}}^{n} E'_i E'_j[G_{Rij}\cos(\delta_i - \delta_j) + B_{Rij}\sin(\delta_i - \delta_j)] \tag{10.53}$$

where n stands for the number of generators. Thus, substituting Equation 10.53 into Equation 10.44, the differential equations for each generator are

$$\frac{d\delta_i}{dt} = \omega_i - \omega_0 \tag{10.54}$$

$$\frac{d\omega_i}{dt} = \frac{\omega_0}{2H_i}\left\{P_{mi} - (E'_i)^2 G_{Rii} - \sum_{\substack{j=1 \\ j\neq i}}^{n} E'_i E'_j[G_{Rij}\cos(\delta_i - \delta_j) + B_{Rij}\sin(\delta_i - \delta_j)] - \frac{D_i}{\omega_0}(\omega_i - \omega_0)\right\} \tag{10.55}$$

In the case of nonlinear load models, the set of nonlinear DAEs cannot be reduced to a set of nonlinear differential Equation 10.38. In such event, the time-domain simulation requires the solution of the nonlinear algebraic Equation 10.49 at each time step. Two network solution methods can be considered in this case: *partitioned solution* and *simultaneous solution* [12]. The selection of the network solution method is associated with the selection of the integration algorithm (explicit or implicit). The partitioned solution approach is used with explicit integration algorithms, whereas the simultaneous solution method is used with implicit integration algorithms.

The partitioned solution first computes the bus voltages and then the state variables. Thus, consider that at time t_{0-} the system is in steady state, which means that derivatives of the state variables are zero; hence, the state variables, bus voltages, and currents are defined by the follwing set of nonlinear equations, obtained by setting $\dot{x} = 0$ in Equation 10.48:

$$f(x_{0-}, \mathcal{U}_{0-}) = 0 \tag{10.56}$$

$$(\mathcal{Y}_{0-})(\mathcal{U}_{0-}) - \mathcal{I}(x_{0-}, \mathcal{U}_{0-}) = 0 \tag{10.57}$$

Assume that a fault occurs at time t_{0+}. Hence, the state variables cannot change instantaneously, that is, $x_{0-} = x_{0+} = x_0$. On the other hand, the network variables change as per the following nonlinear algebraic equations:

$$(\mathcal{Y}_{0+})(\mathcal{U}_{0+}) = \mathcal{I}(x_0, \mathcal{U}_{0+}) \tag{10.58}$$

The bus voltages $\mathcal{U}_{0+}$ are computed iteratively as follows: The value of $\mathcal{I}(x_0, \mathcal{U}_{0-})$ is used to compute the bus voltages by solving the linear set of Equations 10.58, and the resulting bus voltage values are used to update the currents. This process is repeated until the difference of the bus voltages in two consecutive iterations is less than a selected tolerance. Once the bus voltages have been computed, the state variables for $k = 1$ can be determined from the values of the state variables and their derivatives $(x_0, \dot{x}_0)$, based on the chosen explicit integration algorithm, as follows: ·

$$x_1 = x_0 + \Gamma[\dot{x}_0] = x_0 + \Gamma[(x_0, \mathcal{U}_{0+})] \tag{10.59}$$

The simultaneous solution for an implicit integration algorithm computes at the same time the state variables and bus voltages at each time step. The set of nonlinear differential equations is first converted into a set of discrete nonlinear algebraic equations by applying the corresponding implicit integration method, and the combined set of nonlinear algebraic equations is then solved using a Newton-based method. Thus, for the Trapezoidal Rule, the state variables at $k + 1$ are defined by

$$x_{k+1} = x_k + [f(x_k, \mathcal{U}_k) + f(x_{k+1}, \mathcal{U}_{k+1})]\frac{\Delta t}{2} \tag{10.60}$$

which yields a set of nonlinear algebraic equations:

$$h(x_{k+1}, \mathcal{U}_{k+1}) = x_{k+1} - x_k - [f(x_k, \mathcal{U}_k) + f(x_{k+1}, \mathcal{U}_{k+1})]\frac{\Delta t}{2} = 0 \tag{10.61}$$

On the other hand, the set of algebraic equations to be solved at $k + 1$ is

$$g(x_{k+1}, \mathcal{U}_{k+1}) = \mathcal{Y}_{k+1}\mathcal{U}_{k+1} - \mathcal{I}(x_{k+1}, \mathcal{U}_{k+1}) = 0 \tag{10.62}$$

The full set of algebraic equations to be solved is then

$$h(x_{k+1}, \mathcal{U}_{k+1}) = 0 \tag{10.63}$$

$$g(x_{k+1}, \mathcal{U}_{k+1}) = 0 \tag{10.64}$$

The nonlinear algebraic equations 10.63 and 10.64 can be solved using a standard Newton method. Thus, the state variables and bus voltages for an iteration step $\ell + 1$ are obtained by applying

$$\begin{bmatrix} x_{k+1}^{\ell+1} \\ \mathcal{U}_{k+1}^{\ell+1} \end{bmatrix} = \begin{bmatrix} x_{k+1}^{\ell} \\ \mathcal{U}_{k+1}^{\ell} \end{bmatrix} + \begin{bmatrix} \Delta x_{k+1}^{\ell} \\ \Delta \mathcal{U}_{k+1}^{\ell} \end{bmatrix} \tag{10.65}$$

which requires the solution of the following set of linear equations using sparse-matrix techniques:

$$\begin{bmatrix} D_x h|_{k+1} & D_{\mathcal{U}} h|_{k+1} \\ D_x g|_{k+1} & D_{\mathcal{U}} g|_{k+1} \end{bmatrix} \begin{bmatrix} \Delta x_{k+1}^{\ell} \\ \Delta \mathcal{U}_{k+1}^{\ell} \end{bmatrix} = \begin{bmatrix} -h(x_{k+1}^{\ell}, \mathcal{U}_{k+1}^{\ell}) \\ -g(x_{k+1}^{\ell}, \mathcal{U}_{k+1}^{\ell}) \end{bmatrix} \tag{10.66}$$

It should be noted that $D_x h$, $D_{\mathcal{U}} h$, and $D_x g$ are block diagonal matrices, whereas $D_{\mathcal{U}} g$ is the admittance matrix. This particular matrix structure is utilized to more efficiently solve the linear system of Equations 10.66.

EXAMPLE 10.7

Consider the power system depicted in Figure 10.16. The bus and branch data are provided, respectively, in Tables 10.1 and 10.2. The generators' data, with respect to their own MVA bases, are identical for all generators: $H = 3$ s, $D = 2$ pu, and $X' = 0.45$ pu; the transient reactance includes the reactance of the step-up transformer. The loads are modeled as constant admittances. Determine the critical clearing time for a three-phase fault at the sending end of one of the lines between buses 2 and 3, which is cleared by opening the faulted line.

To compute the critical clearing time for the given fault, repeated time-domain simulations are performed for different values of the clearing time until a generator loses synchronism. These time-domain simulations are based on the numerical integration of differential Equations 10.54 and 10.55.

Several calculations must first be performed to obtain the required differential equations. Thus, the generator parameters have to be first converted to a common 100 MVA base; since this is the

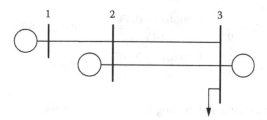

FIGURE 10.16 Single-line diagram of a multimachine three-bus system.

TABLE 10.1
Bus Data of the Multi-Machine System of Figure 10.16

Bus Number	$S_{G\ base}$(MVA)	P_g(MW)	Q_g (Mvar)	P_l (MW)	Q_l (Mvar)	$\mathcal{U}$ (pu)
1	100	80	0.64	0	0	$1.0\angle7.81°$
2	100	80	10.27	0	0	$1.0\angle6.89°$
3	800	640	269.63	800	260	$1.0\angle0°$

TABLE 10.2
Branch Data of the Multi-Machine of Figure 10.16

From (Bus Number)	To (Bus Number)	Circuit	X (pu)
1	2	1	0.02
2	3	1	0.15
2	3	2	0.15

base for the generators at Buses 1 and 2, only the parameters of generator at Bus 3 change base:

$$H = 3 \times \frac{800}{100} = 24 \text{ s}$$

$$D = 2 \times \frac{800}{100} = 16 \text{ pu}$$

$$X' = 0.45 \times \frac{100}{800} = 0.0563 \text{ pu}$$

The current supplied by the generators and the bus voltages behind the transient reactances are then

$$\mathcal{I}_{g1} = \left(\frac{P_{g1} + jQ_{g1}}{U_1}\right)^* = \left(\frac{0.8000 + j0.0064}{1.0\angle 7.81°}\right)^* = 0.8000\angle 7.35°$$

$$\mathcal{I}_{g2} = \left(\frac{P_{g2} + jQ_{g2}}{U_2}\right)^* = \left(\frac{0.8000 + j0.1027}{1.0\angle 6.89°}\right)^* = 0.8066\angle -0.43°$$

$$\mathcal{I}_{g3} = \left(\frac{P_{g3} + jQ_{g3}}{U_3}\right)^* = \left(\frac{6.4000 + j2.6963}{1.0\angle 0°}\right)^* = 6.9448\angle -22.85°$$

$$\mathcal{E}'_1 = \frac{\mathcal{I}_{g1}}{jX'_1} = \frac{0.8000\angle 7.35°}{j0.45} = 1.0655\angle 27.56°$$

$$\mathcal{E}'_2 = \frac{\mathcal{I}_{g2}}{jX'_2} = \frac{0.8066\angle -0.43°}{j0.45} = 1.1064\angle 25.88°$$

$$\mathcal{E}'_3 = \frac{\mathcal{I}_{g3}}{jX'_3} = \frac{6.9448\angle -22.85°}{j0.0563} = 1.2066\angle 17.36°$$

The equivalent admittance of the existing load can then be determined:

$$\mathcal{Y}_{l3} = \frac{P_{l3} - jQ_{l3}}{V_{l3}^2} = \frac{8.0 - j2.6}{1^2} = 8.0 - j2.6$$

The prefault admittance matrix is

$$\mathcal{Y}^{pref} = \begin{bmatrix} -j2.2222 & 0 & 0 & +j2.2222 & 0 & 0 \\ 0 & -j2.2222 & 0 & 0 & +j2.2222 & 0 \\ 0 & 0 & -j17.7778 & 0 & 0 & +j17.7778 \\ +j2.2222 & 0 & 0 & -j52.2222 & +j50.0000 & 0 \\ 0 & +j2.2222 & 0 & +j50.0000 & -j65.5556 & +j13.3333 \\ 0 & 0 & +j17.7778 & 0 & +j13.3333 & 8.0000 - j33.7111 \end{bmatrix}$$

The corresponding fault admittance matrix is simply obtained by introducing a very small impedance (e.g., $Z_f = j10^{-9}$) in the diagonal term corresponding to the self-term of the faulted bus; thus,

$$
\mathcal{Y}^f = \begin{bmatrix}
-j2.2222 & 0 & 0 & +j2.2222 & 0 & 0 \\
0 & -j2.2222 & 0 & 0 & +j2.2222 & 0 \\
0 & 0 & -j17.7778 & 0 & 0 & +j17.7778 \\
+j2.2222 & 0 & 0 & -j52.2222 & +j50.0000 & 0 \\
0 & +j2.2222 & 0 & +j50.0000 & -j10^9 & +j13.3333 \\
0 & 0 & +j17.7778 & 0 & +j13.3333 & 8.0000 - j33.7111
\end{bmatrix}
$$

The postfault period admittance matrix is obtained by disconnecting one of the circuits between buses 2 and 3:

$$
\mathcal{Y}^{posf} = \begin{bmatrix}
-j2.2222 & 0 & 0 & +j2.2222 & 0 & 0 \\
0 & -j2.2222 & 0 & 0 & +j2.2222 & 0 \\
0 & 0 & -j17.7778 & 0 & 0 & +j17.7778 \\
+j2.2222 & 0 & 0 & -j52.2222 & +j50.0000 & 0 \\
0 & +j2.2222 & 0 & +j50.0000 & -j58.8889 & +j6.6667 \\
0 & 0 & +j17.7778 & 0 & +j6.6667 & 8.0000 - j27.0444
\end{bmatrix}
$$

Thus, the admittance matrices reduced with respect to the internal generator buses for the prefault, fault, and postfault periods are, respectively,

$$
\mathcal{Y}_R^{pref} = \begin{bmatrix}
0.0330 - j1.7740 & 0.0345 + j0.3693 & 0.3658 + j1.0818 \\
0.0345 + j0.3693 & 0.0360 - j1.8365 & 0.3821 + j1.1299 \\
0.3658 + j1.0818 & 0.3821 + j1.1299 & 4.0540 - j5.7894
\end{bmatrix}
$$

$$
\mathcal{Y}_R^f = \begin{bmatrix}
0.0000 - j2.1277 & 0.0000 + j0.0000 & 0.0000 + j0.0000 \\
0.0000 + j0.0000 & 0.0000 - j2.2222 & 0.0000 + j0.0000 \\
0.0000 + j0.0000 & 0.0000 + j0.0000 & 2.1062 - j8.9024
\end{bmatrix}
$$

$$
\mathcal{Y}_R^{posf} = \begin{bmatrix}
0.0223 - j1.6525 & 0.0233 + j0.4963 & 0.3086 + j0.8875 \\
0.0233 + j0.4963 & 0.0244 - j1.7038 & 0.3223 + j0.9269 \\
0.3086 + j0.8875 & 0.3223 + j0.9269 & 4.2604 - j5.5238
\end{bmatrix}
$$

With the voltages of the buses behind the transient reactance and the reduced admittance matrices, Equations 10.54 and 10.55 can be solved numerically; the Euler Predictor–Corrector method ttis used here. This requires the initial values of the state variables:

$$
x_0 = [\delta_{G1,0} \quad \delta_{G2,0} \quad \delta_{G3,0} \quad \omega_{G1,0} \quad \omega_{G2,0} \quad \omega_{G3,0}]
$$
$$
= [27.56 \quad 25.88 \quad 17.36 \quad 2 \times \pi \times 50 \quad 2 \times \pi \times 50 \quad 2 \times \pi \times 50]
$$

Observe that the admittance matrices $\mathcal{Y}_R^{pref}$, $\mathcal{Y}_R^f$, and $\mathcal{Y}_R^{posf}$ must be used for each of the corresponding periods during the integration process.

The critical clearing time is determined simulating the fault repeatedly, increasing the clearing time by 1 ms until a generator loses synchronism. Figure 10.17 depicts the time trajectories of the angle differences with respect to δ_3 for clearing times of 372 ms (solid line) and 373 ms (dashed line). The fault is applied at $t = 1$ s to show the steady-state conditions.

It must be noted that the sample system has two areas: an exporting area (Buses 1 and 2), and an importing area (Bus 3). A fault on the circuit that links the exporting and importing areas limits the power transfer between these areas, and results in the acceleration of the generators in the exporting area and in the deceleration of the generators in the importing area. This is clearly observed on the angles differences depicted in Figure 10.17 and cannot be seen on the absolute angles, making these types of plots more informative.

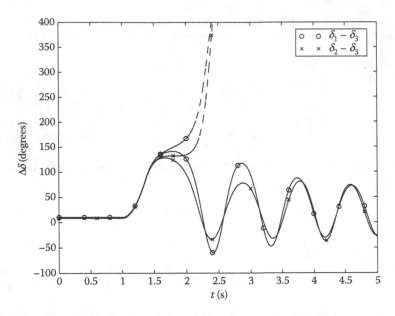

FIGURE 10.17   Time-domain simulation of a multimachine system after a three-phase fault and clearance.

10.2.1.4   Effects of Modeling Detail on Large-Disturbance Stability

So far, the transient stability phenomena has been explained based on a simplified model of the generator, the so-called classical model, which represents the electrical machine with a constant voltage source behind a transient reactance, and assumes that the mechanical power supplied by the turbine is constant. However, more detailed models of the generator represent the synchronous machine in more detail, as well as the excitation system, and the speed governor and turbine systems.

To assess the effect of modeling detail on large-disturbance stability of generators, we consider here a single machine connected to an infinite-bus system (see Example 10.5). The complexity of the modeling is increased progressively: the first model includes a detailed model of the synchronous machine; the second model comprises models of both the synchronous machine and the excitation system (a static excitation system); and the last model includes the turbine model (a steam turbine) also. The parameters of the synchronous machine, excitation, and speed governor and turbine are given in Table 10.3. Details regarding some of these models can be found in Chapter 9 and Appendix C.

TABLE 10.3

Data of Synchronous Machine, Excitation System, and Turbine Models

T'_{d0}	T''_{d0}	T'_{q0}	T''_{q0}	X_d	X_q	X'_d	X''_d	X'_q	X''_q	R_a
8.0	0.03	0.4	0.05	1.8	1.7	0.3	0.55	0.25	0.25	0

K_A	T_A	T_B	T_C	T_R	R_C	X_C	$E_{f\,dmin}$	$E_{f\,dmax}$
200	0	0	0	0.01	0	0	−6	6

K	T_1	T_2	T_3	T_4	T_5	T_6	K_1	K_2	K_3	P_{min}	P_{max}
20	0	0	0.2	0.3	7	0.6	0.3	0.3	0.4	0	0.85

TABLE 10.4

Critical Clearing Times for Several Modeling Options

Model	t_{cri} (ms)
Simplified ($D = 2$ pu)	236
Detailed (machine)	309
Detailed (machine + excitation)	323
Detailed (machine + excitation + turbine)	331

The effect of modeling detail on large-disturbance stability is measured by means of the impact of the various models used on the critical clearing time for a three-phase fault at the sending end of the line connecting the generator to the infinite bus. Table 10.4 compares the critical clearing times for the three modeling options; the critical clearing time obtained for the classical model is also provided for comparison purposes. Observe that as the modeling detail is increased, the critical clearing time increases; in other words, simplified models lead to conservative estimates of the critical clearing times, that is, the actual system is more stable (larger stability regions) than predicted by simplified calculations. The changes in critical clearing time values illustrate the contribution of static excitation and speed governor-turbine systems to transient stability; notice that the contribution of a static excitation system is somewhat higher that the contribution of the speed governor-turbine system. Furthermore and as previously discussed, the results obtained with the detailed representation of the field and damper winding dynamics clearly show the inaccuracy of the over-simplified classical model, in spite of the relatively high-nonzero value of the damping factor D that is typically introduced to try to indirectly represent these dynamics.

Figure 10.18 illustrates the generator response for a three-phase fault clearance at the critical clearing time for the three modeling options under consideration. The fault is applied at $t = 1$ s to depict the

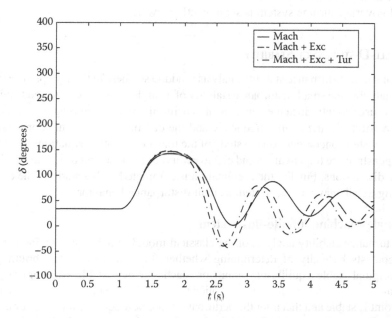

FIGURE 10.18 Generator response for several modeling options when the fault is cleared at the critical clearing time.

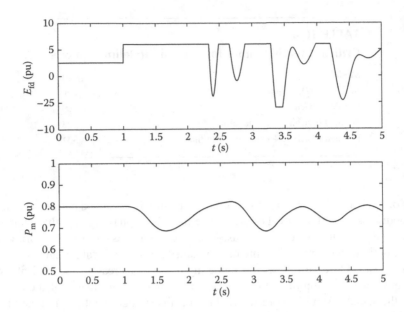

FIGURE 10.19 Time variations of field voltage and mechanical power.

pre- and postfault steady-state conditions. Observe that the generator control systems not only increase the critical clearing times, but also result both in an increase in the amplitude of the first swing and the frequency of the oscillations; in other words, these controls reduce system damping, and are considered to be one of the main reasons for oscillation problems in power systems, as discussed in Section 10.2.2.

Figure 10.19 displays the time variation of the field voltage and the mechanical power supplied by the turbine for a detailed model of the generator and its controls. These show that the excitation system is quite fast, since the field voltage reaches almost instantaneously its ceiling value of 6 pu, and that the speed governor-turbine system is significantly slower.

10.2.2 SMALL-DISTURBANCE STABILITY

The problem of small-disturbance stability analysis is addressed here in three steps. The first step concentrates on analyzing the small-disturbance stability of a single machine connected to an infinite bus. This example is used to introduce the concepts of synchronizing and damping torques, as well as for obtaining expressions for the natural frequency and the damping of the generator's natural oscillations. The second step concentrates on the study of the linear differential equations of a multimachine system; the eigenstructure (eigenvalues and eigenvectors) of the associated state matrix is used as the basis for these discussions. Finally, the presentation is completed with a study of the effect that generator modeling details have on the system's small-disturbance behavior.

10.2.2.1 Single-Machine-Infinite-Bus System

The small-disturbance stability analysis of the classical model of a single machine connected to an infinite bus consists basically of determining whether the generator's equilibrium point comes back to the original stable equilibrium point, or reaches a new stable equilibrium point after a small-disturbance in the mechanical power supplied by the turbine. This study assumes that the initial equilibrium point is stable and that after the perturbation, the new equilibrium point first exists, which might not be the case as discussed in Section 10.3, and second, is stable, which might not be the case when generator controls are considered in the analysis, as discussed at the end of Section 10.2.2. In the case of small disturbances, the nonlinear differential Equations 10.3 and 10.4 that describe the generator dynamic behavior can be linearized around the operating point to study the generator response as follows, based on a Taylor-series expansion:

$$\frac{d\Delta\delta}{dt} = \Delta\omega \tag{10.67}$$

$$\frac{d\Delta\omega}{dt} = \frac{\omega_0}{2H}\left(\Delta P_m - \frac{E'V_\infty}{X_e}\cos\delta_0\Delta\delta - \frac{D}{\omega_0}\Delta\omega\right)$$

$$= \frac{\omega_0}{2H}\left(\Delta P_m - K\Delta\delta - \frac{D}{\omega_0}\Delta\omega\right) \tag{10.68}$$

Observe in Equation 10.68 that, apart from the mechanical torque (represented by ΔP_m), there is a *synchronizing torque* (proportional to the rotor angle, that is, $K\Delta\delta$) and a *damping torque* (proportional to the rotor speed deviation, that is, $D/\omega_0\Delta\omega_0$) applied to the generator's rotor. The constant K is typically referred to as the *synchronizing torque coefficient*.

Equations 10.67 and 10.68 can be also written in matrix form as

$$\begin{bmatrix} \Delta\dot\delta \\ \Delta\dot\omega \end{bmatrix} = \begin{bmatrix} 0 & 1 \\ \frac{-K\omega_0}{2H} & \frac{-D}{2H} \end{bmatrix}\begin{bmatrix} \Delta\delta \\ \Delta\omega \end{bmatrix} + \begin{bmatrix} 0 \\ \frac{\omega_0}{2H} \end{bmatrix}\Delta P_m \tag{10.69}$$

which in "standard" linear system compact form are

$$\Delta\dot x = A\Delta x + b\Delta u \tag{10.70}$$

where A is the state matrix and b is the input vector.

The small-disturbance stability of a generator connected to an infinite bus can be analyzed applying the Laplace transform to the set of linear differential Equations 10.70. Thus, assuming zero initial conditions, one has

$$\Delta x(s) = (sI - A)^{-1}b\Delta u(s) \tag{10.71}$$

The small-disturbance stability of the generator is then determined by the roots of the characteristic equation:

$$\det(sI - A) = 0 \tag{10.72}$$

which results in

$$\det\begin{bmatrix} s & -1 \\ \frac{K\omega_0}{2H} & s + \frac{D}{2H} \end{bmatrix} = s^2 + \frac{D}{2H}s + \frac{K\omega_0}{2H} = 0 \tag{10.73}$$

For the zero damping case, the roots of the characteristic equation 10.73 are

$$s_{1,2} = \pm\sqrt{\frac{-K\omega_0}{2H}} \tag{10.74}$$

Depending on the sign of the synchronizing torque coefficient K, two cases can be discussed. If the synchronizing torque coefficient is positive, the characteristic equation has two pure complex roots and the generator exhibits a permanent oscillatory behavior; this coefficient is positive when the rotor angle δ_0 is between $0°$ and $90°$ (sep's). If the synchronizing torque coefficient is negative, the characteristic equation has two real roots, one positive and another negative; the negative root corresponds to an exponentially decreasing response, whereas the positive root results in an exponentially increasing behavior. The synchronizing torque coefficient is negative when the rotor angle δ_0 is between 90° and 180° (uep's). The case of the coefficient becoming zero at $\delta_0 = 90°$ (saddle-node bifurcation point), which corresponds to a maximum loading point of the system, is discussed in detail in Section 10.3.

In the general case of the damping factor $D > 0$, the comparison of the characteristic Equation 10.73 with the normalized form of a second-order system

$$s^2 + 2\zeta\omega_n s + \omega_n^2 = 0 \tag{10.75}$$

provides the expressions of the natural frequency ω_n and the damping ζ of the generator natural oscillation:

$$\omega_n = \sqrt{\frac{K\omega_0}{2H}} \tag{10.76}$$

$$\zeta = \frac{D}{2H}\frac{1}{2\omega_n} = D\sqrt{\frac{1}{8HK\omega_0}} \tag{10.77}$$

It is interesting to note that the natural frequency is proportional to the inverse of both the square root of the generator inertia H, and the square root of the equivalent reactance that connects the generator to the infinite-bus X_e. On other hand, the damping is directly related to the damping factor D.

EXAMPLE 10.8

Consider the generator of Example 10.5. Determine the roots of the characteristic equation, the natural frequency, and the damping of the generator oscillation for a damping factor $D = 2$ pu referred to the generator MVA base. Determine as well the time trajectory and final value of the internal angle, and the final value of the speed deviation for a 5% reduction in the mechanical power supplied by the turbine.

The synchronizing torque coefficient can be determined from the operating point of the generator as follows:

$$K = \frac{E'V_\infty}{X_e}\cos\delta_0 = 1.7364$$

The roots of the characteristic equation can then be obtained from Equation 10.73:

$$s_{1,2} = -\frac{D}{4H} \pm j\frac{1}{2}\sqrt{4\frac{K\omega_0}{2H} - \left(\frac{D}{2H}\right)^2} = -0.1667 \pm j9.5335$$

and the natural frequency and damping are computed using Equations 10.76 and 10.77

$$\omega_n = 9.5350 \text{ rad/s}$$

$$f_n = \frac{\omega_n}{2\pi} = 1.5175 \text{ Hz}$$

$$\zeta = 0.0175 \text{ pu} = 1.75\%$$

The natural oscillation frequency is within a typical 0.1–2 Hz range for synchronous generators, but the damping ratio of these oscillations is very low compared to other control systems. This damping value would not be admissible in practice, since acceptable values are typically greater than 5%.

From Equation 10.71, the transfer functions between the variation of mechanical power supplied by the turbine and the variation of rotor angle, and the variation of rotor speed are

$$
\begin{bmatrix} \Delta\delta(s) \\ \Delta\omega(s) \end{bmatrix} = \begin{bmatrix} s & -1 \\ \frac{K\omega_0}{2H} & s + \frac{D}{2H} \end{bmatrix}^{-1} \begin{bmatrix} 0 \\ \frac{\omega_0}{2H} \end{bmatrix} \Delta P_m(s)
$$

$$
= \frac{1}{s^2 + \frac{D}{2H}s + \frac{K\omega_0}{2H}} \begin{bmatrix} s + \frac{D}{2H} & 1 \\ \frac{-K\omega_0}{2H} & s \end{bmatrix} \begin{bmatrix} 0 \\ \frac{\omega_0}{2H} \end{bmatrix} \Delta P_m(s)
$$

$$
= \frac{1}{s^2 + \frac{D}{2H}s + \frac{K\omega_0}{2H}} \begin{bmatrix} \frac{\omega_0}{2H} \\ s\frac{\omega_0}{2H} \end{bmatrix} \Delta P_m(s)
$$

From these transfer functions, given that the system is stable, the final values of the angle and speed variation can be obtained by applying the Final Value Theorem assuming that $\Delta P_m(s) = \Delta P_m/s$, with $\Delta P_m = -0.05 \times 0.8\,\text{pu} = -0.04\,\text{pu}$. Thus,

$$
\Delta\delta(\infty) = \lim_{s\to 0} \Delta\delta(s)s = \lim_{s\to 0} \frac{\omega_0}{2H} \frac{1}{s^2 + \frac{D}{2H}s + \frac{K\omega_0}{2H}} \frac{\Delta P_m}{s} s = \frac{\Delta P_m}{K}
$$

$$
= \frac{-0.04}{1.7364} = -0.0230\,\text{rad} = -1.32°
$$

$$
\Delta\omega(\infty) = \lim_{s\to 0} \Delta\omega(s)s = \lim_{s\to 0} \frac{\omega_0}{2H} \frac{s}{s^2 + \frac{D}{2H}s + \frac{K\omega_0}{2H}} \frac{\Delta P_m}{s} s = 0
$$

Observe that the final value of the angle variation is proportional to the magnitude of the step change in mechanical power. This is an estimation based on the linealization of the non-linear differential equations; thus, the final value of the angle assuming a linear model is 23.4172°, whereas its actual final value obtained from the nonlinear model is 23.4240°. The final value of the speed deviation is zero, which confirms that the generator remains in synchronism.

The time variation of the angle deviation due to the step change in mechanical power is obtained computing the inverse Laplace transform of the rotor angle deviation $\Delta\delta(s)$:

$$
\Delta\delta(t) = \mathcal{L}^{-1}\left\{ \frac{\omega_0}{2H} \frac{1}{s^2 + \frac{D}{2H}s + \frac{K\omega_0}{2H}} \frac{\Delta P_m}{s} \right\}
$$

$$
= \frac{\Delta P_m}{K} \mathcal{L}^{-1}\left\{ \frac{\omega_n^2}{s(s^2 + 2\zeta\omega_n + \omega_n^2)} \right\}
$$

$$
= \frac{\Delta P_m}{K}\left[1 - \frac{1}{\sqrt{1-\zeta^2}} e^{-\zeta\omega_n t} \sin\left(\omega_n\sqrt{1-\zeta^2}\,t + \phi\right) \right]
$$

where

$$
\phi = \arctan\frac{\sqrt{1-\zeta^2}}{\zeta} = 1.5533\,\text{rad}
$$

Figure 10.20 shows the time variation of the rotor angle deviation for a 5% reduction in the mechanical power supplied by the turbine. The step change is applied at $t = 1$ s, so that the initial steady-state conditions can be clearly seen.

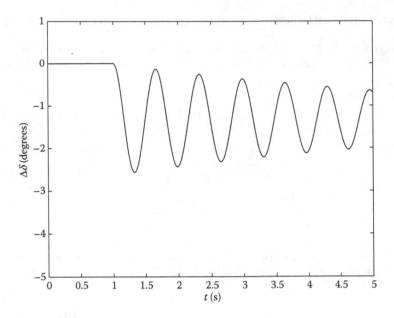

FIGURE 10.20 Rotor angle deviation in Example 10.8.

10.2.2.2 Multi-Machine System

The model of a multimachine power system for small-disturbance stability analysis can be obtained linearizing the nonlinear differential equations of the system around the operating point. If the loads are represented as constant admittances, the linealization of Equations 10.54 and 10.55 yields

$$\frac{d\Delta\delta_i}{dt} = \Delta\omega_i \tag{10.78}$$

$$\frac{d\Delta\omega_i}{dt} = \frac{\omega_0}{2H_i}\left(\Delta P_{mi} - \sum_{j=1}^{n}\left.\frac{\partial P_{ei}}{\partial\delta_j}\right|_0 \Delta\delta_j - \frac{D_i}{\omega_0}\Delta\omega_i\right) \tag{10.79}$$

where

$$\left.\frac{\partial P_{ei}}{\partial\delta_j}\right|_0 = E_i'E_j'[G_{Rij}\sin(\delta_{i_0} - \delta_{j_0}) - B_{Rij}\cos(\delta_{i_0} - \delta_{j_0})] = K_{ij} \tag{10.80}$$

$$\left.\frac{\partial P_{ei}}{\partial\delta_i}\right|_0 = \sum_{\substack{j=1\\j\neq i}}^{n} E_i'E_j'[-G_{Rij}\sin(\delta_{i_0} - \delta_{j_0}) + B_{Rij}\cos(\delta_{i_0} - \delta_{j_0})]$$

$$= -\sum_{\substack{j=1\\j\neq i}}^{n} K_{ij} \tag{10.81}$$

Observe that Equation 10.80 is expressed in terms of a matrix of synchronizing torque coefficients K_{ij}.

If the equations of all generators are written together in matrix form, one obtains

$$
\begin{bmatrix} \Delta\dot{\delta}_1 \\ \vdots \\ \Delta\dot{\delta}_n \\ \hline \Delta\dot{\omega}_1 \\ \vdots \\ \Delta\dot{\omega}_n \end{bmatrix} =
\begin{bmatrix}
0 & \cdots & 0 & 1 & \cdots & 0 \\
\vdots & \ddots & \vdots & \vdots & \ddots & \vdots \\
0 & \cdots & 0 & 0 & \cdots & 1 \\
\hline
-\frac{\omega_0}{2H_1}K_{11} & \cdots & -\frac{\omega_0}{2H_1}K_{1n} & \frac{-D_1}{2H_1} & \cdots & 0 \\
\vdots & \ddots & \vdots & \vdots & \ddots & \vdots \\
-\frac{\omega_0}{2H_n}K_{n1} & \cdots & -\frac{\omega_0}{2H_n}K_{nn} & 0 & \cdots & \frac{-D_n}{2H_n}
\end{bmatrix}
\begin{bmatrix} \Delta\delta_1 \\ \vdots \\ \Delta\delta_n \\ \hline \Delta\omega_1 \\ \vdots \\ \Delta\omega_n \end{bmatrix}
$$

$$
+ \begin{bmatrix}
0 & \cdots & 0 \\
\vdots & \ddots & \vdots \\
0 & \cdots & 0 \\
\hline
\frac{\omega_0}{2H_1} & \cdots & 0 \\
\vdots & \ddots & \vdots \\
0 & \cdots & \frac{\omega_0}{2H_n}
\end{bmatrix}
\begin{bmatrix} \Delta P_{m1} \\ \vdots \\ \Delta P_{mn} \end{bmatrix}
\tag{10.82}
$$

which has the well-known form

$$
\Delta\dot{x} = A\Delta x + B\Delta u
\tag{10.83}
$$

where B is now an input matrix.

The small-disturbance stability analysis of multimachine systems is not performed in practice based on computing the roots of the characteristic equation, given the difficulties of calculating the determinant of a matrix that can be of large dimension if the number of generators is large. Thus, this analysis is typically carried out by determining the analytical solution of the linear system expressed in terms of the exponential of the state matrix A. This analytical solution has two terms: the homogeneous solution (solution to nonzero initial conditions and zero input), and the particular solution (solution to zero initial conditions and nonzero input):

$$
\Delta x = \Delta x_{\mathrm{h}} + \Delta x_{\mathrm{p}} = e^{At}\Delta x(0) + \int_0^t e^{A(t-\tau)}B\Delta u(\tau)\mathrm{d}\tau
\tag{10.84}
$$

where the exponential of the A matrix may be computed using a Taylor-series expansion as follows:

$$
e^{At} = I + \frac{1}{1!}At + \frac{1}{2!}A^2t + \frac{1}{3!}A^3t\cdots
\tag{10.85}
$$

A meaningful approach to the computation of e^{At} is based on the eigenstructure (eigenvalues and eigenvectors) of the state matrix A. Thus, a complex eigenvalue μ_i and right v_i and left w_i eigenvectors of the $N \times N$ matrix A are defined as

$$
Av_i = v_i\mu_i
\tag{10.86}
$$

$$
w_i^{\mathrm{T}}A = \mu_i w_i^{\mathrm{T}}
\tag{10.87}
$$

where $i = 1, \ldots, N$. Both the right and left eigenvectors can be arbitrarily scaled by any given factor; hence, these vectors are typically normalized, so that

$$w_i^T v_i = 1 \tag{10.88}$$

If all N eigenvalues of the state matrix A are distinct, Equations 10.86 through 10.88 can be rewritten as

$$
\begin{aligned}
AV &= V\mu \\
WA &= \mu W \\
WV &= I
\end{aligned}
\tag{10.89}
$$

where

$$V = [\, v_1 \quad \cdots \quad v_N \,], \; W = \begin{bmatrix} w_1^T \\ \vdots \\ w_N^T \end{bmatrix}, \; \mu = \begin{bmatrix} \mu_1 & \cdots & 0 \\ \vdots & \ddots & \vdots \\ 0 & \cdots & \mu_N \end{bmatrix}$$

Therefore, substituting I by VW and A by $V\mu W$ in Equation 10.85, one obtains

$$e^{At} = VW + \frac{1}{1!} V\mu Wt + \frac{1}{2!} V\mu^2 Wt + \frac{1}{3!} V\mu^3 Wt + \cdots \tag{10.90}$$

$$= V\left(I + \frac{1}{1!}\mu t + \frac{1}{2!}\mu^2 t + \frac{1}{3!}\mu^3 t + \cdots\right)W = Ve^{\mu t}W \tag{10.91}$$

Hence, the solution of the set of linear differential Equations 10.83 can be written in the following form:

$$\Delta x = Ve^{\mu t}W\Delta x(0) + \int_0^t Ve^{\mu(t-\tau)}WB\Delta u(\tau)d\tau \tag{10.92}$$

The homogeneous solution can be also written as

$$\Delta x = \sum_{i=1}^{N} \left[w_i^T \Delta x(0)\right] e^{\mu_i t} v_i \tag{10.93}$$

The linear system stability is determined by this homogeneous solution.
   The analysis of Equation 10.93 shows that

- The system evolves according to the N modes defined by their eigenvalues and corresponding right and left eigenvectors.
- A real negative (positive) eigenvalue indicates an exponentially decreasing (increasing) behavior.
- A complex eigenvalue with negative (positive) real part indicates an oscillatory decreasing (increasing) behavior.

- The components of the right eigenvectors measure the contribution of the state variables to the corresponding modes:

$$\Delta x_j = v_{ji}\alpha_i e^{\mu_i t}$$

- The left eigenvectors weigh the initial conditions on the corresponding modes:

$$\alpha_i = \left[w_i^T \Delta x(0) \right]$$

The right eigenvectors are not really appropriate for measuring the participation of a variable in a mode, since they are dimension dependent. A nondimensional measure of the participation of the ith mode in the jth variable is the product of the jth components of right and left eigenvectors corresponding to the ith mode, which is known as the *participation factor* [13]:

$$p_{ji} = w_{ji}v_{ji} \qquad\qquad (10.94)$$

EXAMPLE 10.9

Consider the power system described in Example 10.7. Determine the eigenvalues, right eigenvectors, and participation factors.

The matrix of synchronizing torques coefficients and the state matrix are, respectively,

$$K = \begin{bmatrix} 1.7198 & -0.4341 & -1.2857 \\ -0.4364 & 1.8527 & -1.4163 \\ -1.4522 & -1.5674 & 3.0197 \end{bmatrix}$$

$$A = \begin{bmatrix} 0 & 0 & 0 & 1.0000 & 0 & 0 \\ 0 & 0 & 0 & 0 & 1.0000 & 0 \\ 0 & 0 & 0 & 0 & 0 & 1.0000 \\ -90.0465 & 22.7276 & 67.3188 & -0.3333 & 0 & 0 \\ 22.8521 & -97.0081 & 74.1560 & 0 & -0.3333 & 0 \\ 9.5048 & 10.2589 & -19.7637 & 0 & 0 & -0.3333 \end{bmatrix}$$

The eigenanalysis of the state matrix A yields the following six eigenvalues:

$\mu_{1,2}$	$\mu_{3,4}$	μ_5	μ_6
$-0.1667 \pm j10.8096$	$-0.1667 \pm j9.4824$	-0.3333	0.0000

For this simple model, the number of complex conjugate pairs is equal to the number of generators minus 1; if the system had an infinite bus, the number of complex pairs would be equal to the number of generators for the classical model. The zero eigenvalue is due to the existence of a linear dependent variable and equation, since the system phasors require a reference, that is, the angle of a generator is dependent on the other generator angles; in the presence of an infinite-bus (reference), this zero eigenvalue disappears [14].

The eigenanalysis of the state matrix A also generates the right eigenvectors associated with the eigenvalues. Each of the following eigenvectors are normalized with respect to the component with the largest magnitude:

The components corresponding to the speed deviation of the generators (the last three components) provide the mode shape. The complex pair $\mu_{1,2}$ corresponds to an oscillation with natural frequency 1.7204 Hz, with Generator 2 oscillating against Generators 1 and 3. The complex pair $\mu_{3,4}$ corresponds to an oscillation of natural frequency 1.5092 Hz in which Generators 1 and 2 oscillate against Generator 3. The eigenvalue μ_5 corresponds to an exponentially decreasing response of the three generators with a 3 s time constant.

$v_{1,2}$	$v_{3,4}$	v_5
$+0.0000 \pm j0.0714$	$-0.0019 \mp j0.1054$	$+1.00000$
$-0.0014 \mp j0.0925$	$-0.0013 \mp j0.0754$	$+1.00000$
$+0.0000 \pm j0.0028$	$+0.0004 \pm j0.0253$	$+1.00000$
$-0.7716 \pm j0.0000$	$+1.0000$	-0.33333
$+1.0000$	$+0.7154 \mp j0.0000$	-0.33333
$-0.0301 \mp j0.0000$	$-0.2400 \pm j0.0000$	-0.33333

The participation factors require the computation of the left eigenvectors. In this case, the matrix of left eigenvectors is simply the inverse of the matrix of right eigenvectors. Thus,

$p_{1,2}$	$p_{3,4}$	p_5
$0.1856 \mp j0.0029$	$0.2594 \mp j0.0046$	0.0000
$0.3122 \mp j0.0048$	$0.1336 \mp j0.0023$	0.0000
$0.0022 \mp j0.0000$	$0.1071 \mp j0.0019$	0.0000
$0.1856 \pm j0.0029$	$0.2594 \pm j0.0046$	0.1100
$0.3122 \pm j0.0048$	$0.1336 \pm j0.0023$	0.1084
$0.0022 \pm j0.0000$	$0.1071 \pm j0.0019$	0.7816

The participation factors for the generator angles are identical to the participation factors for the generator speeds. Observe that the oscillation associated with the complex pair $\mu_{1,2}$ is mainly due to Generators 1 and 2 (exporting area); thus, based on the aforementioned right eigenvector analysis, this is a *local oscillation* mode. On the other hand, even though the oscillation associated with the complex pair $\mu_{3,4}$ is still dominated by Generators 1 and 2, Generator 3 (importing area) has a bigger participation in this mode, which combined with the right eigenvector analysis make this an *inter-area oscillation* mode. The response associated with eigenvalue μ_5 is dominated by Generator 3.

10.2.2.3   Effects of Modeling Detail on Small-Disturbance Stability

The small-disturbance stability problem has been explained here considering systems of increasing complexity, starting with a single-machine infinite system and continuing with a multimachine system. In both cases, a simplified classical model of the generator has been used. Hence, in this section, the effect of modeling detail on small-disturbance stability is discussed.

Following the approach used to evaluate the impact of modeling detail on large-disturbance stability, a single generator connected to an infinite bus system is used in this case as well (Example 10.5). Thus, the effect of modeling detail on the natural frequency and damping of the electromechanical oscillations of the generator is analyzed for three modeling options of increasing complexity. The first case considers a detailed model of the synchronous machine. The second case comprises models of both the synchronous machine and the excitation system, assuming a static excitation. The last case adds a steam turbine model to the detailed machine and static excitation system models. The parameters of synchronous machine, excitation system, and speed governor turbine are given in Table 10.3.

The linear model of a generator connected to an infinite bus has 6, 8, and 11 state variables for each of the modeling cases previously mentioned, with as many eigenvalues as there are state variables. Electromechanical oscillations are mainly characterized by the eigenvalue with a natural frequency around 1 Hz and low damping. Using the participation factors, the variables with the greatest contribution to this electromechanical mode are the generators' rotor angles and speeds. Table 10.5 depicts the lowest damped or "critical" electromechanical mode for the three modeling options considered; the eigenvalue of the simplified model is also included here for comparison purposes. Observe that

TABLE 10.5

Electromechanical Eigenvalue in Case of Three Modeling Options

Model	μ	ζ (%)	f_n (Hz)
Simplified ($D = 2$ pu)	$-0.1667 \pm j9.5335$	1.75	1.5175
Detailed (machine)	$-0.6806 \pm j9.1097$	7.45	1.4498
Detailed (machine + excitation)	$-0.3672 \pm j9.5225$	3.85	1.5156
Detailed (machine + excitation + turbine)	$-0.3065 \pm j9.5710$	3.20	1.5233

the natural frequency of this electromechanical mode is hardly affected by modeling detail; the impact of modeling on the damping of this electromechanical mode deserves a more detailed discussion.

If one compares the simplified model with the first detailed model, which models the synchronous machine in detail, it can be observed that the damping significantly increases from the former to the latter. This confirms the contribution of the machine field and damper windings to oscillation damping, as discussed at the end of Section 10.2.1. When the excitation system model is included, the damping of the critical electromechanical mode worsens significantly, decreasing from 7.45% to a relatively small value of 3.85%. The reason for this damping deterioration is the features of the static excitation, that is, a low time constant (10 ms) and high gain (200 pu). When the model of the turbine is incorporated into the system model, the damping of this critical electromechanical mode is not significantly affected, although it does decrease some more. The plots in Figure 10.21 further illustrate these observations, comparing the generator response for a 5% sudden reduction in the mechanical power supplied by the turbine for the three modeling options considered; the disturbance is applied at $t = 1$ s.

10.2.2.4 Undamped Oscillations

As discussed above, the excitation and voltage control systems of the generator negatively affect system damping. This can be further compounded by increasing loading and contingencies, which further stress the system, to the point where undamped oscillations may suddenly appear on the system at

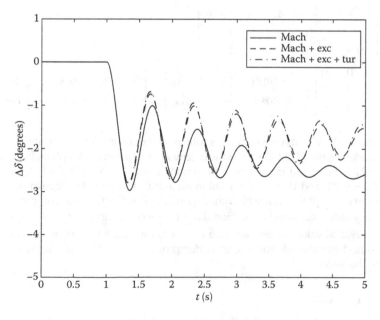

FIGURE 10.21 Comparison of generator response for three modeling options for a 5% sudden reduction in mechanical power.

high loading conditions after contingencies. This phenomenon has been observed in practice and determined to be the main reason for some major system blackouts, as discussed in detail in Section 10.2.4.

The appearance after system contingencies at high loading conditions of these undamped oscillations, whether inter-area (0.1–1 Hz) or local (1–3 Hz), can be directly associated with eigenvalue jumping from the left-half to the right-half of the complex plane, that is, the damping of the critical electromechanical mode becomes negative. This phenomenon can be studied through Hopf bifurcation analysis [15], which is readily explained by means of the example.

EXAMPLE 10.10

In Example 10.1, the DAE equations of the generator-infinite-bus system with AVR yield a linearized model of the form

$$\Delta \dot{x} = \left.\frac{\partial f}{\partial x}\right|_0 \Delta x + \left.\frac{\partial f}{\partial y}\right|_0 \Delta y$$

$$0 = \left.\frac{\partial g}{\partial x}\right|_0 \Delta x + \left.\frac{\partial g}{\partial y}\right|_0 \Delta y$$

at an equilibrium point (x_0, y_0) for a given set of parameter values (p_0, λ_0). This equation can be rewritten in matrix form as:

$$\begin{bmatrix} \Delta \dot{x} \\ 0 \end{bmatrix} = \begin{bmatrix} D_x f|_0 & D_y f|_0 \\ D_x g|_0 & D_y g|_0 \end{bmatrix} \begin{bmatrix} \Delta x \\ \Delta y \end{bmatrix}$$

Hence, by eliminating the algebraic variables Δy, the state matrix can be readily shown to be

$$A = D_x f|_0 - D_y f|_0 D_y g|_0^{-1} D_x g|_0$$

which from Equations 10.3 through 10.7 can be written as:

$$A = \begin{bmatrix} -\frac{D_G}{M} & -\frac{E'_0 V_\infty}{M X_e} \cos \delta_0 & -\frac{V_\infty}{M X_L} \sin \delta_0 \\ 1 & 0 & 0 \\ 0 & 0 & 0 \end{bmatrix}$$
$$- \begin{bmatrix} 0 & 0 \\ 0 & 0 \\ -K_v & 0 \end{bmatrix} \begin{bmatrix} \frac{V_\infty}{X_L} \sin \delta_{10} & \frac{V_\infty V_{10}}{X_L} \cos \delta_{10} \\ -\frac{V_\infty}{X_L} \cos \delta_{10} & \frac{V_\infty V_{10}}{X_L} \sin \delta_{10} \end{bmatrix}^{-1} \begin{bmatrix} 0 & -\frac{E'_0 V_\infty}{X_e} \cos \delta_0 & -\frac{V_\infty}{X_e} \sin \delta_0 \\ 0 & -\frac{E'_0 V_\infty}{X_e} \sin \delta_0 & \frac{V_\infty}{X_e} \cos \delta_0 \end{bmatrix}$$

The state matrix A yields three eigenvalues: one real and a complex conjugate pair (an electromechanical mode). Figure 10.22 illustrates the eigenvalues in the complex plane for all equilibrium points as $\lambda = P_d$ is increased from 0 to a maximum value, for $M = D = 0.1$, $Kv = 1$, $X_L = 0.5$, $X'_G = 0.25$, $X_{Th} = 0.25$, and $V_\infty = V_{1o} = 1$ (all in pu and s). Observe the eigenvalues crossing the imaginary axis at $\mu = \pm j\beta \approx \pm j1.3$, which corresponds to $\lambda = P_d \approx 1.25$ pu. This critical electromechanical mode yields a sustained oscillation (limit cycle) of period $T = \frac{2\pi}{\beta} \approx 4.8$ s, that is, $f \approx 0.2$ Hz, which is a typical value for an inter-area oscillation mode. For loading levels $\lambda = P_d > 1.25$ pu, the electromechanical mode has a negative damping, which can be associated with undamped oscillations in the system.

10.2.3 COUNTERMEASURES

The nature of large-disturbance and small-disturbance stabilities requires different approaches for their improvement.

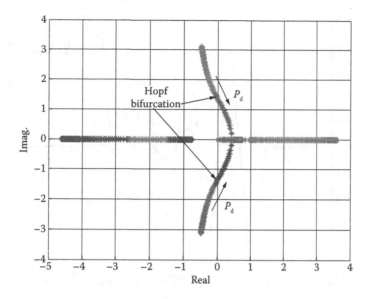

FIGURE 10.22 Eigenvalue loci for the generator with AVR system in Example 10.10.

10.2.3.1 Improvement of Large-Disturbance Stability

Large-disturbance stability can be improved either by reducing the fault clearing times or by increasing the fault critical clearing times, that is, increasing the size of the postfault stability region. The reduction of the fault clearing times requires fast protection systems and circuit breakers, which are given nowadays. The increase of fault critical clearing times, on the other hand, requires more sophisticated approaches, such as increasing the response of the excitation system, which is currently feasible thanks to static excitation systems that have very low time constants and high gains, or reducing the equivalent reactance seen by the generator by means of either low-reactance transformers or series compensation [16].

Another way to increase the critical clearing times is by reducing the acceleration of the generator rotor during the fault. Such reduction can be achieved in steam turbines by reducing the mechanical power supplied by the turbine closing the intercept valves. This is accomplished with *fast valving* systems, which set to zero, for one or two seconds, the mechanical power supplied by the medium and low pressure turbines. Another approach to reduce the generator acceleration consists of connecting a resistive load at the generator terminals; this resistor is referred to as a *braking resistor*.

The above-mentioned methods consider neither generator tripping nor load shedding. However, generator tripping followed by load shedding is an approach widely used to maintain large-disturbance stability in systems where the power transfer between two areas is constrained by large-disturbance stability. In this case, to avoid loss of synchronism after a fault on the line between the two areas, generators in the exporting area are tripped, which leads to load shedding being activated by underfrequency relays as a result of the unbalance between supply and demand [17], as discussed in Section 10.4.

10.2.3.2 Improvement of Small-Disturbance Stability

The main aim of small-disturbance stability improvement is to increase the damping of low-damped or undamped electromechanical oscillations. This can be achieved by applying a braking torque to the generator rotors proportional to their rotor speed, that is, increase system damping; however, direct application of such braking torque is not possible. A number of controllers have been designed to indirectly apply such braking torques, in particular Power System Stabilizers (PSS), Flexible AC

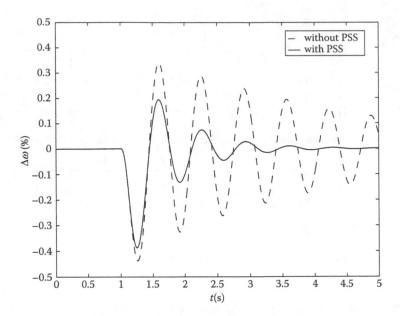

FIGURE 10.23 Comparison of the speed deviation of a generator connected to an infinite bus after a step change on the voltage regulator reference with and without PSS.

Transmission System (FACTS) controllers (especially series-connected controllers), and HVDC link damping controls.

PSSs are the most cost-effective devices for the improvement of small-disturbance stability and are relatively simple supplementary controls of excitation systems. These controllers modulate the reference of the voltage regulator that modulates the field voltage and hence the electric power supplied by the generator. PSSs can use as input signals the rotor speed deviation or the electric power, or a combination of these two, and incorporate lead/lag compensation components to compensate for the phase lag/lead between the input signal and the electric power.

PSS design has attracted much attention from various groups and individuals in both industry and academia [14,18,19]. Figure 10.23 compares the response of a generator connected to an infinite bus after a 10% sudden increase in the reference of the excitation system, with and without PSS. The PSS used in this case is depicted in Figure 10.24 and was designed based on sensitivities of the electromechanical modes with respect to the PSS parameters [20]. The main components of the PSS transfer function are a two-stage phase compensation $(1 + sT_1)/(1 + sT_2) \times (1 + sT_3)/(1 + sT_4)$; the gain K_S; and a high pass or washout filter $(sT_5)/(1 + sT_5)$. The parameters of the stabilizer used here are $K_S = 1.99$ pu, $T_1 = T_3 = 0.17$ s, $T_2 = T_4 = 0.06$ s, and $T_5 = 5$ s.

FACTS are power electronic controllers that have been designed to make the operation of AC transmission systems more flexible [21]. FACTS controllers can be classified into three categories: shunt-, series-, and mixed-connected devices. SVC and STATic synchronous COMpensators (STATCOM) are popular shunt FACTS controllers. The main purpose of these devices is bus voltage control, as discussed in detail in Chapter 9; however, these are also used for oscillation damping by introducing PSSs in their voltage controllers [15]. Instances of series devices are Thyristor Controlled

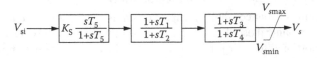

FIGURE 10.24 Model of a power system stabilizer.

Series Compensators (TCSC) and Static Synchronous Series Compensators (SSSC), which are mainly used for inter-area oscillation damping [22]. Unified Power Flow Controllers (UPFC), which are mixed devices that include both a STATCOM and an SSSC with a common DC-link capacitor, may also be used for oscillation damping control. Some of these controllers are discussed in more detail in Chapters 7 and 9.

Converter stations of HVDC links can also contribute to small-disturbance improvement by modulating the power absorbed by the converter stations [23].

10.2.4 Real System Event

The August 10, 1996 collapse of the Western Electricity Coordinating Council (WECC) system is a good practical example of how undamped oscillations can appear and trigger a major system blackout [24]. As discussed in Section 9.1.1, the WECC system comprises the whole North American western interconnection, with various system operators coordinated by WECC, which at the time of the blackout was called the Western System Coordinating Council (WSCC). The day of the blackout, there were high ambient temperatures in the northwest, with the following high power transfers north to south:

- The California-Oregon interties were transferring 4330 MW north to south.
- The Pacific DC intertie was loaded to 2680 MW, north to south as well.
- There was a 2300 MW transfer from British Columbia, Canada.

Between approximately 14:00 and 15:40 p.m., three 500 kV sections of lines from the lower Columbia River to loads in northern Oregon, near the border with Washington, tripped due to tree flashovers. A fourth 500 kV line tripped at 15:47 p.m. due to yet another tree flashover, followed immediately by a 115 kV line trip, triggering a growing 0.23 Hz oscillation (negative damping of about 3%) in the north–south system interconnections. As a result of these undamped oscillations, the system split into four large islands (as planned), leading to more than 7.5 million customers experiencing outages from a few minutes to up to 9 hours, with a total load loss of 30.5 GW. From post-mortem analyses, it was concluded that the main solution to the problem was to retune PSSs at generating units located in New Mexico, as well as add PSSs to units located in southern California.

10.3 VOLTAGE STABILITY

As explained in Section 10.1.2, this section concentrates on discussing voltage stability issues associated with power system operation, which can be better studied using long-term, small-perturbation analysis techniques and tools. Most of the material discussed here can be found in detail in Reference 25.

10.3.1 Voltage Collapse and Maximum Loadability

The inability of the transmission system to supply the grid's demand leads to a voltage collapse problem, that is, the voltages throughout the network collapse, rapidly decreasing below desirable values. This problem is associated with the disappearance of a sep due to a saddle-node or limit-induced bifurcation, typically due to contingencies in the system.

The concepts of saddle-node and limit-induced bifurcations and their association with voltage stability can be readily explained with the help of the generator–load system model illustrated in Figure 10.25. The generator is modeled using the simple second-order classical model discussed in Example 10.1 and utilized throughout Section 10.2, whereas the loads are represented using the following dynamic model:

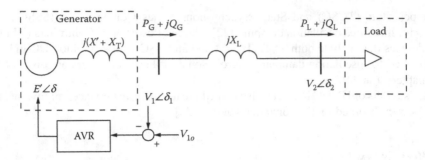

FIGURE 10.25   Generator–load system example.

- A frequency-dependent model for the active power demand:

$$P_L = P_d + D_L \dot{\delta}_2 \Rightarrow \dot{\delta}_2 = \frac{1}{D_L}(P_L - P_d)$$

where P_d corresponds to the steady-state power demand of the load.
- A voltage-dependent model for the reactive power demand:

$$Q_L = Q_d + \tau \dot{V}_2 \Rightarrow \dot{V}_2 = \frac{1}{\tau}(Q_L - kP_d)$$

assuming a constant power factor in steady state, that is, $Q_d = kP_d$.

This load model is needed so that the dynamic phasor model of the generator–load system makes sense; in other words, there would be no proper phasor dynamics to discuss with a static load model, since the generator internal angle would lack a reference with respect to which one may define synchronism in this case. Nevertheless, the static power-flow equations associated are sufficient to study voltage stability issues in this simple system, as well as in more complex ones, as discussed below.

A simpler AVR model than that used in Example 10.1 is used here; thus, the AVR is assumed to react instantaneously to V_1 terminal variations, that is, the AVR gain is infinity. Therefore, assuming that the generator reactive power output is within its limits ($Q_{G_{min}} \leq Q_G \leq Q_{G_{max}}$), the generator–load system equations can be shown to be

$$\dot{\omega} = \frac{1}{M}\left(P_d - \frac{E'V_2}{X_e}\sin\delta - D_G\Delta\omega\right) \tag{10.95}$$

$$\dot{\delta} = \omega - \frac{1}{D_L}\left(\frac{E'V_2}{X_e}\sin\delta - P_d\right) \tag{10.96}$$

$$\dot{V}_2 = \frac{1}{\tau}\left(-\frac{V_2^2}{X_e} + \frac{E'V_2}{X_e}\cos\delta - kP_d\right) \tag{10.97}$$

$$0 = Q_G - \frac{V_{1o}^2}{X_L} + \frac{V_{1o}V_2}{X_L}\cos(\delta_1 - \delta_2) \tag{10.98}$$

$$0 = \frac{V_{1o}V_2}{X_L}\sin(\delta_1 - \delta_2) - \frac{E'V_2}{X_e}\sin\delta \tag{10.99}$$

$$0 = V_2^2\left(\frac{1}{X_L} - \frac{1}{X_e}\right) + \frac{E'V_2}{X_e}\cos\delta - \frac{V_{1o}V_2}{X_L}\cos(\delta_1 - \delta_2) \tag{10.100}$$

where $X_e = X' + X_T + X_L$ and $\delta = \delta_1 - \delta_2$. Observe that these equations are basically DAE equations of the form (Equation 10.1), where $x = [\omega\ \delta\ V_2]^T$, $y = [Q_G\ \delta_1\ \delta_2]^T$, $p = V_{1o}$, and $\lambda = P_d$; $f(\cdot)$ in this case is defined by the right hand side of Equations 10.95 through 10.97, whereas $g(\cdot)$ corresponds to the basic power flow Equations 10.98 through 10.100, defining the algebraic variables y.

If a generator reactive power limit is reached, that is, $Q_G = Q_{G\min/\max}$, the algebraic equations 10.98 through 10.100 change as follows (there is no change in the differential equations):

$$0 = Q_{G\min/\max} - \frac{V_1^2}{X_L} + \frac{V_1 V_2}{X_L}\cos(\delta_1 - \delta_2)$$

$$0 = \frac{V_1 V_2}{X_L}\sin(\delta_1 - \delta_2) - \frac{E' V_2}{X_e}\sin\delta$$

$$0 = V_2^2\left(\frac{1}{X_L} - \frac{1}{X_e}\right) + \frac{E' V_2}{X_e}\cos\delta - \frac{V_1 V_2}{X_L}\cos(\delta_1 - \delta_2)$$

In this case, the generator loses voltage control, which is reflected in the DAE model as follows: $y = [V_1\ \delta_1\ \delta_2]^T$ and $p = Q_{G\min/\max}$; in other words, V_1 becomes a variable, that is, is allowed to change, whereas Q_G is a fixed value.

The power flow equations used to calculate the equilibrium points and associated steady-state variable values $x_0 = [\omega_0\ \delta_0\ V_{2o}]^T$ and $y_0 = [Q_{G0}\ \delta_{1o}\ \delta_{2o}]^T$ are the following, considering that $\omega = \omega_0$ $(\Delta\omega = 0)$ from the equilibrium conditions $\dot\omega = 0$ and $\dot\delta = 0$ in Equations 10.95 and 10.96, respectively:

$$0 = P_d - \frac{E' V_{2o}}{X_e}\sin\delta_0$$

$$0 = kP_d + \frac{V_{2o}^2}{X_e} - \frac{E' V_{2o}}{X_e}\cos\delta_0$$

$$0 = Q_{G0} - \frac{V_{1o}^2}{X_L} + \frac{V_{1o} V_{2o}}{X_L}\cos\delta_0 \qquad (10.101)$$

$$0 = \frac{V_{1o} V_{2o}}{X_L}\sin(\delta_{1o} - \delta_{2o}) - \frac{E' V_{2o}}{X_e}\sin\delta_0$$

$$0 = V_{2o}^2\left(\frac{1}{X_L} - \frac{1}{X_e}\right) + \frac{E' V_{2o}}{X_e}\cos\delta_0 - \frac{V_{1o} V_{2o}}{X_L}\cos(\delta_{1o} - \delta_{2o})$$

The solutions to these power flow equations yield the equilibrium points for the system, which in this case are basically two: one sep and one uep, as per the associated eigenvalues obtained from the state matrix (see Example 10.10):

$$A = D_x f|_0 - D_y f|_0 D_y g|_0^{-1} D_x g|_0$$

Assuming $X_L = 0.5$, $M = D_L = 0.1$, $D_G = \tau = 0.01$, $k = 0.25$, and $V_{1o} = 1$ (all in pu and s), and with the help of a continuation power flow (CPF) program PSAT [26], the power flow solutions yield the PV curve or nose curve (bifurcation diagram) shown in Figure 10.26 for the system without limits. Similarly, a QV curve, depicted in Figure 10.27, can be obtained, where the reactive power Q_s represents the power injected at the load bus (observe that $Q_s \neq Q_G$). The saddle-node bifurcation point SN in this figure corresponds to a point where the state matrix A is singular (one zero eigenvalue); this bifurcation is typically associated with a singular power flow Jacobian. The latter is not always the case, since for more complex dynamic models, the singularity of the state matrix does not necessarily correspond to a singularity of the power flow Jacobian, and vice versa; however, in practice this issue is basically ignored, since system stability models and associated controls are forced to match the

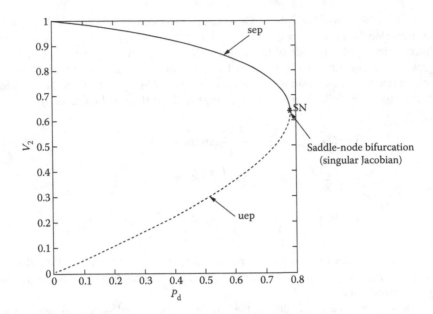

FIGURE 10.26 PV curve (bifurcation diagram) for the generator–load system example.

power flow solutions in steady state. Observe that the SN point corresponds to a maximum value of $\lambda_{max} = P_{d_{max}} \approx 0.78$ pu, which is the reason why this point is also referred to as the maximum loading point or maximum loadability point. For a load greater than $P_{d_{max}}$, there are no power flow solutions, which also means in practice that no steady-state solutions of the system dynamic model exist, regardless of modeling complexity, as discussed above.

If reactive power limits are considered and assuming $Q_{G_{max/min}} = \pm 0.5$ pu, the PV curve depicted in Figure 10.28 is obtained. In this case, the maximum loading or loadability point $\lambda_{max} = P_{d_{max}} \approx 0.65$ pu corresponds to the point where the generator reaches its maximum reactive power limit $Q_G = Q_{G_{max}} = 0.5$ pu and hence loses control of V_1. This point is referred to as a

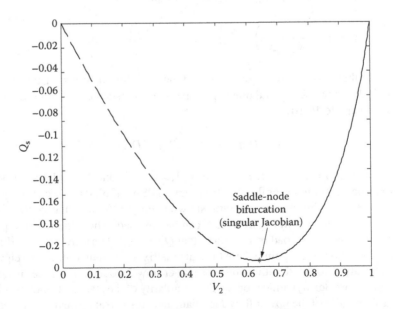

FIGURE 10.27 QV curve for the generator–load system example ($Q_s = -kP_d$).

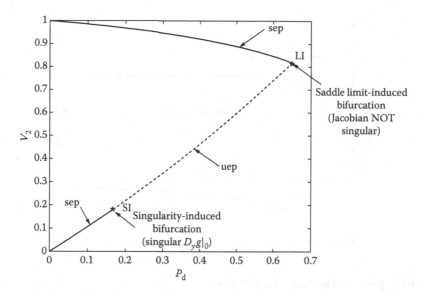

FIGURE 10.28 PV curve (limit-induced bifurcation diagram) for the generator–load system example.

limit-induced static bifurcation point, "beyond" which there are no more power flow solutions due to the limit recovery mechanism of the voltage regulator. Observe that in the low-voltage solution region, there is a point where the Jacobian of the algebraic constraints $D_y g|_0$ becomes singular; this point is referred to as a singularity induced (SI) bifurcation point. An extensive discussion on the latter can be found in Reference 27; however, in practice these types of bifurcations are not very relevant, since they can be readily removed by changing the model of some of the system loads.

A "contingency" in the generator–load system example without reactive power limits is simulated by increasing the line impedance $X_L = 0.5 \rightarrow 0.6$ pu at a loading condition of $\lambda_0 = P_{d0} = 0.7$ pu. Observe the voltage collapse process depicted in Figure 10.29, which is characterized by a slow voltage decay followed by a rapid voltage fall. This collapse is accompanied by an increase in generator

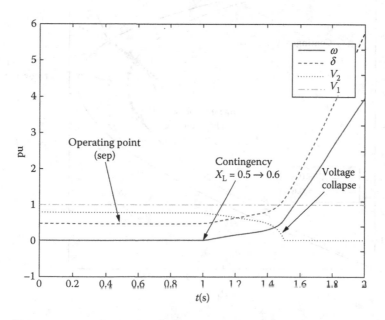

FIGURE 10.29 Voltage collapse time trajectory for the generator–load example due to a contingency.

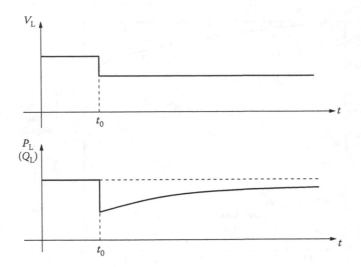

FIGURE 10.30 Typical load recovery characteristics.

angular speed deviation as well as internal angle, which is to be expected, since this is basically an unstable system condition triggered by the lack of a stable operating condition (no power flow solutions).

Figure 10.30 depicts a typical long-term load recovery process that has been observed in actual system loads (e.g., thermostatic loads and load tap changers (LTC) effects), which also applies to the load model used for the generator–load system example discussed here. This load recovery behavior together with the PV curves shown in Figure 10.31 enable explanation of the collapse process depicted in Figure 10.29. Thus, the two PV curves shown in Figure 10.31 correspond to the original

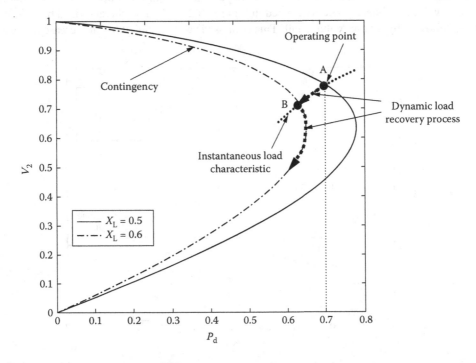

FIGURE 10.31 Collapse process on PV curves.

system and the system after the contingency, without generator limits; observe that the maximum loadability decreases as the transmission system is weakened, which is typically the case, that is, $\lambda_{max} = P_{d_{max}}$ is reduced from approximately 0.78 to 0.65 pu as X_L increases from 0.5 to 0.6 pu. The initial operating point A defined by $\lambda_0 = P_{d0} = 0.7$ pu "disappears" after the contingency, that is, there is no power flow solution for this loading level (no intersection of $P_{d0} = 0.7$ with the corresponding PV curve, since $P_{d0} > P_{d_{max}}$). Initially, the load demand decreases as the load voltage drops immediately after the contingency, as illustrated in Figure 10.30; hence, the system instantaneously moves from point A to B in Figure 10.31, since most loads in practice behave as impedances during the first few moments after sudden voltage changes due to contingencies. However, as the load demand slowly recovers, the voltage continues to drop, typically following the PV curve of the weaker system, and eventually collapses, since there is no power flow solution for this system when the load fully recovers. These observations and behavior also apply to the case when generator limits are considered and apply as well to real systems, in general.

10.3.2 Analysis Techniques

10.3.2.1 Continuation Power Flows

As per the aforementioned discussions, PV curves and maximum loadability points allow to study and detect voltage stability problems. A technique to compute these curves and associated maximum loadabilities is the CPF. This technique traces the solutions z_0 to the power flow equation:

$$G(z, \lambda) = 0$$

as the bifurcation parameter λ changes, where z typically includes the steady-state phasor voltage magnitudes and angles at load buses. The steps are depicted in Figure 10.32 and guarantee convergence when parameterization techniques are used. Thus,

1. *Predictor*: This step is used to compute an approximate value of the power flow solution $z_1 + \Delta z_1$ from an initial solution point z_1, as λ changes from λ_1 to $\lambda_1 + \Delta \lambda_1$. In other words, this step is used to compute Δz_1 and $\Delta \lambda_1$. This can be accomplished by means of one of the following popular techniques:
   a. *Tangent vector method*: This is a good method as it closely follows the PV curves, but is relatively slow. The basic idea in this case is to compute the tangent vector $t = dz/d\lambda$ at the initial solution point, with the tangent vector being defined by the equation

$$D_z G|_1 \underbrace{\frac{dz}{d\lambda}}_{t_1} = -\frac{\partial G}{\partial \lambda}\bigg|_1$$

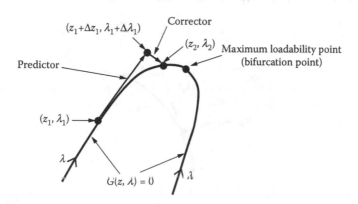

FIGURE 10.32 Continuation power flow steps.

From this equation, which is basically a by-product of a power flow solution, the values of $\Delta\lambda_1$ and Δz_1 can be readily obtained as follows:

$$\Delta\lambda_1 = \frac{k}{\|t_1\|}$$

$$\Delta z_1 = \Delta\lambda_1\, t_1$$

where k is an arbitrary constant used to control the step size ($k = 1$ usually works well). Observe that the value of $\Delta\lambda_1$ decreases as the process approaches the maximum loadability point, since $t_1 \to \infty$ as $\lambda_1 \to \lambda_{\max}$; thus, more solution points are obtained around the maximum loading point, which is a desirable feature of this method. Furthermore, the tangent t defines sensitivities at the associated power flow solution point, which are useful to detect critical load buses where the load voltage magnitudes are the most sensitive to load variations (defined by the largest entries in t associated with load voltage magnitudes), to place reactive power compensation devices (e.g., capacitors). The norm of this vector can also be used as an index to predict proximity to a saddle-node bifurcation [25].

b. *Secant method*: This method is illustrated in Figure 10.33, where one can observe that two power flow solutions z_{1_a} and z_{1_b}, corresponding, respectively, to two loading levels λ_{1_a} and λ_{1_b}, are used to compute Δz_1 and $\Delta\lambda_1$ as follows:

$$\Delta z_1 = k\,(z_{1_a} - z_{1_b})$$

$$\Delta\lambda_1 = k\,(\lambda_{1_a} - \lambda_{1_b})$$

where again, k is an arbitrary constant used to control the step size ($k = 1$ usually works well as well). This method is relatively faster than the tangent method but can lead to numerical problems during the correction step at "sharp corners" (e.g., limits), since, as depicted in Figure 10.34, it may yield a predictor estimate farther from the desired solution (z_2, λ_2) than the tangent vector technique.

2. *Corrector*: The basic idea in this step is to add an extra equation to the power flow equations, that is, solve

$$G(z, \lambda) = 0$$

$$\rho(z, \lambda) = 0$$

for (z, λ) to obtain (z_2, λ_2), using $(z_1 + \Delta z_1, \lambda_1 + \Delta\lambda_1)$ obtained from the predictor step process as the initial guess. These equations can be guaranteed to be nonsingular for the appropriate choice of $\rho(\cdot)$, which is typically defined as follows:

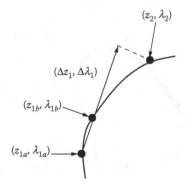

FIGURE 10.33   Secant prediction.

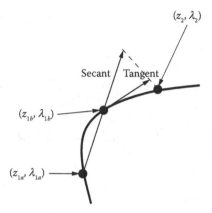

FIGURE 10.34 Problems with secant prediction.

a. *Perpendicular intersection technique*: In this case, $\rho(z, \lambda)$ is chosen based on the perpendicular vector at $(z_1 + \Delta z_1, \lambda_1 + \Delta \lambda_1)$ as depicted in Figure 10.35a. Thus, the perpendicular condition yields the following set of equations:

$$G(z, \lambda) = 0$$
$$(z_1 + \Delta z_1 - z)^T \Delta z_1 + (\lambda_1 + \Delta \lambda_1 - \lambda)\Delta \lambda_1 = 0$$

This method is guaranteed to be nonsingular even in the case where $\lambda_2 = \lambda_{max}$ corresponds to a saddle-node bifurcation point of $G(\cdot)$, which yields a singular Jacobian D_{z}-$G|_2$. However, observe in Figure 10.35b that for a large predictor step, the method may fail to converge; in this case, a simple step cutting technique can be used until convergence is attained (e.g., halving the step, as shown in Figure 10.35b).

b. *Parameterization technique*: A simple parameterization procedure, such as fixing the loading parameter λ, or when close to the maximum loading point where convergence problems may be encountered, fixing one of the system variables $z_i \in z$ (typically a load voltage magnitude), can be used as a corrector as illustrated in Figure 10.36; the latter approach is also referred to as a local parameterization technique, as discussed below. This procedure can be represented by the following equations:

$$G(z, p, \lambda) = 0$$
$$\lambda = \lambda_1 + \Delta \lambda_1 \quad \text{or} \quad z_i = z_{i_1} + \Delta z_{i_1}$$

3. *Parameterization*: This procedure is needed to avoid singularities during the predictor and corrector steps. The most common parameterization methods are as follows:

a. *Local parameterization*: In this case, a variable $z_i \in z$ (typically a load voltage magnitude) is exchanged with the bifurcation parameter λ during the corrector step, that is, the PV curve is "rotated" to eliminate Jacobian singularities. This procedure is depicted in Figure 10.36.

b. *Arc length*: This is an alternative predictor procedure and is based on assuming that the PV curve (bifurcation manifold) is basically an arc of length s. Thus, z and λ are assumed to be functions of s in the following predictor equations:

$$D_z G|_1 \Delta z_1 + \left.\frac{\partial G}{\partial \lambda}\right|_1 \Delta \lambda_1 = 0$$

$$\Delta z_1^T \Delta z_1 + \Delta \lambda_1^2 = k$$

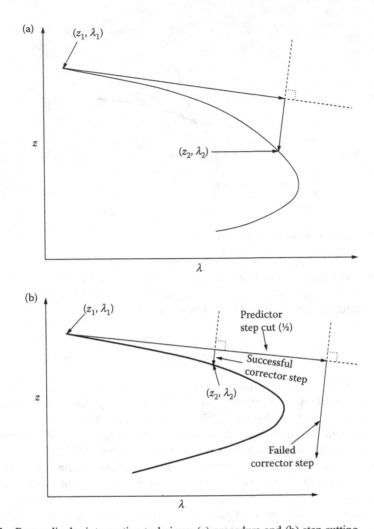

FIGURE 10.35 Perpendicular intersection technique: (a) procedure and (b) step cutting.

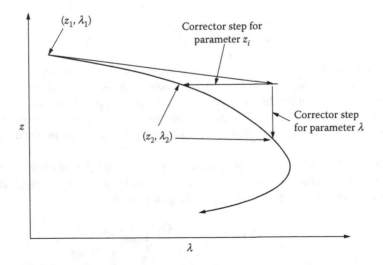

FIGURE 10.36 Parameterization technique.

which are then solved for Δz_1 and $\Delta \lambda_1$, with k being a user-defined parameter to control the step length ($k = 1$ usually works well).

There is no real need for parameterization in practice if a step-cutting technique as the one described above is used.

It is interesting to note that a set of continuous power flow solutions as the system loading is changed evenly, that is, for a constant $\Delta \lambda$, generates the upper side of the PV curves as efficiently as a CPF based on a secant predictor and a parameterization corrector, if a simple step-cutting technique is used. Thus, this technique has been implemented in some popular commercial packages for voltage stability studies, given its simplicity. However, this approach does not allow to compute the lower part (low-voltage solutions) of the PV curve, which may be useful for certain stability analyses.

10.3.2.2 Direct Methods

Techniques to directly compute the maximum loadability point (saddle-node or limit-induced bifurcation) without computing the PV curves are known as direct methods. These methods are based on the fact that maximum loadability problem can be stated, in general, as an optimization problem as follows:

$$
\begin{aligned}
& \text{Max}\lambda \\
& \text{s.t. } G(z, p, \lambda) = 0 \\
& \quad z_{min} \leq z \leq z_{max} \\
& \quad p_{min} \leq p \leq p_{max}
\end{aligned}
\tag{10.102}
$$

where controlled parameters p are now considered, since this optimization approach allows not only to compute the maximum loading point, but also to maximize it if these control parameters are considered as optimization variables. This is one of the main advantages of the optimization approach.

If limits are ignored in Equation 10.102, the KKT conditions of this optimization problem for given values of the control parameters ($p = p_0$) yield the following equations:

$$
\begin{aligned}
G(z, \lambda) &= 0 & \rightarrow & \quad \text{power flow equations} \\
D_z^T G(z, \lambda)w &= 0 & \rightarrow & \quad \text{singularity condition} \\
D_\lambda^T G(z, \lambda)w &= -1 & \rightarrow & \quad w \neq 0 \text{ condition}
\end{aligned}
$$

which are the basic transversality conditions of a saddle-node bifurcation point, with the Lagrange multipliers w corresponding directly to the zero-left eigenvector of a singular power flow Jacobian $D_z G|_0$. These nonlinear equations do not yield a solution for (z, λ, w) if the maximum loading point is a limit-induced bifurcation, since the singularity condition is not met in this case. In this case, the following optimization problem may be solved to obtain the maximum loading point associated with either a saddle-node or a limit-induced bifurcation:

$$
\begin{aligned}
& \text{Max}\lambda \\
& \text{s.t. } F(z, p_0, \lambda) = 0 \\
& \quad z_{min} \leq z \leq z_{max}
\end{aligned}
\tag{10.103}
$$

The solution to Equation 10.103 does not directly correspond to the maximum loadability point obtained using a CPF technique, since the generators' reactive power limits and terminal voltages are handled differently in these optimization models than in the CPF, as discussed in detail in Reference 28. However, by solving these optimization problems, or better yet solving Equation 10.102, one can find the optimal system conditions and control settings that maximize loadability margins and not only the maximum loadability margin for a given operating point; in other words, an optimization approach is significantly more versatile than CPF techniques. For some examples of a variety of optimization models and applications directly associated with voltage stability, the interested reader is referred to References 49–52.

Optimization problems (Equations 10.102 and 10.103) are basically nonconvex, nonlinear programing (NLP) problems; hence, efficient NLP solution techniques such as interior point methods can be used to solve these problems. This approach is usually more efficient than a CPF method to obtain maximum loading points; however, CPFs yield intermediate power flow solution information, thus yielding load voltage and generator reactive power profiles with respect to load variations that can be very useful for analysis and planning purposes.

10.3.2.3 Voltage Stability Indices

Voltage stability indices can also be used to determine proximity to voltage collapse points. These indices are computationally inexpensive compared to determining loadability margins using CPF or direct methods, and do not require making assumptions regarding the direction of generation and load changes; hence, these can be readily used for online applications. However, most indices present highly nonlinear profiles, especially when control limits are reached and are designed to either detect proximity to saddle-node or limit-induced bifurcations, but not both.

Many indices have been proposed, but the most popular ones are the minimum singular value index and the reactive power reserve indices. The former is based on the fact that the system Jacobian becomes singular at the saddle-node bifurcation point; hence, the associated singular value can be used to measure the system proximity to such a bifurcation point. The latter is used to detect proximity to limit-induced bifurcation points, which are typically associated with reactive power limits of generators; thus, by monitoring the amount of generator reactive power reserves, the proximity to these bifurcations can be determined. Both indices are computationally inexpensive and hence can be used for online monitoring applications; however, they are both unreliable for predicting proximity when far from the corresponding bifurcation point, given their highly nonlinear behavior, as illustrated in Figure 10.37. Special linearization techniques may be used to define indices with quasi-linear profiles based on these indices, as explained in detail in Reference 29.

10.3.3 Countermeasures

To avoid voltage stability problems during system operation, PV curves are typically used to define either online or off-line transmission system ATCs, as per Equation 10.8 and as illustrated in

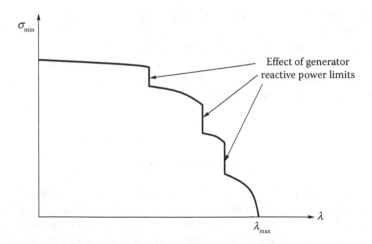

FIGURE 10.37 Typical profile of the minimum singular value index. Reactive power reserve indices show similar profiles.

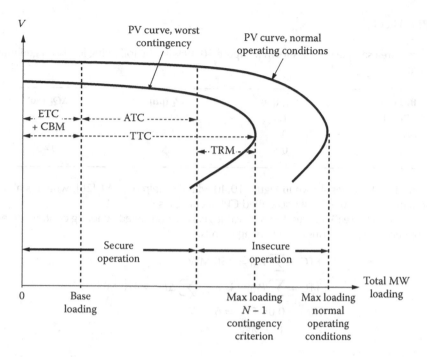

FIGURE 10.38 Computation of ATC using PV curves.

Figure 10.38. In this figure, the TRM is defined as a given distance away from the maximum load-ability point for the worst single contingency ($N - 1$ contingency criterion); for example, for WECC, the TRM is defined as 5% of the TTC for normal operating conditions [30].

The PV curves needed to determine the system's ATC, as per Figure 10.38, are obtained assuming certain loading and dispatch profiles, that is, explicitly defining how the load and generation will change for a given operating horizon. For online applications, this horizon can be as short as every 5 min, as in the case of Ontario; for off-line computations, it can be as short as 1 h. In practice, the ATC values are determined for the system's main transmission corridors, which are then considered during the operation of the system as power transfer limits on these corridors, whose power transfers are continuously monitored. Thus, for ATC computation purposes, load changes are modeled as follows:

$$P_L = P_{L0} + \lambda \Delta P_L$$
$$Q_L = Q_{L0} + \lambda \Delta Q_L$$

where P_{L0} and Q_{L0} define the active and reactive base load powers, respectively, and ΔP_L and ΔQ_L define the changes in active and reactive powers, respectively. On the other hand, generator power dispatch to supply the load changes is defined, with the exception of the slack bus, as

$$P_G = P_{G0} + \lambda \Delta P_G$$

where P_{G0} represents the base power level, and ΔP_G defines the direction of generator dispatch.

The following example illustrates the computation of the ATC using a CPF technique.

EXAMPLE 10.11

For the three-area sample system shown in Figure 10.39 with the following load and generator dispatch data:

Bus name	ΔP_G (pu)	ΔP_L (pu)	ΔQ_L (pu)
AREA 1	1.5	0	0
AREA 2	0	1.5	0.56
AREA 3	0.5	0.5	0.40

one obtains the PV curves shown in Figure 10.40 with the help of PSAT [26], which is a practical implementation of the previously described CPF techniques.

On the basis of these PV curves, the ATC values can be computed, ignoring contingencies without loss of generality and considering voltage limits:

$$ETC = \sum P_{L0} = 350 \text{ MW}$$

$$TTC = \sum P_{L0} + \lambda_{maxv\,min} \sum \Delta P_L = 470 \text{ MW}$$

$$TRM = 0.05 \text{ TTC} = 6 \text{ MW}$$

$$\Rightarrow ATC = 114 \text{ MW}$$

Ignoring voltage limits:

$$TTC = \sum P_{L0} + \lambda_{maxvc} \sum \Delta P_L = 670 \text{ MW}$$

$$TRM = 0.05 \text{ TTC} = 33.5 \text{ MW}$$

$$\Rightarrow ATC = 286.5 \text{ MW}$$

To increase system loadability and thus increase transfer capacities to reduce the possibility of voltage stability problems, the most common control and protections techniques are as follows:

- Increase reactive power supply in the most critical areas, that is, those areas most sensitive to voltage problems; sensitivity studies at the maximum loading conditions are very useful in

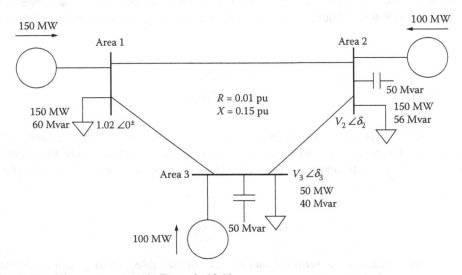

FIGURE 10.39   Three-area system in Example 10.11.

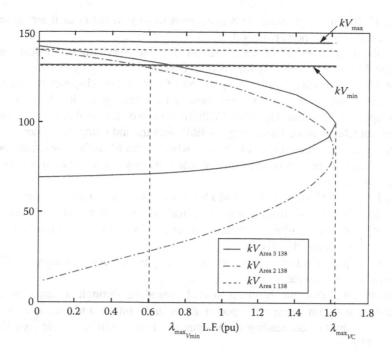

FIGURE 10.40 PV curves for the three-area test system in Example 10.11.

this case [31]. This can be accomplished by introducing shunt compensators such as mechanical switch capacitors (MSC) or STATic COMpensators (SVC and STATCOM) in the area, or by coordinated management of the area reactive power sources by means of secondary voltage control strategies. These control solutions are discussed in detail in Chapter 9.

- Undervoltage relays may be used to reduce load demand on the system, that is, shed load, thus increasing loadability margins to avoid voltage collapse. These relays are protective devices located at subtransmission substations to trip load feeders, and are triggered by long-duration voltage dips on the primary feeders. Their operation is somewhat similar to taps on LTCs, that is, load feeders are tripped in steps and not all at once (e.g., 1%–2% of the substation load is shed at any given time), with time delays in the order of 1–2 min after the voltage falls below certain values, which are typically in the 0.8–0.9 pu range. The larger the voltage dips, the faster and the larger the number of feeders that are tripped.

- Voltage recovery by LTCs may also be blocked during significant voltage dips to reduce demand on the system, since by this mechanism load recovery is slowed down or eliminated all together, thus reducing system demand to avoid voltage collapse.

10.3.4 REAL SYSTEM EVENT

The August 14, 2003 blackout of the northeastern part of the North American grid is a good practical example of voltage stability problems leading to major system failures [32]. The North American eastern interconnection (see Section 9.1.1) was not under particularly stressed conditions, with ambient temperatures and system conditions being relatively normal for that time of the year. The problems started and evolved all around the Cleveland-Akron (critical) area in Ohio, operated by First Energy. The following are the sequence of events that resulted in near 62 GW of load being without service for about 4 h:

- At about 13:30 p.m., a generator in a generation facility in the critical area tripped, thus reducing local reactive power support.
- Between about 14:00 and 15:00 p.m., three 345 kV lines servicing the critical area tripped due to tree flashover at relatively low loading conditions.
- With the third line trip at around 15:00 p.m., NERC's $N-1$ contingency criterion was violated; however, neither the local system operator First Energy nor its "back-up," the Midwest Independent System Operator (MISO), detected the violation of these NERC operating guidelines, since First Energy's EMS servers and computer systems were basically unoperational since around 14:15 p.m., whereas the MISO's state estimator was not converging due to problems with the system topology representation associated with inappropriate monitoring.
- Between 15:00 and 15:40 p.m. more 348 kV lines tripped due to tree flashovers, leaving the 348 kV transmission system servicing the critical area unoperational, and thus leading to line trips and rather low-voltage conditions at the 135 kV level, with voltages reaching values of 0.85 pu at around 16:00 p.m.
- At about 16:05 p.m., the 135 kV system was basically out of service, leaving the Cleveland-Akron area without service.
- Immediately after this, the blackout started spreading through a series of cascading events associated with unexpected power flows and voltage and frequency oscillations throughout the north east, leading in the span of about 7 min to a wide-area blackout at around 16:13 p.m.

It took a total of about 72 h to fully restore service, with the economic losses associated with this event being roughly estimated to be between $7 and $10 billion.

This event is a typical example of system-wide instabilities triggered by lack of enough voltage stability margins and adequate reactive power support in a very specific critical area. In view of the various problems with system operators and operating tools and IT hardware, undervoltage relays and LTC blocking protection mechanisms might have kept most of the system intact. Thus, one of the recommendations of the U.S.–Canada Joint Task Force has been to look at the implementation of these protection schemes throughout the North East.

10.3.5 Fault-Induced Delayed Voltage Recovery

FIDVR refers to the phenomenon in which system voltages remain significant low for a relatively long period of time after a system contingency due to severe faults in the grid, and is related to not enough reactive power compensation (voltage support) and increased current demand from mostly motor loads near the fault [2]. Thus, the problem is typically associated with induction motor loads, such as aggregated air-conditioning compressors, which during low voltage conditions near their terminals increase their current demand, resulting in further depressing the load voltage, especially if the grid feeding the load is not stiff. If the voltage is too low, FIDVR could potentially lead to motor loads stalling and possibly a fast voltage collapse, as depicted in Figure 10.41. This problem can be avoided with faster fault clearing, additional reactive power/voltage support on the low voltage areas, and/or undervoltage protections on the load side, which would trip loads compounding the low voltage conditions.

The load current increase with a terminal voltage dip is related to the characteristics of induction motors and their mechanical loads, which are typically constant or speed dependent torques, as in the case of compressors such as those in residential air conditioners, conveyers, and pumps. Thus, for example, Figure 10.42 illustrates the simulated phase current ($i_{a_{IM}}$) and voltage ($v_{a_{IM}}$) and their corresponding RMS values (I_{IM} and V_{IM}), as seen from the 230 kV high voltage side of the load feeder, of a 100 MW aggregated induction motor (IM) star-up with a speed dependent mechanical load (fan load), followed by a severe voltage dip at the motor load terminals due to a significant

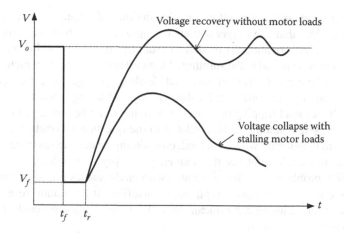

FIGURE 10.41 FIDVR system voltages for a fault at t_f and clearance at t_r with and without motor loads.

contingency on the feeding system; observe the current increase associated with the terminal voltage reduction.

10.4 FREQUENCY STABILITY

As mentioned in Section 10.1.3, frequency stability refers to the ability of a power system to maintain an acceptable frequency after a system disturbance resulting in significant imbalance between generation and load demand [1]. The frequency can only be maintained if the balance between system generation and load can be restored as soon as possible after the disturbance, with as little as possible unintentional tripping of load. If the system goes into a frequency unstable condition, the frequency either drops very fast or sustained frequency swings occur, which can lead to tripping of generators and loads by frequency relays.

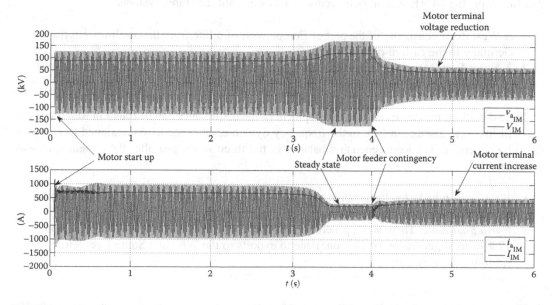

FIGURE 10.42 Simulated aggregated induction motor (IM) load start-up and contingency in the load feeding system.

A severe system disturbance can result in large excursions of frequency, power flows, voltages, and other system variables that can trigger actions not only from controls and protections, but also from other processes that are not normally modeled in standard transient stability and voltage stability studies. Protection actions can lead to the splitting of the system into islands, which may result in significant discrepancies between generation and load, leading to large frequency deviations. These deviations could be too large to cope with for the primary fast acting frequency controls discussed in Section 9.3, and hence load tripping or generator tripping must be employed to stabilize the frequency. In an island, system stability is then related to the question of whether an equilibrium can be reached with minimal unintentional loss of load, considering issues such as boiler dynamics, which are slower processes that could influence the frequency stability of a system.

Frequency stability problems are often associated with inadequacies in equipment responses, poor coordination of control and protection equipment, or insufficient generation reserves. It should be noted that any disturbance causing a significant loss of load or generation could be of concern for the frequency stability of the system.

A frequency instability could either be short term or long term. An example of short-term frequency instability is the formation of an island system with lack of generation and insufficient load shedding, so that the frequency drops very fast, causing a blackout in the system within seconds. If the frequency is not stabilized, generator protections and Volts/Hertz protections might trip more generators, further aggravating the situation. Underfrequency load shedding schemes may also lead to load shedding depending on both the frequency value and the speed of the frequency drop, that is, df/dt. Long-term frequency instability is caused often by complex interactions between boiler/reactor dynamics, steam turbine overspeed control, and associated protections and controls. Such instabilities are developed during time frames from tens of seconds to several minutes.

It is important to mention that abnormal voltage magnitudes could also trigger frequency instabilities or influence their behavior. Thus, low voltages might cause operation of impedance relays leading to system islands. One the other hand, high voltages may cause generator trips as well as lead to increased load levels affecting the load-generation balance.

As an example of how a system could be split into islands with load-generation imbalances leading to significant frequency excursions, the disturbance in the UCTE system of November 4, 2006 is considered here [33]. As a result of a line trip in the northwestern part of Germany, which triggered further line trips, the UCTE system (see Section 9.1.1) was split into three systems:

1. A West Zone consisting of the part of Europe west of a line from the northwestern part of Germany to the eastern part of Slovenia.
2. A North East Zone consisting of the UCTE system northeast of the line mentioned above.
3. A South East Zone consisting of the Balkan states and Greece.

More details on the causes and the development of the disturbance can be found in Reference 33. We will limit the discussion here to the frequency dynamics after the aforementioned three zones (islands) appeared. The load-generation balance in the three zones just after the system split was as follows:

- *West Zone*: It had a total generation of 182,700 MW, with a 8940 MW power imbalance due to missing imports from the East.
- *North East Zone*: There was a generation surplus of more than 10,000 MW (approximately 17% of total generation of this zone) due to exports to the West and South East before the splitting of the system.
- *South East Zone*: There was a total generation of 29,100 MW, and total load of 29,880 MW.

The large imbalance in the West led to a very fast drop of the system frequency from its 50.00 Hz set-point value to about 49 Hz, as illustrated in Figure 10.43. This substantial frequency drop

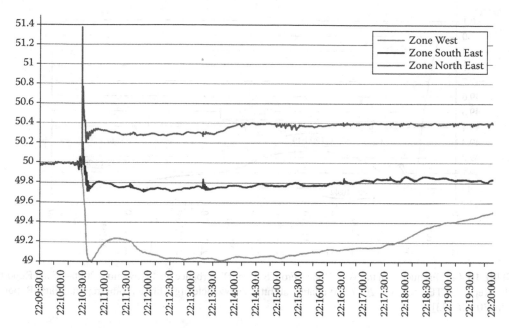

FIGURE 10.43   Frequency recordings after the splitting of the UCTE system of November 4, 2006. (From System Disturbance on 4 November 2006—Final Report. Tech. Rep., UCTE, January 2007. Available at http://www.ucte.org/_library/otherreports/Final-Report-20070130.pdf. [33])

triggered a sequence of events on generators and activation of defense plans, which will be further elaborated below. As a result of the incident, a total of about 17,000 MW of load was shed and 1600 MW of pumps was disconnected. In the North East, the generation surplus caused a rapid frequency increase up to 51.4 Hz, as shown in Figure 10.43, which was reduced to 50.3 Hz by automatic actions, that is, standard and emergency primary frequency control (see Section 9.3.2), and automatic tripping of generating units sensitive to high frequency values (mainly windmills). The total tripping of wind generation is estimated to be 6200 MW, which played a crucial role in decreasing the frequency during the first seconds of the disturbance. In the South East, there was a small power imbalance of about 770 MW; just after the event, a 210 MW generator was tripped, but the frequency was stabilized by the primary frequency controls (see Figure 10.43).

The results of the different events and actions taken in the West Zone are depicted in Figure 10.44, which are as follows:

1. Separation of the Western area from the rest of the UCTE system.
2. Stop of the frequency decrease, mainly due to the action of the defense plan (load shedding and shedding of pumps).
3. Beginning of frequency increase due to primary reserves' action.
4. Frequency recovery to a local maximum of 49.2 Hz.
5. Slow frequency raise, which led back to the normal 50 Hz value at about 22:25 p.m.

In the regions between the indicated points in Figure 10.44, the frequency behavior can be explained as follows:

- *Range 1–2*: The large power imbalance causes a frequency drop of up to 120–150 mHz/s. Moreover, the power imbalance was larger than the primary reserve available in the area. During these first 11 s, pumped-storage units were tripped, but it was not enough to halt the frequency decrease.

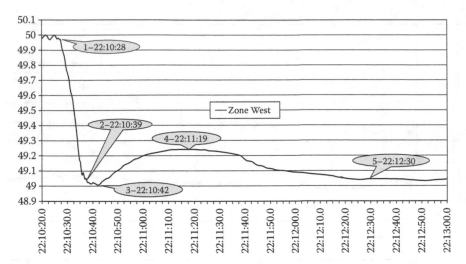

FIGURE 10.44 Frequency in the West Zone. (From System Disturbance on 4 November 2006—Final Report. Tech. Rep., UCTE, January 2007. Available at http://www.ucte.org/_library/otherreports/Final-Report-20070130.pdf. [33])

- *Range 2–3*: During this period, frequency stopped decreasing mainly due to automatic load shedding according to defense plans of the TSOs.
- *Range 3–4*: Frequency started to increase due to the activation of the last primary reserves. A (local) maximum frequency of 49.2 Hz was reached.
- *Range 4–5*: During this time range, the frequency decreased due to the exhaustion of primary reserves, the trip of some generators, and the reenergization of loads by some TSOs.
- *Range 5 and after*: After 22:12 p.m., the frequency was stable and started to increase after 22:15 p.m. to reach a value near 50 Hz at about 22:24 p.m. This was due to manual increase of generation and start of additional generation units by some TSOs.

It is important to highlight the fact that to understand the main reasons for all blackout examples described in this chapter, one needs not only to look at technical issues associated with system stability, but also consider issues such as inappropriate/unexpected operation of protections, lack of coordination among neighboring system operators, and inadequate monitoring and maintenance of the transmission grid. These are patterns that can be detected in most major blackouts worldwide.

10.5 IMPACT OF WIND AND SOLAR GENERATION ON STABILITY

This section describes the impact of different types of RE generators and associated controls on power system stability, based on the material presented in Section 9.5 and complementing it. Thus, voltage stability grid issues related to these kinds of sources are first presented, followed by a discussion on angle stability and then frequency stability, highlighting the main effect on system stability for the various types of RE generators and controls. Finally, some relevant and particular stability problems directly associated with the fast converter control systems of RE generators are reviewed. Several system examples from the technical literature are used to illustrate some of the phenomena explained next.

10.5.1 VOLTAGE STABILITY

As mentioned in Section 9.5, depending the RE generation technology and controls, RE plants would have different types of impact on system voltage stability. Thus, Type A wind-power generators do not provide any significant reactive power support nor voltage control, as discussed in detail in Section 9.5.1; on the contrary, these represent a significant reactive power demand on the system, especially at low voltage conditions. Hence, these kinds of wind-power plants can be detrimental to voltage stability, since these are basically induction machines, which during low voltage conditions after contingencies near the plant terminals (Point of Common Coupling or PCC) increase their current demand, leading to FIDVR problems (see Section 10.3.5), which is why these types of plants must include reactive power compensation equipment (typically capacitor banks) at the PCC and trip during nearby faults.

With the development and deployment of wind-power Type C (Doubly-Fed Induction Generator or DFIG) and now Type D (full-converter interface) generators, as well as SPV plants, which are all capable of reactive power and voltage control, and based on the experience with Type A generators at the time, utilities initially required RE plants to operate in PQ control mode (see Section 9.5.1) at unity power factor (zero reactive power demand) to avoid low voltage problems during normal operation, and trip at low terminal voltage conditions triggered by nearby faults to prevent FIDVR issues during contingencies. However, with the increased penetration of RE sources in the grid replacing traditional power plants, it became apparent that RE generation tripping during faults and their lack of voltage support would lead to other stability problems, in particular frequency stability due to the sudden loss of significant amounts of generated power during sever contingencies, as discussed in more detail in Section 10.5.3, as well as poor voltage support during normal operating and contingency conditions. Therefore, utilities then required that RE plants would remain online and provide reactive power support during faults, defining low-voltage ride-through (LVRT) requirements for these plants, as explained in detail in Section 9.5.2.

With RE becoming a significant part of the generation park in many jurisdictions, utilities are now demanding that RE plants provide similar voltage control capabilities as traditional power plants [4,5], so that proper stability margins can be maintained under all system conditions for secure system operation. Current RE generators and associated controls are certainly capable of providing continuous voltage regulation services from the technical perspective (see Section 9.5.2), especially full-converter RE plants which can be seen as STATCOMs from the a grid voltage regulation perspective, but this requires increased converter ratings, of about 10%–20% depending on the reactive power support requirements (e.g., a converter operating at 0.9 power factor and supplying nominal power, results in an approximately 10% increase in current, and thus a rating increase of about 10%), which do have an effect on RE plant costs, albeit not significant (e.g., considering that the cost of the electrical infrastructure of a wind-power generator is about 10% of the total cost of the system [34], a 10% increase in the electrical system ratings would mean a total cost increase of approximately 1%).

The impact on the PV curves for the benchmark system depicted in Figure 10.45a of the Type C (DFIG) wind plant illustrated in Figure 10.45b, with PQ and voltage controls, is discussed in detail in Reference 35. Figure 10.46 shows the PV curves obtained for this system after a trip of Line 2–4 with an overall system load increase; observe that the system with a DFIG plant with PQ control at unity power factor (Case B) has the worst maximum loadability and voltage profile, whereas the system with the DFIG plant with voltage controls (Case C) is closer to the maximum loadability and voltage profiles of the base system with a synchronous generator (SG) at Bus 2 (Case A), if the Hopf Bifurcation (HB) in the latter case is removed using PSSs, as discussed in Section 10.2.3.2. This is due to plants with voltage control being able to inject more reactive power into the system than plants with PQ controls, but not as much as the traditional generator, since the reactive power injection capabilities of Type C plants are lesser than similarly sized traditional generation plants. However, this would not be the case with full-converter RE plants based on Type D and SPV generators, which if habilitated with voltage controls, are able to inject similar amounts of reactive power than a traditional plant of comparable ratings. Note that RE plants are simply modeled as power injections in the context of the power flows used to obtain PV curves, regulating voltage within reactive power limits for plants with voltage control capabilities; thus, RE plants are simply modeled as traditional PQ or PV generator buses, depending on their controls, in PV-curve calculations. Similar results are reported for

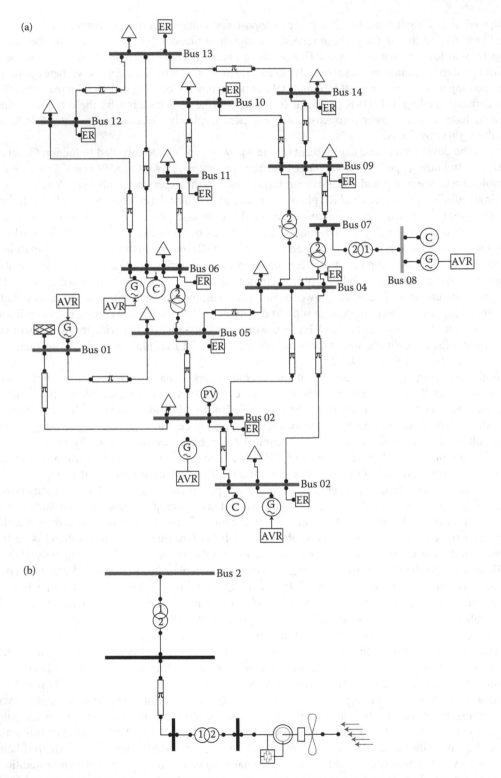

FIGURE 10.45 PSAT diagrams of IEEE 14-bus benchmark system for DFIG stability studies. (a) Base system. (b) DFIG plant replacing the generator at Bus 2. (From Muñoz J. C. and Cañizares C. A., Comparative stability analysis of DFIG-based wind farms and conventional synchronous generators, IEEE-PES Power Systems Conference and Exposition (PSCE), Phoneix, AZ, US, 1-7, March 2011. [35])

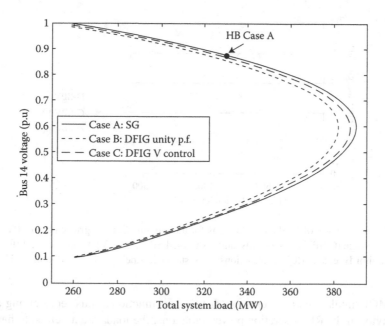

FIGURE 10.46   IEEE 14-bus benchmark system PV curves for different types of generators and controls at Bus 2. (From Muñoz J. C. and Cañizares C. A., Comparative stability analysis of DFIG-based wind farms and conventional synchronous generators, IEEE-PES Power Systems Conference and Exposition (PSCE), Phoenix, AZ, US, 1–7, March 2011. [35])

a full-converter-interfaced wind farm in Reference 36, demonstrating that such plants with voltage controls yield higher maximum system loadabilities and better voltage profiles than with PQ unity-power-factor controls, discussing as well the impact of wind-farm LTCs and converter limits on the PV curves.

Large full-converter SPV plant at the transmission system level, with PQ unity-power-factor and voltage controls and located at an existing SPV plant site, and distributed SPV generators, with PQ controls operating at unity power factor and located at the main load center at the distribution system level, are compared from the voltage stability perspective in a real system in Reference 37. Thus, Figure 10.47 shows the loadability margins of this system for an overall load increase; observe that in this case there is no significant effect on the maximum loadability margins for the different types of controls of the SPV plant at different penetration levels, whereas the distributed SPV generators have a significant positive impact on these margins, since these unload the transmission network by supplying the system demand directly at the load level, thus increasing the systems maximum loadability.

The effect on voltage stability of power variations in RE plants due to wind speed and solar irradiation changes depends on the time scales and types of generator technologies and controls. Thus, as discussed in Reference 38, wind-power generators power outputs do not change significantly in the short term (few seconds), due to the turbine inertia and pitch control, especially in varying-speed Type C and D generators; furthermore, in large plants, aggregation reduces even further this variability. Hence, for these types of wind-power plants, the variability of wind does not really affect the short-term voltage stability of the grid. However, in the long-term, even though the power ramps are relatively slow due to aggregation, as shown in Reference 39, this variations can be significant in the span of hours, thus affecting PV curves. For example, in Reference 40, for a real system with 31% penetration of variable RE generation with PQ unity-power-factor controls, and assuming a 15% margin of variations, which is a realistic value in the span of a couple of hours for system aggregated wind-power plants [39], the bounds of the PV curves shown in Figure 10.48 are obtained using

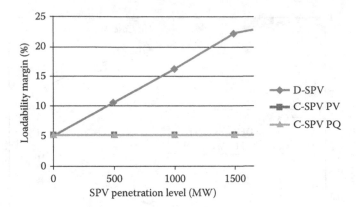

FIGURE 10.47 Comparison of loadability margins for the Ontario-Canada grid with an SPV plant, with PQ (C-SPV PQ) and voltage (C-SPV PV) controls, and distributed SPV (D-SPV) generators at different penetration levels. (From Tamimi B. et al., IEEE Transactions on Sustainable Energy, 4(3), 680–688, 2013. [37])

Mote Carlo (MC) simulations with uniform distribution functions, thus representing a pessimistic statistical behavior of the RE generation power variations; the loads are assumed to have a constant power factor and are increased proportionally to their base values, and similarly for the corresponding dispatch of traditional generators. Observe that the maximum loadability interval resulting from the depicted PV curve bounds is significant (about 1500 MW or 15% of the maximum loading margin), capturing all loadability margins of the associated PV curves for all possible RE generation power variations within the assumed interval.

For SPV plants, the short-term power output variability is more significant than with wind-power generators, as demonstrated in Reference 41, due to the lack of inertia of SPV generators and power control mechanisms, as well as quick changes in solar irradiation levels due to rapid cloud movement, although these variations are ameliorated with aggregation. Therefore, SPV plants with PQ unity-power-factor controls would have a significant effect on short-term voltage stability; however,

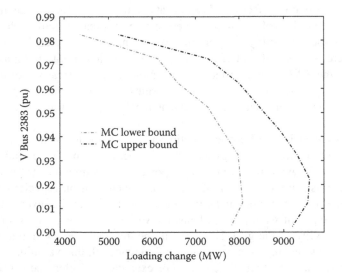

FIGURE 10.48 PV curve boundaries obtained with MC simulations for output power variations of PQ unity-power-factor RE generators introduced in a real 2383-bus system. (From Muñoz J. C. et al., IEEE Transactions on Power Systems, 28(4), 4475–4487, 2013. [40])

with voltage controls, the negative impact associated with variable power output would be significantly reduced, given the voltage regulation provided by such controls. In the long-term, on the other hand, even though SPV plant power ramps are smoothed out by aggregation [41], injected power could vary significantly in the span of a few hours, and thus, like wind-power plants, PV curves and associated maximum loadability margins for PQ unity-power-factor SPV plants would be similar to those illustrated in Figure 10.48, since for PV-curve calculations, both wind-power and SPV plants are represented using the same models.

10.5.2   Angle Stability

As in the case of voltage stability, RE generators effect on angle stability depends on the type of RE generators and associated controls. This section focuses on Type C and D wind-power and full-converter SPV plants, since Type A generators are being/have been effectively replaced by Type C or D generators, due to the lack of controllability and high reactive power demand of these types of plants. The impact on angel stability of these types of plants can be evaluated in terms of small-disturbance angle stability or small angle oscillations, measured by the damping of the system's critical modes, and in terms of the large-disturbance angle stability or system transient stability, which can be measured by the system's critical clearing time (CCT) for sever contingencies in the system. Depending on the RE generation controls, positive or negative effects can be observed, as demonstrated by the examples described next; these effects also depend on the grid itself, the RE plant location in the grid, and the limited or no inertia of these types of plants, which have a more significant and negative impact on frequency stability, as discussed in Section 10.5.3.

Figure 10.46 shows that for a Line 2–4 trip in the benchmark system illustrated in Figure 10.45, the critical-mode damping becomes zero (HB) at a total system load of about 330 MW, and negative for higher loading, for the base system with a traditional generator at Bus 2 and no PSSs (Case A), whereas the system critical damping remains positive for all loading levels for the system with a DFIG plant at Bus 2, for both PQ (Case B) and voltage (Case C) controls, thus eliminating small oscillatory behavior in these cases. This is confirmed by the time-domain simulation of the contingency depicted in Figure 10.49, where the system small oscillatory behavior can only be observed for Case A. From these results, it is evident that RE generation and associated controls, including their tuning, could have a dramatic effect on the critical mode damping, and thus have positive or detrimental effects on the system's small-disturbance angle stability, which is also corroborated by the damping studies for converter-based RE plants presented in Reference 42. This highly unpredictable behavior of the critical-mode damping is evident in the results of the SPV generation stability studies in Reference 37, and illustrated in Figure 10.50, where it is clear that no particular discernable pattern of the system's critical damping can be observed with respect to SPV generation type, control, or penetration levels.

Regarding the impact of RE plants on transient stability, Figure 10.51 illustrates the changes on CCTs as SPV generator types and controls change. Observed that, as expected, the SPV plant with unity-power-factor PQ controls has a negative impact on system transient stability, worsening as penetration levels increase. On the other hand, the SPV plant with voltage control does not negatively affect transient stability regardless of penetration levels in this case, whereas distributed SPV generators improve stability as the penetration level increases, which is to be expected, since as penetration grows, the system demand is reduced. The following general conclusions can be drawn from these results:

- PQ control in RE generators negatively affect transient stability.
- PV control in RE generators may or may not affect transient stability, depending on several factors, such as penetration levels and location of RE generation.
- RE generation near the load is advantageous for transient stability, since it reduces system demand, thus unloading the system making it more stable and hence more secure.

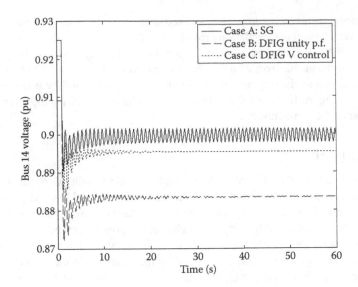

FIGURE 10.49 IEEE 14-bus benchmark time response at a remote bus for a Line 2–4 trip at about 330 MW system loading with different types of generators and controls at Bus 2. (From Muñoz J. C. and Cañizares C. A., Comparative stability analysis of DFIG-based wind farms and conventional synchronous generators, IEEE-PES Power Systems Conference and Exposition (PSCE), Phoenix, AZ, US, 1-7, March 2011. [35])

The variability in power output of RE plants effect on angle stability depends mostly on the short-term power ramps of these plants, since both small- and large-disturbance angle stability are mostly contingent on the system conditions and its response a few seconds before and after a contingency. Hence, the small power ramp changes of wind-power plants in the short-term would not have a significant impact on system angle stability; however, SPV plants relatively faster power ramps in the short-term could affect system angle stability. In that case, multiple simulations would be required to understand the impact of the variability of SPV plant output on angle stability, as in the case of the previously discussed PV-curve bounds, but now applied to critical-mode damping and time trajectories. Various techniques have been proposed to reduce computational burden of these types of studies, as discussed and demonstrated, for example, in References 43,44; however, the "brute force" approach of multiple simulations is the most practical. For example, Figure 10.52 illustrates the

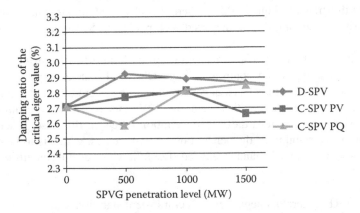

FIGURE 10.50 Comparison of critical-mode damping for the Ontario-Canada grid with an SPV plant, with unity-power-factor PQ (C-SPV PQ) and voltage (C-SPV PV) controls, and distributed SPV (D-SPV) generators at different penetration levels. (From Tamimi B. et al., IEEE Transactions on Sustainable Energy, 4(3), 680–688, 2013. [37])

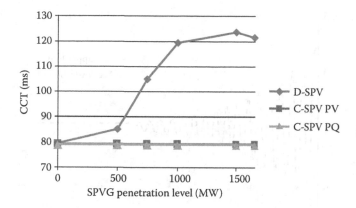

FIGURE 10.51 Comparison of CCTs for the Ontario-Canada grid with an SPV plant, with unity-power-factor PQ (C-SPV PQ) and voltage (C-SPV PV) controls, and distributed SPV (D-SPV) generators at different penetration levels. (From Tamimi B. et al., IEEE Transactions on Sustainable Energy, 4(3), 680–688, 2013. [37])

two-area benchmark system from [18] modified by replacing the standard generators G2 and G4 with DFIG Type C generators with terminal voltage controls; a $\pm 10\%$ variations in the power output of both DFIGs yields the transient stability stable and unstable boundaries depicted in Figure 10.53, which were obtained with MC simulations conservatively assuming a uniform distribution function of the output powers of the RE generators.

10.5.3 FREQUENCY STABILITY

The impact of RE generation plants on frequency stability can be significant, due to the little or no inertia of these kinds of plants, as well as the limited or lack of control of power, as discussed in Section 9.5.3. For example, Figure 10.54 illustrates the frequency responses at a large generator for a fault in the Ontario-Canada system with an SPV plant with unity-power-factor PQ and voltage controls, as well as distributed SPV generators at the largest load center, generating approximately 2 GW of solar power (about 10% penetration level); observe the better frequency response of the system for the case of the distributed SPV generation, as expected, since the system demand is reduced and its overall inertia is not affected by inertia-less SPV generation, whereas the case with the SPV plant shows larger and more sustained frequency oscillations due to the reduced system inertia. This problem can be addressed by introducing power-droop and virtual-inertia controls in RE generators, as demonstrated and explained in detail in Section 9.5.3. It is important to highlight the fact that these

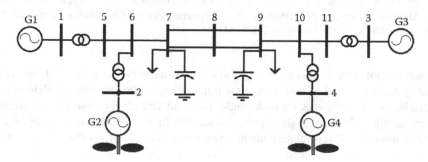

FIGURE 10.52 Modified two-area benchmark system to study the effect on transient stability of power output variations of DIFG generators. (From Muñoz J. C., Affine Arithmetic Based Methods for Power Systems Analysis Considering Intermittent Sources of Power, PhD Thesis, University of Waterloo, 2013. [44])

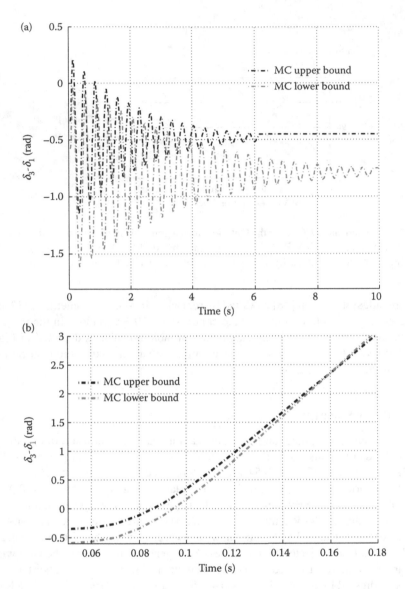

FIGURE 10.53 Transient stability boundaries obtained with Monte Carlo (MC) simulations for a fault and tripping of one of the Lines 8–9 in the test system of Figure 10.52, for a $\pm10\%$ variation on the power output of the wind-power generators. (a) Stable for a 3-cycle fault clearing. (b) Unstable for an 8-cycle fault clearing. (From Muñoz J. C., Affine Arithmetic Based Methods for Power Systems Analysis Considering Intermittent Sources of Power, PhD Thesis, University of Waterloo, 2013. [44])

types of power output controls also affect angle stability, since these controls would affect the plant and thus the systems damping as well as its time trajectories, as demonstrated in Reference 45.

RE generation power variations, for both wind-power and SPV plants, have a larger overall impact on frequency stability than on angle stability, since the time constants of a systems frequency response are in the order of several seconds to a few minutes (typically in the 10 s to 1 min range). In these time ranges, the aggregated SPV plant output variations may not be as significant as in the short-term time scale relevant to angle stability, but may still be relatively large [41]; on the other hand, changes in the wind-power generation outputs may be larger than in the case of the few seconds associated with angle stability phenomena, given the longer time scales in this case [38,39]. Once

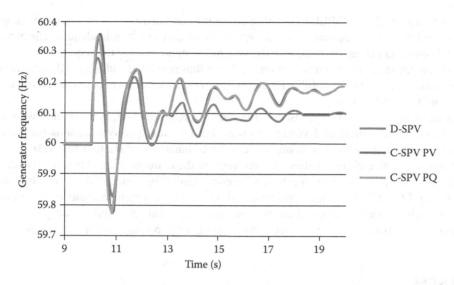

FIGURE 10.54 Comparison of frequency responses at a large generator for a fault in the Ontario-Canada grid with no AGC and an SPV plant, with unity-power-factor PQ (C-SPV PQ) and voltage (C-SPV PV) controls, and distributed SPV (D-SPV) generators at about 10% penetration level. (From Tamimi B. et al., IEEE Transactions on Sustainable Energy, 4(3), 680–688, 2013. [37])

more, to study this issue, the best approach would be a Monte Carlo simulation approach in practice, to determine the systems frequency bounds and change patterns associated with variable RE generation, thus allowing to evaluate the impact of RE plant output changes on the frequency stability of the system. The effect of these changes can be ameliorated with power-droop and virtual-inertia controls, since with these, RE plants are able to control their power output, albeit somewhat limitedly depending on the available RE resources and plant limits; observe that virtual inertia will slow down the plant power ramps and hence result in smoother and hence better system frequency control, as shown in Section 9.5.3.

10.5.4 CONTROL STABILITY

Thus far, this chapter has focused on sinusoidal, fundamental-frequency stability phenomena, which can be explained and studied using fundamental-frequency phasor models and tools, including average converter models that capture the relevant and relatively large time constants of the stability issues discussed so far (in the order of fractions of a Hz to a few Hz). In traditional systems with low penetration of converter-interfaced systems, such as HVDC links, this approach has been sufficient for the most part to understand, define, and describe power system stability, using phasor-domain approximations to study sub-synchronous (tens of Hz but lower than fundamental frequency) and super-synchronous (higher than fundamental frequency) problems such as sub-synchronous resonance [18] and HVDC commutation failures [46], respectively. However, with the growing penetration of converter-interfaced RE generation, which are based on fast-switching voltage source converters (VSCs), such phasor-based models and techniques may not be sufficient for grid stability analyses. Thus, two particular relevant stability issues that have been identified in VSC-interfaced RE generation that require detailed modeling and simulation, including switching processes and controls associated with pulse-width modulation (PWM) control, are PLL synchronization [47] and resonance of VSC filters and controls, referred to as harmonic stability [48].

The problem of PLL synchronization in VSC-based wind-power generators is described in detail in Reference 47, where it is shown that during low voltage conditions at the RE generator terminals, due to nearby faults, the VSC PLL may fail to detect the zero crossings of the system voltage, thus failing to synchronize the generator with the system. This may result in inadequate current-control signals and converter current outputs, with the system failing to provide the required reactive power services during faults, as discussed in Section 9.5.2, and thus possibly resulting in RE generation tripping, which may lead to system stability problems due to the sudden loss of active and reactive power services from these generators.

The issue of harmonic stability, discussed in detail in Reference 48, is associated with interactions of the converter current and voltage controls that lead to super-synchronous oscillations on the converter current and voltage outputs in the order of hundreds of Hz to a few kHz, because of the inductive/capacitive characteristics of converters at these frequencies. These converter output fast oscillations may also appear due to, or be compounded by, interactions with the converter high-frequency LC/LCL filters and capacitances of plant cables, which resonate with the converter controls in their frequency range. This phenomenon may lead to sustained or growing super-synchronous oscillations on the converter voltage and current outputs, which are undesirable.

REFERENCES

1. IEEE-CIGRE, Joint Task Force on Stability Terms, Definitions, Kundur, P., Paserba, J., Ajjarapu, V., Andersson, G., Bose, A., Cañizares, C.A., Hatziargyriou, N., Hill D., Stankovic, A., Taylor, C., Van Cutsem, T., Vittal, V., Definition and classification of power system stability, *IEEE Transactions on Power Systems*, 19(3), 1387–1401, August 2004.
2. A Technical Reference Paper: Fault-Induced Delayed Voltage Recovery, Transmission Issues Subcommittee and System Protection and Control Subcommittee, NERC, June 2009.
3. Available transfer capability definitions and determination. Tech. Rep., NERC, USA, 1996.
4. Gao D. W., Muljadi E., Tian T., Miller M., Wang W. Comparison of standards and technical requirements of grid-connected wind power plants in China and the United States, NREL, NREL/TP-5D00-64225, September 2016.
5. Sourkounis C., Tourou P., Grid Code Requirements for Wind Power Integration in Europe, Conference Papers in Energy, Hindawi Publishing Corp., 1–9, 2013.
6. Rodriguez P., Candela I., Luna A., Control of PV generation systems using the synchronous power controller, IEEE Energy Conversion Congress and Exposition (ECCE), Denver, CO, US, 993–998, September 2013.
7. Fouad A. A., and Vittal V., Power System Transient Stability Using the Transient Energy Function Method, Prentice Hall, 1992.
8. Pavella, M., and Murthy P. G., Transient Stability of Power Systems, John Wiley and Sons, USA, 1994.
9. Pai M. A., Energy Function Analysis for Power System Stability, Kluwer Academic Publishers, USA, 1989.
10. Xue, Y., Wehenkel, L., Belhomme, R., Rousseaux, P., Pavella, M., Euxibie, E., Heilbronn, B., and Lesigne, J.-F., Extended equal area criterion revisited [EHV power systems], *IEEE Transactions on Power Systems*, 7(3), 1012–1022, August 1992.
11. Strang G., Introduction to Applied Mathematics, Wellesley-Cambridge Press, USA, 1986.
12. Power system dynamic analysis, phase I, EPRI Report EL-484, July 1977.
13. Pérez-Arriaga I. J., Verghese G. C., and Schweppe F. C., Selective modal analysis with applications to electric power systems. Part I: Heuristic introduction. Part II: The dynamic stability problem, *IEEE Transactions on Power Apparatus and Systems*, PAS-101(9), 3117–3134, September 1982.
14. Rogers G., Power System Oscillations, Kluwer Academic Publishers, USA, 2000.
15. Mithulananthan, N., Cañizares, C. A., Reeve, J., and Rogers, G. J., Comparison of PSSS, SVC and STATCOM Controllers for damping power system oscillations, *IEEE Transactions on Power Systems*, 18(2), 786–792, May 2003.
16. Anderson P. M., and Farmer R. G., Series Compensation of Power Systems, PBLSH!, USA, 1996.
17. Elmore, W. A. (ed.), Protective Relaying: Theory and Applications, Marcel Dekker, USA, 1994.
18. Kundur P., Power System Stability and Control, McGraw-Hill, USA, 1994.
19. Pagola, F. L., Pérez-Arriaga, I. J., and Verghese, G. C., On Sensitivities, residues and participations, *IEEE Transactions on Power Systems*, PWRS-4(1), 278–285, February 1989.

20. Rouco L., Eigenvalue-based methods for analysis and control of small-signal stability of power systems, IEE Colloquium on Power System Dynamics Stabilisation, University of Warwick, Coventry, England, February 1998.

21. Hingorani N. G., and Gyugyi L., Understanding FACTS, IEEE Press, USA, 2000.

22. Gama C., Brazilian north-south interconnection control-application and operating experience with a TCSC, *Proc. IEEE-PES Summer Meeting*, 2, 1103–1108, July 1999.

23. Smed, T., and Andersson, G., Utilizing HVDC to damp power oscillations, *IEEE Transactions on Power Delivery*, 8(2), 620–627, April 1993.

24. Taylor, C. W., Improving grid behaviour, *IEEE Spectrum*, 36(6), 40–45, June 1999.

25. Cañizares C. A. (ed.) Voltage stability assessment: Concepts, practices and tools, IEEE-PES Power Systems Stability Subcommittee Special Publication, SP101PSS, August 2002.

26. Milano F., An Open Source Power System Analysis Toolbox, *IEEE Transactions on Power Systems*, 20(3), 1199–1206, 2005. Available at http://www.power.uwaterloo.ca/downloads.htm.

27. Venkatasubramanian, V., Schuttler, H., and Zaborszky, J., Dynamics of large constrained nonlinear systems-a taxonomy theory, *Proceedings of the IEEE*, 83(11), 1530–1560, November 1995.

28. Rosehart, W., Roman C., and Schellenberg, A., Optimal power flow with complementarity constraints, *IEEE Transactions on Power Systems*, 20(2), 813–822, May 2005.

29. Cañizares, C. A., Mithulananthan, N., Berizzi, A., and Reeve, J., On the linear profile of indices for the prediction of saddle-node and limit-induced bifurcation points in power systems, *IEEE Transactions on Circuits and Systems-I*, 50(12), 1588–1595, December 2003.

30. Voltage stability criteria, undervoltage load shedding strategy, and reactive power reserve monitoring methodology. Tech. Rep., *WECC*, May 1998. Available at www.wecc.biz.

31. Greene S., Dobson I., and Alvarado F. L., Sensitivity of transfer capability margins with a fast formula, *IEEE Transactions on Power Systems*, 17(1), 34–40, February 2002.

32. U.S.–Canada Power System Outage Task Force, Final report on the August 14, 2003 blackout in the United States and Canada: Causes and recommendations. Tech. Rep., April 2004. Available at https://www.energy.gov/oe/downloads/blackout-2003-final-report-august-14-2003-blackout-united-states-and-canada-causes-and.

33. System Disturbance on 4 November 2006 – Final Report. Tech. Rep., UCTE, January 2007. Available at http://www.ucte.org/_library/otherreports/Final-Report-20070130.pdf.

34. Moné C., Smith A., Maples B., Hand M., 2013 Cost of Wind Energy Review, NREL, NREL/TP-5000-63267, February 2015.

35. Muñoz J. C., Cañizares C. A., Comparative stability analysis of DFIG-based wind farms and conventional synchronous generators, IEEE-PES Power Systems Conference and Exposition (PSCE), Phoenix, AZ, US, 1–7, March 2011.

36. Souxes T., Vournas C., System stability issues involving distributed sources under adverse network conditions, Bulk Power Systems Dynamics and Control X Symposium, IREP, Espinho, Portugal, 1–9, August 2017.

37. Tamimi B., Cañizares C. A., and Bhattacharya K., System stability impact of large-scale and distributed solar photovoltaic generation: The case of Ontario, Canada, *IEEE Transactions on Sustainable Energy*, 4 (3), 680–688, 2013.

38. Goi R., Krstulovic J., and Jakus D., Simulation of aggregate wind farm short-term production variations, *Renewable Energy*, 35, 2602–2609, 2010.

39. Holttinen H., Hourly Wind Power Variations in the Nordic Countries, *Wind Energy*, 8(2), 173–195, 2005.

40. Muñoz J. C., Cañizares C. A., Bhattacharya K., and Vaccaro A., An Affine Arithmetic Based Method for Voltage Stability Assessment of Power Systems with Intermittent Generation Resources, *IEEE Transactions on Power Systems*, 28(4), 4475–4487, 2013.

41. Mills A., Dark Shadows, *IEEE Power and Energy Magazine*, 9(3), 33–41, 2011.

42. Quintero J., Vittal V., Heydt G. T., and Zhang H., The impact of increased penetration of converter control-based generators on power system modes of oscillation, *IEEE Transactions on Power Systems*, 29(5), 2248–2257, 2014.

43. Muñoz J. C., Cañizares C. A., Bhattacharya K., Vaccaro A., Affine arithmetic based methods for voltage and transient stability assessment of power systems with intermittent generation sources, Bulk Power Systems Dynamics and Control IX Symposium, IREP, Rethymnon, Greece, 1–12, August 2013.

44. Muñoz J. C., Affine Arithmetic Based Methods for Power Systems Analysis Considering Intermittent Sources of Power, PhD Thesis, University of Waterloo, 2013.

45. Remon D., Canizares C. A., and Rodriguez P., Impact of 100-MW-scale PV plants with synchronous power controllers on power system stability in northern Chile, *IET Generation, Transmission & Distribution*, 11(11), 2958–2964, 2017.

46. Arrillaga J., Watson N. R., Computer Modeling of Electrical Power Systems, 2nd edition, John Willey, 2001.

47. Göksu Ö., Teodorescu R., Bak C. L., Iov F., Kjaer P. C., Instability of wind turbine converters during current injection to low voltage grid faults and PLL frequency based stability solution, *IEEE Transactions on Power Systems*, 29(4), 1683–1691, 2014.

48. Wang X., Blaabjerg F., and Wu W., Modeling and analysis of harmonic stability in an ac power-electronics-based power system, *IEEE Transactions on Power Electronics*, 29(12), 6421–6432, 2014.

49. Cañizares, C. A., Calculating optimal system parameters to maximize the distance to saddle-node bifurcations, *IEEE Transactions on Circuits and Systems-I*, 45(3), 225–237, March 1998.

50. Rosehart, W., Cañizares C. A., and Quintana, V., Multi-objective optimal power flows to evaluate voltage security costs, *IEEE Transactions on Power Systems*, 18(2), 578–587, May 2003.

51. Milano, F., Cañizares C. A., and Conejo, A. J., Sensitivity-based security-constrained OPF market clearing model, *IEEE Transactions on Power Systems*, 20(4), 2051–2060, November 2005.

52. Kodsi S., and Canizares C. A., Application of a stability-constrained optimal power flow to tuning of oscillation controls in competitive electricity markets, To appear in *IEEE Transactions on Power Systems*, 2007.

11 Three-Phase Power Flow and Harmonic Analysis

Wilsun Xu and Julio García-Mayordomo

CONTENTS

11.1   INTRODUCTION

The proliferation of power quality-sensitive loads in recent years has made power quality one of the major concerns for utility companies, manufacturers, and customers. Power quality refers to the characteristics of the power supply required to make electrical equipment work properly. Two of the important power quality issues are the imbalance of three-phase voltages and the distortion of sinusoidal voltage waveforms in the form of harmonics. In recent years, considerable efforts have been made to improve the management of power quality in power systems. The area of power quality analysis, especially that of three-phase and harmonic power flow algorithms, has experienced significant development. Well-accepted component models, solution techniques, and analysis procedures have been established. TPLF study and network harmonic analysis are becoming an important component of power system analysis and design.

TPLF study is used to determine the level of negative- and zero-sequence components, voltage, or current caused by the presence of unbalanced loads or network configurations in a power system. Harmonic analysis includes the study of waveform distortions caused by nonlinear loads and the determination of harmonic resonance conditions in a network. The first part of this chapter presents representative techniques for TPLF analysis. The second part explains the subject of power system harmonic analysis.

11.2   THREE-PHASE POWER FLOW

The objective of TPLF analysis is to determine the degree of voltage and current unbalance at the fundamental frequency. The unbalances are mainly expressed in terms of negative- and zero-sequence voltages at various locations of a power system. The existence of relative low levels of negative- and zero-sequence voltages in a network represents a good power quality behavior with respect to unbalances.

11.2.1   THE NEED FOR THREE-PHASE POWER FLOW STUDIES

Main low-frequency disturbances, which are produced under normal operation of power systems, can be classified into harmonics, imbalances, and flicker. Unbalances are related with the fundamental frequency and are mainly expressed in terms of negative- and zero-sequence voltages for every busbar of a power system. Hence, the existence of relative low levels of negative- and zero-sequence voltages in a network represents a good power quality behavior with respect to unbalances.

Main unbalance sources in transmission systems are ac arc furnaces [1] and high-speed trains. Moreover, the asymmetry in overhead lines produce a certain degree of unbalance but with an impact lower than the aforementioned unbalanced loads. Unbalances can be reduced by means of static var compensators [2].

On the other hand, the three-phase transformer connections and the more usual configurations of unbalanced loads make easier the propagation of negative-sequence component than the zero component. This explains the fact that most unbalance studies are focused in the generation and propagation of negative-sequence component. The existence of negative-sequence voltages produce overheating in rotating machines and displacements of zero crossings in static power converter, generating noncharacteristic harmonics. To control this negative impact, standards and recommendations set compatibility levels for unbalances. These compatibility levels are expressed in percent as the ratio between negative and positive voltages, and values of 2% (for low- and medium-voltage networks) and 1% (for high-voltage networks) have been proposed [3].

The TPLF is the most suitable tool to perform unbalance studies. This three-phase formulation can be considered as an extension of the well-known single-phase power flow (SPLF), where additional aspects must be included to reproduce different situations where unbalances are present. Moreover, the TPLF can be the first step of a more complex algorithm which is known as three-phase harmonic power flow.

11.2.1.1 State-of-the-Art Review

Today there is a lack of commercial packages for performing a true TPLF for transmission systems. To overcome this limitation, a combination of a conventional single-phase load flow with a conventional three-phase short-circuit program has been recently proposed [4]. However, a more detailed analysis is required to use TPLFs [5,6]. A significant part of these developments are oriented to set fundamental frequency TPLFs to be included as a part of the so "named" three-phase harmonic power flows [7–9], where the iterative process is solved applying the Newton method.

This problem is solved in [7] and [10] by using the classical residual error functions ΔP and ΔQ for each bus and for each phase. The main disadvantage of this procedure is the limited flexibility for representing unbalanced PQ loads. Namely, only wye-grounded PQ loads are considered. Under this condition, it is not possible to represent the most usual three-phase configurations for unbalanced loads in transmissions systems such as arc furnaces and high-speed trains. Arc furnaces must be considered as ungrounded loads with three wires while high-speed trains must be treated as ungrounded single loads with two wires connected between two line conductors [11,12].

Ungrounded PQ loads, in wye and delta connection, are treated in a TPLF algorithm [8]. In this algorithm, the unknowns are the sequence voltages in polar coordinates and the nonlinear equations are formulated in terms of balance currents or residual currents instead of residual powers. The main limitation of this procedure is the representation of the PV buses. They are defined by the positive complex voltage behind the machine's reactance. Therefore, the terminal voltage as well as the active power of the PV machine is not controlled.

A formulation based on current residuals has been recently proposed for single-phase [13] and three-phase [14] configurations. The current residuals are written in rectangular coordinates of the phase quantities. This results in a sparse structure of the Jacobian matrix when Newton method is used. The main drawback of this method is the representation of PV and PQ buses. PQ buses include only grounded PQ loads. On the other hand, PV buses are defined by specifying the active power and the voltage magnitude of each phase. Such procedure could be suitable for distribution networks. However, for transmission systems, the internal structure of the synchronous machines (positive sequence EMFs and positive, negative, and zero-sequence impedances) makes reasonable to specify only the total active power and the terminal voltage magnitude (of positive sequence) since both quantities can be controlled by acting on the magnitude and argument of the EMF (of positive sequence).

The aforementioned limitations of modeling PV and PQ buses are overcome through the method described in [9]. This procedure provides great flexibility in modeling any kind of unbalance at the fundamental frequency. This method not only uses constraints on nodal quantities but also constraints at the branch levels in a full Newton formulation. However, this leads to a very large Jacobian matrix with additional magnitudes considered as unknowns since not only bus voltages are unknowns but also branch currents of PQ loads, branch currents of the PV machines, and complex EMFs of the PV machines are unknowns.

11.2.1.2 Formulation Based on Current Residuals

The objective of this chapter is to present a compact and flexible formulation as described in [9] and [14], respectively, but with the advantages indicated in [15] and [16]. The main characteristics of this formulation are as follows:

1. Full Newton solution for the nonlinear equations based on current balance equations or current residuals.

2. Voltages and currents are expressed in rectangular coordinates and in sequence quantities to improve the sparsity of the Jacobian matrix.
3. Several PQ load configurations and their combinations are treated in a flexible way. Each PQ load is represented by voltage and power controlled current sources.
4. Constraints of PV buses coincide with the proposals of [9] and [17]. Namely, active power and terminal voltages, both of positive sequence, are specified.

These characteristics provide the following advantages:

- In accordance with characteristics 3 and 4, only bus voltages must be treated as unknowns. This leads to a significant reduction of the nonlinear equation number, in comparison with the method proposed in [9], but it maintains the flexibility of representing any kind of unbalance.
- Characteristic 2 involves using a formulation in sequence magnitudes. This formulation coincides with the practical format of most commercial software oriented to SPLFs and short-circuit studies where decoupled positive, negative, and zero-sequence networks are assumed. In this way, the inclusion of an untransposed line between buses k and m implies a coupling between sequences for the offdiagonal terms of the admittance matrix corresponding to buses k and m. However, for any other couple of buses p and q, which does not contain unbalanced lines, the offdiagonal terms, pq, do not present coupling between sequences, but they present coupling between phases. Under these conditions, and assuming a power flow study in a large network where only a subset of lines requires a detailed representation (unbalanced structure), the offdiagonal terms of the bus admittance matrix and the Jacobian matrix will be sparser in terms of symmetrical components. This property reduces the computation effort when a software package is used [18] to deal with the Jacobian matrix.

11.2.2 Bus Constraints

Let us assume a three-phase system with n buses and with the following notation:

- i and j are indices for sequence components 0, 1, and 2.
- r and x are subscripts for real and imaginary parts, respectively.
- $p, q = \{1, ..., n\}$, n being the total number of buses.
- $\mathcal{U}_{qj} = V_{rqj} + jV_{xqj}$ is the voltage at bus q and sequence j.
- $\mathcal{Y}_{pqij} = G_{pqij} + jB_{pqij}$ is the pq element of the nodal admittance matrix for sequences i and j.
- $\mathcal{I}_{pi} = I_{rpi} + jI_{xpi}$ is the current of sequence i leaving bus p. This current is produced by independent ac sources and PQ loads.

The formulation of the power flow assumes that the unknowns are the bus voltages $\mathcal{U}_{pi}$. Hence, there are six unknowns at each three-phase bus p corresponding to the real and imaginary parts of sequence voltages $\mathcal{U}_{p1}, \mathcal{U}_{p2}$, and $\mathcal{U}_{p0}$. It leads to $6n$ total unknowns for a system with n three-phase buses. Therefore, it is necessary to set $6n$ mismatch equations (residual equations) for solving the problem. If there are n_g buses with PV machines, $6n_g$ mismatch equations for PV buses and $6(n - n_g)$ mismatch equations for PQ buses must be formulated.

11.2.2.1 PQ Buses

Currents $\mathcal{I}_{pi}$ of PQ loads at bus p depend on sequence voltages at this bus, and they can be expressed, in a generic form, by

$$\mathcal{I}_{pi} = \mathcal{I}_{pi}(\mathcal{U}_{p1}, \mathcal{U}_{p2}, \mathcal{U}_{p0}, P_p^{\text{sp}}, Q_p^{\text{sp}}) \tag{11.1}$$

where P_p^{sp} and Q_p^{sp} are specified powers. The details of Equation 11.1 will be explained in other sections. With this, the application of Kirchhoff's laws at bus p and sequence i leads to

$$\Delta \mathcal{I}_{pi} = \mathcal{I}_{pi} + \sum_{q=1}^{n} \sum_{j=0}^{2} \mathcal{Y}_{pqij} \mathcal{U}_{qj} \tag{11.2}$$

The residual terms $\Delta \mathcal{I}_{pi}$ are equal to zero in the solution, namely, when the power flow calculations have been completed and therefore the bus voltages $\mathcal{U}_{qj}$ have reached their final values. Under this condition, a "current balance situation" between the currents $\mathcal{I}_{pi}$ of PQ loads and the currents of the linear elements (which form the bus admittance matrix) is satisfied. However, at the beginning of the iterative process, an estimation of the bus voltages is used by which certain discrepancies, $\Delta \mathcal{I}_{pi}$, appear between both currents. These terms $\Delta \mathcal{I}_{pi}$ define the current balance equations (also called as mismatch current equations or residual equations) and they must be forced to zero by means of an iterative process. Equation 11.2 can be expressed in terms of its real and imaginary parts as follows:

$$\begin{bmatrix} \Delta I_{rpi} \\ \Delta I_{xpi} \end{bmatrix} = \sum_{q=1}^{n} \sum_{j=0}^{2} \begin{bmatrix} G_{pqij} & -B_{pqij} \\ B_{pqij} & G_{pqij} \end{bmatrix} \begin{bmatrix} V_{rqj} \\ V_{xqj} \end{bmatrix} + \begin{bmatrix} I_{rpi} \\ I_{xpi} \end{bmatrix} \tag{11.3}$$

Taking into account the three sequences ($i = 0, 1, 2$) and the $(n - n_g)$ PQ buses, $6(n - n_g)$ mismatch equations are obtained from Equation 11.3.

11.2.2.2 PV Buses

The behavior of PV buses is strongly conditioned by the synchronous machines. These machines are represented by means of three decoupled sequence networks. The positive sequence network of a PV machine connected at bus p is formed by an internal EMF $\mathcal{E}_{p1}$ in series connected with an admittance $\mathcal{Y}_{p1}$. Both terms are substituted by the terminal voltage $\mathcal{U}_{p1}$ in this power flow formulation. The negative and zero-sequence networks are represented, respectively, by constant admittances $\mathcal{Y}_{p2}$ and $\mathcal{Y}_{p0}$, which are directly included in the nodal admittance matrix.

From a physical point of view, the internal EMF $\mathcal{E}_{p1}$ is adjusted to control the terminal voltage and to provide a specific active power. This power coincides with the positive sequence power P_{gp1} generated by the PV machine if, as usual, the real parts of admittances $\mathcal{Y}_{p2}$ and $\mathcal{Y}_{p0}$ are neglected. Therefore, PV buses are defined by the magnitude V_{p1}^{sp} of the complex voltage $\mathcal{U}_{p1}$ and the specified power P_{gp1}^{sp}, both of positive sequence, while the complex voltage $\mathcal{U}_{p1}$ is only specified in the particular case that bus p is the slack bus.) With these argumentations, the mismatch equations may be classified as follows:

- $4n_g$ mismatch equations for the negative ($i = 2$) and zero ($i = 0$) sequence currents. These equations present the form indicated in Equation 11.3. The positive sequence equations ΔI_{rp1} and ΔI_{xp1} are not considered for PV buses.
- Two mismatch equations for the slack busbar. In this case, the positive sequence voltage magnitude and phase are specified. It is equivalent to define the real and imaginary parts of complex voltage $\mathcal{U}_{p1}$. With this, the mismatch equations are

$$\Delta V_{rp1} = V_{rp1} - V_{rp1}^{sp}; \quad \Delta V_{xp1} = V_{xp1} - V_{xp1}^{sp} \tag{11.4}$$

- $2n_g - 2$ mismatch equations for the PV busbar. In this case, the positive active power P_{gp1} and the positive sequence voltage magnitude V_{p1} are specified. With this, the mismatch equations are

$$\Delta P_{gp1} = P_{gp1} - P_{gp1}^{sp}; \quad \Delta V_{p1}^2 = (V_{rp1}^2 + V_{xp1}^2) - (V_{p1}^{sp})^2 \tag{11.5}$$

The generated power P_{gp1} and the positive sequence power P_{dp1} demanded by a local PQ load at bus p are related by

$$\frac{1}{3}(P_{gp1} - P_{dp1}) = \text{Re}\left\{ \sum_{q=1}^{n} \sum_{j=0}^{2} \mathcal{Y}_{pq1j}^* \mathcal{U}_{qj}^* \mathcal{U}_{p1} \right\} \tag{11.6}$$

and in terms of real and imaginary parts

$$\frac{1}{3}(P_{gp1} - P_{dp1}) = \sum_{q=1}^{n} \sum_{j=0}^{2} [G_{pq1j}(V_{rp1}V_{rqj} + V_{xp1}V_{xqj}) + B_{pq1j}(V_{xp1}V_{rqj} - V_{rp1}V_{xqj})] \tag{11.7}$$

where the power P_{dp1} is equivalent to

$$\frac{1}{3}P_{dp1} = \text{Re}\left\{ \mathcal{U}_{p1}\mathcal{I}_{p1}^* \right\} = V_{rp1}I_{rp1} + V_{xp1}I_{xp1} \tag{11.8}$$

In a general case, the term P_{dp1} cannot be specified since it depends on the current components I_{rp1} and I_{xp1}. These components and their sensitivities with respect to the voltages can be evaluated after considering the analytical expression (Equation 11.1). It will be discussed in other sections.

11.2.3 NEWTON–RAPHSON SOLUTION

Simultaneous solution of Equations 11.3 through 11.5 is performed by means of the Newton–Raphson (NR) algorithm. It requires an exact evaluation of terms of the Jacobian matrix. If bus p is a PQ bus, the two rows corresponding to bus p and sequence i of the NR algorithm can be expressed by

$$\begin{bmatrix} \Delta I_{rpi} \\ \Delta I_{xpi} \end{bmatrix} = -\sum_{q=1}^{n} \sum_{j=0}^{2} \begin{bmatrix} H_{pqij} & N_{pqij} \\ J_{pqij} & L_{pqij} \end{bmatrix} \begin{bmatrix} \Delta V_{rqj} \\ \Delta V_{xqj} \end{bmatrix} \tag{11.9}$$

where

$$H_{pqij} = \frac{\partial \Delta I_{rpi}}{\partial V_{rqj}} = G_{pqij}; \quad N_{pqij} = \frac{\partial \Delta I_{rpi}}{\partial V_{xqj}} = -B_{pqij} \tag{11.10}$$

$$J_{pqij} = \frac{\partial \Delta I_{xpi}}{\partial V_{rqj}} = B_{pqij}; \quad L_{pqij} = \frac{\partial \Delta I_{xpi}}{\partial V_{xqj}} = G_{pqij} \tag{11.11}$$

For a given value of p and q and for the three sequences ($i, j = 0, 1, 2$), the terms $H, N, J,$ and L of Equation 11.9 form a 6×6 block in the Jacobian matrix. These elements are identical to the corresponding elements of the bus admittance matrix. Therefore, they must not be updated during the iterative process. Moreover, they describe the coupling between different busbars and different sequences. In large power systems, these terms represent an important part of the Jacobian. Therefore, the number of nonzero elements of the Jacobian can be reduced if the admittances are expressed in symmetrical components.

In the particular case corresponding to the diagonal elements ($p = q$), the terms of the Jacobian matrix must include the effect of the PQ loads. Therefore, Equations 11.10 and 11.11 are transformed into

$$H_{ppij} = \frac{\partial \Delta I_{rpi}}{\partial V_{rpj}} = G_{ppij} + \frac{\partial I_{rpi}}{\partial V_{rpj}}; \quad N_{ppij} = \frac{\partial \Delta I_{rpi}}{\partial V_{xpj}} = -B_{ppij} + \frac{\partial I_{rpi}}{\partial V_{xpj}} \tag{11.12}$$

$$J_{ppij} = \frac{\partial \Delta I_{xpi}}{\partial V_{rpj}} = B_{ppij} + \frac{\partial I_{xpi}}{\partial V_{rpj}}; \quad L_{ppij} = \frac{\partial \Delta I_{xpi}}{\partial V_{xpj}} = G_{ppij} + \frac{\partial I_{xpi}}{\partial V_{xpj}} \tag{11.13}$$

The sensitivities of the currents I_{rpi} and I_{xpi} with respect to the voltages V_{rpi} and V_{xpi} depend on the models selected for PQ loads. These sensitivities are functions of the voltages V_{rpi} and V_{xpi}. Details of these terms are described in the following sections.

If bus p is the slack bus, for $i = 1$, the terms ΔI_{rp1} and ΔI_{xp1} of Equation 11.9 must be substituted by ΔV_{rp1} and ΔV_{xp1}, respectively. Hence, according to Equation 11.4, the only nonzero elements H_{pp11} and L_{pp11} of Equation 11.9 are

$$H_{pp11} = \frac{\partial \Delta V_{rp1}}{\partial V_{rp1}} = 1; \quad L_{pp11} = \frac{\partial \Delta V_{xp1}}{\partial V_{xp1}} = 1 \tag{11.14}$$

If bus p is a PV bus, for $i = 1$, the terms ΔI_{rp1} and ΔI_{xp1} of Equation 11.9 must be substituted by ΔV_{p1}^2 and ΔP_{gp1}, respectively. In accordance with Equation 11.5, the contribution of ΔV_{p1}^2 to the nonzero elements H_{pp11} and N_{pp11} is

$$H_{pp11} = \frac{\partial \Delta V_{p1}^2}{\partial V_{rp1}} = 2V_{rp1}; \quad N_{pp11} = \frac{\partial \Delta V_{p1}^2}{\partial V_{xp1}} = 2V_{xp1} \tag{11.15}$$

while the contribution of ΔP_{gp1} to the nonzero elements J_{pq1j} and L_{pq1j} is

$$J_{pq1j} = \frac{\partial \Delta P_{gp1}}{\partial V_{rqj}}; \quad L_{pq1j} = \frac{\partial \Delta P_{gp1}}{\partial V_{xqj}} \tag{11.16}$$

In accordance with Equations 11.5 through 11.8 and Equation 11.16, we have

$$\frac{1}{3}J_{pq1j} = G_{pq1j}V_{rp1} + B_{pq1j}V_{xp1} + \left(\frac{1}{3}\frac{\partial \Delta P_{dp1}}{\partial V_{rpj}} \right) \tag{11.17}$$

$$\frac{1}{3}L_{pq1j} = G_{pq1j}V_{xp1} - B_{pq1j}V_{rp1} + \left(\frac{1}{3}\frac{\partial \Delta P_{dp1}}{\partial V_{xpj}} \right) \tag{11.18}$$

Equations 11.17 and 11.18 are valid in all cases except for $p = q$ and $j = 1$. The terms in brackets in Equations 11.17 and 11.18 are only considered when $p = q$. For $p = q$ and $j = 1$, we have

$$\frac{1}{3}J_{pp11} = \frac{1}{3}\frac{\partial \Delta P_{dp1}}{\partial V_{rp1}} + G_{pp11}V_{rp1} + B_{pp11}V_{xp1} + \sum_{q=1}^{n}\sum_{j=0}^{2}(G_{pq1j}V_{rqj} - B_{pq1j}V_{xqj}) \tag{11.19}$$

$$\frac{1}{3}L_{pp11} = \frac{1}{3}\frac{\partial \Delta P_{dp1}}{\partial V_{xp1}} + G_{pp11}V_{xp1} - B_{pp11}V_{rp1} + \sum_{q=1}^{n}\sum_{j=0}^{2}(G_{pq1j}V_{xqj} + B_{pq1j}V_{rqj}) \tag{11.20}$$

$$\frac{1}{3}\frac{\partial \Delta P_{dp1}}{\partial V_{rpj}} = V_{rp1}\frac{\partial I_{rp1}}{\partial V_{rpj}} + V_{xp1}\frac{\partial I_{xp1}}{\partial V_{rpj}} + (I_{rp1}) \tag{11.21}$$

$$\frac{1}{3}\frac{\partial \Delta P_{dp1}}{\partial V_{xpj}} = V_{rp1}\frac{\partial I_{rp1}}{\partial V_{xpj}} + V_{xp1}\frac{\partial I_{xp1}}{\partial V_{xpj}} + (I_{xp1}) \tag{11.22}$$

The terms in brackets in Equations 11.21 and 11.22 are only considered when $j = 1$. The sensitivities of currents I_{rp1} and I_{xp1} with respect to voltages V_{rpj} and V_{rpj} are calculated in accordance with the description of PQ loads in Section 11.2.4.

On the other hand, it is important to note that, for the negative-($i = 2$) and zero-($i = 0$) sequence components, Equations 11.9 through 11.13 must be applied to slack bus and PV busbars.

This Newton's formulation has been implemented as a program, which is able to read files of massive input data corresponding to power flow and fault analysis used by most Spanish utilities. The program uses a sparse matrix package [18], which takes advantage of the aforementioned structure of the Jacobian matrix. The procedure can be explained in accordance with the following steps:

1. Set initial values for sequence voltages at all busbars.
2. Calculate mismatch current equations for PQ buses, mismatch power and voltage equations for PV buses, and mismatch voltage equations for the slack busbar.
3. Evaluate the terms of the Jacobian matrix.
4. Update sequence voltages at all busbars by means of the Newton algorithm. If all incremental values of the sequence voltages are less than a given tolerance, go to step 5. Otherwise go to step 2.
5. Print the results.

11.2.4 CONSTANT POWER COMPONENTS

Constant impedance loads of any structure, capacitors, and filters are included in the bus admittance matrix. The rest of the shunt elements form the current demand $\mathcal{I}_{pi}$ at bus p of Equation 11.1. To simplify the notation, the subscript p will be omitted from here on. A generalized current demand is presented in Figure 11.1. Here, the existence of six different types of elements acting simultaneously is considered. The simplest element corresponds with the set of independent current sources defined as type 6. Type 6 current sources can be specified or used as a link between a fundamental power flow and an iterative harmonic analysis [8]. The rest of the elements of Figure 11.1 present a nonlinear relationship of the current with the terminal voltage.

The load modeling issue has been explained in Section 7.7 of Chapter 7. The objective of this section is to obtain analytical expressions of these relations for the first five elements in terms of sequence components in rectangular coordinates. The sequence components of voltage $\mathcal{U}_a$ and current $\mathcal{I}_a$ can be expressed by

$$\mathcal{U}_0 = V_{r0} + jV_{x0}; \quad \mathcal{U}_1 = V_{r1} + jV_{x1}; \quad \mathcal{U}_2 = V_{r2} + jV_{x2} \tag{11.23}$$

$$\mathcal{I}_0 = I_{r0} + jI_{x0}; \quad \mathcal{I}_1 = I_{r1} + jI_{x1}; \quad \mathcal{I}_2 = I_{r2} + jI_{x2} \tag{11.24}$$

In accordance with this scheme, the current demand of PQ loads of types 1, 2, 3, and 4 can be expressed by

$$\mathcal{I}_i^{(1)} = f_1(\mathcal{U}_0, \mathcal{U}_1, \mathcal{U}_2, P_{t1}^{sp}, Q_{t1}^{sp}); \quad i = 0, 1, 2 \tag{11.25}$$

$$\mathcal{I}_i^{(2)} = f_2(\mathcal{U}_0, \mathcal{U}_1, \mathcal{U}_2, P_a^{sp}, Q_a^{sp}, P_b^{sp}, Q_b^{sp}, P_c^{sp}, Q_c^{sp}); \quad i = 0, 1, 2 \tag{11.26}$$

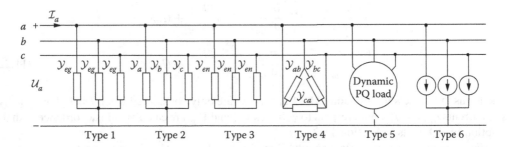

FIGURE 11.1   Generalized current demand at a bus.

$$\mathcal{I}_i^{(3)} = f_3(\mathcal{U}_1, \mathcal{U}_2, P_{t3}^{\text{sp}}, Q_{t3}^{\text{sp}}); \quad i = 1, 2 \tag{11.27}$$

$$\mathcal{I}_i^{(4)} = f_4(\mathcal{U}_1, \mathcal{U}_2, P_{ab}^{\text{sp}}, Q_{ab}^{\text{sp}}, P_{bc}^{\text{sp}}, Q_{bc}^{\text{sp}}, P_{ca}^{\text{sp}}, Q_{ca}^{\text{sp}}); \quad i = 1, 2 \tag{11.28}$$

where the terms P and Q represent the specified active and reactive powers according to the power flow constraints. Specific characteristics of each load type are:

- Load type 1 is grounded and structurally balanced, namely, the admittances $\mathcal{Y}_{eg}$ present the same value for each sequence and each branch. Under this condition, the total three-phase active P_{t1}^{sp} and reactive Q_{t1}^{sp} powers must be only specified. Therefore, current $\mathcal{I}_i^{(1)}$ of sequence i depends on the three sequence voltages and on the parameters P_{t1}^{sp} and Q_{t1}^{sp} as indicated in Equation 11.25.
- Load type 2 is grounded and structurally unbalanced, and shunt admittances are different in each phase. Hence, the active and reactive power of each branch must be specified. Therefore, current $\mathcal{I}_i^{(2)}$ of sequence i depend on the three sequence voltages and on the parameters $P_a^{\text{sp}}, Q_a^{\text{sp}}, P_b^{\text{sp}}, Q_b^{\text{sp}}, P_c^{\text{sp}}, Q_c^{\text{sp}}$ as indicated in Equation 11.26.
- Load type 3 is ungrounded and structurally balanced, namely, the admittances $\mathcal{Y}_{en}$ present the same value for each branch and for positive and negative sequences. Under this condition, the total three-phase active P_{t3}^{sp} and reactive Q_{t3}^{sp} powers must be only specified. Therefore, current $\mathcal{I}_i^{(3)}$ of sequence i depends on the positive and negative sequence voltages and on the parameters P_{t3}^{sp} and Q_{t3}^{sp} as indicated in Equation 11.27.
- Load type 4 is ungrounded and structurally unbalanced, and shunt admittances are different in each branch. Hence, the active and reactive power of each branch must be specified. Therefore, current $\mathcal{I}_i^{(4)}$ of sequence i depend on the positive and negative sequence voltages and on the parameters $P_{ab}^{\text{sp}}, Q_{ab}^{\text{sp}}, P_{bc}^{\text{sp}}, Q_{bc}^{\text{sp}}, P_{ca}^{\text{sp}}, Q_{ca}^{\text{sp}}$ as indicated in Equation 11.28.

It is important to note that a generalized PQ load should include a combination of load types 1, 2, 3, and 4. That means setting 16 values for the aforementioned power parameters at each bus where only six variables (corresponding to the real and the imaginary parts of three-phase voltages) are defined. This explains the fact by which most previous TPLF formulations use only load type 2 (with six power specifications), where the power flow algorithm is defined in terms of power residuals instead of current residuals.

Load type 5 represents the behavior of rotating machine loads with unequal negative- and positive impedances, as in the case of induction or synchronous machines. For this load, it is assumed that the negative $\mathcal{Y}_2$ and zero $\mathcal{Y}_0$ sequence admittances are known and they are included in the bus admittance matrix, namely

$$\mathcal{I}_2^{(5)} = \mathcal{Y}_2 \mathcal{U}_2; \quad \mathcal{I}_0^{(5)} = \mathcal{Y}_0 \mathcal{U}_0 \tag{11.29}$$

The positive sequence admittance is not known and is to be determined from the three-phase active P_1^{sp} and reactive Q_1^{sp} power specifications of positive sequence, or total active P^{sp} and reactive Q^{sp} power specifications (including the three sequences). Therefore, the positive sequence current of load type 5 can by expressed by one of the following equations:

$$\mathcal{I}_1^{(5)} = f_5(\mathcal{U}_1, P_1^{\text{sp}}, Q_1^{\text{sp}}); \quad \mathcal{I}_1^{(5)} = f_5(\mathcal{U}_1, P^{\text{sp}}, Q^{\text{sp}}) \tag{11.30}$$

Equations 11.25 through 11.28 and Equation 11.30 form a set of nonlinear functions which describe the behavior of a PQ load in a general case. Hence, the total current $\mathcal{I}_i$ of Equation 11.1 can be expressed by

$$\mathcal{I}_i = \sum_{s=1}^{6} \mathcal{I}_i^{(s)} \tag{11.31}$$

The sensitivities of the currents with respect to the voltages in Equations 11.12 and 11.13 can be obtained by summing the individual contribution of each element. For instance

$$\frac{\partial I_{ri}}{\partial V_{xj}} = \sum_{s=1}^{5} \frac{\partial I_{ri}^{(s)}}{\partial V_{xj}} \tag{11.32}$$

11.2.4.1 Balanced Loads

Admittances $\mathcal{Y}_{eg}$ and $\mathcal{Y}_{en}$ of load types 1 and 3 of Figure 11.1 are related with powers and voltages as follows:

$$P_{t1}^{sp} - jQ_{t1}^{sp} = \mathcal{Y}_{eg}(V_a^2 + V_b^2 + V_c^2); \quad P_{t3}^{sp} - jQ_{t3}^{sp} = \mathcal{Y}_{en}(V_{an}^2 + V_{bn}^2 + V_{cn}^2) \tag{11.33}$$

where

$$V_{012}^2 = \frac{V_a^2 + V_b^2 + V_c^2}{3} = V_0^2 + V_1^2 + V_2^2; \quad V_{12}^2 = \frac{V_{an}^2 + V_{bn}^2 + V_{cn}^2}{3} = V_1^2 + V_2^2 \tag{11.34}$$

Using Equations 11.33 and 11.34, sequence currents of load types 1 and 3 can be expressed in function of sequence voltages by

$$\mathcal{I}_i^{(1)} = \mathcal{Y}_{eg}\mathcal{U}_i = \frac{P_{t1}^{sp} - jQ_{t1}^{sp}}{3V_{012}^2}\mathcal{U}_i \quad \text{with } i = 0, 1, 2 \tag{11.35}$$

$$\mathcal{I}_i^{(3)} = \mathcal{Y}_{en}\mathcal{U}_i = \frac{P_{t3}^{sp} - jQ_{t3}^{sp}}{3V_{12}^2}\mathcal{U}_i \quad \text{with } i = 1, 2 \tag{11.36}$$

The sensitivities of current $\mathcal{I}_i^{(1)}$ with respect to the sequence voltages provide a block of 36 elements, while for current $\mathcal{I}_i^{(3)}$, only 16 terms are obtained. For instance, the sensitivity of the real part of the positive current $\mathcal{I}_i^{(3)}$ with respect to the imaginary part of the negative-sequence voltage $\mathcal{U}_2$ is given by

$$\frac{\partial \mathcal{I}_{r1}^{(3)}}{\partial V_{x2}} = \frac{-2V_{x2}(P_{t3}^{sp}V_{r1} + Q_{t3}^{sp}V_{x1})}{3V_{12}^4} \tag{11.37}$$

Similar expressions are obtained for the rest of the derivatives. For space reasons, these derivatives are omitted here.

11.2.4.2 Grounded Unbalanced Loads

Load type 2 of Figure 11.1 presents three different phase admittances $\mathcal{Y}_a$, $\mathcal{Y}_b$, and $\mathcal{Y}_c$. These admittances can be expressed as functions of phase specified powers and phase voltages. With this formulation, the phase current is

$$\mathcal{I}_\nu = \mathcal{Y}_\nu\mathcal{U}_\nu = \frac{P_\nu^{sp} - jQ_\nu^{sp}}{V_\nu^2}\mathcal{U}_\nu \quad \text{with} \quad \nu = a, b, c \tag{11.38}$$

The sequence current $\mathcal{I}_i^{(2)}$ ($i = 0, 1, 2$) of load type 2 is obtained by application to $\mathcal{I}_\nu$ of the transformation in symmetrical components. On the other hand, the sensitivities of the phase currents with respect to the phase voltages are

$$L_\nu = \frac{\partial I_{r\nu}}{\partial V_{r\nu}} = -\frac{\partial I_{x\nu}}{\partial V_{x\nu}} = \frac{P_\nu^{sp}(V_{x\nu}^2 - V_{r\nu}^2) - 2Q_\nu^{sp}V_{r\nu}V_{x\nu}}{V_\nu^4} \tag{11.39}$$

$$N_\nu = \frac{\partial I_{r\nu}}{\partial V_{x\nu}} = \frac{\partial I_{x\nu}}{\partial V_{r\nu}} = \frac{Q_\nu^{sp}(V_{r\nu}^2 - V_{x\nu}^2) - 2P_\nu^{sp}V_{r\nu}V_{x\nu}}{V_\nu^4} \tag{11.40}$$

These sensitivities can be grouped inside a matrix $[H_\nu]$, so that

$$[\Delta I_\nu] = [H_\nu][\Delta V_\nu] \tag{11.41}$$

where

$$[\Delta I_\nu] = [\Delta I_{ra}, \Delta I_{xa}, \Delta I_{rb}, \Delta I_{xb}, \Delta I_{rc}, \Delta I_{xc}] \tag{11.42}$$

$$[\Delta U_\nu] = [\Delta V_{ra}, \Delta V_{xa}, \Delta V_{rb}, \Delta V_{xb}, \Delta V_{rc}, \Delta U_{xc}] \tag{11.43}$$

Matrix $[H_\nu]$ is diagonal by blocks of 2×2 elements. For instance, for phase ν we have a block with the following form:

$$\begin{bmatrix} L_\nu & N_\nu \\ N_\nu & -L_\nu \end{bmatrix} \tag{11.44}$$

The application to Equation 11.41 of the transformation in symmetrical components yields

$$[\Delta I_{\nu s}] = [H_{\nu s}][\Delta V_{\nu s}]; \quad [H_{\nu s}] = [T]^{-1}[H_\nu][T] \tag{11.45}$$

$$[\Delta I_{\nu s}] = [\Delta I_{r0}, \Delta I_{x0}, \Delta I_{r1}, \Delta I_{x1}, \Delta I_{r2}, \Delta I_{x2}] \tag{11.46}$$

$$[\Delta U_{\nu s}] = [\Delta V_{r0}, \Delta V_{x0}, \Delta V_{r1}, \Delta V_{x1}, \Delta V_{r2}, \Delta U_{x2}] \tag{11.47}$$

$$[T] = \frac{1}{2} \begin{bmatrix} 2 & 0 & 2 & 0 & 2 & 0 \\ 0 & 2 & 0 & 2 & 0 & 2 \\ 2 & 0 & -1 & \sqrt{3} & -1 & -\sqrt{3} \\ 0 & 2 & -\sqrt{3} & -1 & \sqrt{3} & -1 \\ 2 & 0 & -1 & -\sqrt{3} & -1 & \sqrt{3} \\ 0 & 2 & \sqrt{3} & -1 & -\sqrt{3} & -1 \end{bmatrix} \tag{11.48}$$

Matrix $[H_{\nu s}]$ contains the sensitivities of the sequence currents $I_{ri}^{(2)}$ and $I_{xi}^{(2)}$ with respect to the sequence voltages V_{rj} and V_{xj} for $i = 0,1,2$.

11.2.4.3 Ungrounded Unbalanced Loads

Load type 4 of Figure 11.1 presents three different branch admittances $\mathcal{Y}_{ab}$, $\mathcal{Y}_{bc}$, and $\mathcal{Y}_{ca}$. These admittances can be expressed in function of branch specified powers and branch voltages. With this formulation, the branch current is

$$\mathcal{I}_\mu = \mathcal{Y}_\mu \mathcal{U}_\mu = \frac{P_\mu^{sp} - jQ_\mu^{sp}}{V_\mu^2} U_\mu \quad \text{with} \quad \mu = ab, bc, ca \tag{11.49}$$

$$\mathcal{I}_a = \mathcal{I}_{ab} - \mathcal{I}_{ca}; \quad \mathcal{I}_b = \mathcal{I}_{bc} - \mathcal{I}_{ab}; \quad \mathcal{I}_c = \mathcal{I}_{ca} - \mathcal{I}_{bc} \tag{11.50}$$

The sequence current $\mathcal{I}_i^{(4)}$ ($i = 1, 2$) of load type 4 is obtained after the application to $\mathcal{I}_\nu$ ($\nu = a, b, c$) in Equation 11.50 of the transformation in symmetrical components.

On the other hand, the sensitivities of branch currents $I_{r\mu}$ and $I_{x\mu}$ with respect to branch voltages $V_{r\mu}$ and $V_{x\mu}$ present an identical form as indicated by Equations 11.39 and 11.40 for load type 2. These sensitivities can be grouped inside a matrix $[H_\mu]$, so that

$$[\Delta I_\mu] = [H_\mu][\Delta V_\mu] \tag{11.51}$$

where

$$[\Delta I_\mu] = [\Delta I_{rab}, \Delta I_{xab}, \Delta I_{rbc}, \Delta I_{xbc}, \Delta I_{rca}, \Delta I_{xca}] \tag{11.52}$$

$$[\Delta U_\mu] = [\Delta V_{rab}, \Delta V_{xab}, \Delta V_{rbc}, \Delta V_{xbc}, \Delta V_{rca}, \Delta U_{xca}] \tag{11.53}$$

The application to Equation 11.51 of the transformation in symmetrical components yields

$$[\Delta I_{\mu s}] = [H_{\mu s}][\Delta U_{\mu s}]; \quad [H_{\mu s}] = [T]^{-1}[H_\mu][T] \tag{11.54}$$

where

$$[\Delta I_{\mu s}] = [\Delta I_{rab0}, \Delta I_{xab0}, \Delta I_{rab1}, \Delta I_{xab1}, \Delta I_{rab2}, \Delta I_{xab2}] \tag{11.55}$$

$$[\Delta U_{\mu s}] = [\Delta V_{rab0}, \Delta V_{xab0}, \Delta V_{rab1}, \Delta V_{xab1}, \Delta V_{rab2}, \Delta U_{xab2}] \tag{11.56}$$

Terms ΔV_{rab0} and ΔV_{xab0} are equal to zero since branch voltages do not contain homopolar components. With this, Equation 11.54 can be reduced to

$$[\Delta I'_{\mu s}] = [H'_{\mu s}][\Delta V'_{\mu s}] \tag{11.57}$$

The vectors $[\Delta I'_{\mu s}]$ and $[\Delta V'_{\mu s}]$ coincide, respectively, with those of Equations 11.59 and 11.60 if the homopolar component is eliminated. Matrix $[H'_{\mu s}]$ is a submatrix of $[H_{\mu s}]$ formed by the lower block of 4×4 elements of $[H_{\mu s}]$. On the other hand, the relations (in symmetrical components) between branch magnitudes $\mathcal{U}_{ab}$, $\mathcal{I}_{ab}$ and phase magnitudes $\mathcal{U}_a$, $\mathcal{I}_a$, are given by

$$[\Delta V'_{\mu s}] = [T_{ab}][\Delta V'_{\nu s}]; \quad [\Delta I'_{\mu s}] = \frac{1}{3}[T_{ab}][\Delta I'_{\nu s}] \tag{11.58}$$

$$[T_{ab}] = \frac{\sqrt{3}}{2}\begin{bmatrix} \sqrt{3} & -1 & 0 & 0 \\ 1 & \sqrt{3} & 0 & 0 \\ 0 & 0 & \sqrt{3} & 1 \\ 0 & 0 & -1 & \sqrt{3} \end{bmatrix} \tag{11.59}$$

Vectors $[\Delta I'_{\nu s}]$ and $[\Delta V'_{\nu s}]$ coincide, respectively, with vectors indicated in Equations 11.46 and 11.47 if the homopolar component is eliminated. With this, we have

$$[\Delta I'_{\nu s}] = [H'_{\nu s}][\Delta V'_{\nu s}]; \quad [H'_{\nu s}] = 3[T_{ab}]^{-1}[H'_{\mu s}][T_{ab}] \tag{11.60}$$

Matrix $[H'_{\nu s}]$ contains the sensitivities of the sequence currents $I_{ri}^{(4)}$ and $I_{xi}^{(4)}$ with respect to the sequence voltages V_{rj} and V_{xj} for $i = 1, 2$.

11.2.4.4 Rotating Loads

Load type 5 of Figure 11.1 presents a balanced structure and it can be represented by means of three sequence admittances $\mathcal{Y}_1$, $\mathcal{Y}_2$, and $\mathcal{Y}_0$. Admittances $\mathcal{Y}_2$ and $\mathcal{Y}_0$ have a fixed value and they present a current–voltage relation as indicated in Equation 11.29. However, the admittance $\mathcal{Y}_1$ of positive sequence does not present a fixed value since it depends on voltage and powers (of positive sequence) according to

$$\mathcal{Y}_1 = \frac{P_1 - jQ_1}{3V_1^2} \tag{11.61}$$

Hence, the positive current is

$$\mathcal{I}_1^{(5)} = \frac{P_1 - jQ_1}{3V_1^2}\mathcal{U}_1 \tag{11.62}$$

where

$$P_1 = P^{\text{sp}}; \quad Q_1 = Q^{\text{sp}} \tag{11.63}$$

if the three-phase active P_1^{sp} and reactive Q_1^{sp} powers are specified, or

$$P_1 = P^{\text{sp}} - 3(G_2V_2^2 + G_0V_0^2); \quad Q_1 = Q^{\text{sp}} + 3(B_2V_2^2 + B_0V_0^2) \tag{11.64}$$

if the total active P^{sp} and reactive Q^{sp} powers (including the three sequences) are specified. Terms G_2 and G_0 are the conductances and B_2 and B_0 are the susceptances of admittances $\mathcal{Y}_2$ and $\mathcal{Y}_0$, respectively. With this, the combination of Equations 11.62 and 11.63 or Equation 11.64 provides the positive current of load type 5.

The sensitivities of the positive current with respect to the positive voltage (in terms of real and imaginary parts) are

$$\frac{\partial I_{r1}^{(5)}}{\partial V_{r1}} = -\frac{\partial I_{x1}^{(5)}}{\partial V_{x1}} = \frac{P_1(V_{x1}^2 - V_{r1}^2) - 2Q_1V_{r1}V_{x1}}{3V_1^4} \tag{11.65}$$

$$\frac{\partial I_{r1}^{(5)}}{\partial V_{x1}} = \frac{\partial I_{x1}^{(5)}}{\partial V_{r1}} = \frac{Q_1(V_{r1}^2 - V_{x1}^2) - 2P_1V_{r1}V_{x1}}{3V_1^4} \tag{11.66}$$

If the total active P^{sp} and reactive Q^{sp} powers (including the three sequences) are specified, the sensitivities of the positive current with respect to the negative- and zero-sequence voltages (in terms of real and imaginary parts) must be added, so that

$$\frac{1}{V_{r2}}\frac{\partial I_{r1}^{(5)}}{\partial V_{r2}} = \frac{1}{V_{x2}}\frac{\partial I_{r1}^{(5)}}{\partial V_{x2}} = -2\frac{G_2V_{r1} - B_2V_{x1}}{V_1^2} \tag{11.67}$$

$$\frac{1}{V_{r2}}\frac{\partial I_{x1}^{(5)}}{\partial V_{r2}} = \frac{1}{V_{x2}}\frac{\partial I_{x1}^{(5)}}{\partial V_{x2}} = -2\frac{G_2V_{x1} + B_2V_{r1}}{V_1^2} \tag{11.68}$$

Similar expressions can be derived for the sensitivities of the positive current with respect to the zero-sequence voltage.

11.2.5 REACTIVE POWER LIMITS

A PV machine that operates within the reactive power limits has been indicated in Section 11.2.2. If the reactive power limits ($Q_g^{\min}$, $Q_g^{\max}$) are violated, the PV machine can be considered as load type 5. In accordance with Equation 11.63, load type 5 includes specified powers of positive sequence. If the total active P^{sp} and reactive Q^{sp} powers (including the three sequences) are specified, the effect of negative- and zero-sequence reactive powers of Equation 11.64 is taken into account by means of the susceptances B_2 and B_0, while the conductances G_2 and G_0 are ignored since the resistances of the PV machine are negligible in comparison with the reactances.

The discrimination between PV and PQ operation of a PV machine must be performed at each iteration by means of the generated reactive power Q_{gp} at bus p. If the total active P^{sp} and reactive

Q^{sp} powers (including the three sequences) are specified to represent the PQ operation of the PV machine, the term Q_{gp} is

$$Q_{gp} = Q_{gp1} + 3(B_{p2}V_{p2}^2 + B_{p0}V_{p0}^2) \tag{11.69}$$

where, the effect of the negative- and zero-sequence components on the generated power of PV machine is taken into account. The positive component Q_{gp1} is determined from the local demand and from the reactive power which flows across the network. Therefore, by similar notation to that used in Equation 11.6 and 11.8, the positive sequence component Q_{gp1} can be expressed by

$$\frac{1}{3}Q_{gp1} = \text{Im}\left\{\mathcal{U}_{p1}\mathcal{I}_{p1}^* + \sum_{q=1}^{n}\sum_{j=0}^{2}\mathcal{Y}_{pq1j}^*\mathcal{U}_{qj}^*\mathcal{U}_{p1}\right\} \tag{11.70}$$

If the positive active P_1^{sp} and reactive Q_1^{sp} powers are specified for representing the PQ operation of the PV machine, only the term Q_{gp1} of Equation 11.69 must be evaluated. As in single-phase load flows, the transition from PV to PQ operation is activated if $Q_{gp} > Q_{gp}^{\text{max}}$ or $Q_{gp} < Q_{gp}^{\text{min}}$. For PQ operation, the reactive power specification Q_{gp}^{sp} is equal to Q_{gp}^{max} or Q_{gp}^{min}. The transition from PQ to PV operation is activated if the following conditions are satisfied: $Q_{gp}^{\text{sp}} = Q_{gp}^{\text{max}}$ and $V_{p1} > V_{p1}^{\text{sp}}$ or $Q_{gp}^{\text{sp}} = Q_{gp}^{\text{min}}$ and $V_{p1} < V_{p1}^{\text{sp}}$.

11.2.6 EXAMPLES

This section presents several experiments in which, the performance of the TPLF is pointed out. Power flow studies in networks with 14, 118, and 935 buses are considered.

11.2.6.1 IEEE 14-Bus System

The IEEE 14-bus system is widely known and is used to illustrate several aspects related to power system analysis. Figure 11.2 shows the scheme and Tables 11.1 through 11.3 present the data

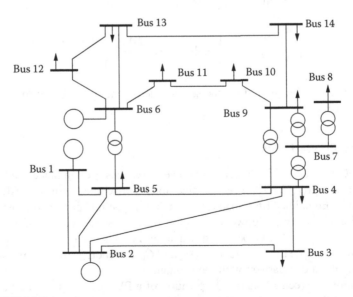

FIGURE 11.2   IEEE 14-bus system.

TABLE 11.1

Bus Data for the IEEE 14-Bus System

Bus	Type	U_{base} (kV)	P_d (MW)	Q_d (Mvar)	V_1 (%)	V_1 (degrees)
1	Osc.	230	0.000	0.000	106.000	0.00
2	PV	230	0.000	0.000	104.500	−4.58
3	PQ	230	80.000	20.000	98.740	−11.14
4	PQ	230	47.790	−3.900	100.962	−9.59
5	PQ	230	7.599	1.599	102.240	−8.40
6	PV	115	0.000	0.000	107.000	−14.64
7	PQ	—	0.000	0.000	98.730	−12.74
8	PQ	13.8	0.000	12.900	96.373	−17.26
9	PQ	115	29.499	16.599	99.152	−14.43
10	PQ	115	9.000	5.799	99.756	−14.75
11	PQ	115	3.501	1.800	102.957	−14.79
12	PQ	115	6.099	1.599	105.099	−15.45
13	PQ	115	13.500	5.799	104.240	−15.41
14	PQ	115	14.901	5.001	100.063	−15.79

corresponding to buses, generators, and branches, respectively, for a base of 100 MVA. Additionally, capacitor bank of 6.33 Mvar is connected at bus 9. SPLF solution is indicated in Table 11.1.

TPLF calculations require additional data corresponding to negative- and zero-sequence data and grounding specifications. This additional information is as follows:

1. All PQ loads will be treated as ungrounded, structurally balanced (type 3 loads as shown in Figure 11.1).
2. Transformer impedances are equal for the three sequences. Moreover, all Y-windings are grounded.
3. The subtransient reactances of Table 11.2 are selected for representing the negative- and zero-sequence reactances of the PV machines.
4. For all transmission lines, a zero-sequence resistance $R_0 = 3R_1$ and zero-sequence reactance $X_0 = 3.5X_1$ are assumed. R_1 and X_1 are indicated in Table 11.3. Namely, all overhead lines are modeled as balanced elements for the base case.
5. For the seven lines which link buses 1, 2, 3, 4, and 5, a zero-sequence susceptance $B_0 = 3.5B_1$ is assumed.

With this system, two experiments are considered to show the imbalance introduced by the geometric configuration of overhead lines and the effect of structurally unbalanced loads.

TABLE 11.2

Data for Generators in the IEEE 14-Bus System

Bus	V_{sp} (%)	P_g (MW)	$X_{subtrans}$ (pu)
1	106.000	261.681	0.25
2	104.500	18.300	0.25
6	107.000	−11.200	0.25

TABLE 11.3

Positive Sequence Data for Lines and Transformers in the IEEE 14-Bus System

Branch	$N_{initial}$	N_{final}	R_1 (pu)	X_1 (pu)	B_1 (pu)	t (pu)	δ_t (degrees)
1	1	2	0.01937	0.05916	0.05279		
2	1	5	0.05402	0.22300	0.04920		
3	2	3	0.04697	0.19794	0.04380		
4	2	4	0.05810	0.17628	0.03740		
5	2	5	0.05693	0.17384	0.03386		
6	3	4	0.06700	0.17099	0.03460		
7	4	5	0.01335	0.04209	0.01280		
8	4	7	0.00000	0.20900	0.00000	1.0000	−30.00
9	4	9	0.00000	0.55618	0.00000	1.0000	0.00
10	5	6	0.00000	0.25020	0.00000	1.0000	0.00
11	6	11	0.09495	0.19887	0.00000		
12	6	12	0.12285	0.25575	0.00000		
13	6	13	0.06613	0.13024	0.00000		
14	7	8	0.00000	0.17615	0.00000	1.0000	−30.00
15	7	9	0.00000	0.11000	0.00000	1.0000	0.00
16	9	10	0.03181	0.08448	0.00000		
17	9	14	0.01270	0.27033	0.00000		
18	10	11	0.08203	0.19202	0.00000		
19	12	13	0.22087	0.19985	0.00000		
20	13	14	0.17089	0.34795	0.00000		

Case 1   Unbalanced Lines

In this case, for the seven lines that link buses 1, 2, 3, 4, and 5, a lumped parameter line model is adopted. It takes into account the geometric layout of the conductors (duplex with layer disposition), skin effect, and earth return path according to Carson equations. Two types of line configurations are considered. Table 11.4 shows the parameters of untransposed lines (types LA-180 and LA-280) for a length of 1 km while Table 11.5 indicates phase conductor and length of each line.

With this input data, and with the hypothesis of perfectly transposed lines, the positive sequence parameters R_1, X_1, and B_1 provided by this model are very similar to the parameters indicated in Table 11.3. However, in this case an untransposed line model is considered. Under this condition,

TABLE 11.4

Parameters of Untransposed Lines with Conductors LA-180 and LA-280, and a Length of 1 km

Phase	Phase Conductor LA-180			Phase Conductor LA-280		
$\alpha\beta$	$R_{\alpha\beta}$ (Ω)	$X_{\alpha\beta}$ (Ω)	$B_{\alpha\beta}$ (μS)	$R_{\alpha\beta}$ (Ω)	$X_{\alpha\beta}$ (Ω)	$B_{\alpha\beta}$ (μS)
aa	0.1893	0.5849	3.4191	0.1494	0.5784	3.4973
ab	0.0875	0.3062	−0.6383	0.0875	0.3062	−0.6662
ac	0.0861	0.2630	−0.2104	0.0861	0.2630	−0.2170
bb	0.1907	0.5841	3.5221	0.1508	0.5777	3.6074
bc	0.0875	0.3062	−0.6383	0.0875	0.3062	−0.6662
cc	0.1893	0.5849	3.4191	0.1494	0.5784	3.4973

TABLE 11.5

Phase Conductor and Length of Each Line

Line	Conductor	Length (km)
Bus 1–Bus 2	LA-180	26.702
Bus 1–Bus 5	LA-280	102.902
Bus 2–Bus 3	LA-280	91.338
Bus 2–Bus 4	LA-180	79.564
Bus 2–Bus 5	LA-180	78.463
Bus 3–Bus 4	LA-180	77.177
Bus 4–Bus 5	LA-180	18.997

these seven lines introduce coupling among sequences, which produce structural imbalances in the system. The results are shown in Table 11.6. The existence of negative- and zero-voltage components is due to the unbalance introduced by these seven lines. In particular, levels of 0.5% for the zero-sequence voltage and 1.22% for the negative sequence appear at bus 3.

Case 2   Unbalanced Loads

In this case, only unbalanced effect of PQ loads is considered. To do that, the PQ load at bus 3 of the base case is substituted by a parallel combination of load types 1, 2, 3, and 4 with the following characteristics:

- Type 1: $P_{t1} = 20\,MW$ and $Q_{t1} = 5\,Mvar$.
- Type 2: $P_a = 20\,MW$ and $Q_a = 5\,Mvar$.
- Type 3: $P_{t3} = 20\,MW$ and $Q_{t3} = 5\,Mvar$.
- Type 4: $P_{ab} = 20\,MW$ and $Q_{ab} = 5\,Mvar$.

The voltages provided by the power flow are indicated in Table 11.7, where high levels of negative- and zero-sequence voltages appear. Zero-sequence voltages are produced by load type 2 while

TABLE 11.6

Power Flow Solution for Case 1

Bus	V_0 (%)	θ_0 (degrees)	V_1 (%)	θ_1 (degrees)	V_2 (%)	θ_2 (degrees)
1	0.260	−16.64	106.000	0.00	0.796	53.49
2	0.033	37.96	104.500	−4.60	0.268	110.13
3	0.507	118.50	98.972	−11.16	1.223	171.17
4	0.238	129.36	100.963	−9.59	0.871	174.72
5	0.173	131.38	102.325	−8.41	0.670	173.27
6	0.092	131.95	107.000	−14.63	0.427	173.04
7	0.122	129.07	98.732	−12.74	0.729	170.79
8	0.000	0.00	96.375	−17.26	0.712	140.79
9	0.137	128.81	99.154	−14.43	0.667	168.10
10	0.129	129.03	99.758	−14.74	0.621	168.16
11	0.111	130.03	102.958	−14.79	0.524	169.73
12	0.095	131.64	105.099	−15.45	0.439	171.63
13	0.098	130.65	104.240	−15.40	0.454	170.15
14	0.121	127.80	100.065	−15.79	0.571	165.39

TABLE 11.7

Power Flow Solution for Case 2

Bus	V_0 (%)	θ_0 (degrees)	V_1 (%)	θ_1 (degrees)	V_2 (%)	θ_2 (degrees)
1	1.224	−125.70	106.000	0.00	3.010	−99.21
2	2.030	−132.45	104.500	−4.61	3.541	−101.73
3	9.782	−141.88	98.174	−11.17	7.311	−107.46
4	2.465	−130.74	100.815	−9.62	4.098	−104.29
5	1.890	−128.07	102.147	−8.43	3.685	−102.71
6	0.996	−127.67	107.000	−14.67	2.226	−104.36
7	1.267	−130.97	98.622	−12.77	3.480	−107.79
8	0.000	0.00	96.265	17.23	3.397	−137.79
9	1.428	−131.18	99.064	−14.46	3.213	−110.14
10	1.351	−130.88	99.684	−14.78	3.016	−109.89
11	1.177	−129.72	102.920	−14.82	2.618	−107.94
12	1.029	−128.02	105.092	−15.49	2.270	−105.82
13	1.058	−128.95	104.228	−15.44	2.326	−107.18
14	1.270	−131.88	100.007	−15.82	2.802	−112.19

negative-sequence voltages are produced due to loads type 2 and type 4. It is important to note the unbalance level at bus 3, where zero- and negative-sequence voltages are 9.8% and 7.3%, respectively. In the rest of the buses, the negative-sequence voltage is higher than 2%, and it presents values higher than the zero-sequence voltage.

If the combined effect of cases 1 and 2 would be studied, zero- and negative-sequence voltages at bus 3 would be 10.2% and 7.6%, respectively. On the other hand, from the comparison of the results of cases 1 and 2, we can state that the imbalance introduced by the nonperfectly transposed overhead lines is lower enough than the effect of structurally unbalanced loads. It partially justifies the use of SPLF studies although the lines are not transposed.

11.2.6.2   IEEE 118-Bus System

The IEEE 118-bus system is also a reference network for power system analysis. In this application, the positive sequence data of the system are specified with a base of 100 MVA and the subtransient reactances of generators are 0.1 pu. All PQ loads are balanced (type 3) with the exception of unbalanced PQ loads (type 4) at buses 11, 60, and 78. They are specified by

- Bus 11: Type 4. $P_{ab} = 70$ MW and $Q_{ab} = 23$ Mvar.
- Bus 60: Type 4. $P_{bc} = 78$ MW and $Q_{bc} = 3$ Mvar.
- Bus 78: Type 4. $P_{ca} = 71$ MW and $Q_{ca} = 26$ Mvar.

Figure 11.3 shows the negative-sequence voltage profile provided by the TPLF. The highest negative voltage level is produced at buses where the unbalanced loads are connected.

11.2.6.3   Large-Scale System

In this case, the Spanish transmission system is considered. It consists of a network with 935 busbars. The unbalance is introduced by phase-to-phase loads corresponding to a high-speed railway. The case includes nine traction substations with a total unbalanced load of 337.22 MW. That means a negative situation where traction substations are heavily loaded. Each substation has two single-phase transformers as shown in Table 11.8. Each transformer is denoted by symbol T and a double subscript. The first subscript indicates the substation number. The transformers are connected alternatively between different phases of the high-voltage network, for instance, *ab*, *bc*, *ca*, to reduce the unbalance level (see Table 11.8). Two cases are studied with these data.

Case 1

The unbalance effect due to the load distribution of Table 11.8 is studied in case 1. Figure 11.4 shows the negative-sequence voltage profile for the 935 busbars. A moderate unbalance level is observed whose highest value is 0.67%.

Case 2

In this case, failures in transformers T_{42} and T_{51} are considered. Under this condition, the loads of these transformers are transferred to transformers T_{41} and T_{52} of the two neighboring substations. With this, the total load of transformers T_{41} and T_{52} is 46.1 MW and 46.3 MW, respectively, providing a significant demand of single-phase loads connected between phases *b* and *c*. The consequence is

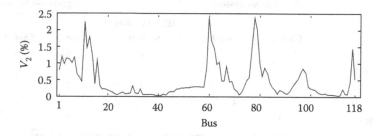

FIGURE 11.3 Negative-sequence voltage profile for the 118-bus system.

TABLE 11.8
Details of Single-Phase Transformers and Unbalanced Loads in the Large-Scale System

Transf	Load (MW)	Connection	Transf	Load (MW)	Connection
T_{11}	17.00	bc	T_{52}	23.15	bc
T_{12}	17.00	ca	T_{61}	17.51	ca
T_{21}	17.60	ab	T_{62}	17.51	ab
T_{22}	17.60	bc	T_{71}	17.60	bc
T_{31}	17.40	ca	T_{72}	17.60	ca
T_{32}	17.40	ab	T_{81}	18.00	ab
T_{41}	23.05	bc	T_{82}	18.00	bc
T_{42}	23.05	ca	T_{91}	17.30	ca
T_{51}	23.15	ab	T_{92}	17.30	ab

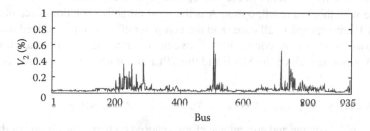

FIGURE 11.4 Negative-sequence voltage profile for the 935-bus system, case 1.

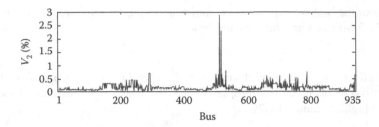

FIGURE 11.5   Negative-sequence voltage profile for the 935-bus system, case 2.

TABLE 11.9
Number of Iterations to Reach the Convergence

	IEEE 14-Bus System			Large-Scale System	
	Case 1	Case 2	IEEE 118-Bus System	Case 1	Case 2
Option A: SPLF	3	3	4	6	6
Option A: TPLF	2	3	3	2	3
Option B: TPLF	4	4	5	7	7
t_A/t_B (%)	≈100	≈100	58	32	36

an important increase of negative-sequence voltage in the system, as shown in Figure 11.5, where the highest value of 2.9% can be observed.

11.2.6.4 Convergence Results

Reactive power limits for PV machines, and local PQ loads at PV buses are considered for the IEEE 118-bus system and the large-scale system. Two strategies have been applied to reach the convergence. Option A is based on using a previous SPLF where the unbalanced PQ loads are handled as balanced PQ loads. The solution of the SPLF provides the initial values for the positive sequence voltages in the three-phase power flow (TPLF). Option B is oriented to use only a TPLF, starting from a flat voltage profile for the positive sequence. The initial values for the negative- and zero-sequence voltages in the TPLF are equal to zero for both options. Solution oscillations and possible divergences, caused by switching between PV and PQ types of buses, are avoided by ignoring the reactive power limits in the first iterations. In this way, the TPLF was executed without considering reactive power limits in the first two iterations for the IEEE 118-bus system and in the first three iterations for the large-scale system.

Table 11.9 shows the number of iterations to reach convergence with mismatches of 10^{-6} pu. The relative computer time (t_A/t_B) between the two options is also reported.

In accordance with this last result, option A is the most attractive solution since only two to three iterations of TPLF are required in all cases and the computer effort can be reduced in the proportion 1/3 in large power systems. A comparison between the amount of computer time needed for options A and B is possible since the SPLF and the TPLF use the same sparse matrix package [18].

11.3   FUNDAMENTALS OF POWER SYSTEM HARMONICS

Distortion of sinusoidal voltage and current waveforms caused by harmonics is one of the major power quality concerns in electric power industry. Waveform distortion is defined as a steady-state deviation from an ideal sinusoidal wave of power frequency. It originates in the nonlinear characteristics of devices and loads in the power system. Typical harmonic sources are variable frequency drives (VFDs) and other power electronics-based equipment. Techniques developed to assess the harmonic impact of nonlinear loads include harmonic power flow and network frequency response analysis.

11.3.1 Definition

A precise method to characterize steady-state periodic waveform distortion is to use the spectral components of the waveform called harmonics. According to the Fourier theory, a periodic function of period T seconds and fundamental frequency $f_0 = 1/T$ (Hz), can be represented by a trigonometric series as [19]

$$f(t) = c_0 + \sum_{h=1}^{\infty} c_h \cos(h\omega_0 t + \phi_h) \tag{11.71}$$

from where the Discrete Fourier Transform (DFT) coefficients can be defined as

$$C_h = c_h \angle \phi_h = \frac{2}{T} \int_{-\frac{T}{2}}^{\frac{T}{2}} f(t) e^{-jh\omega_0 t} dt$$

$$\omega_0 = 2\pi/T$$

We can see that function $f(t)$ has been decomposed into a series of sinusoidal components with different frequencies. The component of $h\omega_0$ is called the hth harmonic of the periodic function. The magnitude of the dc component is c_0. The component with $h = 1$ is called the fundamental frequency component. The coefficients c_h and ϕ_h are known as the hth order harmonic magnitude and phase angle, respectively. In the field of power engineering, function $f(t)$ represents voltage or current at various locations of a network. The resulting Fourier components are called voltage or current harmonics. Figure 11.6 shows the example of a distorted current waveform and its harmonic components.

11.3.2 Electric Quantities Under Harmonic Conditions

On the basis of the Fourier theory described earlier, power system voltage and current waveforms can be represented using harmonic components as follows:

$$v(t) = \sum_{h=1}^{H} v_h(t) = \sum_{h=1}^{H} \sqrt{2} V_h \cos(h\omega_0 t + \phi_h) \tag{11.72}$$

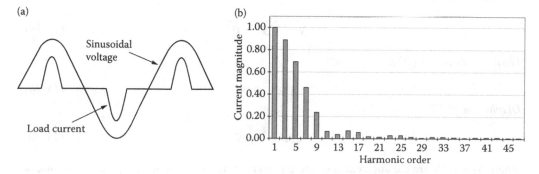

FIGURE 11.6 Waveform distortion and harmonics: (a) voltage and current waveforms and (b) harmonic spectrum of current waveform.

$$i(t) = \sum_{h=1}^{H} i_h(t) = \sum_{h=1}^{H} \sqrt{2}I_h \cos(h\omega_0 t + \delta_h) \qquad (11.73)$$

where H is the highest harmonic order available or of interest, which is normally equal to 50. For practical harmonic analysis, $H = 30 \sim 40$ are generally sufficient. The dc terms are usually neglected due to their extremely small values. The RMS values of the hth order harmonic voltage and current are V_h and I_h, respectively.

The RMS values of $v(t)$ and $i(t)$ are related to the RMS values of the harmonic components, V_h or I_h, as follows:

$$V_{RMS} = \sqrt{\frac{1}{T}\int_0^T v^2(t)dt} = \sqrt{\sum_{h=1}^{H} V_h^2} \qquad (11.74)$$

$$I_{RMS} = \sqrt{\frac{1}{T}\int_0^T i^2(t)dt} = \sqrt{\sum_{h=1}^{H} I_h^2} \qquad (11.75)$$

The average power absorbed by a load experiencing a voltage of $v(t)$ and current of $i(t)$, called active power, can be determined according to its definition and has the following form:

$$P = \frac{1}{T}\int_0^T v(t)i(t)dt = \sum_{h=1}^{H} V_h I_h \cos(\phi_h - \delta_h) = \sum_{h=1}^{H} P_h \qquad (11.76)$$

The equation reveals that the active power is a sum of the active powers produced by each harmonic component. There is no contribution from the voltage at one frequency and the current at another. At present, there is still no consensus on the definitions and physical meanings regarding other power indices such as apparent power and reactive power. A recently published IEEE Std 1459–2000 [20] recommends the following definitions:

Apparent power:

$$S = V_{RMS}I_{RMS} = \sqrt{S_1^2 + S_N^2} \qquad (11.77)$$

Nonactive power:

$$N = \sqrt{S^2 - P^2} \qquad (11.78)$$

Total power factor (PF):

$$PF = P/S \qquad (11.79)$$

Displacement PF:

$$dPF = P_1/S_1 \qquad (11.80)$$

where S_1 and P_1 are the apparent and active powers at the fundamental frequency. The above definitions are established for single-phase quantities. Similar definitions have also been proposed for three-phase quantities in [20].

11.3.3 Indices to Characterize Power System Harmonics

It is not convenient to use a set of DFT coefficients C_h to describe the main characteristics of a distorted waveform. As a result, several indices have been developed to measure the significance of harmonic distortion in a voltage or current waveform. The most commonly used indices are discussed in this subsection.

11.3.3.1 Total and Individual Harmonic Distortion

The total harmonic distortion (THD) index is the most commonly used index to characterize the level of harmonics contained in a waveform:

$$\text{THD}_V = \frac{\sqrt{\sum_{h=2}^{H} V_H^2}}{V_1} \quad \text{or} \quad \text{THD}_I = \frac{\sqrt{\sum_{h=2}^{H} I_H^2}}{I_1} \tag{11.81}$$

The THD index is usually expressed in percentage. It measures the relative magnitude of all harmonic components with respect to that of the fundamental frequency component. For a perfect sinusoidal wave at fundamental frequency, the THD is zero. The voltage THD values shall be $<5\%$ or 10% for most customer interconnection points. The current THD can vary a lot depending on the nature of the current. The amount of a particular harmonic component contained in a waveform can be measured using a similar index called individual harmonic distortion (IHD). IHD for the hth harmonic voltage or current is defined as V_h/V_1 and I_h/I_1, respectively.

11.3.3.2 Total Demand Distortion

There is a problem in applying the THD or IHD indices to current harmonics when the fundamental frequency current approaches zero. The resulting THD or IHD can be very high but they are not necessarily the representatives of the true harmonic levels in a system. As a result, an index called the total demand distortion (TDD) is introduced to characterize the current harmonics as follows [21]:

$$\text{TDD} = \frac{\sqrt{\sum_{h=2}^{H} I_H^2}}{I_L} \tag{11.82}$$

where I_L is the maximum fundamental frequency value of the current. The maximum value is specified as the peak demand current over 15 or 30 min intervals averaged over a 12-month period.

11.3.3.3 Transformer K-Factor

Transformer K-factor is an index used to calculate the derating of standard transformers when harmonic currents are present [22]. The K-factor is defined as

$$K = \frac{\sum_{h=1}^{H} h^2 (I_h/I_1)^2}{\sum_{h=1}^{H} (I_h/I_1)^2} \tag{11.83}$$

The above equation is based on the assumption that the transformer winding losses produced by high-frequency currents are proportional to the square of current frequencies. The K-factor characterizes the extra heat a transformer may produce because of harmonics flowing through it. If there are no harmonics, $K = 1$. A higher K value indicates that transformers with higher harmonic tolerance level are needed to supply the current. Such transformers are called K-rated transformers.

11.3.3.4 V · T and I · T Products

In some cases such as harmonic–telephone interference, the impact of harmonics cannot be characterized using percentage or normalized values alone. The actual levels of harmonics are more useful. The $V \cdot T$ and $I \cdot T$ products are an example. These indices quantify the degree of harmonic-induced noises on telephone circuits and are defined as

$$V \cdot T = \sqrt{\sum_{h=1}^{H} (w_h V_h)^2} \quad \text{or} \quad I \cdot T = \sqrt{\sum_{h=1}^{H} (w_h I_h)^2} \tag{11.84}$$

where w_h is a weighting factor counting for the sensitivity of human ear and telephone audio circuitry at different frequencies. Since telephone noise is caused by harmonic current-induced voltages, the $I \cdot T$ product is the most commonly used index to quantify the potential of harmonic–telephone interference. In addition to the $V \cdot T$ and $I \cdot T$ products, other indices such as TIF (Telephone Influence Factor) and C-message weighted index have also been proposed [21].

11.3.4 Characteristics of Power System Harmonics

Most harmonic-producing loads in power systems cause the same form of distortion to the positive and negative cycles of a waveform. Such waveforms are called half-wave symmetry whose mathematical expression is

$$f(t) = -f(t \pm \frac{T}{2}) \tag{11.85}$$

It can be shown that waveforms of half-wave symmetry do not contain even order harmonics. So, even order harmonics are rare in power system voltage and current waveforms.

Furthermore, three-phase power electronic devices, which may be the most common harmonic sources in power systems, do not produce triple order harmonics under ideal operating conditions due to their circuit topology. As a result, triple harmonics are also rarely encountered in three-phase systems. This leaves 5th, 7th, 11th, 13th, and so on, as the main harmonics of concern in three-phase power systems.

For a three-phase system under balanced conditions, the hth order harmonic voltage of each phase can be expressed as

$$v_{ah}(t) = \sqrt{2}V_h \cos{(h\omega_0 t + \phi_h)} \tag{11.86a}$$

$$v_{bh}(t) = \sqrt{2}V_h \cos{(h\omega_0 t - 2h\pi/3 + \phi_h)} \tag{11.86b}$$

$$v_{ch}(t) = \sqrt{2}V_h \cos{(h\omega_0 t + 2h\pi/3 + \phi_h)} \tag{11.86c}$$

It can be seen that the phase angle differences among the phases are harmonic order-dependent. This results in different phase sequences for different harmonic orders. For example, the fundamental frequency component is positive sequence, the third harmonic is zero sequence, the fifth harmonic is negative sequence, and so on. Table 11.10 summarizes this phenomenon.

The real-life power systems do not operate under ideal conditions. As a result, even order harmonics may be detected. There are also triple harmonics in three-phase systems. A single harmonic may contain all three sequence components. However, the harmonics and their characteristics described in this section will be dominant. So we will more likely encounter fifth negative sequence, seventh positive sequence harmonics, and so on. Harmonic analysis is mainly focused on determining such dominant harmonics.

TABLE 11.10

Sequence Characteristics of Harmonics

Positive Sequence	Negative Sequence	Zero Sequence
$h = 1$	$h = 2$	$h = 3$
$h = 4$	$h = 5$	$h = 6$
$h = 7$	$h = 8$	$h = 9$

In addition to harmonics, a distorted waveform may also contain interharmonics. Interharmonics are those components whose frequencies are not integral multiples of the fundamental frequency. A waveform that contains interharmonics is not periodic on a cycle-by-cycle basis. As a result, the one cycle-based DFT analysis shown earlier is not effective to analyze interharmonics. The subject of interharmonics is beyond the scope of this book.

11.3.5 Network Responses to Harmonic Excitations

The concern on power system harmonics originates from the proliferation of harmonic-producing loads. In a simplified form, such loads can be considered as harmonic current sources. The harmonic currents interact with the network impedances, resulting in voltage harmonics or voltage distortion. The distorted voltages in turn may cause adverse effects on power system equipment and other customers. The primary objective of power system harmonic analysis is to determine the extent and characteristics of harmonic propagation in a power system.

Since a power network is a linear network when harmonic sources are excluded, the harmonic voltage V_h caused by the hth harmonic current injected by a load can be determined according to the following equation:

$$V_h = \mathcal{Z}(h)I_h \tag{11.87}$$

where $\mathcal{Z}(h)$ is the transfer impedance between the node where the voltage harmonic is to be determined and the node where the harmonic current is injected. The above equation shows that the transfer impedance $\mathcal{Z}(h)$ plays an important role in the system response to harmonics. If $\mathcal{Z}(h)$ becomes very large, a large voltage harmonic will result. On the other hand, if $\mathcal{Z}(h)$ is very small, the impact of the harmonic current on that particular node becomes insignificant. In the following, we use two examples to illustrate the responses of power systems under harmonic excitations.

The first example is a harmonic-producing facility that has shunt capacitors installed to compensate its PF. The case is shown in Figure 11.7.

Assume that the supply system can be represented by a Thevenin impedance of $\mathcal{Z}_S = jhX_S$, where h is the harmonic order (or per-unit frequency normalized to the fundamental frequency), the total impedance seen by the harmonic current source can be determined as

$$\mathcal{Z}_{\text{total}} = \mathcal{Z}_S // \mathcal{Z}_C = \frac{jhX_SX_C}{X_C - h^2X_S} \tag{11.88}$$

It can be seen that when $h^2X_S = X_C$, $\mathcal{Z}_{\text{total}}$ approaches infinity and a very high-voltage harmonic may result if the harmonic current has a frequency close to

$$h_R = \sqrt{X_C/X_S} \tag{11.89}$$

The above frequency is called the resonance frequency of the system. In this case, the resonant components X_S and X_C are in parallel. The resulting resonance is called parallel resonance. Parallel

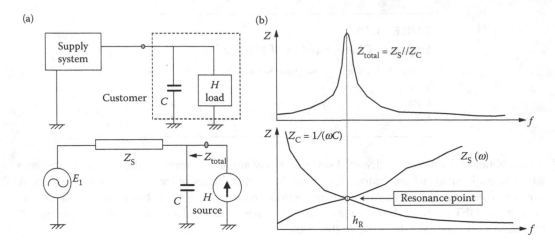

FIGURE 11.7   (a) Equivalent circuit of an example system and (b) its frequency response.

resonance can lead to very high-voltage distortion in power system and high circulating harmonic currents in the resonant components. It is the main cause of harmonic-related equipment damage.

The second example involves inserting an inductor to the shunt capacitor branch as shown in Figure 11.8; the total impedance of the inductor capacitor branch is

$$Z_{LC} = jhX_L - jX_C/h \qquad (11.90)$$

and it will approach zero when $h_R = \sqrt{X_C/X_L}$.

This phenomenon is called series resonance since the resonant components are in series. For the configuration shown in Figure 11.8, we can see that the total impedance seen by the harmonic current source will be zero at the resonant frequency. In other words, the harmonic current injected by the source will be shunted (by-passed) by the series LC branch at that frequency. As a matter of fact, the series LC branch is the simplest form of a shunt harmonic filter.

If we examine the system of Figure 11.8 further, we can see that for frequencies less than h_R, the filter branch is capacitive. We can, therefore, treat it as a capacitor branch. This capacitor branch is in parallel with the system impedance X_S seen from the harmonic current injection point. As a result, a parallel resonance will result at a frequency less than h_R. This situation can be expressed mathematically as

$$h_p X_S = X_{C-eq}(h_p) = \frac{X_C}{h_p} - h_p X_L \qquad (11.91)$$

where h_p denotes the parallel resonance frequency of the composite system. The total impedance as a function of frequency is plotted in Figure 11.8. It can be seen that a shunt filter will result in a parallel

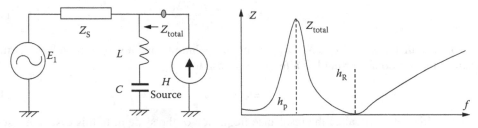

FIGURE 11.8   Effect of series LC branch and frequency responses of Z_{total}.

resonance at h_p and a series resonance at h_R. This demonstrates the complexity of harmonic mitigation using shunt filters. The objective of filter design is to determine the filter component values so that the filter branches can trap various harmonics and, at the same time, avoid parallel resonance at harmonic frequencies.

11.4 POWER SYSTEM HARMONIC ANALYSIS

Power system harmonic analysis is to determine the impact of harmonic-producing loads on a power system. Harmonic analysis has been widely used for system planning, equipment design, troubleshooting, and so on. The following are the most commonly encountered applications of power system harmonic analysis:

- Verifying compliance with harmonic limits.
- Determining harmonic distortion levels for equipment selection.
- Designing harmonic mitigation measures such as harmonic filters.
- Checking if dangerous parallel resonance exist for a given network configuration.

There are two areas of importance in harmonic analysis. The first area is to determine the frequency response characteristics of a network. The results help in verifying whether resonance conditions exist and what changes can be made to mitigate the situation. The second one is to find the actual harmonic voltage or current levels in a network. Such a study is needed, for example, to verify if harmonic limits are complied and to find out the ratings of the equipment to be installed. In this book, we call the first type of harmonic analysis as "Network Frequency Response Analysis" and the second type as "Harmonic Power Flow Analysis." The harmonic power flow analysis needs to define a system operating point at which harmonic distortion levels is to be determined. The operating point is normally defined by the power flow results at the fundamental frequency. The network frequency response analysis, on the other hand, is independent of the power flow results as long as the component impedances are included properly.

Depending on the degree of modeling complexity, harmonic analysis can also be classified into balanced and unbalanced harmonic analysis. Balanced harmonic analysis assumes that the study system is balanced among three phases so single-phase harmonic analysis is sufficient. Harmonic analysis conducted for transmission systems and industry power systems usually belongs to this type. The unbalanced harmonic analysis needs to model all three phases and, sometimes, the neutral conductors of a system. This leads to multiphase harmonic analysis. Examples of multiphase harmonic analysis include the harmonic assessments of unbalanced distribution systems and commercial systems. Owing to space limitation, we concentrate only on balanced harmonic analysis in this book.

Harmonic analysis further consists of two tasks. The first task is to establish high-frequency models for various network components including harmonic-producing devices. The second task is to compute network frequency response or harmonic power flows based on the models. Section 11.4.1 deals with the modeling issues and Sections 11.4.2 and 11.4.3 address the network analysis subjects.

11.4.1 COMPONENT MODELS FOR HARMONIC ANALYSIS

The subject of power system component models has been covered in Chapter 7. In this section, we concentrate only on the modeling issues specific to harmonic analysis. Since harmonics involve frequencies from 50/60 Hz to about 3000 Hz, the models shall take into account the component response characteristics in that frequency range. As most harmonic analysis tools are based on frequency domain algorithms, component models normally have a frequency domain form as well.

11.4.1.1 Linear Network Components

In this chapter, linear network components refer to those components that do not generate harmonics or are not treated as harmonic sources. For example, a transmission line is a linear component that does not produce harmonics. A transformer is a potential harmonic source. But for many harmonic studies, it is treated as a linear component. When harmonics from a transformer are to be investigated, the transformer will be modeled as a harmonic source. The subject of harmonic source modeling is covered in Section 11.4.1.2.

11.4.1.1.1 Overhead Lines and Underground Cables

A transmission line or cable consists of series impedance and shunt admittance. The recommended model for lines and cables is the exact equivalent π-circuit as shown in Figure 11.9. To construct a line/cable model, the unit-length series impedance and shunt admittance parameters are first computed according to the physical arrangement of the conductors. There are line and cable constants programs available to accomplish this task [23]. The series inductance and resistance of the results are frequency-dependent due to the earth return effect and the conductor skin effect. Experience shows that such frequency dependency can be neglected for the majority of harmonic assessment problems. A simple approach is to use the inductance and resistance parameters calculated at the fundamental frequency. An improved approach is to use the parameters calculated at the key harmonic frequency of interest.

It is important, however, to model the long-line effect caused by the shunt admittance. This is because the harmonics are high-frequency phenomena. The shunt admittance, though small at the fundamental frequency, can become quite significant at higher frequencies. The increased admittance could interact with the series impedance of the line/cable, resulting in harmonic resonance. The long-line effect can be easily modeled using an exact equivalent π-circuit. The difference between the exact π-circuit and the commonly used nominal π-circuit is shown in Figure 11.9. In the figure, R_0, X_0, and B_0 are the unit-length impedance and admittance. The line length is ℓ and h is the harmonic order. Note that both models have the same form. The difference is in the values. The exact π-circuit model has included the long-line effect. It can be seen that the exact π-circuit does not make the network solution more complex than the nominal π-circuit.

The main difference between a line and a cable is that a cable generally has a larger shunt capacitance. The long-line effect is therefore more significant. According to experience, for voltages above 10 kV, one should not ignore the shunt admittance for cables longer than 200 m and for lines longer than 1000 m.

11.4.1.1.2 Transformers

A transformer affects harmonic flows through three factors: series impedance, winding connection, and magnetizing branch. For most harmonic studies, a transformer's short-circuit impedance

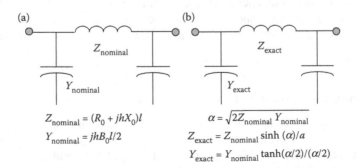

FIGURE 11.9   Equivalent π-circuit models for lines and cables: (a) nominal π-circuit and (b) exact π-circuit.

(Z_{short}) is sufficient to model the series impedance. In reality, the inductance and resistance of the impedance are frequency-dependent because of the interwinding capacitances [23]. Fortunately, the effect of the capacitance becomes significant only at frequencies above 4 kHz for most transformers. So the simple short-circuit-based model is generally sufficient for most cases.

Transformers can produce a $\pm 30°$ phase shift to harmonic voltages and currents, depending on the harmonic order, harmonic sequence, and the transformer winding connection (Y or Δ). This phase-shifting effect could lead to significant harmonic cancelations in a system [24]. It is, therefore, essential to include this effect in the transformer model if there are multiple harmonic sources in the system. A phase-shifter model can be used to represent this effect.

The magnetizing branch of a transformer is a harmonic source. Harmonics are produced due to its nonlinear flux-current (saturation) characteristic. Inclusion of the saturation characteristic is important only when the harmonics generated by a transformer are of concern. If a transformer is not to be treated as a harmonic source, the magnetizing branch can be modeled using its unsaturated magnetizing inductance (Z_{m}). The overall model of a transformer for harmonic analysis has, therefore, the form of Figure 11.10. Note that inclusion of the magnetizing impedance is not important if a transformer is loaded. For an unloaded transformer, however, the branch must be modeled since it may set up a parallel resonance with shunt capacitors in the system. Such a parallel resonance is a significant cause of transformer failure.

11.4.1.1.3 Rotating Machines

This component includes synchronous machines and induction machines. In both machines, the air-gap magnetic field created by the stator harmonics rotates at a speed significantly higher than that of the rotor. The machines, therefore, respond with their short-circuit impedance at harmonic frequencies. In the case of synchronous machines, the impedance is usually taken to be either the negative-sequence impedance or the average of the direct and quadrature subtransient impedances. For induction machines, the impedance is taken to be the locked rotor impedance. The reactance component of the impedance needs to be scaled up according to harmonic frequencies. For both the machines the resistance is frequency-dependent due to conductor skin effects and eddy current losses. The resistance normally increases with frequency in the form of h^α, where h is the harmonic order and α is in the range of 0.5–1.5.

$$Z_{\text{machine}} = R_{\text{short}} h^\alpha + jhX_{\text{short}} \tag{11.92}$$

For the salient-pole synchronous machines, a negative-sequence fundamental frequency current in the stator winding induces a second-order harmonic current in the field winding. The harmonic current can in turn induce a third order harmonic current in the stator. A similar situation arises for the unbalanced harmonic currents in the stator winding. This harmonic conversion mechanism causes a salient-pole synchronous machine to generate harmonic currents. More accurate machine models are available to take into account this effect [25]. Modeling such harmonic generating effects is needed only when there is a significant imbalance in the system and the harmonic from synchronous machine is of primary concern.

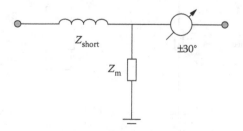

FIGURE 11.10 Equivalent circuit model for transformer.

11.4.1.1.4 Aggregate Loads

Aggregate loads refer to a group of load buses that are treated as one component in harmonic analysis. Typical aggregate loads are distribution feeders seen from a substation bus or a customer plant seen at the point of common coupling. Although such loads typically contain harmonic sources, the main concern is the frequency response characteristics of its equivalent impedance. If the harmonic sources are of concern, the load should not be treated as an aggregate one.

The model for aggregate loads, therefore, has the form of a frequency-dependent impedance. Research has shown that the impedance is not only a function of the individual loads contained in the component but also dependent on the lines or cables connecting the loads. For example, the distribution feeder conductors and shunt capacitors can have a larger impact on the frequency-dependent impedance seen at the feeder terminal than those of the loads connected to the feeder. As a result, it is almost impossible to use a set of general formulae to construct an adequate impedance model for aggregate loads.

In spite of the above considerations, a few models for aggregate loads have been proposed in the past [26]. They are summarized in Equation 11.95 and 11.96 and Figure 11.11. It is important to note that the validity of these models have not been fully verified. For example, model B is derived from measurements taken on a few medium-voltage loads using audio frequency ripple generators [27].

Model A:

$$R_{\text{load}\,h} = \frac{V^2}{P}, \quad X_{\text{load ph}} = h\frac{V^2}{Q} \tag{11.93}$$

Model B:

$$R_{\text{load}\,h} = \frac{V^2}{P}, \quad X_{\text{load ph}} = \frac{hR_{\text{load}\,h}}{6.7(\frac{Q}{P}) - \frac{0.74}{2\pi f_0}}, \quad X_{\text{load sh}} = 0.073hR_{\text{load}\,h} \tag{11.94}$$

where V is the nominal voltage of the load. The components P and Q are active and reactive powers consumed by the load at the fundamental frequency f_0. Since the load is a constant power type at fundamental frequency, the load model has the form shown in Figure 11.11.

11.4.1.1.5 External System

External system for harmonic analysis typically refers to either the utility supply system seen at the point of common coupling from a customer's perspective or the neighboring networks of a utility system under study (Figure 11.12). For the first case, the supply system can be represented as a frequency-dependent harmonic impedance. The impedance is typically determined by performing

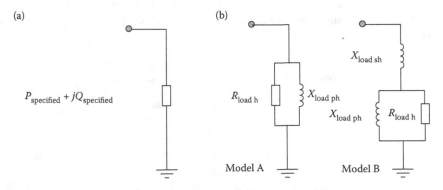

FIGURE 11.11 Model for aggregate linear loads: (a) fundamental and (b) harmonic frequency.

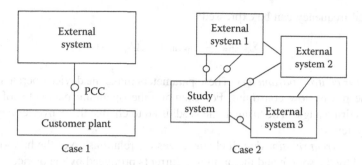

FIGURE 11.12 Two types of external systems.

frequency response calculation on the model of the supply system. Since the number of supply system configurations may need to be investigated, a family of impedance curves could be encountered while dealing with the model of supply systems. If the step-down transformer supplying the customer plant is included as a part of the supply system, it is likely that the transformer impedance will dominate the total equivalent impedance. A much simpler external system model that consists of the supply transformer only could be used for this case.

The external system of the second case can be considered as a multiport network. As a result, they can be represented with a frequency-dependent admittance matrix as follows:

$$\begin{bmatrix} Y_{11}(h) & \cdots & Y_{1k}(h) \\ \vdots & \ddots & \vdots \\ Y_{k1}(h) & \cdots & Y_{kk}(h) \end{bmatrix} \begin{bmatrix} V_1 \\ \vdots \\ V_k \end{bmatrix} = \begin{bmatrix} I_1 \\ \vdots \\ I_k \end{bmatrix} \tag{11.95}$$

where $[\mathcal{V}]$ and $[\mathcal{I}]$ matrices are the voltages and currents at the interface buses. The above $[\mathcal{Y}]$ matrix can also be determined by performing frequency response analysis on the fully modeled external system.

11.4.1.2 Harmonic Sources

There are two main types of harmonic sources. The first type is power electronic devices. These devices produce harmonics due to their waveform switching operations. An example is the variable VFD. The second type produces harmonics due to a nonlinear voltage and current relationship. Examples of the second type are arc furnaces and saturated transformers. Figure 11.13 is a summary of the commonly encountered harmonic sources, organized according to this classification. All harmonic sources have the following three main attributes:

1. *Power flow constraint*: It defines the response of the harmonic-producing devices at the fundamental frequency. For example, the power consumed by the harmonic source at

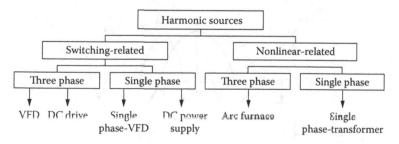

FIGURE 11.13 Classification of commonly encountered harmonic sources.

fundamental frequency can be expressed:

$$S_{\text{specified}} = P_{\text{specified}} + jQ_{\text{specified}} \tag{11.96}$$

2. *Operating (or control) parameters*: These parameters make the device operate in a way that satisfies the power flow constraint. For example, the operating parameter of a converter bridge is its firing angle. The firing angle is adjusted to change the converter power demand from the system.
3. *Voltage and current relationship*: It characterizes the relationship of the harmonic voltages experienced by the source and the harmonic currents produced by the device. The relationship is a function of operating parameters. This is why device operation parameters are singled out in harmonic analysis.

A thyristor-controlled reactor (TCR) is used as an example here to illustrate the above attributes. A TCR is a variable reactor at the fundamental frequency. It is often used as a component of static var compensator (SVC) for reactor power support and voltage control. The structure of TCR and its waveform are shown in Figure 11.14.

The power flow constraint for a TCR in a SVC application is to vary the equivalent TCR reactance so that the SVC can hold its bus voltage to a specified value. It can be expressed as

$$X_1(\alpha) = X_{\text{specified}} \tag{11.97}$$

where X_1 is the fundamental frequency reactance of the TCR and α is the thyristor firing angle. If there is no voltage distortion,

$$X_1(\alpha) = \frac{\pi \omega L}{2(\pi - \sin(2\alpha) - 2\alpha)} \tag{11.98}$$

The operating parameter is the firing angle α in this example. It can be determined, for example, from the above equation. The harmonic voltage and current relationship of the TCR can be determined from the TCR circuit equation

$$L\frac{\mathrm{d}i(t)}{\mathrm{d}t} = v(t), \quad i(t_f) = 0 \tag{11.99}$$

Assume that the voltage has the following form:

$$v(t) = \sum_{h=1}^{H} \sqrt{2}V_h \cos(h\omega t + \phi_h) \tag{11.100}$$

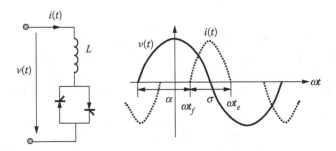

FIGURE 11.14 Thyristor-controlled reactor and its waveforms.

The TCR current can be determined as

$$i(t) = \begin{cases} \sum_{h=1}^{H} \sqrt{2} V_h (h\omega L)^{-1} [\sin(h\omega t + \phi_h) - \sin(h\omega t + \phi_h)], & t_f \leq t \leq t_e \\ 0, & t_0 \leq t \leq t_f \text{ and } t_e \leq t \leq T/2 \end{cases}$$

(11.101)

For a given supply voltage, we can obtain numerical results of the current from the above equation. Transforming the current into frequency domain yields the harmonic current of the TCR. The second example is a nonlinear resistor. A nonlinear resistor may be expressed as

$$i(t) = Gv(t)^r \tag{11.102}$$

where G and r are constants that specify the nonlinear relationship between $v(t)$ and $i(t)$. The above equation defines both the power flow constraint and the harmonic relationship of the device. There is no operating parameter in this case.

Over the past 20 plus years, various models have been proposed to model harmonic sources. In this section, we describe two types of models that are representatives of development in the field. Since there are some commonalties among the models developed for different harmonic sources, the models are presented according to their structures rather than the specific type of harmonic sources.

11.4.1.2.1 Current Source Model

A harmonic source causes harmonic distortion by injecting harmonic currents into the power system. So it is natural to consider harmonic sources as harmonic current sources. In fact, harmonic current source has become the most commonly used model for harmonic source representation. The magnitude and phase of the source can be calculated, for example, from the typical harmonic current spectrum of the device. A commonly used procedure to establish the current source model is as follows:

1. The harmonic-producing load is treated as a constant power load at the fundamental frequency, and the fundamental frequency power flow of the system is solved.
2. The current injected from the load to the system is then calculated and is denoted as $I_1 \angle \theta_1$.
3. The magnitude and phase angle of the harmonic current source representing the load are determined as follows:

$$I_h = I_1 \frac{I_{h-\text{spectrum}}}{I_{1-\text{spectrum}}}, \quad \theta_h = \theta_{h-\text{spectrum}} + h(\theta_1 - \theta_{1-\text{spectrum}}) \tag{11.103}$$

where subscript "spectrum" stands for the typical harmonic current spectrum of the load. This data can be measured or calculated from formulae or time-domain simulations.

The meaning of the magnitude formula is to scale up the typical harmonic current spectrum to match the fundamental frequency power flow result (I_1). The phase formula shifts the spectrum waveform to match the phase angle θ_1. Since the hth harmonic has h-times higher frequency, its phase angle is shifted by h times of the fundamental frequency shift. This spectrum-based current source model is the most common one used in commercial power system harmonic analysis programs. The input data requirement is minimal. Its main disadvantage is inability to model cases involving nontypical operating conditions.

A more advanced current source model is a set of device equations that can be used to compute the current source data for a specific operating condition. For example, device current equation for a TCR is Equation 11.101. For a given $v(t)$ and operating condition defined by parameter α, the current can be determined numerically with that equation. Applying DFT on the current result will yield an operating condition-specific harmonic current that can be used as a current source model.

The most comprehensive current source model is the time-domain simulation-based model. The idea is to simulate the actual power electronic circuit or nonlinear $v(t) \sim i(t)$ relationship of a device in time domain, with a known terminal (harmonic) voltage of the device. The simulation results are then used as harmonic current sources for the network-wide harmonic power flow solution. Since a time-domain simulator can model many details of a device; this model can represent all possible harmonic-producing devices as long as their circuit model and parameters are available.

In summary, the current source model for a harmonic-producing device can be written as follows:

Fundamental frequency power flow constraint:

$$f_1(I_1, V_1) = c \tag{11.104}$$

Harmonic current sources:

$$I_h = f_h(V, \alpha), \quad h = 2, 3, \ldots, H \tag{11.105}$$

where c represents a power flow constraint. For example, the device power consumption is equal to $P_{\text{specified}} + jQ_{\text{specified}}$. The sign V is the known harmonic voltage vector at the device terminal and α represents the device operating parameter. As discussed earlier, Equation 11.105 represents at least three methods of $\mathcal{I}_h$ determination.

11.4.1.2.2 Analytical Device Model

The basic form of this type of model is an analytical expression of the harmonic voltage and current relationship of a device. For example, the model for a nonlinear resistor has the form of

$$i(t) = Gv(t)^r \tag{11.106}$$

and its harmonic domain analytical expression can be established as

$$\sum_{h=-H}^{H} \mathcal{I}_h e^{jh\omega t} = G\left(\sum_{h=-H}^{H} \mathcal{V}_h e^{ptjh\omega t}\right)^r \tag{11.107}$$

where $\mathcal{V}_h$ and $\mathcal{I}_h$ are the harmonic component phasors. It can be seen that, even for a simple nonlinear resistor, the model has become quite complicated. For a six-pulse, three-phase bridge rectifier shown in Figure 11.15, the time-domain expression of its current has the following form [28]:

$$i(t) = A_k + B_k e^{p_k t} + \sum_{h=1}^{H} Y_{h-k} V_h \sin(h\omega t + \phi_h + \delta_{h-k}) \tag{11.108}$$

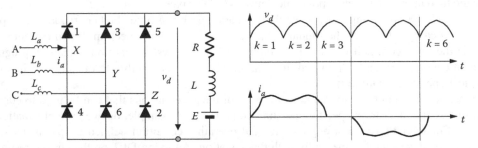

FIGURE 11.15 Six-pulse converter and its waveforms.

where subscript k denotes the kth conduction period of the device within one cycle (see Figure 11.15), The harmonic voltage applied to the rectifier $V_h \angle \phi_h$ and A_k, B_k, p_k, $Y_{h-k} \angle \delta_{h-k}$ are constants that are functions of the rectifier parameters R, L, E, and firing angle α. Expressions of these constants are very complicated. A similar equation has been developed for the TCR in [29].

In summary, this type of model can be generalized as

$$f(v(t), i(t), \alpha) = d \quad \text{in time domain} \tag{11.109}$$

or

$$g(V, \mathcal{I}, \alpha) = d \quad \text{in frequency domain} \tag{11.110}$$

The above model is very similar to the second type current source model described in the previous section. The main motivation of the analytical model, however, is to develop harmonic power flow methods using the Newton algorithm. For a Newton algorithm to work, it is essential to determine the sensitivities of the harmonic currents with respect to its harmonic voltages. The sensitivity results can only be obtained if analytical relationship between the voltage and current is available.

11.4.2 NETWORK FREQUENCY RESPONSE ANALYSIS

Network frequency response analysis is also called frequency scan analysis in some literatures. Its main objective is to identify potential resonance problems in a system. A typical frequency response study involves injecting 1 pu sinusoidal current into the node of interest and the voltage response is calculated [30]. This calculation is repeated for various frequencies of interest. Mathematically speaking, the process is to solve the following network equation at frequency f:

$$[\mathcal{Y}]_f [\mathcal{V}]_f = [\mathcal{I}]_f \tag{11.111}$$

where $[\mathcal{I}]_f$ is the known current vector and $[\mathcal{V}]_f$ is the nodal voltage vector to be solved. For balanced network analysis, only one entry of $[\mathcal{I}]_f$ is nonzero. If multiphase network is involved, a set of positive- or zero-sequence currents may be injected into three phases of a bus, respectively. The results of frequency response analysis are nodal voltages as a function of frequency. Since the injected current is 1 pu, the nodal voltages are essentially the driving point impedance (for the injection node) and the transfer impedances of the network (for other nodes) as shown below

$$
\begin{bmatrix} \mathcal{V}_1 \\ \mathcal{V}_2 \\ \vdots \\ \mathcal{V}_n \end{bmatrix} = [\mathcal{Y}]^{-1} \begin{bmatrix} 1 \\ 0 \\ \vdots \\ 0 \end{bmatrix} = [\mathcal{Z}] \begin{bmatrix} 1 \\ 0 \\ \vdots \\ 0 \end{bmatrix} = \begin{bmatrix} \mathcal{Z}_{11} & \mathcal{Z}_{12} & \cdots & \mathcal{Z}_{1n} \\ \mathcal{Z}_{21} & \mathcal{Z}_{22} & \cdots & \mathcal{Z}_{2n} \\ \vdots & \vdots & \ddots & \vdots \\ \mathcal{Z}_{n1} & \mathcal{Z}_{n2} & \cdots & \mathcal{Z}_{nn} \end{bmatrix} \begin{bmatrix} 1 \\ 0 \\ \vdots \\ 0 \end{bmatrix} = \begin{bmatrix} \mathcal{Z}_{11} \\ \mathcal{Z}_{21} \\ \vdots \\ \mathcal{Z}_{n1} \end{bmatrix} \tag{11.112}
$$

In the above equation, node 1 is assumed as the injection node and the subscript f is omitted. The circuit representation of the frequency response analysis and its sample results are shown in Figure 11.16. From the results, we can see that frequency response curves or the impedances have peaks and valleys at different frequencies. The peaks are parallel resonance points and the valleys are the series resonance points.

Although the frequency response results can identify the existence of harmonic resonance points in a network and reveal their frequencies, more information is needed to solve a resonance problem. For example, answers to the following questions will be very useful:

- Which bus can excite a particular resonance more easily?
- What are the components involved in the resonance?

(a)   (b)

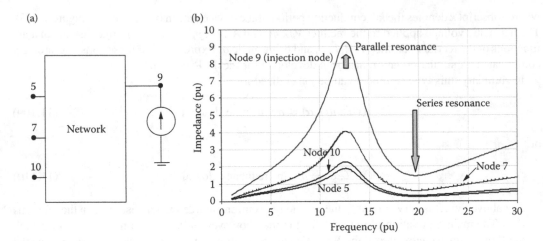

FIGURE 11.16   (a) Network model and (b) frequency response analysis and sample results.

- How far can the resonance propagate in a system?
- Are the resonance phenomena observed at different buses originated from the same cause?

A recently developed method called harmonic resonance mode analysis has the potential to answer the above questions [31]. The basic idea of this method is the following: imagine a system experiences a sharp parallel resonance at frequency f_R based on the frequency scan analysis. It implies that some elements of the voltage vector of Equation 11.111 have large values at f_R, which means that the inverse of the $[\mathcal{Y}(f_R)]$ matrix has large elements. This phenomenon, in turn, is primarily caused by the fact that one of the eigenvalues of the matrix is close to zero. In fact, if the system had no damping at f_R, the matrix would become singular due to one of its eigenvalues becoming zero. To investigate how the $[\mathcal{Y}]_f$ matrix approaches singularity is therefore an attractive way to analyze the problem of harmonic resonance. The well-established theory of eigen-analysis can be readily applied for this purpose. According to the eigen-analysis theory, the $[\mathcal{Y}]$ matrix can be decomposed into the following form:

$$[\mathcal{Y}] = [\mathcal{L}][\Lambda][\mathcal{T}] \tag{11.113}$$

where $[\Lambda]$ is the diagonal eigenvalue matrix, $[\mathcal{L}]$ and $[\mathcal{T}]$ are the left and right eigenvector matrices, respectively. $[\mathcal{L}] = [\mathcal{T}]^{-1}$. The subscript f is omitted here to simplify notation. Substituting Equation 11.113 into Equation 11.111 yields

$$[\mathcal{V}] = [\mathcal{L}][\Lambda]^{-1}[\mathcal{T}][\mathcal{I}] \quad \text{or} \quad [\mathcal{T}][\mathcal{V}] = [\Lambda]^{-1}[\mathcal{T}][\mathcal{I}] \tag{11.114}$$

Defining $[\mathcal{U}] = [\mathcal{T}][\mathcal{V}]$ as the *modal voltage vector* and $[\mathcal{J}] = [\mathcal{T}][\mathcal{I}]$ as the *modal current vector*, respectively, the above equation can be simplified as

$$[\mathcal{U}] = [\Lambda]^{-1}[\mathcal{J}] \quad \text{or} \quad \begin{bmatrix} \mathcal{U}_1 \\ \mathcal{U}_2 \\ \vdots \\ \mathcal{U}_n \end{bmatrix} = \begin{bmatrix} \lambda_1^{-1} & 0 & \cdots & 0 \\ 0 & \lambda_2^{-1} & \cdots & 0 \\ 0 & 0 & \ddots & 0 \\ 0 & 0 & \cdots & \lambda_n^{-1} \end{bmatrix} \begin{bmatrix} \mathcal{J}_1 \\ \mathcal{J}_2 \\ \vdots \\ \mathcal{J}_n \end{bmatrix} \tag{11.115}$$

The inverse of the eigenvalue, λ^{-1}, has the unit of impedance and is named *modal impedance* ($\mathcal{Z}_m$). From Equation 11.113, one can easily see that if $\lambda_1 = 0$ or is very small, a small injection of modal 1 current $\mathcal{J}_1$ will lead to a large modal 1 voltage $\mathcal{U}_1$. On the other hand, the other modal voltages will not be affected since they have no coupling with the mode 1 current. In other words, one can easily identify the location of resonance in the modal domain. The implication is that the resonance actually takes place for a specific mode. It is not related to or caused by a particular bus injection. We therefore call the smallest eigenvalue the *critical mode* of harmonic resonance. This conclusion can be visually demonstrated using Figure 11.17. The first chart shows some of the node voltages of a modified IEEE 14-bus test system when node 9 is injected with 1.0 pu current. The system is modified to have one capacitor so that "one-mode" resonance phenomenon can be seen. It can be seen that most bus voltage experiences a resonance condition. So it is hard to identify the relative significance of each bus in the resonance condition. The eigenvalue chart (Figure 11.17b), on the other hand, shows that only one-mode experiences resonance. So the location of resonance can be easily identified.

The modal current $\mathcal{J}_1$ is a linear projection of the physical currents in the direction of the first eigenvector as follows:

$$\mathcal{J}_1 = \mathcal{T}_{11}\mathcal{I}_1 + \mathcal{T}_{12}\mathcal{I}_2 + \mathcal{T}_{13}\mathcal{I}_3 + \cdots + \mathcal{T}_{1n}\mathcal{I}_n \tag{11.116}$$

It can be seen that if $\mathcal{T}_{13}$ has the largest value, nodal current $\mathcal{I}_3$ will have the largest contribution to the modal 1 current. As a result, bus 3 is the location where the modal 1 resonance can be excited most easily. On the other hand, if $\mathcal{T}_{13} = 0$, current $\mathcal{I}_3$ will not be able to excite the mode no matter how large the current is. The values of the critical eigenvector $[\mathcal{T}_{11}, \mathcal{T}_{12}, \ldots, \mathcal{T}_{1n}]$ can therefore be used to characterize the significance of each nodal current to excite the modal 1 resonance.

The physical nodal voltages are related to the modal voltages by equation $[\mathcal{V}] = [\mathcal{L}][\mathcal{U}]$ as follows:

$$\begin{bmatrix} \mathcal{V}_1 \\ \mathcal{V}_2 \\ \vdots \\ \mathcal{V}_n \end{bmatrix} = \begin{bmatrix} \mathcal{L}_{11} \\ \mathcal{L}_{21} \\ \vdots \\ \mathcal{L}_{n1} \end{bmatrix} \mathcal{U}_1 + \begin{bmatrix} \mathcal{L}_{12} \\ \mathcal{L}_{22} \\ \vdots \\ \mathcal{L}_{n2} \end{bmatrix} \mathcal{U}_2 + \cdots + \begin{bmatrix} \mathcal{L}_{1n} \\ \mathcal{L}_{2n} \\ \vdots \\ \mathcal{L}_{nn} \end{bmatrix} \mathcal{U}_n \approx \begin{bmatrix} \mathcal{L}_{11} \\ \mathcal{L}_{21} \\ \vdots \\ \mathcal{L}_{n1} \end{bmatrix} \mathcal{U}_1 \tag{11.117}$$

The above approximation is possible since $\mathcal{U}_1$ has a value much larger than other modal voltages. This equation reveals that the contribution of the $\mathcal{U}_1$ voltage to the physical voltages can be characterized using the vector $[\mathcal{L}_{11}, \mathcal{L}_{21}, \ldots, \mathcal{L}_{n1}]^{\mathrm{T}}$. If $\mathcal{L}_{31}$ has the largest value, bus 3 will also have the largest value. This implies that bus 3 is the location where the modal 1 resonance can be most easily observed. If $\mathcal{L}_{31} = 0$, the nodal 3 voltage will not be affected by the modal 1 voltage.

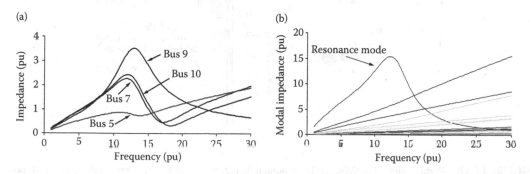

FIGURE 11.17 Frequency responses seen from (a) nodal and (b) modal domains.

In summary, the critical right eigenvector has the potential to characterize the (locational) excitability of the critical mode and the critical left eigenvector can represent the (locational) observability of the critical mode. It is possible to combine the excitability and observability into a single index according to the theory of selective modal analysis [32], as follows:

$$PF_{bm} = \mathcal{L}_{bm}\mathcal{T}_{mb} \tag{11.118}$$

where b is the bus number and m is the mode number. This index characterizes the combined excitability and observability of the critical mode at the same bus. It is called *participation factor* of the node to the critical mode. According to the theory of eigen-analysis, the left and right eigenvector matrices have an inverse relationship, namely, $[\mathcal{L}] = [\mathcal{T}]^{-1}$. Furthermore, if the matrix to be decomposed is symmetric, the eigenvector matrix $[\mathcal{T}]$ will be orthogonal, namely $[\mathcal{L}] = [\mathcal{T}]^{-1} = [\mathcal{T}]^{T}$. Under this condition, the left eigenvector will be equal to the right eigenvector. This is a very important conclusion. It not only simplifies the resonance mode analysis technique but also reveals the unique characteristics of harmonic resonance. Some of the interesting conclusions that can be deducted are as follows:

- The bus that has the highest observability level for a mode is also the one that has the highest excitability level. It implies that if a harmonic current matching the resonance frequency is injected into this bus, the bus will see the highest harmonic voltage level. If the current is injected into a different bus, the harmonic level is likely to be amplified in the system.
- The participation factors are equal to the square of the eigenvectors. As a result, one index, eigenvector or participation factor, is sufficient for resonance mode analysis. The magnitude of the index characterizes how far the resonance will propagate. The bus with the highest participation factor can be considered as the center of the resonance.

The contribution or impact of each network component to a particular resonance can be determined using the sensitivities of modal impedance to component values. Figure 11.18 shows additional sample results of the IEEE 14-bus system. Figure 11.18a plots the participation of each bus in the $f_R = 5.9$ resonance mode. The sizes of the circles are in proportion to the value of participation factors. It can be seen from the figures that this particular resonance involves mainly buses 7, 8, 9, and 10. Figure 11.18b charts the eigen-sensitivity information. The result reveals that the resonance is caused primarily by a shunt capacitor at bus 8 and a series reactance between buses 7 and 8.

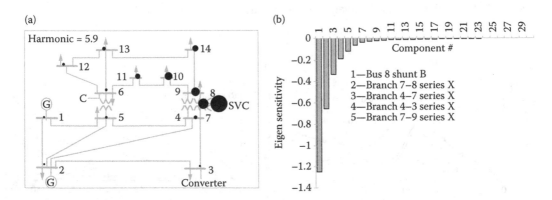

FIGURE 11.18 Participation of each bus and component in a resonance mode: (a) Participation factor of mode $f_R = 5.9$ pu and (b) component sensitivity to mode $f_R = 5.9$ pu.

11.4.3 HARMONIC POWER FLOW ANALYSIS

If one needs to find the harmonic distortion levels for a system, a harmonic power flow analysis needs to be conducted. The main applications of harmonic power flow analysis include, for example, (1) network harmonic distortion assessment, (2) harmonic limit compliance verification, (3) filter performance evaluation, and (4) equipment derating calculation.

11.4.3.1 Formulation of Harmonic Power Flow Problem

A harmonic power flow problem can be formally stated as given an operating condition defined by the fundamental frequency power flow specifications and harmonic source characteristics, the harmonic voltages and currents in various locations of the system are determined. It can be described mathematically as the solution to the following equations:

Linear network equation:

$$[\mathcal{I}]_h = [\mathcal{Y}]_h [\mathcal{V}]_h, \quad h = 1, 3, 5, 7, \ldots, H \tag{11.119}$$

Power flow constraint equations:

$$f_i(\mathcal{V}_i, \mathcal{I}_i) = c_i, \quad i = 1, 2, \ldots, N \tag{11.120}$$

Harmonic source model equations:

$$g_j(\mathcal{V}_j, \mathcal{I}_j, \alpha_j) = d_j, \quad j = 1, 2, \ldots, K \tag{11.121}$$

where $[\mathcal{V}]_h$ and $[\mathcal{I}]_h$ represent the network nodal voltage and current vectors at the hth harmonic, respectively; $[\mathcal{Y}]_h$, the network nodal admittance matrix at the hth harmonic; $\mathcal{V}_i(\mathcal{I}_i)$, the fundamental and harmonic voltages (currents) of node i; c_i, the power flow constraint at node i; $\mathcal{V}_j(\mathcal{I}_j)$, the fundamental and harmonic voltages (currents) of harmonic source j; d_j, the operating constraints of harmonic source j; α_j, the operating parameters of harmonic source j; H, N, and K are the numbers of harmonics, nodes, and harmonic sources, respectively.

An example of power flow constraint equation (for node i) is

$$\mathcal{V}_{i1}\hat{\mathcal{I}}_{i1} = P_{\text{specified}} + jQ_{\text{specified}} \tag{11.122}$$

where subscript 1 stands for fundamental frequency $h = 1$. It is important to note that the above equation does not include harmonic components. In the earlier years of harmonic power flow research, the power flow constraint equation has the following form:

$$\sum_{h=1}^{H} \mathcal{V}_{ih}\hat{\mathcal{I}}_{ih} = S_{\text{specified}} = P_{\text{specified}} + jQ_{\text{specified}} \tag{11.123}$$

which includes all harmonic powers. This formulation has at least one problem—the definition of reactive power is no longer consistent with the recent IEEE standard [20]. Furthermore, it is debatable whether $S_{\text{specified}}$ should include harmonic powers even if a definition exists. It depends on how power demand is metered. In reality, the harmonic contributions to power are very small. It is difficult to justify using the above equation as the equation can make solution algorithms extremely complex. If we exclude the harmonic components in fundamental frequency power flow constraints,

Equation 11.123 can be simplified as

$$f_i(\mathcal{V}_{i1}, \mathcal{I}_{i1}) = c_i, \quad i = 1, 2, \ldots, N \tag{11.124}$$

Examples for Equation 11.121 are the nonlinear resistor Equation 11.102 and converter Equation 11.108 shown in the harmonic source modeling section.

Several numerical solution techniques have been proposed to solve the harmonic power flow equations in the past. The representative techniques are the current source method [26,33], harmonic iteration method [34–37], and the Newton iteration method [28,29,38]. The first two techniques have gained wide acceptance among practicing engineers. Almost all commercial harmonic power flow techniques are based on the current source method. The harmonic iteration method has been used as an improvement to the current source method for cases involving high harmonic voltage distortions. In theory, the Newton iteration method is the most powerful harmonic power flow method in terms of convergence. However, this method has a number of disadvantages. First, it is very difficult and complex to obtain the Jacobian matrices for the harmonic sources. Second, the total Jacobian matrix of the method is huge since it includes all harmonic components. The computational burden can be much larger than the harmonic iteration method. Finally, the method does not have the flexibility to model any harmonic sources. Because of these disadvantages, the Newton method has not been widely used.

11.4.3.1.1 Current Source Method

The foundation of this technique is to model the harmonic sources as known current sources. As discussed in the modeling section, a current source model can be determined using at least three methods. Equation 11.121 of the harmonic power flow equations is thus simplified as follows, which leads to a noniterative solution technique:

$$\mathcal{I}_j = \mathcal{I}_{j-\text{known}}, j = 1, 2, \ldots, K \tag{11.125}$$

1. Solve the fundamental frequency power flow of the network, that is, Equation 11.120. The harmonic-producing loads are modeled as constant power loads as illustrated in Equation 11.96.
2. Determine $\mathcal{I}_{j-\text{known}}$ using one of the three methods described in Section 11.4.1.
3. Calculate the network harmonic voltages and currents by solving Equation 11.119 through 11.121.
4. The results of steps 1 and 3 jointly define the harmonic power flow results of the study system. Harmonic indices such as total harmonic distortions and transformer K factors can be calculated from the results.

As an example, the system shown in Figure 11.19 is solved using the current source method. On the basis of the data provided in the figure, the harmonic impedance of the linear network components are determined as follows:

The branch current $\mathcal{I}$ and bus A voltage can be computed as follows:
Branch current:

$$\mathcal{I} = \frac{\hat{S}_{\text{load}} + \hat{S}_{\text{VFD}} + \hat{S}_{\text{C}}}{\hat{V}_{\text{B}}} = 0.4 - j0.19 = 0.4428\angle - 25.41°$$

Bus A voltage:

$$\mathcal{V}_{\text{A}} = \mathcal{V}_{\text{B}} + (\mathcal{Z}_{\text{T1}} + \mathcal{Z}_{\text{line}} + \mathcal{Z}_{\text{T2}})I = 1.0\angle 0° + (0.01 + j0.17) \cdot (0.4 - j0.19)$$
$$= 1.0384\angle 3.65°$$

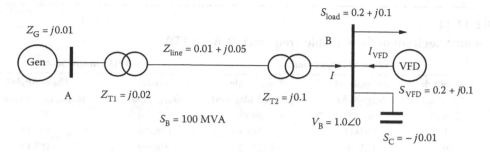

FIGURE 11.19 Single line diagram of the example system.

This establishes the operating point of the system. At harmonic frequencies, the component impedances are

Generator: $\mathcal{Z}_G(h) = j0.01h$

Line and transformer: $\mathcal{Z}_{LT}(h) = 0.01 + j0.17h$

Load: $\mathcal{Z}_{load}(h) = \dfrac{j50h}{5 + j10h}$ (model A is used)

Capacitor: $\mathcal{Z}_C(h) = \dfrac{-j100}{h}$

It can be seen that load Model A of [26] (Equation 11.93) is adopted for the load impedance $\mathcal{Z}_{load}(h)$. This model is not consistent with the impedance definition at the fundamental frequency, which has the form of $\mathcal{Z}_{load}(1) = V^2/(P - jQ)$. An alternative load model that is more consistent with the fundamental frequency model is the following:

$$\mathcal{Z}_{load}(h) = \frac{P}{S^2}V^2 + jh\frac{Q}{S^2}V^2$$

To determine the harmonic current source model for the VFD, we need to calculate the fundamental frequency component of the VFD current:

$$\mathcal{I}_{VFD1} = \left[\frac{\hat{S}_{VFD}}{\hat{V}_B}\right] = 0.2236\angle 153.43°$$

Note that the direction of current is marked in Figure 11.19. Using the above result, the given VFD current spectrum and Equation 11.103, the harmonic current sources representing the VFD are calculated and shown in Table 11.11.

The harmonic voltages can be solved using

$$\begin{bmatrix} \dfrac{1}{\mathcal{Z}_G(h)} + \dfrac{1}{\mathcal{Z}_{LT}(h)} & -\dfrac{1}{\mathcal{Z}_{LT}(h)} \\ -\dfrac{1}{\mathcal{Z}_{LT}(h)} & \dfrac{1}{\mathcal{Z}_{LT}(h)} + \dfrac{1}{\mathcal{Z}_{load}(h)} + \dfrac{1}{\mathcal{Z}_C(h)} \end{bmatrix} \begin{bmatrix} V_A \\ V_B \end{bmatrix} = \begin{bmatrix} 0 \\ I_{VFD} \end{bmatrix}$$

The results are shown in Table 11.12.

Many years of application experiences have shown that this technique is sufficient for many common harmonic analysis tasks. For cases where the harmonic model Equation 11.120 is not available,

TABLE 11.11

Harmonic Spectrum of the Variable Frequency Drive (VFD)

	Given VFD Spectrum			VFD Current Source	
H	Magnitude $(I_{VFDh-Sp})$	Normalized Magnitude	Phase Angle $(\theta_{VFDh-Sp})$ (degrees)	Magnitude (I_{VFDh}) (pu)	Phase Angle $(\theta_{VFDh-Sp})$ (degrees)
1	298.63	1.00000	89.48	0.22361	153.43
3	1.32	0.00442	−58.09	0.00099	133.77
5	69.33	0.23216	−122.92	0.05191	−163.15
7	10.21	0.03419	−136.34	0.00765	−48.66
9	1.37	0.00459	150.75	0.00103	6.34
11	23.42	0.07842	22.19	0.01754	5.69
13	8.79	0.02943	−16.23	0.00658	95.18

TABLE 11.12

Harmonic Voltages at Buses A and B

	Voltage at Bus A			Voltage at Bus B		
H	Magnitude (pu)	% IHD	Phase Angle (degrees)	Magnitude (pu)	% IHD	Phase Angle (degrees)
1	1.03840	—	3.65	1.00000	—	0.00
3	0.00003	0.00	−142.36	0.00053	0.05	−143.43
5	0.00262	0.25	−83.62	0.04712	4.71	−84.26
7	0.00055	0.05	26.18	0.00998	1.00	25.72
9	0.00010	0.01	75.96	0.00178	0.18	75.61
11	0.00216	0.21	69.36	0.03879	3.88	69.07
13	0.00100	0.10	151.94	0.01800	1.80	151.69

the current source method is the only option. If Equation 11.121 is available, it is possible to improve the solution accuracy by solving $\mathcal{I}_j$ from the device equation. In this case, the device voltage contains only the fundamental frequency component solved from the power flow Equation 11.120.

The main disadvantage of the technique is the use of typical or measured harmonic spectrum to represent harmonic-producing devices. It prevents an adequate assessment of cases involving nontypical operating conditions. Such conditions include, for example, partial loading of harmonic-producing devices, excessive harmonic voltage distortions, and unbalanced network conditions. As a result, harmonic iteration methods have been proposed.

11.4.3.1.2 Harmonic Iteration Method

This method is based on the following reasoning: the current source model is obtained with assumed device terminal voltages. The resulting harmonic power flow results are not accurate. The accuracy can be improved if the current source model is updated using the latest harmonic voltage results. This iteration ends until the current sources converge. The following is a more detailed description of the harmonic iteration procedure:

Step 1: Solve the fundamental frequency power flow equation of the system:

$$f_i(\mathcal{V}_{i1}, \mathcal{I}_{i1}) = c_i, \quad i = 1, 2, \ldots, N \tag{11.126}$$

$$[\mathcal{I}]_h = [\mathcal{Y}]_h[\mathcal{V}]_h, \quad h = 1 \tag{11.127}$$

Step 2: Find the currents produced from the harmonic sources by solving the device equation:

$$g_j(\mathcal{V}_j, \mathcal{I}_j, \alpha_j) = d_j, \quad j = 1, 2, \ldots, K \tag{11.128}$$

where $\mathcal{V}_j$ are the device terminal voltages obtained from the previous iteration.

Step 3: Check if the results, $\mathcal{I}_j$, are sufficiently close to those obtained from the previous iteration. If yes, the iteration has converged. Otherwise, go to the next step.

Step 4: Solve for the network harmonic voltages from

$$[\mathcal{I}]_h = [\mathcal{Y}]_h[\mathcal{V}]_h, \quad h = 2, 3, \ldots, H \tag{11.129}$$

where $[\mathcal{I}]_h$ consists of the device harmonic currents solved in step 2. The process is then redirected to step 1.

It can be seen that the main thrust of this iterative method is to update the current source models of the harmonic-producing devices until they accurately represent the device response under the power flow constraints. This general iterative algorithm has a few variations [26,34–36]:

1. The above procedure assumes that the fundamental frequency currents of the nonlinear devices calculated from step 1 will change during the iterative process. As a result, the fundamental frequency power flow equation must be resolved for each iteration. An approximation is to assume that $\mathcal{I}_j$ has no impact on the fundamental frequency power flow results. Consequently, step 1 needs to be performed only once. Although this is an approximation, the resulting accuracy is quite acceptable since it is rare that harmonics will affect fundamental frequency power flow results significantly.

2. Similarly, the device control parameters can be solved from the fundamental frequency results if one assumes that the results will not be affected by harmonic distortions in a meaningful way. This significantly reduces the complexity of the algorithm. If the impact of harmonics on the control parameters must be considered, the control parameter has to be included in the iteration process.

3. There is a range of numerical methods to solve the device equation in step 2. The harmonic source modeling section has shown some methods. The most general method is to simulate the device response in time domain under known supply voltage $\mathcal{V}_j$. This approach is sometimes called hybrid method since it combines time-domain simulation with frequency domain solutions. The hybrid method is the most powerful harmonic power flow solution method since it can model a variety of harmonic source in great detail.

EXAMPLE 11.1

Solve the system shown in Figure 11.20 using the harmonic iteration method. In this case, the harmonic source is a dc drive that is modeled as a voltage-dependent current source. The current source is updated iteratively until convergence is achieved. The solution steps are as follows:

Step 1: Solve the fundamental frequency power flow equation of the system. The dc drive is treated as a constant power load of $\mathcal{S} = 0.2 + j0.1$. The results are

$$\mathcal{V}_A = 1.0384\angle 3.65°, \quad \mathcal{V}_B = 1.0\angle 0°, \quad \text{and} \quad \mathcal{I}_{\text{drive}} = 0.2236\angle 153.43°$$

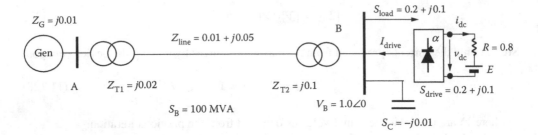

FIGURE 11.20 Single line diagram of the example dc-drive system.

Step 2: Find the operating parameter of the drive—the firing angle and the dc motor voltage. According to [39], the firing angle α can be determined from the drive PF as follows:

$$PF = \frac{3}{\pi} \cos \alpha \qquad (11.130)$$

or

$$\cos \alpha = \frac{\pi PF}{3} = \frac{\pi P_{drive}}{3 S_{drive}} = \frac{2\pi}{3\sqrt{5}} = 0.9366$$

$$\alpha = 20.51°$$

Also according to [39], the dc side voltage can be estimated from

$$V_{dc} = 1.351 V_{ac} \cos \alpha \qquad (11.131)$$

which gives

$$V_{dc} = 1.351 V_B \cos \alpha = 1.351 \times 1.0 \times 0.9366 = 1.2653$$

As a result, the dc motor internal voltage E can be determined as

$$I_{dc} = \frac{P_{drive}}{V_{dc}} = \frac{0.2}{1.2653} = 0.1581 \qquad (11.132)$$

$$E = V_{dc} - I_{dc}R = 1.2653 - 0.1581 \times 0.8 = 1.1388$$

It is important to note that Equation 11.130 through 11.132 are established with the assumption that the dc side current does not contain harmonic components. For this particular example, the dc side inductance is assumed to be 0.1 H.

Step 3: Compute the fundamental frequency and harmonic currents of the dc drive based on the known voltage V_B at the drive terminal, the firing angle α, and the dc motor internal voltage E. There are several ways to determine the currents. The analytical formula 11.108 can be used to compute the numerical values of the drive ac side current waveform. Alternatively, one can use time-domain simulation. The results, obtained using time-domain simulation, are summarized in Table 11.13.

Step 4: Using the dc-drive current as harmonic current sources, the network harmonic voltages are solved from the nodal voltage equation $[\mathcal{V}]_h = [\mathcal{Z}]_h [\mathcal{I}]_h$, where $h = 5, 7, 11, 13, \ldots$ The results are shown in Table 11.14.

Step 5: The bus B voltage now contains harmonic components. It becomes necessary to update the drive harmonic currents. If we assume that the drive control parameters α and E remain unchanged, it is easy to recompute the current under the new drive terminal voltage V_B. This step is the same as step 3 except that voltage V_B now contains harmonic

TABLE 11.13
Bus B Harmonic Voltages and Corresponding Drive Currents

	Bus B Voltage (Drive Input Voltage)		Currents Drawn by the dc Drive	
H	Magnitude (pu)	Phase Angle (degrees)	Magnitude (pu)	Phase Angle (degrees)
1	1.00000	0.00	0.21369	159.62
5	0.00000	0.00	0.04306	257.56
7	0.00000	0.00	0.02991	37.47
11	0.00000	0.00	0.01926	135.42
13	0.00000	0.00	0.01623	274.69

TABLE 11.14
Harmonic Voltages at Buses A and B (First Iteration)

	Voltage at Bus A		Voltage at Bus B	
H	Magnitude (pu)	Phase Angle (degrees)	Magnitude (pu)	Phase Angle (degrees)
1	1.03840	3.65	1.00000	0.00
5	0.00217	−22.92	0.03909	−23.56
7	0.00217	112.31	0.03904	111.85
11	0.00237	−160.91	0.04261	−161.20
13	0.00246	−28.55	0.04437	−28.80

TABLE 11.15
Dc-Drive Current (First Iteration)

H	Magnitude (pu)	Phase Angle (degrees)	ΔI (pu)
1	0.22816	159.62	0.014479
5	0.04595	257.70	0.002889
7	0.03202	37.39	0.002103
11	0.02064	135.50	0.001378
13	0.01732	274.69	0.001093

components. The results are shown in Table 11.15. In the table, ΔI represents the difference between the current (magnitude) obtained in the current step and that obtained in the previous step.

The above approach to compute the drive harmonic current is straightforward and can be repeated for each iteration step. However, it is possible that the resulting drive fundamental frequency current will not meet the power flow constraint of $S_{drive} = 0.2 + j0.1$. Whether the power flow is satisfied can be easily checked as follows:

$$S_{drive\text{-}new} = V_B(-\hat{I}_{drive\text{-}new}) = 0.21 + j0.08$$

The results do confirm that the power drawn by the drive is not exactly equal to the specified power. However, the difference is small. If one must make the drive draw exactly the same amount of power as S_{drive}, the drive parameters α and E must be adjusted. Since Equations 11.130 and 11.132 are no longer accurate due to the presence of harmonics in V_B and I_{dc}, a trial-and-error or iterative process has to be used to compute α and E. The process can be quite complicated. In view of the facts, the $S_{drive-new}$ is very close to S_{drive} and S_{drive} itself is just an approximate operating parameter in many harmonic study cases, there is no practical advantage to produce an exact match. So the simplified approach where α and E are kept constant is adopted.

Step 6: With the newly calculated drive current, the network harmonic voltages are again solved in the same way as step 4. The results are shown in Table 11.16.

Step 7: This step is similar to step 3 and step 5. The new drive harmonic currents are computed and the results are shown in Table 11.17. The column ΔI in the table is used to check the convergence. It can be seen that all harmonic current differences are <0.0005 pu. If 0.0005 pu is the convergence criterion, the harmonic iteration process has converged. The corresponding solution results for bus voltages are shown in Table 11.18.

TABLE 11.16

Harmonic Voltages at Buses A and B (Second Iteration)

	Voltage at Bus A		Voltage at Bus B	
H	Magnitude (pu)	Phase Angle (degrees)	Magnitude (pu)	Phase Angle (degrees)
1	1.03840	3.65	1.00000	0.00
5	0.00232	−22.78	0.04171	−23.42
7	0.00232	112.22	0.04179	111.77
11	0.00254	−160.83	0.04566	−161.12
13	0.00263	−28.56	0.04735	−28.80

TABLE 11.17

Dc-Drive Current (Second Iteration)

H	Magnitude (pu)	Phase Angle (degrees)	ΔI (pu)
1	0.22863	159.62	0.000466
5	0.04614	257.71	0.000192
7	0.03216	37.38	0.000141
11	0.02073	135.50	0.000093
13	0.01739	274.69	0.000072

TABLE 11.18

Harmonic Voltages at Buses A and B (Final Results)

	Voltage at Bus A		Voltage at Bus B	
H	Magnitude (pu)	Phase Angle (degrees)	Magnitude (pu)	Phase Angle (degrees)
1	1.03840	3.65	1.00000	0.00
5	0.00233	−22.77	0.04188	−23.41
7	0.00233	112.22	0.04197	111.76
11	0.00255	−160.83	0.04586	−161.12
13	0.00264	−28.56	0.04755	−28.80

REFERENCES

1. Beites L. F., Mayordomo J. G., Hernández A., Asensi R., Harmonics, interharmonics and unbalances of arc furnaces: A new frequency domain approach, *IEEE Transactions on Power Delivery*, 16(4), 661–668, October 2001.
2. Mayordomo J. G., Izzeddine M., Asensi R., Load and voltage balancing in harmonic power flows by means of static var compensators, *IEEE Transactions on Power Delivery*, 17(3), 761–769, July 2002.
3. Robert A., Marquet J., WG CIGRE/CIRED CCO2 Assessing voltage quality with relation to harmonics, flicker and unbalance. CIGRE, paper 36–203, 1992.
4. Reichelt D., Ecknauer E., Glavitsch H., Estimation of steady-state unbalanced system conditions combining conventional power flow and fault analysis software, *IEEE Transactions on Power Systems*, 11 (1), 422–427, February 1996.
5. Arrillaga J., Bradley D. A., Bodger P. S., *Power System Harmonics*, John Wiley & Sons, New York, 1985.
6. Zhang X. P., Fast three-phase power flow methods, *IEEE Transactions on Power Systems*, 11(3), 1547–1554, August 1996.
7. Smith B. C., Arrillaga J., Power flow constrained harmonic analysis in AC-DC power systems, *IEEE Transactions on Power Systems*, 14(4), 1251–1259, November 1999.
8. Valcárcel M., Mayordomo J. G., Harmonic power flow for unbalanced systems, *IEEE Transactions on Power Delivery*, 8(4), 2052–2059, October 1993.
9. Xu W., Marti J. R., Dommel H. W., A multiphase harmonic power flow solution technique, *IEEE Transactions on Power Systems*, 6(1), 174–182, February 1991.
10. Smith B. C., Arrillaga J., Improved three-phase power flow using phase and sequence components, *IEE Proceedings C*, 145(3), 245–250, May 1998.
11. Mayordomo J. G., López M., Asensi R., Beites L., Rodríguez J. M., A general treatment of traction PWM converters for power flow and Harmonic penetration studies, VIII IEEE International Conference on Harmonics and Quality of Power (ICHQP), Athens (Greece), 685–692, October 1998.
12. Mayordomo J. G., Asensi R., Hernández A., Beites L., Izzeddine M., Iterative harmonic analysis of line side commutated converters used in high speed trains, IX IEEE International Conference on Harmonics and Quality of Power (ICHQP), Orlando (USA), 846–851, October 2000.
13. Da Costa V. M., Martins N., Pereira J. L. R., Developments in the Newton Raphson power flow formulation based on current injections, *IEEE Transactions on Power Systems*, 14(4), 1320–1326, November 1999.
14. García P. A. N., Pereira J. L. R., Carneiro J. R., Da Costa V. M., Martins N., Three-phase power flow calculations using the current injection method, IEEE PES Summer Meeting, 1999.
15. Mayordomo J. G., Izzeddine M., Martínez S., Asensi R., Gómez Expósito A., Xu W., A compact and flexible three-phase power flow based on a full Newton formulation, *IEE Proceedings C*, 149(2), 225–232, March 2002.
16. Mayordomo J. G., Izzeddine M., Martínez S., Asensi R., A contribution for three-phase power flows using the current injection method, IX IEEE International Conference on Harmonics and Quality of Power (ICHQP), Orlando (USA), 295–300, October 2000.
17. Arrillaga J., Smith B., *AC-DC Power System Analysis*, The Institution of Electrical Engineers, London (UK), 1998.
18. Kundert K. S., Sangiovanni-Vincentelli A., *Sparse User's Guide: A Sparse Linear Equation Solver, Version 1.3a*, University of California, Berkeley, April 1988.
19. Oppenheim A. V., Schafer R. W., *Discrete-Time Signal Processing*, Prentice-Hall, Englewood Cliffs, NJ, 1989.
20. IEEE Std. 1459–2000, *IEEE Trial-Use Standard Definitions for the Measurement of Electric Power Quantities Under Sinusoidal, Nonsinusoidal, Balanced, or Unbalanced Conditions*, IEEE, New York, 2000.
21. IEEE Std. 519–1992, *Recommended Practices and Requirements for Harmonic Control in Electric Power Systems*, IEEE, New York, 1993.
22. ANSI/IEEE Standard C57.110-1986, *IEEE Recommended Practice for Establishing Transformer Capability when Supplying Nonsinusoidal Load Currents*, IEEE, New York, 1986.
23. Dommel H. W., *Electromagnetic Transients Program Reference Manual (EMTP Theory Book)*, Dept of Electrical Engineering, University of British Columbia, Canada, August 1986.
24. Derick P., *Power Electronic Converter Harmonics, Multipulse Methods for Clean Power*, IEEE Publication, New York, 1995.

25. Semlyen A., Eggleston J. F., Arrillaga J., Admittance matrix model of a synchronous machine for harmonic analysis, *IEEE Transactions on Power Systems*, PS-2, 833–840, November 1987.

26. IEEE Harmonics Modeling and Simulation Task Force Part I, *IEEE Transactions on Power Delivery*, 11(1), 466–474, January 1996.

27. CIGRE Working Group 36-05, Harmonics, characteristic parameters, methods of study, estimates of existing values in the network, *Electra*, 77, 35–54, July 1981.

28. Xia D., Heydt G. T., Harmonic power flow studies: Part I—formulation and solution, Part II—implementation and practical application, *IEEE Transactions on Power Apparatus and Systems*, PAS-101, 1257–1270, June 1982.

29. Acha E., Rico J.J., Acha S., Madrigal M., Harmonic domain modelling of the three-phase thyristor-controlled reactors by means of switching vectors and discrete convolutions, *IEEE Transactions on Power Delivery*, 11(3), 1678–1684, July 1996.

30. Arrillaga J., Smith B. C., Watson N. R., Wood A. R., *Power System Harmonic Analysis*, John Wiley & Sons, New York, 1997.

31. Xu W., Huang Z., Cui Y., Wang H., Harmonic resonance mode analysis, *IEEE Transactions on Power Delivery*, 20(2), 1182–1190, April 2005.

32. Perez-Arriaga I. J., Verghese G. C., Schweppe F. C., Selective modal analysis with applications to electric power systems, Part I: heuristic introduction, *IEEE Transactions on Power Apparatus and Systems*, pas-101, 9, 3117–3125, September 1982.

33. Pileggi D. J., Chandra N. H., Emanuel A. E., Prediction of harmonic voltages in distribution systems, *IEEE Transactions on Power Apparatus and Systems*, PAS-100, 3, 1307–1315, March 1981.

34. Sharma V., Fleming R. J., Niekamp L., An iterative approach for analysis of harmonic penetration in power transmission networks, *IEEE Transactions on Power Delivery*, 6(4), 1698–1706, October 1991.

35. Dommel H. W., Yan A., Wei S., Harmonics from transformer saturation, *IEEE Transactions on Power Systems*, PWRD-1 2, 209–214, April 1986.

36. Xu W., Drakos J. E., Mansour Y., Chang A., A three-phase converter model for harmonic analysis of HVDC systems, *IEEE Transactions on Power Delivery*, 9(3), 1724–1731, July 1994.

37. Smith B. C., Arrillaga J., Wood A. R., Watson N. R., A review of iterative harmonic analysis for AC-DC power systems, *IEEE Transactions on Power Delivery*, 13(1), 180–185, January 1998.

38. Noda T., Semlyen A., Iravani R., Entirely harmonic domain calculation of multiphase nonsinusoidal steady state, *IEEE Transactions on Power Delivery*, 19(3), 1368–1377, July 2004.

39. Rashid M. H., *Power Electronics*, Prentice-Hall, Englewood Cliffs, NJ, 1993.

12 Electromagnetic Transients Analysis

Juan A. Martínez-Velasco and José R. Martí

CONTENTS

12.1 TRANSIENTS IN POWER SYSTEMS

A transient phenomenon in any type of system can be caused by a change of the operating conditions or of the system configuration. Power system transients can be caused by faults, switching operations, lightning strokes, or load variations. The importance of their study is mainly because of the effects that the disturbances can have on the system performance or to the failures they can cause to power equipment.

Stresses that can damage power equipment are of two types—overcurrents and overvoltages. Overcurrents may damage some power components because of excessive heat dissipation; overvoltages may cause insulation breakdowns (failure through solid insulation) or flashovers (insulation failure through air). Protection against overcurrents is performed by specialized equipment whose operation is aimed at disconnecting the faulted position from the rest of the system by separating the minimum number of power components from the unfaulted sections. Protection against overvoltages can be achieved by selecting an adequate insulation level of power equipment or by installing devices aimed at mitigating voltage stresses. To select an adequate protection against both types of stresses, it is fundamental to know their origin, estimate the most adverse conditions, and calculate the transients they can produce.

Several criteria can be used to classify power system transients. They are as follows:

1. According to their origin, disturbances can be external (lightning strokes) or internal (faults, switching operations, load variations).
2. According to the nature of the physical phenomena, power system transients can be electromagnetic, when it is necessary to analyze the interaction between the (electric) energy stored in capacitors and the (magnetic) energy stored in inductors, or electromechanical, when the analysis involves the interaction between the electric energy stored in circuit elements and the mechanical energy stored in rotating machines.

An accurate calculation of transients in power systems is a very difficult task because of the complexity of the equipment involved and the interaction between components. The solution of most transients is not easy by hand calculation, even for small-size systems. For some cases, one can drastically reduce the size of the equivalent circuit and obtain an equation whose solution can be found in textbooks. For a majority of transients, an accurate, or even an approximate, solution can be obtained only by using a computer.

Transients in power systems were initially analyzed with network analyzers. Since the release of the first digital computers, a significant effort has been dedicated to the development of numerical techniques and simulation tools aimed at solving transients in power systems. Hardware and software developments through years have also motivated the development of more powerful techniques and simulation tools. Computer transients programs have significantly reduced time and cost of simulations, and most transients studies are presently based on the application of a digital computer. The main goal of this chapter is to introduce some of the techniques used to analyze electromagnetic transients in power systems, with emphasis on numerical techniques and the application of a digital computer.

Several techniques have been developed to date for computation of electromagnetic transients in power systems. They can be classified into two groups: time-domain and frequency-domain. Some hybrid approaches (i.e., a combination of both techniques) have been also developed. Among the time-domain solution methods, the most popular one is the algorithm proposed by H. W. Dommel, which is a combination of the trapezoidal rule and the method of characteristics, also known as Bergeron's method. This algorithm was the origin of the electromagnetic transients program (EMTP) [1,2]. This abbreviation is nowadays used to designate a family of simulation tools based on the Dommel's scheme.

A general procedure for transients analysis could consist of the following steps:

1. Collect the information about the origin of the transient phenomenon and data of power equipment that could get involved or affected by the transient.
2. Select the zone of the system that should be included in the model to be analyzed.
3. Choose the best representation for each component included in the study zone.
4. Perform the calculation using one of the techniques developed for transients analysis.
5. Analyze the results.

The tasks that will be carried out in the last step will depend on the main goals of the study. For example, if the goal is to analyze the equipment failure rate caused by overvoltages, the results could force some design improvements in the system. These changes could affect the insulation level of some components or the installation of overvoltage mitigation devices.

The selection of the most adequate representation of a power component in transients studies is not an easy task because of the frequency ranges of the transients that can appear in power systems and the different behavior that a component can have for each frequency range. Section 12.2 is dedicated to analyze this aspect; it presents a short introduction to modeling for transients analysis and to the basic elements that will be included in transient models.

Switch operations are one of the most common causes of electromagnetic transients in power systems. Both the closing and the opening of a switch introduce a change in the system structure which can cause overcurrents and overvoltages. The analysis of switching transients in linear systems can be made by applying the superposition principle. Section 12.3 introduces some fundamental concepts for analysis of switching transients in linear systems.

The performance of power components during a transient phenomenon depends on their physical dimensions and on the main characteristics of the phenomenon. Voltages and currents propagate along conductors with finite velocity, so by default models for electromagnetic transients analysis should consider that electrical parameters are distributed. Only when physical dimensions of those parts of a component affected by a transient are small compared with the wavelength of the main frequencies, a representation based on lumped parameters could be used. The concepts related to wave propagation are introduced in Section 12.4.

Several techniques have been developed for transients analysis. Two of these techniques, the Laplace transform and the Dommel's scheme, are detailed in Section 12.5, which includes several practical examples aimed at illustrating the way in which each technique is usually applied as well as their main advantages and limitations.

Although transient simulation tools are presently applied to a great variety of studies, since the development of the first simulation tools, one of the most important fields has been the calculation of overvoltages in power systems. Section 12.6 summarizes the causes and effects of overvoltages and includes some simple simulation cases of each overvoltage type.

All studies and examples included in those sections are based on single-phase models; they are aimed at introducing electromagnetic transients in power systems. More rigorous multiphase models of overhead transmission lines and insulated cables are presented and discussed in Section 12.7.

12.2 POWER SYSTEM COMPONENTS

12.2.1 INTRODUCTION

The goal of a power system is to satisfy the energy demand of a variety of users by generating, transmitting, and distributing the electric energy. These functions are performed by components whose design and behavior are very complex.

An accurate analysis of most transient phenomena in power systems is a difficult task because of the complexity of power components and the interactions that can occur between them. An introduction to this area such as that aimed in this chapter can be carried out by simplifying models of those components involved in the phenomena to be analyzed. Initially, only single-phase models are used for representing power components in transient phenomena. On the other hand, since only transients of electromagnetic nature are analyzed in this chapter, the representation of mechanical parts is omitted.

This section is dedicated to illustrate the importance that the mathematical model selected for representing power components can have on simulation results, discuss some aspects that have to be accounted for when preparing models for transients calculations, and classify the basic circuit elements that will be used to construct simplified models for electromagnetic transients analysis.

12.2.2 MODELING FOR TRANSIENTS ANALYSIS

Accurate simulation of electromagnetic transients should be based on an adequate representation of power components. The frequency range of transient phenomena in power systems can cover a broad spectrum. An accurate simulation of any electromagnetic transient phenomenon could be based on power component models valid for a frequency range that varies from DC to several MHz. However, a representation valid throughout this range of frequencies is not practically possible for most components. Consider, for instance, the behavior of a transformer during electromagnetic transience. A transformer is a device whose behavior is dominated by magnetic coupling between windings and core saturation when transients are of low or medium frequency, that is, well below the first winding resonance. However, when the transients are caused by high-frequency disturbances (e.g., lightning strokes), then the behavior of the transformer is dominated by stray capacitances and capacitances between windings.

Modeling of power components taking into account the frequency dependence of parameters can be practically made by developing mathematical models that are accurate enough for a specific range of frequencies. Each range of frequencies usually corresponds to some particular transient phenomena [3].

The following aspects are to be considered in digital simulations of electromagnetic transients:

- *Parameters*: Very often only approximated or estimated values are used for some parameters whose influence on the representation of a component can be important or very important. In general, this happens with frequency-dependent parameters in simulations of high-frequency transients, for example, above 100 kHz. It is also important to take into account that some parameters may change because of climatic conditions or be dependent on maintenance.

- *Type of study*: In some studies, the maximum peak voltage is the only information of concern. This maximum usually occurs during the first oscillation after the transient phenomenon starts.
- *The study zone*: The more components the system in study has, the higher the probability of insufficient or wrong modeling. In addition, a very detailed representation of a system will require very long simulation time. Some experience will be, therefore, needed to decide how detailed the study zone should be and how to select models for the most important components.

Guidelines for representation of power components in time-domain digital simulations have been the main subject of several publications [3–5].

12.2.3　Basic Circuit Elements

Power component models for electromagnetic transients analysis will be constructed by using basic circuit elements that can be classified into three categories: sources, passive elements, and switches.

1. *Sources*: They are used to represent power generators and external disturbances that can be the origin of some transients (e.g., lightning strokes). Two types of sources can be distinguished: a voltage source (Thevenin representation) and a current source (Norton representation). The equivalent scheme of a voltage source includes a series impedance, while the equivalent scheme of a current source incorporates a parallel admittance. An ideal behavior for each type of source can be assumed by decreasing to zero the series impedance of a voltage source and the parallel admittance of a current source. A detailed analysis of rotating machines is performed in Appendix C.
2. *Passive elements*: Depending on the transient phenomenon, the behavior of some components can be either linear or nonlinear. The transformer is an example for which either a linear or a saturable model can be required. The representation of linear components will be based on lumped-parameter elements (resistance, inductance, capacitance) and distributed-parameter elements (single-phase lossless line). Figure 12.1 shows their symbols and mathematical models. This list can be expanded by adding other basic elements such as the magnetic coupling or the ideal transformer. If the behavior of a component is nonlinear, then its representation can include nonlinear resistances or saturable inductances.
3. *Switches*: They modify the topology of a network by connecting or disconnecting components although they will be also used to represent faults or short circuits. The behavior of an ideal switch can be summarized as follows: its impedance is infinite when it is opened and zero when it is closed; it can close, if it is originally opened, at any instant regardless of the voltage value at the source side, but it will open, if it is originally closed, only when the current goes through zero. The possibility of opening an ideal switch with a non-zero current can be needed to analyze some phenomena, for example, the *current chopping* phenomenon. Several types of switches can be modeled by using different criteria to determine when they should open or close.

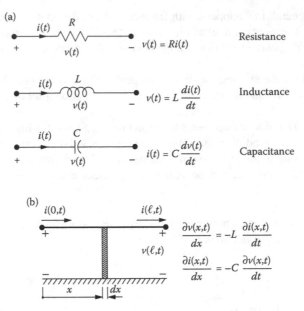

FIGURE 12.1 Basic linear circuit elements: (a) lumped-parameter circuit elements and (b) distributed-parameter single-phase lossless line.

12.3   ANALYSIS OF SWITCHING TRANSIENTS IN LINEAR SYSTEMS

Transients caused by switching operations in linear systems can be analyzed by using the super-position principle. The schemes that result from the application of this principle for analysis of the closing and the opening of a switch are, respectively, shown in Figure 12.2. The details of each case are described in the following paragraphs.

1. Transients caused by the closing of a switch can be analyzed by adding the steady-state sol-ution, which exists before the closing operation, and the transient response of the system which results from short-circuiting voltage sources and open-circuiting current sources to a voltage injected across the switch contacts. Since the voltage across the switch terminals after the operation will be zero, the injected voltage must equal to the voltage that would have existed between switch terminals without closing the switch. Voltages and currents are then calculated by adding the values corresponding to the transient response to the steady-state solution. In some studies, the variable of concern is the current through the switch. To obtain the short-circuit current value, the analysis of the transient response

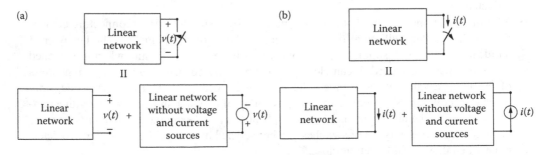

FIGURE 12.2 Application of the superposition principle: (a) closing of a switch and (b) opening of a switch.

will suffice, since the value of this current during steady state, that is, before to the closing operation, is zero (see Figure 12.2a).

2. Transients caused by the opening of a switch can be analyzed by adding the steady-state solution, which exists before the opening operation, and the transient response of the system which results from short-circuiting voltage sources and open-circuiting current sources to a current injected through the switch contacts. Since the current through the switch terminals after the operation will be zero, the injected current must equal to the current that would have existed between switch terminals without closing the switch. When the contacts of a switch start to open, a transient voltage is developed across them. In many transients studies this voltage, known as *transient recovery voltage* (TRV), is the variable of concern. To obtain the TRV waveform, the analysis of the transient response will suffice, since this voltage is zero during steady state, that is, before the opening operation (see Figure 12.2b).

The following example will illustrate how the concepts presented above can be used to analyze transients caused by both types of switching operations.

EXAMPLE 12.1

A fault occurs at the lowest voltage side of a substation. The transient analysis of this case will be based on the single-phase equivalent circuit shown in Figure 12.3. The voltage source in series with resistance R_1 and inductance L_1 represents the Thevenin equivalent of the transmission network seen from the highest voltage side of the transformer; resistance R_2 and inductance L_2 represent the short-circuit impedance of the transformer; finally, the capacitance C represents all cables supplied from the substation. The aim of this example is to analyze, first, the transient overcurrent caused by the short-circuit fault and, second, the transient overvoltage caused upon clearing of the fault.

1. *Closing of the switch*: The circuit of Figure 12.3 is further simplified as shown in Figure 12.4a. The transient current through the switch can be analyzed from the circuit shown in Figure 12.4b, which results upon application of the principle shown in Figure 12.2a.

   Assume that the (phase-to-ground) supply voltage, as seen from the secondary terminals of the transformer, at the time the fault is produced, is

$$e(t) = E_{max} \cos(\omega t + \varphi)$$

where E_{max} is the peak voltage and $f(\omega = 2\pi f)$ is the power frequency of the system.

   The voltage at node 1 before the fault can be deduced from the steady-state solution that exists before the switch operation. If the load is neglected, the voltage can be expressed as follows

$$v_1(t) = V_{max} \cos(\omega t + \varphi - \theta_1)$$

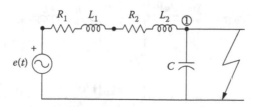

FIGURE 12.3 Simplified scheme for transient analysis.

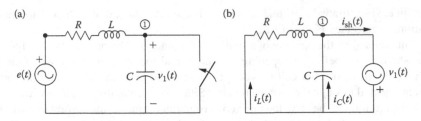

FIGURE 12.4 Equivalent circuit for transient analysis after switch closing.

where

$$V_{max} = E_{max} \frac{1/\omega C}{\sqrt{R^2 + (\omega L - 1/\omega C)^2}}$$

$$\theta_1 = \arctan\left(\frac{\omega L - 1/\omega C}{R}\right) + \pi/2$$

The peak value of $v_1(t)$ can be greater than the peak value of the supply voltage if $\omega^2 LC < 1$, and the resistance value is very low, which are realistic assumptions. Note, in addition, that for a lossless case (i.e., $R = 0$), $\theta_1 = 0$ when $\omega L < 1/\omega C$ and $\theta_1 = \pi$ when $\omega L > 1/\omega C$.

The application of the Kirchhoff's current law at node 1 yields

$$i_{sh}(t) = i_L(t) + i_C(t)$$

The following set of differential equations results from the two branches of the circuit:

$$v_1(t) = L\frac{di_L(t)}{dt} + Ri_L(t)$$

$$i_C(t) = C\frac{dv_1(t)}{dt}$$

Since the expression of $v_1(t)$ is known, the equations can be solved. Remember, on the other hand, that the initial value of the short-circuit current is zero.

2. *Opening of the switch*: The system under fault condition can be represented by the circuit shown in Figure 12.5a. The transient phenomenon caused by the fault clearing will be analyzed by means of the circuit shown in Figure 12.5b, which results upon application of the principle shown in Figure 12.2b. The variable of concern in this case is the TRV, that is, the transient voltage developed across the switch.

If the (phase-to-ground) supply voltage, as seen from the secondary terminals of the transformer, at the time the fault is produced has the same expression as above, the steady-state short-circuit current will be

$$i_{sh}(t) = I_{max} \cos(\omega t + \varphi - \theta_2)$$

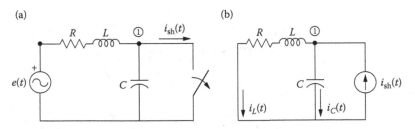

FIGURE 12.5 Equivalent circuit for transient analysis after switch opening.

where

$$I_{max} = \frac{E_{max}}{\sqrt{R^2 + (\omega L)^2}} \qquad \theta_2 = \arctan\left(\frac{\omega L}{R}\right)$$

Note that $\theta_2 = \pi/2$ when losses are neglected; that is, the steady-state short-circuit current is in quadrature and lagging with respect to the voltage that exists across the switch before the opening operation.

The application of the Kirchhoff's current law at node 1 yields the same equation that was obtained in the previous case. Likewise, the relationships between voltages and currents in both branches are the same as above. Therefore, the set of differential equations that was obtained in the previous case is still valid.

Taking into account that

$$i_L(t) = i_{sh}(t) - i_C(t) = i_{sh}(t) - C\frac{dv_1(t)}{dt}$$

the following differential equation results upon substitution of this result in the equation of the first branch

$$v_1(t) = R\left[i_{sh}(t) - C\frac{dv_1(t)}{dt}\right] + L\frac{di_{sh}(t)}{dt} - LC\frac{d^2v_1(t)}{dt^2}$$

After some manipulation, this equation becomes

$$\frac{d^2v_1(t)}{dt^2} + \frac{R}{L}\frac{dv_1(t)}{dt} + \frac{1}{LC}v_1(t) = \frac{R}{LC}i_{sh}(t) + \frac{1}{C}\frac{di_{sh}(t)}{dt}$$

The solution of this case seems to be more complicated although the initial value of the variable of concern, the voltage across the switch, is again zero.

This example was aimed at illustrating how the fundamental concepts presented in this section can be applied to analyze switching transients. These concepts will be used in some examples included in the following sections, but the solutions will be derived using a more elaborated technique for transients analysis.

12.4 WAVE PROPAGATION IN POWER SYSTEMS

12.4.1 INTRODUCTION

The analysis of electromagnetic transients in power components has to consider that the electrical parameters are distributed. During a transient phenomenon, only for conductors whose length is short compared with the wavelengths of the main frequencies of the transient the representation can be based on a lumped-parameter model. This means that the analysis of electromagnetic transients of very high frequency, for example, above 1 MHz, will be based on distributed-parameter models even if conductors are a few meters short.

Voltages and currents in a component represented by a distributed-parameter model do not have the same value over the entire length of the conductor, and they travel with finite velocities whose values depend on the physical characteristics of the component. The analysis of transient phenomena in distributed-parameter models is based on the concept of traveling waves.

The following subsections introduce the concepts associated to wave propagation in power components. The analysis is based on the model of a single-phase lossless distributed-parameter line, which in the rest of this chapter will be also named as ideal line.

The most popular methods for analysis of electromagnetic transients in ideal lines are those developed by Bergeron and Bewley. The Bergeron's method will serve as a basis for the numerical technique presented in this chapter. The method proposed by Bewley, also known as the *lattice diagram*, will be detailed and applied in this section.

12.4.2 SOLUTION OF THE SINGLE-PHASE LOSSLESS LINE EQUATIONS

Figure 12.6 shows the scheme of a single-phase line and the equivalent circuit of a line element. The equations of this circuit can be written as follows:

$$\frac{\partial v(x, t)}{\partial x} = -L\frac{\partial i(x, t)}{\partial t}$$

$$\frac{\partial i(x, t)}{\partial x} = -C\frac{\partial v(x, t)}{\partial t} \qquad (12.1)$$

where L and C are, respectively, the inductance and the capacitance per unit length, and x is the distance with respect to the sending end of the line.

After differentiating with respect the variable x, these equations become

$$\frac{\partial^2 v(x, t)}{\partial x^2} = LC\frac{\partial^2 v(x, t)}{\partial t^2}$$

$$\frac{\partial^2 i(x, t)}{\partial x^2} = LC\frac{\partial^2 i(x, t)}{\partial t^2} \qquad (12.2)$$

The general solution of the voltage equation has the following form:

$$v(x, t) = f_1(x - vt) + f_2(x + vt) \qquad (12.3)$$

where

$$v = \frac{1}{\sqrt{LC}} \qquad (12.4)$$

Both f_1 and f_2 are voltage functions, and v is the so-called *propagation velocity*.

Since the current Equation 12.2 has the same form that the voltage equation, its solution will also have a similar expression. It is, however, possible to obtain a general solution of the current equation based on that for the voltage. This solution could be expressed as follows:

$$i(x, t) = \frac{f_1(x - vt) - f_2(x + vt)}{Z_c} \qquad (12.5)$$

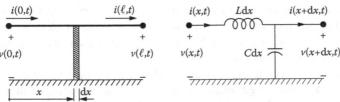

FIGURE 12.6   Scheme of a single-phase lossless line.

where

$$Z_c = \sqrt{\frac{L}{C}} \tag{12.6}$$

is the *surge impedance* of the line. It is a resistance for lossless (ideal) lines.

Note that $f_1(x - vt)$ remains constant if the value of the quantity $(x - vt)$ is also constant. That is, the value of this function will be the same for any combination of x and t such that the above quantity has the same value. The function f_1 represents a voltage traveling wave toward increasing x, while $f_2(x + vt)$ represents a voltage traveling wave toward decreasing x. Both waves are neither distorted nor damped while propagating along the line. The general solution of the voltage and the current at any point along an ideal line is, therefore, constructed by superposition of waves that travel in both directions. The expressions of f_1 and f_2 are determined for a specific case from the boundary and the initial conditions.

From Equations 12.3 and 12.5, one can deduce that the relationship between voltage and current waves is always the surge impedance of the line. However, this relationship can be positive or negative depending on the direction of propagation (see Figure 12.7).

12.4.3 WAVE PROPAGATION AND REFLECTION

The presence of traveling waves that propagate in both directions along a line can be justified by the presence of discontinuity points, that is, points where waves meet a propagation media with different characteristic values (surge impedance and propagation velocity). The following cases will be useful to analyze the physical phenomena that occur when a traveling wave meets a discontinuity point.

- *Line termination*

    Figure 12.8 shows the particular case to be analyzed. Consider that a wave, which propagates along an ideal line, reaches an end where a resistance R_t has been installed.

    Voltage and current for the traveling wave, known as *incident wave*, are related as follows:

$$v_i = Z_c i_i \tag{12.7}$$

where Z_c is the surge impedance of the line along which the incident wave is traveling.

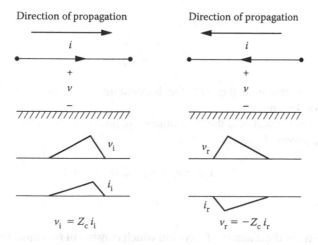

FIGURE 12.7 Relationships between voltage and current waves.

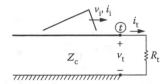

FIGURE 12.8   Line termination.

At the receiving end of the line, the relation between voltage and current is the following one (see Figure 12.8):

$$v_t = R_t i_t \tag{12.8}$$

The surge impedance and the terminal resistance are different, but a mismatch of both voltage and current exists at the discontinuity point; therefore, an adjustment is needed for both variables. A new wave, known as *reflected wave*, is produced when the incident wave reaches the discontinuity point.

Relationships between voltage and current waves at this terminal can be expressed as follows:

$$v_t = v_i + v_r \quad i_t = i_i + i_r \tag{12.9}$$

where subscripts i, r, and t are used to denote incident, reflected, and transmitted waves, respectively.

The reflected wave travels back along the line, far away from the receiving end of the line. Since this wave propagates in an opposite sense to that of the incident wave, voltage and current for this wave are related as follows:

$$v_r = -Z_c i_r \tag{12.10}$$

Substitution of Equations 12.7 through 12.9 into Equation 12.10 yields

$$v_r = r v_i \quad i_r = -r i_i \tag{12.11}$$

where

$$r = \frac{R_t - Z_c}{R_t + Z_c} \tag{12.12}$$

is known as the *reflection coefficient* at the discontinuity point, in this particular case at the receiving end of the line.

Voltage and current waves at the discontinuity point are obtained from the superposition of incident and reflected waves

$$v_t = (1 + r)v_i = t v_i \quad i_t = (1 - r)i_i \tag{12.13}$$

The quantity $t = (1 + r)$ is also known as the *refraction coefficient*.

- *Transition point*

Figure 12.9 shows the diagram of a system which consists of two ideal lines with different surge impedances. Assume that the incident wave travels along Line 1 toward Line 2.

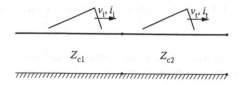

FIGURE 12.9 Transition point.

At the transition (or discontinuity) point, this wave will note a change in the characteristic parameters of the propagation mean. The traveling wave that will pass to Line 2 will depend on the parameters of both lines.

Following a similar reasoning as for the previous case, one can conclude that when the incident wave (v_i, i_i) reaches the transition point, two new waves are generated—a reflected wave (v_r, i_r), which will travel back along Line 1, and a refracted or transmitted wave (v_t, i_t), which will pass and travel along Line 2.

Relationships between voltage and current waves at the transition point can be expressed as follows:

$$v_t = v_i + v_r \quad i_t = i_i + i_r \tag{12.14}$$

On the other hand

$$v_i = Z_{c1} i_i \quad v_r = -Z_{c1} i_r \quad v_t = Z_{c2} i_t \tag{12.15}$$

Substitution of these expressions into the current Equation 12.14 yields

$$v_r = r v_i \quad v_t = (1 + r)v_i = t v_i \tag{12.16}$$

where

$$r = \frac{Z_{c2} - Z_{c1}}{Z_{c2} + Z_{c1}} \quad t = 1 + r = \frac{2Z_{c2}}{Z_{c2} + Z_{c1}} \tag{12.17}$$

r is the reflection coefficient, and t is the refraction coefficient at the transition point.

As for the wave currents, the following results are obtained

$$i_r = -r i_i \quad i_t = (1 - r)i_i \tag{12.18}$$

It is important to keep in mind that the incident wave could have been traveling along Line 2 toward Line 1. In such case, the reflection coefficient would have been

$$r' = \frac{Z_{c1} - Z_{c2}}{Z_{c1} + Z_{c2}} \tag{12.19}$$

One can observe that the expressions of reflection coefficients follow a very simple rule. When an incident wave reaches a discontinuity point, the reflection coefficient is obtained from the equivalent surge impedance that the wave sees at the discontinuity point, Z_{eq}, and the surge impedance of the mean in which the wave is traveling, Z_c, that is,

$$r = \frac{Z_{eq} - Z_c}{Z_{eq} + Z_c} \tag{12.20}$$

The reflection coefficients of the following cases are of interest since they are encountered in many practical cases.

- *Opened line*: The traveling waves that reach an opened terminal meet an infinite impedance or resistance. For this particular case, Equation 12.12 becomes

$$r = 1 \tag{12.21}$$

  Therefore, at an opened terminal

$$v_r = v_i \quad i_r = -i_i \tag{12.22}$$

  from where

$$v_t = v_i + v_r = 2v_i \quad i_t = i_i + i_r = 0 \tag{12.23}$$

  According to these results, when a traveling wave reaches an opened terminal, the voltage wave is doubled, while the current wave is canceled, as expected.
- *Short-circuited line*: The incident waves that reach a short-circuited terminal meet a zero impedance or resistance. In such condition, Equation 12.12 becomes

$$r = -1 \tag{12.24}$$

  Therefore, at a short-circuited terminal

$$v_r = -v_i \quad i_r = i_i \tag{12.25}$$

  from where

$$v_t = v_i + v_r = 0 \quad i_t = i_i + i_r = 2i_i \tag{12.26}$$

  According to these results, when a traveling wave reaches a short-circuited terminal, the voltage wave is canceled, as expected, while the current wave is doubled.
- *Matched line*: A line is matched at one terminal when the incident wave that reaches this terminal meets an impedance or a resistance equal to the surge impedance of the line. The reflection coefficient for this particular case becomes

$$r = 0 \tag{12.27}$$

  Therefore, when a line is matched

$$v_r = 0 \quad i_r = 0 \tag{12.28}$$

  from where

$$v_t = v_i \quad i_t = i_i \tag{12.29}$$

When a traveling wave reaches a matched terminal there will be no reflected wave, which makes sense since the incident wave does not see any change in the characteristic parameters of the propagation mean.

12.4.4 THE LATTICE DIAGRAM

The solution of voltages and currents caused during a given transient phenomenon can be deduced from the boundary and initial conditions of the system under study. The method presented in this subsection obtains the voltage and the current at a given point from the superposition of the traveling waves that are produced at that point after successive reflections. The method is presented with a very simple case for which the results obtained in the previous subsection are used.

The conclusions derived from the cases analyzed above can be summarized as follows:

1. When a line is energized, a voltage and a current start to propagate along the line. The propagation takes place without distortion nor damping and with a finite velocity, being the relationship between voltage and current waves the surge impedance of the line.
2. When a traveling wave meets a discontinuity point, such as another line or a load, it will generate two waves: a reflected wave that will travel back and a transmitted wave that will pass the discontinuity point.

Consider the system shown in Figure 12.10, where a single-phase lossless line is energized from a voltage source. The internal series impedance of the source is a resistance R_0.

The voltage equation at the sending end of the line can be written as follows:

$$e(t) = v(t) + R_0 i(t) \tag{12.30}$$

where $v(t)$ and $i(t)$ are, respectively, the voltage and the current wave that will appear at the sending end of the line and propagate along this line immediately after closing the switch.

Since the relationship between both waves is

$$v(t) = Z_c i(t)$$

the voltage equation becomes

$$e(t) = v(t) + \frac{R_0}{Z_c} v(t) = \left(1 + \frac{R_0}{Z_c}\right) v(t) \tag{12.31}$$

from where

$$v(t) = e(t) \frac{Z_c}{R_0 + Z_c} \tag{12.32}$$

This expression is general; that is, it can be used in all cases where a line is energized from a source and provides the wave voltage that will initially propagate from the sending to the receiving end of the line.

This wave will be reflected back at the receiving end; this reflected wave will encounter a discontinuity point at the sending end, where a new reflected wave will be produced, which will travel to the receiving end. The transient phenomenon will continue with incident and reflected waves propagating between both line ends.

The values of the reflection coefficients at each end of the line are as follows:

- Sending end

$$r_s = \frac{R_0 - Z_c}{R_0 + Z_c}$$

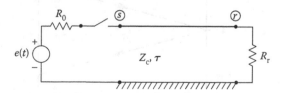

FIGURE 12.10   Energization of an ideal line.

- Receiving end

$$r_r = \frac{R_r - Z_c}{R_r + Z_c}$$

The lattice diagram is a very useful tool to keep track of all traveling waves. The diagram shows the location of the discontinuity points, as well as the incident, the reflected and the transmitted waves at each point. The voltages and currents at one point are determined by the superposition of all traveling waves present at that point.

Assume that τ is the *travel time*, that is, the time that waves needs to travel between both ends of the line shown in Figure 12.10. Then

$$\tau = \frac{\ell}{v} \tag{12.33}$$

where ℓ is the line length, and v is the propagation velocity of waves traveling along this line.

Figure 12.11 depicts the lattice diagram that corresponds to this test system. The diagram shows the traveling waves at each line end. Since the behavior of the line is linear, the term $v(t)$ has been omitted in the diagram; that is, all traveling waves should be multiplied by the voltage wave that starts to travel after the switch is closed.

The following voltage waves arrive to each end of the line:

- Sending end

$$v(t)\varepsilon(t)$$

$$v(t - 2\tau)\varepsilon(t - 2\tau)(r_r + r_s r_r) = v(t - 2\tau)\varepsilon(t - 2\tau)(1 + r_s)r_r \tag{12.34}$$

$$v(t - 4\tau)\varepsilon(t - 4\tau)(r_s r_r^2 + r_s^2 r_r^2) = v(t - 4\tau)\varepsilon(t - 4\tau)(1 + r_s)r_s r_r^2$$

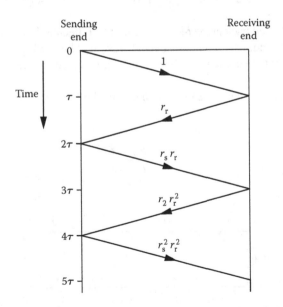

FIGURE 12.11 The lattice diagram.

- Receiving end

$$v(t - \tau)\varepsilon(t - \tau)(1 + r_r)$$

$$v(t - 3\tau)\varepsilon(t - 3\tau)(r_s r_r + r_s r_r^2) = v(t - 3\tau)\varepsilon(t - 3\tau)(1 + r_r)r_s r_r \qquad (12.35)$$

$$v(t - 5\tau)\varepsilon(t - 5\tau)(r_s^2 r_r^2 + r_s^2 r_r^3) = v(t - 5\tau)\varepsilon(t - 5\tau)(1 + r_r)r_s^2 r_r^2$$

where $\varepsilon(t - n\tau)$ is the unit step delayed an interval $n\tau$.

Note that all voltage waves at the sending end include a delay, except the first one.

The following example shows how to use the lattice diagram with a more complex case, which will illustrate the advantages and limitations of this method.

EXAMPLE 12.2

Consider the single-phase system shown in Figure 12.12. It consists of a lossless overhead line and an insulated cable which are energized from a voltage source. Determine the voltages that will result at both ends of each component after closing the switch, if the internal series impedance of the source is a resistance R_0. The model of a lossless-insulated cable is the same that the model of an overhead line; however, surge impedances and propagation velocities can be very different for each component. Assume that the travel time of the line is twice the travel time of the cable.

To analyze this case, it is important to keep in mind that there are three discontinuity points, nodes 1, 2, and 3; therefore, the analysis will be based on the reflection coefficients at each of these points. On the other hand, the reflection coefficient at node 2 will depend on the propagation direction of the incident wave, that is, the reflection coefficient for a wave traveling from the cable to the line will be opposite to the reflection coefficient when the wave travels from the line to the cable. The voltage wave that is generated at node 1 when the switch is closed can be obtained as shown in the previous case

$$v(t) = e(t)\frac{Z_{c1}}{R_0 + Z_{c1}}$$

The reflection coefficients at the discontinuity points are as follows:

- Node 1

$$r_1 = \frac{R_0 - Z_{c1}}{R_0 + Z_{c1}}$$

- Node 2

$$r_{2+} = \frac{Z_{c2} - Z_{c1}}{Z_{c2} + Z_{c1}}$$

$$r_{2-} = \frac{Z_{c1} - Z_{c2}}{Z_{c1} + Z_{c2}}$$

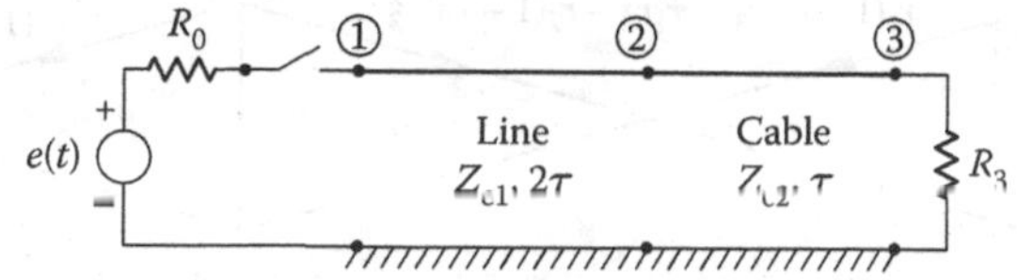

FIGURE 12.12 Example 2: Diagram of the test system.

where r_{2+} is the coefficient when the incident wave travels from the line to the cable, while r_{2-} is the coefficient when the incident wave travels from the cable to the line.

- Node 3

$$r_3 = \frac{R_3 - Z_{c2}}{R_3 + Z_{c2}}$$

Assuming that τ is the travel time for the cable, the lattice diagram of this case is that shown in Figure 12.13. Note that the diagram has been constructed by using the following relationships:

$$r_{2+} = r_2 \quad r_{2-} = -r_2$$

Taking into account that each voltage term incorporates a delay and that each wave included in the diagram has to be multiplied by the voltage wave originated at the time the switch is closed, $v(t)$, the following expressions are obtained for the voltages at all discontinuity points:

- Node 1

$$v_1(t) = v(t) + v(t - 4\tau)\varepsilon(t - 4\tau)(1 + r_1)r_2 + v(t - 6\tau)\varepsilon(t - 6\tau)(1 + r_1)(1 - r_2^2)r_3 + \cdots$$

- Node 2

$$v_2(t) = v(t - 2\tau)\varepsilon(t - 2\tau)(1 + r_2) + v(t - 4\tau)\varepsilon(t - 4\tau)(1 - r_2^2)r_3 + \cdots$$

- Node 3

$$v_3(t) = v(t - 3\tau)\varepsilon(t - 3\tau)(1 + r_2)(1 + r_3) - v(t - 5\tau)\varepsilon(t - 5\tau)(1 + r_2)(1 + r_3)r_2r_3 + \cdots$$

It is evident that the analysis of this case by means of the lattice diagram becomes very complicated after a few reflections. This technique is very useful to understand the physics behind transient phenomena in components with distributed parameters, but its application is limited to those cases in which only the first traveling waves are of interest. The technique is, on the other hand, limited to lossless components, in which waves propagate without distortion and damping.

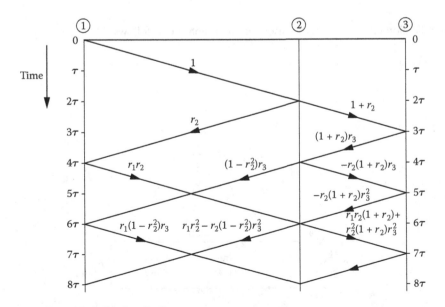

FIGURE 12.13   Example 2: Lattice diagram.

12.5 TECHNIQUES FOR ELECTROMAGNETIC TRANSIENTS ANALYSIS

12.5.1 Introduction

Several techniques can be used to analyze electromagnetic transients in power systems. Although most transients studies are presently based on digital simulation, those techniques applied before the development of the digital computer are still very useful, mainly for educational purposes.

Methods for transients analysis can be classified into two groups: analytical and numerical. The lattice diagram, introduced in the previous section, could be seen as a third category that is very useful when components with distributed parameters are included in the system model.

Before the development of the digital computer, many transients studies were performed with the help of network analyzers. A description of the characteristics of these tools or those of the new real-time digital systems is out of the scope of this chapter.

The following subsections introduce the main principles of each technique and include some examples that will illustrate their advantages and limitations.

12.5.2 Analytical Techniques

12.5.2.1 The Laplace Transform

The performance of a power system during a transient can be described by a set of differential equations that are derived from the differential equations that represent the models of physical components. The simultaneous solution of these equations can be a difficult task. Several techniques have been developed to provide an efficient and a systematic procedure for solving this problem. This chapter presents the application of the Laplace transform. For the application of other transform techniques (e.g., the Fourier transform), see Reference 6.

The Laplace transform of a function of time, $f(t)$, is defined by the following expression:

$$F(s) = \int_0^\infty e^{-st} f(t) \mathrm{d}t \tag{12.36}$$

This integral transforms $f(t)$ into another function whose variable is s, which is a complex variable.

For a detailed description of the Laplace transform, the reader is referred to the specialized literature [6–8].

The following example illustrates the application of the Laplace transform to transients analysis in linear circuits and introduces some concepts that will be used in the subsequent sections.

EXAMPLE 12.3

Consider the case analyzed in Example 12.1, see Figure 12.3. The aim of this example is to analyze the transient voltage across the switch caused upon clearing of the fault by applying the Laplace transform.

The variable of concern (i.e., the recovery voltage) is initially zero, therefore only the transient component of the voltage has to be analyzed. In addition, the driving input will be derived from the steady-state solution of the circuit that exists before the switch operation.

The transient voltage across the switch, caused by the opening operation, is analyzed by means of the circuits shown in Figure 12.14.

The equations of this case are as follows:

$$i_{sh}(t) = i_L(t) + i_C(t)$$

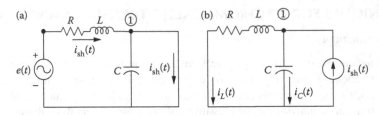

FIGURE 12.14 Analysis after switch opening: (a) equivalent circuit for steady-state analysis prior to switch opening and (b) equivalent circuit for transients analysis after switch opening.

$$v_1(t) = Ri_L(t) + L\frac{di_L(t)}{dt}$$

$$i_C(t) = C\frac{dv_1(t)}{dt}$$

After applying the Laplace transform they become

$$I_{sh}(s) = I_L(s) + I_C(s)$$

$$V_1(s) = RI_L(s) + sLI_L(s)$$

$$I_C(s) = sCV_1(s)$$

since initial values of all variables of the circuit shown in Figure 12.14b are zero.

Solving the second equation for $I_L(s)$, the following expression for $I_{sh}(s)$ is derived upon substitution of the expressions for both $I_L(s)$ and $I_C(s)$ in the first equation

$$I_{sh}(s) = \left(\frac{1}{R + sL} + sC\right)V_1(s)$$

This result can be manipulated to obtain the following form:

$$I_{sh}(s) = \frac{(R + sL)sC + 1}{R + sL} V_1(s) = \frac{s^2 + sa + \omega_0^2}{s + \alpha} CV_1(s)$$

where

$$\alpha = \frac{R}{L} \quad \omega_0 = \frac{1}{\sqrt{LC}}$$

The Laplace transform of the transient voltage across the switch can be now obtained

$$V_1(s) = \frac{I_{sh}(s)}{C} \frac{s + \alpha}{s^2 + sa + \omega_0^2}$$

The expression for $I_{sh}(s)$ in this equation will be derived from the steady-state solution of the circuit shown in Figure 12.14a.

If the switch has an ideal behavior during the opening operation, then it will only open after a zero-crossing of the short-circuit current. Therefore, the time-domain expression for $i_{sh}(t)$, after switch opening, could be written as

$$i_{sh}(t) = I_{max} \cos(\omega t \pm \pi/2) \quad \left(I_{max} = \frac{E_{max}}{\sqrt{R^2 + (\omega L)^2}}\right)$$

Assume that

$$i_{sh}(t) = \frac{E_{max}}{\sqrt{R^2 + (\omega L)^2}} \sin \omega t$$

The Laplace transform of the short-circuit current for analysis of the transient voltage in Figure 12.14b is then

$$I_{sh}(s) = \frac{E_{max}}{\sqrt{R^2 + (\omega L)^2}} \frac{\omega}{s^2 + \omega^2}$$

Substitution of this result in the expression of $V_1(s)$ yields

$$V_1(s) = \frac{E_{max}}{C\sqrt{R^2 + (\omega L)^2}} \frac{\omega}{s^2 + \omega^2} \frac{s + \alpha}{s^2 + s\alpha + \omega_0^2}$$

After some manipulation this equation becomes

$$V_1(s) = \frac{E_{max}}{LC\sqrt{\alpha^2 + \omega^2}} \frac{\omega}{s^2 + \omega^2} \frac{s + \alpha}{s^2 + s\alpha + \omega_0^2} = \omega_0^2 \frac{E_{max}}{\sqrt{\alpha^2 + \omega^2}} \frac{\omega}{s^2 + \omega^2} \frac{s + \alpha}{s^2 + s\alpha + \omega_0^2}$$

The antitransform of this equation is very complex. The analysis can be significantly simplified by assuming $R = 0$, which implies $\alpha = 0$. The Laplace-domain of the transient voltage across the switch is then

$$V_1(s) = \omega_0^2 E_{max} \frac{s}{(s^2 + \omega^2)(s^2 + \omega_0^2)}$$

To obtain the antitransform, the second term of the right-hand side can be rewritten by taking partial fractions

$$V_1(s) = \omega_0^2 E_{max} \frac{1}{\omega_0^2 - \omega^2} \left[\frac{s}{s^2 + \omega^2} - \frac{s}{s^2 + \omega_0^2}\right]$$

Since in real systems $\omega_0 \gg \omega$, the antitransform of this expression can be written as follows:

$$v_1(t) \approx E_{max}(\cos \omega t - \cos \omega_0 t)$$

This result has two similar terms but with a different frequency. With the above relation between frequencies, the TRV will have an undamped sinusoidal waveform with a peak magnitude that doubles that of the AC source voltage.

With a nonzero value of the resistance, the peak value of the TRV would have been lower than that without resistance, and would have shown damped oscillations with a rate of decay that would depend on the time constant of the RL branch.

Consider that the parameters of the equivalent circuit are

$$E_{max} = 1V \quad f = 50\,Hz(\omega = 2\pi 50) \quad L = 175.63\,mH \quad C = 2.0\,\mu F$$

The TRV across the switch will be that depicted in Figure 12.15. These results, deduced with different values of the resistance, confirm the above conclusions. The TRV reaches a peak value

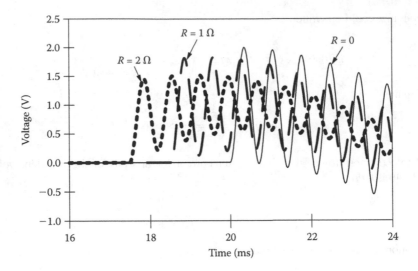

FIGURE 12.15   TRV caused by switch opening.

that doubles that of the AC power supply voltage, and the oscillations remain undamped when the resistance of the circuit is neglected. However, the oscillations decay when the resistance is not neglected, and it is evident that this resistance reduces the peak value of the oscillations.

The results shown in Figure 12.15 were obtained by using a digital computer and simulating the opening of the switch. Note that, by following this approach, the instant at which the switch opens is delayed as the resistance value decreases, and the peak value of the TRV decreases as the resistance value increases.

The study of this example has provided some relationships between voltages and currents in the Laplace-domain that can be used to justify and introduce the equivalents of the basic circuit elements as well as the concepts of operational impedance and operational admittance. A careful analysis of the expressions that result upon application of the Laplace transform would conclude that they are very similar to those that result in steady-state AC analysis using phasors. There is, however, a difference since the initial conditions with which transients begin are important.

The following subsection details the derivation of the Laplace-domain equivalents for basic circuit elements. A methodology for application of these equivalents in transients analysis is presented and applied to some simple test cases.

12.5.2.2   Laplace-Domain Equivalents of Basic Circuit Elements

The application of the Laplace transform to the equations of the basic circuit elements can be used to obtain two types of equivalent circuits for each element—Thevenin and Norton. Only the Norton equivalents are derived in this chapter.

1. *Resistance*: The behavior of a resistance, connected to nodes k and m, during a transient phenomenon is described by the following equation:

$$v_{km}(t) = v_k(t) - v_m(t) = Ri_{km}(t) \qquad (12.37)$$

   The application of the Laplace transform is straightforward

$$V_{km}(s) = V_k(s) - V_m(s) = RI_{km}(s) \qquad (12.38)$$

2. *Inductance*: The equation that represents the behavior of an inductance is the following one:

$$v_{km}(t) = v_k(t) - v_m(t) = L\frac{di_{km}(t)}{dt} \tag{12.39}$$

After applying the Laplace transform, the following form is derived:

$$V_{km}(s) = V_k(s) - V_m(s) = sLI_{km}(s) - Li_{km}(0) \tag{12.40}$$

where $i_{km}(0)$ is the current through the inductance at time $t = 0$.
The solution for the current yields the following expression:

$$I_{km}(s) = \frac{V_{km}(s)}{sL} + \frac{i_{km}(0)}{s} \tag{12.41}$$

which represents the Norton equivalent of an inductance in Laplace-domain, see Figure 12.16.

3. *Capacitance*: The equation that represents the behavior of a capacitance is the following one:

$$i_{km}(t) = C\frac{d}{dt}[v_k(t) - v_m(t)] = C\frac{dv_{km}(t)}{dt} \tag{12.42}$$

After applying the Laplace transform, the following form is derived:

$$I_{km}(s) = sCV_{km}(s) - Cv_{km}(0) \tag{12.43}$$

where $v_{km}(0)$ is the voltage across the capacitance at time $t = 0$.
This equation corresponds to the circuit shown in Figure 12.17, which represents the Norton equivalent of a capacitance in Laplace-domain.

4. *Ideal line*: The scheme and the equations of a single-phase lossless line were presented in Section 12.4. The Laplace transform of Equation 12.1, assuming zero initial conditions,

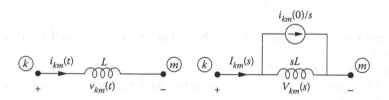

FIGURE 12.16 Laplace-domain equivalent of an inductance.

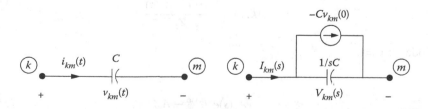

FIGURE 12.17 Laplace-domain equivalent of a capacitance.

provide the following expressions:

$$\frac{\partial V(x, s)}{\partial x} = -sLI(x, s)$$

$$\frac{\partial I(x, s)}{\partial x} = -sCV(x, s)$$

(12.44)

If the equations are differentiated with respect to the variable x, they become

$$\frac{\partial^2 V(x, s)}{\partial x^2} = \beta^2(s)V(x, s)$$

$$\frac{\partial^2 I(x, s)}{\partial x^2} = \beta^2(s)I(x, s)$$

(12.45)

where

$$\beta(s) = s\sqrt{LC}$$

(12.46)

A general solution of the first equation has the following form:

$$V(x, s) = A_1(s)e^{-\beta(s)x} + A_2(s)e^{+\beta(s)x}$$

(12.47)

where $A_1(s)$ and $A_2(s)$ are constants that will be derived from the boundary conditions. The differentiation of this voltage expression gives the following form:

$$\frac{\partial V(x, s)}{\partial x} = -\beta(s)[A_1(s)e^{-\beta(s)x} - A_2(s)e^{+\beta(s)x}]$$

After substituting this expression in Equation 12.44 and solving for the current, the following form is obtained:

$$I(x, s) = -\frac{1}{sL}\frac{\partial V(x, s)}{\partial x} = \frac{[A_1(s)e^{-\beta(s)x} - A_2(s)e^{+\beta(s)x}]}{Z_c}$$

(12.48)

In the most general case, Z_c is an impedance, but for a lossless line it is a resistance, see Equation 12.6.

The solution of Equation 12.44 is obtained from the boundary conditions. If the line is connected to nodes k and m, the following expressions are obtained for each node:

• Node k; $x = 0$

$$V(k, s) = A_1(s) + A_2(s)$$

$$I(k, s) = \frac{A_1(s) - A_2(s)}{Z_c}$$

(12.49)

• Node m; $x = \ell$

$$V(m, s) = A_1(s)e^{-\beta(s)\ell} + A_2(s)e^{+\beta(s)\ell}$$

$$I(m, s) = \frac{A_1(s)e^{-\beta(s)\ell} - A_2(s)e^{+\beta(s)\ell}}{Z_c}$$

(12.50)

From the boundary conditions at node k, one can obtain

$$V(k, s) + Z_c I(k, s) = 2A_1(s)$$
$$V(k, s) - Z_c I(k, s) = 2A_2(s)$$

(12.51)

The solution for $A_1(s)$ and $A_2(s)$ yields

$$A_1(s) = \frac{V(k, s) + Z_c I(k, s)}{2}$$

$$A_2(s) = \frac{V(k, s) - Z_c I(k, s)}{2}$$

(12.52)

From the boundary conditions at node m, one can obtain

$$V(m, s) + Z_c I(m, s) = 2A_1(s)e^{-\beta(s)\ell}$$

$$V(m, s) - Z_c I(m, s) = 2A_2(s)e^{+\beta(s)\ell}$$

(12.53)

The solution for $A_1(s)$ and $A_2(s)$ yields

$$A_1(s) = \frac{V(m, s) + Z_c I(m, s)}{2}e^{+\beta(s)\ell}$$

$$A_2(s) = \frac{V(m, s) - Z_c I(m, s)}{2}e^{-\beta(s)\ell}$$

(12.54)

Before proceeding with the derivation of the line equivalent, the changes in variables and notations shown in Figure 12.18 are introduced. According to this figure

$$V_k(s) = V(k, s) \quad I_{km}(s) = +I(k, s)$$
$$V_m(s) = V(m, s) \quad I_{mk}(s) = -I(m, s)$$

For a lossless line

$$v = \frac{1}{\sqrt{LC}} = \frac{s}{\beta(s)}$$

(12.55)

Therefore,

$$\beta(s)\ell = s\frac{\ell}{v} = s\tau$$

(12.56)

where τ is the travel time.

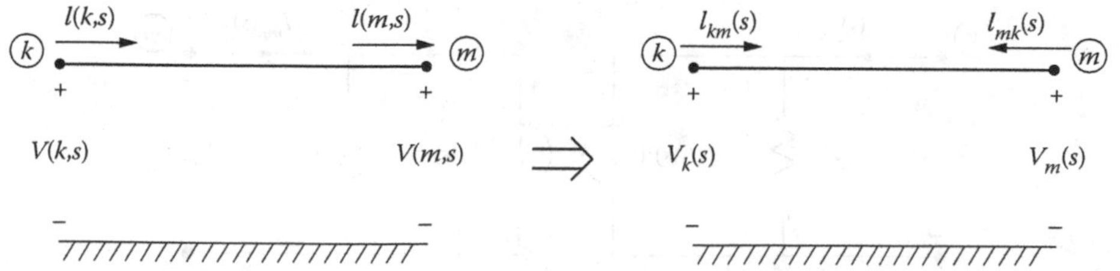

FIGURE 12.18   Change in the notation of the line variables.

After matching the two expressions of $A_1(s)$ and $A_2(s)$ and taking into account the new notations, the following forms are derived:

$$V_k(s) + Z_c I_{km}(s) = [V_m(s) - Z_c I_{mk}(s)]e^{+s\tau}$$
$$V_k(s) - Z_c I_{km}(s) = [V_m(s) + Z_c I_{mk}(s)]e^{-s\tau}$$
(12.57)

Solving for $I_{km}(s)$ from the second equation and for $I_{mk}(s)$ from the first one, the following results are obtained:

$$I_{km}(s) = \frac{V_k(s)}{Z_c} - \left[\frac{V_m(s)}{Z_c} + I_{mk}(s)\right]e^{-s\tau} = \frac{V_k(s)}{Z_c} + I_k(s)$$
$$I_{mk}(s) = \frac{V_m(s)}{Z_c} - \left[\frac{V_k(s)}{Z_c} + I_{km}(s)\right]e^{-s\tau} = \frac{V_m(s)}{Z_c} + I_m(s)$$
(12.58)

These equations correspond to the circuit shown in Figure 12.19, in which the following notations have been used:

$$I_k(s) = -\left[\frac{V_m(s)}{Z_c} + I_{mk}(s)\right]e^{-s\tau}$$
$$I_m(s) = -\left[\frac{V_k(s)}{Z_c} + I_{km}(s)\right]e^{-s\tau}$$
(12.59)

12.5.2.3 Application of the Laplace Transform

The procedure to solve electromagnetic transients by means of the Laplace transform can be summarized as follows:

1. Obtain the equivalent circuit of the test system in the Laplace-domain, taking into account the initial conditions.
2. Use Kirchhoff's laws or any method of circuit analysis (e.g., the method of nodal equations) to obtain the solution of the variables of interest in the Laplace-domain.
3. Apply the inverse Laplace transform to obtain the variables in time-domain.

Although procedures aimed at solving transients in systems of any size have been proposed, this technique is, in general, applied to small-size systems and using a hand procedure. An important limitation is that it can be only applied to linear systems.

The following examples show the application of Laplace-domain circuit equivalents to the analysis of simple linear circuits.

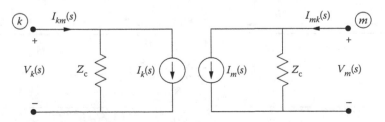

FIGURE 12.19 Laplace-domain equivalent of a single-phase lossless line.

EXAMPLE 12.4

Figure 12.20 shows the scheme of the test system. The goal is to analyze the voltages caused at the receiving end of an ideal line after its energization, when that end is open. Consider that the supply source has a sinusoidal voltage waveform and an internal impedance Z_0. The characteristic parameters of the line are its surge impedance, Z_c, and its travel time, τ.

Figure 12.21 shows the equivalent circuit that results upon substitution of each component by its equivalent in Laplace-domain, once the switch has been closed.

From the expressions of the controlled sources located at each end of an ideal line equivalent circuit, see Equation 12.59, the following equations are deduced:

$$V_1(s) = Z_c[I_{12}(s) - I_1(s)] = Z_c I_{12}(s) + [V_2(s) + Z_c I_{21}(s)]e^{-s\tau}$$
$$V_2(s) = Z_c[I_{21}(s) - I_2(s)] = Z_c I_{21}(s) + [V_1(s) + Z_c I_{12}(s)]e^{-s\tau}$$

Since $I_{21}(s) = 0$, these equations become

$$V_1(s) = Z_c I_{12}(s) + V_2(s)e^{-s\tau} \qquad V_2(s) = [V_1(s) + Z_c I_{12}(s)]e^{-s\tau}$$

The boundary condition at the sending end can be written as follows:

$$E(s) = V_1(s) + Z_0(s)I_{12}(s)$$

Upon substitution of the equation of $V_1(s)$ into this equation

$$E(s) = [Z_0(s) + Z_c]I_{12}(s) + V_2(s)e^{-s\tau}$$

Solving for $I_{12}(s)$ yields

$$I_{12}(s) = \frac{E(s) - V_2(s)e^{-s\tau}}{Z_0(s) + Z_c}$$

Upon substitution of the equation of $V_1(s)$ into the equation of $V_2(s)$

$$V_2(s) = V_2(s)e^{-2s\tau} + 2Z_c I_{12}(s)e^{-s\tau}$$

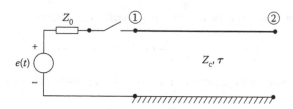

FIGURE 12.20 Diagram of the test system.

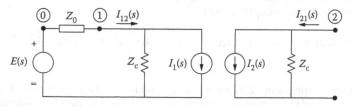

FIGURE 12.21 Laplace-domain circuit for transients analysis.

Further substitution of $I_{12}(s)$ yields

$$V_2(s) = V_2(s)e^{-2s\tau} + 2Z_c\frac{E(s) - V_2(s)e^{-s\tau}}{Z_0(s) + Z_c}e^{-s\tau}$$

Solving for $V_2(s)$ yields

$$V_2(s) = E(s)\frac{2Z_ce^{-s\tau}}{Z_0(s) + Z_c}\frac{1}{1 - \dfrac{Z_0(s) - Z_c}{Z_0(s) + Z_c}e^{-2s\tau}}$$

Since the coefficient of reflection at the sending end of the line is

$$r_s(s) = \frac{Z_0(s) - Z_c}{Z_0(s) + Z_c}$$

the equation can be rewritten as follows:

$$V_2(s) = E(s)\frac{Z_c}{Z_0(s) + Z_c}\frac{2e^{-s\tau}}{1 - r_s(s)e^{-2s\tau}}$$

Note that the two first terms constitute the Laplace transform of the voltage that propagates along the line after energization, $V(s)$, see Subsection 12.4.4 and expression 12.35.

Using $V(s)$ to denote this voltage and taking advantage of the following series expression:

$$\frac{1}{1 - x} \approx 1 + \sum_{n=1}^{\infty} x^n \quad |x| < 1$$

the expression of $V_2(s)$ becomes

$$V_2(s) \approx V(s)2e^{-s\tau}\left[1 + \sum_{n=1}^{\infty}(r_s(s)e^{-2s\tau})^n\right]$$

If $Z_0(s)$ is a resistance, the coefficient of reflection at the sending end of the line will be a real quantity, r_s. The following form is then deduced from the above result:

$$V_2(s) \approx V(s)2e^{-s\tau}[1 + (r_se^{-2s\tau}) + (r_se^{-2s\tau})^2 + (r_se^{-2s\tau})^3 + \cdots]$$

To facilitate the antitransform, the expression is rewritten as follows:

$$V_2(s) \approx 2V(s)[e^{-s\tau} + r_se^{-3s\tau} + r_s^2e^{-5s\tau} + \cdots]$$

If the Laplace transform of the function $v(t)$ is $V(s)$, the antitransform has the following property:

$$\mathcal{L}^{-1}\{V(s)e^{-ns\tau}\} = v(t - n\tau)\varepsilon(t - n\tau)$$

where $\varepsilon(t - n\tau)$ is the unity pulse shifted to the right of interval $n\tau$.

The antitransform is now straightforward

$$v_2(t) \approx 2[v(t - \tau)\varepsilon(t - \tau) + r_sv(t - 3\tau)\varepsilon(t - 3\tau) + r_s^2v(t - 5\tau)\varepsilon(t - 5\tau) + \cdots]$$

Note that this result could have been predicted by using the lattice diagram. In fact, this expression is the same that expression 12.32 with $r_r = 1$. Although the Laplace-domain equivalent of an ideal line can be applied to the analysis of transients in systems with distributed-parameter models, the application of the lattice diagram is in general preferred since it can facilitate the study.

The analysis of the above result is not easy for any source voltage waveform; however, some obvious conclusions can be derived.

- The voltage at the receiving end reaches a value that doubles the voltage of the wave that propagates along the line at the moment the switch is closed. This voltage will, however, depend on the moment at which the energization is made and, therefore, on the wave that starts to propagate.
- Since the voltage wave that propagates along the line depends on the ratio Z_0/Z_c, overvoltages at the receiving end can be controlled by varying the impedance in series with the source.
- In general, $r_s < 0$ and $|r_s| < 1$; therefore, the subsequent waves that arrive to the receiving end will progressively diminish.

EXAMPLE 12.5

Figure 12.22 shows the switch-off of a parallel RLC circuit. The aim of this example is to analyze the TRV across the switch.

Note that the equivalent impedance of the power supply, seen from the load side, has been neglected. The TRV will be then the difference between voltages at switch contacts. Since the voltage at the source side is known and is not altered by the switch operation, only the transient caused at the load side must be analyzed.

Figure 12.23 shows the equivalent circuit that results upon substitution of each component by its equivalent in Laplace-domain, after the switch has been opened.

The equations of this circuit can be written as follows:

$$V_R(s) = V_L(s) = V_C(s) = V(s)$$

$$I_R(s) + I_L(s) + I_C(s) = 0$$

The current equations are

$$I_R(s) = \frac{V(s)}{R}$$

$$I_L(s) = \frac{V(s)}{sL} + \frac{i_L(0)}{s}$$

$$I_C(s) = sCV(s) - Cv_C(0)$$

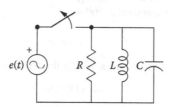

FIGURE 12.22 Single-phase scheme of the test system.

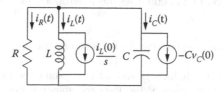

FIGURE 12.23 Scheme for transients analysis.

where $i_L(0)$ and $v_C(0)$ are, respectively, the current through the inductance and the voltage across the capacitance at the time the switch opens.

Substituting these equations into the current equation yields

$$\left[\frac{1}{R} + \frac{1}{sL} + sC\right] V(s) = -\frac{i_L(0)}{s} + Cv_C(0)$$

Solving for $V(s)$

$$V(s) = -\frac{i_L(0)/s}{1/R + 1/sL + sC} + \frac{Cv_C(0)}{1/R + 1/sL + sC}$$

The following form is deduced from this equation:

$$V(s) = -\frac{i_L(0)}{C}\frac{1}{s^2 + s\alpha + \omega_0^2} + v_C(0)\frac{s}{s^2 + s\alpha + \omega_0^2}$$

where

$$\alpha = \frac{1}{RC} \qquad \omega_0^2 = \frac{1}{LC}$$

Assume that Figure 12.23 depicts the single-phase diagram of an unloaded transformer; the L branch represents the unsaturated core inductance, the R branch represents core losses, while the C branch is the winding capacitance. Consider that the power supply is a 50 Hz AC source whose rms phase-to-phase voltage is 25 kV. Assume that the values of the parameters are $L = 2$ H, $C = 60$ nF, and that losses are neglected, so the value of the parallel resistance is infinite.

A simple calculation would conclude that the inductance of the test circuit is dominant; that is, the current supplied by the source is basically the current through the inductance, and it is in quadrature and lagging respect to the voltage. Therefore, at the time the switch opens the current through the inductance is zero, and the voltage across the capacitance is at the peak, being its magnitude that of the supply voltage.

The voltage across the whole RLC circuit would be now given by the following expression:

$$V(s) = v_C(0)\frac{s}{s^2 + s\alpha + \omega_0^2}$$

Therefore, when the influence of the losses is neglected the equation becomes

$$V(s) = v_C(0)\frac{s}{s^2 + \omega_0^2}$$

Since the switch opens when the source voltage is at a peak, this voltage, from the instant at which the switch opens, can be expressed as follows:

$$e(t) = E_{max} \cos \omega t$$

Then $v_C(0) = E_{max}$, and the antitransform of $V(s)$ is the following one:

$$v(t) = E_{max} \cos \omega_0 t$$

The sign in this expression can be either positive or negative since the current across the switch can be canceled when it is decreasing (from positive) or increasing (from negative) to a zero value.

The transient phenomenon that occurs after the interruption of the inductive current can be summarized as follows:

- The capacitance discharges through the inductance. The electric energy stored in this capacitance begins to decrease as the voltage decreases, while the magnetic energy stored in the inductance (which is zero at the instant the switch opens) begins to increase as its current increases. When the capacitance is completely discharged, the energy is stored in the inductance, and a reversal of the above process begins.

- Since losses have been neglected in this study, the transient is not damped, and the oscillations will remain indefinitely. In practice, they will fade away because of losses.
- The frequency of the oscillations originated in the parallel *LC* circuit is the natural frequency of this circuit, which is given by the following expression:

$$f_0 = \frac{\omega_0}{2\pi} = \frac{1}{2\pi\sqrt{LC}}$$

- The TRV (i.e., the voltage across the switch contacts) will be the difference between the source voltage and the voltage developed across the parallel *LC* circuit.

$$v_s(t) = E_{max}[\cos \omega t - \cos \omega_0 t]$$

This voltage has a component at power frequency, 50 Hz, and a component at the natural frequency, f_0, whose value will be usually much higher than that of the power source. As a consequence, the TRV will reach a peak value close to twice the peak value of the source voltage. In addition, this voltage will show a steep front. Both aspects, a high peak value and a steep wave front can be the cause of a dielectric failure of a real breaker.

Figure 12.24 shows the results that correspond to this case. One can observe that the voltage across the transformer, represented here by the parallel *LC* circuit, oscillates at a frequency

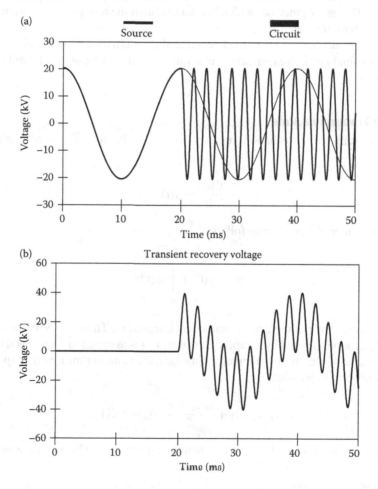

FIGURE 12.24 Interruption of a small inductive current: (a) source and transformer voltages and (b) voltage across the switch.

much higher than that of the source, and the oscillations remain undamped. From the values used in this case, the calculation of the natural frequency for the parallel LC circuit provides a value $f_0 = 459.4\,\text{Hz}$. As predicted, the voltage across the switch reaches a value that doubles that of the source voltage and shows a steeper front.

12.5.3 Numerical Techniques

The Laplace transform has an obvious limitation when it is applied to large systems. The transients analysis of large power systems is usually carried out by using numerical techniques. Time-domain numerical techniques are based on the integration of the differential equations that represent the behavior of a system. Numerical integration is used to transform the differential equations of a circuit element into algebraic equations that involve voltages, currents, and past values. These algebraic equations can be interpreted as representing a resistive companion equivalent of that circuit element. The equations of the whole resistive companion network are assembled using a network technique, for example the nodal admittance equations, and solved as a function of time at discrete instants.

This section presents the principles of the numerical technique proposed by H. W. Dommel [1,2], which is the most popular algorithm presently implemented in simulation tools aimed at solving electromagnetic transients in power systems. This algorithm is based on a combination of the trapezoidal integration, which is used to obtain the resistive companion equivalent of lumped-parameter circuit elements, and the Bergeron's method, which is used to obtain the resistive companion equivalent of distributed-parameter circuit elements.

The following subsections are respectively dedicated to summarize the principles of the trapezoidal integration, to obtain resistive companion equivalents, and to compute the time-domain solution of linear networks.

12.5.3.1 The Trapezoidal Rule

Assume that the differential equation representing the behavior of a circuit element is the following one:

$$\frac{\mathrm{d}y(t)}{\mathrm{d}t} = x(t) \tag{12.60}$$

This equation can be also written as follows:

$$y(t) = y(0) + \int_0^t x(z)\mathrm{d}z \tag{12.61}$$

Figure 12.25 shows the principle of the numerical integration. The area corresponding to a given interval (t_{n-1}, t_n) is approximated by a trapezoid; that is, $x(t)$ is assumed to vary linearly in such interval. Then, if the value of y has been computed at time t_{n-1}, the value at time t_n will be approximated by means of the following expression:

$$y_n = y_{n-1} + \frac{x_{n-1} + x_n}{2}(t_n - t_{n-1}) \tag{12.62}$$

If the computation is carried out using a constant time interval, Δt, also known as *integration time step*, the procedure can be expressed as follows:

$$y(t) = y(t - \Delta t) + \frac{\Delta t}{2}[x(t) + x(t - \Delta t)] \tag{12.63}$$

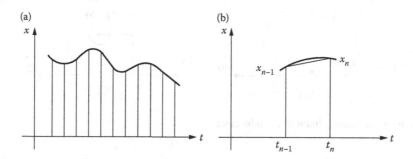

FIGURE 12.25 Application of the trapezoidal rule: (a) primitive function and (b) trapezoidal integration.

12.5.3.2 Companion Equivalents of Basic Circuit Elements

The derivation of the resistive companion equivalents of the basic circuit elements is presented in the following paragraphs.

1. *Resistance*: Since the behavior of a resistance is represented by an algebraic equation

$$v_k(t) - v_m(t) = v_{km}(t) = Ri_{km}(t) \tag{12.64}$$

the application of the trapezoidal rule is not needed, and the companion equivalent is the resistance itself.

2. *Inductance*: The behavior of an inductance is represented by a differential equation

$$v_k(t) - v_m(t) = v_{km}(t) = L\frac{di_{km}(t)}{dt} \tag{12.65}$$

The following expression is deduced from the application of the trapezoidal rule

$$i_{km}(t) = i_{km}(t - \Delta t) + \frac{\Delta t}{2L}[v_{km}(t) + v_{km}(t - \Delta t)] \tag{12.66}$$

Solving for the current yields

$$i_{km}(t) = \frac{\Delta t}{2L}v_{km}(t) + \left[\frac{\Delta t}{2L}v_{km}(t - \Delta t) + i_{km}(t - \Delta t)\right] \tag{12.67}$$

The second term of the right side is known as *history term*. Using the notation

$$I_{km}(t) = \left[\frac{\Delta t}{2L}v_{km}(t - \Delta t) + i_{km}(t - \Delta t)\right] \tag{12.68}$$

the current equation can be written as follows:

$$i_{km}(t) = \frac{\Delta t}{2L}v_{km}(t) + I_{km}(t) \tag{12.69}$$

Figure 12.26 shows the Norton companion equivalent of an inductance in which the history term is defined as a current source.

FIGURE 12.26 Companion circuit of an inductance.

3. *Capacitance*: The behavior of a capacitance is represented by a differential equation

$$i_{km}(t) = C\frac{d}{dt}[v_k(t) - v_m(t)] = C\frac{dv_{km}(t)}{dt} \qquad (12.70)$$

The following expression is deduced from the application of the trapezoidal rule

$$v_{km}(t) = v_{km}(t - \Delta t) + \frac{\Delta t}{2C}[i_{km}(t) + i_{km}(t - \Delta t)] \qquad (12.71)$$

Solving for the current yields

$$i_{km}(t) = \frac{2C}{\Delta t}v_{km}(t) - \left[\frac{2C}{\Delta t}v_{km}(t - \Delta t) + i_{km}(t - \Delta t)\right] \qquad (12.72)$$

As for the inductance, the second term of the right side is known as history term. Using the notation

$$I_{km}(t) = -\left[\frac{2C}{\Delta t}v_{km}(t - \Delta t) + i_{km}(t - \Delta t)\right] \qquad (12.73)$$

the current equation can be written as follows:

$$i_{km}(t) = \frac{2C}{\Delta t}v_{km}(t) + I_{km}(t) \qquad (12.74)$$

Figure 12.27 shows the Norton companion equivalent of a capacitance in which the history term is also defined as a current source.

4. *Ideal line*: As shown in previous sections, the equations of a single-phase lossless line can be expressed as follows:

$$\frac{\partial v(x, t)}{\partial x} = -L\frac{\partial i(x, t)}{\partial t}$$

$$\frac{\partial i(x, t)}{\partial x} = -C\frac{\partial v(x, t)}{\partial t}$$

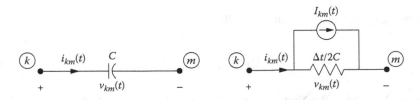

FIGURE 12.27 Companion circuit of a capacitance.

where L and C are, respectively, the inductance and capacitance per unit length, and x is the distance with respect to the sending end of the line.

The general solution of $v(x, t)$ and $i(x, t)$ can be expressed as in Equations 12.3 and 12.5. Upon manipulation of the voltage and current solutions the following form is derived:

$$v(x, t) + Z_c i(x, t) = 2f_1(x - vt) \tag{12.75}$$

where Z_c and v, are respectively, the surge impedance and the propagation velocity.

Note that the form $(v + Z_c i)$ remains constant if the value of the term $(x - vt)$ is also constant. That is, this quantity will be the same for any combination of x and t such that the quantity $(x - vt)$ has the same value.

Assume that τ is the travel time, then the value of $(v + Z_c i)$ at one end of the line will be the same that it was at the other end of an interval of time equal to τ. If the line is connected to nodes k and m, for a wave traveling from k to m the above conclusion can be expressed as follows:

$$v(m, t) + Z_c i(m, t) = v(k, t - \tau) + Z_c i(k, t - \tau) \tag{12.76}$$

Using the change of variables indicated in Figure 12.28

$$v_k(t) = v(k, t) \quad i_{km}(t) = +i(k, t)$$
$$v_m(t) = v(m, t) \quad i_{mk}(t) = -i(m, t)$$

the above relationship becomes

$$v_m(t) - Z_c i_{mk}(t) = v_k(t - \tau) + Z_c i_{km}(t - \tau) \tag{12.77}$$

Following a similar reasoning for waves traveling in the opposite direction, the relationship would be

$$v_k(t) - Z_c i_{km}(t) = v_m(t - \tau) + Z_c i_{mk}(t - \tau) \tag{12.78}$$

Solving for currents at both line ends yields

$$i_{km}(t) = \frac{v_k(t)}{Z_c} - \left[\frac{v_m(t - \tau)}{Z_c} + i_{mk}(t - \tau)\right]$$

$$\tag{12.79}$$

$$i_{mk}(t) = \frac{v_m(t)}{Z_c} - \left[\frac{v_k(t - \tau)}{Z_c} + i_{km}(t - \tau)\right]$$

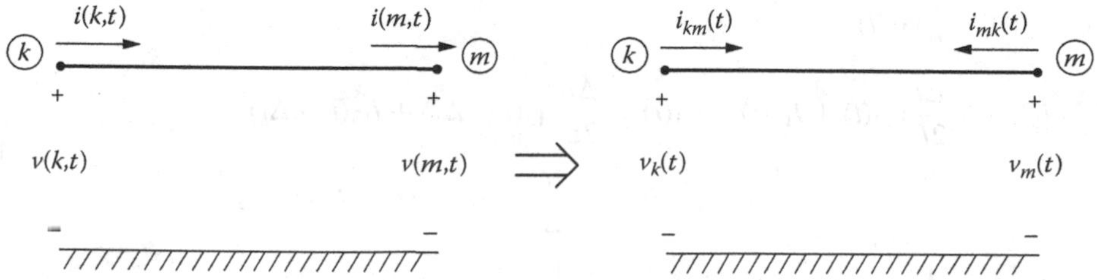

FIGURE 12.28 Change in the notation of line variables.

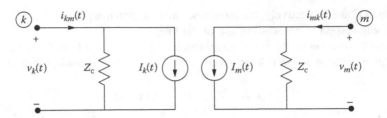

FIGURE 12.29 Companion circuit of a single-phase lossless line.

Using the notation

$$I_k(t) = -\left[\frac{v_m(t-\tau)}{Z_c} + i_{mk}(t-\tau)\right]$$

$$I_m(t) = -\left[\frac{v_k(t-\tau)}{Z_c} + i_{km}(t-\tau)\right]$$

(12.80)

the equations of the line can be written as follows:

$$i_{km}(t) = \frac{v_k(t)}{Z_c} + I_k(t)$$

$$i_{mk}(t) = \frac{v_m(t)}{Z_c} + I_m(t)$$

(12.81)

These equations can be interpreted as the Norton companion equivalent of an ideal line shown in Figure 12.29.

12.5.3.3 Computation of Transients in Linear Networks

Companion equivalents derived above have some obvious advantages: all of them consists of resistances, whose values remain constant if Δt is constant, and current sources, whose values must be updated at any integration time step. Therefore, the solution of a linear network during a transient is the solution of the equations of a purely resistive network whose parameters remain constant during the transient and for which only current source values must be updated at any time step.

Several frameworks can be used to solve the equations of linear resistive networks; only nodal admittance equations are considered in this chapter.

The derivation of the equations of a linear network is illustrated with the case depicted in Figure 12.30, which shows several circuit elements connected to a given node. The application of the Kirchhoff's current law to node 1 gives

$$i_{12}(t) + i_{13}(t) + i_{14}(t) + i_{15}(t) = i_1(t)$$

(12.82)

The following equations are deduced from the companion equivalent of each branch connected to this node:

$$i_{12}(t) = \frac{1}{R}v_{12}(t)$$

$$i_{13}(t) = \frac{\Delta t}{2L}v_{13}(t) + I_{13}(t) \qquad I_{13}(t) = \frac{\Delta t}{2L}v_{13}(t-\Delta t) + i_{13}(t-\Delta t)$$

$$i_{14}(t) = \frac{2C}{\Delta t}v_{14}(t) + I_{14}(t) \qquad I_{14}(t) = -\frac{2C}{\Delta t}v_{14}(t-\Delta t) - i_{14}(t-\Delta t)$$

$$i_{15}(t) = \frac{1}{Z_c}v_1(t) + I_{l1}(t) \qquad I_{l1}(t) = -\frac{1}{Z_c}v_5(t-\tau) - i_{51}(t-\tau)$$

(12.83)

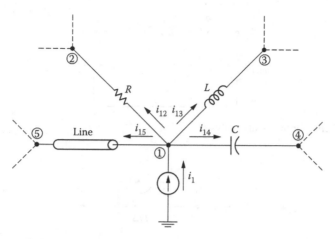

FIGURE 12.30 Generic node of a linear network.

The following equation is obtained upon substitution of Equation 12.83 into Equation 12.82:

$$\left[\frac{1}{R} + \frac{\Delta t}{2L} + \frac{2C}{\Delta t} + \frac{1}{Z_c}\right]v_1(t) - \frac{1}{R}v_2(t) - \frac{\Delta t}{2L}v_3(t) - \frac{2C}{\Delta t}v_4(t) = i_1(t) - I_{13}(t) - I_{14}(t) - I_{l1}(t) \quad (12.84)$$

Using the procedure with all nodes, the equations of a network of any size can be assembled and written as follows:

$$[G][v(t)] = [i(t)] - [I(t)] \quad (12.85)$$

where G is the symmetric nodal conductance matrix, $v(t)$ is the vector of node voltages, $i(t)$ is the vector of current sources, and $I(t)$ is the vector of history terms.

If Δt is constant, only the right-hand side of Equation 12.85 must be updated at any time step since elements of the nodal conductance matrix G remain constant.

In general, some nodes of a power system have known voltages, because voltage sources are connected to them, and the order of the vector of unknown node voltages can be reduced. Assume that A and B are used to denote, respectively, the set of unknown and known node voltages. Nodal admittance equations can be rewritten as follows:

$$\begin{bmatrix} G_{AA} & G_{AB} \\ G_{BA} & G_{BB} \end{bmatrix}\begin{bmatrix} v_A(t) \\ v_B(t) \end{bmatrix} = \begin{bmatrix} i_A(t) \\ i_B(t) \end{bmatrix} - \begin{bmatrix} I_A(t) \\ I_B(t) \end{bmatrix} \quad (12.86)$$

Upon elimination of part B, the following set of equations is obtained:

$$[G_{AA}][v_A(t)] = [i_A(t)] - [I_A(t)] - [G_{AB}][v_B(t)] \quad (12.87)$$

The application of the above technique, based on the companion circuit equivalents, is illustrated with two simple cases.

EXAMPLE 12.6

Figure 12.31 shows the scheme of the test case. It can represent the equivalent circuit for transients analysis of many practical problems in power systems. It can correspond, for instance, to the energization of a capacitor bank, being the power system, seen from the connection point, represented by the AC voltage source in series with its equivalent impedance. The goal is to analyze the

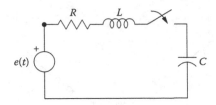

FIGURE 12.31 Scheme for transients analysis.

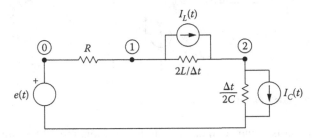

FIGURE 12.32 Companion circuit.

transient phenomenon caused by the closing of the switch for different combinations of the three circuit parameters using the numerical technique described in this section.

The resistive network that results after substituting each circuit element by its companion equivalent is that shown in Figure 12.32. Note that the switch is not included in this representation. Since the initial conditions in all the circuit elements before the closing of the switch are zero, the transient caused by this operation can be analyzed by assuming that the switch does not exist and the voltage source is activated at $t = 0$.

In a series circuit, currents through all elements are the same:

$$i_R(t) = i_L(t) = i_C(t) = i(t)$$

Voltage and current relationships in the three elements are, according to the companion equivalents (see Figures 12.32 and 12.33), as follows:

$$i_R(t) = [v_0(t) - v_1(t)]\frac{1}{R}$$

$$i_L(t) = [v_1(t) - v_2(t)]\frac{\Delta t}{2L} + I_L(t)$$

$$i_C(t) = v_2(t)\frac{2C}{\Delta t} + I_C(t)$$

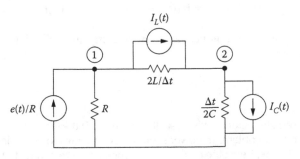

FIGURE 12.33 Modified companion circuit.

where

$$I_L(t) = \frac{\Delta t}{2L}[v_1(t - \Delta) - v_2(t - \Delta)] + i_L(t - \Delta t)$$

$$I_C(t) = -\frac{2C}{\Delta t}v_2(t - \Delta t) - i_C(t - \Delta t)$$

The application of the Kirchhoff's current law to nodes 1 and 2 gives the following set of equations:

$$[v_0(t) - v_1(t)]\frac{1}{R} = [v_1(t) - v_2(t)]\frac{\Delta t}{2L} + I_L(t)$$

$$[v_1(t) - v_2(t)]\frac{\Delta t}{2L} + I_L(t) = v_2(t)\frac{2C}{\Delta t} + I_C(t)$$

Taking into account that the voltage at node 0 is the source voltage, $v_0(t) = e(t)$, the above equations can be expressed in matrix notation as follows:

$$\begin{bmatrix} \frac{1}{R} + \frac{\Delta t}{2L} & -\frac{\Delta t}{2L} \\ -\frac{\Delta t}{2L} & \frac{\Delta t}{2L} + \frac{2C}{\Delta t} \end{bmatrix} \begin{bmatrix} v_1(t) \\ v_2(t) \end{bmatrix} = \begin{bmatrix} e(t)/R - I_L(t) \\ I_L(t) - I_C(t) \end{bmatrix}$$

These equations can be deduced from the application of the admittance nodal equations to the resistive companion equivalent shown in Figure 12.33. This circuit is derived from that shown in Figure 12.32 by substituting the source voltage $e(t)$ in series with parameter R by its Norton equivalent.

Assume that the source voltage value is zero at $t = 0$. The computation of node voltages can be carried out using the following procedure:

1. Solve admittance equations at $t = \Delta t$ without current sources that represent history terms, since the values of these sources are zero.
2. Update $e(t)$ and update history terms [$I_L(t)$ and $I_C(t)$] from the values deduced at the previous step and solve admittance equations.
3. If $t = t_{max}$, the maximum simulation time, stop the procedure; otherwise, continue with step 2.

The computation of the currents can be carried out simultaneously to the computation of node voltages. In this particular case, the common current to the three-series elements can be obtained from any of the equations that relate voltages and currents in the circuit elements.

A detailed analysis of the series RLC circuit can be found in Reference 8. According to the results presented in that book, three different transient responses of the circuit current can be distinguished, depending on the parameter values. Using the following notation:

$$\lambda = \frac{1}{R}\sqrt{\frac{L}{C}}$$

the transient current through the test circuit will be *oscillatory underdamped* if $\lambda > 1/2$, *critically damped* if $\lambda = 1/2$, or *overdamped* if $\lambda < 1/2$.

Consider that the parameters of the test circuit are $L = 2\,\text{mH}$ and $C = 150\,\mu\text{F}$. The resistance value for which the circuit has a critically damped response is $R = 7.3\,\Omega$. Figure 12.34 shows the transient response of the current through the circuit and the voltage across the capacitance with three different values of R. In all cases, the source voltage is a unit step. These results confirm the above conditions and show the influence that the resistance value has on the transient response waveforms and on the peak values. It is evident that the larger the resistance value, the smaller the peak magnitudes of currents and voltages.

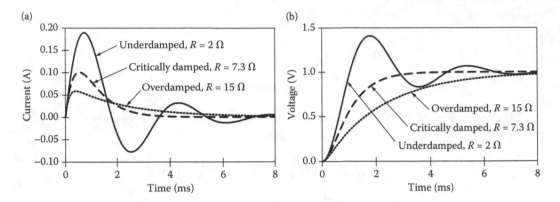

FIGURE 12.34　Transient response of a series RLC circuit: (a) currents and (b) voltages.

EXAMPLE 12.7

Figure 12.35 shows the diagram of the test case that is analyzed in this example. A lightning stroke hits the top of a shielded overhead transmission line tower. The stroke is represented as an ideal current source. This current splits into three parts at the point of impact: two equal parts will travel along the shield wire in opposite directions, the third part will travel toward the bottom of the tower. Reflected waves at the adjacent towers and the footing impedance of the tower will be generated. The superposition of traveling waves may cause an insulator flashover, which is usually known as backflashover.

The scheme that will be used in this study is shown in Figure 12.36. In this simplified scheme, the footing impedance is represented as a constant resistance, R_g, and the effect of the adjacent towers is neglected; that is, shield wire sections at both sides of the point of impact are assumed

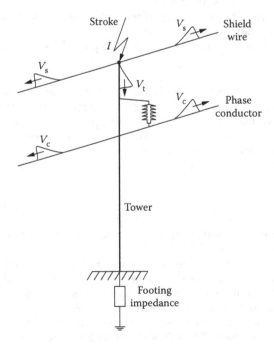

FIGURE 12.35　Diagram of the test system.

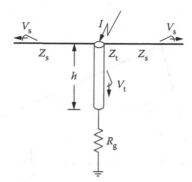

FIGURE 12.36 Simplified scheme of the test system.

to have an infinite length. The tower and the shield wire sections are represented as single-phase ideal lines in which waves propagate at the speed of light. The surge impedances of the tower and the shield wire are, respectively, Z_t and Z_s.

The transient phenomenon caused by the lightning stroke is very complicated since traveling waves along shield wires induce waves on the phase conductors. In addition, phase conductors of actual lines are energized at the time a stroke impacts a tower; power-frequency voltages cannot be always neglected. The presence of the phase conductors is neglected in the scheme shown in Figure 12.36. The goal of this example is to analyze the voltage developed across transmission line towers as a function of the footing resistance and the stroke waveform parameters.

The equivalent circuit can be depicted as in Figure 12.37. The two sections of the shield wire have been joined into a single ideal line, and their surge impedances have been halved. On the other hand, the receiving end of this line, node 3, has been matched. Since no reflected waves are generated at this node, the behavior of this line section reproduces that of an infinite length line.

The lightning stroke is assumed to have a double-ramp waveform as that depicted in Figure 12.38. The characteristic parameters of this waveform are:

- The peak current magnitude, I_p
- The time to crest, t_f, which is the time needed to reach the peak value
- The tail time, t_h, which is the time, measured along the current tail, needed to reach the 50% of the peak value

The nodal admittance equations of the circuit shown in Figure 12.39 can be written as follows:

$$\begin{bmatrix} \dfrac{1}{Z_t}+\dfrac{2}{Z_s} & 0 & 0 \\[2mm] 0 & \dfrac{1}{Z_t}+\dfrac{1}{R_g} & 0 \\[2mm] 0 & 0 & \dfrac{4}{Z_s} \end{bmatrix} \begin{bmatrix} v_1(t) \\[2mm] v_2(t) \\[2mm] v_3(t) \end{bmatrix} = \begin{bmatrix} i_s(t)-I_{t1}(t)-I_{s1}(t) \\[2mm] -I_{t2}(t) \\[2mm] -I_{s2}(t) \end{bmatrix}$$

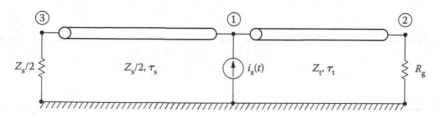

FIGURE 12.37 Simplified equivalent circuit.

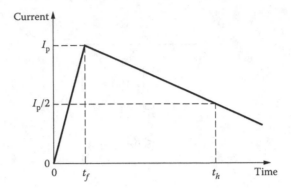

FIGURE 12.38 Lightning stroke current waveform.

where

$$I_{t1}(t) = -\left[\frac{v_2(t - \tau_t)}{Z_t} + i_{21}(t - \tau_t)\right] \qquad I_{t2}(t) = -\left[\frac{v_1(t - \tau_t)}{Z_t} + i_{12}(t - \tau_t)\right]$$

$$I_{s1}(t) = -\left[2\frac{v_3(t - \tau_s)}{Z_s} + i_{31}(t - \tau_s)\right] \qquad I_{s2}(t) = -\left[2\frac{v_1(t - \tau_s)}{Z_s} + i_{13}(t - \tau_s)\right]$$

τ_t and τ_s are the travel times of the line sections that represent the tower and the shield wires, respectively.

The following relationships are also derived from the resistive companion equivalent shown in Figure 12.39:

$$i_{21}(t) = -\frac{v_2(t)}{R_g} \qquad i_{31}(t) = -\frac{v_3(t)}{Z_s/2}$$

Substitution of these expressions into the history terms where these currents are included yields

$$I_{t1}(t) = -v_2(t - \tau_t)\left[\frac{1}{Z_t} - \frac{1}{R_g}\right] \qquad I_{s1}(t) = 0$$

Note that a history term becomes zero. This is an expected result, since the line section that represents the shield wires is matched at the receiving end; therefore, there will not be reflected waves traveling back to the tower from node 3. An interesting conclusion from this result is that the value of the travel time τ_s is unimportant since the voltages and currents originated at node 3 will not affect the rest of the network.

In actual transmission lines, the surge impedance of the tower is about one half of the surge impedance of shield wires

$$Z_t \approx \frac{Z_s}{2}$$

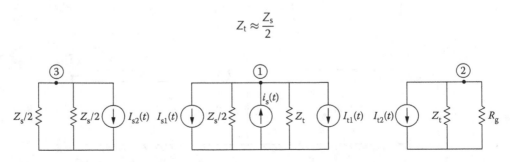

FIGURE 12.39 Resistive companion network.

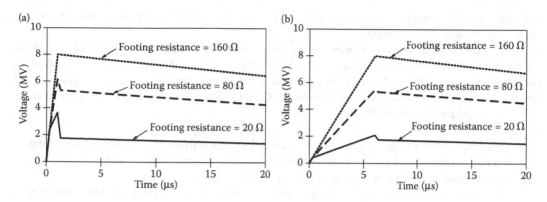

FIGURE 12.40 Voltage at the tower top: (a) stroke waveform 100 kA, 1/50 μs and (b) stroke waveform 100 kA, 6/50 μs.

After inverting the diagonal matrix and taking into account all these results, the equations for numerical calculation of this case become

$$
\begin{bmatrix} v_1(t) \\ v_2(t) \\ v_3(t) \end{bmatrix} = \begin{bmatrix} \dfrac{Z_t}{2} & 0 & 0 \\ 0 & \dfrac{R_g \cdot Z_t}{R_g + Z_t} & 0 \\ 0 & 0 & \dfrac{Z_t}{2} \end{bmatrix} \begin{bmatrix} i_s(t) - I_{t1}(t) \\ -I_{t2}(t) \\ -I_{s2}(t) \end{bmatrix}
$$

The calculation procedure is similar to that detailed in Example 12.6. Since the system is not energized at the time the lightning stroke hits the tower, then

$$
I_{t1}(0) = I_{t2}(0) = I_{s2}(0) = 0
$$

Assume that the values of the tower height and surge impedance are, respectively, $h = 45$ m and $Z_t = 160\,\Omega$. The lightning stroke current waveform is a double ramp, like that shown in Figure 12.38, with $I_p = 100$ kA and $t_h = 50$ μs.

Figure 12.40 depicts results obtained with different values of the footing resistance and two different values of the time to crest, ($t_f = 1$ and $t_f = 6$ μs). These plots show that the voltage developed at the tower top increases with the value of the footing resistance, which becomes an important parameter to achieve a good lightning performance of overhead transmission lines, and decreases as the time to crest of the lightning current increases. The latter effect is more significant with small values of the footing resistance.

12.6 OVERVOLTAGES IN POWER SYSTEMS

12.6.1 INTRODUCTION

An overvoltage is a time-varying voltage stress whose peak value is higher or much higher than the highest value of the rated voltage. Overvoltages can be between a phase conductor and the ground, between phase conductors, or longitudinal (i.e., between terminals that belong to the same phase conductor).

Overvoltage and insulation coordination studies are very important since the information they provide is crucial for power component design and selection of protective devices.

Overvoltages in power systems are presently estimated by means of computers. There are several reasons to use numerical techniques for overvoltage calculations: real-power systems are large and complex, mathematical models for electromagnetic transients calculations are very reliable, digital computations can be very accurate if models and parameters of power components used in computations are adequate and accurate.

A first classification of overvoltages can be made according to their origin: internal (switching operations, faults) or external (lightning strokes) to the power system. A more complete classification considers main overvoltage characteristics: frequency range, duration, peak voltage magnitude, and shape. According to International Electrotechnical Commission (IEC), the following types of overvoltages should be considered [9,10]:

1. *Temporary overvoltages*: Little damped power-frequency overvoltages of relatively long duration (from several milliseconds to several seconds). They can be caused by faults, resonance conditions, load rejection, or a combination of these.
2. *Slow-front overvoltages*: Highly damped transient overvoltages of relatively short duration (from a few milliseconds to a few power-frequency cycles). They can be oscillatory or unidirectional, and their frequency range varies from 2 to 20 kHz. They are usually caused by faults or switching operations.
3. *Fast-front overvoltages*: Transient overvoltages of very short duration (<1 ms). They are highly damped and generally unidirectional. They can be caused by lightning strokes or switching operations.
4. *Very fast-front overvoltages*: Transient overvoltages of very short duration (<1 ms). They can be oscillatory or unidirectional, and their frequency range can vary from 100 kHz to 50 MHz. The most frequent origin, of these overvoltages are faults and switching operations in gas-insulated substations (GIS).

12.6.2 CASE STUDIES

Three illustrative overvoltage studies are presented in this section. Each test case corresponds to a different overvoltage type. All of them have been solved using digital simulation based on the numerical technique detailed in the previous section.

12.6.2.1 Ferroresonance

Ferroresonance is a nonlinear phenomenon that is originated in electric networks with saturable transformers and reactors. In most situations, ferroresonance is a series resonance involving a nonlinear inductance and a capacitance. To understand better this phenomenon, the resonance of a linear series LC circuit is first analyzed. Figure 12.41 shows a series LC circuit energized from an AC voltage source. Assumes that the source has a constant voltage peak magnitude and a variable frequency.

Since the voltages across the inductance and the capacitance are, respectively,

$$V_L = j\omega L I = jX_L I \tag{12.88}$$

$$V_C = \frac{1}{j\omega C} I = -jX_C I \tag{12.89}$$

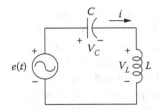

FIGURE 12.41   Linear series LC circuit.

the voltage equation of this circuit can be written as follows:

$$E = V_L + V_C = j(X_L - X_C)I \tag{12.90}$$

Solving for the current yields

$$I = \frac{E}{j(X_L - X_C)} = E \frac{\omega C}{j(\omega^2 LC - 1)} \tag{12.91}$$

The current value will vary if the source frequency varies. Resonance occurs when the circuit impedance $(X_L - X_C)$ is equal to zero. The resonance frequency is obtained from Equation 12.91 and is given by the following expression:

$$\omega_r = \frac{1}{\sqrt{LC}} \tag{12.92}$$

Therefore, when $\omega \to \omega_r$ the current increases toward infinite. In a real case, the system always dissipates energy, and the equivalent resistance of the circuit will limit the current.

Figure 12.42 shows the series LC circuit with a nonlinear inductance. The voltage equation is rewritten as follows:

$$V_L = E - V_C = E - jX_C I \tag{12.93}$$

A justification of the ferroresonance problem that can be originated in this circuit is the simplified voltage–current diagram shown in Figure 12.43. The AC source is represented by its rms voltage, E; V_L and V_C are the rms voltages across the inductance and the capacitance, respectively. The nonlinear inductance is represented by its rms $V - I$ characteristic. The solutions of the circuit correspond to the intersection of V_L and V_C. Note that, depending on the operating point, the voltages across both the inductance and the capacitance can be much larger than the source voltage.

When analyzing this phenomenon it is important to account for the following aspects:

1. Ferroresonance is not strictly a resonance phenomenon since the source frequency is constant. A natural circuit resonance cannot be defined as for a linear circuit, and superposition, as used to obtain the diagram of Figure 12.43, cannot be applied.
2. The current through a saturated reactance is distorted and contains harmonics. This diagram is not correct, since it was derived by assuming that only source frequency voltages and currents do exist in the circuit.
3. The solution depicted in Figure 12.43 gives three possible operating points. A different number of intersections can be obtained with a different capacitance slope. In addition, only operating points 1 and 2 represent stable steady-state solutions.

Although the general requirements for ferroresonance are a voltage source, a saturable inductance, some capacitance and little damping, there are many situations in which this phenomenon can occur.

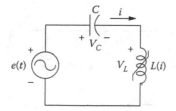

FIGURE 12.42   Nonlinear series LC circuit.

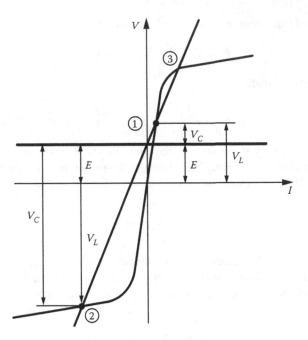

FIGURE 12.43 Voltage–current diagram of a nonlinear series LC circuit.

System conditions that can lead to ferroresonance are ungrounded neutral, small power demand or no load at all, one or two switch poles are open, power supply made through a cable that provides the capacitance.

EXAMPLE 12.8

Figure 12.44 shows the circuit that will be analyzed in this chapter. Note that a resistance representing core losses has been added in parallel with the saturable inductance.

The study of this case can be carried out by considering several approaches for a numerical equivalent of the saturable inductance. Assuming a piecewise linear representation of the saturable inductance, this element is represented at each time step by a linear inductance whose value is selected from the solution obtained at the preceding time step. Since the inductance value is varied during the transient solution, the nodal conductance matrix has to be updated and retriangularized every time the solution moves to another segment of the $\lambda - i$ characteristic. Figure 12.45 shows the companion equivalent circuit that results from assuming that the inductance has a value that is determined from the preceding time step solution.

The voltage and current equations of this circuit are again those used in the previous study.

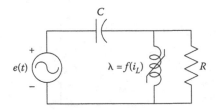

FIGURE 12.44 Ferroresonance analysis of a series LC circuit.

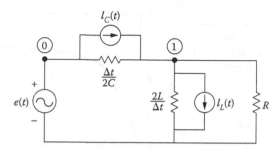

FIGURE 12.45 Companion resistive network—piecewise linear representation.

The application of Kirchhoff's current law at node 1 gives

$$[e(t) - v_1(t)]\frac{2C}{\Delta t} + I_C(t) = v_1(t)\frac{\Delta t}{2L} + I_L(t) + \frac{v_1(t)}{R}$$

Solving for $v_1(t)$ yields

$$\left[\frac{1}{R} + \frac{\Delta t}{2L} + \frac{2C}{\Delta t}\right]v_1(t) = e(t)\frac{2C}{\Delta t} + I_C(t) - I_L(t)$$

From which the following expression is derived:

$$v_1(t) = \left[\frac{2L \times \Delta t}{\Delta t^2 + \Delta t \times 2L/R + 4LC}\right]\left[e(t)\frac{2C}{\Delta t} + I_C(t) - I_L(t)\right]$$

History terms for the capacitance and the inductance are obtained using the equations of the respective companion equivalents.

Note that this algorithm uses the history term of the capacitance and the inductance; therefore, the computation must be simultaneously performed with other variables, namely the currents through the capacitance and the inductance.

As mentioned above, the inductance value must be changed whenever the solution changes to a different slope segment of the saturation curve. This can cause numerical problems since the inductance value used to compute the solution at a given time step can be different from the value that will result once the solution is obtained. Due to this fact, it is recommendable to use a short integration time step, for example, about 1 μs, and represent the $\lambda - i$ characteristic using as many slopes as possible.

The study will be performed by assuming that the supply source is a 50 Hz AC voltage source with a peak value of 20.4 kV, which corresponds to an rms phase-to-phase voltage of 25 kV.

Consider that the magnetizing characteristic of the saturable inductance is that shown in Figure 12.46 and the attached table. It is a voltage–current rms curve that has to be converted to an instantaneous flux–current relationship. The routine saturation available in some transients programs was used. The new magnetizing characteristic is shown in Table 12.1.

The goal of this example is to analyze the influence that the series capacitance and the parallel resistance can have on ferroresonant conditions. The study will be carried out by changing the capacitance value and using a constant parallel resistance whose value will be 40 kΩ. Although ferroresonance can be seen as a steady-state phenomenon, in this example calculations begins with zero values for all variables, including the current sources that represent history terms. Figure 12.47 shows simulation results obtained without and with considering losses in test circuit. Three different values of the series capacitance have been used in both cases. The results were obtained with a piecewise linear representation of the saturable inductance and $\Delta t = 1$ μs.

Voltage (rms value, kV)	Current (rms value, A)
11.44	0.182
13.06	0.266
14.02	0.378
14.64	0.562
15.60	1.204

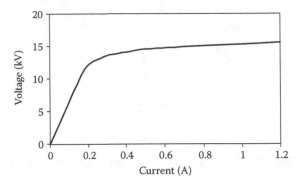

FIGURE 12.46 Magnetizing characteristic of the saturable inductance.

It is evident that large values of both the capacitance and the resistance increase the probability of a ferroresonant scenario. One can also observe that the peak voltage across the saturable inductance increases with the capacitance value. The influence of the parallel resistance is an important result. The resistance value used in this case is rather large compared with what one could find in a more realistic scenario. This means that a small increase of the circuit losses can significantly affect the ferroresonance of this circuit.

12.6.2.2 Capacitor Bank Switching

Various transient problems are associated with capacitor switching since overvoltages and voltage magnification can occur during energization, very large inrush currents can flow after energization, or dielectric reignition and arrester failure can occur after capacitive current interruption. Capacitor switching effects depend on the type of operation (connection, disconnection), the breaker technology, the various parameters of power components involved in the operation, and the conditions with which the action is taken.

In general, maximum overvoltages associated to a normal capacitor bank switching are twice the pre-switch capacitor node voltage, and they are not usually a concern in MV distribution equipment. However, significant overvoltages can occur at remote locations when the operation is made under certain conditions. One of the most serious scenarios is illustrated in Figure 12.48. Magnification of transient overvoltages can occur at the lower voltage capacitor bank location with the energization of a capacitor bank at the MV side of the substation (see Figure 12.48).

The following conditions are required for voltage magnification during capacitor bank energization:

- The reactive power rating of the energized capacitor bank is much greater than that of the capacitor bank installed at the lower voltage

TABLE 12.1

Magnetizing Characteristic

Flux Linkage (Versus)	Current (Peak Value–A)
51.4998	0.25733
58.7901	0.47475
63.1110	0.75493
65.9050	1.30383
70.2259	2.91546

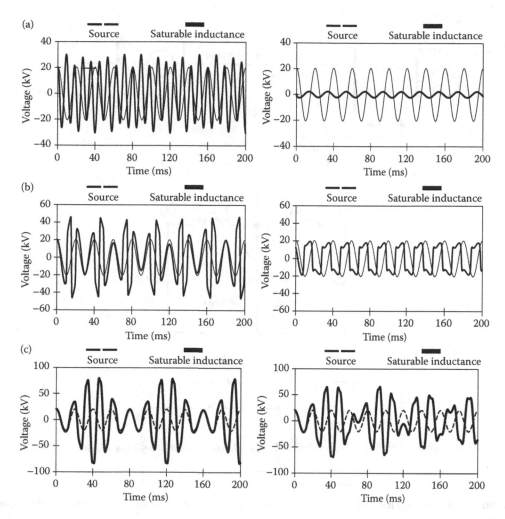

FIGURE 12.47 Simulation results without (left) and with (right) resistance: (a) capacitance $= 0.01\ \mu F$, (b) capacitance $= 0.1\ \mu F$, and (c) capacitance $= 1\ \mu F$.

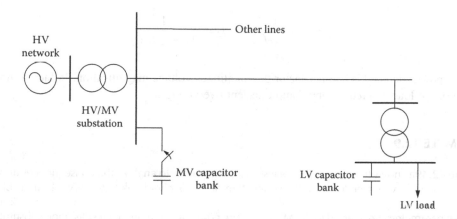

FIGURE 12.48 Connection of a capacitor bank to a distribution network. HV $=$ high voltage; MV $=$ medium voltage; LV $=$ low voltage.

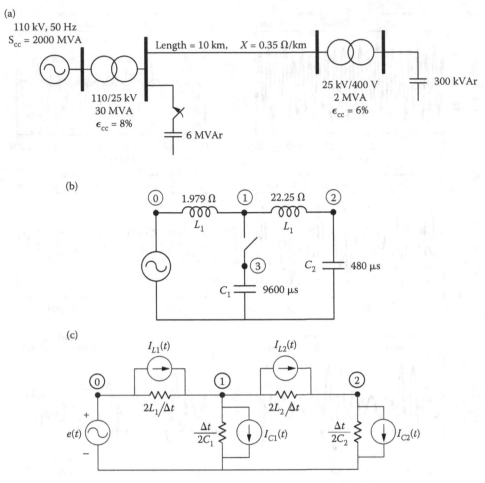

FIGURE 12.49 Voltage magnification: (a) diagram of the test system, (b) simplified scheme, and (c) companion equivalent network.

- The natural frequencies of the system (f_1 and f_2, Figure 12.49b) are nearly equal

$$f_1 = \frac{1}{2\pi \times \sqrt{L_1 C_1}} \qquad f_2 = \frac{1}{2\pi \times \sqrt{L_2 C_2}} \qquad (12.94)$$

A low power demand is another factor that facilitates voltage magnification. In fact, a moderate increase of the load can reduce and damp transient overvoltages.

EXAMPLE 12.9

Figure 12.49a shows the diagram of the test case. The goal is to analyze the consequences at both medium-and low-voltage levels of the connection of a capacitor bank to the MV terminals of the substation. The study will be based on the single-phase equivalent circuit shown in Figure 12.49b, whose parameters are referred to the MV level. The equivalent circuit that results upon substitution of each element by its companion equivalent, once the switch is closed, is depicted in Figure 12.49c.

The application of the Kirchhoff's current law at nodes 1 and 2 yields

$$\frac{\Delta t}{2L_1}[v_0(t) - v_1(t)] + I_{L1}(t) = \frac{2C_1}{\Delta t}v_1(t) + I_{C1}(t) + \frac{\Delta t}{2L_2}[v_1(t) - v_2(t)] + I_{L2}(t)$$

$$\frac{\Delta t}{2L_2}[v_1(t) - v_2(t)] + I_{L2}(t) = \frac{2C_2}{\Delta t}v_2(t) + I_{C2}(t)$$

Since $v_0(t) = e(t)$, the set of equations can be rewritten as follows:

$$\begin{bmatrix} \dfrac{\Delta t}{2L_1} + \dfrac{2C_1}{\Delta t} + \dfrac{\Delta t}{2L_2} & -\dfrac{\Delta t}{2L_2} \\ -\dfrac{\Delta t}{2L_2} & \dfrac{\Delta t}{2L_2} + \dfrac{2C_2}{\Delta t} \end{bmatrix} \begin{bmatrix} v_1(t) \\ v_2(t) \end{bmatrix} = \begin{bmatrix} e(t) \cdot \Delta t/2L_1 \\ 0 \end{bmatrix} - \begin{bmatrix} -I_{L1}(t) + I_{L2}(t) + I_{C1}(t) \\ -I_{L2}(t) + I_{C2}(t) \end{bmatrix}$$

The transient begins with nonzero values for the source currents that represent history terms; that is, for I_{L1}, I_{L2}, and I_{C2}. The initial value of these terms must be obtained from the steady-state solution before the capacitor bank connection.

Transient overvoltages in this case will depend on the value of the source voltage with which the switch is closed. The analysis will be carried out by assuming that the switching operation occurs at the source peak voltage, which will cause the highest overvoltages.

Assume then that the expression of the source voltage to be considered for calculations of the initial conditions is $e(t) = 1 \times \sin\omega t$. The steady-state analysis of the equivalent circuit, see Figure 12.49b, provides the following expression for the current through L_1, L_2, and C_2:

$$i_{L1}(t) = i_{L2}(t) = i_{C2}(t) = 0.486 \cos\omega t \, (\text{mA})$$

The current value at the time the switch is closed corresponds to that obtained with $t = 0$.

The computational procedure to be applied will be similar to that applied to previous examples. Figure 12.50 shows some simulation results. One can observe that the voltage at the node to which the MV capacitor bank is connected drops to zero at the switching time. From these results one can deduce that very high voltages can occur with this operation, and that the highest voltages are produced at the LV side. The presence of losses in a more realistic simulation will not affect too much the peak overvoltages but will produce more damped transient overvoltages.

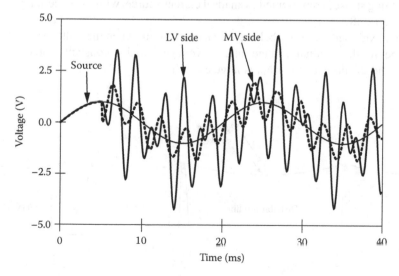

FIGURE 12.50 Simulation results.

Although these results are useful to introduce the voltage magnification problem, they are over-simplified, since calculations were carried out without power demand at the LV side, without including the effect that other distribution lines connected to the MV side of the substation could have, and without considering saturation effects in the distribution transformer model. These effects would have limited transient overvoltages and produced more damped oscillations.

12.6.2.3 Protection against Lightning Overvoltages

The metal oxide surge arrester is a very efficient device for lightning protection of power equipment. Surge arrester is installed in parallel and as close as possible to equipment to be protected. As a first approach one can assume that the overvoltage across terminals of protected equipment is limited to the residual voltage of the arrester. However, some distance between surge arrester and the protected equipment cannot be always avoided. As a consequence, the voltage produced across equipment terminals will be higher than the residual voltage of the surge arrester. This is known as *separation distance effect*.

Figure 12.51 shows the diagram of a substation that is protected at its medium-voltage side by surge arresters. These arresters have been installed at a certain distance from the MV side of the transformer. This example corresponds to a case in which only one feeder is connected to the transformer; more complex situations (i.e., multifeeder substations) can be found in actual systems. The impact of a lightning stroke on the distribution line, at the span close to the transformer, will cause two traveling waves that will propagate along the line in opposite directions from the point of impact. The goal is to analyze the protection provided by the surge arrester to substation equipment, namely to the MV side of the transformer, taking into account that there will be a separation distance between the surge arrester and the transformer.

The study will be performed by using a single-phase equivalent scheme and assuming an idealized behavior of all components of the system:

1. Distribution line sections and leads are represented as single-phase lossless constant distributed-parameter lines.
2. Transformers can be modeled in fast-front transients by their surge capacitance or by the surge impedance of their windings. In this study, the transformer will be represented as an open circuit.
3. The lightning stroke is represented as an ideal current source, with a double-ramp waveform, see Figure 12.38.
4. The metal oxide surge arrester behaves as an open circuit when the voltage across its terminals is below the residual voltage, and its voltage remains constant when this value is reached, and the discharge current is large enough.

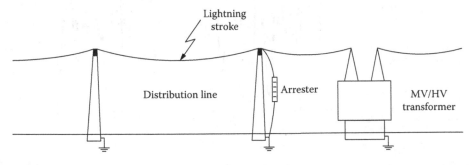

FIGURE 12.51   Diagram of a substation protected by surge arresters.

Figure 12.52 shows the scheme that will be used to analyze the separation distance effect. Note that

- The lightning stroke impacts the distribution line at the midspan, node 0
- The line span, connected to nodes 1 and 2, has been split into two sections and matched at node 1 (i.e., the resistance installed at this point is equal to the surge impedance of the line) to prevent the effect of reflections at this node
- The surge arrester has been installed at node 2
- The transformer, located at node 3, is represented as an open circuit
- The lead section between the surge arrester and the transformer has the same characteristic parameters that the distribution line.

It is assumed that there will not be line flashover. In reality, flashovers can occur between phase conductors at the point of impact, or across insulators. In addition, the effect of arrester leads, that is, connections from arrester terminals to ground and to line conductors, is neglected. The power-frequency voltages are also neglected, so the study will be carried out by assuming that the system is not energized when the lightning stroke impacts the line.

The peak magnitude of any wave that meets the surge arrester is limited to the residual value. As soon as the voltage magnitude exceeds the residual voltage, a relief wave is generated. The addition of this relief wave and the incident wave must equal the wave that propagates beyond the surge arrester. The analysis can be, then, performed by superposing the effects of both the incident and the relief waves.

The initial slope of the stroke current is $S_i = I_{max}/t_f$, see Figure 12.38. Then, traveling waves caused by the impact on the line will have an initial voltage slope S_v given by the following expression:

$$S_v = S_i \frac{Z_c}{2} \tag{12.95}$$

where Z_c is the surge impedance of the distribution line.

The maximum voltage at the transformer is caused when the separation is so long that the voltage at the arrester is already the residual voltage at the time the first reflected wave from the transformer reaches the arrester. Since the maximum voltage of the incident wave that meets the transformer cannot be higher than the residual voltage, the transformer voltage will not overexceed twice the arrester residual voltage.

Assume that the incident wave meets the arrester at $t = 0$, τ is the travel time between the arrester and the transformer, t_a is the time that the arrester needs to reach its residual voltage, and $t_a = 2\tau$. The relief wave starts to travel at the time the arrester voltage reaches the residual value (i.e., at $t = t_a$), and it needs an interval equal to τ to travel along the separation distance. The voltage at the transformer starts to develop an interval equal to τ after the first wave meets the arrester (i.e., at $t = \tau$); however, since the transformer behaves as an open circuit, its voltage slope is twice that of the incident wave voltage, which is the slope of the wave voltage at the arrester. Therefore, when the relief wave meets

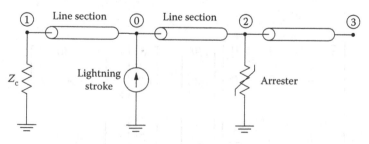

FIGURE 12.52 Scheme of the test system.

the transformer, which occurs at $t = \tau + t_{\mathrm{a}}$, the transformer voltage has just reached twice the residual voltage. From that moment, the effect of the relief wave is to prevent the increase of the transformer voltage.

If the reflected wave from the transformer node meets the surge arrester node at the moment that its voltage reaches the residual voltage, the separation distance fulfills the following condition:

$$V_{\mathrm{res}} = t_{\mathrm{a}} S_v = 2\tau S_v \tag{12.96}$$

from where

$$\tau = \frac{V_{\mathrm{res}}}{2S_v} = \frac{V_{\mathrm{res}}}{Z_{\mathrm{c}} S_i} \tag{12.97}$$

A critical distance is derived from this travel time and from the propagation velocity

$$d_{\mathrm{cri}} = \frac{V_{\mathrm{res}}}{Z_{\mathrm{c}} S_i} \, v \tag{12.98}$$

That is, for any separation distance longer than d_{cri}, the voltage at the transformer will reach twice the residual voltage of the arrester.

EXAMPLE 12.10

The goal is to analyze overvoltages caused at the MV side of the substation transformer of the test system depicted in Figure 12.51, using the representation shown in Figure 12.52. The study will be carried out by assuming that the distribution line does not flashover and neglecting the effect of power system voltages. Consider the following parameters:

Lightning stroke:	Double-ramp current source (I_p)	=30 kA
Overhead line:	Surge impedance (Z_c)	=400 Ω
	Propagation velocity (υ)	=300 m/μs
Surge arrester:	Residual voltage (V_{res})	=100 kV

Two stroke waveforms will be assumed: 8/50 and 2/20 μs. In both cases, the first and the second values indicate the time to crest (t_f) and the tail time (t_h) respectively, see Example 12.7.

Figure 12.53 shows the resistive companion network that results from substituting circuit elements of Figure 12.52 by their companion equivalents. Note that the surge arrester has been represented as a nonlinear resistance R_p.

The application of the Kirchhoff's current law to all nodes yields the following set of algebraic equations:

$$
\begin{bmatrix}
\dfrac{2}{Z_c} & 0 & 0 & 0 \\[2mm]
0 & \dfrac{2}{Z_c} & 0 & 0 \\[2mm]
0 & 0 & \dfrac{2}{Z_c} + \dfrac{1}{R_p} & 0 \\[2mm]
0 & 0 & 0 & \dfrac{1}{Z_c}
\end{bmatrix}
\begin{bmatrix}
v_1(t) \\[2mm] v_0(t) \\[2mm] v_2(t) \\[2mm] v_3(t)
\end{bmatrix}
=
\begin{bmatrix}
0 \\[2mm] i_s(t) \\[2mm] 0 \\[2mm] 0
\end{bmatrix}
-
\begin{bmatrix}
I_{l1r}(t) \\[2mm] I_{l1s}(t) + I_{l2s}(t) \\[2mm] I_{l2r}(t) + I_{l3s}(t) \\[2mm] I_{l3r}(t)
\end{bmatrix}
$$

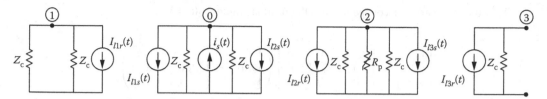

FIGURE 12.53 Equivalent circuit of the test system.

where

$$I_{l1s}(t) = -\left[\frac{v_1(t-\tau_l)}{Z_c} + i_{10}(t-\tau_l)\right] \quad I_{l1r}(t) = -\left[\frac{v_0(t-\tau_l)}{Z_c} + i_{01}(t-\tau_l)\right]$$

$$I_{l2s}(t) = -\left[\frac{v_2(t-\tau_l)}{Z_c} + i_{20}(t-\tau_l)\right] \quad I_{l2r}(t) = -\left[\frac{v_0(t-\tau_l)}{Z_c} + i_{02}(t-\tau_l)\right]$$

$$I_{l3s}(t) = -\left[\frac{v_3(t-\tau_d)}{Z_c} + i_{32}(t-\tau_d)\right] \quad I_{l3r}(t) = -\left[\frac{v_2(t-\tau_d)}{Z_c} + i_{23}(t-\tau_d)\right]$$

being τ_l and τ_d the travel times of the sections that represent, respectively, a half-span of the distribution line and the separation distance between the arrester and the transformer.

The following relationships are derived from the resistive companion equivalent shown in Figure 12.53:

$$i_{10}(t) = -\frac{v_1(t)}{Z_c} \quad i_{32}(t) = 0$$

Substitution of these expressions into the history terms where these currents are included yields

$$I_{l1s}(t) = 0 \quad I_{l3s}(t) = -\frac{v_3(t-\tau_d)}{Z_c}$$

As expected, the history term of the section that represents the half-span that is matched at one end becomes zero. As for in Example 12.7, the value of the travel time of this line section is unimportant since the voltages and currents originated at node 1 will not affect the rest of the network.

The value of the surge arrester resistance during the transient calculation can be derived from its $v - i$ characteristic, which can be approximated by the following expression:

$$i = \text{sign}(v)\text{abs}\left[\beta\left(\frac{v}{V_{res}}\right)^\alpha\right] \tag{12.99}$$

where α and β are surge arrester factors, and V_{res} is the residual voltage.

An ideal behavior of the surge arrester, as that described in the preceding paragraphs, can be obtained by using very high values for parameters α and β; for example, $\alpha \geq 500$ and $\beta \geq 1000$. However, such behavior can be also reproduced by representing the surge arrester as an open switch, when the voltage across it terminals is below the residual voltage, and as a fixed voltage source, with $v = V_{res}$, when the voltage reaches the residual value.

Since the conductance matrix for this case is diagonal, the equations of the resistive companion network can be easily rewritten to obtain voltages. Considering the two possibilities discussed above, the set of equations for each situation will be as follows:

Before the surge arrester voltage reaches the residual value ($R_p \approx \infty$)

$$\begin{bmatrix} v_1(t) \\ v_0(t) \\ v_2(t) \\ v_3(t) \end{bmatrix} = \begin{bmatrix} Z_c/2 & 0 & 0 & 0 \\ 0 & Z_c/2 & 0 & 0 \\ 0 & 0 & Z_c/2 & 0 \\ 0 & 0 & 0 & Z_c \end{bmatrix} \left(\begin{bmatrix} 0 \\ i_s(t) \\ 0 \\ 0 \end{bmatrix} - \begin{bmatrix} I_{l1r}(t) \\ I_{l1s}(t) + I_{l2s}(t) \\ I_{l2r}(t) + I_{l3s}(t) \\ I_{l3r}(t) \end{bmatrix} \right)$$

After the surge arrester voltage reaches the residual value ($v_2(t) = V_{res}$)

$$\begin{bmatrix} v_1(t) \\ v_0(t) \\ v_3(t) \end{bmatrix} = \begin{bmatrix} Z_c/2 & 0 & 0 \\ 0 & Z_c/2 & 0 \\ 0 & 0 & Z_c \end{bmatrix} \left(\begin{bmatrix} 0 \\ i_s(t) \\ 0 \end{bmatrix} - \begin{bmatrix} I_{l1r}(t) \\ I_{l1s}(t) + I_{l2s}(t) \\ I_{l3s}(t) \end{bmatrix} \right)$$

Note that in the second set, the residual voltage $v_2(t) = V_{res}$ is included in the history terms $I_{l2s}(t)$ and $I_{l3r}(t)$. All history terms are initially zero, since power-frequency voltages are neglected; that is, all components are not energized before the lightning stroke impacts the line.

The voltage stresses at the transformer location can be different with different stroke parameters and a different separation distance between arrester and transformer. Figure 12.54 shows simulation results obtained with different values of these parameters.

Some important conclusions can be easily derived from these results:

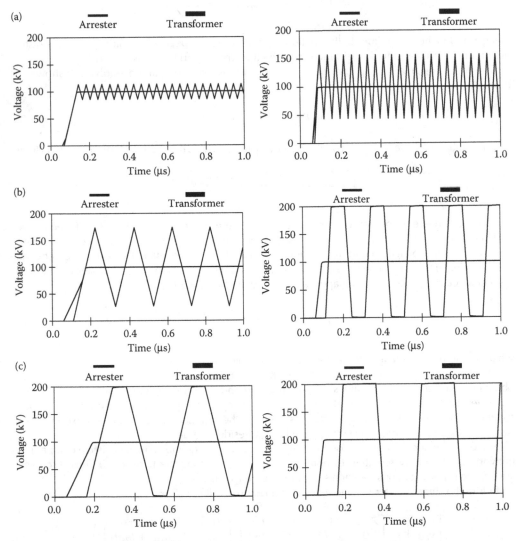

FIGURE 12.54 Stroke waveform: 30 kA, 8/50 μs (left), and 2/20 μs (right): (a) separation distance, $d = 3$ m, (b) separation distance, $d = 12$ m, and (c) separation distance, $d = 30$ m.

- The longer the separation distance between arrester and transformer, the higher the voltage caused at the transformer.
- The steeper the wavefront of the stroke current, I_{max}/t_{max}, the higher the voltage caused at the transformer.
- The maximum overvoltage that can occur at the transformer is twice the residual voltage of the surge arrester.

Upon application of expression 12.98, the following critical distances are obtained for each stroke waveform:

- Waveform: 30 kA, 8/50 µs $\quad d_{cri} = 20$ m
- Waveform: 30 kA, 2/20 µs $\quad d_{cri} = 5$ m

According to these results, the voltage at the transformer location will reach twice the value of the arrester residual voltage with a 8/50 µs waveform in the last case, which is the only one with a separation distance longer than 20 m; however, with a 2/20 µs waveform, except in the first case, for which the separation distance is shorter than 5 m, the voltages at the transformer location are twice the arrester residual voltage.

12.7 MULTIPHASE OVERHEAD LINES AND INSULATED CABLES

12.7.1 INTRODUCTION

AC power transmission systems are three phase. In general, the components of three-phase systems, such as overhead transmission lines, underground cables, transformers, generators, and motors constitute systems of coupled inductances and coupled capacitances involving the phase conductors and the ground return.

The electrical parameters of both overhead lines and insulated cables are obtained by means of formulae that use the line/cable geometry and the properties of the materials that make up the line/cable geometry. The equations that characterize a line or cable are very similar although the behavior of the models and their transient responses can exhibit significant differences. There are also important differences in designs. Both lines and cables are shielded but the reasons to shield a cable are very different from those to shield an overhead line. The large variety of cable designs makes it very difficult to develop and implement a single procedure that could calculate the parameters of any type of cable. The thermal capacity of a cable is a crucial aspect that can justify some design aspects not accounted for in an overhead line. For example, a cross-bonding is used in cables for several reasons, it is a measure that can increase the thermal limit of an underground insulated cable.

A detailed description of procedures for calculating electrical parameters of lines and cables, as well as the manipulation of their equations, and their implementation in a transients tool is beyond the scope of this subsection. Instead, this subsection provides a short introduction to the calculation of electrical parameters of lines and cables, and to the solution methods implemented in transients tools for solving the resulting equations.

12.7.2 OVERHEAD LINES

12.7.2.1 Introduction to Overhead Lines Equations

Figure 12.55 shows a differential section of a three-phase overhead transmission line illustrating the coupling among series inductances and among shunt capacitances.

Despite the apparent complexity of the line section in Figure 12.55, a systematic use of matrix algebra allows the modeling of these structures in a way which resembles the form of the solution for single-phase systems. Care must be exercised, though, when manipulating the equations, in that matrix algebra is not commutative; but other than this, most properties of scalar algebra work as expected when dealing with matrix algebra.

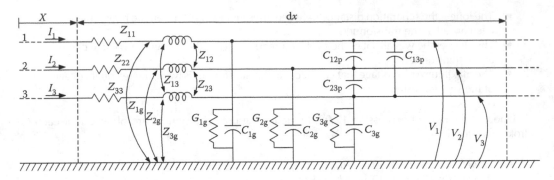

FIGURE 12.55   Differential section of a three-phase transmission line.

With respect to Figure 12.55, the following series impedance and shunt admittance matrices can be defined

$$[Z] = \begin{bmatrix} Z_{11} & Z_{12} & Z_{13} \\ Z_{21} & Z_{22} & Z_{23} \\ Z_{31} & Z_{32} & Z_{33} \end{bmatrix} \quad [Y] = \begin{bmatrix} Y_{11} & Y_{12} & Y_{13} \\ Y_{21} & Y_{22} & Y_{23} \\ Y_{31} & Y_{32} & Y_{33} \end{bmatrix} \tag{12.100}$$

where

$$Z_{ii} = R_i^c + j\omega L_{ii} \tag{12.101}$$

$$Z_{ij} = j\omega L_{ij} \tag{12.102}$$

$$Y_{ii} = G_i + j\omega C_{ii} \tag{12.103}$$

$$Y_{ij} = j\omega C_{ij} \tag{12.104}$$

In general, there can be N conductors (i.e., bundled conductors) per phase and ground wires. For illustration, only three conductors are assumed.

The $[Z]$ and $[Y]$ matrices are symmetrical, that is, $Z_{ij} = Z_{ji}$ and $Y_{ij} = Y_{ji}$.

Notice that in these equations only the identity of the phase conductors 1, 2, and 3 is preserved. The identity of the ground is worked out into the values of the series impedances and shunt admittances, and the original physical system is replaced by an equivalent system of phase conductors over ideal ground. The details of how this is achieved are discussed in the following subsections.

12.7.2.2   Overhead Lines Electrical Parameters

12.7.2.2.1   Introduction

The calculation of both the series impedance and the shunt admittance matrices is presented below. The subsection provides a way to obtain self- and mutual impedances and self- and mutual admittances for conductors above lossy ground. The procedure is applied to three-phase three-conductor overhead lines; details about the approaches to be followed with lines with bundled phases and to account for ground wires in shielded lines are discussed at the end of the subsection. Additional details on the material covered in this section can be found in Section 7.3 of Chapter 7.

12.7.2.2.2   Series Impedance Matrix: Self- and Mutual Impedances for Conductors above Lossy Ground

In Equations 12.101 and 12.102, R_i^c is the resistance of conductor i at frequency ω, L_{ii} is the self-inductance of conductor i at frequency ω when the current is injected in the conductor and returns through the ground, and L_{ij} is the mutual inductance between conductors i and j at frequency ω when the current is injected in the conductor and returns through the ground. The computation of self- and mutual inductances of conductors above ideal ground (resistivity $\rho = 0$) is straightforward using Lord Kelvin's method of images: Imagining the ground as a mirror, one can simply remove the mirror and replace it by the images of the conductors. The normal case, however, is that the ground is not ideal, with a typically used resistivity value of $\rho = 100 \, \Omega.\text{m}$.

For nonideal ground, the equations describing the distortion of the electromagnetic fields can become quite elaborate and Carson's formulas [11] have been traditionally used to calculate these effects. More recently, however, the concept of complex penetration depth [12], which gives a very good approximation to the ground effect in a simple way, has become very popular. There are a number of other factors that introduce uncertainty in the value of the parameters. For example, overhead line conductors will hang or "sag" in between the towers supporting them, thus changing the distances between conductors and ground. This sagging depends on the separation between towers (typically about 300 m in high-voltage transmission systems), on the weight of the conductors (which changes when ice accumulates), and on the temperature of the conductors (which depends on the current and on the outside temperature, for example, whether it is Summer or Winter). In addition, the ground resistivity itself can present large variations as the terrain changes along the length of the line.

In Figure 12.56, conductors 1 and 2 are located at heights h_1 and h_2 above ground. If the ground was ideal, the placement of the image conductors $1'$, $2'$ would be at distances h_1 and h_2 below the ground plane. The method of complex penetration depth makes a correction to the height of the conductors and to the depth of the images by the amount $\bar{p}$, calculated as

$$\bar{p} = \sqrt{\frac{\rho}{j\omega\mu}} \tag{12.105}$$

$\bar{p}$ is the *complex penetration depth* and it is a complex number, ρ is the ground resistivity in $(\Omega.\text{m})$, ω is the frequency in [rad/s], μ is the ground permeability: $\mu = \mu_r\mu_0 \simeq \mu_0$ ($\mu_0 = 4\pi \times 10^{-7}$ in SI units for vacuum). With the heights of the conductors and the depths of the images corrected by $\bar{p}$, the method of images can now be applied as if the ground was ideal. The self- and mutual inductances

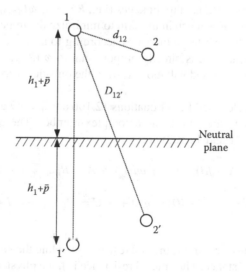

FIGURE 12.56 Method of images for the inductances.

per meter of conductor can be calculated as follows, with $i = 1$ and $j = 2$ in Figure 12.56,

$$L'_{ii} = \frac{\mu}{2\pi} \ln \frac{2(h_i + \bar{p})}{r_{ieq}} = A - jB \tag{12.106}$$

$$L'_{ij} = \frac{\mu}{2\pi} \ln \frac{D_{ij'}}{d_{ij}} = C - jD \tag{12.107}$$

The primes indicate that the inductances have been corrected for the ground effect. The units of L'_{ii} and L'_{ij} are [H/m]. In these equations μ is the magnetic permeability of the medium, and for both air and ground $\mu \simeq \mu_0 = 4\pi \times 10^{-7}$. The distances between conductors are denoted by small case d, while the distances from the conductors to the images are denoted by capital D. The radius of the conductors is denoted by r.

Formulae 12.106 and 12.107 assume that the magnetic field is entirely outside of the conductors and that the conductors' radii are small compared to the distances d_{ij} and D_{ij}. In the actual physical conductors, there is magnetic field both inside and outside the conductors. To account for the internal flux linkages, an *equivalent circular hollow conductor* with a radius $r_{ieq} < r_{iext}$ (r_{iext} is the physical "external" radius) can be used to replace the physical conductor.

The idea is that by having $r_{ieq} < r_{iext}$, the missing internal flux (a hollow conductor has no internal flux) will be compensated for by the flux in the extra space from r_{ieq} and r_{iext}.

When the frequency increases, the current moves toward the outside of the cross section of the conductor, leaving the inside hollow. The higher the frequency, the more the equivalent radius r_{ieq} grows closer to the external radius and, in the limit, for $f \to \infty$, $r_{ieq} = r_{iext}$. The value of geometric mean radius (GMR) commonly found in conductor tables corresponds to $r_{ieq}(50/60 \text{ Hz}) = \text{GMR}$ and includes the corrections for internal flux at power frequency, 50 or 60 Hz, and for the stranding and spiraling of the internal subconductors that make up the conductors.

The concept of skin effect inside a conductor and the concept of complex penetration depth with respect to the surface of the ground have the same physical reason. If the conductor or the ground had zero resistivity, the current would flow in the skin of the conductor and in the surface of the ground to minimize the internal inductance. Elementary current filaments in the center of the conductor or deep inside the ground are linked by more flux than current filaments closer to the surface. Since actual conductors or ground have finite resistivity, some of the current will go inside the conductor or underneath the ground looking for more cross sectional area to minimize the resistance. How important the resistance is with respect to the internal flux linkages (i.e., R versus $\omega L_{\text{internal}}$) will decide how much current should flow inside and how much in the skin to minimize the impedance to the current flow. Since at higher frequencies the resistance increases according to a $\sqrt{\omega}$ law ($R = k\sqrt{\omega}$) (Lord Rayleigh's formula) while the reactance is directly proportional to ω ($X_{\text{internal}} = \omega L_{\text{internal}}$), the current will try to avoid the higher X_{internal} and will move toward the outside, despite having to flow through a reduced cross section.

Inductances L'_{11} and L'_{12} calculated from Equations 12.106 and 12.107 are complex numbers ($A - jB$) and ($C - jD$), respectively, due to $\bar{p}$ being a complex number. The corrected self- and mutual impedances (Equations 12.101 and 12.102) are

$$Z_{ii-corr} = R^c_i + j\omega(A - jB) = (R^c_i + \omega B) + j\omega A = R_{ii-corr} + j\omega L_{ii-corr} \tag{12.108}$$

$$Z_{ij-corr} = j\omega(C - jD) = \omega D + j\omega C = R_{ij-cor} + j\omega L_{ij-corr} \tag{12.109}$$

As a result of $\bar{p}$ being complex, the corrected self impedance $Z_{ii-corr}$ acquires a higher value of resistance (it incorporates the effect of the ground resistivity), while the corrected mutual impedance $Z_{ij-corr}$ acquires a mutual resistance. This mutual resistance is not a physical resistance between conductors, but it is because the currents in the conductors share the lossy ground return. For the line of

Figure 12.55, the series impedance matrix *per each meter* of line is given by

$$\begin{bmatrix} Z_{11} & Z_{12} & Z_{13} \\ Z_{21} & Z_{22} & Z_{23} \\ Z_{31} & Z_{32} & Z_{33} \end{bmatrix} = \begin{bmatrix} Z_{11-corr} & Z_{12-corr} & Z_{13-corr} \\ Z_{12-corr} & Z_{22-corr} & Z_{23-corr} \\ Z_{13-corr} & Z_{23-corr} & Z_{33-corr} \end{bmatrix} \qquad (12.110)$$

12.7.2.2.3 Shunt Admittance Matrix: Self- and Mutual Admittances for Conductors above Ground

In Equation 12.103, the conductances G_i are associated with currents leaking through the insulator chains to ground. For conductors under the *corona* effect, in which the air surrounding the conductor becomes ionized, the value of G_i can be significant. However, corona is normally modeled separately and in what follows noncorona conditions are assumed. Under noncorona conditions, with clean insulators and dry weather, G_i would be zero, and under typical maintenance and average weather, typical values are in the range of 10^{-7} to 10^{-9} S, and can be neglected for most studies.

The terms $j\omega C_{ii}$ in Equation 12.103 and $j\omega C_{ij}$ in Equation 12.104 correspond to "self" and "mutual" capacitances. These mutual capacitance elements can be realized by an equivalent circuit equivalent showing simple capacitors between conductors and between conductors and ground. The value of these simple capacitors can be calculated from the self- and mutual capacitances matrix, as explained later in this subsection.

The starting point for the derivation of the matrix of capacitances is the concept of potential coefficients P in the electrostatic field (Maxwell's coefficients). Consider Figure 12.57. The relationship between the voltages applied to the conductors and the electrical charges needed to produce these voltages can be written as

$$\begin{bmatrix} V_1 \\ V_2 \\ V_3 \end{bmatrix} = \begin{bmatrix} P_{11} & P_{12} & P_{13} \\ P_{21} & P_{22} & P_{23} \\ P_{31} & P_{32} & P_{33} \end{bmatrix} \begin{bmatrix} Q_1 \\ Q_2 \\ Q_3 \end{bmatrix} \qquad (12.111)$$

where P_{ii} and P_{ij} are obtained with the help of Figure 12.58.

Notice that in the case of the electrical field, the electrostatic assumption is followed and the charges on the surface of the conductors and on the surface of the ground, regardless of the frequency and regardless of the resistivity of the conductors or of the ground. As a result, the formulae to calculate the P's in Equation 12.111 use the external radius of the conductors and do not correct for nonideal ground.

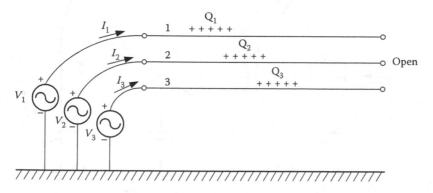

FIGURE 12.57 Currents to ground through the line capacitances.

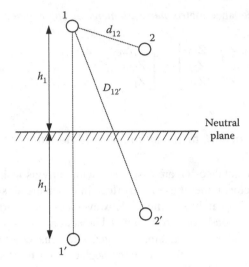

FIGURE 12.58 Method of images for Maxwell's coefficients.

With reference to Figure 12.59, for 1 m of conductors, and with $i = 1$ and $j = 2$,

$$P_{ii} = \frac{1}{2\pi\varepsilon} \ln \frac{2h_i}{r_i} \tag{12.112}$$

$$P_{ij} = \frac{1}{2\pi\varepsilon} \ln \frac{D_{ij'}}{d_{ij}} \tag{12.113}$$

The units of P_{ii} and P_{ij} are (F^{-1} × m). In these equations, ε is the permittivity of the medium, and, in the case of the air and the ground, $\varepsilon \simeq \varepsilon_0$, with $\varepsilon_0 = 8.8542 \times 10^{-12}$ being the permittivity of the

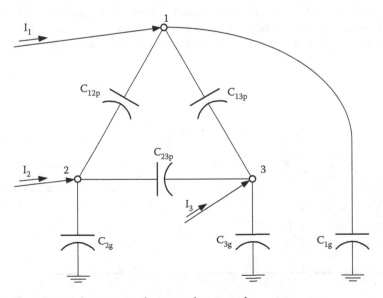

FIGURE 12.59 Capacitances between conductors and to ground.

vacuum. In Equation 12.112, r_i is the actual external radius of conductor i in the case of circular conductors, or an equivalent circular conductor radius for arbitrary shapes.

Once the matrix of Maxwell's coefficients Equation 12.111 has been determined, the physical capacitances of Figure 12.59 can be found.

Starting from Equation 12.111, in compact form,

$$[V] = [P][Q] \tag{12.114}$$

Or, in terms of the current at a given frequency ω,

$$[I] = \frac{d[Q]}{dt} = j\omega[Q] \tag{12.115}$$

Substituting and making the matrices explicit,

$$\begin{bmatrix} V_1 \\ V_2 \\ V_3 \end{bmatrix} = \frac{1}{j\omega} \begin{bmatrix} P_{11} & P_{12} & P_{13} \\ P_{21} & P_{22} & P_{23} \\ P_{31} & P_{32} & P_{33} \end{bmatrix} \begin{bmatrix} I_1 \\ I_2 \\ I_3 \end{bmatrix} \tag{12.116}$$

Solving for the currents,

$$\begin{bmatrix} I_1 \\ I_2 \\ I_3 \end{bmatrix} = j\omega \begin{bmatrix} C_{11} & C_{12} & C_{13} \\ C_{21} & C_{22} & C_{23} \\ C_{31} & C_{32} & C_{33} \end{bmatrix} \begin{bmatrix} V_1 \\ V_2 \\ V_3 \end{bmatrix} \tag{12.117}$$

where the capacitances matrix is the inverse of the matrix of Maxwell's coefficients,

$$\begin{bmatrix} C_{11} & C_{12} & C_{13} \\ C_{21} & C_{22} & C_{23} \\ C_{31} & C_{32} & C_{33} \end{bmatrix} = \begin{bmatrix} P_{11} & P_{12} & P_{13} \\ P_{21} & P_{22} & P_{23} \\ P_{31} & P_{32} & P_{33} \end{bmatrix}^{-1} \tag{12.118}$$

Applying nodal analysis to the circuit of Figure 12.59

$$\begin{bmatrix} I_1 \\ I_2 \\ I_3 \end{bmatrix} = j\omega \begin{bmatrix} (C_{1g} + C_{12p} + C_{13p}) & -C_{12p} & -C_{13p} \\ -C_{12p} & (C_{2g} + C_{12p} + C_{23p}) & -C_{23p} \\ -C_{13p} & -C_{23p} & (C_{3g} + C_{13p} + C_{23p}) \end{bmatrix} \begin{bmatrix} V_1 \\ V_2 \\ V_3 \end{bmatrix} \tag{12.119}$$

Comparing Equation 12.117 with Equation 12.119, it follows that adding all the elements in the first row of Equation 12.117

$$C_{11} + C_{12} + C_{13} = C_{1g} + C_{12p} + C_{13p} - C_{12p} - C_{13p} = C_{1g} \tag{12.120}$$

which is the capacitance from conductor 1 to ground.

Similarly, adding the elements of the second row of Equation 12.117, the capacitance of conductor 2 to ground is determined, and so-and-so-forth. For the mutuals, the relationship is directly

$$C_{12p} = -C_{12} \tag{12.121}$$

and so forth. This process will yield all the physical capacitances between conductors and between conductors and ground in the circuit of Figure 12.59.

For the line of Figure 12.55, neglecting the conductances G, the shunt admittance matrix *per each metre* of line is then given by

$$\begin{bmatrix} Y_{11} & Y_{12} & Y_{13} \\ Y_{21} & Y_{22} & Y_{23} \\ Y_{31} & Y_{32} & Y_{33} \end{bmatrix} = j\omega \begin{bmatrix} C_{11} & C_{12} & C_{13} \\ C_{21} & C_{22} & C_{23} \\ C_{31} & C_{32} & C_{33} \end{bmatrix} \tag{12.122}$$

12.7.2.3 Bundled Phases and Ground Wires

Transmission-level lines use bundled phases and are generally shielded by one or more ground wires to reduce the number of flashovers caused by direct lightning strokes to phase conductors. Bundled phases are used to reduce the electric field strength at the surface of the conductors, and avoid or mitigate the likelihood of corona. Two alternative methods can be used for modeling bundling [2]:

1. The bundled subconductors are replaced with an equivalent single conductor.
   The GMR and equivalent radius of the bundled subconductors are calculated and a single conductor is used to represent the bundle.

$$GMR_{eq} = \sqrt[n]{n \cdot GMR_{cond}^{n-1} \cdot R_b} \tag{12.123}$$

$$R_{eq} = \sqrt[n]{n \cdot R_{cond}^{n-1} \cdot R_b} \tag{12.124}$$

   where n is the number of conductors in bundle, R_b is the radius of bundle, R_{cond} is the radius of conductors, R_{eq} is the radius of equivalent single conductor, GMR_{cond} is the GMR of individual subconductors, and GMR_{eq} is the GMR of equivalent single conductor.

2. All subconductors are represented explicitly and matrix elimination of subconductors is applied.
   When all subconductors are represented explicitly in the series impedance matrix $[Z_{cond}]$ and the matrix of Maxwell's coefficients $[P_{cond}]$, the order of both matrices increases (i.e., for a three-phase line with two subconductors per phase the order of both matrices is 6×6). Since the line model assumes a connection to three phases at each line terminal, an elimination procedure, identical for both matrices, has to be applied to obtain $[Z_{ph}]$ and $[P_{ph}]$. In the case of the matrix of Maxwell's coefficients, the reduction has to be made before calculating the capacitance matrix.

The use of an equivalent single conductor radius and GMR ignores proximity effects and is only valid if the subconductors spacing is much smaller than the spacing between the phases of the line. The option of matrix elimination includes proximity effects and hence is generally more accurate, but the difference with respect to the GMR method is very small.

It is reasonable to assume that the ground wire potential is zero along its length when ground wires are continuous and grounded at each tower and the frequency is below 250 kHz. In such case, a reduction procedure similar to that applied to bundled subconductors is usually carried out for both $[Z_{cond}]$ and $[P_{cond}]$. However, for lightning studies ground wires have to be explicitly modeled since they must be available for lightning overvoltage calculations.

EXAMPLE 12.11

Consider the 345 kV three-phase overhead line shown in Figure 12.60. The line consists of three-phase conductors separated from ground and from each other as indicated in the figure. For

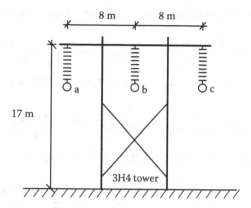

FIGURE 12.60   Example 11: 345 kV transmission line.

simplicity in the example, it is assumed that there are no ground wires. The ground resistivity is $\rho = 100\,\Omega.m$. Each insulator chain consists of 18 plates, each plate is 14.6 cm long. The conductors "sag" between towers is 5 m (the sag is the maximum conductor's hanging distance from the horizontal line due to the conductor's weight). Each phase conductor is of type ACSR 84/19 (i.e., 84 aluminum strands on the outside for current conduction and 19 steel strands in the core for physical strength). The manufacturer's code name for this conductor is "blue bird." From the manufacturer's tables, we find that for 60 Hz and a temperature of operation of 75°C the conductors' resistance is $R_i^c = 0.03386\,\Omega/km$. The tables also provide the conductors' external radius, $r_{i-ext} = 2.238\,cm$, and the geometrical mean radius GMR = 1.79 cm [$= r_{ieq}(60\,Hz)$].

It is desired to obtain both the series impedance matrix of Equation 12.110 and the shunt admittance matrix from Equation 12.122 at a system operating frequency of 60 Hz, as well as the physical capacitances of Figure 12.59.

The series impedances (Equation 12.110) have two components: the series resistances due to the resistivity of the conductors and of the ground return, and the series reactances due to the self- and mutual inductances among conductors and between conductors and ground. The self- and mutual inductances are defined by the geometrical distances with the heights corrected by the complex penetration depth $\bar{p}$ of Equation 12.105. To take into account the effect of the conductors' sags, the physical heights are normally corrected [2] using the following formula:

$$h_{ave} = h_{tower} - \frac{2}{3}sag \qquad (12.125)$$

In the example, considering the length of the insulator chains and the sags,

$$h_a = h_b = h_c = 17 - 0.146 \times 18 - \frac{2}{3}5 = 11.039\,m$$

From Equation 12.105, the complex penetration depth for the given ground resistivity and for 60 Hz is given by

$$\bar{p} = \sqrt{\frac{\rho}{j\omega\mu}} = \sqrt{\frac{100}{j2\pi60 \times 4\pi \times 10^{-7}}} = 459.44\angle{-45°}\,m$$

The self- and mutual inductances can now be calculated from Equations 12.106 and 12.107,

$$L'_{aa} = L'_{bb} = L'_{cc} = \frac{\mu}{2\pi}\ln\frac{2(h_i + \bar{p})}{r_{ieq}} = \frac{4\pi \times 10^{-7}}{2\pi}\ln\frac{2(11.039 + 459.44\angle -45°)}{0.0179}$$

$$= 2.178\angle -4.0476° = 2.1726 - j0.1537\,\mu\text{H/m}$$

$$L'_{ab} = L'_{bc} = \frac{\mu}{2\pi}\ln\frac{D_{ab'}}{d_{ab}} = \frac{4\pi \times 10^{-7}}{2\pi}\ln\frac{\sqrt{8^2 + (2(11.039 + 459.44\angle -45°))^2}}{8}$$

$$= 0.964\angle -9.1718° = 0.9521 - j0.15373\,\mu\text{H/m}$$

$$L'_{ac} = \frac{\mu}{2\pi}\ln\frac{D_{ac'}}{d_{ac}} = \frac{4\pi \times 10^{-7}}{2\pi}\ln\frac{\sqrt{16^2 + (2(11.039 + 459.44\angle -45°))^2}}{16}$$

$$= 0.8279\angle -10.70° = 0.8135 - j0.1537\,\mu\text{H/m}$$

The series impedances can now be calculated from $Z = R + j\omega L$. After multiplying by $j\omega$, the imaginary parts of the complex inductances become resistances that represent the losses due to the ground resistivity while the real parts correspond to the physical inductances,

$$Z_{aa} = Z_{bb} = Z_{cc} = R_a^c + j\omega L'_{aa} = 0.03386 \times 10^{-3} + j2\pi 60\,(2.1726 - j0.1537) \times 10^{-6}\,\Omega/\text{m}$$

$$= 0.0918 + j0.8191\,\Omega/\text{km}$$

$$Z_{ab} = Z_{bc} = j\omega L'_{ab} = j2\pi 60\,(0.9521 - j0.15373) \times 10^{-6}\,\Omega/\text{m}$$

$$= 0.0579 + j0.3589\,\Omega/\text{km}$$

$$Z_{ac} = j\omega L'_{ac} = j2\pi 60\,(0.8135 - j0.1537) \times 10^{-6}\,\Omega/\text{m}$$

$$= 0.0579 + j0.3067\,\Omega/\text{km}$$

The line series impedance matrix is then

$$\begin{bmatrix} Z_{aa} & Z_{ab} & Z_{ac} \\ Z_{ba} & Z_{bb} & Z_{bc} \\ Z_{ca} & Z_{cb} & Z_{cc} \end{bmatrix} = \begin{bmatrix} (0.0918 + j0.8191) & (0.0579 + j0.3589) & (0.0579 + j0.3067) \\ (0.0579 + j0.3589) & (0.0918 + j0.8191) & (0.0579 + j0.3589) \\ (0.0579 + j0.3067) & (0.0579 + j0.3589) & (0.0918 + j0.8191) \end{bmatrix}\,\Omega/\text{km}$$

The Maxwell's potential coefficients are calculated from Equations 12.112 and 12.113

$$P_{aa} = P_{bb} = P_{cc} = \frac{1}{2\pi\varepsilon}\ln\frac{2h_a}{r_a} = \frac{1}{2\pi \times 8.8542 \times 10^{-12}}\ln\frac{2 \times 11.039}{0.02238} = 1.239 \times 10^{11}\,\text{m/F}$$

$$P_{ab} = P_{bc} = P_{ij} = \frac{1}{2\pi\varepsilon}\ln\frac{D_{ab'}}{d_{ab}} = \frac{1}{2\pi \times 8.8542 \times 10^{-12}}\ln\frac{\sqrt{8^2 + (2 \times 11.039)^2}}{8^2} = 0.1936 \times 10^{11}\,\text{m/F}$$

$$P_{ac} = \frac{1}{2\pi\varepsilon}\ln\frac{D_{ac'}}{d_{ac}} = \frac{1}{2\pi \times 8.8542 \times 10^{-12}}\ln\frac{\sqrt{16^2 + (2 \times 11.039)^2}}{16^2} = 0.0958 \times 10^{11}\,\text{m/F}$$

Notice that there is no complex penetration depth correction $\bar{p}$ for the electric field distances. The Maxwell's coefficients matrix is

$$\begin{bmatrix} P_{aa} & P_{ab} & P_{ac} \\ P_{ba} & P_{bb} & P_{bc} \\ P_{ca} & P_{cb} & P_{cc} \end{bmatrix} = \begin{bmatrix} 1.239 & 0.1936 & 0.0958 \\ 0.1936 & 1.239 & 0.1936 \\ 0.0958 & 0.1936 & 1.239 \end{bmatrix} \times 10^{11}\,\text{m/F}$$

Inverting this matrix, the capacitances matrix (Equation 12.118) is obtained,

$$\begin{bmatrix} C_{aa} & C_{ab} & C_{ac} \\ C_{ba} & C_{bb} & C_{bc} \\ C_{ca} & C_{cb} & C_{cc} \end{bmatrix} = [P]^{-1} = \begin{bmatrix} 8.296 & -1.225 & -0.4500 \\ -1.225 & 8.452 & -1.225 \\ -0.4500 & -1.225 & 8.296 \end{bmatrix} \text{pF/m}$$

and neglecting the shunt losses, the line admittance matrix at 60 Hz is given by

$$[Y] = j\omega[C] = j \begin{bmatrix} 3.127 & -0.4620 & -0.1697 \\ -0.4620 & 3.186 & -0.4620 \\ -0.1697 & -0.4620 & 3.127 \end{bmatrix} \times 10^{-6} \text{ S/km}$$

The physical capacitances are (Equations 12.120 and 12.121)

$$C_{ag} = C_{cg} = C_{aa} + C_{ab} + C_{ac} = 8.296 - 1.225 - 0.4500 = 6.621 \text{ pF/m}$$

$$C_{bg} = C_{ba} + C_{bb} + C_{bc} = -1.225 + 8.452 - 1.225 = 6.002 \text{ pF/m}$$

$$C_{abp} = C_{bcp} = -C_{ab} = 1.225 \text{ pF/m}$$

$$C_{acp} = -C_{ac} = 0.4500 \text{ pF/m}$$

12.7.3 Insulated Cables

12.7.3.1 Insulated Cables Designs and Equations

Guidelines for representing insulated cables in transient studies are similar to those proposed for overhead lines. However, the large variety of cable designs makes it very difficult to develop a general computer routine for calculating the parameter of each design. A short description of the main cable designs is provided below.

- *Single core (SC) self-contained cables*: They are coaxial in nature, see Figure 12.61a. The insulation system can be based on extruded insulation (e.g., XLPE) or oil-impregnated paper (fluid-filled or mass-impregnated). The core conductor can be hollow in the case of fluid-filled cables. SC cables for high-voltage applications are always designed with a metallic sheath conductor, which can be made of lead, corrugated aluminum, or copper wires. Such cables are also designed with an inner and an outer semiconducting screen, which are in contact with the core conductor and the sheath conductor, respectively.
- *Three-phase self-contained cables*: They consist of three SC cables which are contained in a common shell. The insulation system of each SC cable can be based on extruded insulation

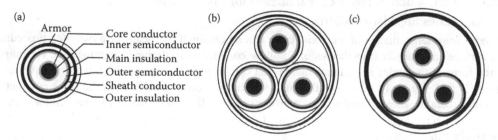

FIGURE 12.61 Some insulated cable designs: (a) single-core cable with armor, (b) three-phase cable design, and (c) pipe-type cable.

or on paper-oil. Most designs can be differentiated into the two designs: (i) one metallic sheath for each *SC* cable, with cables enclosed within metallic pipe (sheath/armor), see Figure 12.61b; (ii) one metallic sheath for each *SC* cable, with cables enclosed within insulating pipe. The space between the *SC* cables and the enclosing pipe is for both designs filled by a composition of insulating materials.

- *Pipe-type cables*: They consist of three *SC* paper cables that are laid within a steel pipe, which is filled with pressurized low-viscosity oil or gas, see Figure 12.61c. Each *SC* cable is fitted with a metallic sheath. The sheaths may be touching each other.

The behavior of a cable can be described by equations similar to those of an overhead line [2,13,14]:

$$[Z(\omega)] = [R(\omega)] + j\omega[L(\omega)] \tag{12.126}$$

$$[Y(\omega)] = [G(\omega)] + j\omega[C(\omega)] \tag{12.127}$$

where $[R]$, $[L]$, $[G]$ and $[C]$ are the cable parameter matrices expressed in per unit length. These quantities are $(n \times n)$ matrices, being n the number of (parallel) conductors of the cable system. The variable ω reflects that these quantities are calculated as function of frequency. Note that a conductance matrix $[G]$ is necessary to represent a cable performance.

The calculation of matrices $[Z]$ and $[Y]$ uses cable geometry and material properties as input parameters. In general, it is necessary to specify:

1. Geometry: location of each conductor (x–y coordinates); inner and outer radii of each conductor; burial depth of the cable system.
2. Material properties: resistivity, ρ, and relative permeability, μ_r, of all conductors (μ_r is unity for all nonmagnetic materials); resistivity, ρ, and relative permeability of the surrounding medium, μ_r; relative permittivity of each insulating material, ε_r.

The current circulating in the core of a cable induces a voltage in the screen of the same cable. If a closed loop is formed then a current circulates in the screen. This increases the system losses and limits the thermal capacity of the cable; therefore, it is important to keep the current in the screen as low as possible. A solution to limit currents and losses is based on the bonding on the cable screens. The techniques implemented to bond screen cables are discussed at the end of this section. The calculation of the series impedance and the shunt admittance matrices for coaxial configurations is presented below. For other designs, see [2,15,16].

12.7.3.2 Calculation of Electrical Parameters for Coaxial Cables

12.7.3.3 Series Impedance Matrix

The series impedance matrix of a coaxial cable can be obtained by means of a two-step procedure. First, surface and transfer impedances of a hollow conductor are derived; then they are rearranged into the form of the series impedance matrix that can be used for describing traveling-wave propagation [17,18]. Figure 12.62 shows the cross section of a coaxial cable with the three conductors (i.e., core, metallic sheath, and armor) and the currents flowing down each one. Some coaxial cables do not have armor. Insulations *A* and *B* are sometimes called bedding and plastic sheath, respectively [2].

Consider a hollow conductor whose inner and outer radii are r_i and r_o, respectively. Figure 12.63 shows its cross section. The inner surface impedance Z_{inn} and the outer surface impedance Z_{out}, both

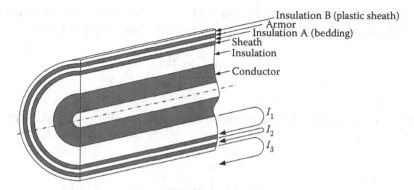

FIGURE 12.62 Single-core self-contained coaxial cable design.

in per unit length (p.u.l.), are given by [17]:

$$Z_{inn} = \frac{\rho m}{2\pi r_i} \frac{I_0(mr_i)K_1(mr_o) + I_1(mr_o)K_0(mr_i)}{I_1(mr_o)K_1(mr_i) - I_1(mr_i)K_1(mr_o)} \quad (12.128)$$

$$Z_{out} = \frac{\rho m}{2\pi r_o} \frac{I_0(mr_o)K_1(mr_i) + I_1(mr_i)K_0(mr_o)}{I_1(mr_o)K_1(mr_i) - I_1(mr_i)K_1(mr_o)} \quad (12.129)$$

where

$$m = \sqrt{\frac{j\omega\mu}{\rho}} \quad (12.130)$$

Notice that m in Equation 12.130 is the inverse of the complex penetration depth in Equation 12.105. As in Equation 12.105, ρ and μ are the resistivity and the permeability of the conductor, respectively, and ω is the frequency. $I_n(\cdot)$ and $K_n(\cdot)$ are the n-th order Modified Bessel Functions of the first and second kind, respectively.

Z_{inn} is the p.u.l impedance of the hollow conductor for the current returning inside the conductor, while Z_{out} is the p.u.l. impedance for the current returning outside the conductor.

The p.u.l. transfer impedance Z_{io} from one surface to the other is calculated as follows [17]:

$$Z_{io} = \frac{\rho}{2\pi r_i r_o} \frac{1}{I_1(mr_o)K_1(mr_i) - I_1(mr_i)K_1(mr_o)} \quad (12.131)$$

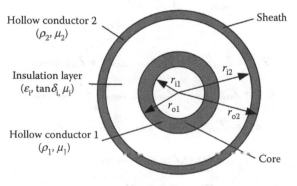

FIGURE 12.63 Cross section of a coaxial cable with a hollow conductor.

The impedance of an insulation layer between two hollow conductors, whose inner and outer radii are, respectively, r_{o1} and r_{i2}, see Figure 12.63, is given by the following expression:

$$Z_{ins} = j\omega \frac{\mu}{2\pi} \ln \frac{r_{i2}}{r_{o1}} \tag{12.132}$$

where μ is the permeability of the insulation.

The ground-return impedance of an underground wire can be calculated by means of the following general expression [19,20]:

$$Z_g = \frac{\rho m^2}{2\pi} \left[K_0(mD_1) - K_0(mD_2) + \int_{-\infty}^{\infty} \frac{e^{-d_2\sqrt{\lambda^2+m^2}}}{|\lambda| + \sqrt{\lambda^2 + m^2}} \cdot e^{j\lambda d_1} \, d\lambda \right] \tag{12.133}$$

where m is given by Equation 12.130 and ρ is the ground resistivity. λ is the integration constant.

The p.u.l. self impedance, Z_g, of a wire placed at a depth of y with radius r is obtained by substituting

$$d_1 = r \quad d_2 = 2y \quad D_1 = r \quad D_2 = \sqrt{r^2 + 4y^2} \tag{12.134}$$

into Equation 12.133.

To obtain the p.u.l. mutual impedance of two wires, $Z_{g,ij}$, placed at depths of y_i and y_j with horizontal separation x_{ij}, substitute

$$d_1 = x_{ij} \quad d_2 = y_i + y_j \quad D_1 = \sqrt{x_{ij}^2 + (y_i - y_j)^2} \quad D_2 = \sqrt{x_{ij}^2 + (y_i + y_j)^2} \tag{12.135}$$

into Equation 12.133.

Consider the coaxial cable shown in Figure 12.62. Assume that I_1 is the current flowing down the core and returning through the sheath, I_2 flows down the sheath and returns through the armor, and I_3 flows down on the armor and its return path is the external ground soil. If V_1, V_2, and V_3 are, respectively, the voltage differences between the core and the sheath, between the sheath and the armor, and between the armor and the ground, the p.u.l. impedance relationships for the three loops shown in Figure 12.62 can be expressed by means of the following impedance matrix [2]:

$$[Z_{loop}] = \begin{bmatrix} Z_{11} & Z_{12} & 0 \\ Z_{21} & Z_{22} & Z_{23} \\ 0 & Z_{32} & Z_{33} \end{bmatrix} \tag{12.136}$$

where

$$Z_{11} = Z_{out(core)} + Z_{ins(core-sheath)} + Z_{inn(sheath)} \tag{12.137}$$

$$Z_{22} = Z_{out(sheath)} + Z_{ins(sheath-armor)} + Z_{inn(armor)} \tag{12.138}$$

$$Z_{33} = Z_{out(armor)} + Z_{ins(armor-ground)} + Z_{g33} \tag{12.139}$$

$$Z_{12} = Z_{21} = -Z_{io(sheath)} \tag{12.140}$$

$$Z_{23} = Z_{32} = -Z_{io(armor)} \tag{12.141}$$

$Z_{inn(conductor)}$, $Z_{out(conductor)}$, and $Z_{io(conductor)}$ are calculated by substituting the inner and outer radii of the conductor into Equations 12.128, 12.129, and 12.131; $Z_{ins(insulator)}$ is calculated by substituting the inner and outer radii of the designated insulator layer into Equation 12.132; $Z_{g,33}$ is the self-ground-return impedance of the armor obtained from Equations 12.133 and 12.134.

Using the following relationships between the loops (i.e., loops 1, 2 and 3) and the conductors (i.e., core, sheath, and armor):

$$V_1 = V_{core} - V_{sheath} \quad V_2 = V_{sheath} - V_{armor} \quad V_3 = V_{armor} \tag{12.142}$$

$$I_1 = I_{core} \quad I_2 = I_{core} + I_{sheath} \quad I_3 = I_{core} + I_{sheath} + I_{armor} \tag{12.143}$$

the following p.u.l. series impedance matrix is obtained for the cable shown in Figure 12.62 when a single coaxial cable is buried alone [2]:

$$[Z_{3\times3}] = \begin{bmatrix} Z_{cc} & Z_{cs} & Z_{ca} \\ Z_{sc} & Z_{ss} & Z_{sa} \\ Z_{ac} & Z_{as} & Z_{aa} \end{bmatrix} \tag{12.144}$$

where

$$Z_{cc} = Z_{11} + 2Z_{12} + Z_{22} + 2Z_{23} + Z_{33} \tag{12.145}$$

$$Z_{cs} = Z_{sc} = Z_{12} + Z_{22} + 2Z_{23} + Z_{33} \tag{12.146}$$

$$Z_{ca} = Z_{ac} = Z_{sa} = Z_{as} = Z_{23} + Z_{33} \tag{12.147}$$

$$Z_{ss} = Z_{22} + 2Z_{23} + Z_{33} \tag{12.148}$$

$$Z_{aa} = Z_{33} \tag{12.149}$$

When two or more parallel coaxial cables are buried together, mutual couplings among the cables must be accounted for. Among the circulating currents I_1, I_2, and I_3, only I_3 has mutual couplings between different cables. Using subscripts a, b and c to denote the phases of the three cables, the p.u.l. series impedance matrix for cable conductors can be expanded into the following form [2]:

$$[Z_{cond}] = [Z_{9\times9}] = \begin{bmatrix} [Z_a] & [Z_{m,ab}] & [Z_{m,ac}] \\ [Z_{m,ba}] & [Z_b] & [Z_{m,bc}] \\ [Z_{m,ca}] & [Z_{m,cb}] & [Z_c] \end{bmatrix} \tag{12.150}$$

where

$$[Z_i] = \begin{bmatrix} Z_{cc,i} & Z_{cs,i} & Z_{ca,i} \\ Z_{sc,i} & Z_{ss,i} & Z_{sa,i} \\ Z_{ac,i} & Z_{as,i} & Z_{aa,i} \end{bmatrix} \quad i = a, b, c \tag{12.151}$$

$$[Z_{m,ij}] = [Z_{m,ji}] = \begin{bmatrix} Z_{g,ij} & Z_{g,ij} & Z_{g,ij} \\ Z_{g,ij} & Z_{g,ij} & Z_{g,ij} \\ Z_{g,ij} & Z_{g,ij} & Z_{g,ij} \end{bmatrix} \quad i, j = a, b, c \tag{12.152}$$

$Z_{g,ab}$, $Z_{g,bc}$, and $Z_{g,ca}$ are the mutual ground-return impedances between the armors of the phases a and b, b and c, and c and a, respectively. These mutual ground-return impedances can be obtained from Equations 12.133 and 12.135.

12.7.3.4 Shunt Admittance Matrix

The p.u.l. admittance matrix for the coaxial cable design depicted in Figures 12.62 and 12.63 when a single cable is buried alone can be obtained as follows:

$$[Y] = \begin{bmatrix} Y_1 & -Y_1 & 0 \\ -Y_1 & Y_1 + Y_2 & -Y_2 \\ 0 & -Y_2 & Y_2 + Y_3 \end{bmatrix} \tag{12.153}$$

where Y_1 is the p.u.l. shunt admittance between core and sheath, Y_2 is the p.u.l. shunt admittance between sheath and armor, and Y_3 is the p.u.l. shunt admittance between armor and ground.

There are no electrostatic couplings between cables when more than two parallel coaxial cables are buried together. Thus, the p.u.l. shunt admittance matrix for a three-phase cable can be expressed as follows:

$$[Y_{cond}] = [Y_{9 \times 9}] = \begin{bmatrix} [Y_a] & 0 & 0 \\ 0 & [Y_b] & 0 \\ 0 & 0 & [Y_c] \end{bmatrix} \tag{12.154}$$

where

$$[Y_i] = \begin{bmatrix} Y_{1i} & -Y_{1i} & 0 \\ -Y_{1i} & Y_{1i} + Y_{2i} & -Y_{2i} \\ 0 & -Y_{2i} & Y_{2i} + Y_{3i} \end{bmatrix} \quad i = a, b, c \tag{12.155}$$

The subscripts a, b, and c denote the phases of the three cables.

If the dielectric losses are considered, the p.u.l. admittances in Equation 12.153 can be obtained from the following expression:

$$Y_i = G_i + j\omega C_i \tag{12.156}$$

where G_i and C_i are the p.u.l. conductance and capacitance of the insulation layer i between two hollow conductors.

The p.u.l. capacitance is given by:

$$C_i = \frac{2\pi\varepsilon_i}{\ln(r_{outi}/r_{inni})} \tag{12.157}$$

where ε_i is the permittivity of the insulation layer, r_{outi} and r_{inni} are, respectively, the outer and the inner radii of the insulation layer. In the configuration shown in Figure 12.63, $r_{outi} = r_{o1}$ and $r_{inni} = r_{i2}$. This calculation has to be applied for any couple of hollow conductors in the cable configuration shown in Figure 12.62.

The value of the conductance G_i is obtained by means of the following form:

$$G_i = \omega C_i \cdot \tan\delta_i \tag{12.158}$$

where $\tan\delta_i$ is the loss factor of insulation i.

The admittance can also be obtained by means of the following single form [2]:

$$Y_i = G_i + j\omega C_i = j\omega \frac{2\pi\varepsilon_0}{\ln(r_{outi}/r_{inni})}(\varepsilon_i' - j\varepsilon_i'') \tag{12.159}$$

where ε_i' is the relative permittivity ($\varepsilon_i = \varepsilon_0\varepsilon_i'$) and ε_i'' accounts for the loss factor of insulation i.

12.7.3.5 Bonding Techniques

The screens (sheaths and armors) of three-phase cables are normally installed in one of three bonding configurations [16,21]:

1. *Single-end bonding*—Grounding of the screens at one end only: This option virtually reduces the currents circulating in the screen to zero, since there is no closed loop. However, it can cause an increase of the screen voltage: one end of the screen is grounded, while a voltage appears at the other end, being the voltage magnitude proportional to the length of the cable. Single-end bonding is limited to short cables.

2. *Both-ends bonding*—Grounding of the screens at both ends: This option reduces the voltage in the screen, which is now close to zero at both ends. However, it provides a closed path for the current in the screen, so both currents and losses are larger than with the previous option. For submarine cables, it is common to adopt solid bonding due to the difficulty in constructing joints offshore. To reduce the loss caused by higher screen currents, the screens of submarine cables are usually designed with a large cross section (i.e., low resistance).

3. *Cross-bonding*—Grounding of the screens at both ends with transposition of the screens: The cable is divided into three sections, called minor sections. The screens are transposed between minor sections and grounded at every third minor section, forming a major section. The minor sections should have a similar length in order to keep the system as balanced as possible and cable may have as many major sections as necessary. The transposition of the screens assures that each screen is exposed to the magnetic field generated by each phase. Assuming a balance system, that is, same magnitude and 120° phase difference between phases, installed in trefoil configuration and that each minor section had exactly the same length, the induced voltages would cancel and the circulating current would be null. Since HV cables are rarely installed in a balanced configuration, it is difficult to achieve such conditions; this option cannot eliminate neither circulating currents nor induced voltages, but it reduces both to low values. Figure 12.64 shows a cross-bonding diagram.

At the distribution level it is normal to use the both-ends bonding technique. Cables longer than 3 km normally adopt cross-bonding to reduce screen currents, suppress screen voltages and increase the thermal capacity of the cable.

The relationships between currents and voltages in a cross-bonded cable are not the same that for cables in which other bonding option has been implemented. In other words, the per unit length series impedance and shunt admittance matrices of a cross-bonded cable are different from those presented above. The methodology to be used for deriving the impedance and the admittance matrices of a cross-bonded cable is beyond the scope of this chapter. A procedure for obtaining the series impedance and shunt admittance matrices of a homogeneously cross-bonded cable is presented in Reference 16.

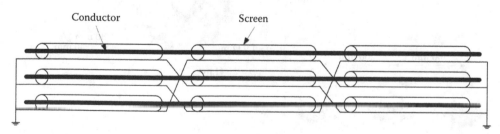

FIGURE 12.64 Example of cross-bonding diagram of a cable.

EXAMPLE 12.12

Consider the underground cable system shown in Figure 12.65. The cable consists of three single-core coaxial cables buried and separated from each other as indicated. Each cable has a core and a sheath, being the sheath ungrounded. The ground resistivity is $\rho = 100\,\Omega.\text{m}$. Assume the permeability of each cable layer is that of vacuum. It is desired to obtain both the series impedance matrix, Equation 12.150, and the shunt admittance matrix, Equation 12.154, at a system operating frequency of 50 Hz.

The series impedance matrix of the three-phase cable system has to be calculated according to Equation 12.150 and has two different types of submatrices: the diagonal submatrices, that correspond to each phase and are calculated according to Equation 12.151, and the off-diagonal submatrices that represent the mutual coupling between phases and are calculated according to Equation 12.152.

For the design depicted in Figure 12.65, the [Z] matrix has the following form:

$$[Z] = \begin{bmatrix} Z_{cc,a} & Z_{cs,a} & Z_{g,ab} & Z_{g,ab} & Z_{g,ac} & Z_{g,ac} \\ Z_{sc,a} & Z_{ss,a} & Z_{g,ab} & Z_{g,ab} & Z_{g,ac} & Z_{g,ac} \\ Z_{g,ba} & Z_{g,ba} & Z_{cc,b} & Z_{cs,b} & Z_{g,bc} & Z_{g,bc} \\ Z_{g,ba} & Z_{g,ba} & Z_{sc,b} & Z_{ss,b} & Z_{g,bc} & Z_{g,bc} \\ Z_{g,ca} & Z_{g,ca} & Z_{g,cb} & Z_{g,cb} & Z_{cc,c} & Z_{cs,c} \\ Z_{g,ca} & Z_{g,ca} & Z_{g,ca} & Z_{g,cb} & Z_{sc,c} & Z_{ss,c} \end{bmatrix}$$

The elements of the diagonal phase submatrices are obtained taking into account the following relationships:

$$Z_{cc,i} = Z_{11,i} + 2Z_{12,i} + Z_{22,i}$$

$$Z_{ss,i} = Z_{22,i}$$

$$Z_{cs,i} = Z_{sc,i} = Z_{12,i} + Z_{22,i}$$

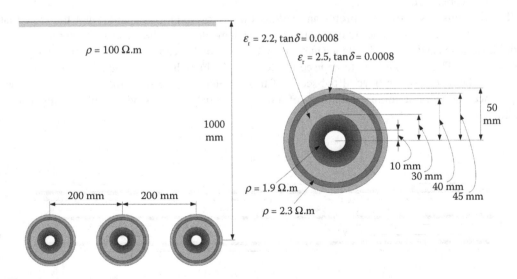

FIGURE 12.65   Example 12: Three-phase underground cable system.

with $i = a, b, c$ and where

$$Z_{11} = Z_{out(core)} + Z_{ins(core-sheath)} + Z_{inn(sheath)}$$

$$Z_{22} = Z_{out(sheath)} + Z_{ins(sheath-ground)} + Z_{g,22}$$

$$Z_{12} = -Z_{io(sheath)}$$

The calculation of each term in these expressions is made by using

- Equation 12.128 for $Z_{inn(sheath)}$,
- Equation 12.129 for $Z_{out(core)}$ and $Z_{out(sheath)}$,
- Equation 12.131 for $Z_{io(sheath)}$,
- Equation 12.132 for $Z_{ins(core-sheath)}$ and $Z_{ins(sheath-ground)}$, and
- Equations 12.133 and 12.134 for $Z_{g,22}$.

The resulting values at 50 Hz are as follows:

$$Z_{cc,a} = Z_{cc,b} = Z_{cc,c} = 0.0613 + j0.6600 \ \Omega/km$$

$$Z_{ss,a} = Z_{ss,b} = Z_{ss,c} = 0.0665 + j0.6269 \ \Omega/km$$

$$Z_{cs,a} = Z_{sc,a} = Z_{cs,b} = Z_{sc,b} = Z_{cs,c} = Z_{sc,c} = 0.0494 + j0.6281 \ \Omega/km$$

The elements of the off-diagonal submatrices (see Equation 12.152) are obtained by means of Equations 12.133 and 12.135

$$Z_{g,ab} = Z_{g,ba} = Z_{g,bc} = Z_{g,cb} = 0.0492 + j0.5308 \ \Omega/km$$

$$Z_{g,ac} = Z_{g,ca} = 0.0492 + j0.4873 \ \Omega/km$$

The series impedance matrix at 50 Hz is then

$$[Z] = \begin{bmatrix} 0.0613 + j0.6600 & 0.0494 + j0.6281 & 0.0492 + j0.5308 & 0.0492 + j0.5308 & 0.0492 + j0.4873 & 0.0492 + j0.4873 \\ 0.0494 + j0.6281 & 0.0665 + j0.6269 & 0.0492 + j0.5308 & 0.0492 + j0.5308 & 0.0492 + j0.4873 & 0.0492 + j0.4873 \\ 0.0492 + j0.5308 & 0.0492 + j0.5308 & 0.0613 + j0.6600 & 0.0494 + j0.6281 & 0.0492 + j0.5308 & 0.0492 + j0.5308 \\ 0.0492 + j0.5308 & 0.0492 + j0.5308 & 0.0494 + j0.6281 & 0.0665 + j0.6269 & 0.0492 + j0.5308 & 0.0492 + j0.5308 \\ 0.0492 + j0.4873 & 0.0492 + j0.4873 & 0.0492 + j0.5308 & 0.0492 + j0.5308 & 0.0613 + j0.6600 & 0.0494 + j0.6281 \\ 0.0492 + j0.4873 & 0.0492 + j0.4873 & 0.0492 + j0.5308 & 0.0492 + j0.5308 & 0.0494 + j0.6281 & 0.0665 + j0.6269 \end{bmatrix} \ \Omega/km$$

For the cable under study, the $[Y]$ matrix has the following form

$$[Y] = \begin{bmatrix} Y_{cs,a} & -Y_{cs,a} & 0 & 0 & 0 & 0 \\ -Y_{cs,a} & Y_{cs,a} + Y_{sg,a} & 0 & 0 & 0 & 0 \\ 0 & 0 & Y_{cs,b} & -Y_{cs,b} & 0 & 0 \\ 0 & 0 & -Y_{cs,b} & Y_{cs,b} + Y_{sg,b} & 0 & 0 \\ 0 & 0 & 0 & 0 & Y_{cs,c} & -Y_{cs,c} \\ 0 & 0 & 0 & 0 & -Y_{cs,c} & Y_{cs,b} + Y_{sg,c} \end{bmatrix}$$

where Y_{cs} are the admittances between cores and sheaths, and Y_{sg} are the admittances between sheaths and ground.

The elements of the admittance matrix are obtained by means of Equation 12.159

$$Y_{cs,a} = Y_{cs,b} = Y_{cs,c} = 0.1070 + j133.7 \times 10^{-6} \text{ S/km}$$

$$Y_{sg,a} = Y_{sg,b} = Y_{sg,c} = 0.3318 + j414.7 \times 10^{-6} \text{ S/km}$$

The admittance matrix at 50 Hz is then given by

$$[Y] = \begin{bmatrix}
0.1070 + j133.7 & -0.1070 - j133.7 & 0 & 0 & 0 & 0 \\
-0.1070 - j133.7 & 0.4388 + j548.4 & 0 & 0 & 0 & 0 \\
0 & 0 & 0.1070 + j133.7 & -0.1070 - j133.7 & 0 & 0 \\
0 & 0 & -0.1070 - j133.7 & 0.4388 + j548.4 & 0 & 0 \\
0 & 0 & 0 & 0 & 0.1070 + j133.7 & -0.1070 - j133.7 \\
0 & 0 & 0 & 0 & -0.1070 - j133.7 & 0.4388 + j548.4
\end{bmatrix}$$
$$\times 10^{-6} \text{ S/km}$$

12.7.4 EQUATIONS OF OVERHEAD LINES AND INSULATED CABLES

12.7.4.1 Introduction

In the frequency-domain, that is, for a given frequency ω and for purely sinusoidal voltages and currents, the voltage and current drops along an elementary line or cable section can be expressed in terms of the series impedance and shunt admittance matrices. The elements of these matrices are calculated according to the methods presented in the previous subsections. Although there are some significant differences in the design of lines and cables, and in the calculation of their corresponding parameters, their voltage and current equations are the same. The rest of this subsection solves the line or cable equations. To simplify the notation, the systems are assumed to have three phases and three conductors.

Consider the three-phase transmission line with phases a, b, c shown in Figure 12.66.

The series voltage drop along a differential length dx is the product of the current times the impedance per unit length $[Z_{ph}]$ times the length dx. Similarly, the shunt current drop along a length dx is the

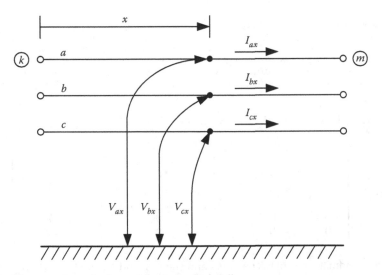

FIGURE 12.66 Phases a, b, c of a three-phase transmission line.

product of the voltage times the admittance per unit length $[Y_{ph}]$ times the length dx. That is,

$$-d[V_{ph}] = dx[Z_{ph}][I_{ph}] \qquad (12.160)$$

$$-d[I_{ph}] = dx[Y_{ph}][V_{ph}] \qquad (12.161)$$

Vectors $[V_{ph}]$ and $[I_{ph}]$ indicate voltages and currents for phases a, b, c at some point x along the line.

$$[V_{ph}] = \begin{bmatrix} V_a(x) \\ V_b(x) \\ V_c(x) \end{bmatrix}; \quad [I_{ph}] = \begin{bmatrix} I_a(x) \\ I_b(x) \\ I_c(x) \end{bmatrix} \qquad (12.162)$$

The sending end of the line is k at $x = 0$, and the receiving end of the line is m at $x = \ell =$ line length.

The impedances matrix $[Z_{ph}]$ is given in units of Ω/m by Equation 12.110 for overhead lines and Equation 12.150 for insulated cables, while the admittances matrix $[Y_{ph}]$ is given in units of S/m by Equation 12.122 for overhead lines and Equation 12.154 for insulated cables.

Dividing these two equations by the distance differential dx gives

$$-\frac{d[V_{ph}]}{dx} = [Z_{ph}][I_{ph}] \qquad (12.163)$$

$$-\frac{d[I_{ph}]}{dx} = [Y_{ph}][V_{ph}] \qquad (12.164)$$

which are the differential equations describing the voltage and current drops along the line.

Taking the second derivative with respect to x, and relating the two equations, gives

$$\frac{d^2[V_{ph}]}{dx^2} = [Z_{ph}][Y_{ph}][V_{ph}] \qquad (12.165)$$

$$\frac{d^2[I_{ph}]}{dx^2} = [Y_{ph}][Z_{ph}][I_{ph}] \qquad (12.166)$$

Equation 12.165 governs the propagation of the voltages along the line, while Equation 12.166 governs the propagation of the currents. Notice that the product of matrices is not commutative, $[Z_{ph}][Y_{ph}] \neq [Y_{ph}][Z_{ph}]$.

The voltage and current propagation equations are linear second-order differential equations with respect to the distance x. As in the case of any ordinary linear second-order equation, the solutions can be expressed as

$$[V_{ph}] = [e^{-x[\gamma_v]}][V_{ph-k}] + [e^{+x[\gamma_v]}][V_{ph-m}] \qquad (12.167)$$

$$[I_{ph}] = [e^{-x[\gamma_i]}][I_{ph-k}] + [e^{+x[\gamma_i]}][I_{ph-m}] \qquad (12.168)$$

with

$$[\gamma_v] = ([Z_{ph}][Y_{ph}])^{1/2}; \quad [\gamma_i] = ([Y_{ph}][Z_{ph}])^{1/2} \qquad (12.169)$$

Equations 12.167 and 12.168 are wave propagation equations. The vector $[V_{ph-k}]$ represents the phase voltages at the sending end of the line k assuming a wave traveling from k towards m but assuming that m is located at infinity (no reflected wave at m); similarly $[V_{ph-m}]$ represents the phase voltages at the receiving end of the line m assuming a wave traveling from m towards k but assuming now that k is located at infinity (no reflected wave at k). Superimposing the waves traveling from k to m and the waves traveling from m to k one gets the combined solution of 12.167. The same meaning applies to the currents.

In principle, Equations 12.167 and 12.168 could be solved directly using functions of matrices. The algebra of functions of matrices, however, can be a bit "tricky" and in the end it relies on the concept of eigenvectors and eigenvalues to evaluate the matrix functions. It is much simpler at this point of the solution to use eigenvalue/eigenvector analysis directly on the defining drop Equations 12.163 and 12.164 and propagation Equations 12.165 and 12.166, as explained next.

12.7.4.2 Modal Decomposition

Eigenvalue and eigenvector analysis (eigenanalysis) provides a simple and powerful tool to study multiphase coupled systems. The pioneering work on this topic was done independently by Wedepohl in 1963 [22] and Hedman in 1965 [23]. Before that time, though, the concept of symmetrical components decomposition, introduced by the Canadian engineer Fortescue in 1918 [24], was already a fundamental concept in power systems analysis. Symmetrical components are a special case of eigenvector decomposition when the system of conductors is perfectly balanced (transposed lines).

The phase voltages can be written in terms of a general base as

$$\begin{bmatrix} V_a \\ V_b \\ V_c \end{bmatrix} = V_1 \begin{bmatrix} u_{11} \\ u_{21} \\ u_{31} \end{bmatrix} + V_2 \begin{bmatrix} u_{12} \\ u_{22} \\ u_{32} \end{bmatrix} + V_3 \begin{bmatrix} u_{13} \\ u_{23} \\ u_{33} \end{bmatrix} = \begin{bmatrix} u_{11} & | & u_{12} & | & u_{13} \\ u_{21} & | & u_{22} & | & u_{23} \\ u_{31} & | & u_{32} & | & u_{33} \end{bmatrix} \begin{bmatrix} V_1 \\ V_2 \\ V_3 \end{bmatrix} \quad (12.170)$$

or,

$$\begin{bmatrix} V_a \\ V_b \\ V_c \end{bmatrix} = \begin{bmatrix} T_v \end{bmatrix} \begin{bmatrix} V_1 \\ V_2 \\ V_3 \end{bmatrix} \quad (12.171)$$

$$[T_v] = \begin{bmatrix} u_{11} & | & u_{12} & | & u_{13} \\ u_{21} & | & u_{22} & | & u_{23} \\ u_{31} & | & u_{32} & | & u_{33} \end{bmatrix} = [u_1 | u_2 | u_3] \quad (12.172)$$

In compact notation,

$$[V_{ph}] = [T_v][V_m] \quad (12.173)$$

Equation 12.173 relates the electrical voltage point expressed in terms of the "old" *phase coordinates* $[V_{ph}]$ with the same electrical point expressed in terms of the "new" *modal coordinates* $[V_m]$. The matrix $[T_v]$ is called the *voltages transformation matrix*.

A similar analysis can be made for the phase currents and one can write

$$[I_{ph}] = [T_i][I_m] \quad (12.174)$$

where $[T_i]$ is the *currents transformation matrix*. In general, matrices $[T_v]$ and $[T_i]$ do not need to be equal.

12.7.4.3 Line and Cable Equations in the Modal Domain

Consider the "change of variables" denoted by Equations 12.173 and 12.174. One can now substitute these transformations into the line equations in phase coordinates, Equations 12.163 through 12.166. For the voltages drop equation,

$$-\frac{d[V_{ph}]}{dx} = [Z_{ph}][I_{ph}] \tag{12.175}$$

$$-[T_v]\frac{d[V_m]}{dx} = [Z_{ph}][T_i][I_m] \tag{12.176}$$

(assuming that $[T_v]$ and $[T_i]$ are independent of x, which is the case for a uniform line).

In terms of modal voltages and currents, it follows that

$$-\frac{d[V_m]}{dx} = ([T_v]^{-1}[Z_{ph}][T_i])[I_m] = [Z_m][I_m] \tag{12.177}$$

where

$$[Z_m] = [T_v]^{-1}[Z_{ph}][T_i] \tag{12.178}$$

is the *modal series impedance matrix*.

Proceeding in a similar way for the currents drop Equation 12.164, one obtains

$$-\frac{d[I_m]}{dx} = ([T_i]^{-1}[Y_{ph}][T_v])[V_m] = [Y_m][I_m] \tag{12.179}$$

where

$$[Y_m] = [T_i]^{-1}[Y_{ph}][T_v] \tag{12.180}$$

is the *modal shunt admittance matrix*. The same variable substitutions can be made in the voltages and currents propagation equations.

The line drop and propagation equations in modal coordinates are summarized next,

$$-\frac{d[V_m]}{dx} = ([T_v]^{-1}[Z_{ph}][T_i])[I_m] = [Z_m][I_m] \tag{12.181}$$

$$-\frac{d[I_m]}{dx} = ([T_i]^{-1}[Y_{ph}][T_v])[V_m] = [Y_m][I_m] \tag{12.182}$$

$$\frac{d^2[V_m]}{dx^2} = [T_v]^{-1}[Z_{ph}][Y_{ph}][T_v][V_m] = [Z_m][Y_m][V_m] \tag{12.183}$$

$$\frac{d^2[I_m]}{dx^2} = [T_i]^{-1}[Y_{ph}][Z_{ph}][T_i][I_m] = [Y_m][Z_m][I_m] \tag{12.184}$$

As will be discussed next, it is possible to find matrices $[T_v]$ and $[T_i]$ so that the coefficient matrices in Equations 12.181 through 12.184 are all diagonal.

12.7.4.4 Eigenvalue/Eigenvector Problem

The premise of the eigenvalue/eigenvector problem is that given a system characterized by a square matrix $[A]$, it is possible to find a transformation matrix $[T_i]$ such that the product

$$[T_i]^{-1}[A][T_i] = \text{diagonal matrix} \tag{12.185}$$

Let the diagonal matrix be $[\Lambda]$, then the objective is to satisfy

$$[T_i]^{-1}[A][T_i] = [\Lambda] \tag{12.186}$$

$$[\Lambda] = \begin{bmatrix} \lambda_1 & 0 & 0 \\ 0 & \lambda_2 & 0 \\ 0 & 0 & \lambda_3 \end{bmatrix} \tag{12.187}$$

The entries λ_i of the diagonal matrix $[\Lambda]$ are the *eigenvalues* of $[A]$.
Let the diagonalizing matrix $[T_i]$ be formed by three column vectors,

$$[T_i] = [u_1|u_2|u_3] \tag{12.188}$$

The vectors u_1, u_2, u_3 which constitute the diagonalizing matrix $[T_i]$ are the *eigenvectors* of $[A]$. From Equation 12.186, multiplying both sides by $[T_i]$,

$$[A][T_i] = [T_i][\Lambda] \tag{12.189}$$

Writing $[T_i]$ in terms of its constituent eigenvectors,

$$[A][u_1|u_2|u_3] = [\lambda_1 u_1|\lambda_2 u_2|\lambda_3 u_3] \tag{12.190}$$

$$[[A]u_1|[A]u_2|[A]u_3] = [\lambda_1 u_1|\lambda_2 u_2|\lambda_3 u_3] \tag{12.191}$$

or,

$$[A]u_i = \lambda_i u_i \tag{12.192}$$

for $i = 1, 2, 3$.

Equation 12.192 is the classic eigenvalue problem. Observe that if u_i in this equation is multiplied by any complex number $\bar{\alpha} = \alpha \angle \theta$, the new vector $\bar{\alpha} u_i$ is also an eigenvector u_i',

$$[A]\bar{\alpha} u_i = \lambda_i \bar{\alpha} u_i = \lambda_i u_i' \tag{12.193}$$

This property means that the length of a given eigenvector can be changed and the eigenvector can be rotated by a given angle and the result will still be an eigenvector. It should also be noted that even though there are multiple choices for the eigenvectors, the eigenvalues λ_i are uniquely defined for a given system matrix $[A]$.

To summarize: The eigenvectors of a linear system's matrix $[A]$ (with any scaling by a complex scalar), arranged in columns in any order, will form transformation matrices $[T_i]$ that will diagonalize the system matrix $[A]$, and will allow the description of an $[N \times N]$ coupled system as N individual decoupled subsystems.

There are a number of standard computer algorithms to compute the eigenvalues and eigenvectors of complex matrices.

12.7.4.5 Decoupled Equations

Assume that the transformation matrix $[T_i]$ in Equation 12.184 is chosen so as to diagonalize the product $[Y_{ph}][Z_{ph}]$. That is, $[T_i]$ is made out of the eigenvectors (Equation 12.188) of the matrix $[A_i] = [Y_{ph}][Z_{ph}]$. Similarly, the transformation matrix $[T_v]$ in Equation 12.183 can be chosen to diagonalize the product $[Z_{ph}][Y_{ph}]$. That is, $[T_v]$ is made out of the eigenvectors (Equation 12.188) of the matrix $[A_v] = [Z_{ph}][Y_{ph}]$.

It can be shown that for the case of a transmission line, if matrices $[T_i]$ and $[T_v]$ are obtained to diagonalize the $[Y_{ph}][Z_{ph}]$ and $[Z_{ph}][Y_{ph}]$ products, then the mixed products $[T_v]^{-1}[Z_{ph}][T_i]$ and $[T_i]^{-1}[Y_{ph}][T_v]$ in the voltage and current drop equations, respectively, will also be diagonal, that is,

$$[T_v]^{-1}[Z_{ph}][T_i] = [Z_m] = \begin{bmatrix} Z_1 & 0 & 0 \\ 0 & Z_2 & 0 \\ 0 & 0 & Z_3 \end{bmatrix} \tag{12.194}$$

$$[T_i]^{-1}[Y_{ph}][T_v] = [Y_m] = \begin{bmatrix} Y_1 & 0 & 0 \\ 0 & Y_2 & 0 \\ 0 & 0 & Y_3 \end{bmatrix} \tag{12.195}$$

With the $[T_v]$ and $[T_i]$ matrices relating the coupled phase quantities and the decoupled modal quantities, the multiphase coupled transmission line can be transformed into an equivalent system of conductors that have no coupling among themselves or to ground.

All the transmission line equations can be written with respect to the equivalent decoupled modal system,

$$\begin{bmatrix} -\dfrac{dV_1}{dx} \\ -\dfrac{dV_2}{dx} \\ -\dfrac{dV_3}{dx} \end{bmatrix} = \begin{bmatrix} Z_1 & 0 & 0 \\ 0 & Z_2 & 0 \\ 0 & 0 & Z_3 \end{bmatrix} \begin{bmatrix} I_1 \\ I_2 \\ I_3 \end{bmatrix}; \quad \begin{bmatrix} -\dfrac{dI_1}{dx} \\ -\dfrac{dI_2}{dx} \\ -\dfrac{dI_3}{dx} \end{bmatrix} = \begin{bmatrix} Y_1 & 0 & 0 \\ 0 & Y_2 & 0 \\ 0 & 0 & Y_3 \end{bmatrix} \begin{bmatrix} V_1 \\ V_2 \\ V_3 \end{bmatrix} \tag{12.196}$$

$$\begin{bmatrix} \dfrac{d^2V_1}{dx^2} \\ \dfrac{d^2V_2}{dx^2} \\ \dfrac{d^2V_3}{dx^2} \end{bmatrix} = \begin{bmatrix} Z_1Y_1 & 0 & 0 \\ 0 & Z_2Y_2 & 0 \\ 0 & 0 & Z_3Y_3 \end{bmatrix} \begin{bmatrix} V_1 \\ V_2 \\ V_3 \end{bmatrix}; \quad \begin{bmatrix} \dfrac{d^2I_1}{dx^2} \\ \dfrac{d^2I_2}{dx^2} \\ \dfrac{d^2I_3}{dx^2} \end{bmatrix} = \begin{bmatrix} Z_1Y_1 & 0 & 0 \\ 0 & Z_2Y_2 & 0 \\ 0 & 0 & Z_3Y_3 \end{bmatrix} \begin{bmatrix} I_1 \\ I_2 \\ I_3 \end{bmatrix}$$

$$\tag{12.197}$$

In the modal equivalent, modes 1, 2, 3 are completely independent from each other and each mode can be fully analyzed as a separate single-phase line. Once the modal voltages and currents are calculated, the corresponding phase voltages and currents can be recovered from Equations 12.173 and 12.174.

12.7.4.6 Symmetrical Components and Other Balanced Transformations

For transposed lines, it is possible to simplify the decoupling problem and assume that the $[Z_{ph}]$ and $[Y_{ph}]$ matrices are balanced. For balanced matrices all elements in the main diagonal positions are equal to each other and all elements in the off-diagonal positions are also equal to each other,

$$[Z_{ph}] = \begin{bmatrix} Z_s & Z_m & Z_m \\ Z_m & Z_s & Z_m \\ Z_m & Z_m & Z_s \end{bmatrix}; \quad [Y_{ph}] = \begin{bmatrix} Y_s & Y_m & Y_m \\ Y_m & Y_s & Y_m \\ Y_m & Y_m & Y_s \end{bmatrix} \quad (12.198)$$

For balanced matrices [A], there are multiple sets of eigenvectors that diagonalize the matrix. The conditions the eigenvectors have to satisfy are the following: (i) the first eigenvector must have all of its components identical; (ii) for the second eigenvector the sum of its elements must be zero; and (iii) for the third eigenvector the sum of its elements must also be zero. The three eigenvectors must satisfy these conditions and also be linearly independent of each other. The following are well-known examples of transformation matrices for balanced system.

12.7.4.7 Symmetrical Components

The classical symmetrical components transformation was introduced by Fortescue in 1918 [24] and diagonalizes balanced matrices using eigenvectors called: zero sequence (modal component 1), positive sequence (modal component 2), and negative sequence (modal component 3). In symmetrical components $[T_v] = [T_i]$ and

$$[T_i] = \begin{bmatrix} 1 & 1 & 1 \\ 1 & a^2 & a \\ 1 & a & a^2 \end{bmatrix}; \quad [T_i]^{-1} = \frac{1}{3}\begin{bmatrix} 1 & 1 & 1 \\ 1 & a & a^2 \\ 1 & a^2 & a \end{bmatrix} \quad (12.199)$$

where $a = 1\angle 120°$. The rule for diagonalizing balanced systems can be easily verified. All elements of the first column are identical, and the elements of the second and third columns, while being independent from each other add up to zero.

12.7.4.8 Karrenbauer's Transformation

Karrenbauer's Transformation is the one used by the EMTP. It was introduced by Karrenbauer in 1967 [2]. In this transformation $[T_v] = [T_i]$ and

$$[T_i] = \begin{bmatrix} 1 & 1 & 1 \\ 1 & -2 & 1 \\ 1 & 1 & -2 \end{bmatrix}; \quad [T_i]^{-1} = \frac{1}{3}\begin{bmatrix} 1 & 1 & 1 \\ 1 & -1 & 0 \\ 1 & 0 & -1 \end{bmatrix} \quad (12.200)$$

Again, the rule that the elements of the first column should be identical and that the elements of the second and third columns add up to zero is satisfied. Karrenbauer's transformation is a minimalistic way of satisfying the balanced matrix conditions, and only uses real numbers, as opposed to the complex coefficients used by symmetrical components. This speeds up the computer calculations.

12.7.4.9 Eigenvalues of Balanced Systems

In the case of balanced systems, the balanced property defines a class of system matrices $[A]_{balanced}$ for which *the eigenvalues have always the same form*, regardless of the particular choice of transformation matrix (Symmetrical Components, Karrenbauer's). For balanced systems, the modal impedances Z_m and admittances Y_m always are

$$[Z_m] = \begin{bmatrix} Z_s + 2Z_m & 0 & 0 \\ 0 & Z_s - Z_m & 0 \\ 0 & 0 & Z_s - Z_m \end{bmatrix} \quad (12.201)$$

$$[Y_m] = \begin{bmatrix} Y_s + 2Y_m & 0 & 0 \\ 0 & Y_s - Y_m & 0 \\ 0 & 0 & Y_s - Y_m \end{bmatrix} \qquad (12.202)$$

The first mode is the zero sequence mode and adds the selves and the mutual, the second and third modes subtract the selves and the mutuals.

EXAMPLE 12.13

Consider the $[Z]$ and $[Y]$ matrices for the transmission line of Figure 12.60, calculated in Example 12.11. Obtain the diagonalizing transformation matrices $[T_v]$ and $[T_i]$ of Equations 12.181 and 12.182, and the modal impedances and admittances of the decoupled equivalent circuit.

1. Using a computer package for eigenvalue/eigenvector extraction, the following results were obtained:

$$[T_v] = \begin{bmatrix} 0.5753 - j0.0034 & -0.7071 - j0.0000 & -0.3861 - j0.0037 \\ 0.5814 + j0.0000 & -0.0000 + j0.0000 & 0.8378 - j0.0000 \\ 0.5753 - j0.0034 & 0.7071 + j0.0000 & -0.3861 - j0.0037 \end{bmatrix}$$

$$[T_i] = \begin{bmatrix} 0.5924 + j0.0000 & -0.7071 - j0.0000 & -0.4111 - j0.0024 \\ 0.5460 + j0.0052 & -0.0000 + j0.0000 & 0.8136 - j0.0000 \\ 0.5924 - j0.0000 & 0.7071 + j0.0000 & -0.4111 - j0.0024 \end{bmatrix}$$

The matrices above are normalized according to the Euclidean norm to make the length of each eigenvector (each column) equal to one. Notice that even though the example line is not balanced, the diagonalizing matrices follow quite closely the conditions of balanced lines, that is, the elements of the first column are very similar to each other and the elements of the second and third columns add up close to zero (this was satisfied exactly for the second column due to the flat line configuration of the example). The flat line configuration also creates a symmetry between the first and last rows.

2. The modal impedances and admittances are given by Equations 12.194 and 12.195. The obtained results are as follows:

$$[Z_m] = \begin{bmatrix} 0.1971 + j1.502 & 0 & 0 \\ 0 & 0.03387 + j0.5124 & 0 \\ 0 & 0 & 0.03439 + j0.4429 \end{bmatrix} \Omega/\text{km}$$

$$[Y_m] = \begin{bmatrix} 0.1709 + j24.19 & 0 & 0 \\ 0 & j32.97 & 0 \\ 0 & 0 & -0.0415 + j37.19 \end{bmatrix} \times 10^{-6}\, \text{S/km}$$

It is left to the reader to verify that even though not exactly the same, the values obtained for $[Z_m]$ and $[Y_m]$ are not too different from the ones for a balanced line (Equations 12.201 and 12.202).

12.7.5 MULTIPHASE LINE AND CABLE MODELS IN THE EMTP

The multiphase line and cable models in the EMTP use modal decomposition to convert between phase and modal domains. Even though phase quantities are used to interface the line/cable model with the rest of the network, the history sources of the models are calculated (updated) in the modal decoupled single-phase equivalents.

Figure 12.67 shows mode 1 of a three-phase line/cable that has been decoupled using the transformation matrices $[T_v]$ and $[T_i]$ discussed above. The EMTP elementary model includes the characteristic impedance for mode 1, Z_{c1}, and history sources $i_{hk1}(t)$ at node k and $i_{hm1}(t)$ at node m. For the full power network, the EMTP solution is performed in terms of phase quantities using the nodal equation

$$[G][v(t)] = [i(t)] \tag{12.203}$$

Consider for example the solution at node m of the line/cable. After the node voltages in phase coordinates have been determined from Equation 12.203, the corresponding modal voltages for the terminals are obtained from

$$\begin{bmatrix} v_{m1}(t) \\ v_{m2}(t) \\ v_{m3}(t) \end{bmatrix} = [T_v]^{-1} \begin{bmatrix} v_{ma}(t) \\ v_{mb}(t) \\ v_{mc}(t) \end{bmatrix} \tag{12.204}$$

These modal node voltages (e.g., $v_{m1}(t)$ in Figure 12.67) are now used to calculate the updated history sources in the single-phase equivalent circuits. Once these sources have been updated they are transferred back to the phase domain,

$$\begin{bmatrix} i_{hma}(t) \\ i_{hmb}(t) \\ i_{hmc}(t) \end{bmatrix} = [T_i] \begin{bmatrix} i_{hm1}(t) \\ i_{hm2}(t) \\ i_{hm2}(t) \end{bmatrix} \tag{12.205}$$

These history current sources are now injected to the right-hand side of the network's nodal solution in Equation 12.203.

The Z_{ci} branches in the modal single-phase circuits are incorporated into the network's $[G]$ matrix after converting them from the modal to the phase quantities. From Equation 12.180,

$$\begin{bmatrix} g_{aa} & g_{ab} & g_{ac} \\ g_{ba} & g_{bb} & g_{bc} \\ g_{ca} & g_{cb} & g_{cc} \end{bmatrix} = [T_i] \begin{bmatrix} \dfrac{1}{Z_{c1}} & 0 & 0 \\ 0 & \dfrac{1}{Z_{c2}} & 0 \\ 0 & 0 & \dfrac{1}{Z_{c3}} \end{bmatrix} [T_v]^{-1} \tag{12.206}$$

The same process explained above for the basic line/cable model is followed in the EMTP for more sophisticated models, like the JMARTI frequency-dependent model [25].

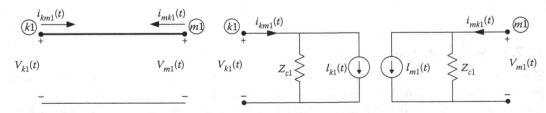

FIGURE 12.67   EMTP line model for mode 1.

REFERENCES

1. Dommel H. W., Digital computer solution of electromagnetic transients in single- and multi-phase networks, *IEEE Trans. on Power Apparatus and Systems*, 88(2), 734–741, 1969.
2. Dommel H. W., *EMTP Theory Book*, Microtran Power System Analysis Corporation, Vancouver, BC, Canada, 2nd Edition, 1996.
3. CIGRE WG 33.02, Guidelines for representation of network elements when calculating transients, 1990.
4. Gole, A. M., Martinez-Velasco, J. A., and Keri, A. J. F. (Eds.), *Modeling and analysis of system transients using digital programs*, IEEE PES Special Publication, TP-133-0, 1999.
5. IEC TR 60071–4, Insulation co-ordination—Part 4: Computational guide to insulation co-ordination and modeling of electrical networks, 2004.
6. Shenkman A. L., *Transient Analysis of Electric Power Circuits Handbook*, Springer, Dordrecht (The Netherlands), 2005.
7. Poularikas A. D., (Ed.), *The Transforms and Applications Handbook*, CRC Press, Boca Raton (FL, USA), 2nd Edition, 2000.
8. Greenwood A., *Electrical Transients in Power Systems*, John Wiley, New York (NY, USA), 2nd Edition, 1991.
9. IEC 60071–1, Insulation co-ordination, *Part 1: Definitions, Principles and Rules*, 8th Edition, 2006.
10. IEC 60071–2, Insulation co-ordination. *Part 2: Application Guide*, 3rd Edition, 1996.
11. Carson J. R., Wave propagation in overhead wires with ground return, *Bell Systems Technical Journal*, 5(4), 539–554, 1926.
12. Deri A., Tevan G., Semlyen A., and Castanheira A., A simplified model for homogeneous and multi-layer earth return, *IEEE Trans. on Power Apparatus and Systems*, 100(8), 3686–3693, 1981.
13. Wedepohl L.M. and Wilcox D.J., Transient analysis of underground power-transmission systems. System-model and wave-propagation characteristics, *Proc. IEE*, 120(2), 253–260, 1973.
14. Ametani A., A general formulation of impedance and admittance of cables, *IEEE Trans. on Power Apparatus and Systems*, 99(3), 902–909, 1980.
15. Gustavsen B., Noda T., Naredo J. L., Uribe F. A., and Martinez-Velasco J. A., Insulated cables, Chapter 3 of Power System Transients. *Parameter Determination*, Martinez-Velasco J. A. (Ed.), CRC Press, Boca Raton (FL, USA), 2009.
16. Ametani A., Nagaoka N., Baba Y., Ohno T., and Yamabuki K., *Power System Transients: Theory and Applications*, Second Edition, CRC Press, Boca Raton (FL, USA), 2016.
17. Schelkunoff S. A., The electromagnetic theory of coaxial transmission lines and cylindrical shields, *Bell System Technical Journal*, 13(4), 532–579, 1934.
18. Rivas R. A. and Marti J. R., Calculation of frequency-dependent parameters of power cables: Matrix partitioning techniques, *IEEE Trans. on Power Delivery*, 17(4), 1085–1092, 2002.
19. Pollaczek F., On the field produced by an infinitely long wire carrying alternating current, *Elektrische Nachrichtentechnik*, 3, 339–359, 1926.
20. Pollaczek F., On the induction effects of a single phase ac line, *Elektrische Nachrichtentechnik*, 4, 18–30, 1927.
21. Faria da Silva F. and Bak C. L., *Electromagnetic Transients in Power Cables*, Springer, London (UK), 2013.
22. Wedepohl L. M., Application of matrix methods to the solution of travelling-wave phenomena in polyphase systems, *Proc. IEE*, 110(2), 2200–2212, 1963.
23. Hedman D. E., Propagation on overhead transmission lines I-Theory of modal analysis, and II-Earth conduction effects and practical results, *IEEE Trans. Power Apparatus and Systems*, 84(3), 200–211, 1965.
24. Fortescue C. L., Method of symmetrical coordinates applied to the solution of polyphase networks, *AIEE*, 37, 1027–1140, 1918.
25. Marti J. R., Accurate modeling of frequency-dependent transmission lines in electromagnetic transient simulations, *IEEE Trans. on Power Apparatus and Systems*, 101(1), 147–157, 1982.

BIBLIOGRAPHY

26. Hase Y., *Handbook of Power Systems Engineering with Power Electronics Applications*, John Wiley, Chichester (West Sussex, UK), 2013.
27. Chowdhuri P., *Electromagnetic Transients in Power Systems*, RSP—John Wiley, Taunton (Somerset, England), 2nd Edition, 2005.

28. Watson N. and Arrillaga J., *Power System Electromagnetic Transients Simulation*, IEE Power and Energy Series, Stevenage (UK), 2003.

29. van der Sluis L., *Transients in Power Systems*, John Wiley, Chichester (West Sussex, UK), 2001.

30. Chowdhuri P. (Section Ed.), Power system transients, included in *Power Systems—The Electric Power Engineering Handbook*, Grigsby L. L. (Ed.), 3rd Edition, CRC Press, Boca Raton (FL, USA), 2012.

31. Dommel H. W., Nonlinear and time-varying elements in digital simulation of electromagnetic transients, *IEEE Trans. on Power Apparatus and Systems*, 90(6), 2561–2567, 1971.

32. Dommel H. W. and Scott Meyer W., Computation of electromagnetic transients, *Proc. of IEEE*, 62(7), 983–993, 1974.

33. Phadke A. (Ed.), *Digital simulation of electrical transient phenomena*, IEEE Tutorial Course, Course Text 81 EHO173-5-PWR, 1981.

34. Martinez-Velasco J. A., (Ed.), *Computer Analysis of Electric Power System Transients*, IEEE Press, Piscataway (NJ, USA), 1997.

35. Dommel H. W., Techniques for analyzing electromagnetic transients, *IEEE Computer Applications in Power*, 10(3), 18–21, 1997.

36. Hileman A. R., *Insulation Coordination for Power Systems*, Marcel Dekker, New York (NY, USA), 1999.

37. Ragaller K., (Ed.), *Surges in High-Voltage Networks*, Plenum Press, New York (NY, USA), 1980.

38. Martinez-Velasco J. A., (Ed.), *Power System Transients. Parameter Determination*, CRC Press, Boca Raton (FL, USA), 2010.

39. Martinez-Velasco J. A., (Ed.), *Transient Analysis of Power Systems. Solution Techniques, Tools and Applications*, John Wiley, Chichester (West Sussex, UK), 2015.

Appendix A: Solution of Linear Equation Systems

Fernando L. Alvarado and Antonio Gómez-Expósito

CONTENTS

A.1 INTRODUCTION

This appendix explains the numerical techniques required for the solution of large sparse systems of linear equations with irregular structure typical of power systems. It includes some more advanced topics such as the solution of modifications relative to a base case and the methods for the determination of selected elements of the inverse matrix.

A.1.1 GAUSSIAN ELIMINATION AND *LU* FACTORIZATION

This section considers the solution of a system of linear equations:

$$Ax = b \tag{A.1}$$

where A is a nonsingular sparse matrix of dimension $n \times n$, and b and x are vectors of dimension n.

In the solution of large power system problems, n is generally very large. For typical power flow and state estimation applications, the number of equations can exceed 5000. However, for optimization applications with intertemporal coupling, this number can be considerably larger. These equations must be solved from three to five times for typical power flow applications to hundreds of times for dynamic equations solutions or for contingency analysis. Thus, it is important to reduce the computational burden.

In most cases, A is structurally symmetric. That is, when an element of A is nonzero, the corresponding transpose element is also nonzero. It is possible to take advantage of this feature for purposes of storage and access of the nonzero elements, since only half the structure needs to be stored. Often, A is also numerically symmetric. This can result in a considerable computational burden reduction provided the symmetry property is retained during the solution process.

The most effective method for solving Equation A.1 is generally Gaussian elimination. The method performs a sequence of elementary transformations on the rows of A until A becomes an upper triangular matrix. These transformations can be (a) normalization of a row, that is, dividing all the elements of the row by the diagonal entry and (b) linear combination of a row with a previously normalized row so as to zero-out a lower triangular entry. The zeroing out of the lower triangular entries can be done by rows or by columns. The vector b is processed as the same time as A. After triangulation, the resulting triangular system of equations can be solved by back substitution, starting with the last row.

In many cases, it is necessary to resolve a set of equations with a single A matrix but with several different independent vectors b. In this case, computation can be reduced by applying the transformations to b by itself, without the need to retriangulate A. This requires that the coefficients used to triangulate A be saved, to use them later as necessary. Because the diagonal entries of the normalized and triangularized matrix A are ones, and because the lower triangular entries of the triangularized matrix A are, by definition, zero, both the diagonal and the lower triangular positions of the original A matrix can be used to store these factors.

There are several refinements and variants of the Gaussian elimination process. A more elegant version of the idea is to use a mostly equivalent process called matrix factorization or triangular decomposition. This method is based on the notion that every nonsingular square matrix can be decomposed as follows:

$$A = L \cdot D \cdot U \tag{A.2}$$

where L and U are lower and upper triangular matrices with unit diagonal, respectively, and D is a diagonal matrix. If A is symmetric, we also have that $L = U^{\mathrm{T}}$.

There are several variants of this factorization process. In some variants, the D matrix is combined with either L or U, yielding a factorization with only two factors:

$$A = L \cdot U \tag{A.3}$$

where either L or U remains a unit triangular matrix.

If A is *positive definite*, D can be factored into two $D^{1/2}$ factors, one factor combined with the unit L lower triangular matrix and the other factor combined with the unit U upper triangular matrix. For a symmetric matrix, this decomposition is called a Choleski decomposition.

Using the LU decomposition idea, the solution requires the following three steps:

1. Factor the matrix A:

$$A = L \cdot U$$

2. Obtain an intermediate vector y from a forward-substitution process:

$$L \cdot y = b \quad \Rightarrow \quad y_1, \ldots, y_n$$

3. Obtain the solution vector x from a back-substitution process:

$$U \cdot x = y \quad \Rightarrow \quad x_n, \ldots, x_1$$

Gaussian elimination in effect combines steps 1 and 2 of this process.

If the matrix A is structurally symmetric, it is possible to access both lower and upper triangular matrix entries using the same index. If this is done, the upper triangular portion of the matrix is accessed by rows and the lower triangular portion is accessed by columns, as shown in Figure A.1.

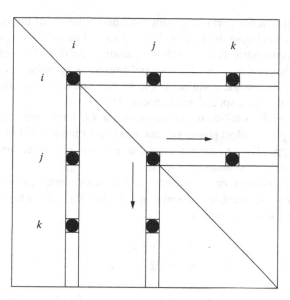

FIGURE A.1   Gaussian elimination sequence for structurally symmetric matrices.

Using the notation suggested in this diagram and assuming the L and U initially contain the entries of A itself, the LU decomposition can be done according to the following pseudo-code:

```
For i = 1, n − 1
    For all j such that u_ij ≠ 0
            u_ij = u_ij/l_ii
    For all j such that u_ij ≠ 0
            l_jj = l_jj − l_ji · u_ij
            For all k > j such that u_ik ≠ 0
                    l_kj = l_kj − l_ki · u_ij
                    u_jk = u_jk − l_ji · u_ik
```

This variant results in a unit upper triangular matrix U. Except for the step involving the normalization of U, the loop follows simultaneously along rows of L and columns of U.

The forward-substitution process is done using

```
For i = 1, n − 1
    b_i = b_i/l_ii
    For all j such that l_ji ≠ 0
            b_j = b_j − l_ji · b_i
b_n = b_n/l_nn
```

And the back substitution is done using

```
For i = n − 1, 1
    For all j such that u_ij ≠ 0
            b_i = b_i − u_ij · b_j
```

The factorization algorithm assumes that none of the diagonal entries is zero at the moment that they are used as a pivot. In theory, to avoid accumulation of unacceptable errors during the factorization process, it is desirable always to choose the largest available absolute value entries as pivots at every stage, even if this requires permuting both rows and columns of the matrix. In practice, it is almost always sufficient to choose the largest absolute-valued entry in the current rows or columns being processed. For unsymmetric matrices, this is a common and desirable technique. For symmetric matrices, a technique that is almost as effective and preserves symmetry is to use 2×2 diagonal blocks as pivots whenever a single pivot proves to be too small. The details of this technique are left to the references in the more specialized literature [1].

For positive-definite or diagonal-dominant matrices, the LU factorization is stable if the diagonal elements are chosen as pivots. Most power system matrices of interest fall in this category. Thus, in the remainder of this appendix we ignore the need to perform row or column permutations of the matrices to preserve numerical stability.

A more important observation for our matrices is that many pivoting operations create new non-zero entries. These new nonzero entries are sometimes called *fills*. To see how fills arise, observe the key steps in the decomposition algorithm:

$$l_{kj} = l_{kj} - l_{ki} \cdot u_{ij}$$
$$u_{jk} = u_{jk} - l_{ji} \cdot u_{ik}$$

If both u_{ij} and u_{ik} are nonzero (and thus also their corresponding lower triangular entries due to our assumption of structural symmetry), then u_{jk} will necessarily be nonzero except for a possible but unlikely fortuitous arithmetic value cancelation. If the u_{ik} entry did not exist (or was zero) in the original matrix, the factors U and L require the creation of a new nonzero entry. For large problems that have not been ordered or numbered in any particular way, the result is that many entries become non-zero, greatly increasing the computational burden. The computational effort for the factorization of a dense matrix is of order n^3. For typical sparse matrices that have been ordered to reduce fill-in, empirical studies have determined that the computational complexity of the factorization step is approximately $n^{1.4}$.

The determination of the optimal (least fill-in) ordering for the factorization of a sparse matrix is a combinatorial problem. This means that, in general, it is not practical to attain an optimal order. Near-optimal heuristic strategies that greatly reduce fill-in have been developed. The fundamental idea of these methods is the observation that a row or column with few entries is less likely to create fill-ins than a denser row or column. A simple ordering of rows/columns in an increasing order of nonzero entries is known as the *a priori* minimum degree algorithm. More sophisticated techniques consider the system graph and continually update the number of nonzero entries in a row/column as the factorization takes place, and order the matrix according to this dynamically updated row/column non-zero entry count. This is the minimum degree algorithm.

Because ordering methods were originally developed in the context of power flow problems [2], they are illustrated in greater detail in the body of Chapter 3.

A.2 FACTORIZATION TREE AND SPARSE VECTORS

Often vector b contains few nonzero entries, that is, not only A but also b are sparse. In the most extreme case, only the ith entry of b, b_i, is nonzero. In this case, we have

$$b = b_i \cdot e_i$$

where e_i is a unit elementary vector, equal to the ith column of the identity matrix. For simplicity, we consider here only this case. We later generalize the presentation to vector with more than one nonzero entry.

If A is a dense (full) matrix, then L and U are also dense. In this case, the forward-substitution process applied to e_i results in a vector y with nonzero entries in all positions y_k, $k \geq i$. The forward-elimination process can start at row i, since all rows up until row $i-1$ involve operations with zero values.

If A is sparse, so are generally L and U provided rows and columns have been properly ordered. Let j correspond to the first nonzero entry in column i of L (or of row i of U, thanks to the assumption of structural symmetry). The forward-substitution algorithm applied to e_i creates a new nonzero entry in row j of y. However, all y entries between row i and row j remain zero. The columns of L in this range do not participate in the forward-substitution process. The idea behind sparse vectors is to develop an automatic process that avoids performing operations involving zero entries [3].

Likewise, the first nonzero entry in column j of L creates another fill-in in the corresponding location of y. If we continue this analysis, we obtain a sequence of columns of L $1, j, ..., n$, where each term in the sequence corresponds to the first nonzero entry in the row defined by the previous term. This sequence is referred to as the factorization path for row i. The path is obtained directly from the data structures used to store the sparse matrix L: it is obtained by accessing the column corresponding to the first off-diagonal entry in each row of i.

Column i of L may contain other nonzero entries for $k > j$. These will in turn create additional fill-ins. What may not be obvious is that all these fill-ins are automatically included when the factorization path is constructed. Thus, it is necessary only to be concerned about the first nonzero entry in each row. This implies that the factorization path for i includes all subsequent columns within row i (in graph terms, all nodes adjacent to i).

The union of all factorization paths for all the rows of the matrix constitutes a tree known as the factorization tree. If a sparse vector contains more than one nonzero entry, the necessary rows for the forward substitution can be obtained by sweeping forward in the factorization tree, starting from the entries corresponding to the initially nonzero entries in b. The factorization tree is an extremely compact way of representing the precedence relations among rows that must be respected during factorization and forward substitution. It is also useful for parallel computation applications, efficient matrix modifications, and system reduction (refer to the subsequent sections and also to Reference [16]).

Consider the following 8×8 matrix:

$$
\begin{array}{c}
\quad\; 1\;\; 2\;\; 3\;\; 4\;\; 5\;\; 6\;\; 7\;\; 8 \\
\begin{array}{c} 1 \\ 2 \\ 3 \\ 4 \\ 5 \\ 6 \\ 7 \\ 8 \end{array}
\left[
\begin{array}{cccccccc}
x & & & & x & & & \\
& x & & & x & & & \\
& & x & x & & x & & \\
& & x & x & & x & x & \\
x & x & & & x & & x & x \\
& & x & x & & x & o & x \\
& & & x & x & o & x & x \\
& & & & x & x & x & x
\end{array}
\right]
\end{array}
$$

where the initial nonzero entries are denoted by "x" and the fill-ins are denoted by "o". The resulting factorization tree is illustrated in Figure A.2.

For this example, a vector with nonzero entries in positions 1 and 2 would involve only rows 5, 7, and 8, which are the rows in the factorization paths for both 1 and 2.

There are applications for which one wishes to determine only some of the entries of the solution vector x. Assume that you are interested only in x_i. If A is dense, all previous unknowns x_k, for $k = i + 1,..., n$, must be calculated. However, if A (and thus U) is sparse, the calculation of x_i requires the knowledge only of those elements of x_k where k belongs to the factorization path of i, this time proceeding upward from the bottom of the tree. This process is called fastback substitution. If one wishes to obtain several variables, it is necessary to sweep the factorization tree forward from all these rows to determine which entries will be required during the fastback-substitution process.

In the previous example, if we wish to obtain x_1, it will be necessary to have previously calculated x_5, x_7, and x_8. No other entries of x are required. If, in addition to x_1, we wanted to calculate x_4, it

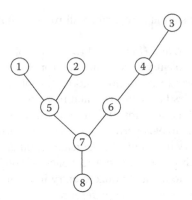

FIGURE A.2   Factorization tree for the eight-node example.

would be necessary to also obtain x_6 according to the factorization tree. Obviously, a small example does not allow us to illustrate the significance of the potential computational savings that can be attained by this process. Even for very large systems, the number of rows involved in the calculation of an entry using sparse vectors typically will not exceed 20 or 30 entries, often fewer.

The fast-forward and fastback-substitution processes just described are sometimes referred to as FF and FB.

The ordering used for the reduction of fill-in during the factorization step affects the shape of the factorization tree and the length of the factorization paths. For purposes of FF/FB, the average length of the factorization path is more important than the number of fills during the factorization process. It is best to have trees with many branches and short paths, even when this results in additional fill-in. Generally (but not always), the same heuristics that lead to a reduction of fill-in in L and U result in acceptable results from the viewpoint of the shape of the factorization tree. Furthermore, because algorithms such as the minimum degree algorithm offer some flexibility in their implementation, it is possible to take advantage of this flexibility to improve the shape of the factorization tree without worsening the fill-in situation [4,5]. An extreme example of incompatibility between the two objectives is the tridiagonal matrix. The simple natural tridiagonal ordering results in a linear tree with maximum dependencies among all nodes. It is possible to reorder a tridiagonal matrix so that some fill-in occurs and the shape of the factorization tree is considerably improved.

A.3   SPARSE INVERSE

The fastest way to obtain arbitrary individual entries of the inverse of a sparse matrix is to use the FF/FB method.

If one wishes to obtain an entire column of the inverse, this can be done with FF operations, followed by ordinary back substitution. For a symmetric matrix, careful ordering of the computations can reduce the computation burden by almost half by computing the entries of the upper triangular portion starting with the last column, and moving back to the first entry. The needed entries from the lower triangular matrix are taken from the previously computed upper triangular matrix entries.

Let sparse inverse denote the subset of matrix inverse entries that correspond to entries that were nonzero in the L and U factors of A. This specific subset is useful in some applications such as short-circuit calculations [17], some state estimation calculations, and contingency analysis. This subset can be obtained efficiently without having to calculate any entries other than those in the subset. The computational effort of this process grows linearly with the number of nonzero entries in the factored matrices.

As is the case for almost all other procedures described in this appendix, the sparse inverse can be obtained by rows or by columns with the same effort. In the "by columns" version, the algorithm conceptually calculates the upper triangle of the sparse inverse, starting with the last column. Each entry in every column is obtained by conventional back substitution. At each step of the algorithm, all the required entries have been previously obtained. As a result of symmetry, the required lower triangular entries are obtained from their corresponding transpose entry.

The Matlab (Matrix Laboratory) pseudo-code for the sparse inverse algorithm for a symmetric matrix is (assuming that L and D are available) as follows:

```
for i=n:-1:1
  [J,I]=find(L(i+1:n,i));
  J=J+i;
  for jj=1:length(J)
    j=J(jj);
    Z(j,i)=-sum(Z(i+1:j,j).*L(i+1:j,i));
    Z(i,j)=Z(j,i);
  end
  Z(i,i)=1/D(i,i)-sum(Z(i+1:n,i).*L(i+1:n,i));
end
```

The critical step involves the interior product of the columns of Z and L. Figure A.3 illustrates the computational sequence for a 20×20 matrix.

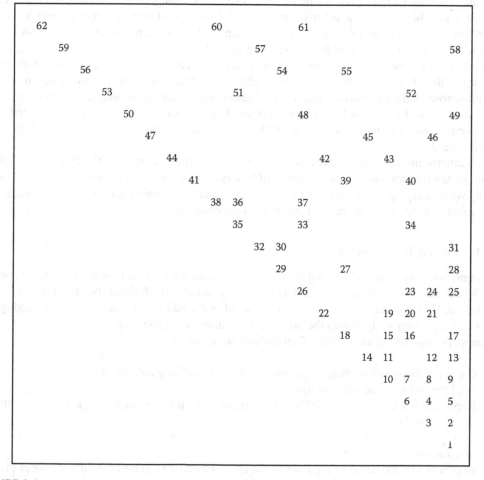

FIGURE A.3 Computational sequence for the computation of the sparse inverse.

The sequence shown is one of many possible sequences. The sparse inverse can be obtained using any sequence that respects the precedence relationships in the factorization tree. Using the same tree, it is possible to extend the computation to include additional entries of the inverse [6]. In almost all cases it is necessary to obtain, in addition to the desired inverse entries, some additional entries as indicated by the factorization tree.

A.4 MODIFICATION OF THE COEFFICIENT MATRIX

The most expensive step when solving a sparse linear system lies generally in the row ordering and factorization of the coefficient matrix. When a previously factorized matrix gets modified, it is normally inefficient to perform its factorization from scratch. Two better alternatives exist to reflect the matrix changes into the solution: (1) modify its LU factors or (2) modify the solution. In this section, the first option is discussed.

Two approaches exist within this category: one that recomputes a selected fraction of the table of factors, another that updates the old values. The changes may refer either to the modification of the numerical values or to the matrix structure or topology. In turn, topology changes may give rise to the creation or elimination of elements. Finally, matrix modifications may be temporary or permanent. Starting from a base case, temporary changes are applied to a limited number of solutions after which the original situation is reverted to. Permanent changes do not revert to the base case and are applied to an indefinite number of subsequent solutions. In general, it is more efficient for permanent changes to modify the solution rather than the factors. However, if the changes must be applied to more than a solution, it may be preferable to modify the factors. The exact point from which an approach is better than the other cannot be established by means of simple rules, as it depends on the particular system being solved, but certain rules of thumb will be provided below.

The key observation is that, when a sparse factorized matrix is modified, only a subset of the elements of the LU factors is affected by the modification. When an element of row k gets modified, only those rows in the factorization path of row k are affected, that is, the same set of rows that would be involved in the FF process for the respective singleton vector. If several rows are modified, then the affected rows are those belonging to the subtree arising by the union of the individual factorization paths.

As formerly indicated, there exist two categories of methods to modify the factors of a sparse matrix: partial refactorization, where a subset of factors is recomputed nearly from scratch, and factor update, where existing factors are simply updated [7]. The latter option tends to be more efficient if the number of modified elements is small, but not necessarily.

A.4.1 Partial Refactorization

For simplicity, only changes affecting the numerical values, not the matrix structure, will be considered. Sometimes, the resulting value after a modification is null, which should be reflected in the sparsity pattern. However, this possibility will be ignored as the additional logic required to modify the matrix structure frequently offsets the possible gains in terms of added sparsity.

The steps involved in the partial refactorization are as follows:

1. Carry out the required changes in rows $i, j, k, \ldots$ of the original matrix.
2. Find the union of the paths of rows $i, j, k, \ldots$.
3. Replace rows $i, j, k, \ldots$ of the factorized matrix with the modified rows $i, j, k, \ldots$ of the original matrix.
4. Replace the remaining rows in the union of paths with the nonmodified rows of the original matrix.
5. Repeat the factorization of the rows in the union of paths keeping in mind the correct precedences among rows.

A possible pseudo-code implementation in Matlab is as follows:

```
while ~isempty(c)
  k=c(1); c=c(2:length(c));
  [iDum,iSet,iVal]=find(A(k,1:k-1));
  while ~isempty(iSet)
    i=iSet(1); iSet=setdiff(iSet,i);
    if ismember(i,cpath),
      A(k,i)=A(k,i)/A(i,i);
    end
    [jDum,jSet,jVal]=find(A(i,i+1:n));
    jSet=jSet+i;
    jSet=unique(intersect(jSet,cpath));
    while ~isempty(jSet)
      j=jSet(1); jSet=setdiff(jSet,j);
      if j<k, iSet=unique(union(iSet,j)); end
      A(k,j)=A(k,j)-A(k,i)*A(i,j);
    end
  end
end
```

The above algorithm replaces the values of matrix A with those of the new factors, but only for those rows contained in path c that should correspond with the composite path of the rows modified in A. This joint path is easily obtained as a subset of the factorization tree.

A.4.2 FACTOR UPDATE

While partial refactorization obtains new factors based on original matrix values, the factor update technique modifies existing factors without referring to the original matrix. Except for rounding off errors, both techniques provide identical results. The difference lies just in the computational cost, as the path involved is identical.

The algorithm can be generalized to any number and type of changes, but the most popular version, discussed here, refers to rank-1 modifications. By repeatedly applying elemental rank-1 modifications, arbitrarily complex modifications can be obtained. The advantages of performing rank-1 modifications are its simplicity, the fact that a single auxiliary vector is needed in the symmetrical case and the fact that in many practical situations only this modification arises.

This method requires that the changes be expressed as a product of matrices. For rank-1 modifications, changes in A can be expressed as

$$A' = A + BrC^T \tag{A.4}$$

where A is $n \times n$ matrix, r is a scalar for rank-1 changes, and B, C are topological n-vectors defining the elements of A affected by the changes. Usually, these vectors are null except for one or two elements taking the values $+1$ or -1.

For a structurally symmetric matrix factorized in the form LDU, the rank-1 factor update algorithm is composed of a preparatory phase followed by the updating itself.

Preparatory phase: It is composed of the following steps:

- Determine the sparse vectors B and C.
- Find the composite path of rows corresponding to the nonnull elements of B and C.
- Let $\beta \leftarrow r$ (the scalar β is modified by the algorithm).

Computation phase: For every row i in the path

1. Prepare the ith row:

$$d_{ii} \leftarrow d_{ii} + B_i\beta C_i$$
$$C_1 \leftarrow \beta B_i$$
$$C_2 \leftarrow \beta C_i$$

2. Process all elements $u_{ij} \neq 0$, $i \neq j$. For each j process all $u_{ik} \neq 0$, $i \neq k$. For those elements do

$$B_j \leftarrow B_j - B_i\ell_{ki}$$
$$C_j \leftarrow C_j - C_i u_{ik}$$
$$u_{ik} \leftarrow u_{ik} - C_1 B_j/d_{ii}$$
$$\ell_{ki} \leftarrow \ell_{ki} - C_2 C_j/d_{ii}$$

3. Update β:

$$\beta \leftarrow \beta - C_1 C_2/d_{ii}$$

During the updating process, fill-in elements in the working vectors B and C take place only in those positions corresponding to the path. Therefore, only those positions should be reset before each updating, which is done on the fly when determining the new path.

The above procedure can be generalized for rank-m modifications by simply letting B and C be $n \times m$ connectivity matrices and replacing the scalar β by an $m \times m$ matrix. The resulting path is then the composite path corresponding to all nonnull elements of B and C. The main difference with respect to a sequence of rank-1 changes is that the updating is completed by sweeping only once the composite path. The arithmetic operations over the factors are, however, identical. Logically, $m \times n$ working matrices, rather than column vectors, are needed.

An advantage of the factor updating process over partial refactorization is that the original matrix, which is not retained in certain applications, is not required.

A.5   REDUCTIONS AND EQUIVALENTS

Another advanced topic of interest is the reduction of sparse matrices. For the discussion that follows, the set of nodes is divided into the retained nodes r and the eliminated ones e. In turn, the set r can be subdivided into frontier or border nodes b and inner nodes i. Border nodes are those connected simultaneously to nodes belonging to set e and i, which means that there is no direct connection between these two sets. The subset i does not play any role in the reduction, but it is frequently manipulated along with set b. Sometimes the set i is empty, the set r being composed only of border nodes.

On the basis of those sets, the nodal equation can be subdivided into

$$\begin{bmatrix} Y_{ee} & Y_{eb} & 0 \\ Y_{be} & Y_{bb} & Y_{bi} \\ 0 & Y_{ib} & Y_{ii} \end{bmatrix} \begin{bmatrix} V_e \\ V_b \\ V_i \end{bmatrix} = \begin{bmatrix} I_e \\ I_b \\ I_i \end{bmatrix} \tag{A.5}$$

Eliminating the variables corresponding to the set e yields

$$(Y_{bb} - Y_{be}Y_{ee}^{-1}Y_{eb})V_b + Y_{bi}V_i = I_b - Y_{be}Y_{ee}^{-1}I_e \tag{A.6}$$

which, using new symbols, can be written as

$$Y'_{bb}V_b + Y_{bi}V_i = I'_b \qquad (A.7)$$

The reduced admittance matrix, Y_{eq}, and the equivalent current injections, I_{eq}, are

$$Y_{eq} = -Y_{be}Y_{ee}^{-1}Y_{eb} \qquad (A.8)$$

$$I_{eq} = -Y_{be}Y_{ee}^{-1}I_e \qquad (A.9)$$

The reduction is not performed by explicitly inverting Y_{ee} but eliminating a bus each time. The reduction is nothing but an ordinary factorization limited to the set e. The operations that eliminate e modify the submatrix Y_{bb}, which is converted into Y_{eq}, as indicated by Equation A.8. If the equivalent injections, I_{eq}, are also needed, they can be obtained by partial forward elimination with the factors of the equivalent, as shown by Equation A.9.

The techniques to compute equivalents greatly differ depending on whether they are large or small. A large equivalent arises when a large portion of the network is to be retained (limited reduction). The notion of adaptive equivalent explains how an equivalent can be obtained even when the original matrix is replaced by its factors. Finally, small equivalents arise when the network is reduced to a few nodes (the Thevenin or Norton equivalent would be the limit case).

A.5.1 LARGE EQUIVALENTS

Usually, the goal of a large equivalent is to reduce the computational cost of repeatedly working with a system, by reducing the retained portion. In this case, it is necessary to consider the fills arising in the retained system, as the elimination of e creates new equivalent branches among the nodes in b. Unless some care is exercised when selecting e (or r), the reduction in the number of buses may be offset by the increment in the number of nonzero elements in the reduced system, leading to no saving at all.

If e represents a single connected network, its elimination will create equivalent branches (that is, pairs of new nonzero elements) between each possible pair of nodes in b. If, on the other hand, e consists of two or more subnetworks, the elimination of each subnetwork will create new elements only between pairs of nodes of b to which such subnetwork is connected. In certain cases, it is possible that by transferring some nodes from set e, initially selected, to set r, the eliminated system gets divided into several disconnected subnetworks. This node readjustment, that reduces the fills in the equivalent, can be used in several ways when reducing sparse matrices [8].

Once e is identified, the fill-in elements in b are fully determined, irrespective of the order in which the nodes of e are eliminated. However, the intermediate storage needed, as well as the computational effort, can be reduced by appropriate techniques.

The number of branches of a large equivalent can be further reduced by discarding those whose admittance is below a threshold, which can be done during or after the reduction process. In the first case, the ordering and reduction processes should be performed simultaneously. The accuracy deterioration originated by this approximation can be controlled to acceptable levels for certain applications. The problem lies in finding the proper balance between accuracy and speed of solution.

A.5.2 ADAPTIVE REDUCTION

Instead of obtaining a single large equivalent for many different problems, it is more efficient in some applications to find equivalents specifically adapted to the need of each individual problem, or even to the different solution steps of the same problem. The adaptive reduction is a procedure to obtain equivalent submatrices of a matrix by recovering its factors [9,10], which is possible because some of the operations needed for the reduction are also carried out during the factorization. Note that, when the factors L, D, and U are multiplied, the original matrix is reconstructed.

A key observation for the adaptive reduction is that the nodes of any factorization path, simple or compound, ending in the last row of the equation system, can be numbered at the end without any implication for the factorization process. This means that, given a compound path, all its nodes can be placed in the lower rightmost part of the matrix without the nonzero pattern and the numerical values of the factors being affected. On the other hand, as formerly explained, the reduction process amounts to a factorization that is stopped when the retained nodes are reached. Both ideas can be combined by letting the set of retained nodes to contain not only the nodes in which we are interested, but also those belonging to the associated joint path. This way, the elimination of external nodes only originates the fills determined by its table of factors. This property can be exploited to selectively recover the original values of a matrix from its factors, modify the values recovered, and refactorize again the recovered portion.

To explain this concept, let us express a matrix in terms of its factors as follows:

$$\begin{bmatrix} Y_{ee} & Y_{eb} \\ Y_{be} & Y_{bb} \end{bmatrix} = \begin{bmatrix} L_{ee} & \\ L_{be} & L_{bb} \end{bmatrix} \begin{bmatrix} U_{ee} & U_{eb} \\ & U_{bb} \end{bmatrix} \tag{A.10}$$

Equating the equivalent of Equation A.8 with the submatrices of Equation A.10 yields

$$\begin{aligned} Y_{be} Y_{ee}^{-1} Y_{eb} &= (L_{be} U_{ee})(L_{ee} U_{ee})^{-1}(L_{ee} U_{eb}) \\ &= L_{be} U_{ee} U_{ee}^{-1} L_{ee}^{-1} L_{ee} U_{eb} \\ &= L_{be} U_{eb} \end{aligned} \tag{A.11}$$

An examination of Equation A.11 shows that the equivalent for the set b can be computed by partial FF operations on the columns of U corresponding to the set e. When handling each of those columns, all operations except those of rows belonging to b can be skipped. In the columns of L used for the partial FF processes, all operations except those corresponding to b can be omitted. In other words, only the operations directly involving the set b in Equation A.5 have to be carried out to calculate the adaptive reduction of a factorized matrix. With the proper logic, an adaptive reduction can be obtained several times faster than a conventional equivalent from scratch.

The above ideas can be implemented in several ways. An equivalent computed through adaptive reduction is always sparse because it contains only fill-in elements that were present in the original factorization. The drawback, which tends to offset this advantage, is that the desired set of nodes has to be augmented with (sometimes many) additional nodes completing the path. Usually, this leads to an equivalent larger than needed.

A.5.3 SMALL EQUIVALENTS

Frequently, the need arises to obtain equivalents composed of very few nodes, and the above methods, oriented to large equivalents, can be very inefficient (e.g., the adaptive reduction may involve hundreds of nodes even if only three are needed). These small equivalents are more efficiently obtained through FF/FB operations on unitary vectors. Such operations generate the inverse equivalent, which must be therefore inverted to obtain the desired equivalent.

Two different schemes exist, the first of which is described below, for a structurally symmetric matrix. The steps required to obtain the 3×3 equivalent for any nodes i, j, and k are as follows:

1. Find the joint path of i, j, and k.
2. Form the unitary vector i, and perform the FF process along the path of i.
3. Perform the FB substitution on the compound path.

4. Save the elements at positions i, j, and k obtained after the FB process. Reset the path for the next iteration.
5. Repeat steps 2–4 for nodes j and k.

The three sets of elements obtained at positions i, j, and k as a result of the three cycles constitute the 3×3 inverse of the equivalent. If this is needed explicitly, it can be obtained by direct inversion, although sometimes its inverse, or factorized inverse, suffices. This example can be easily generalized to equivalents of any size. It is also worth mentioning that, for a symmetric matrix, some operations of the FB processes can be saved by computing only a triangular half of the inverse of the equivalent.

A second possibility (more efficient for symmetric matrices) is as follows:

1. Find the path of i.
2. Form the unitary vector i, and perform the FF process on the path by omitting the diagonal operation. Let F_i be the resulting vector from this operation, which is stored.
3. Divide F_i by d_{ii} to obtain $\tilde{F}_i$.
4. Repeat the steps 1–3 for j and k.
5. Calculate the six different elements of the 3×3 equivalent through the following operations on the sparse vectors obtained in steps 2 and 3:

$$z_{ii} = \tilde{F}_i^T F_i$$
$$z_{ji} = \tilde{F}_j^T F_i \quad z_{jj} = \tilde{F}_j^T F_j$$
$$z_{ki} = \tilde{F}_k^T F_i \quad z_{kj} = \tilde{F}_k^T F_j \quad z_{kk} = \tilde{F}_k^T F_k$$

This scheme is faster for symmetric matrices because the six products among sparse vectors usually require less effort than the three FB solutions of the first scheme, particularly if row ordering oriented to the reduction of the average path length is adopted. As the size of the equivalent grows, the relative advantage of the second approach decreases. This is due to the fact that the cost associated with the vector products grows with the square of the equivalent size, whereas the effort to invert a matrix grows with the cube.

Either of the two schemes is practical up to a certain equivalent size, the approach oriented to large equivalents being more efficient from that point on. Anyway, for a wide range of equivalent sizes the choice is irrelevant. In general, a matrix should be reduced only up to the point in which its sparse nature starts to deteriorate or, alternatively, very small equivalents should be sought.

A.6   COMPENSATION

This is a technique to obtain the solution of a locally modified system based on the original table of factors [11]. The technique is also known as the inverse matrix lemma [12] or rank-1 inverse matrix modifications [13], among other names.

Let us assume that we are interested in the solution of the modified system $(Y + \Delta Y)V = I$, with

$$\Delta Y = BRC^T$$

where R is a scalar or a reduced-rank matrix.

The compensation-based procedure is organized in a preparatory and a solution phase [14]. The preparatory phase consists of finding a small equivalent matrix, Y_{eq}, and a reduced vector, I_{eq}, containing all the modified elements. Changes to the original matrix are instead applied to Y_{eq} and I_{eq}, leading to a new modified matrix Y_{mod} and vector I_{mod}. The solution vector of this small problem is then applied to the reduced unmodified network to obtain

$$\hat{I}_{eq} = Y_{eq} V_{mod} \tag{A.12}$$

and, as a consequence, the small-size compensating vector ΔI_{eq}:

$$\Delta I_{eq} = I_{eq} - \hat{I}_{eq}$$

In turn, this vector is expanded in a full-size sparse compensating vector, ΔI, by completing with zeros the required positions (when appropriate sparsity techniques are employed, this is just a matter of resorting to adequate indices). Finally, ΔI is applied to the original, full-size (unmodified) system. The results are the same as if the network had been modified, a new matrix Y and injection vector I had been built, and the resulting system had been solved.

In the solution phase, the vector ΔI is used to compensate the base-case solution in such a way that network changes are taken into account. In practice, such compensation can be made at three different instants, leading to three variants:

1. *Precompensation*: The vector ΔI is added to the base-case current injection vector. Then, a full system solution is performed with this modified vector to obtain the compensated voltage vector.
2. *Intermediate compensation*: FF procedures are applied to each sparse ΔI vector. These results are added to the intermediate vector obtained after the base-case forward elimination. Finally, the modified solution system is obtained by full backward substitution.
3. *Postcompensation*: By means of FF and full backward processes on ΔI the incremental vector ΔV is obtained, which is added to the base-case solution V to obtain the modified solution.

The most used version is postcompensation, even though intermediate compensation is frequently the fastest.

An interesting version of the preparatory phase expresses the network changes through minimum-rank matrices, which sometimes saves arithmetic operations [15]. However, this formulation is equivalent in many practical cases to the conventional ones, in which the size of the modification agrees with the number of nodes or branches affected.

Let us assume that a set of changes affects buses i, j, and k, and that the independent vector remains unchanged. In the node-oriented set-up the stages described above are

Preparatory phase: It is composed of the following steps:

1. Calculate the small equivalent of nodes i, j, and k, as explained in Section A.5.3, to obtain the 3×3 inverse of the equivalent, Z_{eq} (save the vectors resulting after the FF phase to be later used in the solution phase of the intermediate compensation).
2. Invert Z_{eq} to obtain Y_{eq}, the base-case equivalent.
3. Obtain the base-case equivalent currents, I_{eq}, from $I_{eq} = Y_{eq}V$, where the three values of vector V are those of the base case.
4. Modify Y_{eq} according to the network changes to obtain Y_{mod}.
5. Calculate V_{mod} from $Y_{mod}V_{mod} = I_{eq}$.
6. Obtain $I_{mod} = Y_{eq}V_{mod}$.
7. Find the three compensating currents, ΔI_{eq}, from $\Delta I_{eq} = I_{mod} - I_{eq}$.
8. Express ΔI_{eq} as a full-size sparse vector, ΔI.

Precompensation: The steps for the precompensation are:

a. Add ΔI to the independent vector I previously stored.
b. Perform ordinary forward/backward substitution on the resulting vector to obtain the compensated voltage vector.

Intermediate compensation: The steps for this version are:

a. Multiply each of the three vectors arising after the FF processes of step 1 by their corresponding compensating currents of step 6.
b. Add the three vectors of step (a) to the vector arising after the forward elimination of the base case.
c. Perform the backward substitution to obtain the modified voltages.

Postcompensation: It is composed of the following steps:

a. Perform a FF elimination followed by a full backward substitution on the sparse vector ΔI of step 7, to obtain the vector of compensating voltages.
b. Add the vector computed previously to the base-case voltage vector to obtain the modified voltages.

The inverse of the equivalent provided by step 1 of the preparatory phase can be obtained by means of FF/FB operations on the corresponding connectivity vectors, for each set of changes to the original matrix. In applications where intensive use of compensation is made, it may be more efficient to calculate during the preparatory phase all elements of the inverse that will be needed subsequently. Frequently, these elements are in the sparse inverse.

Explicit computation of the inverse of the equivalent can be avoided by resorting to triangular factorization, which is relevant when the equivalent is large. Changes to the independent (injection) vector, eventually associated to topological modifications, can be taken into account during the preparatory phase by adding them to the base-case currents of the equivalent.

A.7 QR FACTORIZATION AND GIVENS ROTATIONS

Any nonsingular matrix A can be also decomposed as the product of an orthogonal matrix Q and an upper triangular matrix R,* in such a way that $A = QR$. The importance of such a decomposition is the following: by definition, a matrix is orthogonal when $Q^{-1} = Q^T$. Therefore, solving an equation system by means of QR decomposition involves the following steps:

1. Orthogonal factorization of the coefficient matrix:

$$A = QR$$

2. Computation of the intermediate vector y:

$$y = Q^T b$$

3. Solving for the unknown vector x by back substitution:

$$Rx = y \quad \Rightarrow \quad x_n \overset{\rightarrow}{\cdots} x_1$$

The main advantage of using orthogonal matrices is that the resulting solution method is much more stable numerically than the LU decomposition or Gaussian elimination.[†] The disadvantage is that, in general, QR decomposition involves more computation and fill-in than alternative methods.

---

* This matrix R is not in general the same as matrix U in the LU decomposition.
† A second advantage is that QR factorization can be employed with rectangular matrices A in which case the solution x is the one with minimum norm.

The most complex step is by far the factorization itself. There exist several methods to achieve the orthogonal factorization, but the best suited to sparse matrices is the one based on Givens rotations. Givens' method builds a sequence of orthogonal elementary matrices, each one designed to eliminate an element of the lower triangular part of A. To illustrate the method, consider for simplicity a 2×2 matrix. In this case, the Givens rotation matrix has the form

$$A = \begin{bmatrix} a_{11} & a_{12} \\ a_{21} & a_{22} \end{bmatrix}$$

The objective is to eliminate the term a_{21}. The Givens rotation matrix capable of doing this is

$$Q = \begin{bmatrix} \cos\theta & \sin\theta \\ -\sin\theta & \cos\theta \end{bmatrix}$$

In spite of having four terms, this matrix has only a degree of freedom: the angle θ. It is easy to verify that this matrix Q is orthogonal according to its definition. Simply check that

$$\begin{bmatrix} \cos\theta & \sin\theta \\ -\sin\theta & \cos\theta \end{bmatrix} \begin{bmatrix} \cos\theta & -\sin\theta \\ \sin\theta & \cos\theta \end{bmatrix} = \begin{bmatrix} 1 & 0 \\ 0 & 1 \end{bmatrix}$$

for any value of θ. The specific value of θ that eliminates the term a_{21} is

$$\theta = \arctan a_{21}/a_{11}$$

If matrix A is multiplied by matrix Q the result is a triangular matrix. This result generalizes for the elimination of any element a_{ij} of A.

When applied to larger matrices, the elements in the lower triangular part of A should be sequentially eliminated, one by one. The elimination sequence can be by rows or by columns, starting with the first element of the second row and ending with the $n - 1$th element of the last row. Two possible elimination sequences are

$$\begin{bmatrix} x & x & x & x & x \\ 1 & x & x & x & x \\ 2 & 3 & x & x & x \\ 4 & 5 & 6 & x & x \\ 7 & 8 & 9 & 10 & x \end{bmatrix} \quad \text{or} \quad \begin{bmatrix} x & x & x & x & x \\ 1 & x & x & x & x \\ 2 & 5 & x & x & x \\ 3 & 6 & 8 & x & x \\ 4 & 7 & 9 & 10 & x \end{bmatrix}$$

Generally, Givens rotations are implemented row-wise. An alternative method, known as Householder reflections, performs eliminations of entire columns.

The result of the Givens rotations is a sequence of τ matrices Q_k that yields, when applied to matrix A, an upper triangular matrix (τ is the number of nonzero elements, including the so-called intermediate fill-in). It is possible to multiply all these matrices to build a single orthogonal matrix $Q = Q_\tau Q_{\tau-1} \ldots Q_1$. However, this is a wrong strategy for large-scale systems, as the initial sparsity is significantly lost. Each of these matrices Q_k can be represented by three numbers, namely: the coordinates i and j of the eliminated element and the value of the rotation angle θ. Therefore, the τ matrices require only 3τ storage elements. But if they get multiplied, the resulting matrix Q generally has got many more nonzero elements. Apart from this, those elements cannot be represented as elegantly through a rotation angle θ. Therefore, it is not advisable to build the combined matrix Q explicitly, even though it is also an orthogonal matrix.

Another important observation refers to the structure of the triangular matrix R arising from this process. To illustrate this concept and the problem consider the following matrix:

$$A = \begin{bmatrix} a_{11} & a_{12} & \\ a_{21} & a_{22} & a_{23} \\ & a_{32} & a_{33} \end{bmatrix}$$

If a conventional LU factorization (Gaussian elimination) was performed, no fill-in elements would arise. However, when the QR factorization is applied to element a_{21} a fill-in appears at position a_{13}. This happens because multiplying by matrix Q_1 adds a multiple of row 1 to a multiple of row 2 and, at the same time, a multiple of row 2 to a multiple of row 1. As the entry a_{23} is nonzero, this generates a fill-in at position a_{13}. The result is that matrix R tends to have the second "neighbor" structure of matrix A (note, however, that some cancelations are possible under certain conditions).

Therefore, to get better results when performing the QR factorization, one should try to order the columns of A in such a way that a minimum number of fill-ins arise in R. To achieve this goal, a Tinney-2 like method should be applied to reduce the fill-ins of matrix $A^T A$. The product $A^T A$, rather than A, is used for two reasons. First, the fill-ins depend on the second-neighbor structure, (i.e., the structure of $A^T A$), and second, the method is of application this way to rectangular matrices.

Another practical aspect related to QR factorization is the ordering of the rows of A. This ordering has no effect on the fill-ins of R, but it is critical for the number of intermediate fill-ins arising during the factorization process.

Finally, a relevant version of the QR method, known as the hybrid method, should be mentioned (it is also known as the *Corrected SemiNormal Equations* method). If our original factorized equation

$$QRx = b$$

is premultiplied by A^T (taking into account that $A^T = R^T Q^T$), the result is

$$R^T Q^T QRx = A^T b$$

which reduces to

$$R^T Rx = A^T b$$

The conclusion is, first, that there is no need to save the dense matrix Q or its elemental components. Second, once R is obtained, the method reduces to multiply the original matrix by the vector b, followed by forward-elimination and backward-substitution processes using R^T and R, respectively.

REFERENCES

1. Duff, I. S., Erisman, A., and Reid, J., *Direct Methods for Sparse Matrices*, Clarendon Press, Oxford, 1986.
2. Tinney, W. F. and Walker, J. W., Direct solutions of sparse network equations by optimally ordered triangular factorization, *Proceedings IEEE*, 55, 1967, 1801–1809.
3. Tinney, W. F., Brandwajn, V., and Chan, S., Sparse vector methods, *IEEE Transactions on Power Apparatus and Systems*, PAS-104(2), 1985, 295–301.
4. Gómez-Expósito, A. and Franquelo, L. G., Node ordering algorithms for sparse vector method improvement, *IEEE Transactions on Power Systems*, 3(1), 1988, 73–79.
5. Betancourt, R., An efficient heuristic ordering algorithm for partial matrix refactorization, *IEEE Transactions on Power Systems*, 3(3),1988, 1181–1187.
6. Betancourt, R. and Alvarado, F. L., Parallel inversion of sparse matrices, *IEEE Transactions on Power Systems*, 1(1),1986 ,74–81.
7. Chan, S. M. and Brandwajn, V., Partial matrix refactorization, *IEEE Transactions on Power Systems*, 1(1), 1986, 193–200.

8. Tinney, W. F., Powell, W. L., and Peterson, N. M., Sparsity-oriented network reduction, *PICA Proceedings*, 1973, 384–390.
9. Tinney, W. F. and Bright, J. M., Adaptive reductions for power flow equivalents, *IEEE Transactions on Power Systems*, 2(2), 1987, 351–360.
10. Enns, M. K. and Quada, J. J., Sparsity-enhanced network reduction for fault studies, *IEEE Transactions on Power Systems*, 6(2), 1991, 613–621.
11. Alsaç, O., Stott, B., and Tinney, W. F., Sparsity-oriented compensation methods for modified network solutions, *IEEE Transactions on Power Apparatus and Systems*, PAS-102, 1983, 1050–1060.
12. Sage, A. P. and Melsa, J. L., *Estimation Theory with Applications to Communications and Control*, McGraw-Hill, New York, 1971.
13. Householder, A. S., *The Theory of Matrices in Numerical Analysis*, Dover, New York, 1974.
14. Alvarado, F. L., Mong, S. K., and Enns, M. K., A fault program with macros, monitors, and direct compensation in mutual groups, *IEEE Transactions on Power Apparatus and Systems*, PAS-104(5),1985, 1109–1120.
15. van Amerongen, R. A. M., A rank-oriented setup for the compensation algorithm, *IEEE Transactions on Power Systems*, 5(1), 1990, 283–288.
16. Alvarado, F. L., Tinney, W. F., and Enns, M. K., Sparsity in large-scale network computation, *Advances in Electric Power and Energy Conversion System Dynamics and Control*, Leondes, C. T. (ed.), Control and Dynamic Systems, Academic Press, 41, Part 1, London, 1991, 207–272.
17. Takahashi, K., Fagan, J., and Chen, M. S., Formation of a sparse bus impedance matrix and its application to short circuit study, PICA Proceedings, 1973, 63–69.

Appendix B: Mathematical Programing

Antonio J. Conejo

CONTENTS

This appendix describes the fundamentals of mathematical programing, both linear and nonlinear. From a practical viewpoint, it is convenient to use an environment for the formulation and solution of mathematical optimization problems, such as GAMS [1] or AMPL [2], which allows focusing on modeling, not on solution algorithms. Nevertheless, to assess results, it is appropriate to be familiar with the working principles of the optimization algorithms. That is why this appendix provides an overview of practical algorithms to solve optimization problems. Prior to the description of the algorithms, this appendix also provides the theory underlying these algorithms.

    Concerning models based on mathematical programing as well as solution algorithms, the books by Castillo et al. [3] and Conejo et al. [4] might interest the reader.

B.1 LINEAR PROGRAMING

Linear programming allows formulating many problems related to control, operation, planning, economics, and regulations of electric energy systems. On the other hand, the solution of a well-formulated linear programing problem (linear problem) can always be found. Thus, linear programing is a robust modeling technique. Moreover, it is possible to solve very large linear programming problems, involving hundreds of thousands of variables and constraints, using personal computers in seconds. Excellent books on linear programing include Bazaraa et al. [5], Chvátal [6], and Luenberger [7].

B.1.1 FUNDAMENTALS

Consider the linear programing problem

$$
\begin{aligned}
\text{minimize}_{x_1, x_2} \quad & z = -3x_1 - 5x_2 \\
\text{subject to} \quad & 3x_1 + 2x_2 \leq 18 \\
& x_1 \leq 4 \quad x_2 \leq 6 \quad x_1 \geq 0 \quad x_2 \geq 0
\end{aligned}
\tag{B.1}
$$

Figure B.1 depicts the geometry of the feasible region (points that meet constraints) of this problem. This feasible region is a polygon. As shown below, the corners of this polygon are of particular interest.

In problem (B.1), it can be observed that the contour lines of the objective function are straight lines. Note also that the minimizer, feasible point in which the objective function attains its minimum value, coincides with a corner of the feasibility region.

Motivated by the structure of problem (B.1), the canonical form of any linear programing problem is defined as

$$
\begin{aligned}
\text{minimize}_x \quad & z = c^\mathsf{T} x \\
\text{subject to} \quad & Ax \leq b \quad x \geq 0 \quad x \in \mathbb{R}^n
\end{aligned}
\tag{B.2}
$$

where x is the n-dimensional unknown vector, c is the n-dimensional cost vector, A is the $m \times n$ constraint matrix, and b is the m-dimensional right-hand-side resource vector.

Linearity of both the objective function and the constraints confers problem (B.2) a singular structure. Constraint linearity makes the feasible region a convex polyhedron. On the other hand, the linearity of the objective function makes its gradients constant and thus the objective function always increases along certain directions and decreases along the opposite ones. Therefore, the frontier of the feasibility region and particularly its vertices are important. Contour curves abandon the feasible region through just one point (single optimal solution) or through a face of the polyhedron (multiple optimal solutions). It might also happen that contour curves never abandon the feasible region (unbounded problem).

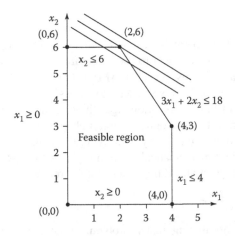

FIGURE B.1 Example of a linear programing problem.

An alternative form of expressing any linear programing problem, denominated standard form, is

$$\text{minimize}_x \quad z = c^T x$$
$$\text{subject to} \quad Ax = b \quad x \geq 0 \quad x \in \mathbb{R}^n \tag{B.3}$$

Without loss of generality vector b is considered nonnegative. A problem formulated in standard form requires m to be smaller than or equal to n ($m \leq n$).

As illustrated through problem (B.1), the feasible region of a linear programing problem is a convex polyhedron. Moreover, the standard form of a linear programing problem allows characterizing algebraically the vertices of this convex polyhedron, which are the candidate points to minimizers.

Since the most common form is the standard form, rules to convert any problem to the standard form are stated below:

1. Constraints of the type greater than or equal to are converted to equality constraints through nonnegative surplus variables (y_i). The constraint $\sum_j a_{ij}x_j \geq b_i$ is equivalent to the two constraints below:

$$\sum_j a_{ij}x_j - y_i = b_i \quad \text{and} \quad y_i \geq 0 \tag{B.4}$$

2. Similarly, constraint $\sum_j a_{ij}x_j \leq b_i$ is equivalent to the two constraints below:

$$\sum_j a_{ij}x_j + y_i = b_i \quad \text{and} \quad y_i \geq 0 \tag{B.5}$$

   In this case, the variable y_i is denominated slack variable.
3. If a variable is free, $-\infty < x_i < \infty$, it can be substituted by the difference of two nonnegative variables, that is,

$$x_i = y_i - z_i \quad y_i \geq 0 \quad z_i \geq 0 \tag{B.6}$$

4. Finally, note that maximizing a given objective function z is equivalent to minimizing its opposite $-z$.

Example (B.1) is written below in standard form (x_3, x_4, and x_5 are slack variables):

$$\text{minimize}_{x_1,x_2,x_3,x_4,x_5} \quad z = -3x_1 - 5x_2$$
$$\text{subject to} \quad x_1 + x_3 = 4$$
$$x_2 + x_4 = 6 \tag{B.7}$$
$$3x_1 + 2x_2 + x_5 = 18$$
$$x_1 \geq 0 \quad x_2 \geq 0 \quad x_3 \geq 0 \quad x_4 \geq 0 \quad x_5 \geq 0$$

Using the standard form of a linear programing problem, consider the partition

$$A = [B|N] \quad x = \begin{bmatrix} x_B \\ x_N \end{bmatrix} \quad c = \begin{bmatrix} c_B \\ c_N \end{bmatrix} \tag{B.8}$$

where B is an m-dimensional full-rank square matrix denominated the basis, N is an $m \times (n-m)$ matrix, x_B is an m-dimensional vector containing so-called basic variables (we say that these variables are in the basis), x_N is an $(n-m)$-dimensional vector containing so-called nonbasic variables (we say that these variables are not in the basis), c_B is an m-dimensional cost vector corresponding to basic variables, and c_N is an $(n-m)$-dimensional cost vector corresponding to nonbasic variables.

We say that vector x is a solution if $Ax = b$, and a feasible solution if $Ax = b$, $x \geq 0$, and a basic feasible solution if $Ax = b$, $x \geq 0$, $x_N = 0$. Observe that a basic solution, feasible or not, can be generated by computing $x_B = B^{-1}b$ and enforcing $x_N = 0$. It is relevant to note that the extreme points of $\{Ax = b, x \geq 0\}$ are indeed basic feasible solutions. This is illustrated below.

We consider again Example (B.1) and obtain all basic feasible solutions. We show below the interest of these solutions, which can be obtained constructing all possible matrices B. The number of matrices B is given by the combination of five columns taken three at a time. All basic feasible solutions are (nonfeasible ones are not provided) as follows:

$$\begin{bmatrix} x_1^* \\ x_2 \\ x_3 \end{bmatrix} = \begin{bmatrix} 1 & 0 & 1 \\ 0 & 1 & 0 \\ 3 & 2 & 0 \end{bmatrix}^{-1} \begin{bmatrix} 4 \\ 6 \\ 18 \end{bmatrix} = \begin{bmatrix} 2 \\ 6 \\ 2 \end{bmatrix} \quad z = -36$$

$$\begin{bmatrix} x_1 \\ x_2 \\ x_4 \end{bmatrix} = \begin{bmatrix} 1 & 0 & 0 \\ 0 & 1 & 1 \\ 3 & 2 & 0 \end{bmatrix}^{-1} \begin{bmatrix} 4 \\ 6 \\ 18 \end{bmatrix} = \begin{bmatrix} 4 \\ 3 \\ 3 \end{bmatrix} \quad z = -27$$

$$\begin{bmatrix} x_1 \\ x_4 \\ x_5 \end{bmatrix} = \begin{bmatrix} 1 & 0 & 0 \\ 0 & 1 & 0 \\ 3 & 0 & 1 \end{bmatrix}^{-1} \begin{bmatrix} 4 \\ 6 \\ 18 \end{bmatrix} = \begin{bmatrix} 4 \\ 6 \\ 6 \end{bmatrix} \quad z = -12$$

$$\begin{bmatrix} x_2 \\ x_3 \\ x_5 \end{bmatrix} = \begin{bmatrix} 0 & 1 & 0 \\ 1 & 0 & 0 \\ 2 & 0 & 1 \end{bmatrix}^{-1} \begin{bmatrix} 4 \\ 6 \\ 18 \end{bmatrix} = \begin{bmatrix} 6 \\ 4 \\ 6 \end{bmatrix} \quad z = -30$$

$$\begin{bmatrix} x_3 \\ x_4 \\ x_5 \end{bmatrix} = \begin{bmatrix} 1 & 0 & 0 \\ 0 & 1 & 0 \\ 0 & 0 & 1 \end{bmatrix}^{-1} \begin{bmatrix} 4 \\ 6 \\ 18 \end{bmatrix} = \begin{bmatrix} 4 \\ 6 \\ 18 \end{bmatrix} \quad z = 0$$

We observe that each basic feasible solution corresponds to a corner of the feasible polygon of Figure B.1. The geometric analysis of this example shows that the minimizer is located at one of these corners. Therefore, we infer that the minimizer corresponds to a basic feasible solution. Generally, it can be shown that this is so. Thus, a possible strategy to locate the minimizer is to obtain all basic feasible solutions and select the one with the smallest objective function value. The fundamental theorem of linear programing, which we do not show (see, e.g., [7] for a formal proof), says precisely so. That is, if a linear programing problem has a solution, this solution is a basic feasible solution.

Finally, note that finding all basic feasible solutions is computationally costly, as an upper bound of the number of basic feasible solutions is $(n!)/(m!(n-m)!)$.

B.1.2 THE SIMPLEX MECHANISM

The algorithm below is denominated revised simplex and allows solving linear programing problems by jumping from basic feasible solution to basic feasible solution while decreasing the objective function. The considered problem is

$$\begin{aligned} \text{minimize}_x \quad & z = c^T x \\ \text{subject to} \quad & Ax = b \quad x \geq 0 \quad x \in \mathbb{R}^n \end{aligned} \tag{B.9}$$

Consider the format below

$$\begin{aligned} \text{minimize}_{x_B, x_N} \quad & z = c_B^T x_B + c_N^T x_N \\ \text{subject to} \quad & B x_B + N x_N = b \\ & x_B \geq 0 \quad x_N \geq 0 \quad x_B \in \mathbb{R}^m \quad x_N \in \mathbb{R}^{n-m} \end{aligned} \tag{B.10}$$

Since matrix B is invertible, we can write $x_B = B^{-1}b - B^{-1}Nx_N$, and the objective function then becomes $z = c_B^T B^{-1}b - c_B^T B^{-1} Nx_N + c_N^T x_N$.

We define below certain vectors and matrices that remain constant as long as the basis B remains constant,

$$\tilde{b} = B^{-1}b \quad \lambda^T = c_B^T B^{-1} \quad Y = B^{-1}N \quad d^T = \lambda^T N - c_N^T \tag{B.11}$$

where $\tilde{b}$ is an m-dimensional column vector, λ is an m-dimensional column vector, Y is an $m \times (n - m)$ matrix, and d is an $(n - m)$-dimensional column vector. Substituting the above matrix and vectors in problem (B.10), we obtain

$$
\begin{aligned}
&\text{minimize}_{x_B, x_N} \quad z = c_B^T \tilde{b} - d^T x_N \\
&\text{subject to} \qquad x_B = \tilde{b} - Yx_N \\
&\qquad\qquad\qquad x_B \geq 0 \quad x_N \geq 0 \quad x_B \in \mathbb{R}^m \quad x_N \in \mathbb{R}^{n-m}
\end{aligned}
\tag{B.12}
$$

where vector d is the so-called reduced cost vector, and λ is a highly relevant sensitivity vector (as analyzed in Section B.1.3.1).

In problem (B.12), the objective function depends only on nonbasic variables and the equality constraints express basic variables as a function of nonbasic variables. This formulation allows switching from one basic feasible solution to another one. This is achieved increasing a nonbasic variable until a basic one reaches zero, which implies that the nonbasic and the basic variables swap status.

The simplex algorithm uses formulation (B.12) for moving among basic feasible solutions with the target of decreasing the objective function value until no further decrease is possible, which implies that the optimal solution has been reached. Observe that:

Observation 1. If the element d_j of the reduced cost vector is positive, then, the objective function value decreases as the nonbasic variable x_{Nj} increases.

Observation 2. For fixed j, if some elements Y_{ij}'s are positive, the corresponding basic variables x_{Bi} decrease provided that the nonbasic variable x_{Nj} increases.

Observation 3. To move from basic feasible solution to basic feasible solution, the nonbasic variable x_{Nj} can be increased until the first basic variable reaches zero, that is, until $\tilde{b}_i - Y_{ij}x_{Nj}$ becomes zero for the first i. The value of x_{Nj} becomes the

$$\text{minimum}_{1 \leq i \leq m} \left\{ \frac{\tilde{b}_i}{Y_{ij}} : Y_{ij} > 0 \right\}$$

while x_{Bi} becomes 0.

Taking into account the above observations, the revised simplex algorithm works as follows:

Step 1. Obtain an initial basic feasible solution. We analyze below in this section how to do that.

Step 2. Check if the current solution is optimal. This is so if all elements of the reduced cost vector are nonpositive. If not, continue.

Step 3. Find out which nonbasic variable should enter the basis. This can be done selecting the nonbasic variable whose corresponding element in the reduced cost vector has the highest positive value. Note that other selection rules are possible.

Step 4. Select the basic variable leaving the basis through the criterion stated in observation 3 above.

Step 5. Build the new basis, get a new basic feasible solution, and continue with step 2.

The simplex algorithm needs an initial feasible solution to start with. An initial basic feasible solution for any linear programing problem can be obtained solving the linear programing problem below, which has a trivial initial basic feasible solution ($y = b$, $x = 0$).

$$\text{minimize}_{x,y} \quad z = \sum_{i=1}^{m} y_i$$

$$\text{subject to} \quad Ax + y = b \quad x \geq 0 \quad y \geq 0 \quad x \in \mathbb{R}^n \quad y \in \mathbb{R}^m \tag{B.13}$$

Solving the problem above is referred to as Phase I of the simplex algorithm, Phase II being the actual solution of the considered problem.

Finally, note that it might happen that the value of a basic variable is zero. We say that such solution is degenerated. In rare cases, degeneracy might lead to a cycling behavior of the simplex algorithm. However, appropriate tricks to avoid such behavior are available. Check, for instance, [7].

B.1.3 SENSITIVITY AND DUALITY

B.1.3.1 Sensitivity

We analyze below the so-called sensitivities. These parameters allow deriving fruitful information from the solution of a linear programing problem. Sensitivities are of high practical interest as they provide information on how the objective function marginally changes as certain parameters of the problem change.

Being B^* the basis of the optimal solution of a linear programing problem, x_B^* the optimal basic variable vector, and z^* the objective function optimal value, then

$$x_B^* = B^{*-1}b \quad \text{and} \quad z^* = c_B^T x_B^* \tag{B.14}$$

We consider marginal changes in the right-hand-side vector b so that the basis remains unchanged, that is, $b^* \rightarrow b^* + \Delta b$, and B^* remains unchanged.

This marginal change in b produces marginal changes in the basic variable vector and in the objective function optimal value $x_B^* \rightarrow x_B^* + \Delta x_B$ and $z^* \rightarrow z^* + \Delta z$.

Taking into account Equation B.14, these marginal changes can be expressed as

$$\Delta x_B = B^{*-1} \Delta b \quad \text{and} \quad \Delta z = c_B^T \Delta x_B \tag{B.15}$$

and

$$\Delta z = c_B^T \Delta x_B = c_B^T B^{*-1} \Delta b. \tag{B.16}$$

Taking into account that $c_B^T B^{*-1}$ has been denoted by λ^{*T}, then $\Delta z = \lambda^{*T} \Delta b$. That is,

$$\lambda_j^* = \frac{\Delta z}{\Delta b_j} \quad \forall j = 1, 2, \ldots, m \tag{B.17}$$

Therefore, λ_j^* is the marginal change of the objective function originated by a marginal change in the right-hand-side vector corresponding with constraint j, provided that the basis remain unchanged.

B.1.3.2 Duality

Duality is a mathematical concept that links two mathematical structures and allows extracting information from one of these structures based on information available from the other one. Duality theory is introduced below through an illustrative example.

An oil factory manufactures artificial oil that contains different amounts of m lubricants. The artificial oil is produced mixing up n natural oils easily found in the market at prices $c_1, c_2,\ldots, c_n$. We denote by a_{ij} the quantity of lubricant i contained in one unit of natural oil j. The minimum amounts of lubricants that should contain the artificial oil are $b_1, b_2,\ldots, b_m$. If $x_1, x_2,\ldots, x_n$ are the unknown quantities of natural oils to be mixed up so that the cost of the artificial oil is minimum and meets lubricant content requirements, the problem below allows computing optimal values for those unknown quantities

$$\text{minimize}_{x_1,\ldots,x_n} \quad z = \sum_{j=1}^{n} c_j x_j$$

$$\text{subject to} \quad \sum_{j=1}^{n} a_{ij} x_j \geq b_i \quad i = 1, 2, \ldots, m \tag{B.18}$$

$$x_j \geq 0 \quad\quad j = 1, 2, \ldots, n$$

On the other hand, we consider that a lubricant factory manufactures the m lubricants required to obtain the artificial oil and sell them in the market at prices $\lambda_1, \lambda_2,\ldots, \lambda_m$. For this business to thrive, the cost of producing any natural oil out of lubricants should be below (or equal to) the market price of that natural oil, that is,

$$\sum_{i=1}^{m} \lambda_i a_{ij} \leq c_j \quad j = 1, 2, \ldots, n \tag{B.19}$$

For the lubricant factory to maximize its profit, the lubricants should be sold at prices $\lambda_1, \lambda_2,\ldots, \lambda_m$ obtained by solving the linear programing problem

$$\text{maximize}_{\lambda_1,\ldots,\lambda_m} \quad z = \sum_{i=1}^{m} \lambda_i b_i$$

$$\text{subject to} \quad \sum_{i=1}^{m} \lambda_i a_{ij} \leq c_j \quad j = 1, 2, \ldots, n \tag{B.20}$$

$$\lambda_i \geq 0 \quad\quad i = 1, 2, \ldots, m$$

The two problems above are dual problems and contain the same components but used differently. Note that they constitute two ways to look at the same reality. These two ways are complementary. Thus, we say that the dual problem of the linear programing problem

$$\text{minimize}_x \quad z = c^{\mathsf{T}} x$$
$$\text{subject to} \quad Ax \geq b \quad x \geq 0 \quad x \in \mathbb{R}^n \tag{B.21}$$

is the linear programing problem

$$\text{maximize}_\lambda \quad z = \lambda^{\mathsf{T}} b$$
$$\text{subject to} \quad \lambda^{\mathsf{T}} A \leq c^{\mathsf{T}} \quad \lambda \geq 0 \quad \lambda \in \mathbb{R}^m \tag{B.22}$$

The duality relates the variables of one problem with the constraints of its dual, following strict rules. For linear programing problems duality is symmetric, that is, the dual of the dual is the primal.

Conversion dual/primal rules are stated below without proof [3]. If, on the other hand, the primal problem objective direction is minimizing, the dual problem objective direction is maximizing. Less than or equal to constraints of the primal problem lead to nonpositive variables in the dual problem, while greater than or equal to constraints in the primal problem lead to nonnegative variables in the dual problem. Equality constraints of the primal problem originate free variables in the dual problem.

If the primal problem objective direction is maximizing, the objective direction of the dual problem is minimizing. Less than or equal to constraints in the primal problem originate nonnegative variables in the dual problem, whereas greater than or equal to constraints in the primal problem lead to nonpositive variables in the dual problem. Equality constraints in the primal problem originate free variables in the dual problem. Additionally, note that the dual problem of the dual problem is the primal problem.

We formulate below four theorems that relate the primal and dual problems.

Consider the asymmetrical couple of dual problems

$$\text{minimize}_x \quad c^T x$$
$$\text{subject to} \quad Ax = b \quad x \geq 0 \quad x \in \mathbb{R}^n \tag{B.23}$$

and

$$\text{maximize}_\lambda \quad \lambda^T b$$
$$\text{subject to} \quad \lambda^T A \leq c^T \quad \lambda \in \mathbb{R}^m \tag{B.24}$$

We say that this couple of linear programing problems is asymmetrical because the primal one only contains equality constraints while the dual one only contains inequality constraints.

The *weak duality theorem* states that if x is a feasible solution for the primal problem (B.23) and λ is a feasible solution for the dual problem (B.24), then $\lambda^T b \leq c^T x$ holds.

The *strong duality theorem* states that if x^* is the minimizer of the primal problem (B.23), then, there exists a vector λ^* that is the maximizer of the dual problem (B.24), and such that $\lambda^{*T} b = c^T x^*$ holds.

Using the strong duality theorem, it is possible to show that the dual variables are sensitivities. This is why λ is used to denote both dual variables and sensitivities.

The optimal value of the primal problem objective function is $z^* = c^T x^* = c_B^T x_B^*$. The vector of optimal basic variables is $x_B^* = B^{*-1} b$. Then, $z^* = c^T x^* = c_B^T B^{*-1} b$. Using the strong duality theorem $(c^T x^* = \lambda^{*T} b)$, we conclude that $\lambda^{*T} = c_B^T B^{*-1}$. That is, dual variables and sensitivities coincide.

Considering a couple of asymmetrical dual linear programing problems, the *complementary slackness theorem* states that

$$\lambda_j^*(A_j x^* - b_j) = 0$$
$$x_i^*(\lambda^{*T} A_i - c_i) = 0 \tag{B.25}$$

Note that the conditions above imply that if a sensitivity is nonnull then the corresponding constraint is binding (it is satisfied as an equality), and conversely, if a constraint is not binding (it is satisfied as an inequality) then the corresponding sensitivity is null.

B.2 MIXED-INTEGER LINEAR PROGRAMING

A mixed-integer linear programing problem is a linear programing problem that includes integer variables. If the integer variables are binary (most common case), the problem is denominated 0/1 mixed-integer linear programing problem. On the other hand, if all variables are integer, the problem is denominated strict mixed-integer linear programing problem.

A mixed-integer linear programing problem is then formulated as

$$\text{minimize}_{x_1,\ldots,x_n} \quad \sum_{j=1}^{n} c_j x_j$$

$$\text{subject to} \quad \sum_{j=1}^{n} a_{ij} x_j = b_i \quad \forall i = 1, 2, \ldots, m \tag{B.26}$$
$$x_j \geq 0 \qquad \forall j = 1, 2, \ldots, n$$
$$x_j \in I \qquad j = 1, 2, \ldots, o$$
$$x_j \in \mathbb{R} \qquad j = o + 1, \ldots, n.$$

Textbooks addressing mixed-integer linear programing problems include Nemhauser and Wolsey [8] and Castillo et al. [4]. Solution techniques are based on branch and bound strategies, and, most recently, on efficacious branch and cut strategies [9].

B.2.1 BRANCH AND BOUND

The branch and bound technique is based on solving a sequence of progressively more constrained linear programing problems. These additional constraints partition the feasibility region into subregions facilitating the solution of the original mixed-integer linear programing problem. Initially, upper and lower bounds of the objective function optimal value are established and, then, through branching strategies, the upper bound is progressively decreased and the lower bound increased as better solutions of the mixed-integer linear programing problem and its relaxations are generated. The gap between these upper and lower bounds provides a measure of the quality of the current solution.

If minimizing, note that a lower bound of the objective function optimal value is obtained relaxing integrality constraints and solving the resulting problem. And, an upper bound for the optimal objective function value is the objective function value corresponding to any solution meeting integrality constraints.

The branch and bound algorithm works as follows:

Step 1. Initializing. Establish an upper bound (∞) and a lower bound ($-\infty$) of the objective function optimal value. Solve the initial problem relaxing integrality constraints. If the solution obtained meets integrality conditions, this solution is the minimizer and the procedure concludes. On the other hand, if the relaxed problem is infeasible, the original problem is infeasible and the procedure concludes. Otherwise, the algorithm continues in step 2.

Step 2. Branching. Selecting a variable that does not meet its integrality constraint, two problems (branches) are generated from the previous problem as indicated below. If the value of variable x_k is "a" (e.g., $x_k = 5.3$), the first branch problem is the original problem incorporating the constraint $x_k \leq [a]$, where $[\cdot]$ denotes the integer part of a (e.g., $x_k \leq 5$), and the second branch problem is the original problem incorporating the constraint $x_k \geq [a] + 1$ (e.g., $x_k \geq 6$). The above problems are added to a list of problems to be processed, either sequentially or in parallel. Observe that this partition procedure completely covers the whole feasible region.

Step 3. Solving. To solve the first problem in the list.

Step 4. Updating and bounds. If the solution of the current problem meets integrality constraints and its objective function value is below the current upper bound, the upper bound is updated and the solution stored as the best solution found so far. If integrality constraints are not satisfied and the objective function value is in between the lower and the upper bounds, the lower bound is updated, and the current problem leads through branching to new problems, which are added to the list of problems to be processed.

Step 5. Cutting. If the solution of the considered problem meets integrality constraints, no further branching is possible and the branch is discarded. If the solution of the considered problem does not meet integrality constraints and the corresponding objective function value is higher than the upper bound, no better solution can be found through further branching and thus the branch receives no further consideration. If the current problem is infeasible, no further branching makes sense and thus the branch is discarded.

Step 6. Optimality? If the list of problems to be processed is empty, the procedure concludes; the minimizer is the current best solution. If, on the other hand, the list is not empty, the procedure continues with step 3.

Observe that branching is stopped for three reasons: (i) problem infeasibility, (ii) meeting integrality constraints, and (iii) the objective function value of a noninteger solution being above the current upper bound. We say that branching is stopped due to infeasibility, integrality, and bounds, respectively.

At branching time, different variables are usually available for branching. How to select the most appropriate one generally requires a thorough knowledge of the problem under study. Additionally, the problems waiting to be processed in the processing list can be considered using a deep-first strategy, a wide-first strategy, or a mixed strategy. Selecting the right strategy is crucial to achieve an efficient solution. Physical knowledge of the system under consideration might lead to select the best processing strategy. A deep-first strategy quickly produces infeasible problems that leads to branch elimination, and generates good lower bounds as these bounds are provided by problems progressively more and more constrained. On the other hand, a wide-first strategy allows processing almost identical problems and that might constitute an advantage. Note finally that any processing strategy can be parallelized.

B.3 NONLINEAR PROGRAMING: OPTIMALITY CONDITIONS

We establish below the conditions to be met by a point to be a local minimizer of a nonlinear programing problem. Note that we characterize below local minimizers, not global ones, and recall that in linear programing, global minimizers are characterized, not local ones. Note also that we work with differentiable and sufficiently smooth functions. A nonlinear programing problem might include, or not, constraints, and the objective function or any constraint should be a nonlinear function. A point is a local minimizer if its corresponding objective function value is lower than the objective function of any point in its neighborhood. This definition is further formalized below. We formulate both necessary and sufficient conditions. Necessary conditions are met not only by minimizers but also by other points, that is, a point meeting necessary conditions is not necessarily a minimizer. Sufficient conditions are met only by minimizers, that is, a point meeting sufficient conditions is a minimizer. However, points not meeting sufficient conditions might be minimizers. We consider first- and second-order conditions. First-order conditions involve just first-order derivatives, while second-order conditions involve both first- and second-order derivatives. Finally, it should be noted that understanding optimality conditions helps to understand how most algorithms to solve nonlinear programing problems work.

The material in this section is covered in detail by Bazaraa et al. [10], Bertsekas [11], Gill et al. [12], and Luenberger [7].

B.3.1 UNCONSTRAINED PROBLEMS

The unconstrained problem is

$$\text{minimize}_x \quad f(x) \atop x \in \mathbb{R}^n \tag{B.27}$$

where $f(x) : \mathbb{R}^n \to \mathbb{R}, f(x) \in C^2$ (up to second-order derivatives continuous).

We define below a local minimizer and a strict local minimizer.

Local minimizer: $x^* \in \mathbb{R}^n$ is a local minimizer of $f(x)$ if there exists an $\varepsilon > 0$, such that $f(x^*) \leq f(x), \forall x \in \mathbb{R}^n$ and $|x - x^*| < \varepsilon$. If $f(x^*) < f(x)$, the local minimizer is said to be strict.

First-order optimality conditions for unconstrained problems are provided below.

Theorem B.1: Let $f(x) : \mathbb{R}^n \to \mathbb{R}$, such that $f(x) \in C^1$, and $x \in \mathbb{R}^n$. If $\nabla f(y) \neq 0$, then y is not a local minimizer.

Proof. A proof sketch is provided for the cases of $\mathbb{R}^1$ and $\mathbb{R}^2$. In $\mathbb{R}^1$ the condition above implies a null derivative, that is, a horizontal tangent line; in $\mathbb{R}^2$ the theorem requires the gradient to be null, that is, a horizontal tangent plane. ∎

Second-order sufficient conditions are provided below.

Theorem B.2: Let $f(x) : \mathbb{R}^n \to \mathbb{R}$, such that $f(x) \in C^2$ and $x^* \in \mathbb{R}^n$. If $\nabla f(x^*) = 0$ and $\nabla^2 f(x^*) > 0$, then x^* is a strict local minimizer of $f(x)$.

Proof. A definite positive Hessian matrix ($\nabla^2 f(x^*) > 0$) at the considered point implies a locally convex geometry (round deep valley). This plus the null gradient condition ($\nabla f(x^*) = 0$) imply a local minimizer. ∎

B.3.2 CONSTRAINED PROBLEMS

A constrained nonlinear programing problem has the general form

$$\begin{aligned} \text{minimize}_x \quad & f(x) \\ \text{subject to} \quad & h(x) = 0 \quad g(x) \leq 0 \quad x \in \mathbb{R}^n \end{aligned} \tag{B.28}$$

where $f(x) : \mathbb{R}^n \to \mathbb{R}, h(x) : \mathbb{R}^n \to \mathbb{R}^m, g(x) : \mathbb{R}^n \to \mathbb{R}^p$, and $m \leq n$.

Functions $h(x)$ and $g(x)$ have the form

$$h(x) = \begin{bmatrix} h_1 x) \\ h_2(x) \\ \vdots \\ h_m(x) \end{bmatrix} \quad \text{and} \quad g(x) = \begin{bmatrix} g_1(x) \\ g_2(x) \\ \vdots \\ g_p(x) \end{bmatrix} \tag{B.29}$$

We consider that

$$f(x) \in C^2 \quad h_i(x) \in C^2 \quad \forall i = 1, 2, \ldots, m \quad g_i(x) \in C^2 \quad \forall i = 1, 2, \ldots, p$$

We define below a local minimizer and a strict local minimizer.

Local minimizer: $x^* \in \mathbb{R}^n$, so that $h(x^*) = 0$ and $g(x^*) \leq 0$, is a local minimizer of $f(x)$ subject to $h(x) = 0$, $g(x) \leq 0$, if there exists an $\varepsilon > 0$ such that $f(x^*) \leq f(x)$, $\forall x \in \mathbb{R}^n$ with $h(x) = 0$, $g(x) \leq 0$ and $|x - x^*| < \varepsilon$. If $f(x^*) < f(x)$ the local minimizer is said to be strict.

Regularity conditions defining regular points are stated below. These conditions are important because they allow characterizing minimizers. We denote by Ω the set of binding inequality constraints.

Regular point: $y \in \mathbb{R}^n$ satisfying $h_i(y) = 0$ $(i = 1, 2, \ldots, m)$, $g_j(y) \leq 0$ $(j = 1, 2, \ldots, p)$ is a regular point of these constraints if the gradients of the binding constraints evaluated at y are linearly independent. That is, $\nabla h_i(y)$, $i = 1, 2, \ldots, m$ and $\nabla g_j(y), j \in \Omega$ are linearly independent.

Regularity allows formulating optimality conditions as stated below. Nonregular points might be minimizers but cannot be characterized through the optimality conditions below.

First-order optimality conditions are given below.

Theorem B.3: KKT conditions. If

1. x^* is a local minimizer of $f(x)$ subject to $h(x) = 0$ and $g(x) \leq 0$
2. x^* is a regular point of $h(x) = 0$ and $g(x) \leq 0$

then

there exist vectors $\lambda \in \mathbb{R}^m$ and $\mu \in \mathbb{R}^p, \mu \geq 0$, such that
$$\nabla f(x^*) + \lambda^T \nabla h(x^*) + \mu^T \nabla g(x^*) = 0$$
$$\mu^T g(x^*) = 0$$

Proof. For the sake of illustration, we simply show the validity of these conditions in $\mathbb{R}^2$ for a problem containing just one inequality constraint. This inequality partitions $\mathbb{R}^2$ into two regions, one feasible and the other infeasible. If the curvature of the constraint and the objective function contours coincide (parallel gradients and in the same direction), the minimizer is in the interior of the feasible region and then $\mu = 0$. On the other hand, if the curvature of the constraint and the objective function contours oppose each other (parallel gradient and different directions), the minimizer lies in the boundary of the feasible region, being the constraint binding, and μ should be positive for the gradient and the objective function and the constraints to be linearly dependents. Since only the two situations above are possible, we conclude that $\mu \geq 0$. ∎

Degenerated constraints are characterized below.

Degenerated constraint: A binding constraint is degenerate if its associate multiplier is null and nondegenerate if otherwise.

In what follows, we consider binding constraints to be nondegenerate. This assumption simplifies the derivations below, and, generally, it is not a practical limitation.

Optimality conditions are formulated in a convenient manner through the Lagrangian function defined below.

Lagrangian: The Lagrangian of the constrained problem (B.28) is defined as $\mathcal{L}(x,\lambda,\mu) = f(x) + \lambda^T h(x) + \mu^T g(x)$.

Using the Lagrangian function and considering binding inequality constraints nondegenerate, the first-order optimality conditions are

$$\nabla_x \mathcal{L}(x, \lambda, \mu) = 0 \quad \nabla_\lambda \mathcal{L}(x, \lambda, \mu) = 0$$
$$\text{If } g_i < 0 \Rightarrow \mu_i = 0 \quad \text{and} \quad \text{if } g_i = 0 \Rightarrow \mu_i > 0 \tag{B.30}$$

The above conditions constitute a system of nonlinear equalities and inequalities. Thus, the solution of a nonlinear system of equalities and inequalities is equivalent to solving an equality/inequality constrained nonlinear programing problem. Finally, note that solving nonlinear systems of equalities and inequalities is not simple.

Second-order sufficient conditions are formulated below.

Theorem B.4: If

1. x^* satisfies $h(x^*) = 0$ and $g(x^*) \leq 0$
2. $\lambda \in \mathbb{R}^m$ and $\mu \geq 0 \in \mathbb{R}^p$ satisfy

$$\nabla f(x^*) + \lambda^T \nabla h(x^*) + \mu^T \nabla g(x^*) = 0$$
$$\mu^T g(x^*) = 0$$

3. $\nabla^2 f(x^*) + \sum_{i=1}^{m} \lambda_i \nabla^2 h_i(x^*) + \sum_{j \in \Omega} \mu_j \nabla^2 g_j(x^*) > 0$ on the subspace $\{y : \nabla h(x^*)^T y = 0,$ $\nabla g_j(x^*)^T y = 0 \; \forall j \in, \Omega\}$, where Ω is the set of indices of binding inequality constraints

then

- x^* is a strict local minimizer of $f(x)$ subject to $h(x) = 0$ and $g(x) \le 0$.

∎

The proof of the conditions above can be found, for instance, in [7] or [12].
Multiplier vectors λ and μ are sensitivity parameters as shown below.

Theorem B.5: Consider the nonlinear mathematical programing problem

$$\begin{aligned}
\text{minimize}_x \quad & f(x) \\
\text{subject to} \quad & h(x) = c \quad g(x) \le d
\end{aligned} \tag{B.31}$$

such that $f(x) \in C^2$, $h_i(x) \in C^2$, $(i = 1, 2, \dots, m)$, and $g_j(x) \in C^2$, $(j = 1, 2, \dots, p)$.

For $c = 0$ and $d = 0$, let x^* be such that $h(x^*) = 0$ and $g(x^*) \le 0$ and regular, and consider that x^*, λ^*, and μ^* meet second-order sufficiency conditions for a strict local minimizer and that no binding constraint is degenerate.

Then, $\forall c \in \mathbb{R}^m$ and $\forall d \in \mathbb{R}^p$ in a neighborhood of $(0,0)$, the solution of problem (B.31), x, can be parameterized as a function of c and d, that is, $x(c, d)$ and the equalities below hold.

$$\nabla_c f(x(c,d))]_{(0,0)} = -\lambda^{*T}$$

$$\nabla_d f(x(c,d))]_{(0,0)} = -\mu^{*T}$$

Note that $x(0, 0) = x^*$.

∎

The proof of the theorem above can be found, for instance, in [7] or [12].

The results above show that λ_i/μ_i provides the change of the objective function as a result of a change in the right-hand-side of element c_i/d_i. Actually, λ_i/μ_i are dual variables.

B.4 UNCONSTRAINED PROBLEMS: SOLUTION METHODS

This section reviews algorithms to solve the unconstrained nonlinear programing problem

$$\begin{aligned}
\text{minimize}_x \quad & f(x) \\
\text{subject to} \quad & x \in \mathbb{R}^n
\end{aligned} \tag{B.32}$$

where $f(x) : \mathbb{R}^n \to \mathbb{R}$ and $f(x) \in C^2$.

The material in this section can be extended through the textbooks by Bazaraa et al. [10], Gill et al. [12], and Luenberger [7].

From a practical viewpoint, we briefly describe the following methods: steepest-descent, Newton, quasi-Newton, conjugate directions, and coordinate descent. Unconstrained solution methods are also important to solve constrained problems.

The general working mechanism of these algorithms is as follows: (i) start at an initial point, (ii) identify a descent direction, and (iii) determine the advance over the descent direction so that the objective function value decreases. This process is repeated until no improvement in the objective function can be achieved. Finally, observe that this procedure locates a local minimum.

B.4.1 Steepest-Descent Method

Consider an initial point x_0, the minus-gradient as a descent direction, and an advance step α, that is, $x_\alpha = x_0 - \alpha \nabla f(x_0)$, $\alpha > 0$, then $x_\alpha - x_0 = -\alpha \nabla f(x_0)$.

Using Taylor expansion, the objective function at x_α becomes

$$f(x_\alpha) \approx f(x_0) + \nabla f(x_0)^T(x_\alpha - x_0) = f(x_0) - \nabla f(x_0)^T \alpha \nabla f(x_0)$$
$$= f(x_0) - \alpha[\nabla f(x_0)]^2 \tag{B.33}$$

For α small enough, if $\nabla f(x_0) \neq 0$ we conclude that $f(x_\alpha) < f(x_0)$, that is, the minus-gradient is a descent direction.

Next, we consider descent directions different from the minus-gradient. Consider the step $x_\alpha = x_0 + \alpha d$, such that (i) $\alpha \geq 0$, (ii) $\nabla f(x_0) \neq 0$, (iii) $d \in \mathbb{R}^n$, and (iv) $\nabla f(x_0)^T d < 0$. Using Taylor expansion and evaluating the objective function at x_α we obtain

$$f(x_\alpha) \approx f(x_0) + \nabla f(x_0)^T(x_\alpha - x_0)$$
$$= f(x_0) + \alpha \nabla f(x_0)^T d \tag{B.34}$$

and since $\nabla f(x_0)^T d < 0$ we conclude that $f(x_\alpha) < f(x_0)$ for α small enough.

It is also relevant to consider the step $x_\alpha = x_0 - \alpha D \nabla f(x_0)$, such that (i) $\alpha \geq 0$ and (ii) D is a definite positive matrix. Using Taylor expansion and evaluating the objective function at x_α we get

$$f(x_\alpha) \approx f(x_0) + \nabla f(x_0)^T(x_\alpha - x_0)$$
$$= f(x_0) - \alpha \nabla f(x_0)^T D \nabla f(x_0) \tag{B.35}$$

and since $D > 0$ and $\nabla f(x_0)^T D \nabla f(x_0) > 0$, then $f(x_\alpha) < f(x_0)$ for α small enough.

Therefore, the steps below are descent steps:

Steepest-descent iteration 1: $x_{k+1} = x_k + \alpha_k d_k$, $k = 1, 2, \ldots$, where (i) $\alpha_k \geq 0$, (ii) $\nabla f(x_k)^T d_k < 0$ if $\nabla f(x_k) \neq 0$, and (iii) $d_k = 0$ if $\nabla f(x_k) = 0$.

Steepest-descent iteration 2: $x_{k+1} = x_k - \alpha_k D_k \nabla f(x_k)$, $k = 1, 2, \ldots$, where (i) $\alpha_k \geq 0$ and (ii) D is a positive definite matrix. For D_k equal to the identity matrix, the algorithm above is the so-called steepest-descent algorithm.

Steepest-descent methods work as follows:

Step 1. Select a descent direction, d_k or $-D \nabla f(x_k)$.
Step 2. Select a search step α_k such that the objective function value decreases.

Alternatively to using the minus-gradient direction is selecting a positive definite diagonal matrix D_k to modify that minus-gradient direction. Each diagonal element of D_k modifies specifically each gradient component, and the resulting descent direction might be particularly effective.

The search step is obtained by either using a line search or a rule that guarantees a large enough descent. A line search works as stated below.

Line search: Once fixed x_k and d_k, we need to find α_k being the argument that minimizes the function $\phi(\alpha)$, where $\phi(\alpha) = f(x_k + \alpha d_k)$ with x_k and d_k fixed.

Observe that $\alpha \in \mathbb{R}$, and, therefore, the search is in one dimension.

Possible search techniques include: (i) Fibonacci search, (ii) golden search, (iii) quadratic function fitting, and (iv) cubic function fitting. These methods are described, for instance, in [7].

Alternatively, search rules are easy to implement and computationally cheap. These rules guarantee that the advance is neither too small nor too large. The Armijo and Goldstein rules can be found in [7].

Although laborious, it is not complicated to show that a sequence generated by a steepest-descent algorithm converges to a local minimizer under not very restrictive conditions. That is, convergence is guaranteed and the convergence rate is linear in the following sense. A sequence $\{x_k\}$ generated by a steepest-descent method meets

$$\lim_{k \to \infty} \frac{\|x_{k+1} - x^*\|}{\|x_k - x^*\|} = \frac{1}{a} \quad a > 0 \tag{B.36}$$

That is, the distance to the minimizer at iteration $k+1$ is a time smaller than the distance to the minimizer at iteration k.

Additionally, it can be proved that

$$f(x_{k+1}) - f(x^*) \le \left(\frac{L-l}{L+l}\right)^2 [f(x_k) - f(x^*)] \tag{B.37}$$

where L and l are, respectively, the largest and smallest eigenvalues of the Hessian matrix of the objective function evaluated at the minimizer. Thus, the objective function error at iteration $k+1$ is smaller than or equal to the objective function error at iteration k times a coefficient that depends on the eigenvalues of the Hessian of the objective function at the minimizer. If all eigenvalues are equal ($L = l$), the error vanishes in one step. In $\mathbb{R}^2$, the above implies a sphere-shaped geometry and the one-step error elimination becomes intuitive. If, on the other hand, $L \gg l$, the error decreases from iteration to iteration but by a very small quantity. In $\mathbb{R}^2$, the above implies a narrow ellipsoid-shaped geometry. It can be concluded that it is convenient to pursue a sphere-shaped geometry around the minimizer. However, no specific procedures to achieve this are available. Nevertheless, simple and generally efficacious rules are available.

A reasonable possibility, although not always efficacious, is to perform a change of variables. But how? A promising alternative is to carry out the change of variables in such a manner that all components of the minimizer are of the same order of magnitude. If each component x_j of x is bounded as $a_j \le x_j \le b_j$, the change of variables

$$y_j = \frac{2x_j}{b_j - a_j} - \frac{a_j + b_j}{b_j - a_j} \quad \forall j \tag{B.38}$$

makes each variable y_j to be between -1 and 1. This variable transformation is generally efficacious and advisable if no further information is available.

An alternative variable transformation is

$$y_j = \frac{x_j - a_j}{b_j - a_j} \quad \forall j \tag{B.39}$$

which makes each variable y_j to be between 0 and 1.

The steepest-descent iterative procedure is stopped once one or several of the conditions below are satisfied:

Condition 1. The objective function value does not decrease sufficiently, that is,

$$\frac{|f(x_{k+1}) - f(x_k)|}{|f(x_{k+1})| + 1} < \epsilon_1$$

Condition 2. The solution does not change sufficiently, that is,

$$\frac{\|x_{k+1} - x_k\|}{\|x_{k+1}\| + 1} \leq \epsilon_2$$

Condition 3. The gradient at the considered solution is close enough to zero, that is,

$$\|\nabla f(x_{k+1})\| < \epsilon_3$$

B.4.2 NEWTON AND QUASI-NEWTON METHODS

A second-order Taylor expansion of the objective function value at x_k is

$$f(x) \approx f(x_k) + \nabla f(x_k)^T \Delta x + \frac{1}{2} \Delta x^T \nabla^2 f(x_k) \Delta x \tag{B.40}$$

The trick is to determine Δx so that the quadratic expression above attains its minimizer. The derivative with respect to Δx of Equation B.40 is made equal zero. Thus,

$$\nabla f(x_k) + \nabla^2 f(x_k) \Delta x = 0 \quad \text{and} \quad \Delta x = -[\nabla^2 f(x_k)]^{-1} \nabla f(x_k) \tag{B.41}$$

Therefore, the Newton iteration is

$$x_{k+1} = x_k - [\nabla^2 f(x_k)]^{-1} \nabla f(x_k) \quad k = 1, 2, \ldots. \tag{B.42}$$

The observations below are in order:

Observation 1. Around a local minimizer the Hessian matrix is definite positive and the Newton method is well defined.

Observation 2. Outside the proximity of a local minimizer, the Hessian matrix might not be positive definite or could even be singular.

Observation 3. The Newton method exhibits quadratic convergence in the sense below. A Newton-generated sequence $\{x_k\}$ meets

$$\lim_{k \to \infty} \frac{\|x_{k+1} - x^*\|}{\|x_k - x^*\|^2} = \frac{1}{b} \quad b > 0 \tag{B.43}$$

That is, the distance to the minimizer at iteration $k + 1$ is b times smaller than the squared distance to the minimizer at iteration k.

The conclusion from the above properties is that Newton method is fast if it converges, but it might not converge.

Approximate variants of Newton method include the following:

1. Factorization of the Hessian not at every iteration but just every several iterations, and using available factors until they are updated.
2. Numerical calculation of the Hessian matrix.
3. Using a Hessian matrix containing just diagonal elements (considering zero the off-diagonal element).

Quasi-Newton methods have the same structure as Newton methods, but the inverse of the Hessian matrix is progressively approximated. These methods differ just in the way they approximate the inverse of the Hessian matrix. We sketch below several quasi-Newton methods.

Davidon–Fletcher–Powell quasi-Newton method works as follows:

Step 1. Consider an arbitrary positive definite matrix H_0 and an initial point x_0. Set $k = 0$.

Step 2. Set $g_k = \nabla f(x_k)$ and $d_k = -H_k g_k$.

Step 3. Minimize $f(x_k + \alpha d_k)$ with respect to $\alpha > 0$ to obtain $x_{k+1} = x_k + \alpha_k d_k$, $p_k = \alpha_k d_k$, and also $g_{k+1} = \nabla f(x_{k+1})$. If convergence is attained (use one or several of the stopping rules stated in Section B.4.1), stop; otherwise, continue.

Step 4. Set $q_k = g_{k+1} - g_k$, and compute

$$H_{k+1} = H_k + \frac{p_k p_k^{\mathrm{T}}}{p_k^{\mathrm{T}} q_k} - \frac{H_k q_k q_k^{\mathrm{T}} H_k}{q_k^{\mathrm{T}} H_k q_k} \tag{B.44}$$

Update k and continue with step 2.

It should be noted that the inverse of the Hessian matrix (H_k) is approximated using information pertaining to gradients in successive iterations.

Broyden–Fletcher–Goldfarb–Shanno quasi-Newton method is similar to the above method, but updating the inverse Hessian as

$$H_{k+1} = H_k + \left(1 + \frac{q_k^{\mathrm{T}} H_k q_k}{q_k^{\mathrm{T}} p_k}\right) \frac{p_k p_k^{\mathrm{T}}}{p_k^{\mathrm{T}} q_k} - \frac{p_k q_k^{\mathrm{T}} H_k + H_k q_k p_k^{\mathrm{T}}}{q_k^{\mathrm{T}} p_k} \tag{B.45}$$

If we enforce $H_k = I$ in the expression above, the resulting method does not require storing H_k and its behavior is generally good enough (memoryless quasi-Newton method).

Additionally, it can be shown that the Hessian updating below provides a good scaling for the inverse Hessian (self-scaling quasi-Newton method).

$$H_{k+1} = \left(H_k - \frac{H_k q_k q_k^{\mathrm{T}} H_k}{q_k^{\mathrm{T}} H_k q_k}\right) \frac{p_k^{\mathrm{T}} q_k}{q_k^{\mathrm{T}} H_k q_k} + \frac{p_k p_k^{\mathrm{T}}}{p_k^{\mathrm{T}} q_k} \tag{B.46}$$

Finally, it should be noted that the constructing procedures of the inverse Hessian provided above guarantee that this matrix is positive definite. It should also be noted that the line search in step 3 should be precise to adequately approximate the inverse Hessian.

B.4.3 Conjugate Directions Methods

Conjugate direction methods (see [7] and [12]) are based on extrapolating properties exhibited by quadratic functions. For a quadratic function $f(x) = x^{\mathrm{T}} A x + B x + C$ with A symmetric and definite positive, there exists a set of independent directions (conjugate directions) that, if successively used as descent directions, allows finding the minimizer in a finite number of steps.

To derive the iterative procedure of a conjugate direction method is complex but the actual structure of the iteration is simple. Fletcher–Reeves conjugate direction algorithm works as follows:

Step 1. Consider a point x_0, compute $g_0 = \nabla f(x_0)$ and set $d_0 = -g_0$.

Step 2. For $k = 0, 1, \ldots, n - 1$ carry out the steps below:

i. If convergence is attained (use one or several of the stopping rules stated in Section B.4.1), stop; otherwise, continue.

ii. Set $x_{k+1} = x_k + \alpha_k d_k$ where α_k minimizes in α the function $f(x_k + \alpha d_k)$.
iii. Calculate $g_{k+1} = \nabla f(x_{k+1})$.
iv. While $k \neq n - 1$, do $d_{k+1} = -g_{k+1} + \beta_k d_k$ where

$$\beta_k = \frac{g_{k+1}^{\mathrm{T}} g_{k+1}}{g_k^{\mathrm{T}} g_k} \tag{B.47}$$

Step 3. Replace x_0 with x_n and go to step 2.

Polak–Ribiere conjugate direction method is similar to the above method, but updating parameter β_k as

$$\beta_k = \frac{(g_{k+1} - g_k)^{\mathrm{T}} g_{k+1}}{g_k^{\mathrm{T}} g_k} \tag{B.48}$$

Updating rule (B.48) performs usually better than rule (B.47).

B.4.4 COORDINATE DESCENT METHOD

Methods that do not use derivatives generally converge slowly and their convergence is not guaranteed. These methods are of interest if the derivatives of the objective function are not available. One of such methods is the coordinate descent method that works as follows:

Step 1. Consider a solution x_0.
Step 2. Improve iteratively $f(x_0)$ solving for $i = 1, 2, \ldots, n$

$$\text{minimize}_{x_i} \quad f(x_1, x_2, \ldots, x_n) \tag{B.49}$$

Step 3. Repeat the above step until convergence, or until a large enough number of iterations has been carried out.

Note that not to compute the gradient is computationally an advantage but not being able to use the rich gradient information is a disadvantage.

Note also that a coordinate descent method is a descent-direction method whose successive search directions are single-variable directions.

B.5 CONSTRAINED PROBLEMS: SOLUTION METHODS

This section considers computational procedures to solve the problem

$$\begin{aligned} \text{minimize}_x \quad & f(x) \\ \text{subject to} \quad & x \in \mathcal{S} \end{aligned} \tag{B.50}$$

where $f(x) : \mathbb{R}^n \rightarrow \mathbb{R}, f(x) \in C^2$, and $\mathcal{S} \subset \mathbb{R}^n$.

Set $\mathcal{S}$ represents equality and inequality constraints and bounds on variables. For equality constraints only, the set $\mathcal{S}$ has the form $\mathcal{S} = \{x : h_i(x) = 0, \forall i = 1, 2, \ldots, m\}$. For inequality constraints only, the set $\mathcal{S}$ has the form $\mathcal{S} = \{x : g_i(x) \leq 0, \forall i = 1, 2, \ldots, p\}$.

Additional information on solution methods for constrained nonlinear programing problems can be found in Bazaraa et al. [10], Gill et al. [12], Luenberger [7], and Bertsekas [11].

Among the many techniques available to solve constrained nonlinear programing problems, this appendix considers penalty and barrier methods, augmented Lagrangian methods, and the primal–dual interior point method.

B.5.1 PENALTY AND BARRIER METHODS

B.5.1.1 Penalty Methods

The constrained mathematical programing problem

$$\text{minimize}_x \quad f(x)$$
$$\text{subject to} \quad x \in S \tag{B.51}$$

is substituted by the unconstrained problem below, which incorporates into the objective function a penalty term,

$$\text{minimize}_x \quad f(x) + c\, P(x) \tag{B.52}$$

where $c \in \mathbb{R}$, $c > 0$; $P(x) : \mathbb{R}^n \to \mathbb{R}$ so that $P(x)$ is a continuous function, $P(x) \geq 0 \; \forall x \in \mathbb{R}^n$ and $P(x) = 0$ if and only if $x \in S$.

In the case of just equality constraints, $S = \{x : h_i(x) = 0, \forall i = 1, 2, \ldots, m\}$, and the penalty function $P(x)$ might have the form $P(x) = 1/2 \sum_{i=1}^{m} (h_i(x))^2$. In the case of just inequality constraints, $S = \{x : g_i(x) \leq 0, \forall i = 1, 2, \ldots, p\}$, and the penalty function $P(x)$ might have the form $P(x) = 1/2 \sum_{i=1}^{p} (\text{maximum}\,[0, g_i(x)])^2$.

Note that the search for the minimizer occurs outside the feasible region, which is approached as $c \to \infty$.

Penalty methods work as follows:

Step 1. Set a sequence $\{c_k\}$, $k = 1, 2, \ldots$, so that $c_k > 0$ and $c_{k+1} > c_k$.

Step 2. Build the penalized function $q(c_k, x) = f(x) + c_k P(x)$.

Step 3. For each k solve the problem $\text{minimize}_x \, q\,(c_k, x)$ and obtain x_k. Obtain a new penalty parameter (from the sequence described in step 1), and use x_k as the starting point in step 2. Stop once one or several of the stopping rules stated in Section B.4.1 are satisfied.

It can be easily shown that the limit point of a sequence of points $\{x_k\}$ generated by a penalty procedure is a solution of the original problem (B.51).

Penalty parameters and Lagrange multiplies are closely related as stated below. Consider the problems

$$\text{minimize}_x \quad f(x)$$
$$\text{subject to} \quad h(x) = 0 \tag{B.53}$$

and

$$\text{minimize}_x \quad f(x) + c_k\, \gamma(h(x)) \tag{B.54}$$

where $h(x) : \mathbb{R}^n \to \mathbb{R}^m$ and $\gamma(h) : \mathbb{R}^m \to \mathbb{R}$ is the penalty function.

The first-order optimality condition of problem (B.54) at x_k is

$$\nabla f(x_k) + c_k \nabla_h \gamma(h(x_k)) \nabla h(x_k) = 0 \tag{B.55}$$

and the first-order optimality condition of problem (B.53) at x_k is

$$\nabla f(x_k) + \lambda_k^T \nabla h(x_k) = 0 \tag{B.56}$$

Imposing that both conditions must hold at the minimizer implies

$$\lambda_k^T = c_k \nabla_h \gamma(h(x_k)) \tag{B.57}$$

Note that $x_k \to x^*$ leads to $\lambda_k \to \lambda^*$, where x^* is a regular point of $h(x)$ and a solution of problem (B.53) and λ^* is the multiplier vector associated with that solution.

Inequality constrained problems are considered next.
Consider the problems

$$\begin{aligned} \text{minimize}_x \quad & f(x) \\ \text{subject to} \quad & g(x) \le 0 \end{aligned} \tag{B.58}$$

and

$$\text{minimize}_x \quad f(x) + c_k \gamma(g(x)) \tag{B.59}$$

where $g(x) : \mathbb{R}^n \to \mathbb{R}^m$ and $\gamma(g) : \mathbb{R}^m \to \mathbb{R}$ is the penalty function.
The first-order optimality condition of problem (B.59) at x_k is

$$\nabla f(x_k) + c_k \nabla_g \gamma(g(x_k)) \nabla g(x_k) = 0 \tag{B.60}$$

and the first-order optimality conditions of problem (B.58) at x_k are

$$\begin{aligned} \nabla f(x_k) + \mu_k^\mathrm{T} \nabla g(x_k) = 0 \quad & \mu_k \ge 0 \\ \mu_k^\mathrm{T} g(x_k) = 0 & \end{aligned} \tag{B.61}$$

At the minimizer, it should hold

$$\mu_k^\mathrm{T} = c_k \nabla_g \gamma(g(x_k)) \quad \mu_k \ge 0 \tag{B.62}$$

Note that as $x_k \to x^*$ leads to $\mu_k \to \mu^*$, where x^* is a regular point of $g(x)$ and a solution of problem B.58 and μ^* is the multiplier vector associated with that solution.

Note also that nonbinding constraints have null multipliers.

The structure of the Hessian matrix is analyzed below. Consider the penalized objective function

$$q(c,x) = f(x) + c\gamma(e(x)) \tag{B.63}$$

where $e(x)$ includes both equality and inequality constraints.

The gradient of $q(c, x)$ is

$$\nabla q(c,x) = \nabla f(x) + c \nabla_e \gamma(e(x)) \nabla e(x) \tag{B.64}$$

and the Hessian matrix of $q(c, x)$ is

$$\nabla^2 q(c,x) = \nabla^2 f(x) + c\nabla_e \gamma(e(x)) \nabla^2 e(x) + c\nabla e(x)^\mathrm{T} \nabla_e^2 \gamma(e(x)) \nabla e(x) \tag{B.65}$$

Denominating $\nabla^2 \mathcal{L}(x) = \nabla^2 f(x) + \lambda^\mathrm{T} \nabla^2 e(x)$, where $\lambda^\mathrm{T} = c \nabla_e \gamma(e(x))$, the Hessian becomes

$$\nabla^2 q(c,x) = \nabla^2 \mathcal{L}(x) + c\nabla e(x)^\mathrm{T} \nabla_e^2 \gamma(e(x)) \nabla e(x) \tag{B.66}$$

The observations below are in order:

Observation 1. Matrix $\nabla^2 \mathcal{L}(x)$ approximates the Hessian of the Lagrangian function of problem (B.51) at the minimizer as x_k gets closer to x^*. Therefore, neither the limit matrix nor its condition number depend on c.

Observation 2. Matrix $[c\nabla e(x)^\mathrm{T} \nabla_e^2 \gamma(e(x)) \nabla e(x)]$ approaches infinity as c approaches infinity.

Therefore, as c increases, the Hessian matrix (Equation B.66) becomes worse and worse conditioned.

B.5.1.2   Barrier Methods

To use a barrier method, the feasible region of the considered problem needs to be defined through inequalities.

The constrained problem

$$\text{minimize}_x \quad f(x)$$
$$\text{subject to} \quad x \in \mathcal{S} \tag{B.67}$$

can be substituted by the unconstrained problem below, which incorporates into the objective function a barrier term,

$$\text{minimize}_x \quad f(x) + \frac{1}{c}B(x) \tag{B.68}$$

where $c \in \mathbb{R}$; $c > 0$; $B(x) : \mathbb{R}^n \to \mathbb{R}$ is continuous and $B(x) > 0 \ \forall x \in \mathbb{R}^n$, and $B(x) \to \infty$ if x approaches the feasibility boundary of $\mathcal{S}$.

For just inequality constraints, the barrier function $B(x)$ might take the form

$$B(x) = -\sum_{i=1}^{p} \frac{1}{g_i(x)} \tag{B.69}$$

Barrier methods work as follows:

Step 1. Establish a sequence $\{c_k\}$, $k = 1, 2, \ldots$, so that $c_k > 0$, $c_{k+1} > c_k$.
Step 2. Build the barrier function $r(c_k, x) = f(x) + 1/(c_k B(x))$.
Step 3. Solve the problem

$$\text{minimize}_x \quad r(c_k, x)$$
$$\text{subject to} \quad x \in \text{interior of } \mathcal{S} \tag{B.70}$$

and obtain x_k. The procedure concludes once one or several of the stopping criteria stated in Section B.4.1 are satisfied. Otherwise, the algorithm continues obtaining a new barrier parameter c_k and using x_k as the starting point in step 2.

It is easy to prove that the limit point of a sequence $\{x_k\}$ generated by a barrier method is a solution of problem (B.67).

Barrier penalty parameters and Lagrange multipliers are closely related. Consider the problems

$$\text{minimize}_x \quad f(x)$$
$$\text{subject to} \quad g(x) \le 0 \tag{B.71}$$

and

$$\text{minimize}_x \quad f(x) + \frac{1}{c_k}\eta(g(x)) \tag{B.72}$$

where $g(x) : \mathbb{R}^n \to \mathbb{R}^p$ and $\eta(g) : \mathbb{R}^p \to \mathbb{R}$ is the barrier function.

The first-order optimality condition of problem (B.72) at x_k is

$$\nabla f(x_k) + \frac{1}{c_k}\nabla_g \eta(g(x_k))\nabla g(x_k) = 0 \tag{B.73}$$

and the first-order optimality conditions of problem (B.71) at x_k are

$$\nabla f(x_k) + \mu_k^T \nabla g(x_k) = 0 \quad \mu_k \ge 0$$
$$\mu_k^T g(x_k) = 0 \tag{B.74}$$

Requiring that both conditions are satisfied at the minimizer implies

$$\mu_k^{\mathrm{T}} = \frac{1}{c_k} \nabla_g \eta(g(x_k)) \quad \mu_k \geq 0 \tag{B.75}$$

Note that $x_k \to x^*$ leads to $\mu_k \to \mu^*$, where x^* is a regular point of $g(x)$ and a solution of problem (B.71) and μ^* is the vector of Lagrangian multipliers associated with that solution. Note also that non-binding constraints have null multipliers.

It is simple to prove that the Hessian of the objective function incorporating a barrier function suffers a similar ill-conditioning as an objective function that incorporates a penalty term.

B.5.2 AUGMENTED LAGRANGIAN METHODS

The augmented Lagrangian method or method of the multipliers is considered next. For simplicity, only equality constrained problems are considered. The extension to both equality and inequality constrained problems is described, for instance, in [13]. Consider the problem

$$\begin{aligned} &\text{minimize}_x \quad f(x) \\ &\text{subject to} \quad h(x) = 0 \end{aligned} \tag{B.76}$$

The augmented Lagrangian function is defined as

$$\mathcal{L}(x,\lambda) = f(x) + \lambda^{\mathrm{T}} h(x) + \frac{1}{2} c[h(x)]^2 \tag{B.77}$$

where $c \in \mathbb{R}$ is a positive constant.

Consider also the equality constrained optimization problem

$$\begin{aligned} &\text{minimize}_x \quad f(x) + \lambda^{\mathrm{T}} h(x) \\ &\text{subject to} \quad h(x) = 0 \end{aligned} \tag{B.78}$$

Observe that the quadratic penalty function plus the objective function of problem (B.78) is the augmented Lagrangian of problem (B.76).

Additionally, the equality constrained problem below is considered

$$\begin{aligned} &\text{minimize}_x \quad f(x) + \frac{1}{2} c[h(x)]^2 \\ &\text{subject to} \quad h(x) = 0 \end{aligned} \tag{B.79}$$

Observe that the Lagrangian function of problem (B.79) is the augmented Lagrangian of problem (B.76).

On the basis of the above observations, the ingredients of the augmented Lagrangian method are described below.

The first-order optimality condition of problem (B.76) is

$$\nabla f(x) + \lambda^{*\mathrm{T}} \nabla h(x) = 0 \tag{B.80}$$

where λ^* is the optimal Lagrange multiplier vector.

On the other hand, consider the equality constrained problem

$$\begin{aligned} &\text{minimize}_x \quad f(x) + \lambda_k^{\mathrm{T}} h(x) \\ &\text{subject to} \quad h(x) = 0 \end{aligned} \tag{B.81}$$

The first-order optimality condition for problem (B.81) is

$$\nabla f(x) + \lambda_k^{\mathrm{T}} \nabla h(x) + \bar{\lambda}^{\mathrm{T}} \nabla h(x) = 0 \tag{B.82}$$

where $\bar{\lambda}$ is the optimal Lagrange multiplier vector.

Note that the solutions of the two problems previously formulated should be identical. Considering the first-order optimality conditions for both of them, we conclude that $\bar{\lambda} = \lambda^* - \lambda_k$.

On the other hand, the penalized problem associated with problem (B.81) is

$$\text{minimize}_x \quad f(x) + \lambda_k^T h(x) + \frac{1}{2} c[h(x)]^2 \tag{B.83}$$

The optimal multiplier vector of this problem can be analytically obtained (see Equation B.57) and is

$$\bar{\lambda} = c\nabla_h \left[\frac{1}{2}(h(x))^2 \right] = c\, h(x_k) \tag{B.84}$$

where x_k is the solution of the penalized problem (B.83).

The above Lagrange multiplier vectors are related as follows:

$$\bar{\lambda} = \lambda^* - \lambda_k \quad \text{and} \quad \bar{\lambda} = c\, h(x_k) \tag{B.85}$$

It can be concluded that $\lambda^* = \lambda_k + c\, h(x_k)$, and an appropriate updating rule for λ^* is

$$\lambda_{k+1} = \lambda_k + c\, h(x_k) \tag{B.86}$$

Therefore, the augmented Lagrangian method to solve problem B.76 works as follows:

Step 1. Consider an appropriate penalty constant c and initialize the multiplier vector λ_k.
Step 2. Compute x_k, solving the unconstrained problem:

$$\text{minimize}_x \quad f(x) + \lambda_k^T h(x) + \frac{1}{2} c[h(x)]^2$$

Step 3. Update the multiplier vector as $\lambda_{k+1} \leftarrow \lambda_k + c\, h(x_k)$.
Step 4. Repeat steps 2 and 3 until convergence in λ is attained.

It can be proved that for c large enough, the above procedure converges to a solution of problem (B.76). On the other hand, c should be small enough to avoid ill-conditioning. It is worth mentioning that the penalty parameter c can be increased with the iterations to facilitate convergence.

More sophisticated multiplier updating procedures that incorporate second derivative information are available, see [12] and [13]. The augmented Lagrangian method can be extended to consider inequality constraints as illustrated in [4] and [13].

B.5.3 PRIMAL–DUAL INTERIOR POINT METHODS

Interior point methods were first developed to tackle linear programing problems. However, these algorithms can successfully solve general convex nonlinear problems. The primal–dual interior point algorithm for linear programing problems is developed below. Additional information is available in [14].

Consider the problem

$$\begin{aligned} \text{minimize}_x \quad & c^T x \\ \text{subject to} \quad & Ax = b \quad x \geq 0 \end{aligned} \tag{B.87}$$

where $x, c \in \mathbb{R}^n$, $b \in \mathbb{R}^m$, and $A \in \mathbb{R}^m \times \mathbb{R}^n$.

The dual problem of problem (B.87) is

$$\begin{aligned} \text{maximize}_y \quad & b^T y \\ \text{subject to} \quad & A^T y \leq c \end{aligned} \tag{B.88}$$

Converting inequality constraints into equalities using slack variables, problem (B.88) above becomes

$$\text{maximize}_{y,z} \quad b^T y$$
$$\text{subject to} \quad A^T y + z = c \quad z \geq 0 \tag{B.89}$$

The dual variables are $y \in \mathbb{R}^m$ and $z \in \mathbb{R}^n$.

To eliminate nonnegativity constraint, logarithmic barriers are used, that is,

$$\text{maximize}_{y,z} \quad b^T y + \mu \sum_{j=1}^{n} \ln z_j \tag{B.90}$$
$$\text{subject to} \quad A^T y + z = c$$

The interior point method relies on solving different instances of problem (B.90) for increasing values of the parameter μ. This parameter is defined so that $\mu_0 > \mu_1 > \mu_2 > \cdots > \mu_\infty = 0$. The following result shows the interest of the above algorithm.

Parameter sequence $\{\mu_k\}$ results in a sequence of problems B.90, whose solutions, $\{x_k\}$, progressively approach the solution of the original problem (B.89). This result is shown, for instance, in [14].

The Lagrangian of problem (B.90) is

$$\mathcal{L}(x, y, z, \mu) = b^T y + \mu \sum_{j=1}^{m} \ln z_j - x^T (A^T y + z - c) \tag{B.91}$$

The first-order optimality conditions for problem (B.90) are

$$\nabla_x \mathcal{L}(\cdot) = A^T y + z - c = 0 \quad \nabla_y \mathcal{L}(\cdot) = Ax - b = 0 \quad \nabla_z \mathcal{L}(\cdot) = XZe - \mu e = 0 \tag{B.92}$$

where

$$X = \text{diag}(x_1, x_2, \ldots, x_n) \quad Z = \text{diag}(z_1, z_2, \ldots, z_n) \quad e = (1, 1, \ldots, 1)^T \tag{B.93}$$

Note that the dimension of e is $n \times 1$.

To solve by Newton the nonlinear system of Equation B.92, x, y, and z are substituted by $x + \Delta x$, $y + \Delta y$, and $z + \Delta z$, respectively. Ignoring second-order terms, system (B.92) becomes

$$Z \Delta x + X \Delta z = \mu e - XZe$$
$$A \Delta x = 0 \tag{B.94}$$
$$A^T \Delta y + \Delta z = 0$$

Primal and dual search directions are obtained solving system (B.94) for Δy, Δz, and Δx, that is,

$$\Delta y = -(AXZ^{-1}A^T)^{-1} AZ^{-1} v(\mu)$$
$$\Delta z = -A^T \Delta y \tag{B.95}$$
$$\Delta x = Z^{-1} v(\mu) - XZ^{-1} \Delta z$$

where $v(\mu) = \mu e - XZe$.

The Newton iteration is then

$$x^{k+1} = x^k + \alpha_p \Delta x$$
$$y^{k+1} = y^k + \alpha_d \Delta y \tag{B.96}$$
$$z^{k+1} = z^k + \alpha_d \Delta z$$

where $0 \leq \alpha_p \leq 1$, $0 \leq \alpha_d \leq 1$ are primal and dual search steps, respectively.

The primal search step concerns variable x, while the dual one concerns variables y and z. The selection of α_p and α_d is carried out so that x and z remain positive (y does not need to be positive). To enforce positiveness, parameter σ ($0 < \sigma < 1$) is used. Thus,

$$\alpha_x = \text{minimum}\left\{\frac{-x_i}{\Delta x_i} \text{ so that } \Delta x_i \leq -\delta\right\} \quad \alpha_p = \text{minimum}\{1, \sigma\alpha_x\} \tag{B.97}$$

$$\alpha_z = \text{minimum}\left\{\frac{-z_j}{\Delta z_j} \text{ so that } \Delta z_j \leq -\delta\right\} \quad \alpha_d = \text{minimum}\{1, \sigma\alpha_z\} \tag{B.98}$$

where δ is a tolerance parameter (example, $\delta = 0.0001$) and σ is generally equal to 0.99995.

The above adjustments are important if Δx_i and Δz_j are large enough.

The duality gap is the difference between the objective function value of the primal problem and the objective function value of the dual one. For feasible values x and y, the duality gap is $c^T x - b^T y$. This duality gap is an appropriate metric of proximity to optimality.

If the available initial solution is not feasible, Newton system (B.94) becomes

$$Z\,\Delta x + X\,\Delta z = \mu e - XZe$$
$$A\,\Delta x = b - Ax \tag{B.99}$$
$$A^T\Delta y + \Delta z = c - A^T y - z$$

whose solution is

$$\Delta y = -(AXZ^{-1}A^T)^{-1}(AZ^{-1}v(\mu) - AXZ^{-1}r_d - r_p)$$
$$\Delta z = -A^T\Delta y + r_d \tag{B.100}$$
$$\Delta x = Z^{-1}v(\mu) - XZ^{-1}\Delta z$$

where

$$r_p = b - Ax, \quad r_d = c - A^T y - z \tag{B.101}$$

are primal and dual residuals, respectively.

If the initial solution is infeasible, both feasibility and optimality are achieved simultaneously. That is, the residuals and the duality gap approach zero simultaneously.

Barrier penalty parameter μ can be updated as explained below. Equation B.92 leads to $\mu = z^T x/n$. This updating equations is modified to avoid a quick approximation to the frontier of the feasible region and to achieve a centered descent path through parameter ρ. Thus,

$$\mu = \rho\frac{z^T x}{n} \tag{B.102}$$

Values for ρ are experimentally obtained. A reasonable selection as proposed in [15] is

$$\rho = \begin{cases} 0.1 & \text{if } c^T x > b^T y \\ 0.2 & \text{if } c^T x < b^T y \end{cases} \tag{B.103}$$

A reasonable stoping criterion is to achieve a small enough duality gap, that is,

$$\frac{|c^T x^k - b^T y^k|}{\max\{1, |c^T x^k|\}} \leq \varepsilon \tag{B.104}$$

where ε is a per unit tolerance parameter. The numerator of the above expression is the duality gap, whereas the denominator is the maximum between 1 and the absolute value of the primal objective function. This stopping criterion avoids division by zero.

The extension of the above results to linear programing problems that include upper and lower bounds on variables is easily achieved. See, for instance, [15].

The primal–dual interior point algorithm works as follows:

Step 1. Set $k = 0$ and select an initial solution (x^0, y^0, μ_0) and a tolerance parameter ε. This initial solution can be primal/dual feasible or not.

Step 2. Compute the vector of slack dual variables $z^k = c - A^T y^k$, and build the diagonal matrices $X^k = \text{diag}(x_1^k, x_2^k, \ldots, x_n^k)$ and $Z^k = \text{diag}(z_1^k, z_2^k, \ldots, z_n^k)$.

Step 3. Calculate vector $v(\mu_k) = \mu_k e - X^k Z^k e$, and residual vectors $r_p = Ax - b$ and $r_d = c - A^T y - z$. If the current solution is primal and dual feasible, residual vectors are zero.

Step 4. Calculate search directions Δy, Δz, and Δx, solving the corresponding Newton system. Use Equation B.100.

Step 5. Calculate the steps α_p and α_d. Use Equations B.97 and B.98.

Step 6. Update primal and dual variables using Equation B.96.

Step 7. If the duality gap is small enough (Equation B.104), stop; a solution has been found within a level of accuracy of ε. Otherwise, set $k \leftarrow k + 1$ and continue with step 8 below.

Step 8. Update the barrier parameter μ_k (Equations B.102 and B.103) and continue with step 2.

REFERENCES

1. Brooke, A., Kendrick, D., Meeraus, A., and Raman, R., *GAMS A User's Guide*, GAMS Development Corporation, Washington, 2005.
2. Fourer, R., Gay, D. M., and Kernighan, B. W., *AMPL: A Modeling Language for Mathematical Programming*, Second Edition, Duxbury Press/Brooks/Cole Publishing Company, Boston, 2002.
3. Castillo, E., Conejo, A. J., Pedregal, P., García, R., and Alguacil, N., *Building and Solving Mathematical Programming Models in Engineering and Science*, John Wiley & Sons, New York, 2001.
4. Conejo, A. J., Castillo, E., Mínguez, R., and García Bertrand, R., *Decomposition Techniques in Mathematical Programming; Engineering and Science Applications*, Springer, New York, 2006.
5. Bazaraa, M. S., Jarvis, J. J., and Sherali, H. D., *Linear Programming and Network Flows*, Second Edition, John Wiley and Sons, New York, 1990.
6. Chvátal, V., *Linear Programming*, W. H. Freeman and Company, New York, 1983.
7. Luenberger, D. G., *Linear and Nonlinear Programming*, Second Edition, Addison-Wesley Publishing Company, Reading, MA, 1984.
8. Nemhauser, G. L., and Wolsey, L. A., *Integer and Combinatorial Optimization*, John Wiley and Sons, New York, 1988.
9. Bixby, R. E., Solving real-world linear problems: A decade and more of progress, *Operantions Research*, 50(1), 3–15.
10. Bazaraa, M. S., Sherali, H. O., and Shetty, C. M., *Nonlinear Programming: Theory and Algorithms*, Second Edition, John Wiley and Sons, New York, 1993.
11. Bertsekas, D. P., *Nonlinear Programming*, Athena Scientific, Belmont, MA, 1995.
12. Gill, P. E., Murray, W., and Wright, M. H., *Practical Optimization*, Academic Press Inc., London, 1981.
13. Bertsekas, D. P., *Constrained Optimization and Lagrange Multiplier Methods*, Academic Press Inc., New York, 1982.
14. Wright, S. J., *Primal-Dual Interior-Point Methods*, Siam, PA, 1997.
15. Vanderbei, R. J., *Linear Programming—Foundations and Extensions*, Second Edition, Kluwer Academic Publishers, Boston, MA, 2001.

Appendix C: Dynamic Models of Electric Machines

Luis Rouco

CONTENTS

C.1 INTRODUCTION

This appendix provides dynamic models of electric machines for power system stability studies. Models of both induction and synchronous machines, including simplified models for stability and fault analysis are presented and discussed.

The appendix starts with the model of electric machines' rotor dynamics. The electromagnetic model of the induction machine is then presented and discussed, based on the space vector approach. The model of the synchronous machine is presented as an extension of the induction machine model, incorporating the representation of the field winding.

C.2 ROTOR DYNAMICS MODEL OF ELECTRIC MACHINES

The equation that describes the dynamics of the rotor of an electric machine is the motion equation of a rigid body. In power system stability studies, the rigid body comprises not only the rotor of the electric machine but also the rotor of the generators' prime mover or the motors' mechanical load. For other studies, such as subsynchronous resonance, each rotating mass is represented individually.

Assuming that the electric machine is a generator, the motion equation of the generator rotor is

$$J\frac{d\Omega}{dt} = T_\mathrm{m} - T_\mathrm{e} - T_\mathrm{d} = T_\mathrm{m} - T_\mathrm{e} - K_\mathrm{d}(\Omega - \Omega_0) \tag{C.1}$$

where J is the inertia momentum of the rotating masses expressed in Nm/s^2; Ω is the mechanical angular speed expressed in rad/s; T_m is the mechanical torque expressed in Nm; T_e is the electrical torque expressed in Nm; T_d is the damping torque expressed in Nm, which is assumed to be proportional to the speed deviation from the synchronous speed; K_d is the damping coefficient expressed in Nm/s; Ω_0 is the synchronous mechanical angular speed expressed in rad/s.

Equation C.1 can be expressed in per unit (pu) by dividing it by the base torque T_base:

$$\frac{J}{T_\text{base}}\frac{d\Omega}{dt} = \frac{T_\text{m}}{T_\text{base}} - \frac{T_\text{e}}{T_\text{base}} - \frac{K_\text{d}}{T_\text{base}}(\Omega - \Omega_0) \tag{C.2}$$

where the base torque can be expressed in terms of the apparent power base S_base and the mechanical angular speed base Ω_base (the synchronism speed Ω_0) as follows:

$$T_\text{base} = \frac{S_\text{base}}{\Omega_\text{base}} = \frac{S_\text{base}}{\Omega_0}$$

Hence, Equation C.2 becomes

$$\frac{J\Omega_0^2}{S_\text{base}\Omega_0}\frac{d\Omega}{dt} = \tau_\text{m} - \tau_\text{e} - \frac{K_\text{d}\Omega_0^2}{S_\text{base}\Omega_0}(\Omega - \Omega_0) \tag{C.3}$$

where τ_m and τ_e are, respectively, the mechanic and electric torques in pu.

In terms of the inertia constant H and the damping factor D, Equation C.3 can be written as

$$\frac{2H}{\Omega_0}\frac{d\Omega}{dt} = \tau_\text{m} - \tau_\text{e} - \frac{D}{\Omega_0}(\Omega - \Omega_0) \tag{C.4}$$

where the inertia constant H is defined as the pu rotating kinetic energy at base speed with respect to the base power, that is,

$$H = \frac{E_\text{c}}{S_\text{base}} = \frac{\frac{1}{2}J\Omega_0^2}{S_\text{base}}$$

The units of H are s, since the rotating kinetic energy units are Ws and the base power is in VA. On the other hand, the damping factor D is defined as

$$D = \frac{K_\text{a}\Omega_0^2}{S_\text{base}}$$

If the mechanical angular velocities Ω and Ω_0 are expressed in terms of the electrical angular velocities ω and ω_0, respectively, and the number of poles p ($\omega = p/2\Omega$, and similarly for ω_0), Equation C.4 becomes

$$\frac{2H}{\omega_0}\frac{d\omega}{dt} = \tau_\text{m} - \tau_\text{e} - \frac{D}{\omega_0}(\omega - \omega_0) \tag{C.5}$$

C.3   DYNAMIC MODELS OF INDUCTION MACHINES

C.3.1   EQUATIONS IN PHASE VARIABLES

Figure C.1 depicts the equivalent windings of each phase of a two-pole, wound-rotor, three-phase induction machine, with the rotor rotating at ω_r. At any given time, the angle between the magnetic axes of the stator phase a and rotor phase a is defined as θ. Figure C.2 illustrates the currents and

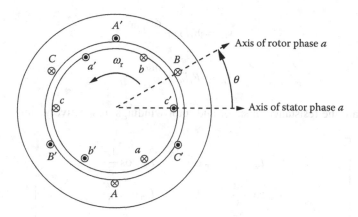

FIGURE C.1   Induction machine windings.

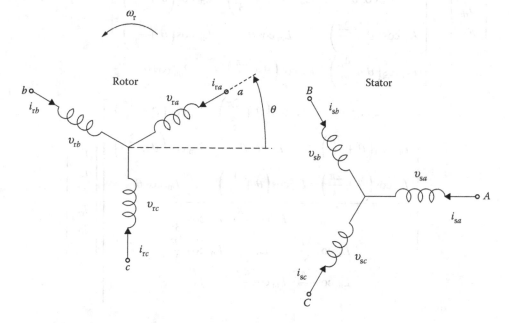

FIGURE C.2   Induction machine stator and rotor circuits.

voltages of each winding, assuming motor operation and a Y connection for both stator and rotor windings. Applying Kirchoff's voltage law to each winding results in

$$v_{sa} = R_s i_{sa} + \frac{d\psi_{sa}}{dt} \tag{C.6}$$

$$v_{sb} = R_s i_{sb} + \frac{d\psi_{sb}}{dt} \tag{C.7}$$

$$v_{sc} = R_s i_{sc} + \frac{d\psi_{sc}}{dt} \tag{C.8}$$

$$v_{ra} = R_r i_{ra} + \frac{d\psi_{ra}}{dt} \tag{C.9}$$

$$v_{rb} = R_r i_{rb} + \frac{d\psi_{rb}}{dt} \tag{C.10}$$

$$v_{rc} = R_r i_{rc} + \frac{d\psi_{rc}}{dt} \tag{C.11}$$

where R_s and R_r are the resistance of stator and rotor windings, respectively, and

$$
\begin{bmatrix} \psi_{sa} \\ \psi_{sb} \\ \psi_{sc} \\ \psi_{ra} \\ \psi_{rb} \\ \psi_{rc} \end{bmatrix} =
\left[
\begin{array}{ccc|ccc}
L_{ss} & L_m \cos\frac{2\pi}{3} & L_m \cos\frac{4\pi}{3} & L_m \cos\theta & L_m \cos\left(\theta + \frac{2\pi}{3}\right) & L_m \cos\left(\theta + \frac{4\pi}{3}\right) \\
L_m \cos\frac{4\pi}{3} & L_{ss} & L_m \cos\frac{2\pi}{3} & L_m \cos\left(\theta + \frac{4\pi}{3}\right) & L_m \cos\theta & L_m \cos\left(\theta + \frac{2\pi}{3}\right) \\
L_m \cos\frac{2\pi}{3} & L_m \cos\frac{4\pi}{3} & L_{ss} & L_m \cos\left(\theta + \frac{2\pi}{3}\right) & L_m \cos\left(\theta + \frac{4\pi}{3}\right) & L_m \cos\theta \\
\hline
L_m \cos\theta & L_m \cos\left(\theta + \frac{4\pi}{3}\right) & L_m \cos\left(\theta + \frac{2\pi}{3}\right) & L_{rr} & L_m \cos\frac{2\pi}{3} & L_m \cos\frac{4\pi}{3} \\
L_m \cos\left(\theta + \frac{2\pi}{3}\right) & L_m \cos\theta & L_m \cos\left(\theta + \frac{4\pi}{3}\right) & L_m \cos\frac{4\pi}{3} & L_{rr} & L_m \cos\frac{2\pi}{3} \\
L_m \cos\left(\theta + \frac{4\pi}{3}\right) & L_m \cos\left(\theta + \frac{2\pi}{3}\right) & L_m \cos\theta & L_m \cos\frac{2\pi}{3} & L_m \cos\frac{4\pi}{3} & L_{rr}
\end{array}
\right]
\begin{bmatrix} i_{sa} \\ i_{sb} \\ i_{sc} \\ i_{ra} \\ i_{rb} \\ i_{rc} \end{bmatrix}
\tag{C.12}
$$

with L_m representing the mutual (magnetizing) inductance of both stator and rotor windings, and L_{ss} and L_{rr} representing the stator and the rotor self-inductances, respectively, which result from the sum of the corresponding leakage and magnetizing inductances:

$$L_{ss} = L_s + L_m$$
$$L_{rr} = L_r + L_m$$

C.3.2 Equations in Complex Form

Equations C.6 through C.11 are a set of linear, time-varying differential equations, due to the variation of the angle θ with time ($d\theta/dt = \omega_r$). These equations can be converted into a set of linear, time-invariant differential equations applying an appropriate variable transformation, that is, Park's transformation, which is the approach followed in classical books on power system dynamics. In this appendix, we use a transformation more commonly used in modern textbooks on vector control of electric machines [1,2].

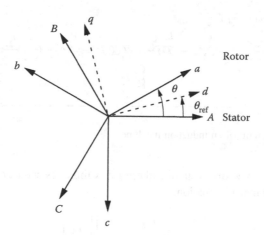

FIGURE C.3 Reference frame of an induction machine.

The transformation used here is characterized by the fact that the resulting variables, the so-called space vectors, are complex magnitudes. Thus, a three-phase system of either stator or rotor variables (voltages, currents, and fluxes) can be represented, respectively, by space vectors with respect to a reference frame (see Figure C.3) according to

$$\vec{x}_s = ke^{j\theta_{ref}}\left(x_{sa} + x_{sb}e^{j\frac{2\pi}{3}} + x_{sb}e^{j\frac{4\pi}{3}}\right) \tag{C.13}$$

$$\vec{x}_r = ke^{j(\theta_{ref}-\theta)}\left(x_{ra} + x_{rb}e^{j\frac{2\pi}{3}} + x_{rb}e^{j\frac{4\pi}{3}}\right) \tag{C.14}$$

where k is a scaling factor.

If Equations C.6 through C.8 are multiplied, respectively, by $ke^{-j\theta_{ref}}$, $ke^{-j\theta_{ref}}e^{j\frac{2\pi}{3}}$, and $ke^{-j\theta_{ref}}e^{j\frac{4\pi}{3}}$, and then added up, the following complex equation is obtained:

$$\begin{aligned}\vec{v}_s &= R_s\vec{i}_s + e^{-j\theta_{ref}}\frac{d\left(e^{j\theta_{ref}}\vec{\psi}_s\right)}{dt} \\ &= R_s\vec{i}_s + \frac{d\vec{\psi}_s}{dt} + j\frac{d\theta_{ref}}{dt}\vec{\psi}_s\end{aligned} \tag{C.15}$$

This equation shows two components of the induced stator-winding voltages: the speed voltages $j\frac{d\theta_{ref}}{dt}\vec{\psi}_s$, and the "transformer" voltages $\frac{d\vec{\psi}_s}{dt}$. On the other hand, if Equations C.9 through C.11 are multiplied, respectively, by $ke^{-j(\theta_{ref}-\theta)}$, $ke^{-j(\theta_{ref}-\theta)}e^{j\frac{2\pi}{3}}$, and $ke^{-j(\theta_{ref}-\theta)}e^{j\frac{4\pi}{3}}$, and then added up, the following complex equation is obtained for the rotor voltages:

$$\begin{aligned}\vec{v}_r &= R_r\vec{i}_r + e^{-j(\theta_{ref}-\theta)}\frac{d\left(e^{-j(\theta_{ref}-\theta)}\vec{\psi}_r\right)}{dt} \\ &= R_r\vec{i}_r + \frac{d\vec{\psi}_r}{dt} + j\frac{d(\theta_{ref}-\theta)}{dt}\vec{\psi}_r\end{aligned} \tag{C.16}$$

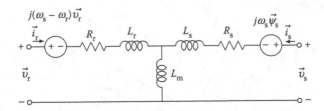

FIGURE C.4 Equivalent circuit of an induction machine.

Expressions of the space vectors of stator and rotor flux linkages can be obtained in a similar way, resulting in the following matrix equation:

$$
\begin{bmatrix} \vec{\psi}_s \\ \vec{\psi}_r \end{bmatrix} = \begin{bmatrix} L_{ss} & L_m \\ L_m & L_{rr} \end{bmatrix} \begin{bmatrix} \vec{i}_s \\ \vec{i}_r \end{bmatrix}
\tag{C.17}
$$

Assuming that the reference frame is rotating at synchronous speed, that is, $\frac{d\theta_{ref}}{dt} = \omega_s$, Equations C.15 through C.17 can be represented as the equivalent circuit depicted in Figure C.4. This circuit can also be represented as the two circuits shown in Figure C.5, one for the d-axis and another for the q-axis.

C.3.3 Electromagnetic Torque

The rotor output power expressed in terms of the space vectors of rotor voltage and current is

$$
P_r = -\mathrm{Re}\left\{ \vec{v}_r \vec{i}_r^* \right\}
\tag{C.18}
$$

If the expression of $\vec{v}_r$ in Equation C.16 is incorporated into this equation, it results in

$$
P_r = -\mathrm{Re}\left\{ R_r \vec{i}_r \vec{i}_r^* + \frac{d\vec{\psi}_r}{dt} \vec{i}_r^* + j(\omega_s - \omega_r)\vec{\psi}_r \vec{i}_r^* \right\}
\tag{C.19}
$$

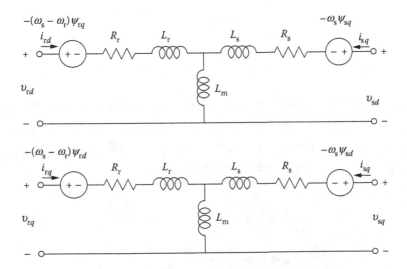

FIGURE C.5 Equivalent circuits of an induction machine for the d- and q-axes.

TABLE C.1
Eigenvalues and Participation Factors for the Detailed Fifth-Order Linear Model of an Induction Machine

	$\mu_{1,2}$ $-1.2839 \pm j314.1109$	$\mu_{2,3}$ $-7.0598 \pm j12.3228$	μ_5 -11.2618
ψ_{sd}	**0.5001**	0.0000	0.0002
ψ_{sq}	**0.5001**	0.0001	0.0000
ψ_{rd}	0.0001	**0.6002**	0.0959
ψ_{rd}	0.0001	0.0550	**1.0290**
$s = \omega_s - \omega_r$	0.0000	0.5535	0.0670

Three terms can be identified here: the rotor losses $-R_r \vec{i}_r^2$, the variation of the magnetic energy $\mathrm{Re}\{-d\vec{\psi}_r/dt\,\vec{i}_r^*\}$, and the electromagnetic power $\mathrm{Re}\{-j(\omega_s - \omega_r)\vec{\psi}_r\,\vec{i}_r^*\}$. The electromagnetic torque can be simply obtained dividing the electromagnetic power by the rotor speed minus the speed of the rotating frame, that is,

$$\tau_e = \frac{P_e}{\omega_s - \omega_r} = -\mathrm{Re}\left\{j\vec{\psi}_r\,\vec{i}_r^*\right\} = -\mathrm{Im}\left\{\vec{\psi}_r^*\,\vec{i}_r\right\} \tag{C.20}$$

C.3.4   MODEL FOR POWER SYSTEM STABILITY STUDIES

The model of an induction machine for power system stability studies neglects the dynamics of the stator transients; in other words, the derivative of the stator flux space vector is assumed to be equal to zero. This assumption can be justified by means of an eigenvalue analysis of the linearized equations of the detailed model of an induction generator connected to an infinite bus. Table C.1 shows the eigenvalues and corresponding participation factors (see Chapter 10) obtained for this system model.* Two distinct dynamic patterns can be observed here, as highlighted in boldface [3]: one close to the fundamental-frequency associated with the stator fluxes and characterized by eigenvalues $\mu_{1,2}$, and another in the frequency range of stability studies and characterized by eigenvalues $\mu_{3,4}$ and μ_5; the complex pair $\mu_{3,4}$ corresponds to the rotor dynamics in the d-axis (excitation flux and slip s), whereas the real μ_5 corresponds to the rotor dynamics in the q-axis. Observe the little effect of the stator fluxes on the rotor variables and vice versa.

The aforementioned assumption yields the following set of differential-algebraic equations (DAE) for the rotor dynamics and the electromagnetic model of the induction machine (third-order approximate model):

$$\frac{d\omega_r}{dt} = \frac{\omega_0}{2H}(\tau_e - \tau_m) \tag{C.21}$$

$$\frac{d\vec{\psi}_r}{dt} = -R_r \vec{i}_r - j(\omega_s - \omega_r)\vec{\psi}_r \tag{C.22}$$

$$0 = -\vec{v}_s + R_s \vec{i}_s + j\omega_s \vec{\psi}_s \tag{C.23}$$

---

* The machine is assumed to be operating at rated conditions, with parameters: $R_s = 0.001$ pu, $L_s = 0.10$ pu, $L_m = 4$ pu, $R_r = 0.01$ pu, $L_r = 0.15$ pu, $H = 3$ s.

$$\begin{bmatrix} 0 \\ 0 \end{bmatrix} = \begin{bmatrix} \vec{\psi}_s \\ \vec{\psi}_r \end{bmatrix} + \begin{bmatrix} L_{ss} & L_m \\ L_m & L_{rr} \end{bmatrix} \begin{bmatrix} \vec{i}_s \\ \vec{i}_r \end{bmatrix}$$ (C.24)

$$0 = -\tau_e - \text{Im}\{-\vec{\psi}_r^* \vec{i}_r\}$$ (C.25)

The incorporation of an induction machine model into a power system stability simulation tool requires considering the stator voltages and currents as "interface" variables. Thus, Equations C.22 through C.25 can be rewritten in a more compact form as

$$\dot{x} = f\left(x, z, u, \vec{v}_s\right)$$ (C.26)

$$0 = g\left(x, z, u, \vec{v}_s\right)$$ (C.27)

$$\vec{i}_s = h\left(x, z, u, \vec{v}_s\right)$$ (C.28)

where

$$x^T = [\omega_r \ \vec{\psi}_r]$$

$$z^T = [\vec{i}_r \ \tau_e]$$

$$u^T = \tau_m$$

Figure C.6 shows a comparison of the dynamic response of the detailed fifth-order and the reduced third-order models of an induction machine operating as a generator for a 100-ms three-phase solid fault at its terminals. Machine slip and electromagnetic torque are displayed; the solid line corresponds to the fifth-order model, whereas the dash line corresponds to the third-order model. Observe that the reduced model does not contain the fundamental-frequency oscillations that the detailed model shows; however, the general dynamic response of the machine is adequately approximated by the simplified model.

C.4 DYNAMIC MODELS OF SYNCHRONOUS MACHINES

C.4.1 EQUATIONS IN d- AND q-AXES VARIABLES

The synchronous machine model is developed here from the aforementioned induction machine model. This is possible since a synchronous machine contains only an additional winding in the rotor with respect to the induction machine, that is, the field winding, which is located in the d-axis; the damper windings are basically the short-circuited windings on the machine's rotor. Thus, the induction machine model represented by the d- and q-axes equivalent circuits in Figure C.5 are modified assuming

- The reference frame is considered to be on the rotor, that is, $d\theta_{ref}/dt = \omega_r$. Hence, no speed voltages are induced in the rotor.
- A new branch is added to the d-axis equivalent circuit corresponding to the field winding.
- The reluctance of the magnetic paths on the d- and q-axes are different due to pole saliency. Therefore, different values of the magnetizing inductance L_m are used in the d- and q-axes equivalent circuits, that is, L_{md} and L_{mq}.

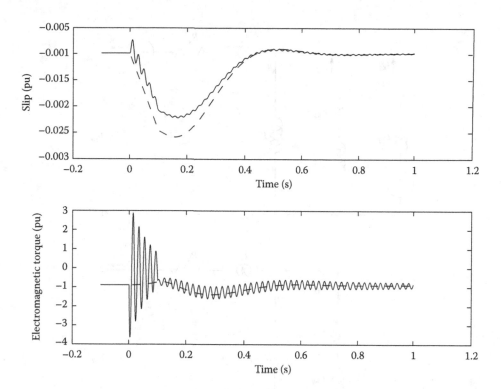

FIGURE C.6 Comparison of the detailed fifth-order and reduced third-order models of an induction machine for a 100-ms three-phase solid fault at its terminals.

- The resistance and the leakage inductances of the rotor windings are different on the d- and q-axes.

Figure C.7 shows the equivalent circuits of the synchronous machine for both d- and q-axes; these circuits result from the previous considerations, assuming generator operation. From these equivalent circuits, the voltage equations of the synchronous machine model are

$$v_{sd} = -R_s i_{sd} + \frac{d\psi_{sd}}{dt} - \omega_r \psi_q \tag{C.29}$$

$$v_{sq} = -R_s i_{sq} + \frac{d\psi_{sq}}{dt} + \omega_r \psi_d \tag{C.30}$$

$$e_{fd} = R_{fd} i_{fd} + \frac{d\psi_{fd}}{dt} \tag{C.31}$$

$$0 = R_{kd} i_{kd} + \frac{d\psi_{kd}}{dt} \tag{C.32}$$

$$0 = R_{kq} i_{kq} + \frac{d\psi_{kq}}{dt} \tag{C.33}$$

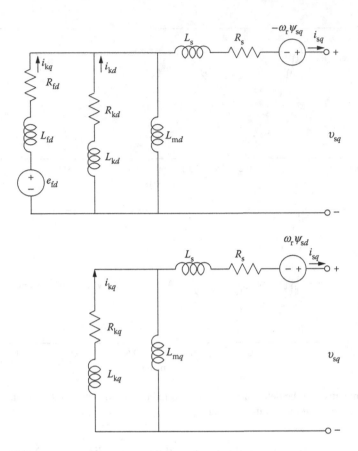

FIGURE C.7   Equivalent circuits of a synchronous machine for the *d*- and *q*-axes.

where

$$
\begin{bmatrix}
\psi_{sd} \\
\psi_{fd} \\
\psi_{kd} \\
\psi_{sq} \\
\psi_{kq}
\end{bmatrix}
=
\begin{bmatrix}
L_s + L_{md} & L_{md} & L_{md} & 0 & 0 \\
L_{md} & L_{fd} + L_{md} & L_{md} & 0 & 0 \\
L_{md} & L_{md} & L_{kd} + L_{md} & 0 & 0 \\
0 & 0 & 0 & L_s + L_{mq} & L_{mq} \\
0 & 0 & 0 & L_{mq} & L_{kq} + L_{mq}
\end{bmatrix}
\begin{bmatrix}
-i_{sd} \\
i_{fd} \\
i_{kd} \\
-i_{sq} \\
i_{kq}
\end{bmatrix}
\tag{C.34}
$$

On this detailed model of the synchronous machine, the field and the damper rotor windings are represented as two circuits on the *d*- and *q*-axes. Damper windings can be represented with different degrees of modeling detail. In the IEEE Standard [4], a number of alternatives are discussed, depending on the number of circuits represented on each axis; the model presented here corresponds to Model 2.1 in this Standard.

The detailed models presented here for both induction and synchronous machines are in terms of the circuit parameters. The circuit parameters of induction machines are usually provided by manufacturers. However, manufacturers of synchronous machines usually provide the transient and subtransient reactances and time constants instead of leakage and magnetizing reactances and resistances. Section C.5 presents and discusses the two simplified models of the synchronous machine from which these transient and subtransient reactances can be obtained. The interested reader can further refer to [5] for the expressions of the time constants.

C.4.2   ELECTROMAGNETIC TORQUE

The expression of the electromagnetic torque of the induction machine has been presented above in terms of rotor quantities. In contrast, the expression of the electromagnetic torque of the synchronous machine is given here in terms of stator quantities. Thus, the stator input power expressed in terms of the space vectors of stator voltage and current can be written as

$$
\begin{aligned}
P_s &= \mathrm{Re}\left\{\vec{v}_s \vec{i}_s^*\right\} \\
&= \mathrm{Re}\left\{-R_s \vec{i}_s \vec{i}_s^* + j\omega_r \vec{\psi}_s \vec{i}_s^* + \frac{d\vec{\psi}_s}{dt}\vec{i}_s^*\right\}
\end{aligned}
\tag{C.35}
$$

The electromagnetic power term can be easily identified in this equation; hence, the electromagnetic torque is computed dividing this power by the rotor speed:

$$
\tau_e = \frac{\mathrm{Re}\{j\omega_r \vec{\psi}_s \vec{i}_s^*\}}{\omega_r} = \mathrm{Im}\{\vec{\psi}_s \vec{i}_s^*\}
\tag{C.36}
$$

C.4.3   ROTOR ANGLE

The rotor angle δ_r is defined as the angle between the terminal voltage phasor and the q-axis. Hence, the components of the terminal voltage in the d- and q-axes rotating at rotor speed are

$$
v_{sd} = v_s \sin\delta_r
\tag{C.37}
$$

$$
v_{sq} = v_s \cos\delta_r
\tag{C.38}
$$

The rotor angle is related to the rotor speed as follows:

$$
\frac{d\delta_r}{dt} = \omega - \omega_0
\tag{C.39}
$$

C.4.4   MODEL FOR POWER SYSTEM STABILITY STUDIES

As in the case of the induction machine, the model of a synchronous machine for power system stability also neglects the dynamics of the stator transients. This assumption can be justified as well by means of an eigenvalue analysis of the linear set of equations corresponding to the detailed model of a synchronous generator connected to an infinite bus. The participation factors and eigenvalues shown in Table C.2 also present two distinct dynamic patterns:* one close to fundamental-frequency associated with the stator fluxes and characterized by eigenvalues $\mu_{1,2}$, and another in the frequency range of stability studies and characterized by eigenvalues $\mu_4 - \mu_7$; the complex pair $\mu_{5,6}$ corresponds to the rotor dynamics, and the real eigenvalues μ_4 and μ_7 are associated with the dynamics of the field and the damper windings.

If the rotor speed deviations with respect to the synchronous speed are assumed small, which is typically the case in stability studies, the speed voltages $j\omega_r \vec{\psi}_s$ can be approximated by $j\omega_s \vec{\psi}_s$. From this and the abovementioned assumptions, a fifth-order approximate model represented by

---

* The machine operates at unity power factor and rated conditions, with parameters: $R_s = 0.005$ pu, $L_s = 0.1$ pu, $L_d = 1.05$ pu, $L_q = 0.7$ pu, $L'_d = 0.35$ pu, $L''_d = 0.25$ pu, $L''_q = 0.3$ pu, $T'_{d0} = 5$ s, $T''_{d0} = 0.03$ s, $T''_{q0} = 0.05$ s, $H = 3$ s.

TABLE C.2
Eigenvalues and Participation Factors for the Detailed Seventh-Order Linear Model of a Synchronous Machine

	$\mu_{1,2}$ $-5.7189 \pm j313.8063$	μ_3 -45.1916	μ_4 -46.6577	$\mu_{5,6}$ $-0.8950 \pm j9.4084$	μ_7 -0.4011
ψ_{sd}	**0.5009**	0.0042	0.0001	0.0080	0.0003
ψ_{sq}	**0.5010**	0.0003	0.0000	0.0058	0.0007
ψ_{fd}	0.0000	0.0002	0.0005	0.0104	**0.9961**
ψ_{kd}	0.0004	0.2136	**0.8023**	0.0198	0.0028
ψ_{kq}	0.0007	**0.8611**	0.1948	0.0682	0.0029
ω_r	0.0000	0.0369	0.0017	**0.5192**	0.0009
δ_r	0.0008	0.0341	0.0006	**0.5198**	0.0020

the following equations can be obtained:

$$\frac{d\delta_r}{dt} = \omega_r - \omega_0 \tag{C.40}$$

$$\frac{d\omega_r}{dt} = \frac{\omega_0}{2H}(\tau_m - \tau_e) \tag{C.41}$$

$$\frac{d\psi_{fd}}{dt} = -R_{fd}i_{fd} + e_{fd} \tag{C.42}$$

$$\frac{d\psi_{kd}}{dt} = -R_{kd}i_{kd} \tag{C.43}$$

$$\frac{d\psi_{kq}}{dt} = -R_{kq}i_{kq} \tag{C.44}$$

$$0 = -v_d + R_s i_d - \omega_s \psi_q \tag{C.45}$$

$$0 = -v_q + R_s i_q + \omega_s \psi_d \tag{C.46}$$

$$\begin{bmatrix} 0 \\ 0 \\ 0 \\ 0 \\ 0 \end{bmatrix} = \begin{bmatrix} \psi_{sd} \\ \psi_{fd} \\ \psi_{kd} \\ \psi_{sq} \\ \psi_{kq} \end{bmatrix}$$

$$+ \begin{bmatrix} L_s + L_{md} & L_{md} & L_{md} & 0 & 0 \\ L_{md} & L_{fd} + L_{md} & L_{md} & 0 & 0 \\ L_{md} & L_{md} & L_{kd} + L_{md} & 0 & 0 \\ 0 & 0 & 0 & L_s + L_{mq} & L_{mq} \\ 0 & 0 & 0 & L_{mq} & L_{kq} + L_{mq} \end{bmatrix} \begin{bmatrix} -i_{sd} \\ i_{fd} \\ i_{kd} \\ -i_{sq} \\ i_{kq} \end{bmatrix} \tag{C.47}$$

$$0 = -\tau_e + \psi_{sd}i_{sq} - \psi_{sq}i_{sd} \tag{C.48}$$

All variables are referred to a reference frame rotating at synchronous speed. Stator voltage and current components in the rotating d- and q-axes reference frame are then transformed to a synchronously rotating frame R–I (complex phasor plane) using the rotor angle δ_r, which is the angle between the real q-axis and the R-axis. This transformation is accomplished by means of the following equations:

$$\begin{bmatrix} V_R \\ V_I \end{bmatrix} = \begin{bmatrix} \sin \delta_r & \cos \delta_r \\ -\cos \delta_r & \sin \delta_r \end{bmatrix} \begin{bmatrix} v_{sd} \\ v_{sq} \end{bmatrix} \tag{C.49}$$

$$\begin{bmatrix} I_R \\ I_I \end{bmatrix} = \begin{bmatrix} \sin \delta_r & \cos \delta_r \\ -\cos \delta_r & \sin \delta_r \end{bmatrix} \begin{bmatrix} i_{sd} \\ i_{sq} \end{bmatrix} \tag{C.50}$$

The aforementioned machine equation can be written in a DAE form as follows:

$$\dot{x} = f(x, z, u, V) \tag{C.51}$$

$$0 = g(x, z, u, V) \tag{C.52}$$

$$I = h(x, z, u, V) \tag{C.53}$$

where

$$x^T = \begin{bmatrix} \delta & \omega & \psi_{fd} & \psi_{kd} & \psi_{kq} \end{bmatrix}$$
$$z^T = \begin{bmatrix} \psi_{sd} & \psi_{sq} & i_{sd} & i_{fd} & i_{kd} & i_{sq} & i_{kq} & v_{sd} & v_{sq} & \tau_e \end{bmatrix}$$
$$u^T = \begin{bmatrix} \tau_m & e_{fd} \end{bmatrix}$$
$$V^T = \begin{bmatrix} V_R & V_I \end{bmatrix}$$
$$I^T = \begin{bmatrix} I_R & I_I \end{bmatrix}$$

Figure C.8 illustrates a comparison of the dynamic response of the detailed seventh-order and reduced fifth-order models of a synchronous machine working as generator for a 100-ms three-phase solid fault at its terminals. Machine speed and electromagnetic torque are displayed; the solid line corresponds to the seventh-order model, whereas the dash line depicts the fifth-order model response. Observe that the more detailed model contains fundamental-frequency oscillations that are not found in the reduced model; however, the latter adequately approximates the overall dynamic trend of the machine.

C.5 SYNCHRONOUS MACHINE SIMPLIFIED MODELS

This section discusses the *subtransient* and *transient* models of the synchronous machine, which are often used in stability studies. Both models assume that rotor fluxes are constant. The subtransient model considers two circuits on each axis, whereas the transient model assumes one circuit on each axis; the subtransient model corresponds to Model 2.2 in the IEEE Standard [4], while the transient model corresponds to Model 1.1.

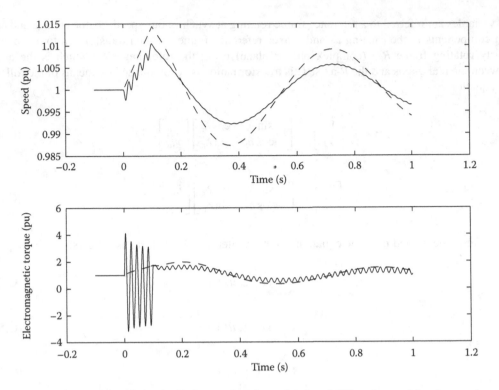

FIGURE C.8 Comparison of the detailed seventh-order and reduced fifth-order models of a synchronous machine for a 100-ms three-phase solid fault at its terminals.

C.5.1 SUBTRANSIENT MODEL

The stator equations of the synchronous machine in the d- and q-axes are

$$v_{sd} = -R_s i_{sd} - \omega_s \psi_{sq} \tag{C.54}$$

$$v_{sq} = -R_s i_{sq} + \omega_s \psi_{sd} \tag{C.55}$$

Assuming that the rotor fluxes remain constant, the following expressions of the stator fluxes in the d- and q-axes can be obtained from the equivalent circuits depicted in Figure C.9:

$$\psi_{sd} = -\left(L_s + \frac{1}{\frac{1}{L_{md}} + \frac{1}{L_{fd}} + \frac{1}{L_{kd}}}\right) i_{sd}$$
$$-\frac{1}{\frac{1}{L_{md}} + \frac{1}{L_{fd}} + \frac{1}{L_{kd}}}\left(\frac{\psi_{fd}}{L_{fd}} + \frac{\psi_{kd}}{L_{kd}}\right) \tag{C.56}$$

$$\psi_{sq} = -\left(L_s + \frac{1}{\frac{1}{L_{mq}} + \frac{1}{L_{kq1}} + \frac{1}{L_{kq2}}}\right) i_{sq}$$
$$-\frac{1}{\frac{1}{L_{mq}} + \frac{1}{L_{kq1}} + \frac{1}{L_{kq2}}}\left(\frac{\psi_{kq1}}{L_{kq1}} + \frac{\psi_{kq2}}{L_{kq2}}\right) \tag{C.57}$$

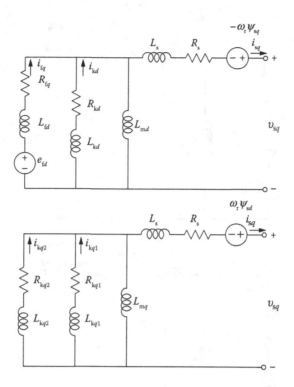

FIGURE C.9 Equivalent circuits of the d- and q-axes subtransient model of a synchronous machine.

If Equations C.56 and C.57 are substituted in Equations C.54 and C.55, one obtains

$$v_{sd} = -R_s i_{sd} + X_q'' i_{sd} + E_d'' \tag{C.58}$$

$$v_{sq} = -R_s i_{sq} - X_d'' i_{sq} + E_q'' \tag{C.59}$$

where X_d'' and X_q'' are, respectively, the subtransient reactances on the d- and q-axes. If rotor saliency is neglected, that is, $X_d'' = X_q''$, Equations C.58 and C.59 can be rewritten as the single complex equation:

$$\mathcal{E}_s'' = (R_s + jX'')\mathcal{I}_s + \mathcal{U}_s \tag{C.60}$$

where X'' is the subtransient reactance, and $\mathcal{E}_s''$ is the voltage behind the subtransient impedance.

Subtransient models are typically used to represent synchronous machines in short-circuit studies.

C.5.2 TRANSIENT MODEL

Assuming that the rotor fluxes remain constant, expressions of the stator fluxes in the d- and q-axes can be obtained from the equivalent circuits of Figure C.10 as follows:

$$\psi_{sd} = -\left(L_s + \frac{1}{\frac{1}{L_{md}} + \frac{1}{L_{fd}}}\right) i_{sd} - \frac{1}{\frac{1}{L_{md}} + \frac{1}{L_{fd}}} \frac{\psi_{fd}}{L_{fd}} \tag{C.61}$$

$$\psi_{sq} = \left(L_s + \frac{1}{\frac{1}{L_{mq}} + \frac{1}{L_{kq}}}\right) i_{sq} + \frac{1}{\frac{1}{L_{mq}} + \frac{1}{L_{kq}}} \frac{\psi_{kq}}{L_{kq}} \tag{C.62}$$

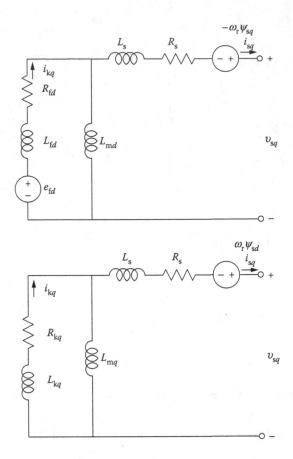

FIGURE C.10 Equivalent circuits of the d- and q-axes transient model of a synchronous machine.

If Equations C.61 and C.62 are substituted in Equations C.54 and C.55, the following expressions can be obtained:

$$v_{sd} = -R_s i_{sd} - X_q' i_{sq} + E_d' \tag{C.63}$$

$$v_{sq} = -R_s i_{sq} + X_d' i_{sd} + E_q' \tag{C.64}$$

where X_d' and X_q' are, respectively, the transient reactances on the d- and q-axes. If rotor saliency is neglected, that is, $X_d' = X_q'$, Equations C.63 and C.64 yield the following single complex equation:

$$\mathcal{E}_s' = (R_s + jX')\mathcal{I}_s + \mathcal{U}_s \tag{C.65}$$

where X' is the transient reactance, and $\mathcal{E}_s'$ is the voltage behind the transient impedance.

Transient models are used to represent synchronous machines in simplified power system stability studies.

REFERENCES

1. Leonard W., *Control of Electric Drives*, Springer-Verlag, Berlin, 1996.
2. Novotny D. W., and Lipo T. A., *Vector Control and Dynamics of AC Drives*, Oxford University Press, Oxford, 1996.

3. Pérez-Arriaga I. J., Verghese G. C., and Schweppe F. C., Selective modal analysis with applications to electric power systems. Part I: Heuristic introduction. Part II: The dynamic stability problem, *IEEE Transactions on Power Apparatus and Systems*, PAS-101(9), 3117–3134, September 1982.
4. IEEE Power Engineering Society, *IEEE Guide for Synchronous Generator Modeling Practices in Stability Studies*, IEEE Std. 1110–1991, 1991.
5. Kundur P., *Power System Stability and Control*, New York, McGraw-Hill, 1994.

Index

Printed in the United States
by Baker & Taylor Publishing Services

Printed in the United States
by Baker & Taylor Publisher Services